4th International Conference on the "Hydrology of the Great Rivers of Africa"

Hydrology of Large River Basins of Africa

Proceedings of the International Association of Hydrological Sciences (PIAHS) Volume 384, 2021

Cotonou, Benin
16 – 20 November 2021

Editors:

E. Amoussou
G. Mahé
O. Amrouni
A. Bodian
C. Cudennec
S. Dietrich
D. J. Kodja
W. E. Vissin

ISBN: 978-1-7138-4503-4

Printed from e-media with permission by:

Curran Associates, Inc.
57 Morehouse Lane
Red Hook, NY 12571

Some format issues inherent in the e-media version may also appear in this print version.

Printed with permission by Curran Associates, Inc. (2022)

International Association of Hydrological Sciences
Centre for Ecology and Hydrology
Wallingford, Oxfordshire OX10 8BB
United Kingdom

Phone: 441 491 692 442
Fax: 441 491 692 448

http://iahs.info/

Additional copies of this publication are available from:

Curran Associates, Inc.
57 Morehouse Lane
Red Hook, NY 12571 USA
Phone: 845-758-0400
Fax: 845-758-2633
Email: curran@proceedings.com
Web: www.proceedings.com

TABLE OF CONTENTS

SECTION 2: HYDROLOGICAL REGIMES AND CHANGES

SECTION 3: HYDROLOGICAL SECURITY

Proc. IAHS, 384, 1–4, 2021
https://doi.org/10.5194/piahs-384-1-2021

Preface: Hydrology of Large River Basins in Africa

Ernest Amoussou[1,2,3,4], **Gil Mahe**[3,4,5], **Oula Amrouni**[3,6], **Ansoumana Bodian**[7], **Christophe Cudennec**[4,8], **Stephan Dietrich**[3,9], **Domiho Japhet Kodja**[2], **and Expédit Wilfrid Vissin**[2]

[1]Université de Parakou, Parakou, Bénin
[2]Laboratoire Pierre Pagney: Climate, Water, Ecosystems and Development (LACEEDE), Université d'Abomey-Calavi, Cotonou, Bénin
[3]FRIEND-Water program, Intergovernmental Hydrological Program, UNESCO/Flow Regimes from International, Experimental and Network Data, Paris, France
[4]International Association of Hydrological Sciences, Wallingford, UK
[5]HSM, Université Montpellier, IRD, CNRS, IMT, Case MSE, 163 rue Auguste Broussonnet, 34090, Montpellier, France
[6]Institut National des Sciences et Technologies de la Mer, Salammbô, Université de Carthage, Tunis, Tunisie
[7]Laboratoire Leïdi "Dynamique des territoires et développement", Université Gaston Berger (UGB), BP 234-Saint-Louis, Sénégal
[8]UMR SAS, Institut Agro, INRAE, Rennes, France
[9]International Centre for Water Resources and Global Change, Federal Institute of Hydrology, Koblenz, Germany

Correspondence: Ernest Amoussou (ernestamoussou@gmail.com) and Gil Mahe (gil.mahe@ird.fr)

Published: 16 November 2021

FRIEND-Water is a flagship program of the UNESCO Intergovernmental Hydrological Program. Since it started in 1985 the International Association of Hydrological Sciences (IAHS) gave its scientific support to the every 4 years global FRIEND conferences and some other occasional issues, which resulted in the publication of previously called "redbooks" (available online at https://iahs.info, last access: 18 October 2021), and now *Proceedings of IAHS* (*PIAHS*). This collaboration between FRIEND-Water and IAHS is very important to keep developing the visibility and attractiveness of hydrological and related sciences in all countries of the World, particularly in developing countries. This *PIAHS* issue publishes articles presented at the 4th International Conference on the Hydrology of Large Rivers Basins of Africa, organized by hydrologists of Benin. The co-editors of this issue come from Benin, Senegal, Tunisia, Germany and France, which shows the strong South-North relationships between institutions and colleagues. From more than 230 abstracts, 132 papers were submitted to the *PIAHS* issue, and 56 were eventually accepted after a double peer-reviewing process involving 122 reviewers from 21 countries, listed below.

This conference is the fourth of a series started in 2015 in Tunisia, followed by a second edition in Senegal in 2016, and a third one in Algeria in 2018. Scheduled in 2020 according to the 2-years pace, it has been delayed due to the covid crisis, but is organized mainly with physical presence in November 2021 in Cotonou, Benin. As usual this conference is the occasion to organize training courses, side-event workshops, and for the first time a *PIAHS* volume. This cycle of conferences also refers to the World Large Rivers Initiative launched by UNESCO in 2014 to which it contributes for the continent. The scientific coverage is comprehensive across the global theme of large rivers; ranging from observation, characterisation and modelling of various processes and settings, to the analysis of regimes, changes and security issues. Articles published in that issue, are hence much related to most of the 10 International Commissions of IAHS, and to cross-cutting and agenda-setting initiatives: innovation and assimilation in hydrometry and the data-service value chain (Tauro et al., 2018; Cudennec et al., 2020; in consistence with Pecora and Lins, 2020 and Dixon et al., 2021), Prediction in Ungauged Basins – PUB (Hrachowitz et al., 2013), Hydrological change – Panta Rhei (Montanari et al., 2013;

McMillan et al., 2016) and related Water security (Young et al., 2015; Ceola et al., 2016) and nexus approach (Liu et al., 2017; Connor et al., 2020; Heal et al., 2021), and socio-hydrology and citizen engagement (Di Baldassarre et al., 2019; Nardi et al., 2021). The clustering of the articles in three sections captures that alignment through a progression from hydrological dynamics, to regimes and changes, and to security. The range of African Large Rivers addressed across these articles is a richness as such, with an ultimate value for comparative hydrology; whereas some local studies are also included, which are valuable for documenting knowledge and models.

This open access (as all *PIAHS* volumes) bilingual issue is of particular interest and value for a large community of authors and readers in Africa, in terms of knowledge capitalization and sharing, both at the regional and international levels, within academia and in the science-policy bridging – as advocated by learned societies and UNESCO in terms of Open Science and public good (Cudennec and Hubert, 2008; Hughes et al., 2014; Quinn et al., 2018; Boulton et al., 2021; UNESCO, 2021). Leading authors come from 15 African countries, Canada, France and UK. Beyond these figures, this issue shows the great expectation of African researchers to integrate international institutional activities and share common scientific reflections. This *PIAHS* volume, as a contribution to the FRIEND-Water program, shortly follows the one published in 2020 out of the Beijing Global FRIEND-Water conference in 2018 (Yu et al., 2020). The next Global FRIEND-Water conference is scheduled to be held in Dakar, Senegal in March 2023. The contribution of the African scientific community to IAHS activities is strongly increasing (Hughes et al., 2015; Croke and Jewitt, 2018; Makurira et al., 2018), which is also visible by the very large registration of African colleagues to the ongoing 23 UPHs in Africa initiative (Blöschl et al., 2019), whose results should be integrated to the global 23 UPHs discussions scheduled in Montpellier IAHS XIth scientific assembly in May–June 2022.

Acknowledgements. Many contributors helped funding this issue and the conference, Institutions: UNESCO Intergovernmental Hydrological Program, the International Centre for Water Resources and Global Change of Koblenz, Germany, IAHS including its SYSTA granting mechanism, the Research Institute for Development – IRD France, the World Meteorological Organization, l'Agence Universitaire de la Francophonie, le CRASTE-LF Centre de Recherche Africain des Sciences et Technologies de l'Espace en Langue Française Maroc, Ministère de l'Enseignement Supérieur et de la Recherche Scientifique du Bénin, Ministère de l'Eau et des Mines du Bénin, Ministère du Cadre de Vie et du Développement Durable du Bénin, Ministère de l'Agriculture et de l'Alimentation du Bénin, Ambassade de France au Bénin, Ambassade de France au Congo, Agence Française de Développement, Académie Nationale des Sciences Arts et Lettres du Bénin, AHR Foundation Bénin; and Universities and Laboratories: University of Parakou Bénin, University of Abomey-Calavi Bénin, Institut de Géographie de l'Aménagement du Territoire et de l'Environnement, Institut National de l'Eau, Laboratoire Pierre Pagney Climat, Ecosystème, Eau et Développement, Laboratoire GéoSciences Environnement Toulouse France, Laboratoire HydroSciences Montpellier France. The following hydrologists have contributed as reviewers: Abderrahmane Khedimallah, Abderrazak Bouanani, Larbi Djabri, Ahmed Kettab, Amin Zettam, Abdelhadi Ammari, Ayoub Zeroual, Belkacem Bekkoussa, Faiza Hallouz, Hind Meddi, Mohamed Meddi, Ayoub Zeroual, Sabrina Taibi, Samir Toumi, Sid-Ahmed Bouguerra, from Algeria; Dawit Berhane from Australia; Firmin Adandedji, Constant Houndenou, K. Olive Adissoda, Fidèle Ahehehinnou Yedo, Gervais Atchade, Djigbo Félicien Badou, Eliezer Biao, Michel Boko, Aymar Bossa, Cyr Gervais Etene, Eric A. Alamou, Moussa Gibigaye, Jean Hounkpe, Domiho Japhet Kodja, Olivier Koudamiloro, Hervé Koumassi, Maurice Ahouansou, Brice Mongbo, Pierre I. Akponikpe, Maïeul G. Quenum, S. Henri Totin Vodounon, Expédit Wilfrid Vissin, Ibouraïma Yabi, Josué Zandagba, Salou Zoulkif, Wilfrid Expédit Vissin, Ernest Amoussou, Rafiath Bamisso, N. P. Maximilien Boko, from Benin; Yetchékpo Patrick Gbohoui, Issoufou Ouedraogo, Yacouba Yira, Augustin Yameogo from Burkina-Faso; Wilfried Pokam, Jules Rémy Ndam, Rodric Merime Nonki, Elisabeth Fita Dassou from Cameroon; Yangbo Chen from China; Martin Kabantu from Congo; François Assemian, Modeste Kacou, Kouakou Lazare Kouassi, Auguste Kouassi Kouame, Amani Michel Kouassi, Djibril Dabissi Noufe, Salomon Obahoundje, Zamblé Armand Tra Bi, Vami Nguessan, Mouamé Y. Morton, Eric Pascal Zahiri from Côte D'Ivoire; Negash Tessema from Ethiopia; Alain Dezetter, Andrew Ogilvie, Benjamin Polh, Pascal Breil, Charles Perrin, Christophe Bouvier, Christophe Brachet, Christophe Cudennec, Daniel Loudiere, Olivier Delaigue, Didier Orange, Arona Diedhiou, Emmanuelle Cadot, Gil Mahe, Jean-Louis Perrin, Jean Emmanuel Paturel, Sandra Rome, Alban Delavenne, Sylvain Bigot, Manuella Grippa, Yves Tramblay, Valentin Wendling, Théo Vischel, Telesphore Brou, Etienne Guilpart, Vazken Andreassian from France; Felix Opdehipt, Jörg Dietrich, Stephan Dietrich from Germany; Lhoussaine Bouchaou from Morocco; Nicolas Vigaud from USA; Olusegun Adeaga, Isaiah Akoteyon from Nigeria; Cyriaque Rufin Nguimalet from Central African Republic; Mamadou Lamine Mbaye, Soussou Sambou, Raymond Malou, Amadou Abou Sy, Abdoulaye Faty, Ibrahima Mall, Lamine Diop, Gayane Faye, Assoumana Bodian from Senegal; Lamboni Batablinle, Benoît Koubodana from Togo; Sihem Benabdallah, Noureddine Gaaloul, Majid Mathlouthi, Oula Amrouni from Tunisia; Nilay Dogulu from Turkey; Declan Conway, Bastien Dieppois, Maria Uberia, Abdou Khouakhi from United Kingdom.

References

Blöschl, G., Bierkens, M. F. P., Chambel, A., Cudennec, C., Destouni, G., Fiori, A., Kirchner, J. W., McDonnell, J. J., Savenije, H. H. G., Sivapalan, M., Stump, C., Toth, E., Volpi, E., Carr, G., Lupton, C., Salinas, J., Széles, B., Viglione, A., Aksoy, H., Allen, S. T., Amin, A., Andréassian, V., Arheimer, B., Aryal, S. K., Baker, V., Bardsley, E., Barendrecht, M. H., Bartosova, A., Batelaan, O., Berghuijs, W. R., Beven, K., Blume, T., Bogaard, T., Borges de Amorim, P., Böttcher, M. E., Boulet, G., Breinl, K., Brilly, M., Brocca, L., Buytaert, W., Castellarin,

A., Castelletti, A., Chen, X., Chen, Y., Chen, Y., Chifflard, P., Claps, P., Clark, M. P., Collins, A. L., Croke, B., Dathe, A., David, P. C., de Barros, F. P. J., de Rooij, G., Di Baldassarre, G., Driscoll, J. M., Duethmann, D., Dwivedi, R., Eris, E., Farmer, W. H., Feiccabrino, J., Ferguson, G., Ferrari, E., Ferraris, S., Fersch, B., Finger, D., Foglia, L., Fowler, K., Gartsman, B., Gascoin, S., Gaume, E., Gelfan, A., Geris, J., Gharari, S., Gleeson, T., Glendell, M., Gonzalez Bevacqua, A., González-Dugo, M. P., Grimaldi, S., Gupta, A. B., Guse, B., Han, D., Hannah, D., Harpold, A., Haun, S., Heal, K., Helfricht, K., Herrnegger, M., Hipsey, M., Hlaváciková, H., Hohmann, C., Holko, L., Hopkinson, C., Hrachowitz, M., Illangasekare, T. H., Inam, A., Innocente, C., Istanbulluoglu, E., Jarihani, B., Kalantari, Z., Kalvans, A., Khanal, S., Khatami, S., Kiesel, J., Kirkby, M., Knoben, W., Kochanek, K., Kohnová, S., Kolechkina, A., Krause, S., Kreamer, D., Kreibich, H., Kunstmann, H., Lange, H., Liberato, M. L. R., Lindquist, E., Link, T., Liu, J., Loucks, D.P., Luce, C., Mahé, G., Makarieva, O., Malard, J., Mashtayeva, S., Maskey, S., Mas-Pla, J., Mavrova-Guirguinova, M., Mazzoleni, M., Mernild, S., Misstear, B. D., Montanari, A., Müller-Thomy, H., Nabizadeh, A., Nardi, F., Neale, C., Nesterova, N., Nurtaev, B., Odongo, V. O., Panda, S., Pande, S., Pang, Z., Papacharalampous, G., Perrin, C., Pfister, L., Pimentel, R., Polo, M. J., Post, D., Prieto Sierra, C., Ramos, M.-H., Renner, M., Reynolds, J. E., Ridolfi E., Rigon, R., Riva, M., Robertson, D.E., Rosso, R., Roy, T., Sá, J. H. M., Salvadori, G., Sandells, M., Schaefli, B., Schumann, A., Scolobig, A., Seibert, J., Servat, E., Shafiei, M., Sharma, A., Sidibe, M., Sidle, R. C., Skaugen, T., Smith, H., Spiessl, S. M., Stein, L., Steinsland, I., Strasser, U., Su, B., Szolgay, J., Tarboton, D., Tauro, F., Thirel, G., Tian, F., Tong, R., Tussupova, K., Tyralis, H., Uijlenhoet, R., van Beek, R., van der Ent, R.J., van der Ploeg, M., Van Loon, A.F., van Meerveld, I., van Nooijen, R., van Oel, P. R., Vidal, J.-P., von Freyberg, J., Vorogushyn, S., Wachniew, P., Wade, A. J., Ward, P., Westerberg, I. K., White, C., Wood, E. F., Woods, R., Xu, Z., Yilmaz, K. K., and Zhang, Y.: Twenty-three unsolved problems in hydrology (UPH) – a community perspective, Hydrolog. Sci. J., 64, 1141–1158, https://doi.org/10.1080/02626667.2019.1620507, 2019.

Boulton, G., Cudennec, C, Castle, D., Dogulu, N., Haliwell, J., Hélein, F., Maras, P., Rossini, M., Sanchez, J. A., Sud, M., and Vigen J.: ISC Statement on Open Science and the UNESCO initiative, available at: https://council.science/current/news/open-science-and-the-unesco-initiative/, last access: 5 October 2021.

Ceola, S., Montanari, A., Krueger, T., Dyer, F., Kreibich, H. ,Westerberg, I., Carr, G., Cudennec, C., Elshorbagy, A., Savenije, H., Van Der Zaag, P., Rosbjerg, D., Aksoy, H., Viola, F., Petrucci, G., MacLeod, K., Croke, B., Ganora, D., Hermans, L., Polo, M.J., Xu, Z., Borga, M., Helmschrot, J., Toth, E., Ranzi, R., Castellarin, A., Hurford, A., Brilly, M., Viglione, A., Blöschl, G., Sivapalan, M., Domeneghetti, A., Marinelli, A., and Di Baldassarre, G.: Adaptation of water resources systems to changing society and environment – A statement by the International Association of Hydrological Sciences, Hydrolog. Sci. J., 61, 2803–2817, https://doi.org/10.1080/02626667.2016.1230674, 2016.

Connor, R., Timmerman, J., Uhlenbrook, S., and Koncagül, E.: Chapter 9 – Water-Climate-Energy-Food-Environment Nexus, The United Nations World Water Development Report 2020 "Water and Climate Change", 118–125, ISBN 978-92-3-100371-4, available at: https://en.unesco.org/themes/water-security/wwap/wwdr/2020 (last access: 5 October 2021), 2020.

Croke, B. and Jewitt, G.: Editorial: Special Issue on Water security and the food-water-energy nexus: drivers, responses and feedbacks at local to global scales, Proc. IAHS, 376, 1–1, https://doi.org/10.5194/piahs-376-1-2018, 2018.

Cudennec, C. and Hubert, P.: Multi-objective role of HSJ in processing and disseminating hydrological knowledge, Hydrolog. Sci. J., 53, 485–487, https://doi.org/10.1623/hysj.53.2.485, 2008.

Cudennec, C., Lins, H., Uhlenbrook, S., and Arheimer, B.: Towards FAIR and SQUARE hydrological data, Hydrolog. Sci. J., 65, 681–682, https://doi.org/10.1080/02626667.2020.1739397, 2020.

Di Baldassarre, G., Sivapalan, M., Rusca, M., Cudennec, C., Garcia, M., Kreibich, H., Konar, M., Mondino, E., Mård, J., Pande, S., Sanderson, M. R., Tian, F., Viglione, A., Wei, J., Wei, Y., Yu, D. J., Srinivasan, V., and Blöschl, G.: Socio-hydrology: Scientific challenges in addressing the sustainable development goals, Water Resour. Res., 55, 6327–6355, https://doi.org/10.1029/2018WR023901, 2019.

Dixon, H., Sandström, S., Cudennec, C., Lins, H.F., Abrate, T., Bérod, D., Chernov, I., Ravalitera, N., Sighomnou, D., and Teichert, F.: Intergovernmental cooperation for hydrometry – what, why, how?, Hydrolog. Sci. J., 1–15, https://doi.org/10.1080/02626667.2020.1764569, 2021.

Heal, K. V., Bartosova, A., Hipsey, M. R., Chen, X., Buytaert, W., Li, H.-Y., McGrane, S. J., Gupta, A. B., and Cudennec C.: Water quality: the missing dimension of water in the water–energy–food nexus, Hydrolog. Sci. J., 66, 745–758, https://doi.org/10.1080/02626667.2020.1859114, 2021.

Hrachowitz, M., Savenije, H. H. G., Blöschl, G., McDonnell, J. J., Sivapalan, M., Pomeroy, J. W., Arheimer, B., Blume, T., Clark, M. P., Ehret, U., Fenicia, F., Freer, J. E., Gelfan, A., Gupta, H. V., Hughes, D. A., Hut, R. W., Montanari, A., Pande, S., Tetzlaff, D., Troch, P. A., Uhlenbrook, S., Wagener, T., Winsemius, H.C., Woods, R.A., Zehe, E., and Cudennec, C.: A decade of Predictions in Ungauged Basins (PUB) – a review, Hydrolog. Sci. J., 58, 1198–1255, https://doi.org/10.1080/02626667.2013.803183, 2013.

Hughes, D. A., Heal, K. V., and Leduc, C.: Improving the visibility of hydrological sciences from developing countries, Hydrolog. Sci. J., 59, 1627–1635, https://doi.org/10.1080/02626667.2014.938653, 2014.

Hughes, D. A., Jewitt, G., Mahe, G., Mazvimavi, D., and Stisen, S.: A review of hydrological sciences research in Africa over the last decade, Hydrolog. Sci. J., 60, 1865–1879, https://doi.org/10.1080/02626667.2015.1072276, 2015.

Liu, J., Yang, H., Cudennec, C., Gain, A. K., Hoff, H., Lawford, R., Qi, J., de Strasser, L., Yillia, P. T., and Zheng, C.: Challenges in operationalizing the water-energy-food nexus, Hydrolog. Sci. J., 62, 1714–1720, https://doi.org/10.1080/02626667.2017.1353695, 2017.

Makurira, H., Mazvimavi, D., Kapangaziwiri, E., Kileshye Onema, J.-M., and Gumindoga, W.: Preface: Understanding spatio-temporal variability of water resources and the implications for IWRM in semi-arid eastern and southern Africa, Proc. IAHS, 378, 1–1, https://doi.org/10.5194/piahs-378-1-2018, 2018.

McMillan, H., Montanari, A., Cudennec, C., Savenije, H., Kreibich, H., Krueger, T., Liu, J., Mejia, A., Van Loon, A., Aksoy, H., D, Baldassarre, G., Huang, Y., Mazvimavi, D., Rogger, M., Sivakumar, B., Bibikova, T., Castellarin, A., Chen, Y., Finger, D., Gelfan, A., Hannah, D. M., Hoekstra, A. Y., Li, H., Maskey, S., Mathevet, T., Mijic, A., Acuña, A. P., Polo, M. J., Rosales, V., Smith, P., Viglione, A., Srinivasan, V., Toth, E., van Nooyen, R., and Xia, J.: Panta Rhei 2013–2015: Global perspectives on hydrology, society and change, Hydrolog. Sci. J., 61, 1174–1191, https://doi.org/10.1080/02626667.2016.1159308, 2016.

Montanari, A., Young, G., Savenije, H. H. G., Hughes, D., Wagener, T., Ren, L.L., Koutsoyiannis, D., Cudennec, C., Toth, E., Grimaldi, S., Blöschl, G., Sivapalan, M., Beven, K., Gupta, H., Hipsey, M., Schaefli, B., Arheimer, B., Boegh, E., Schymanski, S. J., Di, Baldassarre, G., Yu, B., Hubert, P., Huang, Y., Schumann, A., Post, D. A., Srinivasan, V., Harman, C., Thompson, S., Rogger, M., Viglione, A., McMillan, H., Characklis, G., Pang, Z., and Belyaev, V.: "Panta Rhei – Everything Flows": Change in hydrology and society – The IAHS Scientific Decade 2013–2022, Hydrolog. Sci. J., 58, 1256–1275, https://doi.org/10.1080/02626667.2013.809088, 2013.

Nardi, F., Cudennec, C., Abrate, T., Allouch, C., Annis, A., Assumpção, T., Aubert, A. H., Bérod, D., Braccini, A. M., Buytaert, W., Dasgupta, A., Hannah, D.M., Mazzoleni, M., Polo, M. J., Sæbø, Ø., Seibert, J., Tauro, F., Teichert, F., Teutonico, R., Uhlenbrook, S., Wahrmann Vargas, C. and Grimaldi, S.: Citizens AND HYdrology (CANDHY): conceptualizing a transdisciplinary framework for citizen science addressing hydrological challenges, Hydrolog. Sci. J., https://doi.org/10.1080/02626667.2020.1849707, in press, 2021.

Pecora, S. and Lins, H. F.: E-monitoring the nature of water, Hydrolog. Sci. J., 65, 683–698, https://doi.org/10.1080/02626667.2020.1724296, 2020.

Quinn, N., Blöschl, G., Bárdossy, A., Castellarin, A., Clark, M., Cudennec, C., Koutsoyiannis, D., Lall, U., Lichner, L., Parajka, J., Peters-Lidard, C. D., Sander, G., Savenije, H., Smettem, K., Vereecken, H., Viglione, A., Willems, P., Wood, A., Woods, R., Xu, C.-Y., and Zehe, E.: Joint editorial: Invigorating hydrological research through journal publications, Hydrol. Earth Syst. Sci., 22, 5735–5739, https://doi.org/10.5194/hess-22-5735-2018, 2018.

Tauro, F., Selker, J., Van De Giesen, N., Abrate, T., Uijlenhoet, R., Porfiri, M., Manfreda, S., Caylor, K., Moramarco, T., Benveniste, J., Ciraolo, G., Estes, L., Domeneghetti, A., Perks, M. T., Corbari, C., Rabiei, E., Ravazzani, G., Bogena, H., Harfouche, A., Broccai, L., Maltese, A., Wickert, A., Tarpanelli, A., Good, S., Lopez Alcala, J. M., Petroselli, A., Cudennec, C., Blume, T., Hut, R., and Grimaldi, S.: Measurements and observations in the XXI century (MOXXI): innovation and multi-disciplinarity to sense the hydrological cycle, Hydrolog. Sci. J., 63, 169–196, https://doi.org/10.1080/02626667.2017.1420191, 2018.

UNESCO: Draft text of the UNESCO recommendation on Open Science, available at: https://en.unesco.org/science-sustainable-future/open-science/recommendation, last access: 5 October 2021.

Young, G., Demuth, S., Mishra, A., and Cudennec, C.: Hydrological sciences and water security: An overview, Proc. IAHS, 366, 1–9, https://doi.org/10.5194/piahs-366-1-2015, 2015.

Yu, Z., Lu, C., Cai, J., Yu, D., Mahe, G., Mishra, A., Cudennec, C., Van Lanen, H. A. J., Orange, D., and Amani, A.: Preface: Hydrological processes and water security in a changing world, Proc. IAHS, 383, 3–4, https://doi.org/10.5194/piahs-383-3-2020, 2020.

Proc. IAHS, 384, 5–18, 2021
https://doi.org/10.5194/piahs-384-5-2021

Open Access PIAHS

The UNESCO FRIEND-Water program: accelerates, shares and transfers knowledge and innovation in hydrology across the world in the frame of the Intergovernmental Hydrological Program (IHP)

Gil Mahé[1], Gamal Abdo[2], Ernest Amoussou[3], Telesphore Brou[4], Stephan Dietrich[5], Ahmed El Tayeb[6], Henny van Lanen[7], Mohamed Meddi[8], Anil Mishra[9], Didier Orange[10], Thi Phuong Quynh Le[11], Raphael Tshimanga[12], Patrick Valimba[13], Santiago Yepez[14], Andrew Ogilvie[15], and Oula Amrouni[16]

[1]HSM, Université Montpellier, IRD, MSE, 300 Av. E. Jeanbrau, 34090 Montpellier, France
[2]Water Research Center, University of Khartoum, Khartoum, Sudan
[3]Department of Geography and Spatial Planning, University of Parakou, - Parakou, Benin
[4]Dept. de Géographie, Laboratory OIES (Océan Indien: Espaces et Sociétés), Université de La Reunion, 15 avenue René Cassin, CS 92003, 97744 Saint Denis CEDEX 9, La Réunion
[5]International Centre for Water Resources and Global Change, Federal Institute of Hydrology, P.O. Box 200253, 56002 Koblenz, Germany
[6]The National Water Research Center Ministry of Water Resources, Irrigation and Electricty, Khartoum, Sudan
[7]Hydrology and Quantitative Water Management Group, Wageningen University, Droevendaalsesteeg 4, 6708 PB Wageningen, the Netherlands
[8]National Superior School of Hydraulics, Blida, Algeria
[9]UNESCO, Division of Water Sciences, 7, place de Fontenoy, 75352 Paris 07 SP, France
[10]Eco&Sols, Université Montpellier, IRD, INRAe, CIRAD, SupAgro, LMI iESOL, Centre ISRA-IRD Bel Air, Dakar, Sénégal
[11]Institute of Natural Product Chemistry, Graduate University of Science and Technology, Vietnam Academy of Science and Technology, 18 Hoang Quoc Viet, Cau Giay, Hanoi, Vietnam
[12]Congo Basin Water Resources Research Centre/CRREBaC, Univ. of Kinshasa, P.O. Box 117 KIN XI, Kinshasa, Congo
[13]Department of Water Resources Engineering, College of Engineering and Technology, University of Dar es Salaam, P.O. Box 35131, Dar es Salaam, Tanzania
[14]Departamento Manejo de Bosques y Medio Ambiente, Facultad de Ciencias Forestales, Universidad de Concepción – UdeC, Concepción 407374, Chile
[15]G-EAU, IRD, AgroParisTech, BRGM, CIRAD, INRAe, SupAgro Montpellier, Montpellier, France
[16]National Institute for Marine Sciences and Technologies, Carthage Salammbô, Tunis, Tunisia

Correspondence: Gil Mahé (gil.mahe@ird.fr)

Published: 16 November 2021

Abstract. The FRIEND-Water program (FWP) is the oldest and the most transverse program (i.e. Flagship Initiative) within the Hydrological Intergovernmental Program (IHP) from UNESCO. FRIEND means "Flow Regimes from International and Experimental Network Data". The FWP is dedicated to allow large communities of hydrologists and associated disciplines to collaborate across borders on common shared data and scientific topics which are addressed through large world regions. The program has evolved in its 35 years of existence. There are 8 large regions in the world which gather tenths of hundreds of researchers, and organize events following several research topics defined according to each region priorities. The FWP is chaired by a scientific committee gathering regional coordinators and thematic experts. Each region gives its research priorities which evolve according to the projections given by the member States during the IHP council every year. The future activities of the FWP are defined by the upcoming IHP IX program and with the support of the newly created Montpellier UNESCO Category II Center ICIREWARD, acting as FWP Secretariat.

1 Introduction

At the start in 1985 the purpose of the FWP was to allow researchers from different countries and regions to conduct studies on shared river basins, with the support of national hydrological services which agreed to share data with the UNESCO FRIEND-Water groups. Over time, most hydrological services stopped sharing data and this is the current challenge to reactivate the data exchange through new platform facilities. The FRIEND-Water regional groups develop studies on research topics specific to their regions. However, some topics are common to several regions such as floods and low flows/drought, snow and glacier hydrology, erosion/sedimentation and water quality. The ecohydrology theme is of increasing interest among the different regions. Its development within the FWP is in close link with the Ecohydrology program of UNESCO IHP. A special session on ecohydrology has been organized during the last Global FRIEND-Water conference in Beijing in November 2018, and further joint meeting are scheduled. In addition to the Ecohydrology programme, there are several other thematic IHP programmes and initiatives such as International Flood Initiative (IFI), International Sediment Initiative (ISI), International Drought Initiative (IDI), World large River Initiative (WLRI) and UNESCO's programme for Water and Development Information for Arid Lands – a Global Network (G-WADI). It is important that FRIEND-Water should positions itself in a way that maximises complementarities and synergies and minimizes duplication with other IHP initiatives and programmes. The activities of the program are developed through different meetings, workshops or conferences, together with training sessions and field trips. Many of the communications presented during the workshops are published in special issues of regular indexed journals and conference proceedings of IAHS, the International Association of Hydrological Sciences, with which FRIEND-Water is closely linked since the start. Furthermore, FRIEND-Water intends to promote the dialogue between the scientific community, the user community and policy makers in studies related in the context of SDGs and Paris Agreement. Ongoing research addresses groundwater, in particular shortage under dry environmental conditions. The Groundwater Drought Initiative (GDI) has been established across Europe. Large-scale studies explored how groundwater pumping, which is vitally important for irrigation and associated food security, decreases streamflow globally, and estimate where and when environmentally critical stream flows will no longer be sustained. Furthermore, hydrological knowledge is more and more used to study impacts of natural hazards on socio-economic and environmental conditions. There is special attention to the relationship between droughts and/or floods and people. Classical bottom-up approaches which include vulnerability are combined with top-down approaches to change from reactive to pro-active risk management. In this context, study of multi-hazards (coinciding and cascading) is a focal area that aims at development of Early Warning Systems, including understanding, identification, monitoring and forecasting. Other studies deal with the processes of transfer and transformation of matter within the large global hydrological systems. With emphasis on rivers, lakes and reservoirs, trying to understand the balance of sediments and their role in primary production, and in particular water quality. Currently, the different groups of the FWP are relying on new space technologies to observe the spatial and temporal dynamics of these processes. A particular case is the use of satellite images as a tool to densify and extend in situ observations which were traditionally measured in a punctual manner at hydrological stations. This new type of approach, applied at different scales, allows balances to be made which help to understand the natural variability of the systems, quantifying the transfer processes and determining possible impacts, driven either by climatic or by anthropogenic causes. Based on the thematic scientific feedback FRIEND-Water should also provide feedback to address hydrology and human nature interactions in the context of complex water management problems.

Moreover, more studies are ongoing which use spatial climate (gridded) input data into hydrological models to assess suitability of generated streamflow for rainfed agriculture, engineering designs of hydraulic and flood protection structures and ecological conservation. The emphasis is to understand suitability of global climate data in real life applications. This data increasingly becomes available to patch gaps in observational climate records and useful in large ungauged catchments/basins but needs verification in different areas as some is derived from coarse networks of observational stations particularly in Africa.

During the current phase of the FRIEND-Water program (2018–2022) (Fig. 1), it has been recommended to all regional groups to promote training sessions and workshops on the theme of spatial hydrology, which is a research field that can bring future major improvements in hydrological sciences, closely related to the processing and share of huge hydrological databases. It has been also recommended to develop the information about the program to a large public, including scholars and civil society and to closely fit with

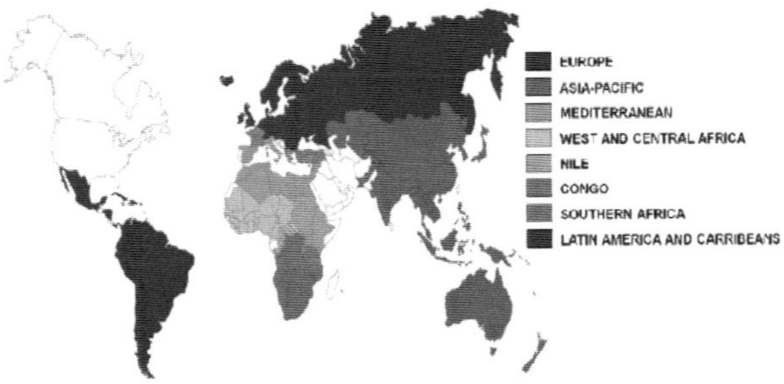

Figure 1. The 8 regions of the FRIEND-Water program during the phase 2018–2022.

the orientations of the Strategic Plan for the IHP-IX that was adopted by the 24th Session of the Intergovernmental Council of the UNESCO IHP (UNESCO, 2021).

2 Regional FRIEND-Water cooperations

2.1 AMIGO FRIEND-Water

Currently, the main challenge will be to implement a standard framework for assessing the conditions and evolution of rivers, considering hydrogeomorphological and ecological properties, which should be a priority for all future river management works. Recently, we have started to apply the River Styles framework to carry out hydrogeomorphological studies with an approach adapted to the local reality, focusing on the development of protocols and procedures (many of them automated using GIS/RS techniques) to assess the condition of rivers in South America and the Caribbean (Gallay et al., 2019). This would allow prioritizing rivers in terms of their conservation and restoration, as well as orienting sustainable management strategies in them, where the hydrogeomorphological factor is sometimes incorporated in a limited way in most of the engineering projects developed along the riverbanks (Nardini et al., 2020). This approach must necessarily take into account in a comprehensive manner the different morphologies and varieties of rivers and also the particular nature of human disturbances (e.g. water diversions, reservoirs, gravel mining, urban sprawl, forestry activities, among others) and their distribution across regions.

South American region has a strong vulnerability face to climate change and it affects not only rainfall regimes and temperature extremes, but also water resources systems. Many countries experiment a serious concern with glacier retreat since last decades caused by a regional warming. In general, Andean mountain catchments are also being affected by anthropogenic interventions, affecting water resources availability and water security (Laqui et al., 2018; Yepez et al., 2017, 2018). In Andean context, some hydrological processes are still unknown and misunderstood because of data scarcity, increasing the gap between decision makers and develop. Beside, this alteration of the hydrological cycle is exacerbated with large scale climatic phenomenon as El Niño, producing droughts or floods, being more remarkable on both sides of the Andes, over Pacific coast and Amazonian regions (Rau et al., 2018). Therefore, researches are expected to provide reliable databases, scientific evidence and communicate to improve water management practices.

The Andes of South America and their surrounding basins are regions sensitive to climate change and tectonic events. Feedback between climate, tectonics and anthropogenic activities has a profound impact on the current landscape. Additionally, such processes have a significant impact on the biodiversity of species. Our investigations use thermochronology and quantitative geomorphology techniques to model the landscape evolution and analyze the influence of these interactions on biodiversity. A combined effect of increased tectonic, climatic and anthropogenic activities produces an increase in the sedimentation rates of rivers. Through numerical modeling, thermochronology and cosmogenic nuclides it is possible estimate long-term erosion rates and compare them with short term (or current) erosion rates in order to discriminate the events that generate these increases (Ferreira Araújo et al., 2017).

Land-use changes lead to disturbances on weathering processes and nutrient cycles. Chemical weathering of continental rocks plays a central role in regulating the carbon cycle and the Earth's climate, as well, in the supplying of essential nutrients, such as N, P and trace metals, for the ecosystems. However, man-induce changes on land use and hydrological features associated with natural resources exploitation, water extraction, damming and urbanization can alter water-rock interaction times, flow regimes, sediment transport (Carretier et al., 2018; Laraque et al., 2019), which in turn, affect natural weathering regime and elemental biogeochemical cycles. Human activity can also impact weathering via acid rain. Burning of fossil fuels releases nitrogen and sulfur compounds that later form nitric and sulfuric acids in rainwater. These are stronger acids than carbonic acid and they accelerate rock dissolution and negatively impact upon ecosystems. Our study aims to understand how climate change and anthropic actions can modify weathering processes at local and global magnitudes, at short and large scale-times, and impact aquatic ecosystems (Moquet et al., 2018).

2.2 CONGO FRIEND-Water

As referred to the study of Alsdorf et al. (2016), the Congo River basin although being the second largest river in the world by the discharge, it remains largely understudied. It is thus very important to strengthen the research and knowledge capacity in this region of the world, based on the emerging new research teams over the many countries that share the Congo River (Beighley et al., 2019; Trigg and Tshimanga, 2020). The Congo River Basin exhibits tremendous heterogeneities, out of which it emerges as an intricate system where complexity will vary consistently over time and space.

Increased complexity in the absence of adequate knowledge will always result in increased uncertainties. One way of simplifying this complexity is through an understanding of organisational relationships of the landscape features, which is termed here as catchment classification (Tshimanga et al., 2021). This is one of the major tasks to start with at the international level, in order to develop further multidisciplinary studies on the basis of this global knowledge shared among all teams. During the past 12 months, there has been a first attempt to establish a framework of catchment classification using an a priori approach. Various methods, including GIS and multivariate analyses, were combined to define the subbasin units and group them in homogeneous regions. A set of 1740 catchment units were partitioned across the whole Congo River Basin, based on a set of comprehensive criteria, including the natural break of the elevation gradient (199 units), inclusion of socio-economic and anthropogenic systems (204 units), and water management units based on traditional (socio-hydrology) nomenclature of the rivers within the basin (1337 units). The identified catchment units were used to assess existing datasets of the basin physical properties, necessary to derive descriptors of the catchments characteristics. Assessment of relationships between physical basin attributes and the catchments through PCA showed that 15 variables out of 26 initially selected contained the information necessary to explain the spatial distribution and grouping of the sub-basins. These 15 variables were included in a cluster analysis using unsupervised classification (hierarchical agglomerative clustering), which identified 11 groups of homogenous catchments based on the use of the environmental descriptors derived from the available physical basin property datasets. This constitutes therefore an a priori clustering of the catchment classification framework for the CRB and should be considered as an initial phase of the iterative process of catchment classification. This a priori clustering will be further used to derive functional relationships of the catchments, necessary to enable hydrological prediction and water management in the basin. The framework is therefore an efficient tool that will help synthesize our understanding of the interaction between climate variability, catchment characteristics (e.g. vegetation, soils, topography) (Mushi et al., 2019) and the resulting hydrological response. It is also a crucial step in improving predictions in ungauged basins, and the new insights of the Congo river hydrology obtained from the most recent field campaigns will help improve the accuracy of the satellite derived hydrological data (Carr et al., 2019). Within FRIEND-Water-UNESCO, we intend to promote the use of this framework to address the challenge of generalizing knowledge derived from local observations and understanding of the catchment response characteristics, as well as the underlying process controls; to provide guidance for measurement and modelling, and constraints for predictions in ungauged basins and to estimate impacts of environmental changes.

2.3 EURO FRIEND-Water

Closely inter-related key issues and challenges to large-scale hydrology were identified to improve prediction of hydrological systems behavior (surface/sub-surface) on large spatial and time scales (i.e. continental – multidecadal) under changing environmental conditions (Massei et al., 2018; Kingston et al., 2020). Large-scale hydrology depends on interactions across the ocean-atmosphere-vegetation/soil/aquifer system. Detection of climate indices and large-scale patterns directly from hydrological data (Renard and Thyer, 2019) is required. Characterizing non-stationarity and time-scale dependence of large-scale relationships also appears fundamental (Dieppois et al., 2013). Other remaining challenges to be studied are how climate system models are capable of representing large-scale relationships at different temporal scales, or how to assess whether model performance in simulating hydroclimatic variables results from realistic large-scale climate processes, or from other compensating biases (Dieppois et al., 2019). Improved knowledge on large-scale relationships would also lead to more skilful hydrological forecasts (Lavers et al., 2017).

As catchment characteristics play a dominant role in filtering out climate variability on river flow, including hydrological extremes (Van Lanen et al., 2013), efforts need to focus on improving representation of key hydrological processes on the large scale, very often oversimplified in large-scale models. For instance, lateral re-distribution by hillslope processes, landscape (Laizé and Hannah, 2010) or subsurface heterogeneities (Hartmann et al., 2017), or the role of groundwater on large-scale hydrological response (Sidibe et al., 2019) needs to be accounted for.

The last decade, substantial progress has been on the detection and projections of hydrological extremes, including drought (e.g. Stahl et al., 2012; Prudhomme et al., 2014). Emerging topics have been identified, such as flash droughts, multi-year droughts (where these were not reported before), cold-climates' droughts, anthropogenically-enhanced droughts (Teuling et al., 2016).

Another major challenge that remains is the attribution of the causes and processes leading to low flow (events), streamflow droughts, and intermittent flow. These necessities improved knowledge on how weather anomalies eventually end up as hydrological drought (groundwater, streamflow), i.e. how climate and catchment controls, and human drivers affect drought development and termination (Van Loon et al., 2016). Groundwater, as an important catchment control, requires more attention (Stoelzle et al., 2014). A step forward is the pan-European Groundwater Drought Initiative (GDI[1]) that recently has been established. Furthermore, attribution calls for further development of hydrological models that dynamically consider changing environmental conditions im-

[1] https://www2.bgs.ac.uk/groundwater/waterResources/groundwaterDroughtInitiative/home.html, last access: 6 January 2021)

plying focus on more process-based rather than a calibration-based modelling approach.

Early Warning Systems (EWS), incl. monitoring and forecasting, will be further developed to mitigate and adapt to hydrological extremes and to enable rapid pro-active risk-management rather than a re-active approach. In this context, a critical appraisal of indices and indicators will be done. Furthermore, hydrological knowledge is more and more used to study impacts of drought on socio-economic and environmental conditions (Stahl et al., 2016; Blauhut et al., 2015; Sutanto et al., 2019a). Classical bottom-up approaches, which use hazard, exposure and vulnerability as a basis are combined with top-down approaches to change from reactive to pro-active risk management. In this context, study of multi-hazards (coinciding and cascading) is becoming a focal area (Sutanto et al., 2019b) in the further development of EWS.

Transnational hydroclimatic data are fundamental to investigate hydrological extremes and large-scale variation in hydrology. Flow data from European river will be regularly updated, as part of further development of information systems the Global Runoff Data Centre (GRDC) and other relevant data centres and observational networks of the Global Terrestrial Network for Hydrology (GTN-H)[2] are involved in. We will encourage hydrometeorological services to provide these data. Large-scale hydrological data rescue (Le Gros et al., 2015) needs continuation, as well as, reconstruction of long-term, high-resolution climate datasets (Devers et al., 2020), and use of remote sensing products (e.g. GRACE). Compatibility between, or merging of, national-scale datasets (Caillouet et al., 2019) is planned to be further investigated to contribute to large-scale databases.

2.4 FRIEND-Water AP

In Asia Pacific region, water resources in many countries are affected by climate change (Yang et al., 2019). Increasing dry season duration, longer and more serious droughts (Wei et al., 2019), could cause water shortage occurred on a large scale and reduce water quality, affect strongly water security and sustainable development. Besides, rising sea levels lead to flooding in the delta and coastal areas, increasing saline intrusion in estuaries and coastal plains and causing erosion and sedimentation to upset the natural and ecological balance (Li et al., 2018; Xi et al., 2019), seriously affecting coastal areas. Thus, researches are hoped to provide useful information and solutions related to ensuring water security in the context of clear climate change today.

The downstream of international rivers, is often suffering from water-related disasters (Try et al., 2018), facing water scarcity and declining water resources due to overexploitation, thus water resources in the Asian countries are hiding many unsustainable factors for development (Kondolf et al., 2018). Water scarcity and water shortage will increase sharply compared to today and most river basins are in a state of stress on water use (Hiraga et al., 2018), especially during the dry season. Water tensions will be severe in key economic basins (Kositgittiwong et al., 2017). Thus, one important topic in Asia is how to reduce the stress on water use, especially during the dry season?

Due to rapid increase of population, industrial and expanded agricultural production, pollution of surface water has been increased in both level and scale (Zhang et al., 2018; Nguyen et al., 2019). Water is available in many places but cannot be used due to its poor quality. Especially, surface water and ecohydrosystem in many urban areas, and industrial parks has been seriously affected due to the untreated wastewater or incompletely treated wastewater. In some countries in Asia Pacific, lack of large investment in industrial and urban wastewater treatment, no effective mechanisms and policies to reduce water pollution, over use of fertilizers and chemicals in the agricultural sector … have accelerated pollution level in this region. Thus, the development of researches related to minimize water degradation is firstly needed (Le et al., 2017; Uk et al., 2018).

Rapid increase in water demand accelerates the groundwater exploitation, leading to over-exploited. Groundwater level in some areas has been continuously reduced and more important, groundwater quality is at high risk of pollution (e.g. As) (Hoa et al., 2020). Thus, studies should be focused on providing information concerning reasonable exploitation of groundwater and protection of both quantity and quality of groundwater.

2.5 FRIEND-Water AOC

In West and Central Africa, as in other regions of the world, studies have shown a growing risk of extreme events (excess rain and temperature, lack or scarcity of rain, etc.) with the advent of global warming of the planet, but which is not without consequences for the capacities of production systems and water resources (Sidibe et al., 2018, 2020; Amoussou et al., 2020). Thus, we are witnessing the destruction of natural and human ecosystems and the loss of biological diversity by overflowing waters (flooding), very marked drying up of beds and anthropogenic pressure. Changes in precipitation and temperature, as well as non-linear effects on humidity, evapotranspiration and soil, affect the quantity and quality of water resources, agriculture, fishing and livestock (Amoussou, 2010; IPCC, 2013) and therefore on the life of the local population.

This climate change/variability coupled with demographic change and land degradation affects the functioning of the terms of the hydrological balance and the frequency of extreme events and therefore the water levels in the lakes (Zare et al., 2017), the hydraulicity downstream of the structures,

[2]GTN-H is a federation of global water data centres, inter alia the GRDC and GEMS/Water Data Centre and operates under the auspices of WMO and the Global Climate Observing System GCOS, http://www.gtn-h.info (last access: 6 January 2021)

the modes of resource use in the developed sites, the dams, the safety of river water and the feminine management of the resource. To this end, it is important to accentuate studies on hydrological standards, intensity, duration and frequency (IDF) of storms, cold or heat waves on the one hand, the increase of the extent of regions affected by drought and unexpected floods, with unevenly distributed consequences for many regions of the world, particularly in developing countries (Parkes et al., 2019), and especially in Sahel (Descroix et al., 2018). In order to overcome these hydrological catastrophic events, the Friend AOC network with the support of the Research Institute for Development (IRD) has written and submitted a GDRI project entitled: "Hydrological Standards in West Africa and Power station for a better design of hydraulic structures",

Hydrological networks quality has been decreasing continuously since decades in the region, despite very huge support from international organization. Thus spatial hydrology development and use are of utmost importance in the region, and the first studies since some years are very encouraging (Bogning et al., 2020). It must be associated with a significant increase of data sharing with large communities of researchers (Dieulin et al., 2019).

In addition, in the lagoon system of the coastal regions, the alternation of high and low water generates sometimes a predominance of the sea current with a strong sediment transit, and sometimes a predominance of the river current with addition and deposition of sediment on the shore. The mechanism of flood transport of sediments from developed basins occurs from upstream to downstream with the deposit of fine particles at the entrance to the floodplains. In addition, in periods of high water, the river current which feeds the lagoon systems increases and exceeds in force the coastal or lagoon current, there is therefore an increase in the solid charge and deposit of coarse sands due to the weakening of the current. For this purpose, there is an elevation of the bottom of the bed on one side and the digging of the other side related to the reinforcement of the current. All of this creates disturbances with the rising tide, causing flocculation and rapid deposition of sediment in the lagoon system, which causes the decline in fishery productivity of water bodies.

The socio-economic and ecological importance of the coastal zones of West and Central Africa in general and lagoon in particular (Adeaga et al., 2017) in a context of environmental crisis leads to a reflection on the functioning, the evolution as well as the management of the use of resources of lagoon spaces. Thus, the animation of this new theme with that of eco-hydrology is necessary to better characterize the hydrological and sedimentary dynamics of the environment in the context of global changes, and sea-level rise, already impacting most coastal areas around the world.

2.6 FRIEND-Water SA

Monitoring of water surfaces and floods by remote sensing Radar is a very important issues for hydrological studies. In the case of Mozambique and Tanzania in the semi-arid rural areas of southern Africa where the availability of water is a major issue for agriculture, the monitoring of surface water is particularly important (André et al., 2016; Mzava et al., 2019), particularly in ungauged watersheds. In addition, large cities, due to the weakness of water facilities, are particularly vulnerable to large-scale floods. Radar images are particularly well suited to the seasonal observation of water surfaces in tropical environments because the Radar signal is freed from cloudiness, which on the contrary constitutes a major limitation of the use of optical images in these environments. In addition, Radar images are very robust for detecting surfaces in water (source, etc.), the signal backscattering being very low (almost zero). On a Radar image, the pixel-waters are those whose values are less than 0.05.

The possibilities offered by the European Space Agency (Copernicus) thanks to its Sentinel 1 sensors make it possible to have at least one image per month. Sentinel 1 (Radar) monthly time series of images can thus be created. Management of hydro agricultural resources in ungauged watersheds: the case of the Kilombero watershed in Tanzania (Johansson and Abdi, 2020). The Kilombero watershed is a territory which potential in terms of water and agricultural resources make it a national issue. Indeed, the Tanzanian government wants to make these vast flood plains the granary of Tanzania through an agricultural irrigation project. At the same time, the Tanzanian government plans to build a large hydroelectric dam, downstream from the Kilombero watershed, which will mobilize almost 66 % of the water resources in the watershed. The management of water in this watershed could therefore arise in terms of competing for access to water between these two projects of national importance. Consideration should be given so that agricultural production in the plain upstream also allows optimal production of hydroelectricity downstream from the watershed. In addition, to better manage water resources, it is important to have a good hydro-environmental and socio-economic database. However, since the 1980s, no hydro-environmental and socio-economic measures have been taken on this 35 000 km^2 watershed. The first step will be to create a database (hydrological, climatic, land use, agricultural, socio-demographic and tourist) in order to better understand the functioning of this territory, which constitutes a national economic issue. In a second step, we will model the environmental dynamics from the database thus created. From an operational perspective, the database will assist in the management of water resources in the Kilombero watershed.

A second research direction which is of major importance is the impact of climate variability and human pressures on surface conditions. A follow-up of the relationships between climatic variability and vegetation must be first analyzed to understand the natural dynamics of the landscape (Brou et al., 2018; Brou, 2019). This study is based on the assumption of an increasing vulnerability of forest ecosystems due to changes in the climate regime recorded during the last thirty years.

To conduct this study, it is important to update the knowledge of climate parameters, in particular the amount of rain and its distribution throughout the year. Equally important is the study of the trend and frequency of climate data for the longest series of observations available on the watershed. The study of the sensitivity of vegetation to climatic variations is carried out by means of a rain/NDVI co-variation analysis. Several parameters will make it possible to follow the interannual seasonal variability of the vegetation: the start and end dates, the duration, the amplitude, the maximum of the vegetative season (Boussougou et al., 2017). To better understand local dynamics, Sentinel 2 images should be used. The comparison of images from several dates will allow to assess, by site, the levels of transformation of landscapes in relation to the degree of soil degradation.

The use of spatial data represents an important skill to strengthen the capacities of students in the monitoring and management of water resources at better spatio-temporal resolutions. In the context of global environmental changes, this training in geomatics is intended, ultimately, to strengthen the capacities of master and doctorate students in hydrology and geography in the field of monitoring water resources through the use of data spatial. This geomatics training is part of a favorable regional and international context marked by the availability of a large quantity of Sentinel, Landsat satellite images in particular. In addition, the development of GPS makes it possible to collect terrain data with more precision. The objective is to allow users in environmental sciences, environmental managers or developers, to acquire the conceptual bases, methods of use and processing of remote sensing data. Applications in the field of water resources and coastal zones must be developed.

2.7 MEDFRIEND-Water

The Mediterranean basin is characterized by limited and unevenly distributed water resources and a considerable development of anthropogenic activities. Hydro-climatic projections predict a decrease in average annual rainfall and runoff in this region of the world. Moreover, water demands have doubled since the 1950s and are expected to increase with the expansion of irrigated areas and urban areas. For the Mediterranean basin, according to climate projections, the temperature will increase by 2 to 4 °C and rainfall will decrease by 4 % to 30 % by 2050. This decrease in precipitation will cause a decrease in flow for all four seasons. As a result of these climate changes, extreme peak flows are expected to increase dramatically. Hydrological droughts will worsen in summer, mainly in the Mediterranean region. The most extreme events are expected to occur in summer, with extreme droughts in the Mediterranean region. The Mediterranean countries are expected to face increasing water shortages. These projections also lead to an increase in the number of days per year when river flows will be below the minimum in the Mediterranean regions. Climate projections show a decrease in groundwater resources in the Mediterranean basin. Overexploitation of groundwater in Mediterranean countries will lead to alarming levels of groundwater and increased pumping costs. Also, soil water stress conditions will drastically reduce agricultural production yields. Water consumption is already higher than water availability and is expected to be critical for global warming. In the countries on the southern shore of the Mediterranean and Spain, the situation of water resources will become more unsustainable in the future, which will have a negative impact on the availability of water resources to ensure sustainable development in these countries.

Thus, the major research topics to deal with in priority in the Mediterranean area are:

- Extreme events or hydrological extremes and water-related disasters (Tramblay and Somot, 2018; Hadour et al., 2020);
- Erosion and transport of solids (Hallouz et al., 2018; Kotti et al., 2018);
- Integrated Water Resources Management (Marignani et al., 2017; Khater et al., 2019);
- Groundwater and global changes (Kuper et al., 2017; Dubois et al., 2018; Erostate et al., 2020);
- Ecohydrology (Amrouni et al., 2019; Ben Moussa et al., 2019);
- Water in agriculture (Venot et al., 2017; Kuper et al., 2018; Reca-Cardena and Lopez-Luque, 2018).

Transversal topics will be of high importance for instance water in the city, education to water questions, participatory science, transition toward a more ecological approach of the water cycle and use. It will be developed through improved connections with all the sectors of the society, and the development of new tools to communicate and disseminate knowledge, mainly web-based.

A high-level challenge is also to improve the co-development of research projects between countries of the Mediterranean and SubSaharan countries, and the Water family of UNESCO is a very good support to this.

2.8 NILE FRIEND-Water

The Water Research Centre, University of Khartoum is a multidisciplinary research and capacity building center established by the University of Khartoum in 2009 with the aim of conducting multi-disciplinary research and development studies addressing the challenges related to water and environmental management in Sudan and the region (Salih, 2018; Basheer et al., 2018). One of the top priority research areas is trans-boundary water management in the Nile basin. Demand for water throughout the basin is constantly growing due to economic development and population growth. Following this rapidly increasing demand, riparian countries have acknowledged the need for joint planning and management. Numerous studies have demonstrated that the river has great potential which, through sustainable and cooperative management and development, could greatly benefit all the basin countries. A research area of specific interest to the Water Research Centre is the Water Energy Food (WEF) Nexus investment and management in the basin. The WEF supply pressures, the growing understanding of the interlink ages between the three resources, and the failure of fragmented management of the different WEF sectors have led to the emergence of such Water-Energy-Food nexus thinking. Although WEF in stressed trans-boundary Basins such as the Nile are subject to high pressure due to competition between riparian countries over the three resources, a systematic understanding of how cooperation between riparian countries impacts the WEF nexus is still lacking (Wheeler et al., 2018; Basheer and Elagib, 2019). Another trans-boundary management challenge is the coordinated operation and management of new water-related infrastructure across the basin's international borders. This is particularly true given that water rights are highly politicized and decisions of exploitation are often guided by ambitious national development plans and complex political dynamics. A good example is the Grand Ethiopian Renaissance Dam (GERD) on the Blue Nile which makes cooperative management between the countries of Ethiopia, Sudan, and Egypt an urgent matter. To enhance confidence building and enhance cooperation in the Nile Basin the Water Research Centre has developed a strategy for training and capacity building in Water Diplomacy and Cooperation (Salih, 2019). A number of workshops have been conducted in which participants from the Nile Basin Countries from relevant disciplines were brought together and trained on trans-boundary water governance and institutions, water laws and conventions, concepts of conflict and conflict resolution mechanisms, negotiations and mediation concepts, role of media in water diplomacy, the role of women in water diplomacy and the role in science in supporting negotiations and water cooperation. An important research area in WRC is the study of hydrological extremes. Sudan, being situated on the eastern edge of the African Sahel, is a well-documented region in facing the problems of climatic change/variability and rainfall fluctuations creating severe floods and drought scenarios (Abdo, 2019). Over the past 50 years, Sudan has encountered a number of high flash floods and droughts that have caused widespread damage on a massive scale to infrastructures, agriculture, health and educational sectors, and loss human and animal lives. Other research areas the Centre is involved in include irrigation management and water use efficiency, water harvesting and management of groundwater recharge, urban water management, echo-hydrology, water quality management and groundwater assessment, development and management (Basheer et al., 2019).

In 2014 the WRC was promoted to a Centre of Excellence in Water Research by the Ministry of Higher Education and scientific Research. In 2017 WRC was selected as one of the AU-NEPAD Centers of excellence in water science and technology in Africa and it is now hosting the secretariat of group of centres in central and Eastern Africa. Over the years, WRC has developed scientific collaboration with many universities, research institutions, specialized networks and programs nationally, regionally and internationally. These include UNESCO IHP, Global G-Wadi (Water and Development Information), African G-Wadi, Arab G-Wadi, International Flood Initiative (IFI), International Sediment Initiative (ISI), World Bank-Nile Basin Initiative NBI/, Eastern Nile Technical Regional Office (ENTRO), UNESCO-IHE, Delft the Netherlands, UNECEF, Cologne University of Science and Technology, Germany, Imperial College of Science and Technology, London, UK, Oxford University, Oxford, UK as well as many Regional Universities including Addis Ababa University, Nairobi University, Kenyatta University, Dar es Salam University, Cairo University, and Ain Shams University. WRC has participated in the two phases of the FRIEND-Water/NILE project and produced a number of papers in the flood frequency, low flow frequency, watershed management and rainfall runoff modeling components.

3 Future FRIEND-Water program developments

3.1 Ecohydrology

Ecohydrology and multiscale studies are a key interest for the future FWP. As more and more well known, the stationarity is dead. As more and more known, the Earth is a non-linear complex system. Recent discoveries on the importance of living things and their environmental feedbacks, on the role of homeostatic processes and breakpoints confirm that global warming is only part of the current ecological crisis which is also manifested by the loss of biodiversity, the loss of soil fertility, the water pollution but also the drastic increase in diseases that affect the entire living world. The current COVID-19 pandemic is just one more dramatic example! The water cycle is at the center of all of these interlocking systems (Orange and Breil, 2018). Also it becomes obvious to hydrologists that it is essential to cross the scales of spatial studies, from large to small watersheds, and vice versa. Understanding the why of a hydroclimatic rupture and its consequences will therefore require interdisciplinary perspectives. It then appears that the study methods promoted by the ecohydrology must also belong to the research corpus of hydrologists.

Ecohydrology is among the key words of the future FWP directions. Then, related to the goal of the IHP9 to achieve Sustainable Water Management (SWM), the FWP will get some interests to incorporate ecohydrology among its key tools. In this context, significant challenges of scientific fields appear to be addressed:

- developing dynamic hydrological models based on changing environmental conditions (e.g. soil-water dynamics) (Hang et al., 2018),
- climate change and sustainable ecological transition (e.g. urban-rural association),
- nature-based solutions as a rising awareness towards environment (e.g. REUSE-Soil-Biodiversity nexus).
- rainwater management in urban areas (e.g. flashfloods).

Ecohydrology is among the key words of the future FWP directions. Then, related to the goal of the IHP9 to achieve Sustainable Water Management (SWM), the FWP will get some interests to incorporate the ecohydrology among its key tools. Active partnership between EcoHydrology theme and the future FWP is scheduled through the opening of demosites. Since 2011 UNESCO-IHP promotes the establishment of various demonstration sites around the world to apply Ecohydrology solutions in watersheds at all scales. Each demonstration site aims to apply Ecohydrology principles and solutions to solve both social and environmental issues (Riaux et al., 2020). These solution-oriented projects and methodologies are shown in a web-platform (http://ecohydrology-ihp.org/demosites/#demosites, last access: 6 January 2021) to disseminate the acquired information to UNESCO's Member States and the general public. This information is dedicated to allow to applicate SWM in other locations.

Then, it is scheduled to propose the opening of 2 joint EH-FRIEND-Water demosites, one in Tunisia dealing with water quality remediation in highly polluted environments (around Bizerte), in relationship with surface and groundwater hydrology; another one in West Africa in Senegal (Roupsard et al., 2018; Faye et al., 2020), called Dakar'2030, dealing with the combination of rainwaters and wastewaters management in the Region of Dakar to water and re-green soils against soils and sea pollution.

3.2 Databases

Regarding the accessibility of data and open access data, the accent should be put on how to connect directly to existing web database, as well as to develop API to deal with connectivity issues. In some countries, even if there is open data, challenges refer to the infrastructure, as communication networks focus on mobile phone; the facility to upload data is not always there. It is necessary to address this, incl. the question of communication software. The world water data centres federated within the GTN-H (inter alia the Global Runoff Data Centre (GRDC), the GEMS/Water Data Centre for Water Quality (GWDC) the Global Groundwater Monitoring Network (GGMN) operated by IGRAC) mostly operates under auspices of UN organization such as the WMO, UNEP or UNESCO and can guarantee long-term operational services to the scientific communities and local stakeholders worldwide since decades. These data centres should be recognized and considered for data rescue due to their long-term provision of homogenized and quality-assured in-situ observations. Vice versa FRIEND-Water communities could support the data centres in raising awareness of difficulties in data acquisition as well as lags in open data policies and technical requirements (e.g. APIs, provision of standards). Research is also necessary for link in-situ with satellite-born observations. Also, here FRIEND-Water could play a crucial role. Third action involves support to projects or initiatives with focus on surface water, river. Groundwater must not be forgotten, as well as interface areas like wetlands and coastal lagoons, as those play important role in the emission of GHGs. In addition, scaling should be addressed, as well spatial analysis of water data and other water-related data (uses, energy, virtual vs real, raw vs. reclaimed ...), at various scales, appropriate to the targets of communication. Implementation of experimental basins or watersheds is an interesting concept; in order to enable benchmarking, and use a methodology that can tackle multidisciplinary factors and approaches. Additionally, not only environmental settings, geopolitical settings, historical background, as well as demography and socio economic situations, but also water stress (considering surface and groundwater) should clearly be determined. Exchange of experiments, of data treatment, and also solutions could be organized through UNESCO Water family. Sharing data must become an open space in the entire UN Water family.

4 Conclusion: toward the implementation of the IHP IX program

The FWP is the most well know program from the IHP across the world. Since 36 years its success comes as well from its very large field of interest covering all topics of hydrology, than from its capacity to gather lot of people to share data, methods, tools and results in local to international events supported by the UNESCO Water Division. The FWP has developed a close relationship with the International Association of Hydrological Sciences-IAHS, which gives a scientific background to the program, and supports the publication of proceedings for each global conference. The FWP has changed over time, to follow the global directions given by the member States in each phase of the IHP program. The next IHP IX phase program will foster actions towards five main objectives: Scientific research and innovation; Water Education in the Fourth Industrial Revolution including sustainability; Bridging the data-knowledge gap; Inclusive water management under conditions of global change; Water Governance based on science for mitigation, adaptation and resilience. FWP, as the only transversal Water program of the IHP, will be able to support these five main objectives, both by launching regional initiatives towards a better integration of the multidimensional aspects of the water sciences including developed links with education, sustainability, governance, and so on, and by keeping a tight link with the IAHS scientific most recent outcomes and initiatives, as for instance the recent FRIEND-Water African Initiative (Mahe et al., 2020) to bring regional inputs to the 23 UPH – Unsolved Problems in Hydrology – (Bloschl et al., 2019).

Recently the FWP has been given the support of a Secretariat based in the UNESCO category 2 Center of Montpellier ICIREWARD. The next global FWP conference is scheduled in Dakar, Senegal in March 2023.

Data availability. No data sets were used in this article.

Author contributions. All autors contributed equally to this study, each one bringing the text for its region and or topic. GM prepared the frame of the paper and supervised the edition process of the paper.

Competing interests. The contact author has declared that neither they nor their co-authors have any competing interests.

Disclaimer. Publisher's note: Copernicus Publications remains neutral with regard to jurisdictional claims in published maps and institutional affiliations.

Special issue statement. This article is part of the special issue "Hydrology of Large River Basins of Africa". It is a result of the 4th International Conference on the "Hydrology of the Great Rivers of Africa", Cotonou, Benin, 13–20 November 2021.

References

Abdo, G.: Water management challenges in Sudan and how to meet them, Proceedings of the 7th Arab Forum for Scientific Research and Sustainable Development Amman – Jordan, 17–19 December, 2019.

Adeaga, O., Mahé, G., Dieulin, C., Elbaz-Poulichet, F., Rouché, N., Seidel, J. L., and Servat, E.: Quality of water resources in the Niger basin and in the region of Lagos (Nigeria), B. Geogr., 13, 51–60, https://doi.org/10.1515/bgeo-2017-0013, 2017.

Alsdorf, D., Beighley, E., Laraque, A., Lee, H., Tshimanga, R., O'Loughlin, F., Mahe, G., Dinga, B., Moukandi, G., and Spencer, R. G. M.: Opportunities for Hydrologic Research in the Congo Basin, Rev. Geophys., 54, 378–409, https://doi.org/10.1002/2016RG000517, 2016.

Amoussou, E.: Variabilité pluviométrique et dynamique hydrosédimentaire du bassin versant du complexe fluvio-lagunaire Mono-Ahémé-Couffo (Afrique de l'Ouest), Thèse de Doctorat unique, Université de Bourgogne, Dijon, France, CRCCNRS UMR5210, 313 pp., 2010.

Amoussou, E., Awoye, H., Totin Vodounon, H. S., Obahoundje, S., Camberlin, P., Diedhiou, A., Kouadio, K., Mahe, G., Houndenou, C., and Boko, M.: Climate and extreme rainfall events in the Mono river basin (West Africa): investigating future changes with regional climate models, Water, 12, 833, https://doi.org/10.3390/w12030833, 2020.

Amrouni, O., Sánchez, A., Khélifi, N., BenMoussa, T., Chiarella, D., Mahé, G., Abdeljaouad, S., and McLaren, P.: Sensitivity assessment of the deltaic coast of Medjerda based on fine-grained sediment dynamics, Gulf of Tunis (Western Mediterranean), J. Coast. Conserv., 23, 571–587, https://doi.org/10.1007/s11852-019-00687-x, 2019.

André, M., Brou, Y. T., and Bigot, S.: Spatio-temporal variability of rainfall and impacts on vegetation in Reunion Island, 32nd Annual Conference of the South African Society for Atmospheric Sciences (SASAS), At Cape Town (South Africa), 30 October–2 November 2016, Cape Town, South Africa, 2016.

Basheer, M. and Elagib, N. A.: Temporal analysis of water-energy nexus indicators for hydropower generation and water pumping in the Lower Blue Nile Basin, J. Hydrol., 578, 124085, https://doi.org/10.1016/j.jhydrol.2019.124085, 2019.

Basheer, M., Wheeler, K., Ribbe, L., Abdo, G., and Zagona, E.: Quantifying and evaluating the impacts of cooperation in transboundary river basins on the Water-Energy-Food nexus: The Blue Nile Basin, Sci. Total Environ., 630, 1309–1323, 2018.

Basheer, M., Sulieman, R., and Ribbe, L.: Exploring management approaches for water and energy in the data-scarce Tekeze-Atbara Basin under hydrologic uncertainty, Int. J. Wat. Res. Dev., 37, 182–207, https://doi.org/10.1080/07900627.2019.1591941, 2019.

Beighley, E., Tshimanga, R. M., and N'Kaya, M. G.: Establishing science campaigns in sub-Saharan Africa, Eos, 100, https://doi.org/10.1029/2019EO117249, 2019.

Ben Moussa, T., Amrouni, O., Hzami, A., Dezileau, L., Mahe, G., Condomines, M., and Saadi, A.: Progradation and retrogradation of the Medjerda delta during the 20th century (Tunisia, Western Mediterranean), Comptes Rendus Geosciences, 341, 340–350, https://doi.org/10.1016/J.Crte.2018.10.004, 2019.

Blauhut, V., Gudmundsson, L., and Stahl, K.: Towards pan-european drought risk maps: quantifying the link between drought indices and reported drought impacts, Environ. Res. Lett., 10, 014008, https://doi.org/10.1088/1748-9326/10/1/014008, 2015.

Blöschl, G., Bierkens, M. F. P., Chambel, A., Cudennec, C., Destouni, G., Fiori, A., Kirchner, J. W., McDonnell, J. J., Savenije, H. H. G., Sivapalan, M., Stumpp, C., Toth, E., Volpi, E., Carr, G., Lupton, C., Salinas, J., Széles, B., Viglione, A., Aksoy, H., Allen, S. T., Amin, A., Andréassian, V., Arheimer, B., Aryal, S. K., Baker, V., Bardsley, E., Barendrecht, M. H., Bartosova, A., Batelaan, O., Berghuijs, W. R., Beven, K., Blume, T., Bogaard, T., Borges de Amorim, P., Böttcher, M. E., Boulet, G., Breinl, K., Brilly, M., Brocca, L., Buytaert, W., Castellarin, A., Castelletti, A., Chen, X., Chen, Y., Chen, Y., Chifflard, P., Claps, P., Clark, M. P., Collins, A. L., Croke, B., Dathe, A.,

David, P. C., de Barros, F. P. J., de Rooij, G., Di Baldassarre, G., Driscoll, J. M., Duethmann, D., Dwivedi, R., Eris, E., Farmer, W. H., Feiccabrino, J., Ferguson, G., Ferrari, E., Ferraris, S., Fersch, B., Finger, D., Foglia, L., Fowler, K., Gartsman, B., Gascoin, S., Gaume, E., Gelfan, A., Geris, J., Gharari, S., Gleeson, T., Glendell, M., Gonzalez Bevacqua, A., González-Dugo, M. P., Grimaldi, S., Gupta, A. B., Guse, B., Han, D., Hannah, D., Harpold, A., Haun, S., Heal, K., Helfricht, K., Herrnegger, M., Hipsey, M., Hlavácikovà, H., Hohmann, C., Holko, L., Hopkinson, C., Hrachowitz, M., Illangasekare, T. H., Inam, A., Innocente, C., Istanbulluoglu, E., Jarihani, B., Kalantari, Z., Kalvans, A., Khanal, S., Khatami, S., Kiesel, J., Kirkby, M., Knoben, W., Kochanek, K., Kohnová, S., Kolechkina, A., Krause, S., Kreamer, D., Kreibich, H., Kunstmann, H., Lange, H., Liberato, M. L. R., Lindquist, E., Link, T., Liu, J., Loucks, D. P., Luce, C., Mahé, G., Makarieva, O., Malard, J., Mashtayeva, S., Maskey, S., Mas-Pla, J., Mavrova-Guirguinova, M., Mazzoleni, M., Mernild, S., Misstear, B. D., Montanari, A., Müller-Thomy, H., Nabizadeh, A., Nardi, F., Neale, C., Nesterova, N., Nurtaev, B., Odongo, V. O., Panda, S., Pande, S., Pang, Z., Papacharalampous, G., Perrin, C., Pfister, L., Pimentel, R., Polo, M. J., Post, D., Prieto Sierra, C., Ramos, M.-H., Renner, M., Reynolds, J. E., Ridolfi, E., Rigon, R., Riva, M., Robertson, D. E., Rosso, R., Roy, T., Sá, J. H. M., Salvadori, G., Sandells, M., Schaefli, B., Schumann, A., Scolobig, A., Seibert, J., Servat, E., Shafiei, M., Sharma, A., Sidibe, M., Sidle, R. C., Skaugen, T., Smith, H., Spiessl, S. M., Stein, L., Steinsland, I., Strasser, U., Su, B., Szolgay, J., Tarboton, D., Tauro, F., Thirel, G., Tian, F., Tong, R., Tussupova, K., Tyralis, H., Uijlenhoet, R., van Beek, R., van der Ent, R. J., van der Ploeg, M., Van Loon, A. F., van Meerveld, I., van Nooijen, R., van Oel, P. R., Vidal, J.-P., von Freyberg, J., Vorogushyn, S., Wachniew, P., Wade, A. J., Ward, P., Westerberg, I. K., White, C., Wood, E. F., Woods, R., Xu, Z., Yilmaz, K. K., and Zhang, Y.: Twenty-Three Unsolved Problems In Hydrology (UPH) – A Community Perspective, Hydrol. Sci. J., 64, 1141–1158, https://doi.org/10.1080/02626667.2019.1620507, 2019.

Bogning, S., Frappart, F., Paris, A., Blarel, F., Niño, F., Saux Picart, S., Lanet, P., Seyler, F., Mahé, G., Onguene, R., Bricquet, J. P., Etame, J., Paiz, M. C., and Braun, J. J.: Hydro-climatology study of the Ogooué River basin using hydrological modeling and satellite altimetry, Adv. Space Res., 68, 672–690, https://doi.org/10.1016/j.asr.2020.03.045, 2020.

Boussougou, G., Brou, Y. T., and Mohamed, I.: Changements de la couverture forestière dans l'île d'Anjouan entre 1995 et 2014, Proceedings of the Spatial Analysis and GEOmatics conference, SAGEO 2015, 194–207, 2017.

Brou, Y. T.: Coviability of the social and ecological systems in Réunion Island's National Park: Climate variability, wildfires, and the vulnerability of biodiversity, in: Co-viability of Social and Ecological Systems, reconnect Human with Biosphere, edited by: Barière, O. and Behnassi, M., Edition Springer, 2, 187–202, https://doi.org/10.1007/978-3-319-78497-7, 2019.

Brou, Y. T., Vololonirainy, R., Boussougou, G., Raharinjanahary, R., Bigot, S., Dumas, S., Ramboarison, R., Razanaka, S., Parany, L., Rakotoarijaona, M., Herve, D., and Rakotoarijaona, J. R.: Pressions agro-démographiques et état de l'occupation des sols dans l'espace protégé d'Ankarafantsika: bilan cartographique et paysager, Carnet de Recherche de l'Océan Indien, Université no. 2, 1–13, 2018.

Caillouet, L., Vidal, J.-P., Sauquet, E., Graff, B., and Soubeyroux, J.-M.: SCOPE Climate: a 142-year daily high-resolution ensemble meteorological reconstruction dataset over France, Earth Syst. Sci. Data, 11, 241–260, https://doi.org/10.5194/essd-11-241-2019, 2019.

Carr, A. B., Trigg, M. A., Tshimanga, R. M., Borman, J. D., and Smith, M. W.: Greater Water Surface Variability Revealed by New Congo River Field Data: Implications for Satellite Altimetry Measurements of Large Rivers, Geophys. Res. Lett., 46, 8093–8101, https://doi.org/10.1029/2019GL083720, 2019.

Carretier, S., Tolorza, V., Regard, V., Aguilar, V., Bermudez, M.A., Martinod., J., Guyot, J. L., Herail, G., and Riquelme, R.: Review of erosion dynamics along the major N-S climatic gradient in Chile and perspectives, Geomorphology, 300, 45–68, https://doi.org/10.1016/j.geomorph.2017.10.016, 2018.

Descroix, L., Guichard, F., Grippa, M., Lambert, L., Panthou, J., Mahe, G., Gal, L., Dardel, C., Quantin, G., Kergoat, L., Bouaita, Y., Hiernaux, P., Vischel, T., Pellarin, T., Faty, B., Wilcox, C., Malam Abdou, M., Mammadou, I., Bandervaere, J. P., Diongue-Niang, A., Ndiaye, O., Sane, Y., Dacosta, H., Gosset, M., Casse, C., Sultan, B., Barry, A., Amogu, O., Nka Nnomo, B., Barry, A., and Paturel, J. E.: Surface hydrology evolution in the Sahelo-Sudanian stripe: An updated review, Water, 10, 748, https://doi.org/10.3390/w10060748, 2018.

Devers, A., Vidal, J.-P., Lauvernet, C., Graff, B., and Vannier, O.: A framework for high-resolution meteorological surface reanalysis through offline data assimilation in an ensemble of downscaled reconstructions, Q. J. Roy. Meteor. Soc., 146, 153–173, https://doi.org/10.1002/qj.3663, 2020.

Dieppois, B., Durand, A., Fournier, M., and Massei, N.: Links between multidecadal and interdecadal climatic oscillations in the North Atlantic and regional climate variability of northern France and England since the 17th century, J. Geophys. Res.-Atmos., 118, 4359–4372, 2013.

Dieppois, B., Pohl, B., Crétat, J., Eden, J., Sidibe, M., New, M., Rouault, M., and Lawler, D.: Southern African summer-rainfall variability, and its teleconnections, on interannual to interdecadal timescales in CMIP5 models, Clim. Dynam., 53, 3505–3527, https://doi.org/10.1007/s00382-019-04720-5, 2019.

Dieulin, C., Mahe, G., Paturel, J. E., Ejjiyar, S., Tramblay, Y., Rouche, N., and El Mansouri, B.: A new 60-year 1940–1999 monthly gridded rainfall data set for Africa, Water, 11, 387, https://doi.org/10.3390/w11020387, 2019.

Dubois, C., Stoffner, F., Kalia, A. C., Sandner, M., Labiadh, M., and Mimouni, M.: COPERNICUS SENTINEL-2 data for the determination of groundwater withdrawal in the Maghreb region, ISPRS Ann. Photogramm. Remote Sens. Spatial Inf. Sci., IV-1, 37–44, https://doi.org/10.5194/isprs-annals-IV-1-37-2018, 2018.

Erostate, M., Huneau, F., Garel, E., Ghiotti, S., Vystavna, Y., Garrido, M., and Pasqualini, V.: Groundwater dependent ecosystems in coastal Mediterranean regions: Characterization, challenges and management for their protection, Water Res., 172, 115461, https://doi.org/10.1016/j.watres.2019.115461, 2020.

Faye, W., Orange, D., Roupsard, O., Do, F. C., Jourdan, C., Fall, A. N., Kane, A., N'diénor, M., Angulo-Jaramillo, R., Lassabatère, L., Di Prima, S., Sauvage, S., and Sanchez-Perez, J.: Ecohydrology and agroforestry for water use efficiency on sahelian agrosystem: the Sine River Basin, In: 4th International conference on Hydrology of African Large River Basins, FRIEND-AOC, Cotonou, Bénin, 6 pp., 24–28 November 2020.

Ferreira Araújo, D., Peres, L. G. M., Yepez, S., Mulholland, D. S., Machado, W., Tonha, M., and Garnier, J.: Assessing man-induced environmental changes in the Sepetiba Bay (Southeastern Brazil) with geochemical and satellite data, Comptes Rendus Geoscience, 349, 290–298, 2017.

Gallay, M., Martinez, J. M., Mora, A., Castellano, B., Yepez, S., Cochonneau, G., Alfonso, J. A., Carrera, J. M., Lopez, J. L., and Laraque, A.: Assessing Orinoco river sediment discharge trend using MODIS satellite images, J. South Am. Earth Sci., 91, 320–331, https://doi.org/10.1016/j.jsames.2019.01.010, 2019.

Hadour, A., Meddi, M., and Mahe, G.: Impact of climate change on the hydrology of North-Western Algeria, J. Hydrol. Reg. Stud., 28, https://doi.org/10.1016/j.ejrh.2020.100671, 2020.

Hallouz, F., Meddi, M., Mahe, G., Toumi, S., and Ali Rahamni, S. E.: Erosion, sedimentation and climate change on the Wadi Mina at the Sidi M'hamed Ben Aouda dam, Algeria, Water, 10, 895, https://doi.org/10.3390/w10070895, 2018.

Hang, H. T. T., Thuy, D. T., Kien, N. T., Le, T. P. Q., Minh, L. T. N., Duc, T. A., Anh, L. H., Cuong, H. T., Kim, D. D., Nemery, J., Orange, D., and Klein, J.: Impact of anthropogenic activities on water quality and plankton communities in the Day River (Red River Delta, Vietnam), Environ. Monit. Assess., 190, 67, https://doi.org/10.1007/s10661-017-6435-z, 2018.

Hartmann, A., Gleeson, T., Wada, Y., and Wagener, T.: Enhanced groundwater recharge rates and altered recharge sensitivity to climate variability through subsurface heterogeneity, P. Natl. Acad. Sci. USA, 114, 2842–2847, https://doi.org/10.1073/pnas.1614941114, 2017.

Hiraga, Y., Kazama, S., Ekkawatpanit, C., and Touge, Y.: Impact of reclamation on the environment of the lower Mekong river basin, J. Hydrol., 18, 143–155, 2018.

Hoa, N. T. Q., Anh, H. Q., Tue, N. M., Trung, N. T., Da, L. N., Van Quy, T., Huong, N. T. A., Suzuki, G., Takahashi, S., Tanabe, S., Thuy, P. C., Dau, P. T., Viet, P. H., and Thuyen, L. H.: Soil and sediment contamination by unsubstituted and methylated polycyclic aromatic hydrocarbons in an informal e-waste recycling area, northern Vietnam: occurrence, source apportionment, and risk assessment, Sci. Total Environ., 709, 135852, https://doi.org/10.1016/j.scitotenv.2019.135852, 2020.

IPCC: Climate Change 2013: The Physical Science Basis. Contribution of Working Group I to the Fifth Assessment Report of the Intergovernmental Panel on Climate Change, edited by: Stocker, T. F., Qin, D., Plattner, G.-K., Tignor, M., Allen, S. K., Boschung, J., Nauels, A., Xia, Y., Bex, V., and Midgley, P. M., Cambridge University Press, Cambridge, United Kingdom and New York, NY, USA, 1535 pp., https://doi.org/10.1017/CBO9781107415324, 2013.

Johansson, E. L. and Abdi, A. M.: Mapping and quantifying perceptions of environmental change in Kilombero Valley, Tanzania, Ambio, 49, 557–568, 2020.

Khater, C., Fenianos, J., Brouillet, D., and Arnaud, N.: Bridging Ecological Rehabilitation and Socio Cognitive Psychology, International Journal of Environmental Sciences & Natural Resources, Juniper Publishers Inc., 17, 2, 39–42, 2019.

Kingston, D. G., Massei, N., Dieppois, B., Hannah, D. M., Hartman, A., Lavers, D. A., and Vidal, J. P.: Moving beyond the catchment scale: Value and opportunities in large-scale hydrology to understand our changing world, Hydrol. Process., 34, 2292–2298, https://doi.org/10.1002/hyp.13729, 2020.

Kondolf, G., Schmitt, M., Rafael, J. P., Carling, P., Darby, S., Arias, M., Bizzi, S., Castelletti, A., Cochrane, T. A., Gibson, S., Kummu, M., Oeurng, C., Rubin, Z., and Wild, T.: Changing sediment budget of the Mekong: Cumulative threats and management strategies for a large river basin, Sci. Total Environ., 1, 625, 114–134, 2018.

Kositgittiwong, D., Ekkawatpanit, C., Chiawyonsin, S., Petpongpan, C., and Ekkphisutsuntorn, P.: Development of wetland water management in Bueng Takreng, Thailand, Wetland Research, 7, 25–31, 2017.

Kotti, F., Dezileau, L., Mahe, G., Habaieb, H., Benabdallah, S., Bentkaya, M., Calvez, R., and Dieulin, C.: The impact of dams and climate on the evolution of the sediment loads to the sea by the Medjerda River using a paleo-hydrological approach, J. Afr. Earth Sci., 142, 226–233, https://doi.org/10.1016/j.jafrearsci.2017.10.003, 2018.

Kuper, M., Leduc, C., Massuel, S., and Bouarfa, S.: Topical collection: Groundwater-based agriculture in the Mediterranean, Hydrogeol. J., 25, 1525–1528, https://doi.org/10.1007/s10040-017-1598-9, 2017.

Kuper, M., Naouri, M., and Hartani, T.: Agriculture in motion: Intertwined geographical and socio-professional mobilities in North Africa's groundwater economy, Meltem, 4, 33–45, https://doi.org/10.32325/iaad.2018.31, 2018.

Laizé, C. and Hannah, D. M.: Modification of climate–river flow associations by basin properties, J. Hydrol., 389, 186–204, https://doi.org/10.1016/j.jhydrol.2010.05.048, 2010.

Laqui, W., Zubieta, R., Rau, P., Mejia, A., Lavado, W., and Ingol, E.: Can artificial neural networks estimate potential evapotranspiration in Peruvian highlands?, Model. Earth Syst. Environ., 5, 1911–1924, 2018.

Laraque A., Lopez, J. L., Yepez, S., and Georgescu, P.: Water and Sediment Budget of Casiquiare Channel Linking Orinoco and Amazon Catchments, Venezuela, Water, 11, 2068, https://doi.org/10.3390/w11102068, 2019.

Lavers, D.A., Zsoter, E., Richardson, D. S., and Pappenberger, F.: An assessment of the ECMWF extreme forecast index for water vapor transport during boreal winter, Weather Forecast., 32, 1667–1674, https://doi.org/10.1175/WAF-D-17-0073.1, 2017.

Le, T. P. Q., Dao, V. N., Rochelle-Newall, E., Garnier, J., Lu, X. X., Billen, G., Duong, T. T., Ho, C. T., Etcheber, H., Nguyen, T. M. H., Nguyen, T. B. N., Nguyen, B. N., Le, N. D., and Long, P. Q.: Total organic carbon fluxes of the Red River system (Vietnam), Earth Surf. Proc. Landf., 42, 1329–1341, https://doi.org/10.1002/esp.4107, 2017.

Le Gros, C., Sauquet, E., Lang, M., Achard, A. L., Leblois, E., and Biton, B.: The hydrological yearbooks published by the Société Hydrotechnique de France: A valuable source of information on hydrology in France, La Houille Blanche, 4, 66–77, https://doi.org/10.1051/lhb/20150048, 2015.

Li, D. F., Lu, X. X., Yang, X., Chen, L., and Lin, L.: Sediment load responses to climate variation and cascade reservoirs in the Yangtze River: A case study of the Jinsha River, Geomorphology, 322, 41–52, 2018.

Mahe, G., Ogilvie, A., Servat, E., and Ardoin-Bardin, S.: 23 UPH in Africa, gathering the contributions of the African hydrologists to the 23 Unsolved Problems in Hydrology, of the IAHS, available at: https://www.researchgate.net/project/23-UPH-in-Africa (last access: 6 January 2021), 2020.

Marignani, M., Bruschi, D., Astiaso Garcia, D., Frondoni, R., Carli, E., Pinna, M. S., Cumo, F., Gugliermetti, F., Saatkamp, A., Doxa, A., Queller, E. M., Chaieb, M., Bou Dagher-Kharrat, M., El Zein, R., El Jeitani, S., Khater, C., Mansour, S., Al-Shami, A., Harik, G., Alameddine, I., El-Fadel, M., and Blasi, C.: Identification and prioritization of areas with high environmental risk in Mediterranean coastal areas: a flexible approach, Sci. Total Environ., 590–591, 566–578, https://doi.org/10.1016/j.scitotenv.2017.02.221, 2017.

Massei, N., Kingston, D. G., Hannah, D. M., Vidal, J. P., Dieppois, B., Fossa, M., Hartmann, A., Lavers, D. A., and Laignel, B.: Understanding and predicting large-scale hydrological variability in a changing environment, 8Th Global FRIEND Water conference "Hydrological Processes and Water Security in a Changing World", Beijing, China, 6–9 November 2018.

Moquet, J. S., Guyot, J. L., Morera, S., Crave, A., Rau, P., Vauchel, P., Lagane, C., Sondag, F., Lavado, C. W., Pombosa, R., and Martinez, J. M.: Temporal variability and annual budget of inorganic dissolved matter in Andean Pacific Rivers located along a climate gradient from northern Ecuador to southern Peru, C.R. Geo. 350, 76–87, https://doi.org/10.1016/j.crte.2017.11.002, 2018.

Mushi, C. A., Ndomba, P. M., Trigg, M. A., Tshimanga, R. M., and Mtalo, F.: Assessment of basin-scale soil erosion within the Congo River Basin: A review, Catena, 178, 64–76, https://doi.org/10.1016/j.catena.2019.02.030, 2019.

Mzava, P., Nobert, J., and Valimba, P.: Land Cover Change Detection in the Urban Catchments of Dar es Salaam, Tanzania using Remote Sensing and GIS Techniques, Tanzania J. Sci., 45, 315–329, 2019.

Nardini, A., Yepez, S., Zuniga, L., Gualtieri, C., and Bejarno, M. D.: A Computer Aided Approach for River Styles – Inspired Characterization of Large Basins: The Magdalena River (Colombia), Water, 12, 1147, https://doi.org/10.3390/w12041147, 2020.

Nguyen, T. X., Nguyen, B. T., Tran, H. T. T., Mai, H., Duong, T. T., and Bach, Q. V.: Seasonal, spatial variation, and potential sources of organochlorine Pesticides in Water and Sediment in the Lower Reaches of the Dong Nai River System in Vietnam, Archives of Environmental Contamination and Toxicology, 77, 514–526, 2019.

Orange, D. and Breil, P.: Integrated watershed management including eco-hydrology and socio-hydrology: Ecological and societal based approaches for watershed management, Introduction text of Session 6 for: Hydrological Processes and Water Security in a Changing World, FRIEND-Water 2018: 8th Global FRIEND-Water Conference, Beijing, China, 6–9 November 2018, 1 pp., 2018.

Parkes, B., Cronin, J., Dessens, O., and Sultan, B.: Climate change in Africa: costs of mitigating heat stress, Clim. Change, 154, 461–476, 2019.

Prudhomme, C., Giuntoli, I., Robinson, E. L., Clark, D. B., Arnell, N. W., Dankers, R., Fekete, B. M., Franssen, W., Gerten, D., Gosling, S. N., Hagemann, S., Hannah, D. M., Kim, H., Masaki, Y., Satoh, Y., Stacke, T., Wada, Y., and Wisser, D.: Hydrological droughts in the 21st century, hotspots and uncertainties from a global multimodel ensemble experiment, P. Natl. Acad. Sci., 111, 3262–3267, https://doi.org/10.1073/pnas.1222473110, 2014.

Rau, P., Bourrel, L., Labat, D., Frappart, F., Ruelland, D., Lavado, W., Dewitte, B., and Felipe, O.: Hydroclimatic change disparity of Peruvian Pacific drainage catchments, Theor. Appl. Climatol., 134, 139–153, 2018.

Reca-Cardena, J. and Lopez-Luque, R.: Chapter 9 – Design Principles of Photovoltaic Irrigation Systems, Advances in Renewable Energies and Power Technologies, Solar and Wind Energies, 1, 295–333, https://doi.org/10.1016/B978-0-12-812959-3.00009-5, 2018.

Renard, B. and Thyer, M.: Revealing hidden climate indices from the occurrence of hydrologic extremes, Water Resour. Res., 55, 7662–7768, https://doi.org/10.1029/2019WR024951, 2019.

Riaux, J., Orange, D., Ogilvie, A., Leduc, C., Mohamed, A. S., Ali, H. M., Sidatt, Z. A., and Wesselink, A.: Une démarche sociohydrologique pour renouveler les savoirs sur le fleuve Sénégal, Colloque International PATEO/UASZ : Vulnérabilité des sociétés et des milieux côtiers et estuariens d'Afrique de l'Ouest, UASZ, Ziguinchor, Sénégal, 19–22 November 2019, Coll. Livres UASZ, 6 pp., 2020.

Roupsard, O., Jourdan, C., Cournac, L., Ndour, B. Y., Tall, L., Chapuis-Lardy, L., Clermont-Dauphin, C., Orange, D., Do, F., Kergoat, L., Le Maire, G., Van der Meersche, K., Timouk, F., Grippa, M., Rocheteau, A., Duthoit, M., Chotte, J. L., and Laclau, J. P.: "Faidherbia-Flux", a new highly instrumented collaborative Observatory in a semi-arid agro-sylvo-pastoral system of Western Africa (Niakhar-Senegal), Int. Conf. ANAEE, session: challenges and opportunities of long-term ecosystem experiments in forests, Functional Ecology, Nancy, 10–13 December 2018, 1 pp., 2018.

Salih, A.: Water Resources in Sudan, Book, Water Research Center, Univ. of Khartoum, Sudan, 2018.

Salih, A.: Coping with water scarcity in the Arab Region, Proceedings of the 7th Arab Forum for Scientific Research and Sustainable Development Amman – Jordan, 17–19 December 2019.

Sidibe, M., Dieppois, B., Mahe, G., Paturel, J. E., Amoussou, E., Anifowose, B., and Lawler, D.: Trend And Variability In A New, Reconstructed Streamflow Dataset For West And Central Africa, And Climatic Interactions, 1950–2005, J. Hydrol., 561, 478–493, https://doi.org/10.1016/j.jhydrol.2018.04.024, 2018.

Sidibe, M., Dieppois, B., Eden, J., Mahé, G., Paturel, J. E., Amoussou, E. Anifowose, B., and Lawler, D.: Interannual to multi-decadal streamflow variability in west and Central Africa: Interactions with catchment properties and large-scale climate variability, Glob. Planet. Change, 177, 141–156, https://doi.org/10.1016/j.gloplacha.2019.04.003, 2019.

Sidibe, M., Dieppois, B., Eden, J., Mahe, G., Paturel, J. E., Amoussou, E., Anifowose, B., Van de Wiel, M., and Lawler, D.: Impacts of near-term climate change on hydrological systems in West and Central Africa, Clim. Dynam., 54, 2041–2070, https://doi.org/10.1007/s00382-019-05102-7, 2020.

Stahl, K., Tallaksen, L. M., Hannaford, J., and van Lanen, H. A. J.: Filling the white space on maps of European runoff trends: estimates from a multi-model ensemble, Hydrol. Earth Syst. Sci., 16, 2035–2047, https://doi.org/10.5194/hess-16-2035-2012, 2012.

Stahl, K., Kohn, I., Blauhut, V., Urquijo, J., De Stefano, L., Acácio, V., Dias, S., Stagge, J. H., Tallaksen, L. M., Kampragou, E., Van Loon, A. F., Barker, L. J., Melsen, L. A., Bifulco, C., Musolino, D., de Carli, A., Massarutto, A., Assimacopoulos, D., and Van Lanen, H. A. J.: Impacts of European drought events: insights from an international database of text-based reports, Nat. Hazards Earth Syst. Sci., 16, 801–819, https://doi.org/10.5194/nhess-16-801-2016, 2016.

Stoelzle, M., Stahl, K., Morhard, A., and Weiler, M.: Streamflow sensitivity to drought scenarios in catchments with different geology, Geophys. Res. Lett., 41, 6174–6183, https://doi.org/10.1002/2014GL061344, 2014.

Sutanto, S. J., van der Weert, M., Wanders, N., Blauhut, V., and Van Lanen, H. A. J.: Moving from drought hazard to impact forecasts, Nat. Commun., 10, 4945, https://doi.org/10.1038/s41467-019-12840-z, 2019a.

Sutanto, S. J., Vitolo, C., Di Napoli, C., D'Andrea, M., and Van Lanen, H. A. J.: Heatwaves, droughts, and fires: exploring compound and cascading dry hazards at the pan-European scale, Environ. Int., 134, 105276, https://doi.org/10.1016/j.envint.2019.105276, 2019b.

Teuling, G. A. J., Tallaksen, L. M., Uijlenhoet, R., Hannah, D. M., Sheffield, J., Svoboda, M., Verbeiren, B., Wagener, T., Rangecroft, S., Wanders, N., and Van Lanen, H. A. J.: Drought in the Anthropocene, Nat. Geosci., 9, 89–91, https://doi.org/10.1038/ngeo2646, 2016.

Tramblay, Y. and Somot, S.: Future evolution of extreme precipitation in the Mediterranean, Clim. Change, 151, 289–302, 2018.

Trigg, A. M. and Tshimanga, R. M.: Capacity Building in the Congo Basin: Rich Resources Requiring Sustainable Development, One Earth 2, 2020 Elsevier Inc, https://doi.org/10.1016/j.oneear.2020.02.008, 2020.

Try, S., Lee, G., Yu, W., Oeurng, C., and Jang, C.: Large-scale Flood Inundation Modeling in the Mekong River Basin, J. Hydrol. Eng., 23, 05018011, https://doi.org/10.1186/s40645-020-00353-z, 2018.

Tshimanga, R. M., Bola, G., Kabuya, P., Nkaba, L., Neal, J., Hawker, L., Trigg, A. M., Bates, P., Hughes, A. D., Laraque, A., Woods, R., and Wagener, T.: Towards a framework of catchment classification for hydrologic predictions and water resources management in the ungauged basin of the Congo River: An a priori approach, AGU Monograph, in press, 2021.

Uk, S., Chihiro, Y., Sokly, S., Try, S., Heejun, Y., Oeurng, C., Shanghang, L., and Seingheng, H.: Tonle Sap Lake: Current Status and Important Research Directions for Environmental Management, Lakes Reservoirs, 23, 1–13, 2018.

UNESCO: Science for a Water Secure World in a Changing Environment, Draft IHP-IX Strategic Plan, The ninth phase of the Intergovernmental Hydrological Programme 2022–2029, Paris, 4 June 2021, available at: https://en.unesco.org/sites/default/files/ihp-ic-xxiv-ref.1_ihp-ix_strategic_plan.pdf (last access: 6 January 2021), 2021.

Van Lanen, H. A. J., Wanders, N., Tallaksen, L. M., and Van Loon, A. F.: Hydrological drought across the world: impact of climate and physical catchment structure, Hydrol. Earth Syst. Sci., 17, 1715–1732, https://doi.org/10.5194/hess-17-1715-2013, 2013.

Van Loon, A. F., Gleeson, T., Clark, J., Van Dijk, A. I. J. M., Stahl, K., Hannaford, J., Di Baldassarre, G., Teuling, A. J., Tallaksen, L. M., Uijlenhoet, R., Hannah, D. M., Sheffield, J., Svoboda, M., Verbeiren, B., Wagener, T., Rangecroft, S., Wanders, N., and Van Lanen, H. A. J.: Drought in the Anthropocene, Nat. Geosci., 9, 89–91, https://doi.org/10.1038/ngeo2646, 2016.

Venot, J. P., Kuper, M., and Zwarteveen, M.: Drip Irrigation: Untold Stories of Efficiency, Innovation & Development, Earthscan, Routledge, London, 2017.

Wei, X., Sauvage, S., Le, T. P. Q., Ouillon, S., Orange, D., Vu, D. V., and Sanchez-Perez, J.-M.: A modelling approach to diagnose the impacts of global changes on discharge and suspended sediment concentration within the Red River basin, Water, 11, 958, https://doi.org/10.3390/w11050958, 2019.

Wheeler, K., Hall, J. W., Abdo, G., Dadson, S. J., Kasprzyk, J. R., Smith, R., and Zagona, E. A.: Exploring Cooperative Transboundary River Management Strategies for the Eastern Nile Basin, Water Resour. Res., 54, 9224–9254, 2018.

Xi, W., Sauvage, S., Le, T. P. Q., Ouillon, S., Orange, D., Vinh, V. D., and Sanchez Perez, J. M.: A modelling approach to diagnose the impacts of global changes on discharge and suspended sediment concentration within the Red River basin, Water, 11, 958, https://doi.org/10.3390/w11050958, 2019.

Yang, X. K., Lu, X. X., Park, E., and Tarolli, P.: Impacts of Climate Change on Lake Fluctuations in the Hindu Kush-Himalaya-Tibetan Plateau, Remote Sens., 11, 1082, https://doi.org/10.3390/rs11091082, 2019.

Yepez, S., Laraque, A., Martinez, J. M., Sa, J. D., Carrera, J. M., Castellanos, B., Gallay, M., and Lopez, J. L.: Retrieval of suspended sediment concentrations using Landsat-8 OLI satellite images in the Orinoco River (Venezuela), Comptes Rendus Geoscience, 350, 20–30, https://doi.org/10.1016/j.crte.2017.08.004, 2018.

Yepez, S. P., Laraque, A., Gualtieri, C., Christophoul, F., Marchan, C., Castellanos, B., Azocar, J. M., Lopez, J. L., and Alfonso, J.: Morphodynamic change analysis of bedforms in the Lower Orinoco River, Venezuela, Proc. IAHS, 377, 41–50, https://doi.org/10.5194/piahs-377-41-2018, 2018.

Zare, A., Barbier, B., Bologo Traore, M., Diarra, A., Mahe, G., and Paturel, J. E.: Climate Forecast perception and needs in Wetlands: A Case Study of the Inner Niger Delta in Mali, Wetlands, 37, 913–923, https://doi.org/10.1007/s13157-017-0926-0, 2017.

Zhang, S. R., Bai, Y. J., Wen, X., Ding, A., and Zhi, J. H.: Seasonal and downstream alterations of dissolved organic matter and dissolved inorganic ions in a human-impacted mountainous tributary of the Yellow River, China, Environ. Sci. Pollut. Res., 25, 17967–17979, 2018.

Proc. IAHS, 384, 19–23, 2021
https://doi.org/10.5194/piahs-384-19-2021

Open Access PIAHS

Improvement of river network representation in Africa: application of an approach based on digital elevation data and environmental characteristics

Axel P. Belemtougri[1], **Agnès Ducharne**[2], **and Harouna Karambiri**[1]

[1]Laboratoire Eaux HydroSystèmes et Agriculture (LEHSA), Institut International d'Ingénierie de l'Eau et de l'Environnement (2iE), Ouagadougou, Burkina Faso

[2]Milieux Environnementaux, Transferts et Interactions dans les hydrosystèmes et les Sols (METIS), Sorbonne Université, CNRS, EPHE, Paris, France

Correspondence: Axel P. Belemtougri (axelbelemtougri@gmail.com)

Published: 16 November 2021

Abstract. In many continental databases representing rivers across Africa, information on rivers characteristics (length, width, intermittent or perennial) is often incomplete, and small streams are largely underrepresented. Fortunately, the use of topographic information from digital elevation models (DEMs) provides an opportunity to have more detailed information on rivers, particularly small streams in broad areas. A common approach to extract streams from DEM is to consider as a stream all DEM cells that drain at least a certain upstream surface, commonly known as the minimum contributing area (A_{min}). This parameter (A_{min}) is generally defined uniformly over large areas and this independently of the very variable climate and landscape conditions so that the generated streams have a rather uniform spatial distribution. To address this issue, using a 3″ (approx. 90 m × 90 m) DEM of Burkina Faso, the relationship between A_{min} and the observed drainage density ($\mathrm{Dd_{obs}}$) was established with satisfactory performance ($r^2 = 0.86$). In ongoing work, the functional relationship between the observed drainage densities ($\mathrm{Dd_{obs}}$) and environmental variables (lithology, climate, geology, vegetation cover) should allow for the establishment of the relationships between A_{min} and the environmental variables. This relationship allows for spatially variable A_{min} values depending on landscape characteristics. Before extracting river networks in Africa, the next step will be to validate or update these relationships in several countries.

1 Introduction

The accurate representation of river network recently, serves as a decision aid for land managers for efficient monitoring and exploitation of the water resource (Heine et al., 2004; Ouellet Dallaire et al., 2019). Models and many studies of hydrologic processes are sensitive to the accurate representation and density of streams (Yan et al., 2020; Brakebill et al., 2011). However, in many continental databases representing rivers across Africa, information on river's characteristics is often incomplete, and small streams are largely underrepresented (Lehner et al., 2008). National hydrographic databases contain reliable hydrographic information in this sense, either localized or nationwide. However, accessibility to these national hydrographic databases remains difficult in Africa because of the cost required and the reluctance of some countries to share this information.

Fortunately, the exploitation of topographic information from digital elevation models (DEM) provides a major opportunity to obtain detailed river descriptions over large areas (Liu et al., 2005). With the aid of Geographic Information Systems (GIS), river extraction usually involves defining a threshold river pixel number (N_{p}) which is a value from which a DEM pixel is considered a river. The upstream area drained by this number of pixels (N_{p}) corresponds to minimum contributing area (A_{min}) (O'Callaghan and Mark, 1984). A low A_{min} threshold increases the length of the extracted stream and vice versa. The choice of A_{min} value, therefore, condition the length of extracted rivers and influence the drainage density (Dd) which can be defined as the ratio between the total length of rivers in a watershed and its surface (Horton, 1945). Generally, for stream extraction $A_{\min}$ is defined rather arbitrarily and uniformly over large areas due to lack of information, resulting in the generated streams having a rather uniform spatial distribution and often not corresponding to those observed in the field (Passalacqua et al., 2010; Papageorgaki and Nalbantis, 2017). This is the example of the Hydro1k database (US Geological Survey, 2000) and AQUAMAPS (FAO, 2014), which proposes a mapping of rivers extracted from a 30″ (approx. 1 km × 1 km) and 15″ (approx. 500 m × 500 m) DEM for a constant $A_{\min}$ value of 1000 and 100 km^2, respectively.

In a natural environment, however, drainage density varies spatially as a function of multiple environmental variables (climate, lithology, slope, vegetation cover) and to capture this variability, there is a need to stress the use of spatially variable $A_{\min}$ (Tucker et al., 2001; Da Ros and Borga, 1997). Recently, Schneider et al. (2017) proposed an approach to generate a series of variable $A_{\min}$ values as a function of environmental variables. These $A_{\min}$ values are calibrated on different landscape types (constituted of the spatial intersection of Lithology, Climate, and Slope) with a 500 m DEM from HydroSHEDS database to match observed hydrographic information from national hydrographic databases of France and Australia. The calibrated $A_{\min}$ values were extrapolated at the global scale, given the local type of landscape, such as to extract a global hydrographic network, called LCS. Despite satisfactory validation in Brazil and USA against the national hydrographic databases, this LCS network shows difficulties in reproducing very high observed drainage densities (Schneider et al., 2017). The LCS network, for example, partially captures the spatial variability of the drainage density of the national hydrographic network of Burkina Faso, which is probably due to the coarse scale of the lithology layer used.

Based on the methodology used by Schneider et al. (2017), our objective is to use the accuracy of local information collected from national geographic institutes and widely available climatic, landscape, DEM information to establish relationships between the minimum contributing area ($A_{\min}$) and environmental variables (lithology, climate, geology, vegetation cover) to constitute a spatial model. A first step in setting up the spatial model is to define a relationship between the observed drainage densities (Dd_{obs}) from a national hydrographic database and calibrated minimum contributing areas ($A_{\min}$) in Burkina Faso. The same process is applied in Benin and France to evaluate the effect of the scale of a national hydrographic database in the relationship between $A_{\min}$ and Dd_{obs}. The second step will be to define relationships between the observed drainage densities (Dd_{obs}) and environmental variables (lithology, climate, geology, vegetation cover), which should ultimately allow us to establish relationships between $A_{\min}$ and the environmental variables. This study covers the first step of the model calibration in Burkina Faso, and the other steps are ongoing work.

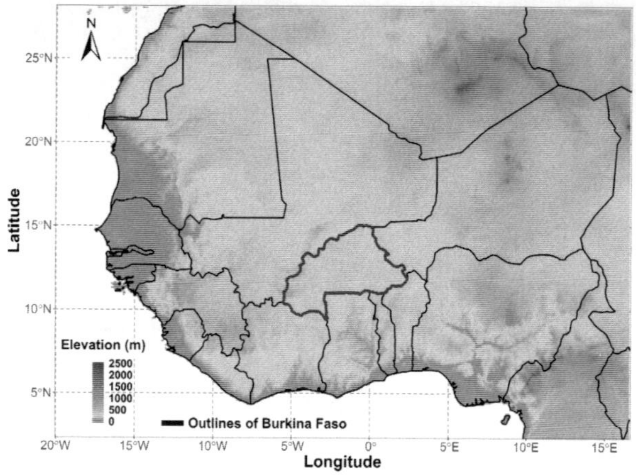

Figure 1. Location of Burkina Faso in West Africa. The borders of the country are shown in dark blue.

2 Materials and methods

2.1 Study area

For the first step of the model calibration, we are interested in Burkina Faso, a landlocked country of 274 200 km^2 located in West Africa and covers quite varied climatic zones with an annual rainfall gradient between South and North (1200 to less than 600 mm yr^{-1} respectively) (PANA, 2007). The values of altitudes are mainly between 250 and 350 m, with a maximum altitude around 750 m making the relief relatively flat (Fig. 1).

2.2 Description of data

The DEM used in this study was extracted from the HydroSHEDS database (Lehner et al., 2008), which is the most widely used database currently for hydrological applications at a resolution of 3″ (about 90 m × 90 m), i.e. the smallest resolution proposed by the database. The reference hydrographic database (national hydrographic database) taken into account in Burkina-Faso is the BNDT available at the scale of 1 : 200 000 and distributed by the IGB (Institut Géographique du Burkina). The reference hydrographic database considered in France and Benin are CARTHAGE and BDI, respectively, available at the same scale of 1 : 50 000, and distributed by the IGN (Institut Géographique National) France and IGN Bénin.

2.3 Methods

All maps and spatial processing are made using ArcGIS software (10.6.1) and the graphics using R software ver-

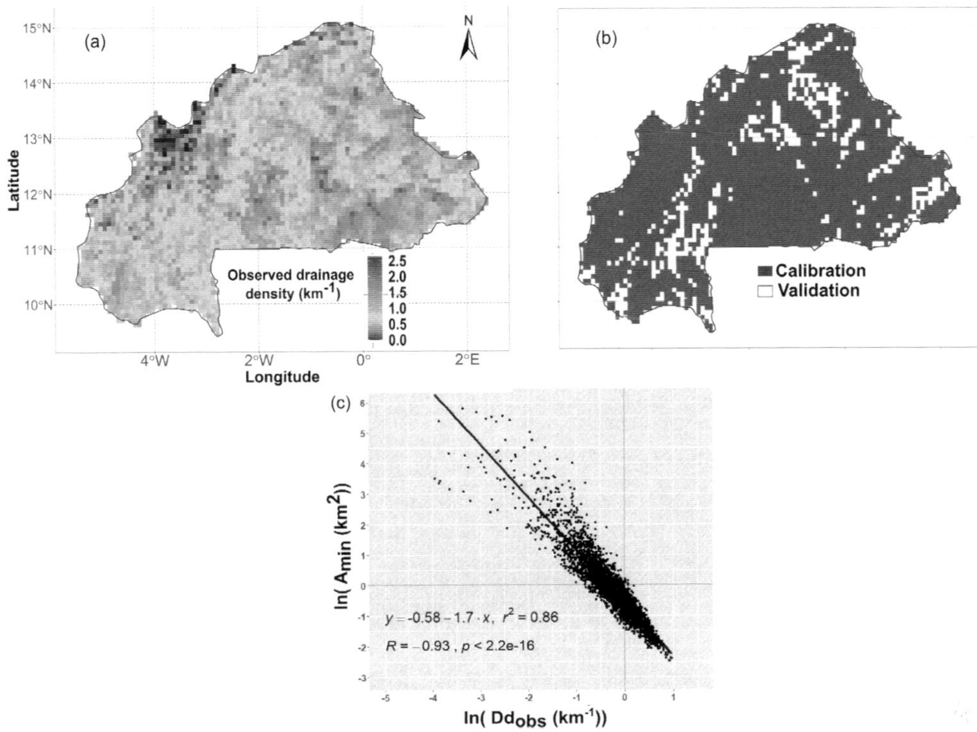

Figure 2. Results of the first stage of model implementation in Burkina Faso. **(a)** Representation of the observed drainage density in 5′ grid cells. **(b)** Calibration and validation areas for the relationship between A_{min} and Dd_{obs}. **(c)** Relation between the calibrated values of A_{min} and Dd_{obs} values in 5′ grid cells, with the axes in logarithm.

sion 3.6.1. The ArcGIS tool has been used to extract 400 river networks in Burkina Faso using the DEM from which the direction and accumulation grids are also progressively extracted. To span a large range of drainage densities, the number of pixels defined to extract these networks (N_p) ranges between 1 to 40 000, corresponding to A_{min} between 0.0081–324 km^2.

The territory of Burkina Faso has been divided into 5′ grid cells (about 64 km^2 at the equator) for this analysis. For each of the 400 extracted networks, the drainage density is calculated as the total river length (km) inside a 5′ grid cell, divided by the surface area of the grid cell (km^2). The same procedure is used to calculate the observed drainage density (Dd_{obs}) using the river network information contained in the reference hydrographic database (BNDT) of Burkina Faso (Fig. 2a). For each grid cell, the calibrated A_{min} value is the one defining the extracted network (among the 400 possibilities) that best approximates the observed drainage density (Dd_{obs}), based on the BNDT river layer, in this grid cell. Thus, the selected A_{min} is the one minimizing the difference in drainage density between the extracted networks and the observed BNDT network in the grid cell. We randomly selected several grid cells that were excluded from the calibration process (Fig. 2b) and that represents 14 % of the territory. It is on these grid cells that validation of the relationship established between calibrated A_{min} and Dd_{obs} was carried out. The same calibration process was applied in Benin and France to evaluate the effect of the scale of hydrographic database in the relationship between A_{min} and Dd_{obs}.

3 Results and discussions

The relationship between A_{min} and Dd_{obs}, based on the first step spatial model in Burkina, the calibrated values of A_{min} (Fig. 2c) obtained vary between 0.084–318 km^2. The minimum value obtained (0.084 km^2) is much lower than those used in the continental databases (Hydro1k and Aquamaps at constant A_{min} values of 1000 and 100 km^2, respectively, LCS with variable values falls between 0.3–193 km^2). This may reflect consideration of higher drainage densities in Burkina compared to the values obtained from the databases mentioned above, given that the lower A_{min} becomes, the higher Dd_{obs} will.

The results of this first stage of the model implementation allowed us to establish a relationship (Fig. 2c) with very satisfactory performances (coefficient of determination $r^2 = 0.86$, correlation coefficient $R = -0.93$, p value $< 2.2 \times 10^{-16}$):

$$A_{min} = 0.56 \cdot Dd_{obs}^{-1.7} \tag{1}$$

Table 1. Performance of validation of the first step of the model.

Statistical parameters	Dd_{obs}	Dd_{sim}
Mean	1.01	0.99
Standard deviation	0.42	0.41
Correlation coefficient	1	0.93
RMSE (Root Mean Square Error)	0	0.15

Table 2. Relationship between A_{min} and Dd_{obs} in different countries and at different scales.

Country	Equations	r^2	R	Scale
Bénin	$A_{min} = 0.41 \cdot Dd^{-1.8}$	0.88	−0.94	1 : 50 000
France	$A_{min} = 0.53 \cdot Dd^{-1.7}$	0.82	−0.91	1 : 50 000
Burkina Faso	$A_{min} = 0.56 \cdot Dd^{-1.7}$	0.86	−0.93	1 : 200 000

with A_{min} (km^2) and Dd_{obs} (km^{-1}). Equation (1) shows a decrease in the minimum contributing area (A_{min}) as a function of an increase in drainage density, in the form of a power relation as mentioned in some studies such as Moglen et al. (1998) and Colombo et al. (2007). It should allow from the observed drainage density to find the corresponding A_{min}, a necessary parameter for stream extraction. Validation of Eq. (1) on the remaining 14 % of the territory (Fig. 2b) allowed us to extract streams with calculated drainage densities (Dd_{sim}) almost similar to those observed (Dd_{obs}), with a correlation coefficient of 0.93 and an RMSE of 0.15 (Table 1).

There is a similarity in the equations between A_{min} as a function of Dd_{obs} at different scales and in different countries with r^2 performances between 0.82 and 0.86 (Table 2).

Likely, this relationship does not depend on the scale of the hydrographic database. Calibration of the hydrographic database at different scales in other countries would help to confirm this hypothesis.

4 Conclusions

This study proposes a method to develop a model of minimum contributing areas (A_{min}) as a function of environmental variables (lithology, precipitation, evapotranspiration, runoff, slope, vegetation cover). The first step presented here leads to a very performant relationship between A_{min} and Dd_{obs} in Burkina Faso. This relationship may even be independent from the scale of the hydrographic database. However, this quality needs confirmation by additional calibration tests in other countries with different scales of hydrographic databases. If validated in other countries with different climates, lithology, and drainage density, this first relationship should allow us to locally estimate the best A_{min} from the observed drainage densities without having to extract 400 different hydrographic networks for example. However, the final goal is to characterize A_{min} in areas lacking hydrographic information, where observed drainage densities are therefore not available. To this end, we plan to establish a second empirical relationship, linking observed drainage densities and environmental variables. Once calibrated in Burkina Faso, this relationship between Dd_{obs} and environmental variables can be combined to the first one between A_{min} and Dd_{obs}, such as to directly infer A_{min} from environmental variables. For validation, we plan to apply this spatial model of A_{min} to extract the rivers of Benin, Kenya, South Africa, and Madagascar, where hydrographic networks have been collected from national institutes. These countries were chosen because of the ease of access to the data and the different types of climate they cover. Upon satisfactory validation, this two-step spatial model of A_{min} could be applied to the entire African continent to extract more accurate hydrographic networks than the current continental databases, thus making it possible to support water management in regions where hydrographic observations are missing or limited.

Data availability. DEM data are freely available at https://doi.org/10.1029/2008EO100001 (Lehner et al., 2008). The hydrographic reference networks of Burkina Faso and Bénin are available on request from the agencies (Institut géographique du Burkina Faso – IGB, Institut géographique national du Bénin – IGN Bénin). The hydrographic network of France is freely available on the website: https://www.data.gouv.fr/fr/datasets/cours-deau-metropole-2016-bd-carthage/ (BD Carthage, 2016).

Author contributions. APB did the conceptualization, methodology, formal analysis, writing – original draft. The co-authors contributed as follows: AD participated in Conceptualization, Methodology, writing – review and editing, supervision; HK participated in conceptualization, methodology, writing – review and editing, supervision.

Competing interests. The contact author has declared that neither they nor their co-authors have any competing interests.

Disclaimer. Publisher's note: Copernicus Publications remains neutral with regard to jurisdictional claims in published maps and institutional affiliations.

Special issue statement. This article is part of the special issue "Hydrology of Large River Basins of Africa". It is a result of the 4th International Conference on the "Hydrology of the Great Rivers of Africa", Cotonou, Benin, 13–20 November 2021.

Financial support. This research has been supported by the African Development Bank Group (project no. P-Z1-IA0-013, grant no. 2100155032824).

References

BD Carthage: Cours d'eau – Métropole 2016 – , available at: https://www.data.gouv.fr/fr/datasets/cours-deau-metropole-2016-bd-carthage/, last access: 14 April 2021.

Brakebill, J. W., Wolock, D. M., and Terziotti, S.: Digital Hydrologic Networks Supporting Applications Related to Spatially Referenced Regression Modeling 1, J. Am. Water Resour. Assoc., 47, 916–932, 2011.

Colombo, R., Vogt, J. V., Soille, P., Paracchini, M. L., and de Jager, A.: Deriving river networks and catchments at the European scale from medium resolution digital elevation data, Catena, 70, 296–305, https://doi.org/10.1016/j.catena.2006.10.001, 2007.

Da Ros, D. and Borga, M.: Use of digital elevation model data for the derivation of the geomorphological instantaneous unit hydrograph, Hydrol. Process., 11, 13–33, 1997.

FAO: AQUAMAPS Global spatial database on water and agriculture, Food and Agriculture Organization of the United Nations, available at: https://data.apps.fao.org/aquamaps/ (last access: 21 April 2021), 2014.

Heine, R. A., Lant, C. L., and Sengupta, R. R.: Development and Comparison of Approaches for Automated Mapping of Stream Channel Networks, Ann. Assoc. Am. Geogr., 94, 477–490, https://doi.org/10.1111/j.1467-8306.2004.00409.x, 2004.

Horton, R. E.: Erosional development of streams and their drainage basins; hydrophysical approach to quantitative morphology, Geol. Soc. Am. Bull., 56, 275–370, 1945.

Lehner, B., Verdin, K., and Jarvis, A.: New Global Hydrography Derived From Spaceborne Elevation Data, EOS Trans. Am. Geophys. Union [data set], 89, 93, https://doi.org/10.1029/2008EO100001, 2008.

Liu, X., Peterson, J., and Zhang, Z.: High-resolution DEM generated from LiDAR data for water resource management, in: Proceedings of the International Congress on Modelling and Simulation (MODSIM05), December 2005, Australia, 1402–1408, 2005.

Moglen, G. E., Eltahir, E. A. B., and Bras, R. L.: On the sensitivity of drainage density to climate change, Water Resour. Res., 34, 855–862, https://doi.org/10.1029/97WR02709, 1998.

O'Callaghan, J. F. and Mark, D. M.: The extraction of drainage networks from digital elevation data, Comput. Vis. Graph. Image Process., 28, 323–344, 1984.

Ouellet Dallaire, C., Lehner, B., Sayre, R., and Thieme, M.: A multidisciplinary framework to derive global river reach classifications at high spatial resolution, Environ. Res. Lett., 14, 024003, https://doi.org/10.1088/1748-9326/aad8e9, 2019.

PANA: Programme National d'adaptation de la variabilité et aux changements climatiques, Ministère de l'environnement et du cadre du vie, Burkina Faso, 2007.

Papageorgaki, I. and Nalbantis, I.: Definition of critical support area revisited, European Water, 57, 273–278, 2017.

Passalacqua, P., Do Trung, T., Foufoula-Georgiou, E., Sapiro, G., and Dietrich, W. E.: A geometric framework for channel network extraction from lidar: Nonlinear diffusion and geodesic paths, J. Geophys. Res.-Earth, 115, F01002, https://doi.org/10.1029/2009JF001254, 2010.

Schneider, A., Jost, A., Coulon, C., Silvestre, M., Théry, S., and Ducharne, A.: Global-scale river network extraction based on high-resolution topography and constrained by lithology, climate, slope, and observed drainage density: Global-Scale River Network Extraction, Geophys. Res. Lett., 44, 2773–2781, https://doi.org/10.1002/2016GL071844, 2017.

Tucker, G. E., Catani, F., Rinaldo, A., and Bras, R. L.: Statistical analysis of drainage density from digital terrain data, Geomorphology, 36, 187–202, https://doi.org/10.1016/S0169-555X(00)00056-8, 2001.

US Geological Survey: HYDRO1k Elevation Derivative Database, available at: https://www.usgs.gov/centers/eros/science/usgs-eros-archive-digital-elevation-hydro1k?qt-science_center_objects=0#qt-science_center_objects (last access: 21 April 2021), 2000.

Yan, Y., Lidberg, W., Tenenbaum, D. E., and Pilesjö, P.: The accuracy of drainage network delineation as a function of environmental factors: A case study in Central and Northern Sweden, Hydrol. Process., 34, 5489–5504, 2020.

Proc. IAHS, 384, 25–29, 2021
https://doi.org/10.5194/piahs-384-25-2021

Open Access PIAHS

Hydrology of Large River Basins of Africa

Estimation bayésienne des courbes de tarage et des incertitudes associées : application de la méthode BaRatin au Congo à Brazzaville

Jérôme Le Coz[1], Guy D. Moukandi N'kaya[2], Jean-Pierre Bricquet[3], Alain Laraque[4], and Benjamin Renard[1]

[1]INRAE, UR RiverLy, Lyon, France
[2]LMEI, CUSI/ENSP, Université Marien N'gouabi, Brazzaville, République du Congo
[3]HSM, IRD, CNRS, UM, Montpellier, France
[4]GET, CNRS/IRD/UPS, Toulouse, France

Correspondence: Jérôme Le Coz (jerome.lecoz@inrae.fr)

Published: 16 November 2021

Résumé. L'inférence bayésienne est une approche intéressante pour estimer les courbes de tarage hauteur-débit des stations hydrométriques et les incertitudes associées car elle permet de croiser les jaugeages et la connaissance a priori des contrôles hydrauliques dans un cadre probabiliste. Utilisable par tous à travers un logiciel gratuit et simple d'utilisation, la méthode BaRatin est ici illustrée avec une application récente à l'analyse de la courbe de tarage du fleuve Congo à l'échelle de Brazzaville. Par rapport aux ajustements conventionnels de courbes de tarage, l'approche bayésienne est basée sur des hypothèses explicites sur les contrôles hydrauliques et fournit une décomposition des sources d'incertitude sur les débits. Ceci facilite la justification et la révision future de la courbe de tarage, et permet une utilisation éclairée des données hydrométriques. La série des débits avec incertitudes du Congo à Brazzaville depuis 1902 ainsi calculée présente des incertitudes très réduites.

1 Introduction

Pour la plupart des cours d'eau disposant d'une station hydrométrique, la série temporelle de débit (ou hydrogramme) est établie en appliquant un modèle hauteur-débit (ou courbe de tarage) à la série temporelle des hauteurs d'eau (ou limnigramme) mesurée en continu à la station. La construction et l'estimation de cette courbe de tarage est un problème complexe qui nécessite de combiner des mesures (appelées jaugeages) ponctuelles et incertaines du couple hauteur-débit avec l'expertise (également incertaine) de l'hydromètre sur les contrôles hydrauliques de sa station. Ces contrôles sont les caractéristiques physiques du chenal ou de la section critique qui déterminent la hauteur d'eau au point de mesure en fonction du débit du cours d'eau. L'inférence bayésienne est une approche intéressante pour ce type de problème car elle permet de croiser les deux sources d'information incertaine dans un cadre probabiliste pour estimer la courbe de tarage et les incertitudes associées. Plusieurs méthodes basées sur le principe bayésien ont été proposées ces dernières années (Kiang et al., 2018), dont la méthode BaRatin (pour Bayesian Rating curve, Le Coz et al., 2014). A travers le logiciel gratuit BaRatinAGE en versions anglaise et française, BaRatin est maintenant utilisé par différents services hydrométriques opérationnels et différents groupes de recherche ou d'ingénierie à travers le monde. La méthode est ici illustrée avec une application récente à l'analyse de la courbe de tarage du fleuve Congo à l'échelle de Brazzaville, plus grande station hydrométrique du continent africain par son bassin versant et par son débit moyen.

2 Matériel et méthode

2.1 Le Congo à Brazzaville-Kinshasa

Le fleuve Congo est le deuxième de la planète et le premier du continent africain par la superficie de son bassin versant (près de $3{,}8 \times 10^6\ km^2$) et par son débit moyen inter-

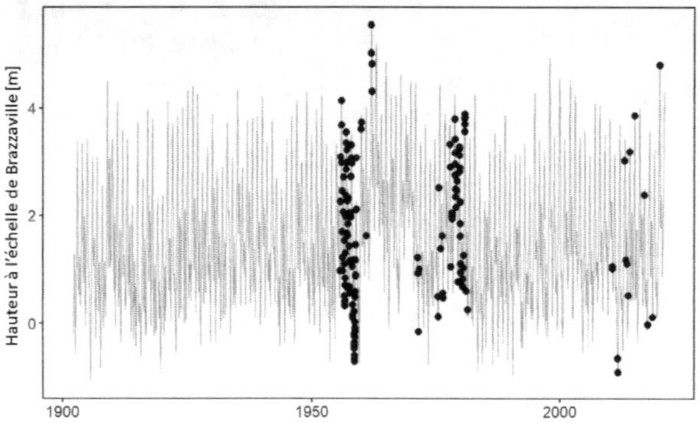

Figure 1. Série de hauteurs d'eau du Congo à l'échelle de Brazzaville et les jaugeages disponibles (points bleus).

annuel (40 500 m^3/s pour la période 1902–2020). Ce fleuve long de 4700 km présente dans sa partie aval un régime équatorial bimodal, avec sa crue principale en décembre et sa secondaire en mai (Alsdorf et al., 2016). C'est au niveau du bief qui sépare les capitales Brazzaville et Kinshasa sur le pourtour du Pool Malebo qu'ont débuté les mesures hydrologiques sur le fleuve Congo, avec des relevés quotidiens de hauteur d'eau depuis fin 1902 jusqu'à aujourd'hui (Devroey, 1957 ; Mateba 22, 1957). Environ dix kilomètres en aval, le Pool Malébo se termine avec les barres rocheuses des rapides du Djoué, au début des chutes de Livingstone. Grâce à un important travail documentaire et l'accès à la base de données SIEREM de l'IRD, les hauteurs d'eau ainsi qu'un riche ensemble de jaugeages historiques (Van Ganse, 1959 ; Lempicka, 1973 ; Molinier, 1979 ; Moukandi N'kaya et al., 2020) ont pu être réunis pour cette étude de la courbe de tarage du Congo à l'échelle de Brazzaville (Fig. 1). En rapportant les relevés de hauteur d'eau faits à Kinshasa à l'échelle de Brazzaville via une régression linéaire, on obtient une longue série hydrométrique sur une station couvrant près de 98 % de la superficie du bassin versant du Congo. L'amplitude des hauteurs d'eau relevées entre 1902 et 2020 va de −1 à 5,55 m. Les incertitudes de mesure (au niveau de probabilité de 95 %) des relevés de hauteur d'eau ont été supposées égales à ±5 cm après le 1er janvier 1955 (lecture directe de l'échelle de Brazzaville) et à ±10 cm avant, pour tenir compte des erreurs dues au transfert des cotes lues à l'échelle de Kinshasa vers celle de Brazzaville.

2.2 Jaugeages et incertitudes

Sur les périodes 1955–1962 et 1971–1981, 139 jaugeages au moulinet sont disponibles. Ils ont été complétés à partir de 2010 grâce à 15 autres mesurés par ADCP dans le cadre d'une coopération internationale. Ils couvrent tout le cycle hydrologique ainsi que la quasi-totalité du marnage observé sur les 120 dernières années (Fig. 1). Eu égard aux conditions de mesure très difficiles sur le Congo avec les moyens de jaugeage de l'époque, les incertitudes de mesure (au niveau de probabilité de 95 %) ont été supposées égales à ±15 % pour l'ensemble des jaugeages au moulinet sauf pour les quatre plus élevés, réalisés dans des conditions de mesure extrêmes (débits supérieurs à 70 000 m^3/s), pour lesquels l'incertitude est estimée à ±20 %. L'incertitude des jaugeages par ADCP réalisés à partir de 2010 est estimée à ±10 %. La méthode de détection des détarages proposée par Darienzo et al. (2021), basée sur une segmentation des résidus jaugeages/courbe de tarage tenant compte des incertitudes, a été appliquée à l'ensemble du jeu de jaugeages. Aucun détarage n'est détecté.

2.3 Méthode BaRatin

La méthode bayésienne BaRatin (Le Coz et al., 2014) permet la construction des courbes de tarage hauteur-débit avec estimation de l'incertitude, en combinant la connaissance a priori sur les contrôles hydrauliques et le contenu d'information des jaugeages incertains. L'équation de la courbe de tarage est dérivée de la combinaison de fonctions puissance pour chacun des contrôles supposés pour le site. L'utilisateur définit également les distributions de probabilité a priori des paramètres physiques de cette équation hauteur-débit, « a priori » signifiant sans utiliser les jaugeages utilisés dans l'estimation bayésienne. Une telle estimation bayésienne est basée sur l'échantillonnage par méthode Monte Carlo par Chaîne de Markov (MCMC) de la distribution a posteriori des paramètres de la courbe de tarage, inférée à partir du théorème de Bayes. L'hydrogramme avec incertitudes est calculé à partir des réalisations de la courbe de tarage et du limnigramme avec incertitudes selon la méthode de propagation décrite par Horner et al. (2018).

3 Résultats

3.1 Définition des a priori hydrauliques

Le modèle de courbe de tarage du Congo à Brazzaville est bâti sur l'hypothèse d'un unique contrôle par déversoir rectangulaire représentant les premiers rapides du Djoué, à l'aval immédiat du Pool Malebo. L'équation donnant le débit Q (en m^3/s) en fonction de la hauteur d'eau h (en m) est donc :

$$Q = C\sqrt{2g}B(h-h_0)^n, \tag{1}$$

où C est le coefficient de débit du déversoir, g l'accélération de la gravité, B la largeur déversante du déversoir, h_0 la cote moyenne de la crête du déversoir, et n un exposant. Les a priori sur les différents paramètres sont exprimés sous forme de distributions gaussiennes (les incertitudes indiquées ci-après correspondent à deux fois l'écart-type, soit un niveau de probabilité de 95 %). Le coefficient $C\sqrt{2g}B$ est estimé comme un seul paramètre du modèle. La valeur centrale et l'incertitude de ce coefficient sont calculées par propagation selon la méthode du GUM (JCGM 100, 2008). Les a priori sont les suivants : $C = 0{,}4 \pm 0{,}1$ (valeur

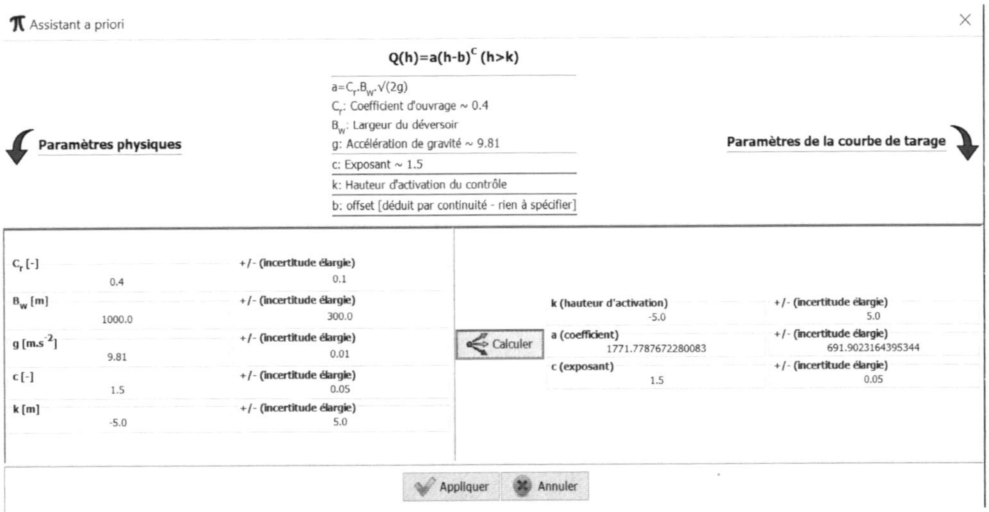

Figure 2. Définition a priori des paramètres du contrôle (déversoir rectangulaire) de la station du fleuve Congo à Brazzaville, via l'interface graphique du logiciel BaRatinAGE.

typique d'un déversoir rectangulaire), $g = 9{,}81 \pm 0{,}01$ m/s^2, $B = 1000 \pm 300$ m (estimé par vue satellite Google Earth, au resserrement le plus étroit en aval du Pool Malebo et présentant un dénivelé brusque, identifié comme la section de contrôle dite des rapides du Djoué), $h_0 = -5 \pm 5$ m (estimé grossièrement d'après les plus basses valeurs du limnigramme), et $n = 1{,}5 \pm 0{,}05$ (exposant théorique pour un déversoir rectangulaire).

L'interface graphique du logiciel BaRatinAGE, d'utilisation simple, permet à l'opérateur de spécifier cette connaissance a priori des contrôles hydrauliques (Fig. 2). Cette information est traduite par le logiciel sous forme d'une équation de courbe de tarage et de distributions de probabilité a priori des paramètres de cette équation. Dans l'exemple présenté ici, le Congo à Brazzaville, le modèle de courbe de tarage est très simple car réduit à un seul contrôle hydraulique pour toute la gamme de hauteurs d'eau possibles. BaRatin permet aussi de modéliser une courbe de tarage en considérant un ensemble de contrôles par chenal (loi de Manning-Strickler) ou par section critique (lois de déversoir ou d'orifice) qui s'ajoutent ou se succèdent les uns aux autres sur des segments de hauteur d'eau.

3.2 Estimation de la courbe de tarage

En combinant l'information a priori et la vraisemblance des observations (les jaugeages munis de leurs incertitudes), le théorème de Bayes permet de calculer (à une constante près) la distribution a posteriori des paramètres de la courbe de tarage, qui est échantillonnée par méthode Monte Carlo par Chaîne de Markov (MCMC). Les résultats sont probabilistes : les échantillons permettent de calculer l'enveloppe d'incertitude au niveau de probabilité souhaité (95 % par défaut). L'incertitude totale combine une incertitude dite paramétrique, traduisant l'incertitude d'estimation des paramètres de l'équation de la courbe de tarage, et une incertitude dite restante (ou structurelle) nécessaire à réconcilier les jaugeages et la courbe de tarage. Cette dernière composante d'incertitude peut traduire les limites du modèle de courbe de tarage pour représenter les contrôles hydrauliques réels de la station et leur éventuelle évolution dans le temps (détarages, influence aval, végétation ...). Parmi le grand nombre d'échantillons, on retient la courbe de tarage « MaxPost » calculée avec le jeu de paramètres échantillonné qui maximise la densité a posteriori. Pour le Congo à Brazzaville (Fig. 3), cette courbe de tarage « MaxPost » vaut :

$$Q = 1806(h + 6{,}29)^{1{,}50} \qquad (2)$$

Dans la gamme jaugée ainsi que pour les très basses eaux, la courbe de tarage BaRatin est très proche de la courbe de tarage historique ORSTOM établie à partir des jaugeages au moulinet sur la période 1971–1979. On constate néanmoins qu'au-dessus de $h = 4{,}2$ m et $Q = 70\,000$ m^3/s, les quatre jaugeages de la crue de 1961–1962 ainsi que le jaugeage ADCP de la forte crue de décembre 2019 sont alignés environ 10 % au-dessus de la courbe de tarage BaRatin. Cette dernière reste statistiquement cohérente avec l'ensemble des jaugeages disponibles et leurs incertitudes propres. Cependant, la pertinence des hypothèses hydrauliques faites sur le contrôle (déversoir unique, sans élargissement en forte crue) jouant en partie sur le haut de la courbe méritera d'être évaluée à l'avenir, à l'aide de futurs jaugeages ADCP en crue. L'incertitude de la courbe de tarage du Congo à Brazzaville (Fig. 3) est extrêmement faible (inférieure à 3 % environ) sur la gamme jaugée (entre 30 000 et 80 000 m^3/s), les parties extrapolées en hautes et basses eaux étant très réduites. L'incer-

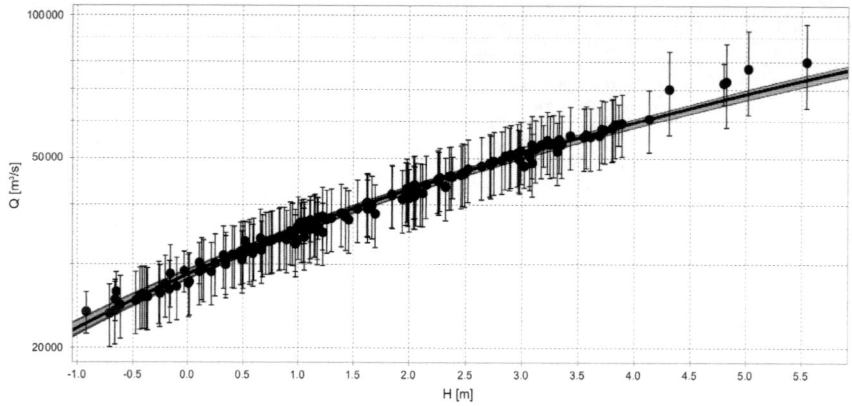

Figure 3. Jaugeages avec incertitudes du Congo à Brazzaville et courbe de tarage avec incertitudes obtenue par la méthode BaRatin.

titude sur les hautes eaux est toutefois peut-être sous-estimée si l'hypothèse d'un contrôle unique se révélait fausse.

3.3 Estimation de la série de débit

Selon la méthode exposée par Horner et al. (2018), chaque échantillon de courbe de tarage est croisé avec un échantillon de hauteur d'eau, issu de l'incertitude spécifiée pour le limnigramme, pour calculer un échantillon de débit. On calcule ainsi l'hydrogramme assorti de son incertitude. L'incertitude des débits calculés pour le Congo à Brazzaville entre 1902 et 2020 varie entre 2 % et 5 % avec une moyenne et une médiane de 3 %. Le budget d'incertitude, c'est-à-dire la contribution relative des sources d'incertitude à l'incertitude totale, peut différer selon les périodes considérées. Pour la forte crue de 1961–1962, l'incertitude paramétrique (plage en rose sur la Fig. 4a) domine largement les incertitudes structurelle et limnimétrique. Avant 1955, en raison du transfert des cotes de l'échelle de Kinshasa vers celle de Brazzaville, l'incertitude due aux relevés de hauteur d'eau (plage en jaune sur la Fig. 4b, exemple de l'étiage sévère de 1905) est plus importante qu'après 1955 et domine le budget d'incertitude, en particulier pour les basses eaux. Mais le contrôle par les rapides du Djoué présentant une excellente sensibilité sur toute la gamme de débit, les incertitudes dues aux relevés de hauteur d'eau n'entraînent que de faibles incertitudes sur l'hydrogramme.

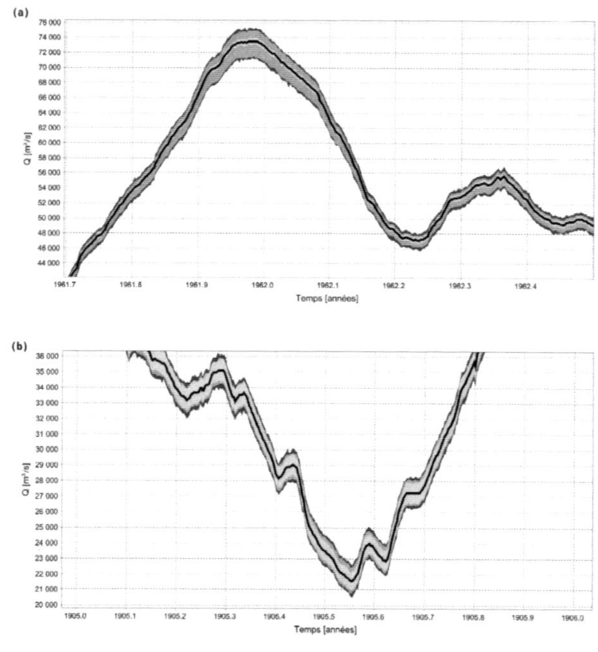

Figure 4. Hydrogrammes avec incertitudes du Congo à Brazzaville calculés par BaRatin pour **(a)** la plus forte crue enregistrée (1961–1962) et **(b)** l'étiage sévère de 1905. Les enveloppes d'incertitude à 95 % correspondent à la combinaison des incertitudes du limnigramme (en jaune), et des incertitudes paramétrique (en rose) et structurelle (en rouge) de la courbe de tarage.

4 Conclusion

La méthode bayésienne BaRatin permet d'établir la courbe de tarage et l'hydrogramme munis de leurs incertitudes d'une station hydrométrique fonctionnant avec une relation hauteur-débit univoque. Par rapport aux ajustements conventionnels de courbes de tarage, l'approche bayésienne est basée sur des hypothèses explicites sur les contrôles hydrauliques et fournit une décomposition des sources d'incertitude sur les débits. Cette méthode est ici illustrée avec l'application à la station du Congo à Brazzaville. Grâce au bassin de tranquillisation constitué par le Pool Malébo et par le contrôle stable et sensible des rapides du Djoué, la courbe de tarage et la longue série de débits depuis 1902 présentent des incertitudes très réduites. De futurs jaugeages restent néanmoins nécessaires, notamment en hautes eaux pour confirmer la pertinence du modèle hydraulique par contrôle unique, et en basses eaux pour confirmer l'absence de détarage significatif.

Disponibilité du code. Le logiciel BaRatinAGE est disponible sur simple demande à baratin.dev@lists.irstea.fr.

Disponibilité des données. Les données sont disponibles sur demande auprès des auteurs.

Collaborateurs. JLC a effectué les calculs et rédigé le manuscrit. GDMN'k, JPB et AL ont réuni et critiqué les données, expertisé le site de mesure et relu et corrigé les résultats. BR a contribué à la méthode de calcul, et vérifié les résultats et l'article.

Intérêts concurrents. Les auteurs déclarent qu'ils n'ont aucun conflit d'intérêts.

Clause de non-responsabilité. Publisher's note : Copernicus Publications remains neutral with regard to jurisdictional claims in published maps and institutional affiliations.

Déclaration du numéro spécial. This article is part of the special issue "Hydrology of Large River Basins of Africa". It is a result of the 4th International Conference on the "Hydrology of the Great Rivers of Africa", Cotonou, Benin, 13–20 November 2021.

Remerciements. Les producteurs historiques et actuels des données des stations hydrométriques de Kinshasa et Brazzaville sont chaleureusement remerciés.

Références

Alsdorf, D., Beighley, D., Laraque, A., Lee, H., Tshimanga, R., O'Loughlin, F., Mahé, G., Dinga, B., Moukandi N'kaya, G., and Spencer, R. G. M. : Opportunities for hydrologic research in the Congo Basin, Rev. Geophys., 54, 378–409, https://doi.org/10.1002/2016RG000517, 2016.

Darienzo, M., Le Coz, J., Renard, B., and Lang, M. : Detection of stage-discharge rating shifts using gaugings : a recursive segmentation procedure accounting for observational and model uncertainties, Water Resour. Res., 57, e2020WR028607, https://doi.org/10.1029/2020WR028607, 2021.

Devroey, E. J. : Annuaire hydrologique du Congo belge et du Ruanda-Urundi, Académie royale des Sciences coloniales, Classe des Sciences Techniques, Mémoires in-8°, Nouvelle série, Tome X, fasc. 1, Bruxelles, 1957.

Horner, I., Renard, B., Le Coz, J., Branger, F., McMillan, H. K., and Pierrefeu, G. : Impact of stage measurement errors on streamflow uncertainty, Water Resour. Res., 54, 1952–1976, 2018.

Joint Committee for Guides in Metrology (JCGM 100) : 2008 : Evaluation of measurement data – Guide to the expression of uncertainty in measurement (GUM), JCGM (Joint Committee for Guides in Metrology), BIPM, Sèvres, France, 2008.

Kiang, J. E., Gazoorian, C., McMillan, H., Coxon, G., Le Coz, J., Westerberg, I., Belleville, A., Sevrez, D., Sikorsk, A. E., Petersen-Øverleir, A., Reitan, T., Freer, J., Renard, B., Mansanarez, V., and Mason, R. : A comparison of methods for streamflow uncertainty estimation, Water Resour. Res., 54, 7149–7176, 2018.

Le Coz, J., Renard, B., Bonnifait, L., Branger, F., and Le Boursicaud, R. : Combining hydraulic knowledge and uncertain gaugings in the estimation of hydrometric rating curves : a Bayesian approach, J. Hydrol., 509, 573–587, 2014.

Lempicka, M. : Bilan hydrique du bassin du fleuve Zaïre. I : Ecoulement du bassin 1950–1959, Office National de la Recherche et du Développement, Kinshasa, République Démocratique du Congo, 1973.

Mateba 22 : Observations Limnimétriques : Kinshasa-Matadi-Boma, Inventaire 1903–1983 Navigabilité du bief maritime du fleuve Zaïre, Laboratoire de Recherches Hydrauliques, Borgerhout, Chatelet, Royaume de Belgique, 1984.

Molinier, M. : Note sur les débits et la qualité des eaux du Congo à Brazzaville, Cah, ORSTOM Série Hydrol., 16, 55–66, 1979.

Moukandi N'kaya, G. D., Orange, D., Bayonne Padou, S. M., Datok, P., and Laraque, A. : Temporal variability of sediments, dissolved solids and dissolved organic matter fluxes in the Congo River at Brazzaville/Kinshasa, Geosciences, 10, 341, https://doi.org/10.3390/geosciences10090341, 2020.

Van Ganse, R. : Les débits du fleuve Congo à Léopoldville et à Inga, Bull. Séanc. Acad. r. Sci. colon., Bruxelles, nouvelle série, 5, 737–763, 1959.

Proc. IAHS, 384, 31–35, 2021
https://doi.org/10.5194/piahs-384-31-2021

Open Access
PIAHS

Validation of the altimetry-based water levels from Sentinel-3A and B in the Inner Niger Delta

Adama Telly Diepkilé[1], Flavien Egon[2], Fabien Blarel[3], Eric Mougin[3], and Frédéric Frappart[2]

[1]DER Math-Informatique, FST/USTTB, Bamako, BP:E3206, Mali
[2]LEGOS, UMR 5566, University of Toulouse, CNES/CNRS/IRD/UPS. OMP, Toulouse, France
[3]GET, UMR 5563, University of Toulouse, CNRS/IRD/UPS. OMP, Toulouse, France

Correspondence: Adama Telly Diepkilé (adama.diepkile@usherbrooke.ca)

Published: 16 November 2021

Abstract. The comprehension of water level fluctuations and the sustainability of the Inner Niger River Delta (IND) is a major concern for the scientific community, but also for the local population. Located in the centre of Mali, the heart of the Sahel, the delta is characterised by a floodable area of more than 32 000 km^2 during the rainy season, which contributes very strongly to the vitality of local ecosystem, and is consequently classified as a Ramsar site under the international Convention for Wetlands. In addition, the Delta acts as an environmental and socio-economic development barometer for the entire sub-region. Nowadays, we can observe an increasing fragility of the delta due to climate change, desertification and human activities, and justifies the need for permanent monitoring. The present study is based on the recent successes of radar altimetry, originally designed to monitor the dynamics topography of the ocean, and now very frequently used to retrieve inland water levels, of lakes, rivers, and wetlands. Previous studies evaluated the performances of several radar altimetry missions including Low Resolution Mode (LRM) (Topex-Poseidon, Jason-1/2/3, ERS-2, ENVISAT, and SARAL, and Synthetic Aperture Radar (SAR) Sentinel-3A missions for water level retrievals over 1992–2017. More than 50 times series of water levels were build at the crossing between water bodies and Sentinel-3A and 3B over 2016–2020. Twenty-four comparisons between in-situ and altimetry-based time-series of water levels were achieved over the IND. RMSE generally lower than 0.7 m and r higher than 0.9 were obtained.

1 Introduction

Radar altimetry (RA) has been experiencing numerous inovations in terms of acquisition mode (from Low Resolution Mode – LRM – to Synthetic Aperture Radar – SAR –, and even Interferometry SAR – InSAR) and data processing over all types of Earth surfaces (Abdalla et al., 2021). Owing to their availability over almost 30 years, since the launch of Topex/Poseidon and ERS-1 missions in and 1991 and 1992, respectively, the two first missions to provide high accuracy measurements (Stammer and Cazenave, 2017), RA data are increasingly used, over land, to globally monitor the water levels of rivers, lakes and reserveroirs, wetlands and floodplains (Birkett et al., 2011; Crétaux et al., 2017). To ensure the confidence in the reliability of RA-based water levels, it is necessary to determine their quality through comparisons against in-situ measurements. In these operations are routinely performed over the ocean at different calibrration/validation sites (e.g., Bonnefond et al., 2011; Mertikas et al., 2018), few of these facilities are available over land, except at Lake Issykkul, in Kyrgyzstan (Central Asia) (Crétaux et al., 2009). Sentinel-3A and 3B are two of the most recently launched RA missions. They operate in SAR and use the Open-Loop (OL) or Digital Elevation Model (DEM) tracking mode designed to reduce the loss of tracking over hilly areas (Biancamaria et al., 2017; Taburet et al., 2020). Very few studies evaluated the accuracy of both Sentinel-3A and 3B over inland water bodies, and mostly over lakes (Frappart et al., 2021; Kittel et al., 2021; Shu et al., 2021).

Owing the availability of a large number of in-situ measurements, the IND has often been chosen as a study area for validating RA-based water levels (Frappart et al., 2015;

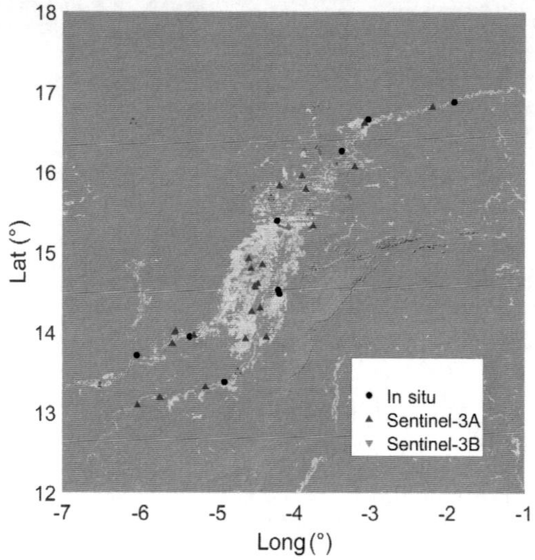

Figure 1. The Niger Inner Delta in Mali. Flooded areas appear in blue. They are made available by http://floodobservatory.colorado.edu/ (last access: 1 June 2021). Locations of in-situ gauges, and VS from Sentinel-3A and B are represented using black dots, red and green triangles, respectively.

Goita and Diepkile, 2012; Normandin et al., 2018). In this study, RA-based water levels defined at the cross-sections of Sentinel-3A and 3B ground-tracks and river and floodplains in the IND were compared to in-situ water stages.

2 Study area and datasets

2.1 The Inner Niger Delta

The study area the inner delta of the Niger River, located in the centre of the Sahelian zone of West Africa, precisely in Mali (Fig. 1a). It extends between latitudes 13 to 17° N, and longitudes 2 to 7° W (Fig. 1). Its floodable area is estimated at 32 000 km^2 (Goita and Diepkile, 2012; Mahé et al., 2009; Seiler et al., 2009; Zwarts et al., 2005). The Delta presents a rather particular ecological environment by the fact that it separates two zones with very precarious climatic conditions: on the one hand, the desert Sahara dominated by sand dunes, and on the other hand, the Sahel which is characterized by recurrent droughts.

2.2 Radar altimetry data from Sentinel-3A and 3B

Sentinel-3 was developed by the European Space Agency (ESA) in the framework of the COPERNICUS program. Two satellites are already operating: Sentinal-3A and 3B launched on 16 February 2016 and 25 April 2018, respectively. They are orbiting at 814.5 km altitude on a 98.65° inclination sun-synchronous orbit with a 27 d repeat period and an equatorial ground-track spacing of $\sim$ 105 km. Sentinel-3A and 3B were placed on the same orbit with a phase difference of 180°. The satellites payloads is composed of SRAL (SAR Radar ALtimeter), a dual-frequency SAR altimeter operating at Ku (13.575 GHz) and C (5.41 GHz), bands, a Microwave Radiometer (MWR) sensor for wet path delay correction over the ocean, and a triple system for Precise Orbit Determination (POD) including a GPS receiver, a LRA and a DORIS instrument (Donlon et al., 2012). The data used in this study are the ones necessary to compute water levels over land (see (Frappart et al., 2021) and Sect. 3). They are made available at: http://ctoh.legos.obs-mip.fr (last access: 15 May 2021).

2.3 In-situ water levels

Records from 10 gauge stations located in IND (see Fig. 1) were provided by the Malian hydrological service, Direction Nationale de l'Hydraulique (DNH). In this study, we used the time series of water levels from February 2016 to December 2020.

3 Methodology

Time-series of water level were produced at the cross-section between an altimeter ground-track and a water body using the the Altimetry Time-Series (AlTiS) software (Frappart et al., 2021). From the Geophysical Data Records (GDR) which contain the along-track altimetry data, AlTiS computed the altimeter height (h) as described in (Crétaux et al., 2017) and Eq. (1):

$$h = H - R - \Delta R_{\text{ion}} - \Delta R_{\text{dry}} - \Delta R_{\text{wet}} - \Delta R_{\text{sol}} - \Delta R_{\text{pol}} - N \quad (1)$$

where H is the altitude of the satellite on its orbit, R the range or distance between the satellite and the surface, ΔR_{ion}, ΔR_{dry}, ΔR_{wet}, ΔR_{sol}, ΔR_{pol}, are the corrections to apply to the range to account for the delay introduced by the ionosphere, the troposphere (dry and wet components), and the effects of the solid Earth and pole tides, and N is the geoid model. Following Frappart et al., 2006), the ranges derived from Offset of Center of Gravity (OCOG) (Wingham et al., 1986) were used as they were found to provide more accurate water levels.

Then, the valid height values were manually selected through AlTiS Graphical User Interface (GUI). Once the valid data are selected, the time-series of water levels is generated computing the median of the heights for each altimetry cycle (i.e., every 27 d for Sentinel-3A and B).

The altimetry-based time-series of water levels generated over the rivers in the Inner Niger Delta were compared to in-situ water stages using the Root-Mean-Square Error (RMSE), Relative (ratio of the RMSE to the mean annual amplitude of the water level at the station multiplied by 100) RMSE (RRMSE), and the Pearson correlation coefficient (r).

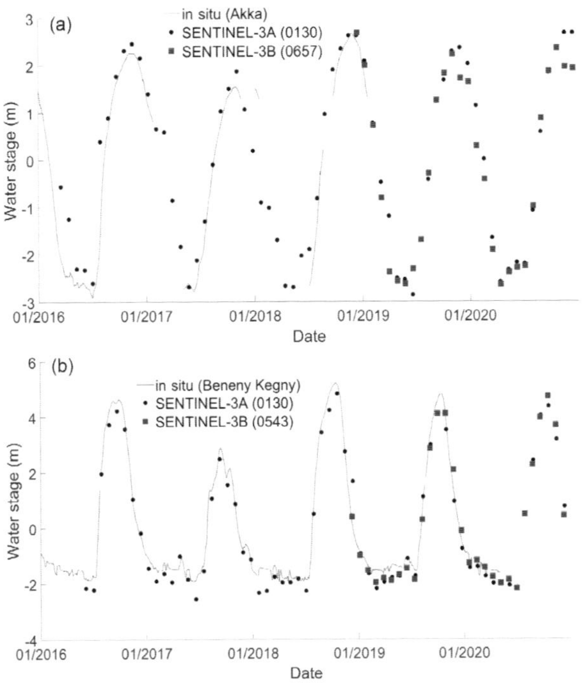

Figure 2. Time-series of water levels from the in-situ gauge stations (blue) of Beneny Kegny **(a)** and Akka **(b)**, Sentinel 3A (black dots) and 3B (red squares).

4 Results

Several tenths of RA-based time-series of water levels were generated over the IND (rivers and floodplains) using Sentinel-3A and 3B data: 28/24 on the Sentinel-3A/3B RA ground-tracks (Fig. 1). They offer a dense network of RA Virtual Stations (VS) were time-series of water levels are estimated. Two examples of comparisons of in-situ and RA-based time-series of water level are presented in Fig. 2. The first one to a cross-section of Sentinel-3A and 3B located on the Bani River, a south-west tributary to the Niger River, at (4.74° W, 13.51° N), the second one one to a cross-section of Sentinel-3A and 3B located on the Niger River at (4.27° W, 15.46° N), on the nort central part of the IND. Comparisons were made with gauge records of Beneny Kegny (4.917° W, 13.383° N) and Akka (4.233° W, 15.400° N) stations, respectively. The distance to the Sentinel-3 SV is 28 and 8 km, respectively.

RA-based water levels exhibit similar temporal variations as the in-situ gauge stations, with a well-marked seasonal cycle that can reach 6 m (Fig. 2). RMSE of 0.35 m/0.30 m, and r of 0.99/0.99 were found at Beneny Kegny Akka for Sentinel-3A and 3 B, respectively, and RMSE of 0.45 m and r of 0.98 at Akka for Sentinel-3A. As too few in-situ measurements were collected or made available for the Akka gauge station during the period of operation of Sentinel-3B, we were unable to estimate RMSE, RRSME and r for Sentinel-3B. Nev-

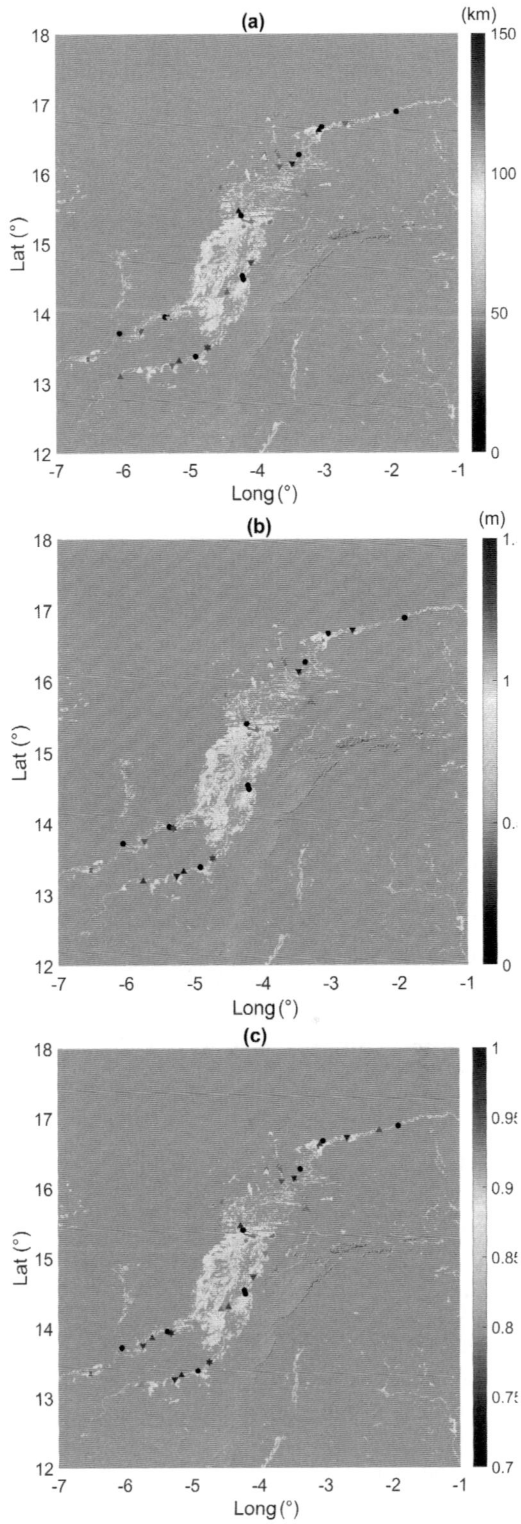

Figure 3. Comparisons between in-situ and altimetry- based water levels for Snetinel-3A (upward-pointing triangles) and 3B (downward-pointing triangles) over the IND: **(a)** distance between in in-situ stations and VS, **(b)** RMSE (m), **(c)** R.

ertheless, as RA-based water levels from Sentinel-3A and Sentinel-3B exxhibit very similar temporal variations and taking into account the good agreement between in-situ and Sentinel-3A-based water levels, similar perfomances can be expected from Sentinel-3B based water stages. This latter example brings into focus the strong interest of radar altimetry to provide a continuous monitoring of water levels in case in-situ gauge stations stopped operating for a reason or another.

The same evaluation parameters (distance between the VS and the in-situ station, RMSE, RRMSE and R) were estimated for 12 Sentinel-3A VS and 12 Sentinel-3B VS. They are presented on Fig. 3, except the RRMSE not shown here. Comparisons were made on distances between the in-situ station and the VS ranging from 7 to 126 km (Fig. 3a). The RMSE range from 0.15 to 1.39 m, with 13 values over 0.5 m (Fig. 3b). If these values can be considered quite high, the corresponding RRMSE are all lower or equal 20 %, and 19 RRMSE values out of 24 are below 15 %. Correlations are all higher or equal to 0.79 (Fig. 3c). Twenty one out of 24 *r* values are above 0.9, and even, 16 of them are higher than 0.95, confirming the good accuracy of Sentinel-3 based water stages reported in previous studies over rivers (e.g., Bogning et al., 2018; Kittel et al., 2021; Normandin et al., 2018). The high RMSE values can be accounted for the distances between the in-situ gauges and the SV. If the temporal variations are likely to exhibit a similar pattern over distances of several tenths of kilometers, the seasonal amplitude can be affected by changes in depth and width of the river along of the river network. As a consequence, the correlation can still present high values when the RMSE increases. This situation is what is observed here: high correlation values were estimated, slightly decreasing as the distance increases, but RMSE generally increases with the distance (e.g., RMSE of 0.93 and 1.39 m were obtained fro distances of 126 and 93 km, respectively, for corresponding *r* of 0.79 and 0.91).

5 Conclusions

An extensive assessment of the quality of the data acquired by Sentinel-3A and 3B was achieved over the IND over the period 2016–2020. Fifty two VS were defined over the rivers and the floodplains using the newly developed AlTiS software, and 24 comparisons against in-situ water levels were made. An overall very good agreement with r values ranging from 0.79 to 1, but generally over 0.90 or 0.95, and RRMSE lower than 20 %, and, most of the times, lower tha 15 %. The high RMSE obtained for some VS (RMSE > 0.50 m) are, most of the time, due to the long distance, above 50 km, between the in-situ gauge stations and the VS. Owing to the density of the in-situ gauge stations over the IND and the rapid availability of the water stages, The Inner Niger Delta is a good site for the validation of the recently launched (e.g., Sentinel-6/Jason-CS in November 2020) or to be launched missions (e.g., Surface Water and Ocean Topography – SWOT in fall 2022).

Code availability. AlTiS is an Open Source project under CeCill license (IDDN certification: IDDN.FR.010.0121234.000.R.X.2020.041.30000 – https://www.iddn.org/cgi-iddn/certificat.cgi?IDDN.FR.010.0121234.000.R.X.2020.041.30000, last access: 3 June 2021, CTOH, 2020). AlTiS software is provided by CTOH along with the radar altimetry data using the following request form: http://ctoh.legos.obs-mip.fr/applications/land_surfaces/altimetric_data/altis (last access: 4 July 2021, Normandin et al., 2018).

Data availability. Radar altimetry data are made freely available by CTOH. They can be obtained using the following request form: http://ctoh.legos.obs-mip.fr/applications/land_surfaces/altimetric_data/altis (last access: 4 July 2021, Normandin et al., 2018).

Author contributions. ATD has made the In-situ data collection for the IND. He has also performed the statistical comparisons, and contributed to the writing of the manuscript. FE has performed the data processing of radar altimetry data from Sentinel-3. FB has extracted radar altimetry data. EM has contributed to the manuscript writing. FF has made the statistical comparisons. He has produced the figures, and carried out the writing of the manuscript.

Competing interests. The contact author has declared that neither they nor their co-authors have any competing interests.

Disclaimer. Publisher's note: Copernicus Publications remains neutral with regard to jurisdictional claims in published maps and institutional affiliations.

Special issue statement. This article is part of the special issue "Hydrology of Large River Basins of Africa". It is a result of the 4th International Conference on the "Hydrology of the Great Rivers of Africa", Cotonou, Benin, 13–20 November 2021.

References

Abdalla, S., Kolachina, A., Adusumilli, S., Bhowmick, A., Alou-Font, E., Amarouche, L., Andersen, O. B., Naeije, M. C., Simons, W. J. F., et al.: Altimetry for the future: Building on 25 years of progress, Adv. Space Res., 68, 319–363, https://doi.org/10.1016/j.asr.2021.01.022, 2021.

Biancamaria, S., Frappart, F., Leleu, A.-S. S., Marieu, V., Blumstein, D., Desjonquères, J.-D., Boy, F., Sottolichio, A., and Valle-Levinson, A.: Satellite radar altimetry water elevations performance over a 200 m wide river: Evaluation over the Garonne River, Adv. Sp. Res., 59, 128–146, https://doi.org/10.1016/j.asr.2016.10.008, 2017.

Birkett, C., Reynolds, C., Beckley, B., and Doorn, B.: From research to operations: The USDA global reservoir and lake monitor, in Coastal Altimetry, Springer Berlin Heidelberg, Berlin, Heidelberg, 19–50, 2011.

Bogning, S., Frappart, F., Blarel, F., Niño, F., Mahé, G., Bricquet, J. P., Seyler, F., Onguéné, R., Etamé, J., Paiz, M. C., and Braun, J. J.: Monitoring water levels and discharges using radar altimetry in an ungauged river basin: The case of the Ogooué, Remote Sens., 10, 350, https://doi.org/10.3390/rs10020350, 2018.

Bonnefond, P., Haines, B. J., and Watson, C.: In situ Absolute Calibration and Validation: A Link from Coastal to Open-Ocean Altimetry, in Coastal Altimetry, Springer Berlin Heidelberg, Berlin, Heidelberg, 259–296, 2011.

CTOH: Centre de Topographie des Océans et l'Hydrosphère, AlTiS – Altimetric Time Series Software, [code] available at: https://www.iddn.org/cgi-iddn/certificat.cgi?IDDN.FR.010.0121234.000.R.X.2020.041.30000 (last access: 4 July 2021) 2020.

Crétaux, J.-F., Nielsen, K., Frappart, F., Papa, F., Calmant, S., and Benveniste, J.: Hydrological applications of satellite altimetry: rivers, lakes, man-made reservoirs, inundated areas, in: Satellite Altimetry Over Oceans and Land Surfaces, edited by: Stammer, D. and Cazenave, A., CRC Press., 459–504, 2017.

Crétaux, J. F., Calmant, S., Romanovski, V., Shabunin, A., Lyard, F., Bergé-Nguyen, M., Cazenave, A., Hernandez, F. and Perosanz, F.: An absolute calibration site for radar altimeters in the continental domain: Lake Issykkul in Central Asia, J. Geod., 83, 723–735, https://doi.org/10.1007/s00190-008-0289-7, 2009.

Donlon, C., Berruti, B., Buongiorno, A., Ferreira, M. H., Féménias, P., Frerick, J., Goryl, P., Klein, U., Laur, H., Mavrocordatos, C., Nieke, J., Rebhan, H., Seitz, B., Stroede, J., and Sciarra, R.: The Global Monitoring for Environment and Security (GMES) Sentinel-3 mission, Remote Sens. Environ., 120, 37–57, https://doi.org/10.1016/j.rse.2011.07.024, 2012.

Frappart, F., Calmant, S., Cauhopé, M., Seyler, F., and Cazenave, A.: Preliminary results of ENVISAT RA-2-derived water levels validation over the Amazon basin, Remote Sens. Environ., 100, 252–264, https://doi.org/10.1016/j.rse.2005.10.027, 2006.

Frappart, F., Fatras, C., Mougin, E., Marieu, V., Diepkilé, A. T., Blarel, F., and Borderies, P.: Radar altimetry backscattering signatures at Ka, Ku, C, and S bands over West Africa, Phys. Chem. Earth, 83, 96–110, https://doi.org/10.1016/j.pce.2015.05.001, 2015.

Frappart, F., Blarel, F., Fayad, I., Bergé-Nguyen, M., Crétaux, J.-F., Shu, S., Schregenberger, J., and Baghdadi, N.: Evaluation of the Performances of Radar and Lidar Altimetry Missions for Water Level Retrievals in Mountainous Environment: The Case of the Swiss Lakes, Remote Sens., 13, 2196, https://doi.org/10.3390/rs13112196, 2021.

Goita, K. and Diepkile, A. T.: Radar altimetry of water level variability in the Inner Delta of Niger River, in 2012 IEEE International Geoscience and Remote Sensing Symposium, IEEE., 5262–5265, 2012.

Kittel, C. M. M., Jiang, L., Tøttrup, C., and Bauer-Gottwein, P.: Sentinel-3 radar altimetry for river monitoring – a catchment-scale evaluation of satellite water surface elevation from Sentinel-3A and Sentinel-3B, Hydrol. Earth Syst. Sci., 25, 333–357, https://doi.org/10.5194/hess-25-333-2021, 2021.

Mahé, G., Bamba, F., Soumaguel, A., Orange, D., and Olivry, J. C.: Water losses in the inner delta of the River Niger: water balance and flooded area, Hydrol. Process., 23, 3157–3160, https://doi.org/10.1002/hyp.7389, 2009.

Mertikas, S. P., Donlon, C., Féménias, P., Mavrocordatos, C., Galanakis, D., Tripolitsiotis, A., Frantzis, X., Tziavos, I. N., Vergos, G. and Guinle, T.: Fifteen years of Cal/Val service to reference altimetry missions: Calibration of satellite altimetry at the permanent facilities in Gavdos and Crete, Greece, Remote Sens., 10, 1557, https://doi.org/10.3390/rs10101557, 2018.

Normandin, C., Frappart, F., Diepkilé, A. T., Marieu, V., Mougin, E., Blarel, F., Lubac, B., Braquet, N., and Ba, A.: Evolution of the performances of radar altimetry missions from ERS-2 to Sentinel-3A over the Inner Niger Delta, [code, data set], Remote Sens., 10, 833, https://doi.org/10.3390/rs10060833, 2018.

Seiler, R., Schmidt, J., Diallo, O., and Csaplovics, E.: Flood monitoring in a semi-arid environment using spatially high resolution radar and optical data, J. Environ. Manage., 90, 2121–2129, https://doi.org/10.1016/j.jenvman.2007.07.035, 2009.

Shu, S., Liu, H., Beck, R. A., Frappart, F., Korhonen, J., Lan, M., Xu, M., Yang, B., and Huang, Y.: Evaluation of historic and operational satellite radar altimetry missions for constructing consistent long-term lake water level records, Hydrol. Earth Syst. Sci., 25, 1643–1670, https://doi.org/10.5194/hess-25-1643-2021, 2021.

Stammer, D. and Cazenave, A.: Satellite altimetry over oceans and land surfaces, Taylor & Francis, Boca Raton, FL., 2017.

Taburet, N., Zawadzki, L., Vayre, M., Blumstein, D., Le Gac, S., Boy, F., Raynal, M., Labroue, S., Crétaux, J.-F., and Femenias, P.: S3MPC: Improvement on Inland Water Tracking and Water Level Monitoring from the OLTC Onboard Sentinel-3 Altimeters, Remote Sens., 12, 3055, https://doi.org/10.3390/rs12183055, 2020.

Wingham, D. J., Rapley, C. G., and Griffiths, H.: New Techniques in Satellite Altimeter Tracking Systems, Proc. IGARSS Symp. Zurich, (SEPTEMBER 1986), 1339–1344, 1986.

Zwarts, L., Beuering, V. B., Kone, B., and Wymenga, E.: The Niger , a lifeline: effective water management in the Upper Niger Basin., 2005.

Proc. IAHS, 384, 37–41, 2021
https://doi.org/10.5194/piahs-384-37-2021

Suivi hydrologique par altimétrie spatiale dans le bassin du Congo

Christophe Brachet[1], Alice Andral[2], Georges Gulemvuga Guzanga[3], Blaise Léandre Tondo[3], Pierre-Olivier Malaterre[4], and Sébastien Legrand[5]

[1]OiEau, Paris, France
[2]CNES, Toulouse, France
[3]CICOS, Kinshasa, RD Congo
[4]G-EAU, Univ Montpellier, AgroParisTech, BRGM, CIRAD, IRD, INRAE, Institut Agro, Montpellier, France
[5]CNR, Lyon, France

Correspondence: Christophe Brachet (c.brachet@oieau.fr)

Published: 16 November 2021

Résumé. Le satellite SWOT sera lancé fin 2022 par le Centre National d'Etudes Spatiales (CNES) français et la National Aeronautics and Space Administration (NASA) américaine. L'altimétrie spatiale permet de compléter les données hydrométriques in situ à travers l'établissement de « stations virtuelles », au croisement de la trace au sol du satellite avec un cours d'eau. SWOT améliorera encore la couverture des zones observées ainsi que la précision grâce à une technologie innovante. Un projet d'appui à la Commission Internationale du bassin Congo-Oubangui-Sangha (CICOS) développé depuis 2016 sur financement de l'Agence Française de Développement (AFD) et facilité par l'Office International de l'Eau (OiEau) permet de promouvoir l'hydrologie spatiale à travers un groupe d'institutions françaises, en appui à la CICOS. Diverses activités ont été développées dont la fourniture d'une base de données spatiales par l'Institut de Recherche pour le Développement (IRD), ainsi que la comparaison avec des données in situ. Une méthodologie innovante a été proposée par l'Institut national de recherche pour l'agriculture, l'alimentation et l'environnement (INRAE) pour passer des altitudes aux débits aux stations virtuelles. Des applications spécifiques ont par ailleurs été développées par la Compagnie Nationale du Rhône (CNR) pour l'hydroélectricité et la navigation fluviale.

1 Introduction

Depuis plusieurs décennies, on constate un fort déclin des réseaux hydrométriques dans le monde et en particulier en Afrique. Le Congo, plus grand fleuve d'Afrique, n'échappe pas à ce constat. Mais plusieurs innovations ou évolutions permettent de surmonter ces manques : l'OMM a par exemple développé HydroHub (Dixon et al., 2020) ou encore WHOS (WMO Hydrological Observing System – Pecora et Lins, 2020), et des approches multidisciplinaires sont promues (Tauro et al., 2018). L'altimétrie spatiale, qui consiste à mesurer les hauteurs d'eau par satellite est une de ces innovations majeures.

Sept entités françaises se sont ainsi regroupées au sein d'un groupe de travail pour valoriser l'altimétrie spatiale dans le bassin du Congo et proposer des applications opérationnelles.

Après avoir rappelé le contexte, le présent article présente le projet d'appui du groupe de travail sur le bassin du Congo ainsi que les principaux résultats obtenus : augmentation de la base de données de stations virtuelles sur le Congo, étude comparative des données spatiales et in situ au droit d'un même site, transformation des altitudes en débits et utilisation pour des applications opérationnelles (navigation fluviale et hydroélectricité).

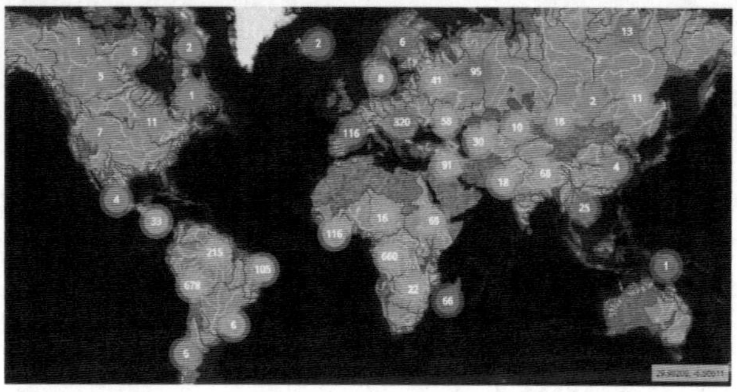

Figure 1. Stations virtuelles disponibles sur Hydroweb pour le suivi des lacs et rivières.

2 Contexte

2.1 L'altimétrie spatiale

L'altimétrie spatiale repose sur les satellites équipés d'un appareil radar (altimètre). Les mesures sont réalisées au droit de « stations virtuelles », qui correspondent aux intersections de la trajectoire du satellite et des cours d'eau.

Les mesures de hauteur d'eau par altimétrie sont accessibles sur le site Hydroweb (http://hydroweb.theia-land.fr, la date du dernier accès : 22 octobre 2021). Hydroweb est une base de données de variations des niveaux d'eau sur les lacs et les fleuves du monde à partir de l'altimétrie satellitaire, qui a été progressivement mise en place depuis 2003 (Crétaux et al., 2011). Aujourd'hui, ce sont plus de 11 000 stations virtuelles qui sont disponibles partout dans le monde (Fig. 1).

L'altimétrie spatiale permet ainsi de compléter le réseau de mesure in situ afin de densifier les données et d'assurer un meilleur suivi hydrologique (Calmant et Seyler, 2006 ; Calmant et al., 2008).

2.2 Le groupe de travail sur l'hydrologie spatiale

Le Groupe de travail sur l'hydrologie spatiale a été créé en 2014 puis consolidé par un accord signé lors de la COP22 à Marrakech. Il regroupe les institutions françaises suivantes : le CNES, l'IRD, l'AFD, l'OiEau (animateur du groupe), l'INRAE, BRLi et CNR.

Les travaux du groupe s'intéressent à la valorisation des données existantes d'altimétrie spatiale (en particulier avec les missions Jason et Sentinel3A) pour des applications opérationnelles sur des bassins versants, et à la préparation des applications aval pour le futur satellite SWOT. Il permet également de capitaliser les acquis et avancées obtenues.

2.3 La CICOS et le bassin du Congo

Le bassin du Congo est le second bassin au monde après celui de l'Amazone avec une superficie de $3{,}7 \times 10^6\ km^2$ et un débit moyen de $41\,000\ m^3\ s^{-1}$.

La CICOS est une organisation intergouvernementale regroupant l'Angola, le Cameroun, la Centrafrique, le Congo, la République Démocratique du Congo et le Gabon. Elle a été créée en 1999 à travers l'Accord créant la CICOS et instituant un Régime Fluvial Uniforme (CICOS, 1999), complété en 2007 (CICOS, 2007) par un mandat de Gestion Intégrée des Ressources en Eau du bassin du Congo. En décembre 2015 lors de la COP 21 à Paris, une convention a été signée entre la CICOS et l'AFD afin de renforcer le suivi hydrologique dans le bassin du Congo en tenant compte de l'adaptation aux changements climatiques.

3 Le projet de suivi hydrologique du bassin du Congo

3.1 Objectifs et activités du projet

Le groupe de travail a choisi le bassin du Congo comme bassin pilote avec les objectifs spécifiques suivants : (1) l'établissement d'une base de données de stations virtuelles, (2) l'étude de la conversion des élévations altimétriques en débits, (3) l'élaboration d'un système d'information hydrologique et (4) les applications sur l'hydroélectricité et la navigation.

3.2 Etablissement d'une base de données de stations virtuelles

Le projet a permis de mettre en œuvre près de 550 stations virtuelles sur l'ensemble du bassin du Congo (CICOS, 2016b, c). La calibration d'une station virtuelle a été réalisée avec une station in situ (Maluku-Trechot) installée au droit d'une trace de satellite. La précision estimée de la mesure altimétrique est ainsi de 14 cm (Fig. 2). Cette précision sera encore accrue avec SWOT, dont les incertitudes seront moindres.

3.3 Passage des altitudes aux débits

Le passage des altitudes aux débits est l'un des chantiers importants en cours au sein du groupe. Diverses méthodologies ont été développées ces dernières années dans le cadre du projet SWOT et du DAWG (Discharge Algorithm Working Group) pour estimer les débits des fleuves à partir des futures données SWOT et de manière plus générale, à partir de données issues des mesures satellitaires altimétriques et de masques d'eau (Durand et al., 2016 ; Gejadze et Malaterre, 2016, 2017 ; Oubanas et al., 2018a, b).

La méthodologie décrite dans ces articles Oubanas et al. (2018a, b) et dénommée SIC4DVar fonctionne avec des données de type purement SWOT (hauteur, largeur, pente), et est donc adaptée à un contexte non-jaugé sans aucune donnée in situ complémentaire. Si des données complémentaires sont disponibles, la méthode peut les intégrer (bathymétrie, frottements, autres données satellitaires, etc.)

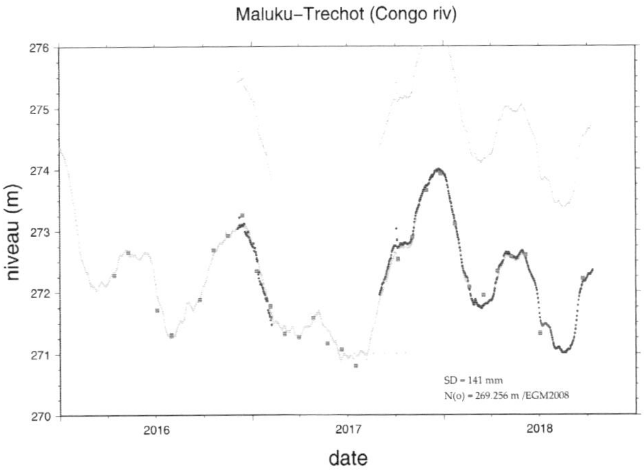

Figure 2. Comparaison entre les hauteurs d'eau estimées par altimétrie avec l'altimètre Sentinel 3 et les stations de Maluku-Trechot et de Brazzaville situées de part et d'autre de la trace altimétrique.

Des étapes de cette méthodologie ont été appliquées au fleuve Congo et sur l'Oubangui pour estimer les débits à partir des données altimétriques disponibles depuis 2002 grâce à diverses missions : Envisat, Saral, Jason 2 et 3, Sentinel 3A. Les données sont utilisées dans un algorithme d'inversion de l'équation de Manning-Strickler pour calculer une bathymétrie, ainsi que pour positionner et dimensionner les chutes qui sont détectées automatiquement.

Le modèle hydraulique généré avec la bathymétrie et les frottements obtenus permet ensuite différentes exploitations en simulation. Une exploitation du modèle est de générer des courbes de tarage $Q(Z)$ par des séries de calculs en régime permanent pour différents débits couvrant les gammes réalistes observées sur le terrain. Une fois ces courbes de tarage générées, elles peuvent être exploitées de manière opérationnelle et en temps réel pour calculer des débits à partir des cotes altimétriques satellitaires.

Même si de nombreuses étapes sont perfectibles et en cours de tests, les résultats obtenus et validés sur des données in situ sont très encourageants, par exemple sur l'Oubangui avec une très bonne reconstitution de la $Q(Z)$ in situ (Root-Mean-Square Error RMSE = 3 %), sur le Congo avec une très bonne reconstitution de la bathymétrie dans les zones où des relevés ADCP étaient disponibles.

3.4 Applications pour l'hydroélectricité et la navigation sur le bassin du Congo

3.4.1 Navigation

La Sangha, affluent de rive droite transfrontalier du fleuve Congo, draine un bassin versant de 213 000 km^2. Le transport fluvial est un enjeu majeur sur cette rivière, qui n'est pas navigable toute l'année. La hauteur d'eau lue quotidiennement à l'échelle limnimétrique d'Ouesso par le GIE-SCEVN (Groupement d'Intérêt Économique – Service Commun d'Entretien des Voies Navigables pour les deux États Congo et Centrafrique), est utilisée comme indicateur de navigabilité sur le fleuve Sangha.

Un modèle opérationnel de prévision des hauteurs d'eau à Ouesso au pas de temps journalier a ainsi été développé afin d'anticiper les arrêts de navigation ou d'adapter le chargement des bateaux. Ce modèle statistique, calé sur des données allant jusqu'à 2012, utilise les données de hauteur d'eau à Ouesso, ainsi que les données de précipitations GPCP (Adler et al., 2018).

L'ajout des données d'altimétrie spatiale dans le modèle a été étudié. L'apport des données a été mesuré par l'évolution du R^2 ajusté du modèle avec et sans les stations virtuelles sur la même période. Les résultats sont contrastés, mais d'autres types de modélisation (comme la modélisation distribuée avec assimilation des données spatiales), pourraient s'avérer plus adaptées aux données spatiales.

3.4.2 Hydroélectricité

L'hydroélectricité est par ailleurs un des enjeux identifiés dans le cadre du SDAGE de la CICOS (Schéma Directeur d'Aménagement et de Gestion des Eaux, CICOS, 2016a), dans un contexte où les populations locales ont un accès limité à l'électricité.

CNR a déterminé les tronçons les plus propices au développement de l'hydroélectricité dans les bassins du Congo et de l'Ogooué. La méthodologie a été établie à partir d'un indicateur de puissance hydroélectrique, calculée comme le produit du débit et de la pente du cours d'eau. Le débit a été reconstitué par similitude des bassins versants, à partir des débits des stations in situ et des données de précipitations du Climatic Research Unit (University of East Anglia). La pente entre deux stations consécutives est calculée comme la différence entre l'altitude moyenne aux stations virtuelles.

Des tronçons avec un potentiel hydroélectrique intéressant ont ainsi été identifiés, et correspondent à des zones de fort débit et/ou de forte pente. A l'inverse, des zones à plus faible potentiel sont apparues, telles que la Cuvette Centrale. La confrontation de ces résultats avec les projets hydroélectriques déjà identifiés sur le bassin confirme l'intérêt de la méthodologie comme outil d'aide à la décision permettant à la CICOS d'identifier les tronçons les plus intéressants et d'étudier ultérieurement des sites à équiper, en coopération avec les Etats. Néanmoins certaines limitations sont apparues, comme le manque de données dans les zones de montagne ou sur les cours d'eau de moindre largeur. L'utilisation d'un MNT suffisamment précis pourrait constituer une alternative à l'utilisation de l'altimétrie.

4 Le programme SWOT : présentation et attentes

Le programme satellitaire SWOT (Surface Water and Ocean Topography) du CNES et de la NASA est un pro-

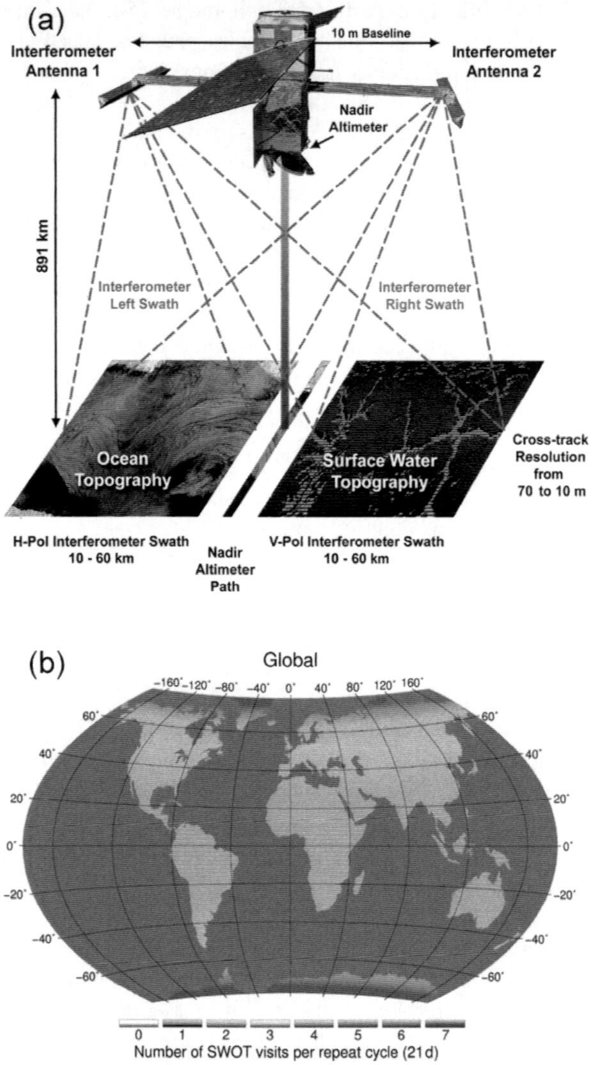

Figure 3. (**a**) Principes de la mesure SWOT (en haut). (**b**) Couverture spatiale et temporelle de SWOT (nombre de revisites de SWOT par cycle de 21 jours) (en bas).

jet de satellite d'observation de la Terre qui fournira à l'horizon 2023 et pour la première fois les variations spatio-temporelles des niveaux d'eau à l'échelle globale des grands fleuves, lacs et rivières et les niveaux des océans. Par rapport aux missions précédentes, SWOT propose une nouvelle technologie non encore testée depuis l'espace, l'altimétrie à large fauchée (Biancamaria et al., 2016, cf. Fig. 3).

L'altimétrie à large fauchée permettra d'obtenir des données continues, sur des tronçons de cours d'eau d'au moins 100 m de large, et non plus des données ponctuelles. Il est prévu que les données soient mises à disposition par biefs d'environ 10 km de long, ainsi qu'à des résolutions plus fines (ex. : sections tous les 200 m et même pixels de quelques dizaines de m^2). La précision attendue de SWOT d'environ 8 cm à l'échelle des biefs. La technologie à large fauchée permettra d'obtenir une couverture globale des altitudes, des plans d'eau et de pente, ce qui constituera un progrès par rapport à l'altimétrie conventionnelle.

5 Conclusion

L'altimétrie spatiale se présente comme une technologie innovante de mesure hydrologique au côté du réseau classique des stations in situ. La densification du réseau de mesure par les stations virtuelles sur le bassin du Congo, grâce aux différentes missions satellitaires altimétriques existantes, est un réel atout dans le suivi hydrologique de ce grand bassin transfrontalier et l'amélioration de la connaissance de ses ressources en eau.

Les appuis du groupe de travail sur l'hydrologie spatiale à la CICOS et l'arrivée prochaine du satellite SWOT et ses produits avals, permettront l'amélioration de la gestion et de la prise de décision quant à l'utilisation des ressources en eau du bassin du Congo (navigation fluviale, hydroélectricité, forêts inondées et biodiversité...) et d'autres grands bassins transfrontaliers (Niger et Sénégal en particulier), toujours en complémentarité des observations de terrain qui restent indispensables.

Disponibilité du code. Le code des logiciels utilisés n'est pas du domaine public. Il est la propriété de la CICOS qui représente ses 6 Etats membres.

Disponibilité des données. Les données utilisées ne sont pas du domaine public. Elles sont la propriété de la CICOS qui représente ses 6 Etats membres.

Collaborateurs. AA du Centre National d'Etudes Spatiales a mis en œuvre avec les collègues de l'Institut de Recherche pour le Développement les activités du projet relatives à l'altimétrie spatiale. POM de l'Institut national de recherche pour l'agriculture, l'alimentation et l'environnement a plus spécifiquement développé l'outil de Passage des altitudes aux débits. SL de la Compagnie Nationale du Rhône a réalisé les applications pour l'hydroélectricité et la navigation sur le bassin du Congo. GGG et BLT de la Commission Internationale du bassin Congo-Oubangui-Sangha ont contribué à l'ensemble des activités, en tant que maître d'ouvrage. CB de l'Office International de l'Eau a préparé et coordonné la rédaction de l'article avec la contribution de tous les co-auteurs.

Intérêts concurrents. Les auteurs déclarent qu'ils n'ont aucun conflit d'intérêts.

Clause de non-responsabilité. Publisher's note : Copernicus Publications remains neutral with regard to jurisdictional claims in published maps and institutional affiliations.

Déclaration du numéro spécial. This article is part of the special issue "Hydrology of Large River Basins of Africa". It is a result of the 4th International Conference on the "Hydrology of the Great Rivers of Africa", Cotonou, Benin, 13–20 November 2021.

Remerciements. Nos vifs remerciements vont à Gil Mahé de l'Institut de Recherche pour le Développement.

Financement. This research has been supported by the Agence Française de Développement (grant no. CZZ2054).

Références

Adler, R., Wang, J., Sapiano, M., Huffman, G., Bolvin, D., Nelkin, E., and NOAA CDR Program : Global Precipitation Climatology Project Climate Data Record, Version 1.3 (Daily) Precipitation, NOAA National Centers for Environmental Information, https://doi.org/10.7289/V5RX998Z, 2018.

Biancamaria, S., Lettenmaier, D. P., and Pavelsky, T. M. : The SWOT mission and its capabilities for land hydrology, Surv. Geophys., 37, 307–337, https://doi.org/10.1007/s10712-015-9346-y, 2016.

Calmant, S. and Seyler, F. : Continental surface waters from satellite altimetry, C. R. Geosci., 338, 1113–1122, https://doi.org/10.1016/j.crte.2006.05, 2006.

Calmant, S., Seyler, F., and Cretaux, J.-F. : Monitoring Continental Waters by Satellite Altimetry, Surv. Geophys., 29, 247–269, https://doi.org/10.1007/s10712-008-9051-1, 2008.

CICOS : Accord instituant un régime fluvial uniforme et créant la CICOS, accessible à l'adresse suivante : https://www.cicos.int/wp-content/uploads/Accord-Instituant-un-R%C3%A9gime-Fluvial-Uniforme-et-cr%C3%A9ant-la-CICOS...…-2.pdf (la date du dernier accès : 22 octobre 2021), 1999.

CICOS : Additif à l'Accord instituant un régime fluvial uniforme et créant la CICOS, accessible à l'adresse suivante : https://www.cicos.int/wp-content/uploads/Additif_a_l_Accord_Instituant_un_Regime_Fluvial_Uniforme_et_creant_la_CICOS.pdf (la date du dernier accès : 22 octobre 2021), 2007.

CICOS : "Schéma Directeur d'Aménagement et de Gestion des Eaux de la CICOS. Programme de Mesures 2016–2020", 161 pp., accessible à l'adresse suivante : https://www.cicos.int/publications/cicos-sdage-programme-des-mesures/ (la date du dernier accès : 22 octobre 2021), 2016a.

CICOS : Rapport sur l'amélioration et la validation de données altimétriques spatiales sur le bassin du Congo, CICOS, Kinshasa, RD Congo, 45 pp., 2016b.

CICOS : Rapport sur la mise à disposition de données altimétriques satellitaires à la CICOS, CICOS, Kinshasa, RD Congo, 17 pp., 2016c.

Cretaux, J. F., Jelinski, W., Calmant, S., Kouraev, A., Vuglinski, V., Berge Nguyen, M., Gennero, M-C., Nino, F., Abarca Del Rio, R., Cazenave, A., and Maisongrande, P. : SOLS : A lake database to monitor in the Near Real Time water level and storage variations from remote sensing data, Adv. Space Res., 47, 1497–1507, https://doi.org/10.1016/j.asr.2011.01.004, 2011.

Dixon, H., Sandström, S., Cudennec C., Lins H., Abrate T., Bérod D., Chernov I., Ravalitera N., Sighomnou D., and Teichert F. : Intergovernmental cooperation for hydrometry – what, why, how ?, Hydrolog. Sci. J., Special Issue : Hydrological data : opportunities and barriers, https://doi.org/10.1080/02626667.2020.1764569, 2020.

Durand, M., Gleason, C., Garambois, P.-A., Bjerklie, D., Smith, L., Roux, H., Rodriguez, E., Bates, P., Pavelsky, T., and Monnier, J. : An intercomparison of remote sensing river discharge estimation algorithms from measurements of river height, width, and slope, Water Resour. Res., 52, 4527–4549, https://doi.org/10.1002/2015WR018434, 2016.

Gejadze, I. Y. and Malaterre, P.-O. : Design of the control set in the framework of variational data assimilation, J. Comput. Phys., 325, 358–379, https://doi.org/10.1016/j.jcp.2016.08.02, 2016.

Gejadze, I. Y. and Malaterre, P.-O. : Discharge estimation under uncertainty using variational methods with application to the full Saint-Venant hydraulic network model, Int. J. Numer. Meth. Fl., 83, 405–430, https://doi.org/10.1002/fld.4273, 2017.

Oubanas, H., Gejadze, I. Y., Malaterre, P.-O., and Mercier, F. : River discharge estimation from synthetic SWOT-type observations using variational data assimilation and the full Saint-Venant model, J. Hydrol., 559, 638–647, https://doi.org/10.1016/j.jhydrol.2018.02.004, 2018a.

Oubanas, H., Gejadze, I. Y., Malaterre, P.-O., Durand, M., Frasson, R., Wui, R., and Domeneghetti, A. : Discharge estimation in ungauged basins through variational data assimilation : The potential of the SWOT mission, Water Resour. Res., 54, 2405–2423, https://doi.org/10.1002/2017WR021735, 2018b.

Pecora, S. and Lins, H. : E-monitoring the nature of water, Hydrolog. Sci. J., 65, 683–698, https://doi.org/10.1080/02626667.2020.1724296, 2020.

Tauro, F., Selker, J., van de Giesen, N., Abrate, T., Uijlenhoet, R., Porfiri, M., Manfreda, S., Caylor, K., Moramarco, T., Benveniste, J., Ciraolo, G., Estes, L., Domeneghetti, A., Perks, M., Corbari, C., Rabiei, E., Ravazzani, G., Bogena, H., Harfouche, A., Brocca, L., Maltese, A., Wickert, A., Tarpanelli, A., Good, S., Lopez Alcala, J. M., Petroselli, A., Cudennec, C., Blume, T., Hut, R., and Grimaldi, S. : Measurements and observations in the XXI century : innovation and multi-disciplinarity to sense the hydrological cycle, Hydrolog. Sci. J., 63, 169–196, https://doi.org/10.1080/02626667.2017.1420191, 2018.

Proc. IAHS, 384, 43–48, 2021
https://doi.org/10.5194/piahs-384-43-2021

Open Access PIAHS

Studying recent hydroclimatic variability in Madagascar despite deficient measurement networks: use of CHIRPS and GRACE data at the scale of the Mahajunga province

Sylvain Bigot[1], **Dominique Dumas**[2], **Télesphore Y. Brou**[3], **Rivo Ramboarison**[4], **Samuel Razanaka**[5], **and Nathalie Philippon**[1]

[1]UGA – Institut des Géosciences de l'Environnement (IGE), UMR 5001, 38000 Grenoble, France
[2]Université Lyon 3 – Environnement Ville Société (EVS), UMR 5600, 69362 Lyon, France
[3]Université de La Réunion – Laboratoire OIES, Océan Indien: Espace et Sociétés, 97744 Saint-Denis, France
[4]Cabinet d'études BEST-IDEES, Antananarivo, Madagascar
[5]Centre National de Recherche en Environnement (CNRE), Antananarivo, Madagascar

Correspondence: Sylvain Bigot (sylvain.bigot@univ-grenoble-alpes.fr)

Published: 16 November 2021

Abstract. Given the lack of in situ hydroclimatic measurements and networks in Madagascar, the GRACE (2003–2016) spatial gravimetry data, combined with other satellite data such as CHIRPS rainfall estimates or fire monitoring using GFED products, make it possible to establish an interannual assessment of certain climatic and environmental covariations at the northwest scale of the country. The results show a negative trend in continental rainfall and water content, especially after 2007, but also a time lag in the linear variations and trends of the Water Equivalent Height as well as the number of detected fires (variable indirectly measuring the pressure of deforestation by slash and burn agriculture).

1 Introduction and environmental context

The intensification of hydroclimatic risks requires the Malagasy authorities to adapt their national rural and urban development strategies and to be able to implement resilience policies. Especially since, at the same time, the island has been affected for several decades by intense deforestation which fundamentally modifies bio-geophysical and ecosystemic processes. Depending on these threats, Madagascar therefore urgently needs reliable hydroclimatic diagnostics for its many national socio-environmental issues (e.g., agriculture, water resources, biodiversity, and health situation). Despite the objectives of strengthening hydrometeorological measurement services and an adequate institutional framework in Madagascar, the national network and infrastructure remain deficient. With a surface area of 587 295 km^2 (i.e. the 5th largest island in the world), the country currently has only 28 operational meteorological stations listed by the WMO. The hydrological network is in a similar state of obsolescence, and the flows of the large Madagascar rivers are no longer monitored continuously. The national hydroclimatic balances are generally based on imprecise, old (Chaperon et al., 1993) and anisotropic estimates that do not allow for a effective study of regional or intra-seasonal variations, which are decisive for the many Madagascar environmental problems (Bigot et al., 2018).

An alternative solution for studying hydroclimatic variability is the use of satellite estimates. The data used here to analyze rainfall variations and soil water storage are CHIRPS (Climate Hazards Group InfraRed Precipitation with Station data) and GRACE (Gravity Recovery And Climate Experiment) data. The study focuses on the Boeny region (Fig. 1), in the province of Mahajanga (in the northwest of the country) where knowledge of the seasonal evolution of water resources is lacking, but crucial for socio-economic purposes. Given the water resources in the Betsiboka River (605 km, about 49 000 km^2), the largest river in Madagascar, this region is considered the 'water tower' of the agricultural plain of Marovoay (located just downstream), the second largest rice growing area in Madagascar. But intense deforestation

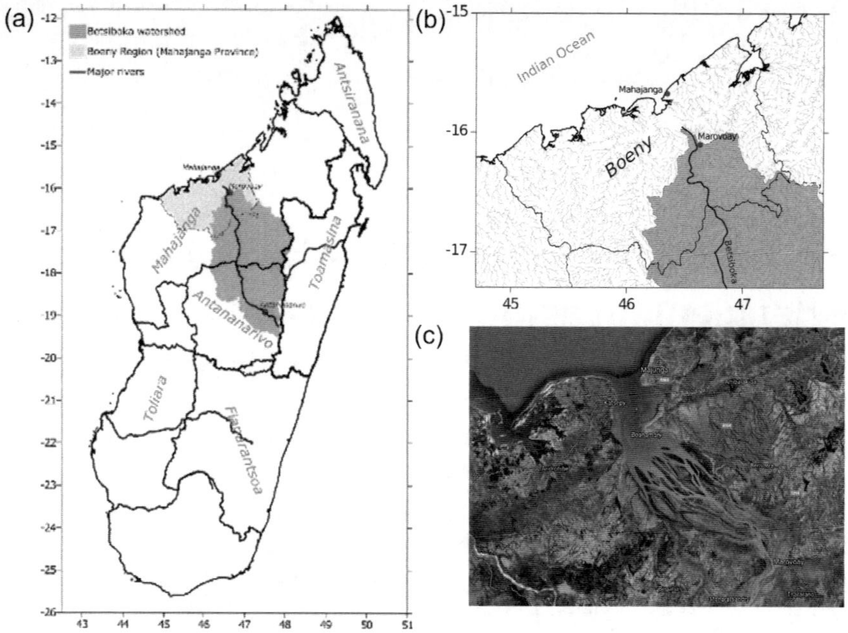

Figure 1. **(a)** Location of the Betsiboka river catchment area (in blue) and the Boeny region (in grey). **(b)** Zoom on the Boeny region and the regional hydrographic network (data from HydroRIVERS 1.0 database). **(c)** Zoom on the Betsiboka River delta and Bombetoka Bay (TerraMetric image 2020, from © Google Maps).

of the dry forests has considerably modified the water supply by accentuating the runoff during the rainy season, and by decreasing the regulating effect of the soils on the runoff.

The soils, deprived of the protection of the forest cover, are subject to very high erosion, which causes silting of the rice plots, and in turn contributes to deforestation for access to other arable land. This river catchment contributes to laterite and sediment transport from the uplands to its mouth in Bombetoka Bay and the Mozambique Channel (Fig. 1), with an estimated mean annual flow of $271\,m^3\,s^{-1}$ and seasonal variations ranging from 400 to $4500\,m^3\,s^{-1}$ at the watershed scale (Duvail et al., 2017), given the uncertainties due to the lack of recent in situ observations. Interannual classifications from remote sensing show that the total area occupied by the delta wetland was about $200\,km^2$ in 1973, increasing to $765\,km^2$ in 1989 and $1600\,km^2$ in 2000, an increase of 800 % (Raharimahefa and Kusky, 2010).

The main objective of the study is to compare the interannual variations of precipitation analyzed with CHIRPS data to the hydrological time series of GRACE over the period 2003–2016. The challenge is to better define their correlation, which would eventually allow to use them to better determine and calibrate some uses of hydro- or agro-climatic modeling that are currently unavailable or deficient in terms of in situ data in Madagascar.

A secondary objective is also to analyze the time lag between rainfall–hydrological variations analyzed through CHIRPS and GRACE data, and the number of fires detected by satellite (which is a direct indicator of deforestation).

2 The data analyzed and the methodology used

2.1 GRACE data (2003–2016)

2.1.1 Origin and interest in hydroclimatology

The GRACE satellite mission is a joint project between the National Aeronautic and Space Administration (NASA, USA) and the Deutschen Zentrum für Luftund Raumfahrt (DLR, Germany): http://grace.jpl.nasa.gov (last access: 29 November 2019). Launched into orbit on 17 March 2002, GRACE consists of two twin co-orbital and microwave telemetry interconnected satellites, flying in nearly identical polar orbits with an inclination of 89.5° and at an altitude of 500 km, with the instrumental binomial approximately 220 km apart (NASA, 2004). The GRACE mission ended on 12 October 2017 due to a power outage. The new GRACE-FO (Follow-On) mission was launched on 22 May 2018 to continue and enhance these acquisitions. The satellites fly over the entire Earth's surface in about 30 d, providing monthly estimates of a global gravity model with a surface spatial resolution of about 380 km.

The accuracy of continental water heights is about 0.7 cm for a drainage area of $400\,000\,km^2$, and 0.3 cm for one of $4\,000\,000\,km^2$ (Swenson et al., 2003). Errors were estimated to be about 40 mm at the equator, decreasing to 15 mm in the polar regions. The main variations in gravity signals are due to changes in the distribution of water, snow and ice stored on Earth. On a continental scale, the analysis of water stocks and the derivation of water balances via GRACE provide a holistic view of total flows within a basin, as it integrates the flow of rivers, but also surface or deep water inflows and outflows. The regionalized experiments identified at the tropical scale are, for example, balances for the Congo, Amazon, Nile and Volta basins, these very large catchment areas being well adapted to the accuracy of GRACE signatures (Ahmed et al., 2014). Work has shown that once filtered, temporal variations in GRACE data acquired over central Africa and up to 10° S are essentially controlled by components of the hydrological cycle, especially runoff and recharge (Ahmed et al., 2014). GRACE estimates of land water storage now provide a new assimilation component for the simulation of continental surface models, and can be used, in conjunction with other remote sensing data, to develop operational hydroclimatic services (Dai, 2011; Frappart and Ramillien, 2018; Tapley et al., 2019).

2.1.2 The choice of GRACE Mascon products

Provided free by various distribution organizations: the Goddard Space Flight Center (GSFC: https://earth.gsfc.nasa.gov/geo/, last access: 29 November 2019), the Jet Propulsion Laboratory (JPL: https://grace.jpl.nasa.gov/, last access: 29 November 2019), the Center for Space Research (CSR: http://www2.csr.utexas.edu/grace/, last access: 29 November 2019) or the German Research Centre for Geosciences (GFZ: http://gravis.gfz-potsdam.de/home, last access: 29 November 2019), there are level 1 products (measured geophysical harmonics) and level 2 advanced solutions derived from the raw measurements (gravity fields and products expressed in geoid height or Water Equivalent Height – WEH). The data are used either with the initial spatial resolution of the spherical harmonic solutions, or with regionalized solutions that allow a better estimation of the variance of the water heights without being affected by spectral truncation effects and allow a better geographical localization of the hydrological patterns. In particular, there is the so-called "Mascon" (for "mass concentration") approach, which recalculates discrete cells from the original data, each with its own gravity signal (Watkins et al., 2015).

However, even if the spatial resolution of the GRACE Mascon data is better (at 1° of arc resolution), their fundamental temporal definition is finally close to that defined with the original GRACE data due to the spatial autocorrelation.

2.2 The rainfall estimates CHIRPS

Rainfall variability is analyzed using CHIRPS (Climate Hazards Group InfraRed Precipitation with Station data) data available on a quasi-planetary scale, starting in 1981 with a daily or 10 d resolution. They integrate both satellite data (0.05° resolution) and time series of in situ reference stations (Funk et al., 2015). Although not obviously as reliable a measurement as in situ precipitation, and despite the uncertainties inherent in these satellite products, the performance of CHIRPS data is excellent to study environmental processes (Dembélé and Zwart, 2016). The estimates obtained are sufficiently accurate to study tropical rainfall patterns and their intra-seasonal and interannual variability over a multidecadal temporal depth, particularly in regions with poor meteorological networks, as is typically the case in Madagascar (Bigot et al., 2018).

2.3 Main methodological points

The methodology used is based on a classical descriptive statistical approach, which seeks to identify synchronous and asynchronous linear relationships between the different indicators selected. Two steps are used:

1. the analysis of linear correlations between CHIRPS rainfall series and GRACE hydrological series (WEH: Water Equivalent Height) at Boeny Region scale between January 2003 and December 2016, and from standardized values;

2. the analysis of linear co-variations between these hydroclimatic series and time series of some environmental indicators: the PDSI (Palmer Drought Severity Index; from TerraClimate data) and regional fire count (from Global Fire Emissions Database data); the main objective is to be able to determine if there is a lag between hydroclimatic variations and those theoretically directly related to the risk of drought or fire intensity.

3 Results

At the global scale, GRACE trends suggest an increase in water storage at high and low latitudes, and rather a decrease at mid-latitudes with drying trends (Tapley et al., 2019). Analysis of the GRACE Mascon (2002–2016) time series at the scale of Madagascar shows a rather different and opposite trend to that of much of Africa, with a generalised decline in Water Equivalent Height – WEH (Fig. 2). The strongest downward trends ($> 0.6\,\mathrm{cm\,yr^{-1}}$) are observed on the eastern facade, plains and central reliefs. The upper catchment area of the Betsiboka shows a trend of $-0.62\,\mathrm{cm\,yr^{-1}}$ of the WEH, with values decreasing towards the downstream basin to reach $-0.36\,\mathrm{cm\,yr^{-1}}$.

In Madagascar these hydrological trends are attributed to natural variations, anthropogenic climate change, and human water management and land use practices. Rather, increased precipitation and/or decreased temperatures explain an increase in total water levels (due to reservoir storage and decreased evaporation), and vice versa. However, anthropogenic practices can modulate these variations depending

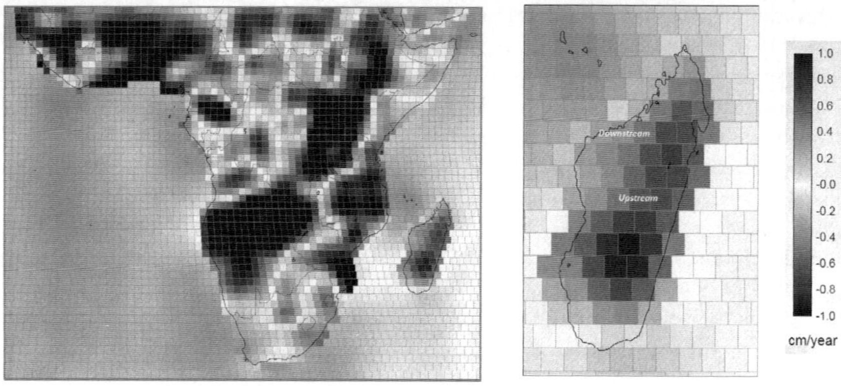

Figure 2. Trend (in cm yr^{-1}) of the Water Equivalent Height (WEH) at the spatial resolution of GRACE Mascon cells ($\sim 1°$) over the 2002–2016 period on the African continent scale (left panel) and for a zoom on Madagascar (with the location of the upstream and downstream regions of Betsiboka); adapted from Goddard Space Flight Center and Colorado Center for Astrodynamics Research data.

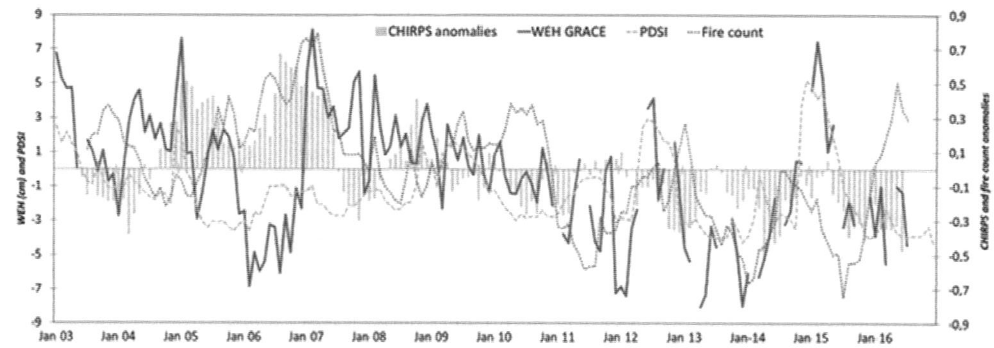

Figure 3. Covariations at Boeny Region scale between January 2003 and December 2016 of rainfall anomalies (CHIRPS data, reduced centred values), Water Equivalent Height (WEH, in cm; GRACE Mascon data), PDSI (Palmer Drought Severity Index; TerraClimate data) and regional fire count (Global Fire Emissions Database data).

on climate variability, or even reverse the expected evolutions. Goel (2006), for example, showed at the scale of the Congo Basin that the decrease in groundwater levels in a context of stable rainfall would be mainly explained by changes in land use and deforestation that favour runoff to the detriment of infiltration and recharge.

At the scale of the Boeny region, several hydroclimatic and land-use related covariations can be discriminated, by studying synchronous or asynchronous linear correlations (Fig. 3):

- The CHIRPS time series show a significant negative trend in regional rainfall, especially after 2007 (Bigot et al., 2019), but without having been able to analyze at the same time the hydrological variations.

- The negative trend in continental water resources transcribed by the GRACE 2003–2016 data is also strong and with a visible break around 2007; but several interannual trends correspond to ocean–atmospheric variations imposed by the Indian Ocean Dipole Oscillation (Bigot et al., 2020). For example, the three positive events in 2006, 2012 and 2015 coincide with a significant increase in precipitation at the scale of the study area, and an increase in gravimetrically estimated water heights in the ground.

- The Palmer Drought Severity Index (PDSI) from TerraClimate data, based on a rainfall/temperature calculation, reproduces the main variations observed with GRACE, but is much less accurate because it does not take into account continental and regional components (Tapley et al., 2019).

- As observed in other regional studies and explained by the response of continental mechanisms, there is an increase in linear correlation with a 5 to 6 month lag between rainfall and the variation in moisture and ground storage observed with GRACE (Fig. 4).

- The interannual evolution of the fire counts, a direct index of agricultural activities, deforestation and land-use changes in the region (Bigot et al., 2019), is calculated over the same period for the study area using satellite data from the Global Fire Emissions Database (GFED)

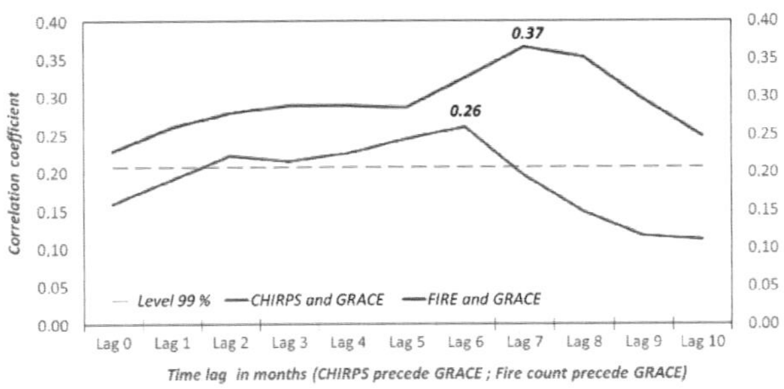

Figure 4. Synchronous linear correlation (lag 0) and up to an 10-month time lag between CHIRPS precipitation and WEH GRACE monthly variations on the 2002–2016 period (in blue) and between fire count and WEH GRACE (in red); the 99 % significance level is indicated by a dashed line.

portal. The interannual covariation between rainfall anomalies and fire count is strong but more complex than that evident on a seasonal scale, and the succession between the rainy season and the wildfire season. The most significant correlation ($R = 0.37$) occurs when fires precede hydrological changes by 7 months (Fig. 4), which here seems to suggest that a change in soil moisture content could be the consequence of the number of fires (and thus deforestation and land-use changes) recorded earlier in the year.

This confirms certain elements of Maina et al. (2012) on Madagascar which shows that there is a significant covariance between forest cover, anthropogenic pressure, river flow, sediment load and interannual and seasonal coastal environmental indicators. With this regional example, it is therefore clear that the GRACE data, apart from their evident contribution in hydroclimatology, could also bring some diagnoses in the sensitivity and evolution of Malagasy forest ecosystems.

However, due to the very short analysis period here (14 years), the time links highlighted on a monthly scale must be put into perspective, as they do not have sufficient statistical robustness to be able to determine and validate a stable rainfall/WEH/PDSI/fires empirical model. This would require, on the one hand, the analysis of rainfall over a longer period (e.g. with the CHIRPS data available since 1979); but also, on the other hand, the availability of updated in situ rainfall and hydrological series in Madagascar, in order to validate these remote sensing products.

4 Conclusion

The contribution of GRACE data allows a more detailed study of hydroclimatic and environmental variations because Madagascar has almost no in situ monitoring network. On the one hand, they allow monitoring and therefore also modelling, the asynchronous consequences of rainfall variations on regional water resources. But more broadly, the GRACE data allow for integrated studies of the relationships between water resource variability and land use change influenced by deforestation and fire use. Of course, a large part of the variance of the components of the water cycle also remains imposed by certain ocean-atmosphere variations in the Indian Ocean or global changes such as the increase in temperatures that impose flow, evapotranspiration, water storage and the seasonal intensity of vegetation fires.

Code availability. All software codes can be requested from the contact author.

Data availability. GRACE land data are available at https://doi.org/10.5067/TELND-3AC64 (Landerer, 2021) and Landerer and Swenson (2012), supported by the NASA MEaSUREs Program.

Author contributions. SB conceived of the presented idea and performed the computations and figures. DD and TYB verified the analytical methods and contributed to the final manuscript. All authors provided critical feedback and helped shape the research, analysis and manuscript.

Competing interests. The contact author has declared that neither they nor their co-authors have any competing interests.

Special issue statement. This article is part of the special issue "Hydrology of Large River Basins of Africa". It is a result of the 4th International Conference on the "Hydrology of the Great Rivers of Africa", Cotonou, Benin, 13–20 November 2021.

References

Ahmed, M., Sultan, M., Wahr, J., and Yan, E.: The use of GRACE data to monitor natural and anthropogenic induced variations in water availability across Africa, Earth-Sci. Rev., 136, 289–300, 2014.

Bigot, S., Dumas, D., Brou, Y. T., Ramboarison, R., Razanaka, S., Descroix, L., Philippon, N., and Gwiazdzinski, L.: La difficulté du monitoring climatique à une échelle fine dans une région fortement impactée par la déforestation tropicale: la forêt sèche du parc national d'Ankarafantsika (Madagascar), Publications de l'Association Internationale de Climatologie, 31, 356–361, 2018.

Bigot, S., Dumas, D., Brou, T., Curt, T., and Razanaka, S.: Feux de végétation et conditions pluviométriques à Madagascar (2001–

2017), Publications de l'Association Internationale de Climatologie, 32, 169–174, 2019.

Bigot, S., Dumas, D., Brou, T., Ramboarison, R., Razanaka, S., and Philippon N.: Etude hydroclimatique dans le nord-ouest de Madagascar: apport des données satellitaires GRACE, Publications de l'Association Internationale de Climatologie, 33, 121–126, 2020.

Chaperon, P., Danloux, J., and Ferry, L.: Fleuves et rivières de Madagascar, in: Monographies hydrologiques, 10, ORSTOM, Paris, France, p. 874, 1993.

Dai, A.: Characteristics and trends in various forms of the Palmer Drought Severity Index during 1900–2008, J. Geophys. Res., 116, 1–26, 2011.

Dembélé, M. and Zwart, S. J.: Evaluation and comparison of satellite-based rainfall products in Burkina Faso, West Africa, Int. J. Remote Sens., 37, 3995–4014, 2016.

Duvail, S., Hamerlynck, O., Paron, P., Hervé, D., Nyingi, W. D., and Leone, M.: The changing hydro-ecological dynamics of rivers and deltas of the Western Indian Ocean: Anthropogenic and environmental drivers, local adaptation and policy response, C. R. Geosci., 349, 269–279, 2017.

Frappart, D. and Ramillien, G.: Monitoring groundwater storage changes using the Gravity Recovery and Climate Experiment (GRACE) Satellite Mission: A review, Remote Sens., 10, 829–853, https://doi.org/10.3390/rs10060829, 2018.

Funk, C., Peterson, P., Landsfeld, M., Pedreros, D., Verdin, J., Shukla, S., Husak, G., Rowland, J., Harrison, L., Hoell, A., and Michaelsen, J.: The climate hazards infrared precipitation with stations – a new environmental record for monitoring extremes, Scient. Data, 2, 150066, https://doi.org/10.1038/sdata.2015.66, 2015.

Goel, P.: Water pollution: Causes, Effects and Control, 2nd Edn., New Age International Publisher, New Delhi, 432 pp., 2006.

Landerer, F.: TELLUS_GRAC_L3_CSR_RL06_LND_v04, Ver. RL06 v04, PO.DAAC [data set], CA, USA, https://doi.org/10.5067/TELND-3AC64, 2021.

Landerer, F. W. and Swenson, S. C.: Accuracy of scaled GRACE terrestrial water storage estimates, Water Resour. Res., 48, W04531, https://doi.org/10.1029/2011WR011453, 2012.

Maina, J., de Moel, H., Vermaat, J. E., Bruggemann, J. H., Guillaume, M. M., Grove, C. A., Madin, J. S., Mertz-Kraus, R., and Zinke, J.: Linking coral river runoff proxies with climate variability, hydrology and land-use in Madagascar catchments, Mar. Pollut. Bull., 64, 2047–2059, 2012.

NASA: CALIPSO, CloudSat, GRACES: Sciences writers's guide – A guide for reporters to understand the mission and purpose of NASA's satellites, sciences mission directorate, NP-2005-8-709-GSFC, NASA, Greenbelt, MD, USA, 44 pp., 2004.

Raharimahefa, T. and Kusky, T. M.: Environmental monitoring of Bombetoka Bay and the Betsiboka Estuary, Madagascar, using multi-temporal satellite data, J. Earth Sci., 21, 210–226, 2010.

Swenson, S., Wahr, J., and Milly, P. C. D.: Estimated accuracies of regional water storage variations inferred from the Gravity Recovery and Climate Experiment (GRACE), Water Resour. Res., 39, 1223, https://doi.org/10.1029/2002WR001808, 2003.

Tapley, B. D., Watkins, M. M., Flechtner, F., Reigber, C., Bettadpur, S., Rodell, M., Sasgen, I., Famiglietti, J. S., Landerer, F. W., Chambers, D. P., Reager, J. T., Gardner, A. S., Save, H., Ivins, E. R., Swenson, S. C., Boening, C., Dahle, C., Wiese, D. N., Dobslaw, H., Tamisiea, M. E., and Velicogna, I.: Contributions of GRACE to understanding climate change, Nat. Clim. Change, 358, 358–369, 2019.

Watkins, M. M., Wiese, D. N., Yuan, D.-N., Boening, C., and Landerer, F. W.: Improved methods for observing Earth's time variable mass distribution with GRACE using spherical cap mascons, J. Geophys. Res.-Solid, 120, 2648–2671, https://doi.org/10.1002/2014JB011547, 2015.

Proc. IAHS, 384, 49–56, 2021
https://doi.org/10.5194/piahs-384-49-2021

Open Access
PIAHS

Hydrology of Large River Basins of Africa

Estimation par approche inverse d'un champ de transmissivité sur l'ensemble de l'aquifère du Continental Terminal (CT) d'Abidjan

Kouamé Auguste Kouassi[1], William Francis Kouassi[1], Oi Mangoua Jules Mangoua[2], Philippe Ackerer[3], Gountôh Aristide Douagui[1], and Issiaka Savané[1]

[1]UFR Sciences et Gestion de l'Environnement, Laboratoire Geosciences et Environnement, Université NANGUI ABROGOUA, 02 BP 801 Abidjan 02, Côte d'Ivoire
[2]UFR Environnement, Laboratoire des Sciences et Technologies de l'Environnement, Université Jean Lorougnon Guedé Daloa, BP 150 Daloa, Côte d'Ivoire
[3]Département « Laboratoire d'Hydrologie et de Géochimie de Strasbourg », Université de Strasbourg, Strasbourg, France

Correspondence: Kouamé Auguste Kouassi (kouameauguste@yahoo.fr)

Published: 16 November 2021

Résumé. La caractérisation hydraulique des systèmes aquifères est importante pour le développement de scénarios d'exploitation et de stratégies de gestion des eaux souterraines. En particulier dans les aquifères lithologiquement hétérogènes, il ne faut pas négliger les variations de transmissivité (T) à l'échelle locale. Les valeurs de transmissivité à l'échelle locale sont habituellement dérivées d'essais de pompage, mais dans la plupart des cas, leur nombre et leur disponibilité sont plutôt limités. De même, la mesure directe de la transmissivité sur l'ensemble d'un aquifère s'avère onéreux et quasi-impossible techniquement. Dans de telles situations, la modélisation hydrodynamique par approche inverse se présente comme la solution appropriée. Dans cet article, le champ de transmissivité réel de l'aquifère du Continental Terminal (CT) d'Abidjan est recherché par une paramétrisation multi-échelle qui permet de contourner le problème de changement d'échelle et de déterminer ce paramètre hydrodynamique sur l'ensemble de la nappe. Cette modélisation hydrogéologique a été réalisée à partir des modèles conceptuels Est-Ouest et Nord-Sud qui montrent que le CT est un aquifère constitué de deux couches principales n3 et n4, respectivement formées de sables grossiers de 90 m d'épaisseur et de sables fins à moyen de 80 m d'épaisseur. Le calage du modèle bicouche s'est effectué en régime transitoire de 12 pas de temps mensuels avec des données piézométriques de 24 piézomètres, des données de production de 71 forages et des valeurs de transmissivité connues de 40 forages dans un maillage élément fini constitué de 534 mailles et 320 noeuds. Ainsi, ce modèle de la nappe du CT a identifié une structure de dimension 153 nœuds comme la structure la plus proche de celle de l'aquifère du CT. Les valeurs de T variant entre $5{,}4.10^{-5}$ et $1\ m^2\ s^{-1}$ du champ de transmissivité associé à cette dimension optimale, ont été comparées aux valeurs de T d'autres champs de transmissivités publiées dans des études en Afrique et dans le monde. Ces valeurs identifiées sont plausibles et présentes une bonne structure dans l'ensemble.

Published by Copernicus Publications on behalf of the International Association of Hydrological Sciences.

1 Introduction

Les nombreuses sollicitations de l'eau ont pour conséquence de diminuer la réserve d'eau disponible partout dans le monde (De Lacour, 2012). Abidjan, la plus grande ville de Côte d'Ivoire n'échappe pas à cette réalité mondiale. Elle représente moins de 1 % du territoire national mais produit et consomme à elle seule près de 70 % de l'eau potable du pays (DHH, 2001 ; SODECI, 2017). Abidjan est essentiellement alimentée en eau potable à partir des eaux souterraines de la nappe du Continental Terminal (CT). Les fortes augmentations de la demande en eau d'Abidjan (SOGREAH, 1997 ; Kouassi, 2013 ; SODECI, 2017) avec très peu de recharge ont mis à rude épreuve ses eaux souterraines, ce qui s'est traduit par une baisse générale des niveaux statiques (Ahoussi, 2008 ; Kouassi, 2013). Face à la pression démographique que connait Abidjan et au besoin en eau croissant de sa population nous nous sommes interrogés sur la capacité de la nappe d'Abidjan à produire suffisamment d'eau pour satisfaire ces multiples besoins. Cette question trouve sa réponse dans la connaissance de la transmissivité (T) sur l'ensemble de l'aquifère du Continental Terminal. En effet, T est le principal paramètre hydraulique qui contrôle les prélèvements des eaux souterraines. La méthode la plus fiable pour estimer la transmissivité d'un aquifère est d'effectuer une analyse d'essai de pompage ponctuel (Mjemah et al., 2009). Toutefois, cette approche de mesure directe de la transmissivité sur l'ensemble de la nappe nécessite des ressources techniques et financières énormes qu'il est pratiquement et économiquement impossible de satisfaire (Kouassi, 1999). Dans de tels cas, la modélisation hydrodynamique via une approche inverse qui consiste à rechercher les valeurs inconnues des propriétés de l'aquifère (Transmissivité T) en utilisant les variables d'état de celui-ci (charge hydraulique h) connues aux points de mesure, s'impose comme la solution idoine. L'objectif de ce travail est d'estimer un champ de transmissivité sur l'ensemble de l'aquifère du Continental Terminal d'Abidjan. Ce champ renseignera donc les gestionnaires de la nappe d'Abidjan sur la disponibilité et les potentialités de captage de la ressource en eau dans cet aquifère. L'identification de ce champ a été possible grâce à un modèle numérique réalisée à partir du code HPP-GMS basé sur la méthode des éléments finis de Galerkin développée autour d'éléments triangulaires.

2 Methodologies

Dans cette étude, la détermination du champ de transmissivité existant de l'aquifère du Continental Terminal s'est construite autour de trois principales étapes que sont la caractérisation physique de l'aquifère (modèle conceptuel), la conceptualisation du modèle numérique de l'aquifère (maillage, données d'entrées, conditions aux limites etc. ...) et la recherche de la structure et de la dimension la plus proche du milieu naturel suite à une série de simulations des charges par la méthode inverse de paramétrisation multi-échelle.

2.1 Caractérisation physique et modèle numérique de l'aquifère du CT

La caractérisation physique de l'aquifère du CT s'est résumé par des modèles conceptuels servant à simplifier la structure réelle et complexe de l'aquifère en des schémas géologiques fonctionnels et par des hypothèses de fonctionnement établis. Ainsi, le modèle conceptuel de l'aquifère du CT réalisé à partir de 185 fiches techniques de forages, présente une structure bicouche dont l'épaisseur totale varie entre 30 et 160 m dans la direction Est-Ouest et varie également entre 30 et 120 m dans la direction Nord-Sud (Fig. 1).

Cet ensemble ne forme qu'un seul aquifère libre constitué majoritairement sables fins à moyens et de sable grossier avec une épaisseur moyenne captante de 90 m. Les valeurs de transmissivités ponctuelles déduites des essais de pompages de 40 forages irregulièrement repartis, oscillent entre les gammes 10^{-6} et 10^{-1} $m^2 s^{-1}$. Le modèle numérique de la nappe du CT d'Abidjan est en maillage éléments finis comportant 534 mailles pour 320 nœuds (Fig. 1). Les conditions aux limites se résument à des potentiels imposés aux frontières Nord et Sud, à des flux imposés nuls aux limites Ouest et Nord-Est dictés par la perpendicularité des lignes de courants aux equipotentielles et à des flux imposés non nuls entrant à la frontière Sud-Est. La prise en compte des échanges entre la nappe du Continental Terminal et ses rivières ne concerne que les rivières Gbangbo, Banco et Bété. Ces échanges mettent à contribution le niveau de la nappe, la cote du fond de la rivière et la cote de sa surface libre, et un coefficient d'échange entre les deux milieux.

2.2 Estimation de la transmissivité

En considérant l'équation de diffusivité, l'estimation des transmissivités se résument à la minimisation de la fonction objectif sur les charges et sur les transmissivités. La fonction objectif, sert de critère pour déterminer la meilleure solution à un problème d'optimisation. Concrètement, elle indique le degré de rapprochement entre les simulations du modèle et les observations de l'état du système. Pour un domaine de calcul de 534 mailles, seulement 40 mailles se sont vues attribuées une valeur de transmissivité ponctuelle issu d'une analyse d'essai de pompage. Pour réaliser le modèle hydrodynamique, les données piézométriques sont fournies par les mesures effectuées sur 24 piézomètres reparties sur la nappe du CT sur une période de 1 an réparties en 12 pas de temps mensuel. Les sollicitations de la nappe se résument essentiellement à la production journalière des 71 forages fonctionnels des neuf (9) champs captants SODECI d'Abidjan à l'exception des forages du champ captant du plateau qui est fermé.

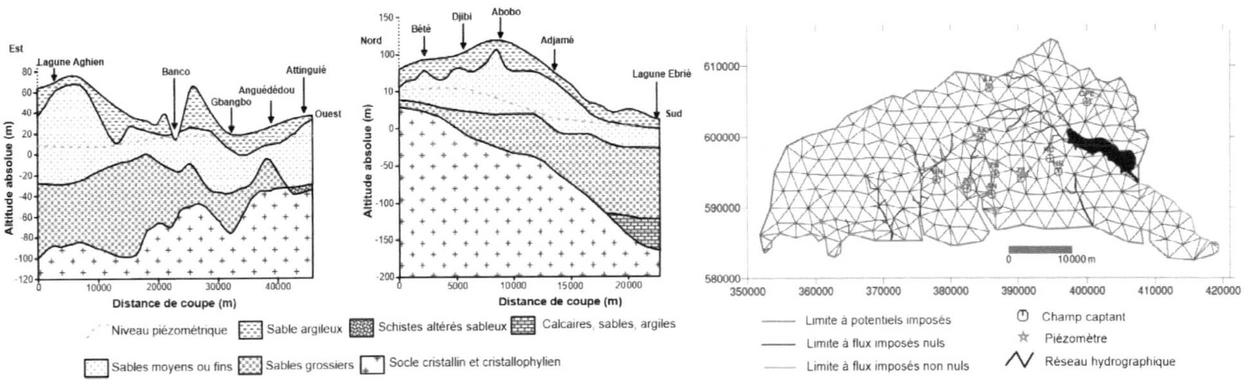

Figure 1. Schéma conceptuel hydrogéologique Est-Ouest et Nord-Sud et Conceptualisation du modèle numérique de l'aquifère du Continental Terminal d'Abidjan.

2.2.1 Paramétrisation multi-échelle

La détermination du champ de transmissivités réel nécessite, la recherche de la structure et de la dimension la plus proche du milieu naturel. Afin d'atteindre cet objectif, nous adoptons la méthode de paramétrisation multi-échelle non automatique dont l'idée est d'utiliser dès le départ plusieurs structures différentes pour générer un ensemble de solutions possibles. Cette approche consiste à représenter la structure du milieu par des structures de plus en plus élaborées, au cours de l'identification des paramètres du modèle. Cette démarche se traduit par une augmentation de la dimension du modèle de paramétrisation, par affinement du maillage, d'une étape à la suivante (Kouassi, 2010). C'est donc dans cet état d'esprit que l'identification des transmissivités du modèle de la nappe d'Abidjan s'est effectuée autour de huit structures de base dénommées structure 100, structure 200 jusqu'à structure 800. La proposition de ces structures de départ a été guidée par l'exploitation simultanée de plusieurs informations sur le milieu physique (transmissivités ponctuelles, débits spécifiques, données géologiques, occupation du sol etc. ...), sans oublier la connaissance générale que nous avons du secteur à partir des études antérieures (Loroux, 1978; SOGREAH, 1996; Kouassi et al., 2010; Kouassi, 2013). Ainsi à l'aide de ces informations nous avons raffiné ou non certaines zones du domaine d'étude pour générer les différentes structures. Chaque structure est par la suite constituée d'une série de maillage de paramétrisation de dimension graduellement croissante à tester. L'ensemble des maillages de paramétrisation à tester des huit structures confondus est de 115 maillages.

La procédure d'identification multi-échelle non automatique démarre par un maillage de grande dimension, composé de quatre mailles et de cinq nœuds couvrant tout le domaine de calcul (Ackerer et al., 1996). Après cette première étape, nous enchaînons directement avec la série de maillages des huits structures de base prédéfinis grâce à l'état de connaissance de l'aquifère (Fig. 2).

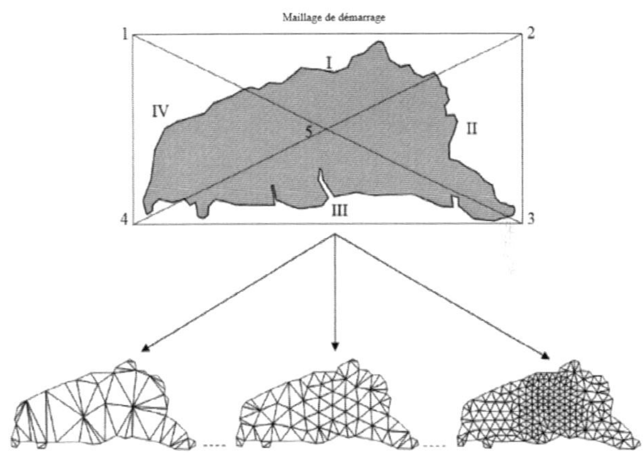

Figure 2. Procédure de la paramétrisation multi-échelle non automatique.

2.2.2 Calage et validation du modèle

Le calage du modèle vise à reproduire au mieux le fonctionnement hydrodynamique de la nappe en ajustant les différents paramètres que l'on estime pouvoir faire varier tout en restant dans la gamme des valeurs réalistes (Leduc, 2005). Pour améliorer les résultats de la simulation, en plus de la transmissivité, les coefficients d'échange nappe-rivières, les flux, la surface libres des rivières et la porosité ont été ajustés.

Une fois le calage lancé, on observe qu'au fur et à mesure que le nombre de paramètres augmente, la fonction objectif diminue. Elle diminue très rapidement au début et se stabilise autour d'une valeur au-delà d'une certaine dimension de la paramétrisation (nombre de paramètres). Tous les maillages de paramétrisation qui ont donné ces résultats satisfaisants sur la valeur de la fonction objectif peuvent être retenus comme approximations possibles de la structure du milieu étudié. Dans ce cas, le seul critère retenu est la valeur de la fonction objectif. On aurait alors une multitude de solutions pour un problème qui en réalité n'en a qu'une seule puisque la structure réelle est unique. Pour limiter les effets de l'inévitable non-unicité de la solution du problème inverse, il est donc indispensable de trouver un critère autre que la fonction objectif. Ce critère servira à évaluer la qualité du jeu de paramètres associé à chaque modèle de paramétrisation et à réduire le nombre de solution au problème inverse. Pour apprécier la qualité de chaque maillage de paramétrisation, nous déterminons donc deux critères de performance.

Le premier critère est celui classique sur les charges, qui indique dans quelle mesure le modèle a pu simuler l'état du système sur la base des données disponibles. Ce critère désignant erreur de modèle est noté Erreur_H (Eq. 1).

$$\text{Erreur\_H} = \sqrt{J(P)/N \cdot N_\text{p}} \qquad (1)$$

Avec :

$$J(P) = \sum_{n=0}^{N-1} \left\| H_\text{c}^{n+1} - H_\text{m}^{n+1} \right\|^2$$

H_c : Vecteur des charges simulées par le modèle aux points d'observations, H_m : Vecteur des charges mesurées sur le terrain aux points d'observations, N : Nombre de pas de temps, N_p : Nombre de points d'observation de la charge sur le terrain.

Erreur_H représente la moyenne sur tous les points de mesure et à tous les pas de temps, de l'écart entre les charges simulées et les charges mesurées.

Le deuxième critère porte sur la qualité des paramètres estimés. Ainsi, le critère Erreur_T qui représente la moyenne des écarts relatifs quadratiques entre les transmissivités identifiées et les transmissivités connues, prend la forme de l'équation (2) suivante :

$$\text{Erreur\_T} = 100 \cdot \sqrt{\frac{1}{N_\text{e}} \sum\nolimits_{j=1}^{N_\text{e}} \left(\frac{Tc_j - Tm_j}{Tm_j} \right)^2} \qquad (2)$$

Avec : Tc_j : Transmissivité identifiée sur l'élément j du maillage de calcul ; Tm_j : Transmissivité du champ connu pour l'élément j, N_e : Nombre total d'éléments du maillage de calcul.

Dès lors que les calages successifs sont réalisés, nous disposons pour chaque maillage de paramétrisation d'une valeur de Erreur_H et d'une valeur de Erreur_T. Ces deux critères Erreur_H et Erreur_T sont alors représentés dans un même graphe en fonction de la dimension de la paramétrisation. La représentation de Erreur_H et de Erreur_T permet dans la mesure du possible, de sélectionner la structure et la dimension optimale du modèle de paramétrisation qui convient pour le calage du modèle (Kouassi, 1999). A l'analyse des courbes de Erreur_H et Erreur_T, nous couplerons les informations recueillies aux résultats de la comparaison entre les charges simulées et les charges observées. Pour encore réduire le nombre de solutions de l'identification des transmissivités et également servir de facteur de validation

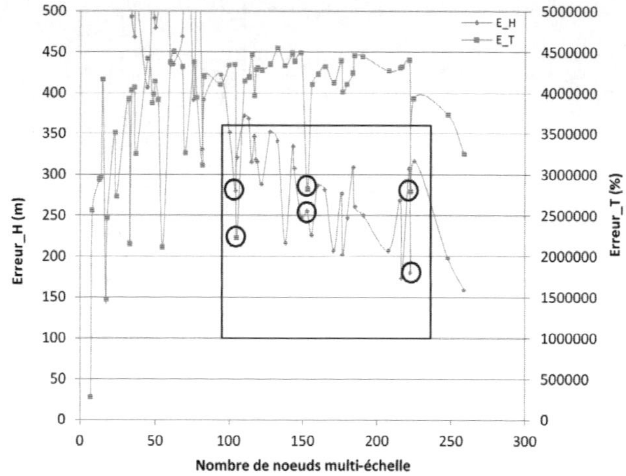

Figure 3. Représentation locale de la fraction de Erreur_H stable et de Erreur_T en fonction du nombre de paramètres multi-échelle.

du champ de transmissivité identifié, des valeurs de transmissivité n'ayant pas servi dans la mise en place du modèle vont être positionnées sur les probables champs de transmissivité solutions du problème inverse. L'objectif de superposer les transmissivités connues sur les champ transmissivités calculées est de rechercher le champ de transmissivité identifié qui reproduit le mieux les valeurs réelles des transmissivités de la nappe d'Abidjan.

3 Resultats et discussions

3.1 Identification du champ de transmissivité

L'analyse des valeurs de Erreur_H et de Erreur_T comprises entre 0 et 5.10^6 issus des différents calages réalisés a permis de choisir trois maillages (maillages 105 ; 153 et 223 nœuds de la structure 700) de paramétrisation pour identifier les paramètres du modèle (Fig. 3). Les valeurs de transmissivités estimées oscillent entre $5{,}4.10^{-5}$ et $1\,\text{m}^2\,\text{s}^{-1}$. Parmi ces trois maillages présélectionnés (Fig. 4), nous ne pouvons pas d'ores et déjà choisir l'un d'entre eux pour représenter les paramètres du modèle car le minimum de E_H n'est pas distinguable de manière incontestée. Ainsi, il nous faut des informations qualitatives complémentaires.

3.2 Détermination de la structure et de la dimension optimale du modèle

L'analyse des courbes de Erreur_H et de Erreur_T en fonction du nombre de paramètres a permis de disposer dans un premier temps d'un ensemble de maillages solutions possibles du problème inverse et ensuite de réduire cet ensemble de solutions à trois maillages proches de la structure vraie. Il a été cependant impossible de déterminer une structure et une dimension unique proche de la structure réelle de la

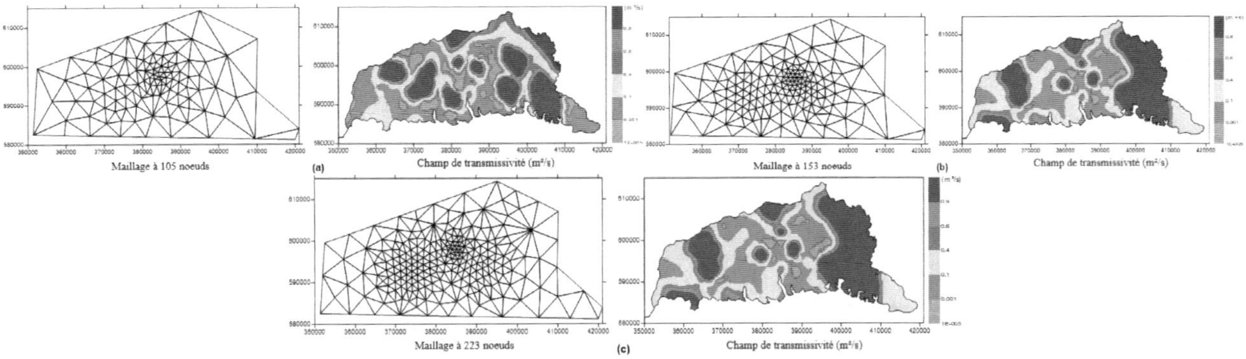

Figure 4. Maillages de paramétrisation et champs de transmissivités associés : **(a)** Maillage de 105 nœuds et son champ de transmissivité ; **(b)** Maillage de 153 nœuds et son champ de transmissivité ; **(c)** Maillage de 223 nœuds et son champ de transmissivité.

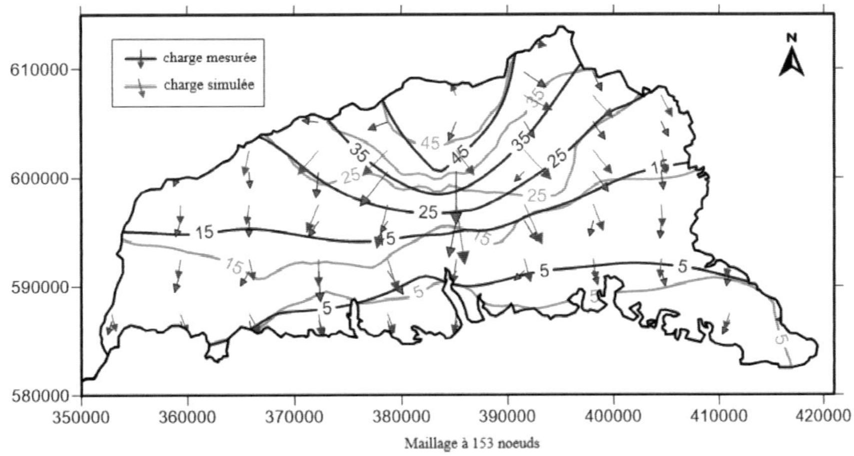

Figure 5. Comparaison des charges mesurées et simulées en juin 2005 dans le maillage de 153 nœuds

nappe parmi ces trois maillages. En regardant dans le détail la portion de la courbe E_H solution du problème inverse précédemment identifiés, il n'est toujours pas possible de départager concrètement les trois champs de transmissivités déterminés et de faire un choix unique. En l'absence d'autres critères de différenciation de ces maillages indiqués, la comparaison des charges simulées associées à chaque maillage de paramétrisation, aux charges mesurées aux points d'observation, est exploitée. Dans l'ensemble, les trois maillages de paramétrisation identifiés restituent quasi fidèlement le sens d'écoulement général de la nappe. Toutefois, il faut noter que les lignes piézométriques issues des charges calculées du maillage de 153 nœuds se rapprochent plus des lignes de références (Fig. 5). La comparaison des charges observées et des charges calculées a permis de déterminer la dimension de 153 nœuds de la structure 700 comme étant la dimension optimale du modèle de la nappe du Continental Terminal.

Au terme de nos investigations, nous proposons le champ de transmissivité associés au maillage de 153 nœuds comme la structure la plus proche de la structure réelle de l'aquifère du Continental Terminal d'Abidjan (Fig. 4b).

Validation des transmissivités

La validation du champs identifié montre que les transmissivités mesurées du champ captant d'Anguédédou appartenant aux gammes de 10^{-3} et 10^{-2} m^2 s^{-1} sont situées dans la classe de transmissivité de 10^{-3} à 10^{-1} m^2 s^{-1} du champ de transmissivité identifié calculés à l'issu des simulations (Fig. 6). A la fin de ce test de validation, nous pouvons donc déduire que le champ de transmissivité issu du maillage de 153 nœuds à la Fig. 7 représente au mieux la structure vraie de l'aquifère du Continental Terminal d'Abidjan.

3.3 Discussion

L'analyse des critères E_H et E_T à la suite des séries de calage a permis de déterminer les maillages à 105, 153 et 223 nœuds de la structure 700 comme solutions du problème inverse. Comme cela est très souvent le cas dans de nom-

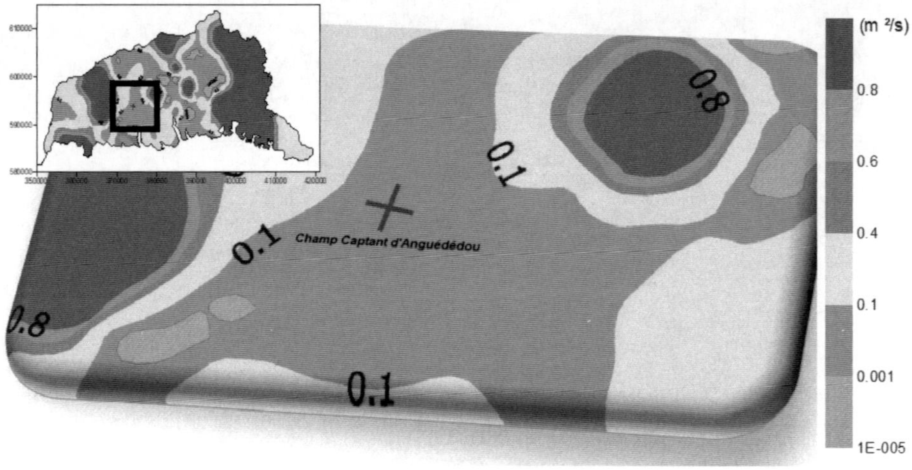

Figure 6. Positionnement du champ captant d'Anguédédou sur le champ de transmissivité identifié.

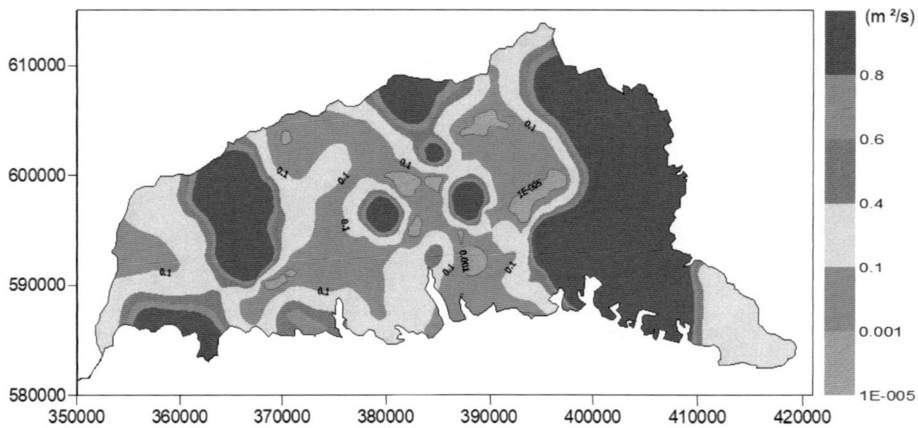

Figure 7. Champ de transmissivité identifié comme la structure la plus proche de la structure de l'aquifère du Continental Terminal d'Abidjan.

breux domaines de géosciences, nous sommes typiquement en présence d'un problème inverse mal posé car il n'admet pas de solution unique et stable. Alors, le seul moyen de réduire ces incertitudes serait d'ajouter de l'information soit sous forme de piézométries observées, soit sous forme d'essais de pompages. Les valeurs des fonctions objectifs E_H et E_T comprises entre 0 et 5. 10^6 sont restées élevées comme dans les travaux de Lehmann (1996) où la fonction objectif $O(p)$ est parfois allée jusqu'à 10^9. Les raisons des fonctions objectif élevées, sont probablement liées à la complexité du milieu d'étude naturel et au nombre insuffisant des points de mesures réelles. Les champs de transmissivités calculées, associés aux trois maillages sélectionnés présentent une bonne structure dans l'ensemble et cela en conformité avec l'hydrodynamique de la nappe du CT. Ces transmissivités estimées ($5{,}4.10^{-5} \leq T < 1\,\mathrm{m^2\,s^{-1}}$) s'inscrivent en partie dans la gamme des transmissivités mesurées dans plusieurs aquifères sédimentaires continentaux ou côtiers de l'Afrique et de celle mesurée par d'autres auteurs (Loroux, 1978 ; Scet Ivoire et al., 1980) sur le bassin sédimentaire côtier ivoirien. Ainsi, les valeurs de transmissivités estimées sur la majeure partie de l'aquifère du CT d'Abidjan variant entre 10^{-1} et $1\,\mathrm{m^2\,s^{-1}}$ sont possibles. Cette démonstration permet également de montrer que l'aquifère du CT est puissant et très productif par endroits. Les séries de calage ont permis de choisir le maillage de 153 nœuds comme la dimension optimale du modèle. Le champ de transmissivités associés à cette dimension optimale est défini comme la structure la plus proche de la structure vraie de l'aquifère du CT. En conséquence, le champ de transmissivité identifié pourra servir de guide dans l'implantation des forages à gros débits sur la nappe d'Abidjan et en retour le taux de réussite de ces implantations permettra d'évaluer le niveau de crédibilité du champ identifié.

4 Conclusions

La modélisation de l'écoulement de la nappe du Continental Terminal a permis de mettre en place un modèle numérique d'écoulement constitué d'un maillage de 534 mailles et de 320 nœuds. Par ailleurs, l'identification des paramètres autour de huit structures de départ s'est soldée par le choix de la structure 700 de dimension 153 nœuds comme la structure la plus proche de la structure vraie. Le champ de transmissivité associé à ce maillage présente de bonnes valeurs d'ensemble variant entre $5{,}4.10^{-5} \leq T < 1\,\mathrm{m}^2\,\mathrm{s}^{-1}$ dont l'essentiel des valeurs oscillent entre 10^{-3} et $1\,\mathrm{m}^2\,\mathrm{s}^{-1}$. Ce champ de transmissivité estimé sur l'aquifère du Continental Terminal d'Abidjan renseigne sur les énormes potentialités d'implantation des forages et le captage de la ressource en eau sur l'ensemble de cet aquifère.

Disponibilité du code. Le code de modelisation dont les réferences suivent : HPP-GMS plus Hydrology Pre/Post Processor V 4.50 © Copyright 1996, 1997 MSB All rights reserved. nous été offert gracieusement par le Directeur de Recherche CNRS Philippe Ackerer du Laboratoire d'hydrologie et de géochimie de Strasbourg (UMR 7517).

Disponibilité des données. En plus des données recoltées sur le terrain, les autres données utilisées ces travaux sont accessibles au public et ont été collectées auprès des structures étatiques et privés ci-dessous : la Direction de l'Hydraulique Humaine (DHH), le Centre de cartographie et de télédetection (CCT), l'entreprise Forage Rationnel Construction (FORACO), la Société de Distribution d'Eau de Côte d'Ivoire (SODECI), et la Société de développement Minier (SODEMI).

Collaborateurs. KAK : Organisation et exécution des travaux de terrain, collecte de données, caractérisation de l'aquifère par un modèle conceptuel, conceptualisation du modèle numérique et rédaction de l'article. WFK : Appui à la réalisation du modèle conceptuel et du modèle numérique de l'aquifère et correction de l'article. NB : Co-auteur décédé le mardi 19 octobre 2021 Oi Mangoua Jules Mangoua : Participation aux travaux de terrain, appui à la réalisation du modèle conceptuel et correction de l'article. PA : Mise à disposition du code HPP-GMS plus V 4.50, appui à l'usage du code et à la réalisation du modèle numérique de l'aquifère. GAD : Participation aux travaux de terrain et correction de l'article. IS : Orientation et correction de l'article.

Intérêts concurrents. Les auteurs déclarent qu'ils n'ont aucun conflit d'intérêts.

Clause de non-responsabilité. Publisher's note : Copernicus Publications remains neutral with regard to jurisdictional claims in published maps and institutional affiliations.

Déclaration du numéro spécial. This article is part of the special issue "Hydrology of Large River Basins of Africa". It is a result of the 4th International Conference on the "Hydrology of the Great Rivers of Africa", Cotonou, Benin, 13–20 November 2021.

Remerciements. Mes remerciements vont l'endroit du Laboratoire Géosciences et Environnement de l'Université NANGUI ABROGOUA (Côte d'Ivoire), du Laboratoire d'Hydrologie et géochimie de Strasbourg (France) et des structures telles que la DHH, le CCT, la SODECI et l'ONEP pour leur appui logistique dans la réalisation de ce travail. Au Dr William Francis KOUASSI, je tiens à dire particulièrement merci de m'avoir initié à la modélisation hydrogéologique, transmis l'abnégation et la hargne au travail. Fauché le 19 octobre 2021 dans la plénitude de ton art des sciences de l'eau, jamais je n'oublierai ton humanité et ta sollicitude pour ton prochain cher Maître. Repose en paix.

Références

Ackerer, P. H., Siegel, P., and Blaschke, A. P. : Inverse problem applied to groundwater flow and transport equations using a downscaling parameterization, Proc. of "Computational Methods in Water Resources XI", Cancun, Mexique, 22–26 July, edited by : Aldama, A., Aparicio, J., Brebbia, C., Gray, W. G., Herrera, I., and Pinder, G., Computational Mechanics Publ., 321–328, 1996.

Ahoussi, K. E. : Evaluation quantitative et qualitative des ressources en eau dans le Sud de la Côte d'Ivoire. Application de l'hydrochimie et des isotopes de l'environnement à l'étude des aquifères continus et discontinus de la région d'Abidjan-Agboville, Thèse unique de l'université de Cocody, Côte d'Ivoire, 283 pp., 2008.

De Lacour, G. : Les nappes phréatiques mondialement à la baisse, le journal de l'environnement, 26755, accessible à l'adresse suivante : http://www.journaldelenvironnement.net/article/les-nappes-phreatiques-mondialement-a-la-baisse, la date du dernier accès : 8 décembre 2012.

DHH (Direction de l'Hydraulique Humaine) : Hydraulique Humaine en Côte d'Ivoire, Ministère des Infrastructure Economiques, Abidjan, 66 pp., 2001.

Kouassi, K. A. : Modélisation hydrodynamique en milieu poreux saturé par approche inverse via une paramétrisation multiéchelle : Cas de l'aquifère du Continental Terminal d'Abidjan (Côte d'Ivoire), thèse unique en sciences et gestion de l'environnement de l'université Nangui Abrogoua, 266 pp., 2013.

Kouassi, W. F. : Estimation des paramètres hydrogéologique par approche inverse, Mise en œuvre d'une approche multi-échelle sur un secteur de la nappe d'Alsace, Thèse Université Louis Pasteur-Institut de Mécanique des Fluides, UMR 7507 UPL-CNRS, Strasbourg, 165 pp., 1999.

Kouassi, K. A., Kouassi, F. W., Goula, B. T. A., Kouame, K. I., Dibi, B., and Savane, I. : Conceptual model of Ivorian sedimentary costal basin : case of Abidjan Continental Terminal aquifer, Eurp. J. Sci. Res., 44, 400–419, 2010.

Leduc, C. : Modélisation hydrogéologique Orsay 2004–2005, Notes de cours de modélisation numérique en hydrogéologie, 16 pp., 2005.

Lehmann, F. : Hydrodynamique en milieux poreux hétérogènes non saturés : identification des paramètres par approche inverse,

Thèse Université Louis Pasteur – Institut de Mécanique des Fluides, URA CNRS 854, Strasbourg, 191 pp., 1996.

Loroux, B. F. E. : Contribution à l'étude hydrogéologique du bassin sédimentaire côtier de Côte d'Ivoire, Thèse de Doctorat de 3ème cycle, Université de Bordeaux I, France, 93 pp., 1978.

Mjemah, I. C., Van Camp, M., and Walraevens, K. : Groundwater exploitation and hydraulic parameter estimation for Quaternary aquifer in Dar-es-Salaam, Tanzania, J. Afr. Earth Sci., 55, 134–146, 2009.

Scet Ivoire, Sodeci, Holfelder, Scet International : Alimentation en eau potable d'Abidjan. Plan directeur et étude de la nappe et sa protection contre la pollution. Etude hydrogéologique de la nappe d'Abidjan : ressource et exploitation optimale de la nappe. Rapports no. 2 et no. 3, République de Côte d'Ivoire, Ministère des travaux publics, des transports, de la construction et de l'urbanisme : Direction Centrale de l'Hydraulique, 44 pp., 1980.

SODECI : Rapport d'Activités 2017, Société de Distribution d'Eau de Côte d'Ivoire, 60 pp., 2017.

SOGREAH : Etude de la gestion et de la protection de la nappe assurant l'alimentation en eau potable d'Abidjan, Etude sur modèle mathématique, Rapport de phase 1 et 2, République de Côte d'Ivoire, Ministère des infrastructures Economiques, Direction et Contrôle des Grands Travaux (DGCTX), 22 pp., 1997.

Proc. IAHS, 384, 57–62, 2021
https://doi.org/10.5194/piahs-384-57-2021

Open Access PIAHS

Hydrological Modelling of the Mono River Basin at Athiémé

Eliézer Iboukoun Biao[1,4], Ezéchiel Obada[1,4], Eric Adéchina Alamou[1,4], Josué Esdras Zandagba[1], Amédée Chabi[1], Ernest Amoussou[2], Julien Adounkpe[3], and Abel Afouda[1,3]

[1]Laboratory of Applied Hydrology (LHA), University of Abomey-Calavi, 01 BP 4521, Abomey-Calavi, Benin
[2]Laboratoire Pierre PAGNEY, Climat, Eau, Ecosystème et Développement (LACEEDE), University of Abomey-Calavi, Abomey-Calavi, Benin
[3]West African Science Service Center on Climate Change and Adapted Land Use, GRP Water Resources, University of Abomey-Calavi, Abomey-Calavi, Benin
[4]Laboratoire de Géosciences, de l'Environnement et Applications (LaGEA), National University of Sciences, Technology, Engineering and Mathematics, Abomey, Benin

Correspondence: Eliézer Iboukoun Biao (biaoeliezer@yahoo.fr)

Published: 16 November 2021

Abstract. The objective of this study is to model the Mono River basin at Athiémé using stochastic approach for a better knowledge of the hydrological functioning of the basin. Data used in this study consist of observed precipitation and temperature data over the period 1961–2012 and future projection data from two regional climate models (HIRHAM5 and REMO) over the period 2016–2100. Simulation of the river discharge was made using ModHyPMA, GR4J, HBV, AWBM models and uncertainties analysis were performed by a stochastic approach. Results showed that the different rainfall-runoff models used reproduce well the observed hydrographs. However, the multi-modelling approach has improved the performance of the individual models. The Hermite orthogonal polynomials of order 4 are well suited for the prediction of flood flows in this basin. This stochastic modeling approach allowed us to deduce that extreme events would therefore increase in the middle of the century under RCP8.5 scenario and towards the end of the century under RCP4.5 scenario.

1 Introduction

Water resource management and catchment analysis are crucial aspects of the twenty-first century in hydrological and environmental sciences (Chalkias et al., 2016). Benin Republic has an aquatic ecosystem, supplied and varied, represented from downstream to upstream by a significant maritime component, a system of lagoons and lakes and a hydrographic network in which fit the Lake Ahémé and its channels (Amoussou, 2010). This hydrographic basin – lagoon – lake and maritime system continuum is very sensitive to climate forcing. Nowadays, the lake Ahémé is facing a real problem of sedimentation. The Mono River is one of the main water inlets of this lake. Several hydrological studies have already been conducted in the Mono River basin (Amoussou et al., 2014; Lawin et al., 2019; Koubodana et al., 2019; Amoussou et al., 2020). However, none of these studies have investigated the multi-modelling approach, as well as performing uncertainty analysis using stochastic approach. The objective of this study is therefore to model the Mono River basin at Athiémé using stochastic approach for a better knowledge of the hydrological functioning of the basin.

2 Materials and methods

2.1 Study area

The Mono River basin at Athiémé occupies an area of 21 500 km^2 shared between two West-African countries, Togo and Benin. Specifically, it is located between the latitudes 06°16′ and 09°20′ N, and the longitudes 0°42′ and 2°25′ E (Fig. 1). It hosts the Nangbéto hydropower dam built in 1987 and utilized by the two countries. The river serves as

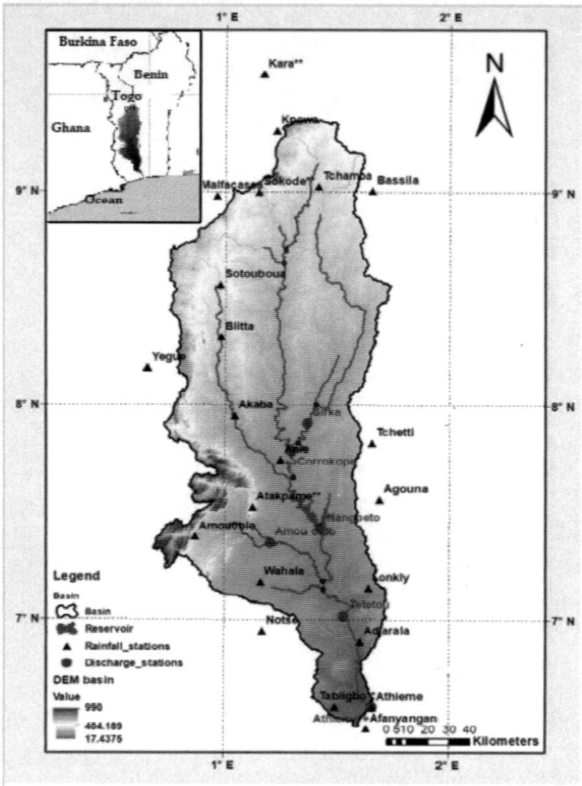

Figure 1. Study area.

natural border between the two countries in the southern part. The climate is tropical (two rainy seasons and two dry seasons) downstream and subequatorial (one rainy season and one dry season) upstream (Lawin et al., 2019). This river basin is patterned in the south by floodplains and plateaus, and higher landforms in the north and north-west, e.g., the Atakora Mountains with a height of 800 m and their southern extensions that are the Togo mountains (Amoussou et al., 2020).

2.2 Methods

2.2.1 Observed and Regional Climate Models (RCM) data used

Observed meteorological data (daily rainfall data, temperature data) and daily discharge data were provided, respectively, by the Benin Meteorological Department, i.e., Météo-Benin and the National Directorate of Water (DG-Eau). The observed data are considered for the period 1961–2012 (good compromise, taking into account the length of the data available in the different stations). The historical and future projections (RCP4.5 and RCP8.5 scenarios) rainfall data of two RCM (HIRHAM5 and REMO) were obtained from CORDEX Africa project (http://www.cordex.org, last access: 3 December 2020). These two RCM were chosen to test their ability to reproduce the observed rainfall data in Mono catchment for impact studies. The reference period selected for examining the rainfall bias was the period 1961–2005. For future projections, the RCP4.5 and RCP8.5 scenarios are considered over the period 2016 à 2100.

2.2.2 Data processing

Preprocessing of hydro-climatic data: This involves reconstructing the missing hydro-climatic data and carrying out the spatialisation by using the ordinary kriging method.

Bias correction: ISI-MIP method (Hempel et al., 2013) was used for the correction of bias contained in the temperature and rainfall data from the investigated RCM.

Calculation of Potential Evapotranspiration (PET): The calculation of daily PET by the formula of Oudin et al. (2005), which only uses as input the temperature, was retained for hydrological modeling. This ETP formula is simple and efficient and allows to obtain better results in terms of flow simulation.

Hydrological modelling: In this study, several hydrological models were used: ModHyPMA, GR4J, HBV and AWBM (Table 1). These models have been used successfully in previous studies (Alamou et al., 2010; Amoussou, 2010; Amoussou, 2015; Biao et al., 2016). They will be calibrated over the period 1961–1965 and validated, firstly, over the period 1971–1975 (period before the establishment of the Nangbéto dam) and secondly over the period 1988–1992 (period after the establishment of the dam in 1987). This was done in order to consider the effect of the operation of the dam on the flow simulations at the Athiémé outlet. Ensembles of these models were also considered because the average of the outputs of the different models could also reduce the uncertainties related to the models structure. In order to evaluate the model performance for the calibration and validation, the coefficient of model efficiency CE (Nash and Sutcliffe, 1970) and the coefficient of determination R^2 were used.

Uncertainty analysis: stochastic modeling of flood flows: Hydrological processes are natural phenomena having not only a dynamic aspect but also a stochastic character. Thus, any analysis of hydrological data should seek to take into account both stochastic and dynamic aspects. In addition, precipitation at the entrance to the basin is the result of complex physical phenomena that give it a stochastic behavior. The stochastic formulation of ModHyPMA by methods based on the use of stochastic differential equations (SDE) allows to better take into account the uncertainties in the rainfall-runoff modeling. The SDE that describes the Mono basin at Athiémé can be written in the form of Eq. (1) (Alamou, 2011; Biao et al., 2016):

$$dQ(t) = f(Q,t)dt + G(Q,t)dW(t) \tag{1}$$

 https://doi.org/10.5194/piahs-384-57-2021

Table 1. Characteristics of the rainfall-runoff models used.

Models	Structure	Parameters	Reservoirs	Input	References
ModHyPMA	Physically	2		P, PET + Q	Afouda et al. (2010)
GR4J	Conceptual	4	2	P, PET + Q	Edijatno Nascimento et al. (1999)
HBV	Conceptual	9	3	P, PET + Q	Bergström (1973)
AWBM	Conceptual	7	3	P, PET + Q	Boughton (1993)

where $[dW(t), t \in T]$ is a Brownian motion process. One of the advantages of SDE is the use that can be made of its associated Fokker-Planck equation (FPE). The FPE is a general probabilistic approach to describe the dynamics of various stochastic systems. It models the time evolution of the probability distribution in a system under uncertainty. It can be written in the form of Eq. (2):

$$\frac{\partial P(Q,t)}{\partial t} = -\frac{\partial}{\partial Q}\left[f(Q,t)P(Q,t)\right] + \frac{1}{2}\frac{\partial^2}{\partial Q^2}\left[2G^2(Q,t)P(Q,t)\right] \quad (2)$$

where $f(Q,t)$ et $2G^2(Q,t)$ represent respectively the mean and the variance. Hermite polynomial expansion to approximate probability distribution of flood flows (Afouda and Alamou, 2010; Alamou, 2011; Biao et al., 2016) is given in the form of Eq. (3):

$$P(u,t) = \frac{1}{\sqrt{2\pi}}e^{-\frac{u^2}{2}}\left[1+\sum\nolimits_{i=1}^{n} A_i H_{i+1}(u)\right] \quad (3)$$

where $A_1 = 0$; $A_2 = \frac{\mu_3}{3!}$; $A_3 = \frac{\mu_4 - 3}{4!}$; etc.; $H_n(u) = (-1)^n e^{\frac{u^2}{2}} \frac{d^n}{du^n}\left(e^{-\frac{u^2}{2}}\right)$; $u = \frac{Q - E[Q]}{\sigma_Q}$ and μ represent the centered moments of the variable u. Based on the above expressions of the expansion coefficients, the time series of the observed or simulated maximum daily mean flows are used to make calculations of $P(u,t)$ for the first four orders. From this approximation of probability distribution of flood flows, the values of the 2, 5, 10, 20, 50 and 100 years return period were determined.

3 Results and discussion

3.1 Hydrological modelling

Figures 2 and 3 and Table 2 show the results of the rainfall-runoff simulation in the Mono river basin at Athiémé. A closer look at Figs. 2 and 3 reveals an underestimation of flood peaks by the different models. This confirms the regulatory role of dams in accordance with the conclusions of Payan (2007). In addition, the multi-model approach has improved the performance of individual models. Indeed, the performance in calibration, in validation 1 and in validation 2 of the ensemble of HBV, ModHyPMA and GR4J, as well as

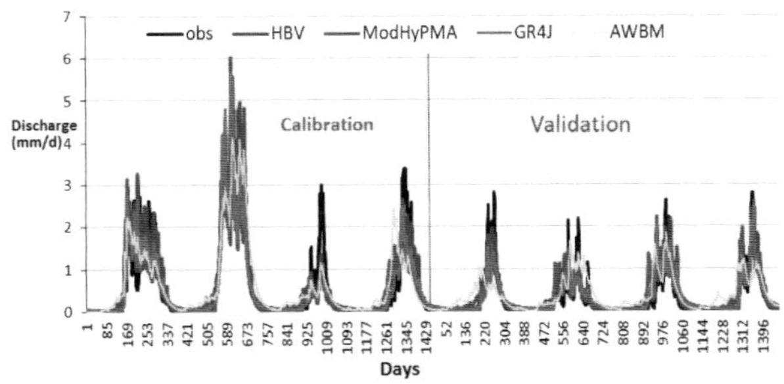

Figure 2. Calibration and validation of the investigated models in the Mono basin at Athiémé.

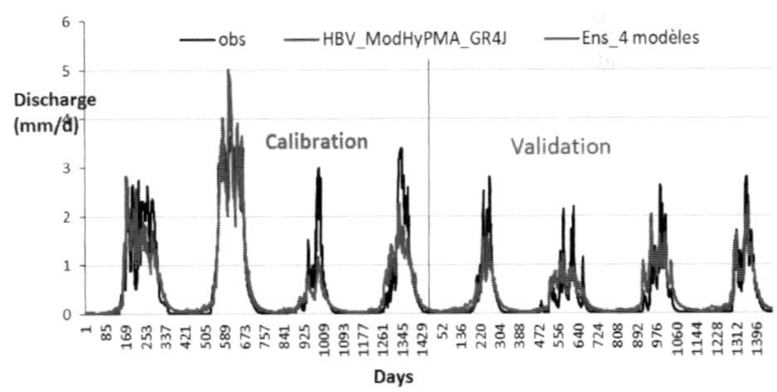

Figure 3. Calibration and validation of the investigated ensemble models in the Mono basin at Athiémé.

the ensemble of the 4 rainfall-runoff models are better than the performance of the models taken individually. However, the validation of the models after the impoundment of the dam is less good than the first validation. This can be explained by the influence of the water releases at the dam.

In order to better take into account the uncertainties, stochastic modeling of flood flows in the Mono basin at Athiémé is crucial.

3.2 Stochastic modelling of flood flows

Figure 4 shows the results of the fitting of the simulated flood flows with approximations based on Hermite orthogonal polynomials. From this figure, it can be deduced that it is the probability distribution based on the Hermite orthogonal polynomials of order 4 which best fits the observation data and those simulated in the Mono basin at Athiémé.

Table 2. Performance of the rainfall-runoff.

	Calibration			Validation_1			Validation_2		
Model	Nashroot	Nash	R^2	Nashroot	Nash	R^2	Nashroot	Nash	R^2
HBV	0.87	0.78	0.82	0.80	0.79	0.80	0.21	0.41	0.64
ModHyPMA	0.83	0.75	0.79	0.72	0.70	0.71	0.51	0.50	0.55
GR4J	0.87	0.78	0.79	0.82	0.81	0.83	0.51	0.49	0.62
AWBM	0.75	0.72	0.72	0.37	0.36	0.41	0.48	0.61	0.64
Ens_HBV_ModHyPMA_GR4J	0.88	0.82	0.83	0.81	0.79	0.81	0.52	0.55	0.64
Ens_of 4_models	0.86	0.82	0.82	0.74	0.75	0.78	0.55	0.58	0.65

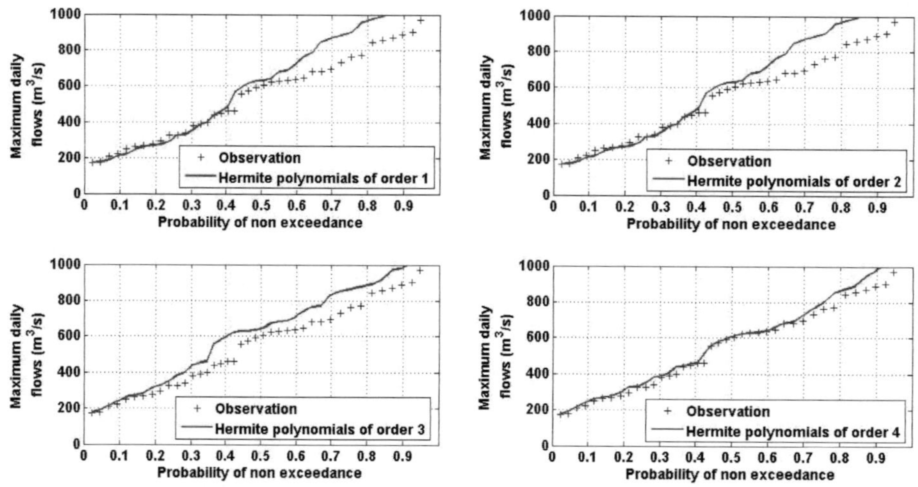

Figure 4. Fitting curve of the probability distribution based on Hermite polynomials for the simulated maximum daily flows in the Mono basin at Athiémé.

The results also showed that the simulated maximum flows, using the observed rainfall data in the basin, are underestimated for the 2 and 5 years return periods, whereas these flows are overestimated for the 100 years return period, in comparison with the observed maximum flows. However, using historical data from the REMO and from the ensemble REMO and HIRHAM5, the relative differences between the values of the simulated flows and those of observed flows are not too high, compared to those obtained with HIRHAM5 alone. These differences vary from 18 % (for $T = 100$ years return period) to 21 % (for $T = 2$ years return period). Therefore, REMO and the ensemble REMO and HIRHAM5 can be used for long-term simulations.

3.3 Frequency analysis of projected flows

The results obtained from the REMO climate model project a decrease change of flood flows, under the two RCP4.5 and RCP8.5 scenarios. However, the ensemble HIRHAM5 and REMO project an increase change for the 5, 10, 20, 50 and 100 years return periods in the Mono basin at Athiémé under the two scenarios. In comparison with the observed maximum flows, an increase ranging between 22 % (for $T =$ 5 years return period) and 24 % (for $T = 100$ years return period) could be obtained. These results allow us to deduce that the extreme events could therefore increase in the middle of the century under RCP8.5 scenario and towards the end of the century under RCP4.5 scenario.

The traditional approach for assessing the probability of outflow of a given frequency has been to pick a storm pattern, choose a runoff model and set the parameters with the best available estimate (for example in Amoussou, 2015). However, one of the most important advantages of the SDE is the associated FPE, which allows one to directly derive the time-varying probabilities associated with the outflow. The stochastic approach used in the present study enables us to take into account both the stochastic and dynamic aspects of the hydrological phenomenon.

4 Conclusions

The main contribution of this paper was to model the Mono River basin at Athiémé using stochastic approach for a better knowledge of the hydrological functioning of the basin. The different hydrological models used in this study reproduce well the observed hydrographs in the Mono basin at Athiémé. However, the multi-model approach has improved the performance of the individual models. Stochastic modeling of flood flows shows that the intensity and frequency of flows are likely to be more pronounced under RCP4.5 and RCP8.5 scenarios. The study concluded that climate change will modify the level of floods in the Mono basin at Athiémé. The main advantage of the SDE approach is that it provides a physically transparent and mathematically tractable description of the stochastic dynamics, indicating how uncertainty in input precipitation and environmental parameters (potential evapotranspiration, temperature) affects the uncertainty in model output. Although the use of the SDE and the associated FPE as proposed in this paper can become more complex, the potential benefits in the areas of decision making, data collection and value of information are of promising importance.

Code availability. The code supporting the findings of this study is available from the corresponding author on request.

Data availability. The data supporting the findings of this study are available from the corresponding author on request.

Author contributions. EIB, EO, JEZ, and AC designed the methodological framework and contributed to the entire strategic and conceptual framework of the study. EIB performed the simulations, analyzed the results and prepared the manuscript under the supervision of EAA, EA, JA, and AA.

Competing interests. The contact author has declared that neither they nor their co-authors have any competing interests.

Disclaimer. Publisher's note: Copernicus Publications remains neutral with regard to jurisdictional claims in published maps and institutional affiliations.

Special issue statement. This article is part of the special issue "Hydrology of Large River Basins of Africa". It is a result of the 4th International Conference on the "Hydrology of the Great Rivers of Africa", Cotonou, Benin, 13–20 November 2021.

References

Afouda, A. and Alamou, E.: Modèle Hydrologique basé sur le Principe de moindre action (MODHYPMA), Annales des Sciences Agronomiques du Bénin, 13, 23–45, 2010.

Alamou, E.: Application du principe de moindre action a la modélisation pluie – débit, Thèse de Doctorat, CIPMA – chaire UNESCO, Université d' Abomey – Calavi, 231 pp., 2011.

Amoussou, E.: Variabilité pluviométrique et dynamique hydro-sédimentaire du bassin- versant du complexe fluvio-lagunaire Mono-Ahémé-Couffo (Afrique de l'Ouest), Thèse de Doctorat unique, Université de Bourgogne, Dijon, CRC – CNRS – UMR5210, 313 pp., 2010.

Amoussou, E.: Analyse hydrométéorologique des crues dans le bassin-versant du Mono en Afrique de l'Ouest avec un modèle conceptuel pluie-débit, FMSH-WP-2015-90, April 2015.

Amoussou, E., Tramblay, Y., Totin, H. S. V., Mahé, G., and Camberlin, P.: Dynamique et modélisation des crues dans le bassin du Mono à Nangbéto (Togo/Bénin), Hydrol. Sci. J., 59, 2060–2071, https://doi.org/10.1080/02626667.2013.87101, 2014.

Amoussou, E., Awoye, H., Henri S. Vodounon, T., Obahoundje, S., Camberlin, P., Diedhiou, A., Kouadio, K., Mahé, G., Houndénou, C., and Boko, M.: Climate and Extreme Rainfall Events in the Mono River Basin (West Africa): Investigating Future Changes with Regional Climate Models, Water, 12, 833 pp., https://doi.org/10.3390/w12030833, 2020.

Bergström, S. and Forsman, A.: Development of a conceptual deterministic rainfall-runoff model, Nord. Hydrol., 14, 147–170, 1973.

Biao, I. E., Alamou, A. E., and Afouda, A.: Improving rainfall-runoff modelling through the control of uncertainties under increasing climate variability in the Oueme River basin (Benin, West Africa), Hydrol. Sci. J., 61, 2902–2915, 2016.

Boughton, W. C.: A hydrograph-based model for estimating the water yield of ungauged catchments. In: Proceedings of the 1993 Hydrology and Water Resources Conference, Institution of Engineers, Australia, National Conference Publication no. 93, 317–324, 1993.

Chalkias, C., Stathopoulos, N., Kalogeropoulos, K., and Karymbalis, E.: Applied Hydrological Modeling with the Use of Geoinformatics: Theory and Practice, Empirical Modeling and Its Applications, Mamun Habib, IntechOpen, https://doi.org/10.5772/62824, 2016.

Edijatno Nascimento, N. O., Yang, X., Makhlouf, Z., and Michel, C.: GR3J: a daily watershed model with three free parameters, Hydrol. Sci. J., 44, 263–277, 1999.

Hempel, S., Frieler, K., Warszawski, L., Schewe, J., and Piontek, F.: A trend-preserving bias correction – the ISI-MIP approach, Earth Syst. Dynam., 4, 219–236, https://doi.org/10.5194/esd-4-219-2013, 2013.

Koubodana, H. D., Diekkrüger, B., Näschen, K., Adounkpe, J., and Atchonouglo, K.: Impact of the Accuracy of Land Cover Data sets on the Accuracy of Land Cover Change Scenarios in the Mono River Basin, Togo, West Africa, Int. J. Adv. Remote Sens. GIS, 8, 3073–3095, 2019.

Lawin, A. E., Houngué, N. R., Biaou, C. A., and Badou, D. F.: Statistical Analysis of Recent and Future Rainfall and Temperature Variability in the Mono River Watershed (Benin, Togo), Climate, 7, 1, https://doi.org/10.3390/cli7010008, 2019.

Nash, J. E. and Sutcliffe, J. V.: River flow forecasting through conceptual models: Part I – A discussion of principles, J. Hydrol., 10, 282–290, https://doi.org/10.1016/0022-1694(70)90255-6, 1970.

Oudin, L., Hervieu, F., Michel, C., Perrin, C., Andréassian, V., Anctil, F., and Loumagne, C.: Which potential evapotranspiration input for a lumped rainfall–runoff model?: Part 2 – Towards a simple and efficient potential evapotranspiration model for rainfall–runoff modelling, J. Hydrol., 303, 290–306, https://doi.org/10.1016/j.jhydrol.2004.08.026, 2005.

Payan, J.-L.: Prise en compte de barrages-réservoirs dans un modèle global pluie-débit, Thèse de Doctorat de L'ENGREF, CEMAGREF, ANTONY, Paris, 257 pp., 2007.

Proc. IAHS, 384, 63–68, 2021
https://doi.org/10.5194/piahs-384-63-2021

Open Access PIAHS

Surface runoff prediction and comparison using IHACRES and GR4J lumped models in the Mono catchment, West Africa

Houteta Djan'na Koubodana[1,2,3], Kossi Atchonouglo[3], Julien G. Adounkpe[2], Ernest Amoussou[4,5], Domiho Japhet Kodja[5], Dambré Koungbanane[6], Koba Yaovi Afoudji[6], Yao Lombo[1], and Kossi E. Kpemoua[1]

[1]Institut Togolais de Recherche Agronomique, P.O. Box 1163, Lomé, Togo
[2]West Africa Science Service Centre on Climate change and Adapted Land Use, WASCAL – Climate Change and Water Resources, University of Abomey Calavi, 03 BP 526 Cotonou, Benin
[3]Faculty of Sciences, University of Lomé, P.O. Box 1515, Lomé, Togo
[4]Département de Géographie et Aménagement du Territoire, Université de Parakou, Parakou, Benin
[5]Laboratory Pierre PAGNEY, Climate, Water, Ecosystem and Development (LACEEDE), 03 BP1122 Cotonou, Benin
[6]Department of Geography, University of Lomé, P.O. Box 1515, Lomé, Togo

Correspondence: Houteta Djan'na Koubodana (koubo2014@gmail.com)

Published: 16 November 2021

Abstract. This study aims to assess simulated surface runoff before and after dam construction in the Mono catchment (West Africa) using two lumped models: GR4J (Rural Engineering with 4 Daily Parameters) and IHACRES (Identification of unit Hydrographs and Component flows from Rainfall, Evapotranspiration and Stream data) over two different periods (1964–1986 and 1988–2010). Daily rainfall, mean temperature, evapotranspiration and discharge in situ data were collected for the period 1964–2010. After the model's initialization, calibration and validation; performances analysis have been carried out using multi-objectives functions developed in R software (version 3.5.3). The results indicate that statistical metrics such as the coefficient of determination (R^2), the Kling–Gupta Efficiency (KGE), the Nash–Sutcliffe coefficient (NSE) and the Percent of Bias (PBIAS) provide satisfactory insights over the first period of simulation (1964–1986) and low performances over the second period of simulation (1988–2010). In particular, IHACRES model underestimates extreme high runoff of Mono catchment between 1964 and 1986. Conversely, GR4J model overestimates extreme high runoff and has been found to be better for runoff prediction of the river only between 1964 and 1986. Moreover, the study deduced that the robustness of runoff simulation between 1964 and 1986 is better than between 1988 and 2010. Therefore, the weakness of simulated runoff between 1988 and 2010 was certainly due to dam management in the catchment. The study suggests that land cover changes impacts, soil proprieties and climate may also affect surface runoff in the catchment.

1 Introduction

Hydrological modelling is a veritable challenge in West Africa as in many regions in the world owing to limited data access and high percent of gaps in these data. Consequently, climate change impacts on water resource and management and surface runoff prediction become difficult (Komi et al., 2016). Furthermore, dam management, industrialization, irrigation and human activities influences hydrology processes in a watershed (Thompson et al., 2014). For example, land use and land cover play an important role in runoff generation and ground water process (Obahoundje et al., 2018). For example, in Mono catchment Koubodana et al. (2019) have mentioned that forest and savannah have decreased in the benefit of cropland from 1975 to 2013. In the same, catchment the studies such as Amoussou et al. (2012), Kissi et al. (2015), Lawin et al. (2019) and Ntajal et al. (2017) have concluded that the source of high discharge causing flooding in the downstream is not only due to climate change impacts but also to the regulation of the Nangbéto dam, land use and the social factors of the communities living in this catchment. Therefore, it is important to investigate discharge simulation over the catchment using different lumped models and also by considering land cover change constant for the period before Nangbéto dam construction and the period after dam installation in the catchment.

2 Geographic location of the study area

The study is conducted in Mono catchment located in West Africa (Fig. 1). It is a main river in Togo, which is shared with Benin country in its last kilometers in the south. This river is located between 06°16′ and 9°20′ N latitude and 0°42′ and 1°40′ E longitude. With a perimeter of 872 092 km, the catchment covers from Alédjo to Athiémé an area of 22 013.14 km^2 and with 88 % of it area in Togo country and the rest (12 %) in Benin (PCCP, 2008). Its length of 308 773 km, Mono catchment has it source in Alédjo mountains in North of Benin before throwing in Atlantic Ocean by the lagoon system. The elevation of the catchment is range from 12 to 948 m (https://srtm.csi.cgiar.org/, last access: 3 June 2021). The watershed shelters the biggest dam of Nangbéto that produce 20 % of total hydroelectricity used by the two countries (Houessou, 2016). The climate is a subequatorial from 6 to 8° N and with two rainy seasons and two dry seasons. It totals 1200 to 1500 mm yr^{-1} in the mountainous area of the South-West and only, 800 to 1000 mm yr^{-1} on the coastal zone. From 8 to 10° N the climate is tropical humid with one rainy season and one dry season (1000 to 1200 mm yr^{-1}). In the winter months (December to March), there is an anti-cyclonic high-pressure area Centered over the Sahara (Koubodana et al., 2020).

It drives the Harmattan, a desiccating, dusty wind that blows rather persistently from the northeast, drying out landscapes all the way to the coast. However, the hydrograph has one peak that indicates that river discharge is mostly controlled by upstream tributaries. The mean annual temperature ranges from 22 to 30 °C (CILSS, 2016; Speth et al., 2010). The population within the Mono catchment is around 3.5 million inhabitants (https://www.oieau.fr/, last access: 3 June 2021). The mains socio economic activities are agriculture, trade, fisheries and livestock husbandry.

3 Datasets and methods

3.1 Data used

Daily historical hydro-climatic time series including observation river discharge, rainfall, mean temperature and evapotranspiration collected at the gauge stations are used for this analysis. The datasets were collected at Direction General de l'Eau et Assainissement du Togo (DGEA-Togo), Direction General de l'Eau (DGEau-Benin), General Direction Générale de la Météorologie Nationale du Togo (DGMN-Togo) and Agence Nationale de la Météorologie du Bénin (METEO-Benin) and cover the period from 1964 to 2010. These data sets were initially used for quality control and initial analysis over the catchment by Amoussou et al. (2012) and Koubodana et al. (2020).

Table 1. Calibration and validation periods.

Sub-period	Warm up period	Calibration period	Validation period
Pre-dam	1 Jan to 31 Dec 1964	1 Jan 1965 to 31 Dec 1968	1 Jan 1969 to 31 Dec 1973
Post-dam	1 Jan to 31 Dec 1988	1 Jan 1989 to 31 Dec 1992	1 Jan 1993 to 31 Dec 1997

3.2 Methods

The mean average of rainfall at Athiémé have been computed by using Thiensen polygons methods of the 21 gauges stations (Fig. 1) in Quantum Geographic Information System (QGIS) (Koubodana et al., 2020; Thiessen, 1911).

The study period was divided first into two-sub period as rainfall breakpoint in 1987 (Koubodana et al., 2020) the predam installation period (1964–1986) and the post-dam installation period (1988–2010) (Table 1).

IHACRES model was used under Hydromad package in R and GR4J model under AirGR package in R. One year was used as model warm-up period in each sub period (Perrin et al., 2003). Calibration was performed for 4 years in each sub period and the same parameters from calibration were used for validation within 5 years after a reasonable compared models sensitivity analysis (Badjana et al., 2017). Four and five years are used with equal length of dry and wet year for calibration and validation periods. The sensitive analysis of IHACRES and GR4J models leading to select optimization parameters and ranges was based on the literature of the past studies done in the region and on Sobol method sensitivity analysis (Badjana et al., 2017; Kodja et al., 2018; Oyerinde et al., 2016).

Many multi-objective functions criterion were adopted in order to evaluate the model parameters optimization: coefficient of determination (R^2) (Dougherty et al., 2000), the Nash–Sutcliffe coefficient (NSE) (Nash and Sutcliffe, 1970), Kling–Gupta Efficiency (KGE) (Gupta et al., 2009) and Percent Bias (PBIAS) (Gupta et al., 2009). These criteria were used after calibration and validation in the catchment and for each sub period (Adeaga et al., 2012; Vansteenkiste et al., 2014).

 https://doi.org/10.5194/piahs-384-63-2021

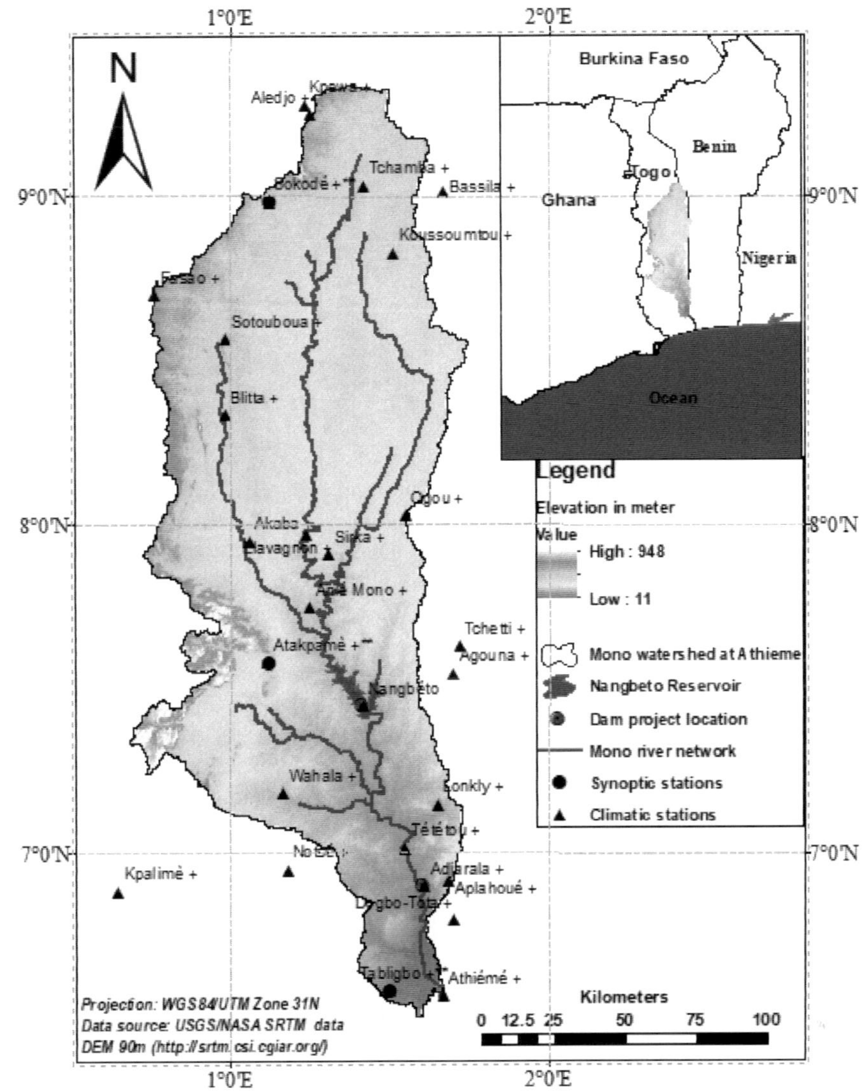

Figure 1. Location of the Mono catchment.

4 Results and discussion

4.1 IHACRES and GR4J model model's performance and uncertainties

GR4J and IHACRES model performance analysis, uncertainties and optimal values are showing in Table 2. The performances values of such as KGE, NSE, R^2 and PBIAS after daily calibration and validation are summarized in Table 2. The acceptable values of KGE, NSE and R^2 are in bolt in the table below.

IHACRES model over the pre-dam have reliable value of KGE, NSE and R^2 higher than 0.5 whereas the PBIAS are minimized in the same period. During the second period of post-dam period, KGE, NSE and R^2 are only acceptable on the calibration (1989–1992) period while poor during the validation period (1993–1997) model performances criterions are insignificant. The parameters after optimization of the model respect the range at 95 % significant level as the value plotted in Table 2.

4.2 IHACRES and GR4J model simulated and observed runoff evolution dynamic

The graphs of Fig. 2 indicate the evolution of daily areal rainfall, simulated and observed runoff over calibration and validation periods both for the two models. The result shows that the observed and simulated runoff values predicted by GR4J and IHACRES models follow exactly the areal rainfall patterns but present some small differences. The GR4J model overestimates discharges whereas IHACRES models

Table 2. Model's performance – uncertainty and optimized parameters. Acceptable performance values are in bold.

GR4J: criteria	KGE	NSE	R^2	PBIAS [%]
Calibration 1965–1968	**0.94**	**0.88**	**0.83**	**13.38**
Validation 1969–1973	**0.82**	**0.80**	**0.90**	20.03
Calibration 1989–1992	**0.70**	0.58	0.33	30.11
Validation 1993–1997	0.38	0.28	0.39	**−3.99**
Parameters	X_1	X_2	X_3	X_4
Period 1965–1973	290.04	−2.18	41.26	7.40
Period 1989–1997	432.68	−4.64	307.96	1.03
IHACRES: criteria	KGE	NSE	R^2	PBIAS [%]
Calibration 1965–1968	0.41	**0.65**	**0.90**	**1.85**
Validation 1969–1973	0.55	**0.90**	**0.75**	−15.71
Calibration 1989–1992	0.36	0.32	**0.74**	**4.10**
Validation 1993–1997	**0.80**	0.26	**0.63**	38.50
Parameters	t_w	τ_s	τ_q	f
Period 1965–1973	58.53	71.3	1.98	0.36
Period 1989–1997	100.00	82	2.10	5.86

Note: In GR4J model, X_1 is maximum capacity of production store (mm), X_2 is the groundwater exchange coefficient (mm); X_3 is the 1 d ahead maximum capacity of routing store (mm) and X_4 is the time peak ordinate of hydrograph unit (day) (Perrin et al., 2003). IHACRES model parameters: t_w is the drying rate reference temperature [%], τ_s is the time constant for slow flow stor (day), τ_q is the time constant for quick flow stor (day) and f is temperature to potential evapotranspiration conversion factor (Andrews et al., 2011).

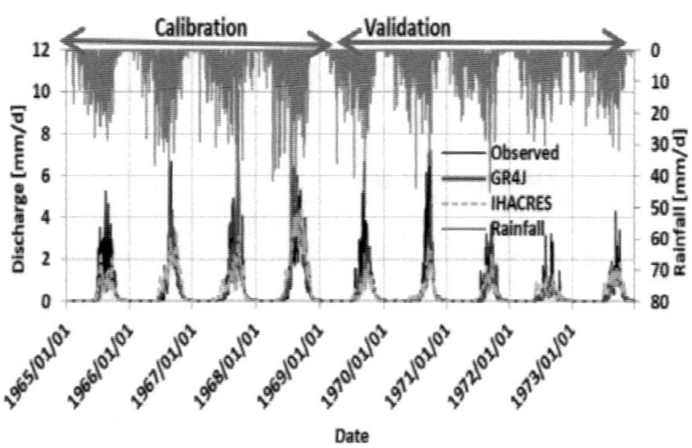

Figure 2. Models observed and simulated discharge between 1965 and 1973.

underestimates over the period of calibration (1965–1968) and period of validation (1969–1973) as shown in Fig. 2.

The two models accurately simulated base flow in the catchment. The lowest runoff periods (1971, 1972 and 1973) which can be explained by the period of drought and the periods of floods (1967 and 1958) in Togo and in West Africa (Koubodana et al., 2020; Oguntunde et al., 2006). The models simulated well base flow over between 1964 and 1986. IHACRES model during pre-dam simulation presents reliable value of KGE, NSE and R^2 coefficients higher than 0.5 whereas and PBIAS are minimized in the same period.

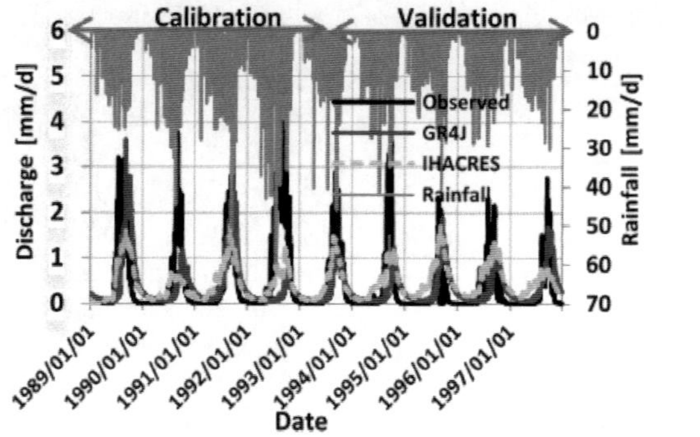

Figure 3. Models observed and simulated discharge between 1988 and 1997.

During the second period of post-dam simulation the coefficients KGE, NSE and PBIAS are only acceptable on the calibration (1989–1992) period while poor during the validation period (1993–1997) as displayed in Fig. 3.

The model performances criterions are significant only for R^2 criteria. The parameters after optimization of the model respect the range at 95 % significant level as the value plotted in Table 2. According to the results, during the pre-dam period predicted runoff and areal rainfall dynamic are similar and the simulated runoff from IHACRES and GR4J models matches with observed runoff during calibration (1965–1968) and validation (1969–1973) periods with R^2, NSE and KGE coefficient between 0.6 and 0.87. The performances of GR4J models are seen to be reliable and robust than IHACRES model between 1964 and 1986. This can be explained by model parameterization, assumptions and number of the parameters in the models. Indeed GR4J model has four parameters while IHACRES model has more than four parameters (Andrews et al., 2011; Perrin et al., 2001). Difference between the hydrograph of the two sub periods (1964–1986) and (1988–2010) maybe due to many factors responsible of runoff generation in tropical zone. According to Dubreuil (1985), the generation of runoff and subsurface runoff depend on spatial variation of rainfall, soil surface formations, texture and structure, vegetation cover, geological layers, drainage density, weathered material and climate and even on Nangbéto dam management. Other factor is the Nangbéto dam management which affects surface runoff dynamic in the downstream of the reservoir and particularly at Athiémé gauge stations.

5 Conclusion

This study has assessed Mono catchment surface runoff using two rainfall–runoff models over the period before dam installation (1964–1986) and after dam installation (1988–2010). The lumped rainfall–runoff IHACRES and GR4J models were successfully calibrated and validated over the two sub-periods using multi-objectives functions. The results show that at daily time step, GR4J model overestimates the observation whereas IHACRES model underestimates observations during pre-dam installation period (1964–1986). However, over post-dam installation period (1988–2010), there is a weak prediction performance of observation with a shift by GR4J model while IHACRES model is underestimating observations. Study has deduced that runoff generation is not linear to rainfall variability in the catchment and maybe influences by anthropogenic and natural factors. Dam management has influences on surface runoff after the year 1987 and the catchment response depends strongly on the data used for the model used parameterization before or after Nangbéto dam installation in 1987. Study suggested a semi-distributed model for an accurate simulation of water components in the catchment for further analysis.

Data availability. All climatic and hydrological data used in the research were collected from Togolese and Beninese Meteorological and hydrological services. The data can be obtained from DGMN-Togo, DGEA-Togo, METEO-Benin and DGEau.

Author contributions. HDK, KA, JGA and EA initiated the concept, HDK collected, analyzed the data and wrote the first manuscript. DJK, DK, KYA, YL and KEP contributed in the manuscript and KA, JGA and EA supervised the work.

Competing interests. The contact author has declared that neither they nor their co-authors have any competing interests.

Disclaimer. Publisher's note: Copernicus Publications remains neutral with regard to jurisdictional claims in published maps and institutional affiliations.

Special issue statement. This article is part of the special issue "Hydrology of Large River Basins of Africa". It is a result of the 4th International Conference on the "Hydrology of the Great Rivers of Africa", Cotonou, Benin, 13–20 November 2021.

Acknowledgements. The authors would like to thank the German Ministry of Education and Research (BMBF), IAHS SYSTA award funding to participate at the "4th International Conference on the Hydrology of African Large River Basins", Academy of Sciences–Islamic Development Bank (TWAS–IsDB) Collaborative Programmes 2020/2021 for the support and Togolese and Beninese Meteorological and hydrological services for data provision.

Financial support. This research has been supported by the Bundesministerium für Bildung und Forschung – BMBF (grant no. Batch-3) and World Academy of Sciences–Islamic Development Bank (TWAS–IsDB) Collaborative Programmes 2020/2021.

References

Adeaga, O., Mahe, G., Dieulin, C., Elbaz-Poulichet, F., Rouche, N., Seide, J. L., and Servat, E.: Rainfall–Runoff simulation in part of lower Niger basin, J. Environ. Sci. Eng. B, 1, 812–819, 2012.

Amoussou, E., Camberlin, P., and Mahé, G.: Impact de la variabilité climatique et du barrage Nangbéto sur l'hydrologie du système Mono-Couffo (Afrique de l'Ouest) l'hydrologie du système Mono-Couffo (Afrique de l'Ouest), Hydrolog. Sci. J., 57, 805–817, https://doi.org/10.1080/02626667.2011.643799, 2012.

Andrews, F. T., Croke, B. F. W., and Jakeman, A. J.: An open software environment for hydrological model assessment and development, Environ. Model. Softw., 26, 1171–1185, https://doi.org/10.1016/j.envsoft.2011.04.006, 2011.

Badjana, H. M., Fink, M., Helmschrot, J., Diekkrüger, B., Kralisch, S., Afouda, A. A., and Wala, K.: Hydrological system analysis and modelling of the Kara River basin (West Africa) using a lumped metric conceptual model, Hydrolog. Sci. J., 62, 1094–1113, https://doi.org/10.1080/02626667.2017.1307571, 2017.

CILSS: Landscapes of West Africa – A Window on a Changing World, Garretson, USA, 2016.

Dougherty, E. R., Kim, S., and Chen, Y.: Coefficient of determination in nonlinear signal processing, Sig. Process., 80, 2219–2235, 2000.

Dubreuil, P. L.: Review of Field Observations of Runoiff generation in the Tropics, J. Hydrol., 80, 237–264, 1985.

Gupta, H. V., Kling, H., Yilmaz, K. K., and Martinez, G. F.: Decomposition of the mean squared error and NSE performance criteria: Implications for improving hydrological modelling, J. Hydrol., 377, 80–91, https://doi.org/10.1016/j.jhydrol.2009.08.003, 2009.

Houessou, S.: Les inondataions et les risques prévisionnels liés aux barrages hydroéléctriques dans la basse vallée du Mono, These de doctorant, Soutenu publiquement le 11 Octobre 2016, Universite d'Abomey Calavi, Benin, 2016.

Kissi, A. E., Abbey, G. A., Agboka, K., and Egbendewe, A.: Quantitative Assessment of Vulnerability to Flood Hazards in Downstream Area of Mono Basin, South-Eastern Togo: Yoto District, J. Geogr. Inf. Syst., 7, 607–619, https://doi.org/10.4236/jgis.2015.76049, 2015.

Kodja, D. J., Mahé, G., Amoussou, E., Boko, M., and Paturel, J.-E.: Assessment of the Performance of Rainfall–Runoff Model GR4J to Simulate Streamflow in Ouémé Watershed at Bonou's outlet (West Africa), Preprints, 2018030090, 18 pp., https://doi.org/10.20944/PREPRINTS201803.0090.V1, 2018.

Komi, K., Amisigo, B. A., and Diekkrüger, B.: Integrated Flood Risk Assessment of Rural Communities in the Oti River Basin, West Africa, Hydrology, 3, 42, https://doi.org/10.3390/hydrology3040042, 2016.

Koubodana, H. D., Diekkrüger, B., Näschen, K., Adounkpe, J., and Atchonouglo, K.: Impact of the Accuracy of Land Cover Data Sets on the Accuracy of Land Cover Change Scenarios in the Mono River Basin, Togo, West Africa, Int. J. Adv. Remote Sens. GIS, 8, 3073–3095, https://doi.org/10.23953/cloud.ijarsg.422, 2019.

Koubodana, H. D., Adounkpe, J., Tall, M., Amoussou, E., Atchonouglo, K., and Mumtaz, M.: Trend Analysis of Hydroclimatic Historical Data and Future Scenarios of Climate Extreme Indices over Mono River Basin in West Africa, Am. J. Rural Dev., 8, 37–52, 2020.

Lawin, A. E., Hounguè, N. R., Biaou, C. A., and Badou, D. F.: Statistical Analysis of Recent and Future Rainfall and Temperature Variability in the Mono River Watershed (Benin, Togo), Climate, 7, 8, https://doi.org/10.3390/cli7010008, 2019.

Nash, J. E. and Sutcliffe, J. V.: River flow forecasting through conceptual models part I – A discussion of principles, J. Hydrol., 10, 282–290, 1970.

Ntajal, J., Lamptey, B. L., Mahamadou, I. B., and Nyarko, B. K.: Flood Disaster Risk Mapping in the Lower Mono River Basin in Togo, West, Int. J. Disast. Risk Reduct., 23, 93–103, https://doi.org/10.1016/j.ijdrr.2017.03.015, 2017.

Obahoundje, S., Diedhiou, A., Ofosu, E. A., Anquetin, S., François, B., Adounkpe, J., Amoussou, E., and Kouame, Y. M.: Assessment of Spatio-Temporal Changes of Land Use and Land Cover over South-Western African Basins and Their Relations with Variations of Discharges, Hydrology, 5, 56, https://doi.org/10.3390/hydrology5040056, 2018.

Oguntunde, P. G., Friesen, J., Van De Giesen, N., and Savenije, H. H. G.: Hydroclimatology of the Volta River Basin in West Africa: Trends and variability from 1901 to 2002, Phys. Chem. Earth Pt. A/B/C, 31, 1180–1188, https://doi.org/10.1016/j.pce.2006.02.062, 2006.

Oyerinde, G. T., Wisser, D., Hountondji, F. C. C., Odofin, A. J., Lawin, A. E., Afouda, A. A., and Diekkrüger, B.: Quantifying Uncertainties in Modeling Climate Change Impacts on Hydropower Production, Climate, 4, 1–15, https://doi.org/10.3390/cli4030034, 2016.

PCCP: PROGRAMME PCCP – From Potential Conflict to Cooperation Potential: cas du basin du Mono (Togo-Benin), Lomé, Togo, 2008.

Perrin, C., Michel, C., and Andre, V.: Does a large number of parameters enhance model performance? Comparative assessment of common catchment model structures on 429 catchments, J. Hydrol., 242, 275–301, 2001.

Perrin, C., Michel, C., and Andre, V.: Improvement of a parsimonious model for streamflow simulation, J. Hydrol., 279, 275–289, https://doi.org/10.1016/S0022-1694(03)00225-7, 2003.

Speth, P., Christoph, M., and Diekkrüger, B.: Impacts of Global Change on the Hydrological Cycle in West and Northwest Africa, Springer, Berlin, Heidelberg, https://doi.org/10.1007/978-3-642-12957-5_1, 2010.

Thiessen, A. H.: Precipitation averages for large areas, Mon. Weather Rev., 39, 1082–1089, 1911.

Thompson, J. R., Green, A. J., and Kingston, D. G.: Potential evapotranspiration-related uncertainty in climate change impacts on river flow: An assessment for the Mekong River basin, J. Hydrol., 510, 259–279, https://doi.org/10.1016/j.jhydrol.2013.12.010, 2014.

Vansteenkiste, T., Tavakoli, M., Van Steenbergen, N., De Smedt, F., Batelaan, O., Pereira, F., and Willems, P.: Intercomparison of five lumped and distributed models for catchment runoff and extreme flow simulation, J. Hydrol., 511, 335–349, https://doi.org/10.1016/j.jhydrol.2014.01.050, 2014.

Proc. IAHS, 384, 69–74, 2021
https://doi.org/10.5194/piahs-384-69-2021

Open Access
PIAHS

Hydrology of Large River Basins of Africa

Challenges of groundwater pollution and management in transboundary basins at the African scale

Issoufou Ouedraogo[1,2] **and Marnik Vanclooster**[3]

[1]Ecole Supérieure d'Ingénierie (ESI), Université de Fada N'Gourma, P.O. Box 54, Fada N'Gourma, Burkina Faso
[2]Laboratoire Géosciences et Environnement (LaGE), Ecole Doctorale Sciences et Technologies, Université Joseph KI-ZERBO, P.O. Box 7021, Ouagadougou, Burkina Faso
[3]Earth and Life Institute, Université catholique de Louvain, Croix du Sud 2, P.O. Box 2, 1348 Louvain-la-Neuve, Belgium

Correspondence: Issoufou Ouedraogo (ouedraogo.issoufou03@gmail.com)

Published: 16 November 2021

Abstract. We address the significant knowledge gap for groundwater pollution in Africa methods by assessing groundwater pollution risk at the African scale. To do so, we compiled the most recent continental-scale information on soil, land use, geology, hydrogeology, and topography in a Geographical Information System at the resolution of 15×15 km and the $1:60\,000\,000$ scale. We produced a vulnerability map by using the generic DRASTIC vulnerability indicator. This map revealed that groundwater is highly vulnerable in Central and West Africa groundwater basins, where the water table is shallow. In addition, very low vulnerability classes are found in the large sedimentary basins of Africa deserts where groundwater is situated in very deep aquifers. The generic groundwater pollution risk map is obtained by overlaying the DRASTIC vulnerability indicator with current land use. The northern, central, and western parts of the African continent are dominated by high vulnerability classes and very strongly related to water table depths and the development of agricultural activities. Given the availability of data, we concentrate first on nitrate vulnerability mapping. To this end, groundwater nitrate contamination data are compiled in literature using meta-analysis technic and used to calibrate as well linear and nonlinear statistical models; the latter performing much better as compared to simple linear statistical models. This study will help to raise awareness of the manager's International Basin Authorities or Transboundary Basin Organizations in Africa and in particular on transboundary groundwater pollution issues.

1 Introduction

Groundwater is an important water resource for meeting the various water demands in Africa. It is a crucial natural resource supporting the socio-economic development of the African continent; however, it is subjected to many pressures. To define sustainable water resources management plans at the continental scale, assessments of groundwater resources and associated pressures are strongly needed (Hasiniaina et al., 2010; Lapworth et al., 2017). Several studies have already been conducted to improve our understanding of African groundwater systems. The review by Xu and Usher (2006) was one of the most important qualitative works providing insights into common issues with regards to groundwater pollution in Africa at a local/regional scale (e.g. Nigeria, Cameroon, Ethiopia). Notwithstanding the compilation in this book and the recent progress in monitoring technology based for instance on remote sensing technology, no study has intended to assess the pan-African scale groundwater vulnerability for pollution. Assessing groundwater vulnerability for pollution at a large scale is particularly important for monitoring progress in sustainable development goals, such as the target on water issues. In this context, assessing the groundwater vulnerability for pollution is important for designing efficient regional-scale groundwater management and protection strategies.

According to Xu and Usher (2006), the degradation of groundwater is the most serious water resource problem in Africa. The two main threats are overexploitation and contamination (MacDonald et al., 2013). Nitrate is a common chemical contaminant of groundwater and the level of contamination has also increased in many African aquifers (Spalding and Exner, 1993; Puckett et al., 2011). Nitrate ingestion has been linked to methemoglobinemia, adverse reproductive outcomes, and specific cancers (Ward et al., 2005). In addition, nitrate is often a proxy of other possible pollutants of groundwater. Nitrate contamination is therefore very informative for overall groundwater quality. Nitrate contamination of groundwater is; however, a space-time variable property and the level of contamination depends on many space-time variable environmental and anthropogenic attributes.

The prediction and mapping of nitrate contamination pose, therefore, formidable technical and scientific challenges. Statistical models can be used to explain the spatial distribution of observed nitrate concentration in terms of available environmental and anthropogenic attributes, or to discriminate sources of contamination (Nolan and Hitt, 2006). For example, Bauder et al. (1993) investigated the major controlling factors for nitrate contamination of groundwater in agricultural areas using multiple linear regression analysis with land use, climate, soil characteristics, and cultivation types as independent variables. Furthermore, nonlinear random forest techniques have been applied to model nitrate in an unconsolidated aquifer in southern Spain (Rodriguez-Galiano et al., 2014). Norouz et al. (2016) used also a random forest method to determine the Malekan Aquifer vulnerability with several variables such as variables related to the DRASTIC method. Most statistical models used in such a context uses linear regression, nonlinear regression, logistic regression or Bayesian approaches to extract the variables that explain nitrate contamination (Mair and El-kadi, 2013; Mattern et al., 2012). If properly identified and validated, the statistical models can be used to predict nitrate contamination, and map possible contamination at the regional scale. Such maps are extremely valuable for designing transboundary water resources management and protection programs.

The purposes of the study were to improve our understanding of groundwater quality issues at the African scale to support the monitoring of the UN Sustainable Development Goals and in particular Goal 6 "Clean water and Sanitation". In this regard, we mapped firstly the vulnerability of groundwater to pollution at the continental scale. Secondly, we tried to identify the environmental factors contributing to the nitrate contamination of groundwater. We used the power of meta-analysis, to build pan-African database of groundwater pollution by nitrates. Then we relied on the meta-database to develop an exploratory linear statistical model of nitrate degradation of groundwater in terms of driving attributes available in the best available database. Finally, we used the meta-database to develop a nonlinear statistical method, and outputs were compared to those derived from a statistical linear model.

Table 1. Weight settings for DRASTIC variables (Aller et al., 1987).

Symbol	Variable	Weight
D_{w}	Depth to Water	5
R_{w}	Net Recharge	4
A_{w}	Aquifer media	3
S_{w}	Soil type	2
T_{w}	Topography	1
I_{w}	Impact of vadose zone	5
C_{w}	Hydraulic Conductivity	3
LU	Land use/LandCover	5

2 Data and methods

In this study, groundwater vulnerability was evaluated using the DRASTIC method, a system typically used in groundwater assessment. The acronym DRASTIC corresponds to the initials of the seven variables that drives vulnerability as defined according to Aller et al. (1987). To implement the DRASTIC index methodology at the pan-African scale, a best available database of environmental factors in various spatial resolutions for the continent was established at the resolution of 15×15 km and the $1:60000000$ scale. This GIS database was related to landcover/land use, soil type, hydrogeology, topography (Ouedraogo et al., 2016). Spatial data were processed in ArcGIS10.3™, QGIS™ 2.2 and Matlab™. The DRASTIC vulnerability index was calculated by the addition of the different products (rating × weight of the corresponding variable), using the following the Eq. (1):

$$\mathrm{DI} = D_{\mathrm{w}}D_{\mathrm{r}} + R_{\mathrm{w}}R_{\mathrm{r}} + A_{\mathrm{w}}A_{\mathrm{r}} + S_{\mathrm{w}}S_{\mathrm{r}} + T_{\mathrm{w}}T_{\mathrm{r}} + I_{\mathrm{w}}I_{\mathrm{r}} + C_{\mathrm{w}}C_{\mathrm{r}} \quad (1)$$

where DI, is the DRASTIC index; D, R, A, S, T, I, and C are the seven variables, as defined in Table 1; and the subscripts "r", and "w" are the corresponding rating and weights. The weights indicate the relative importance of each DRASTIC variable with respect to the other variables. These weights are constant (Ehteshami et al., 1991). The ratings used are in Ouedraogo et al. (2016).

The degree of groundwater vulnerability for pollution was evaluated by combined DRASTIC index and land use Eq. (2):

$$\text{Risk index} = \text{DRASTIC} \times \text{LU} \quad (2)$$

In addition to the pan-African hydrogeological geodatabase, and using the meta-analysis, we have analysed the distribution of nitrate concentration at the continental scale.

This analysis was based on studies reported in the literature because of the scarcity of groundwater monitoring data. Thus, 250 publications dealing with groundwater pollution were retrieved from the literature, mainly from books and the internet Web of Sciences (mainly Scopus™). The data were filtered using the following criteria: (i) the publication should explicitly report on nitrate concentrations in groundwater; (ii) we excluded older studies before 1999 since the intensity of human activities is expected to be significantly different after 1999. The dependent variable is the natural log of sampled groundwater nitrate concentration. By using the database collected at the African scale (Ouedraogo and Vanclooster, 2016), the average mean nitrate concentration is 54.85 mg/L, the standard deviation is 89.91 mg/L, and the median concentration is 27.58 mg/L. The log transformation reduces the influence of very high nitrate values (up to 648 mg/L) on model predictions. The nitrate studies were separated into two groups: 80 % of the dataset for building the model (training dataset) and 20 % of the data set for the validation of the model (tested dataset) (Ouedraogo et al., 2018). Furthermore, we used "Random Forest Regression (RFR)" to explain and predict groundwater nitrate contamination at the continental scale and its performance was compared with Multiple Linear Regression (MLR) techniques. A detailed description of the mathematical formulation of the Random Forest model is found in Breiman (2001), Liaw and Wiener (2002). The R statistical software was used for the modelling.

3 Results and discussion

3.1 Mapping output

We produce the first groundwater vulnerability map at the African scale by using the generic DRASTIC index indicator (Fig. 1). We classified this index into 5 classes, ranging from very low to very high index. This map reveals that groundwater is highly vulnerable in Central and West Africa groundwater basins. The low depth of groundwater in these regions and the high recharge explains this highest vulnerability. In addition, very low vulnerability classes are found in the large sedimentary basins of the African deserts where groundwater is situated in very deep aquifers. The transboundary basins in northern, central, and western parts of the African continent are dominated by high vulnerability classes and are very strongly related to water table depths and the development of agricultural activities. Also, Figure 1 gives an overlaying of Transboundary aquifers (TBAs) defined by Altchenko and Villholth (2013). This figure could be used as in example for transboundary aquifers management (International Network of Basin Organizations – INBO, Global Water Partnership – GWP) in Africa.

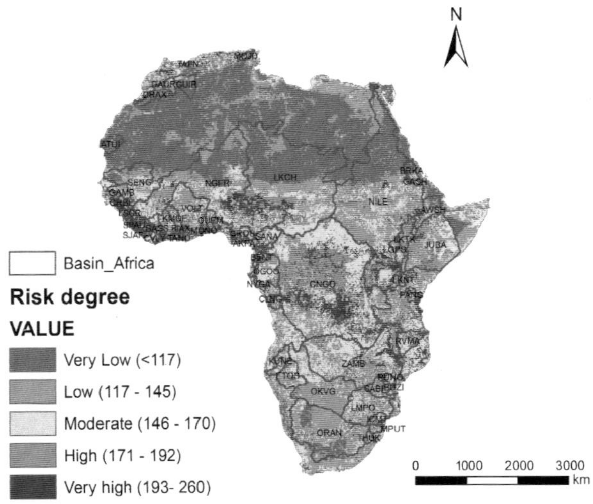

Figure 1. Groundwater pollution risk map under different transboundary aquifers.

3.2 Statistical modelling results

Using the stepwise MLR method firstly, we were able to select only those parameters which have a significant impact on the nitrate concentration values. A Student statistic t test was used to check the statistical significance (with p-values < 0.10) of variables in the final model. The MLR model revealed that the four final explanatory variables yielded an R^2 of 0.64 of variables in the final model indicating that 64 % of the variation of the observed log-transformed means nitrate concentration at the pan African scale is explained by the model. The four variables are depth to groundwater, recharge, aquifer type, and population density. The residual standard error is 0.95 (ln (mg/L)). A plot of predicted versus the observed log nitrate (ln NO_3) concentration in Fig. 2a for the training data, indicates the MLR model shows an acceptable fit with the observed data.

Secondly, with the machine learning technique, the percentage of variance explained in the final model of Random Forest Regression (RFR) is 97.9 % (mean of squared residuals = 0.0407 ln (mg/L)). Figure 2b shows a plot of the predicted versus the observed log-transformed nitrate concentration values for the training data, based on only 80 % of observations. The variable importance plot is a critical output of the random forest algorithm. Figure 3 shows the ranking of the relative importance of environmental attributes on groundwater nitrate occurrence. Higher values of a percent increase in mean squared error (MSE) indicate higher importance.

The comparison between the results shows that RFR has a much higher predictive ability compared to MLR. The better performance of RFR is due to the ability to handle nonlinear relationships between nitrate pollution and explaining factors. The most important explanatory variable with

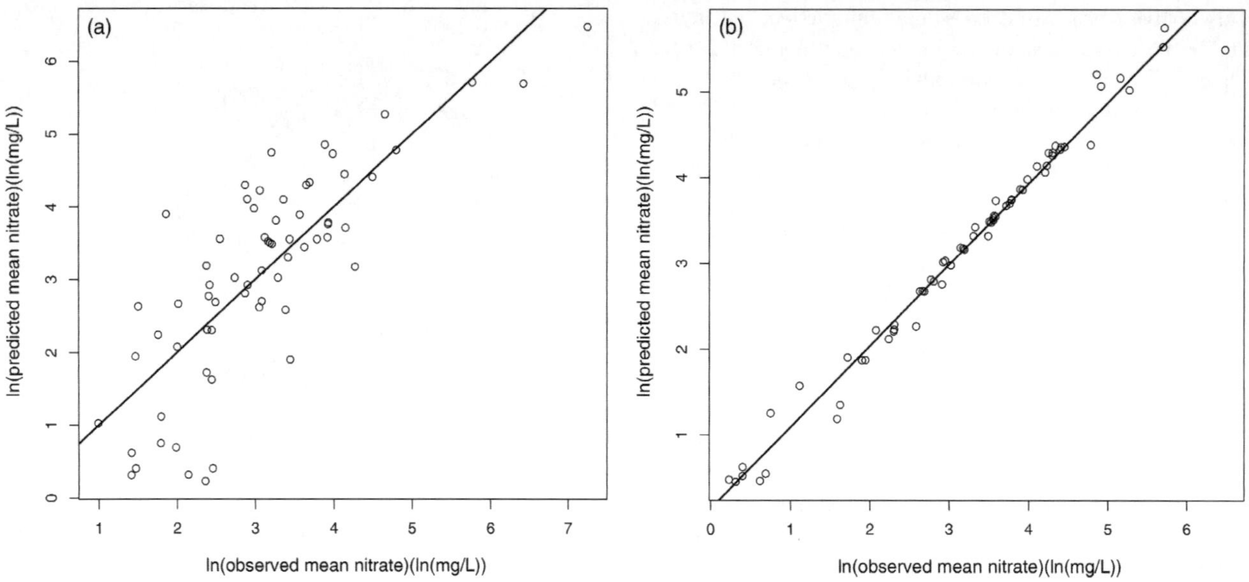

Figure 2. Comparison of observed and predicted log (ln) nitrate concentration on training dataset: **(a)** Linear regression and **(b)** Random Forest Regression.

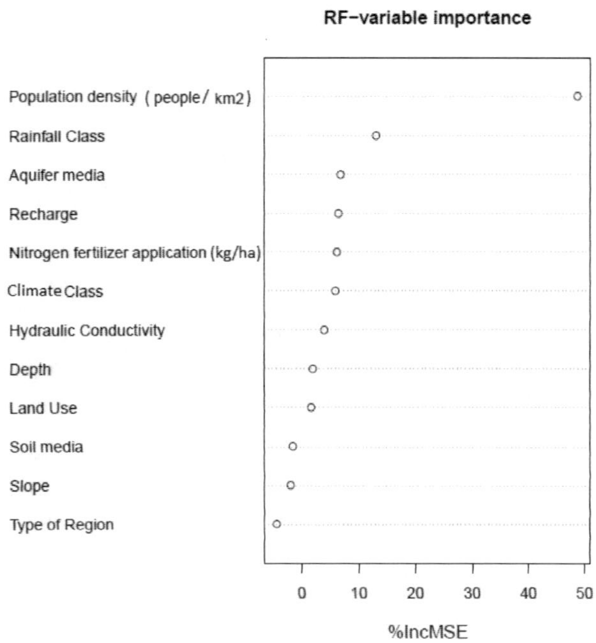

Figure 3. Variable importance section in Random Forest (RF) regression.

the strongest influence on nitrate concentration at the pan-African scale identified in both models was the population density (urban areas, agricultural activity). This is most likely related to the lack of sanitation in major parts of the African continent and the development of small-scale farming systems in peri-urban environments. A strong influence of the population density on groundwater nitrate contamination is consistent with a previous United Nations Environment Program (UNEP) study (e.g., Nolan and Hitt, 2006; Mattern et al., 2009; UNEP/DEWA, 2014; Kihumba et al., 2016; Ouedraogo et al., 2016, 2018). This study at the pan African scale could prompt national or international authorities to foster targeted local investigations because nowadays, many international agencies investigate transboundary basins aquifers in Africa (e.g., Observatoire du Sahara et du Sahel, Intergovernmental Authority for Development, Global Water Partnership, African Ministerial Council on Water African Union).

Despite several limitations in this study, due to the DRASTIC method, the scale at which data were gathered, the spatial resolution of mapping, the uncertainty of spatial data, the information gained from the analysis of historical groundwater quality data focusing on nitrate contamination, the main lesson of the study is to remind that shallow groundwater poses pollution problems in Africa. The map that was designed in this study can increase awareness of citizens and leaders in areas where groundwater pollution is likely to be significant. In addition, they could prompt national or international authorities to foster targeted local investigations. In fact, environmental management needs to be operatively performed at regional and local scales, but investment policies can be addressed at continental (transboundary water management) or even global scales by international agencies and authorities (e.g., Food and Agriculture Organization, United Nations Environment Programme/United Nations Educational, Scientific and Cultural Organization, International Network of Basin Organizations (INBO), and African Ministerial Council on Water African Union.

4 Conclusions

The results of the study have shed light on the pollution problem of shallow groundwater at the pan-African scale.

We demonstrated the unambiguous link between population density and groundwater nitrate contamination. We showed that the Random Forest machine learning techniques are promising for modelling groundwater degradation at the African scale because of its ability to provide meaningful analysis of nonlinear and complex relationships such as those found in hydrogeological studies. This paves the way for creating a water quality map at the continent scale of Africa. The study represents an important step toward developing tools that will allow us to accurately predict the distribution of nitrate contamination in groundwater in the context of climate change.

Code availability. The vulnerability map was implemented using ARC-GIS ESRI software. The statistical models were calibrated in R-software. The ARC-GIS modelmaker and R-scripts to process the data can be obtained by writing to the corresponding author.

Data availability. The spatial data used to map the vulnerability is obtained using public available data as described in the paper. The N literature data base can be obtained by writing to the corresponding authors.

Author contributions. The two authors developed the methodology. IO prepared the datasets, performed the data analysis, wrote the manuscript. MV reviewed the manuscript. The two authors have read and agreed to the published version of the manuscript.

Competing interests. The contact author has declared that neither they nor their co-author has any competing interests.

Disclaimer. Publisher's note: Copernicus Publications remains neutral with regard to jurisdictional claims in published maps and institutional affiliations.

Special issue statement. This article is part of the special issue "Hydrology of Large River Basins of Africa". It is a result of the 4th International Conference on the "Hydrology of the Great Rivers of Africa", Cotonou, Benin, 13–20 November 2021.

Acknowledgements. GIS shapefiles to generate generic attributes were obtained from different sources through the World and also online. In this regard, special thanks go to Tom Gleeson, Petra Döll, Nils Moosdoorf, and Patricia Trambauer. This study has been supported by Université catholique de Louvain through the "Fonds Spécial de Recherche" and Islamic Development Bank Merit Scholarship Program for PhD thesis. We thank two anonymous reviewers for their constructive comments.

References

Aller, L., Bennet, T., Leher, J. H., Petty, R. J., and Hackett, G.: DRASTIC: a standardized system for evaluating ground water pollution potential using hydrogeological settings, US Environmental Proctection Agency, Ada Oklahoma 74820, EPA 600/2-87-035, 662 pp., 1987.

Altchenko, Y. and Villholth, K. G.: Transboundary aquifer mapping and management in Africa: a harmonised approach, Hydrogeol. J., 21, 1497–1517, https://doi.org/10.1007/s10040-013-1002-3, 2013.

Bauder, J. W., Sinclair, K. N., and Lund, R. E.: Physiographic and land use characteristics associated with nitrate nitrogen-nitrogen in Montana groundwater, J. Environ. Qual., 22, 255–262, https://doi.org/10.2134/jeq1993.00472425002200020004x, 1993.

Breiman, L.: Random forests, Mach. Learn., 45, 5–32, https://doi.org/10.1023/A:1010933404324, 2001.

Ehteshami, M., Peralta, R. C., Eisele, H., Dear, H., and Tindall, T.: Assessing groundwater pesticide contamination to groundwater: a rapid approach, Ground Water, 29, 862–868, https://doi.org/10.1111/j.1745-6584.1991.tb00573.x, 1991.

Hasiniaina, F., Zhou, J., and Guoyi, Z. L.: Regional assessment of groundwater vulnerability in Tamtsag basin, Mongolia using drastic model, J. Am. Sci., 6, 65–78, 2010.

Kihumba, A. M., Longo, J. N., and Vanclooster, M.: Modelling nitrate pollution pressure using a multivariate statistical approach: the case of Kinshasa groundwater body, Democratic Republic of Congo, Hydrogeol. J., 24, 425–437, 2016.

Lapworth, D. J., Nkhuwa, D. C. W., Okotto-Okotto, J., Pedley, S., Stuart, M. E., Tijani, M. N., and Wright, J. J. H. J.: Urban groundwater quality in sub-Saharan Africa: current status and implications for water security and public health, Hydrogeol. J., 25, 1093–1116, https://doi.org/10.1007/s10040-016-1516-6, 2017.

Liaw, A. and Wiener, M.: Classification and regression by random forest, Vol. 2/3, December 2002, ISSN 1609-3631, available at: https://cogns.northwestern.edu/cbmg/LiawAndWiener2002.pdf (last access: 9 April 2021), 2002.

MacDonald, A., M., R. Taylor, G., and Bonsor, H. C. (Eds.): Groundwater in Africa – is there sufficient water to support the intensification of agriculture from "Land Grabs", Hand book of land and water grabs in Africa, 376–383, 2013.

Mair, A. and El-Kadi, A. I.: Logistic regression modeling to assess groundwater vulnerability to contamination in Hawaii, USA, J. Contam. Hydrol., 153, 1–23, https://doi.org/10.1016/j.jconhyd.2013.07.004, 2013.

Mattern, S., Fasbender, D., and Vanclooster, M.: Discriminating sources of nitrate pollution in a sandy aquifer, J. Hydrol., 376, 275–284, https://doi.org/10.1016/j.jhydrol.2009.07.039, 2009.

Mattern, S., Raouafi, W., Bogaert, P., Fasbender, D., and Vanclooster, M.: Bayesian Data Fusion (BDF) of monitoring data with a statistical groundwater contamination model to map groundwater quality at the regional scale, J. Water Resour. Protect., 4, 929–943, https://doi.org/10.4236/jwarp.2012, http://hdl.handle.net/2078.1/114949, 2012.

Nolan, B. T. and Hitt, K. J.: Vulnerability of shallow groundwater and drinking-water wells to nitrate in the United States, Environ. Sci. Technol., 40, 7834–7840, https://doi.org/10.1021/es060911u, 2006.

Norouz, H., Asghari Moghaddam Negar, A. M., and Attaallah, N.: Determining vulnerable areas of malekan plain aquifer for nitrate, using random forest method, J. Environ. Stud. Winter, 41, 923–942, 2016.

Ouedraogo, I. and Vanclooster, M.: A meta-analysis and statistical modelling of nitrates in groundwater at the African scale, Hydrol. Earth Syst. Sci., 20, 2353–2381, https://doi.org/10.5194/hess-20-2353-2016, 2016.

Ouedraogo, I., Defourny, P., and Vanclooster, M.: Mapping the groundwater Vulnerability for pollution at the pan-African scale, Sci. Total Environ., 544, 939–953, https://doi.org/10.1016/j.scitotenv.2015.11.135, 2016.

Ouedraogo, I., Defourny, P., and Vanclooster, M.: Application of random forest regression and comparison of its performance to multiple linear regression in modeling groundwater nitrate concentration at the African continent scale, Hydrogeol. J., 27, 1081–1098, https://doi.org/10.1007/s10040-018-1900-5, 2018.

Puckett, L. J., Tesoriero, A. J., and Dubrovsky, N. M.: Nitrogen contamination of surficial aquifers – a growing legacy, Environ. Sci. Technol., 45, 839–844, https://doi.org/10.1021/es1038358, 2011.

Rodriguez-Galiano, V., Mendes, M. P., Garcia-Soldado, M. J., Chica-Olmo, M., and Ribeiro, L.: Predictive modeling of groundwater nitrate pollution using Random Forest and multisource variables related to intrinsic and specific vulnerability: a case study in an agricultural setting (Southern Spain), Sci. Total Environ., 476/477, 189–206, https://doi.org/10.1016/j.scitotenv.2014.01.001, 2014.

Spalding, R. F. and Exner, M. E.: Occurrence of nitrate in groundwater – a review, J. Environ. Qual., 22, 392–402, https://doi.org/10.2134/jeq1993.00472425002200030002x 1993.

UNEP/DEWA: Sanitation and Groundwater Protection – a UNEP Perspective UNEP/DEWA, 18 pp., available at: http://www.bgr.bund.de/EN/Themen/Wasser/Veranstaltungen/symp_sanitat-gwprotect/present_mmayi_pdf.pdf?_blob=_publicationFile&v=2 (last access: 14 August 2019), 2014.

Ward, M. H., DeKok, T. M., Levallois, P., Brender, J., Gulis, G., Nolan, B. T., and VanDerslice, J.: Workgroup Report: Drinking-Water Nitrate and Health – Recent Findings and Research Needs, Environ. Health Perspect., 113, 1607–1614, https://doi.org/10.1289/ehp.8043, 2005.

Xu, Y. and Usher, B.: Groundwater pollution in Africa, Taylor & Francis/Balkema, the Netherlands, 353 pp., 2006.

Proc. IAHS, 384, 75–80, 2021
https://doi.org/10.5194/piahs-384-75-2021

Assessment of hydrological factors affecting shallow groundwater chemistry in the most urbanized area of the coastal sedimentary basin of Togo

Kossitse Venyo Akpataku[1,a], **Masamaeya D. T. Gnazou**[1], **Yao Tse Agbefu Nomesi**[2], **Phintè Nambo**[1], **Komi Doni**[1], **Limam Moctar Bawa**[1], **and Gbandi Djaneye-Boundjou**[1]

[1]Laboratory of Applied Hydrology and Environment, University of Lomé, Togo, BP 1515, Lomé, Togo
[2]Agency of Water and Sanitation for Africa, Office of Togo, BP 3689, Lomé, Togo
[a]current address: Department of Chemistry, University of Kara, Togo, BP 404, Kara, Togo

Correspondence: Kossitse Venyo Akpataku (avenyo@gmail.com)

Published: 16 November 2021

Abstract. The project to reduce environmental and health risks related to the activity of boreholes water sellers in Lomé (in French PRRESAF) was implemented to contribute to the guarantee of public health. One of its objectives is to improve the hydrological understanding of the shallow Continental Terminal (CT) aquifer tapped by water sellers. It is a vital water resource in the coastal sedimentary extending from Ghana to Nigeria. The results presented here are part of this project and aim at assessing the hydrological factors controlling CT groundwater quality and chemistry using principal component analysis (PCA), hierarchical cluster analysis (HCA), and GIS. High-resolution sampling campaigns for major ions water quality analyses were performed to improve the accuracy of factors assessment. The results showed that groundwater is predominantly acidic with varying degrees of mineralization ranging from very soft to brackish waters. The integration of multivariate results to GIS allows classifying samples into five main groups. They revealed the probable recharge and discharge areas, associated hydrochemical evolution along flow paths, and areas under the high impact of land-use dynamics, wastewaters/sewage infiltration, and interactions with surface water systems. This study appears useful for managing boreholes' water sellers' activity relative to the integrity of groundwater resources and public health.

1 Introduction

The coastal town of Lomé is located on the sedimentary basin of the Gulf of Guinea, namely the Keta basin that extends from Ghana to Nigeria. Like in other African cities, there is a galloping growth of the population with a rate of 6.1 % per year and a tremendous socio-economic development these last years (BCR/DGSN/MPRPDAT, 2011; MPT-Togo, 2013). Unfortunately, sanitation and drinking water supply systems have not kept pace with the city's expansion. Septic tanks and sumps are not built according to the rules, and uncollected wastes become uncontrolled rubbish dumps. In general, domestic and industrial wastewaters are not treated before being released into the environment. Runoff from these environments is drained into natural storm basins, some developed as artificial storm pools and connected to the lagoon system (MPT-Togo, 2013; Ayah et al., 2015). All these features constitute potential sources of high contamination of the Continental Terminal shallow aquifer, the primary and overexploited resource for drinking water in Lomé (Alfa-Sika Mande et al., 2012; Gnazou et al., 2017). In addition to formal exploitation by the Togolese Water Company (TdE) which cannot cover the population's needs, the past decade has witnessed a proliferation of informal private boreholes. Despite regulatory texts, water's physicochemical or bacteriological quality is not usually insured before human consumption. In this context, the project for the reduction of environmental and health risks related to the activity of sellers of boreholes water in the city of Lomé (in French PRRESAF) was implemented in order to contribute to the guarantee of public health. One of its objectives is to improve the hydrological understanding of the shallow Continental Terminal (CT) aquifer tapped by water sellers. As part of the project mentioned above, this study aims at assessing the relative importance of hydrological factors affecting the CT shallow groundwater chemistry. A multivariate statistical analysis approach integrated with a Geographical Information System (GIS) was applied to a dataset obtained from the physicochemical characterization of groundwater.

2 Study area

Lomé and its suburban areas lay between the latitudes 6°08′ and 6°18′ N and the longitudes 1°5′ and 1°23′ E (Fig. 1). It covers an area of around 360 km^2 with 1.6 million inhabitants in 2010 (BCR/DGSCN/MPRPDAT, 2011).

The study area consists of the Plateau of Agoenyive (20 to 67 m a.s.l.) from the Central part toward the North, the narrow littoral cordon in the South, and the alluvial plain of Zio in the East with an altitude lower than 20 m a.s.l. The climate is subequatorial Guinean type, characterized by two rainy seasons (April to July and September to October) and two dry seasons (July to August and November to March) with an average rainfall of 900 mm/year. The drainage network is part of the lake Togo Basin in connection with the Zio River and the lagoon system of Lomé. The geological context of the study area integrates that of the post-Paleozoic sedimentary series of the Gulf of Guinea (Ghana-Togo-Benin-Nigeria) (Sylvain et al., 1986; Da Costa et al., 2013). The Geological north-south cross-section of Cretaceous-Quaternary sediments is presented in Fig. 1. The sedimentary basin counts four aquifers. They are linked to Quaternary formations, the sands of the Continental Terminal, the limestone and sand of the Palaeocene, and the sand and sandstone of the Maastrichtian (Gnazou et al., 2017). The Continental Terminal (CT) aquifer is the main one exploited by boreholes water sellers and private boreholes for domestic purposes. With an NW-SE groundwater flow direction, the average transmissivity of CT aquifer is around 5 to 8×10^{-3} m^2/s (Gnazou et al., 2017). The mean depth of informal water sellers boreholes is around 30 m below ground level (m b.g.l.). In the south and east parts, the CT aquifer is influenced by the Quaternary sediments aquifer, where the average depth of boreholes and wells is 10 m with a static level close to ground level.

3 Methods

Nighty six boreholes were monitored in 2015 and 2016 for physicochemical characterization. Parameters such as temperature, pH, electrical conductivity, TDS, and dissolved oxygen are measured in situ using a portable multimeter (WTW, Instruments). Major and minor ions analyses were performed using the French Association of Standardization (AFNOR) methods (Rodier et al., 2009). Results are validated for ionic balance between −10 % and +10 %. The factor analysis based on the Principal Components (PCA) and the hierarchical cluster analysis ACH based on the Ward aggregation method and the measurement of similarity by Euclidean distances are performed as described by Cloutier et al. (2008) using STATISTICA software. ArcGIS software was used for spatial analysis and holistic interpretations.

4 Results and discussions

The pH values ranged between 4 and 8, with an average of 5.8 ± 0.7 indicating the predominance of acidic water types. The electrical conductivity values vary between 234 and 6250 µs/cm, and the corresponding TDS values range from 180 to 4738 mg/L with averages of 1523 ± 1132 µs/cm and 1150 ± 863 mg/L respectively. No significant differences ($p > 0.05$) were observed between the concentrations of chemical species during sampling periods suggesting no significant short-term variation of groundwater chemistry during the two hydrological years. The Durov Diagram presented in Fig. 2a shows that most of the samples exhibit Na-Cl water types.

The principal component analysis (PCA) allows extracting three (3) variables which explain 77 % of the total variance in the dataset (Table 1). Factor 1 accounts for 45 % of the total variance with high positive loadings (> 0.7) in TDS, Ca^{2+}, Mg^{2+}, Na^+, Cl^-, and SO_4^2. Since Cl^- and SO_4^{2-} are ubiquitous in the saline hydro system (Appelo and Postma, 2005), Factor 1 is defined as salinity. The predominance of saline groundwater samples suggests that the conjunction of factors such as dissolution of salts originated from sea spray and marine aerosols deposition, seawater intrusion, evaporation, mixing with brackish water from storm pools, and the lagoon system, along with anthropogenic contamination. They exert substantial control over the overall mineralization of groundwater.

Factor 2 accounts for 20 % of the total variance with high positive loadings in Fe_{total}, pH, and HCO_3^- and a strong negative loading for the NO_3^-. Iron concertation high as 4.4 mg/L was found in the lower Paleocene aquifer and correlated with turbidity, probably due to release from confining clay particles (Gnazou et al., 2011). In this study, concentrations high as 5 mg/L are found in the upper shallow CT aquifer. Their association with pH and HCO_3^- suggests a reductive dissolution of iron oxides by organic matter in acidic water following the Reaction (1) (Appelo and Postma, 2005).

$$CH_2O + 4FeOOH + 7H^+ \rightarrow 4Fe^{2+} + HCO_3^- + 6H_2O \quad (R1)$$

This reductive condition can explain the negative loading for NO_3^- which exists under oxic conditions. Factor 2 is then defined as the redox factor. In general, reduction conditions become important with depth, but they depend on recharge regime and land-use change (Lee et al., 2008).

Factor 3 explains 12 % of the variance with negative loadings in dissolved oxygen (O_2) and positive loadings in pH, HCO_3^-, NH_4^+ and K^+, suggesting high consumption of dissolved oxygen through the oxidation of

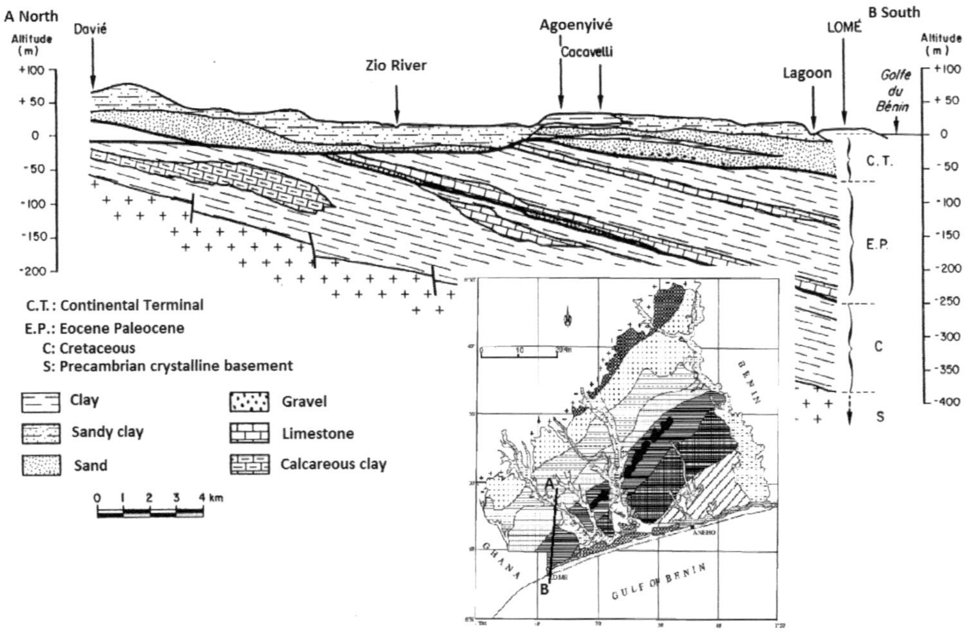

Figure 1. Geological north-south cross-section of Cretaceous-Quaternary sediments (modified after BRGM, 1982 and Sylvain et al., 1986; in Helstrup, 2006 and Gnazou et al., 2011).

Table 1. Results of the factor analysis. Bold values represent loadings > 0.7.

	pH	TDS	Ca^{2+}	Mg^{2+}	Na^+	K^+	HCO_3^-	Cl^-	SO_4^{2-}	NO_3^-	Fe_{tot}	NH_4^+	O_2	Eigenvalues	% variance	% cumul
Factor 1	0.03	**0.97**	**0.82**	**0.93**	**0.91**	0.5	0.17	**0.95**	**0.73**	0.05	−0.07	0.22	−0.01	5.8	45	45
Factor 2	**0.65**	−0.07	0.2	0.07	−0.21	0.06	**0.62**	−0.16	0.06	**−0.84**	**0.67**	−0.02	0.27	2.5	20	65
Factor 3	**0.62**	0.18	0.27	0.1	0.11	**0.7**	**0.68**	−0.03	0.4	0.18	−0.03	**0.73**	**−0.71**	1.5	12	77

anthropogenic organic matter. High NH_4^+ concentrations (> 30 mg/L) are measured, indicating contamination from pit latrines, sewage, septic tank, and dumpsites as in other Sub-Saharan big cities where ammonium concentrations up to 60 mg/L were determined (Lapworth et al., 2017). Cation exchange could occur between NH_4^+ released in aqueous phases, and K^+ fixed on aquifer materials (Appelo and Postma, 2005). Therefore, Factor 3 is defined as indirect recharge from wastewaters.

The ACH allows distinguishing five (5) main samples groups which characteristics are presented in Fig. 2. Except for three boreholes (P56, P73, and P75), the remaining points maintained their membership in the ACH groups, irrespective of the sampling campaigns. The groups obtained can then strongly describe the spatial evolution of water chemistry. The graphical representation of the median chemical composition of groundwater in each group is presented through the Schoeller diagram (Fig. 3b). There is a gradual increase of TDS and ions contents following G5 < G3 < G2 < G4 < G1. The segments connecting major ions concentrations are quite parallel. The similar slope suggests similar geochemical processes, while the hydrochemical evolution along flow paths may explain the gradual increase of TDS. The spatial distribution pattern of groups (Fig. 2e) shows that the main direction from Group 5 to Group 1 corresponds to groundwater flow in the NW-SE direction. The segments connecting NO_3^-, DO, and NH_4^+ are not parallel, suggesting that the impacts of anthropogenic contamination and land-use change on groundwater chemistry are not uniform in the study area. Group 5 samples (median TDS of 348 mg/L) with very low factorial scores are located in the surrounding agriculture areas with isolated residences. These samples indicate zones where groundwater chemistry is controlled by a copious infiltration of rainwater, sea spray, and dissolution of surface salts. Groups 2, 3, and 4 samples represent part of the groundwater system under intermediate evolution along flow paths with gradual increase of salinity and variable degree of NO_3^- contamination. Compared to Group 5, Group 2 samples are highly influenced by anthropogenic contamination due to increased population density (Fig. 2e). Group 4 samples are located in the historical urbanized areas of the city (Fig. 2e). These samples with very low DO and high NH_4^+ recorded high scores for Factor 3 (Fig. 2d), indicating low direct recharge from precipitation

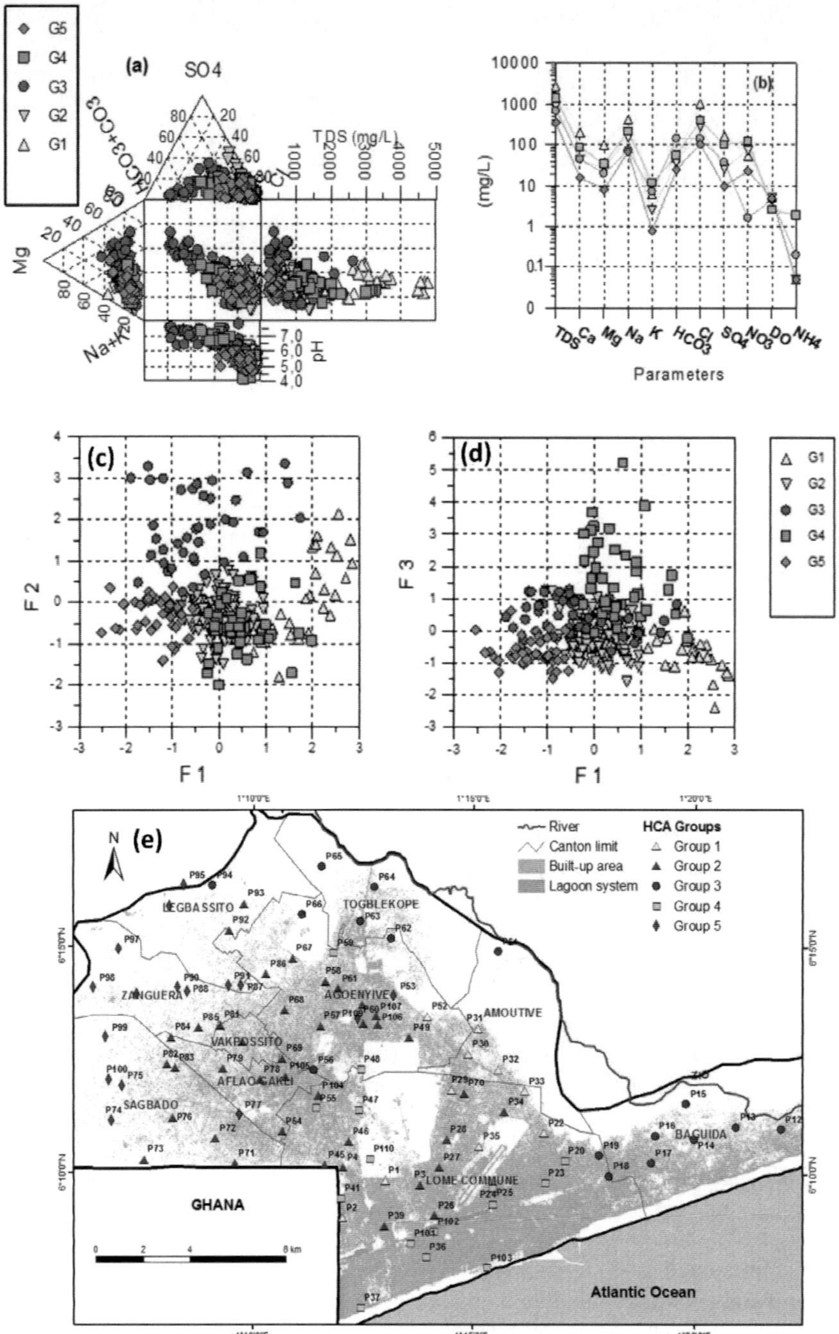

Figure 2. (**a**) Durov diagram of samples and (**b**) Schoeller diagram of the median chemical composition of each CHA group (**b, c**) Factorial scores for samples, (**e**) Spatial distribution of groups of samples. The built-up area is extracted from a SPOT-6 2015-12-08 satellite image (Copyright © AIRBUS Defence & Space).

against an important infiltration of wastewaters from septic tanks and sewer and mixing with the polluted water from the lagoon system.

Low direct recharge may be explained by soil compactness due to urbanization and the high density of the population. Group 3 samples with the highest pH values and lowest NO_3^- concentration display between Na-Cl and Ca-Mg-HCO_3 water types end members (Fig. 2a). They recorded high scores for Factor 2, suggesting iron oxides and nitrate reduction by organic matter due to the lithology's detrital and sand clayey characteristics (Da Costa et al., 2013). Moreover, the localization of most of these samples in or near the flooding ground of Zio River (Fig. 2e) may suggest interactions with freshwater from Zio River and the upper Quaternary sediments along with denitrification processes. Group 1 samples recorded the highest salinity and present Na-Cl and Ca/Mg-Cl water types. Na-Cl water type with advanced hydrochemical evolution flow toward the central part of the alluvial plain of Zio River where Na^+ ions in water are exchanged with Ca^{2+} and Mg^{2+} fixed on sediments. Reverse cation exchange process were highlighted by the points displaying along the line of slope −1 in the diagram (Na^+-Cl^-) vs (Ca^{2+} + Mg^{2+}-HCO_3^--SO_4^{2-}).

5 Conclusions

This study showed that shallow groundwater in the coastal city of Lome is predominantly acidic and Na-Cl water types with mineralization ranging from soft waters to brackish ones. The PCA allowed identifying three main factors accounting for nearly 77 % of the total variance in the water chemistry dataset. The first factor is salinity and results from marine aerosol, sea spray, sea intrusion, contamination from saline lagoon system, increasing salts dissolution along flow paths, and 140 anthropogenic contamination. The second factor related to redox conditions depends on lithology, recharge regime, and land use. The third factor, defined as indirect recharge from wastewaters, has revealed the impact of sanitation conditions on groundwater chemistry in the historic central part of the city. The hierarchical clusters analysis allows dividing samples into five main groups according to the hydrochemical evolution in the NW-SE direction of groundwater flow. In the Zio River alluvial plain, groundwater evolves toward Na-Cl / HCO_3 and Ca-Mg-HCO_3 water types due to mixing with fresh water from the surface and upper Quaternary sediments. These results appear helpful for understanding the evolution of the shallow groundwater chemistry in the coastal sedimentary basin of Togo.

Data availability. Data can be obtained by contacting the authors.

Author contributions. KVA, MDTG, LMB, and GDB conceived and designed the research; All authors contributed to research, activities, and paper writing

Competing interests. The contact author has declared that neither they nor their co-authors have any competing interests.

Disclaimer. Publisher's note: Copernicus Publications remains neutral with regard to jurisdictional claims in published maps and institutional affiliations.

Special issue statement. This article is part of the special issue "Hydrology of Large River Basins of Africa". It is a result of the 4th International Conference on the "Hydrology of the Great Rivers of Africa", Cotonou, Benin, 13–20 November 2021.

Acknowledgements. The Authors thank Sivapalan Young Scientists Travel Awards (SYSTA), IAHS, and the organizers of Friend Water's 4th International Conference on the Hydrology of the Great Rivers of Africa

Financial support. This research has been supported by the Delegation of the European Union to Togo and the Agency of Water and Sanitation for Africa, Office of Togo (grant no. DCI-NSAPVD/2013/308-062).

References

Alfa-Sika Mande, S.-L., Liu, M., Djaneye-Boundjou, G., Liu, F., Bawa, M. L., and Chen, H.: Nitrate in drinking water: A major polluting component of groundwater in gulf region aquifers, south of Togo, Int. J. Phys. Sci., 7, 144–152, https://doi.org/10.5897/ijps11.874, 2012.

Appelo, C. A. J. and Postma, D.: Geochemistry, groundwater and pollution, 2nd Edn., CRC Press, Taylor & Francis Group, Boca Raton London New York, 2005.

Ayah, M., Grybos, M., Tampo, L., Bawa, L. M., Bril, H., and Djaneye-Boundjou, G.: Qualité et pollution des eaux d'un hydrosystème littoral tropical: cas du système lagunaire de Lomé, Togo, Eur. Sci. J., 11, 95–119, 2015.

BCR/DGSCN/MPRPDAT: Quatrième Recensement Général de la Population et de l'Habitat (RGPH4): Résultats définitifs, Document Gouvernemental, Ref. TGO_2010_PHC_v01_M, 57 pp., available at: https://catalog.ihsn.org//catalog/4097/download/55189 (last access: 5 November 2021), 2011.

BRGM: Surveillance piézométrique et chimique des nappes du Bassin Sédimentaire Côtier du Togo. Synthèse des connaissances hydrogéologiques. Bureau de Recherces Géologiques et Minières, Republique Togolaise, Ministère des Travaux Publics, des Mines, de l'Energie, et des Ressources Hydrauliques, Direction de l'Hydraulique et de l'Energie, Orléans Cedex, Rapport BRGM Direction de l'Hydraulique et l'Energie/MTPMERH. CI-04554, 1982.

Cloutier, V., Lefebvre, R., Therrien, R., and Savard, M. M.: Multivariate statistical analysis of geochemical data as indicative of the hydrogeochemical evolution of groundwater in a sedimentary rock aquifer system, J. Hydrol., 353, 294–313, https://doi.org/10.1016/j.jhydrol.2008.02.015, 2008.

Da Costa, P. Y., Affaton, P., Salaj, J., Johnson, A. K. C., and Seddoh, K.: Biozonation des formations sédimentaires du bassin côtier du Togo (Afrique de l'Ouest), Rev. Ivoir Sci. Technol., 21–22, 45–73, 2013.

Gnazou, M. D. T., Bawa, L. M., Banton, O., and Djanéyé-Boundjou, G.: Hydrogeochemical characterization of the coastal Paleocene aquifer of Togo (West Africa), Int. J. Water Resour. Environ. Eng., 3, 10–29, 2011.

Gnazou, M. D. T., Sabi, B. E., Lavalade, J. L., Schwartz, J., Akakpo, W., and Tozo, A.: Multilayered aquifer modeling in the coastal sedimentary basin of Togo, J. Afr. Earth Sci., 125, 42–58, 2017.

Helstrup, T: Environmental isotopic and hydrochemical characteristics of groundwater from the Cretaceous-Eocene limestone aquifer in the Keta Basin, Ghana, and the Coastal Sedimentary Basin of Togo, PhD Thesis, University of Copenhagen, 2006.

Lapworth, D. J., Nkhuwa, D. C. W., Okotto-Okotto, J., Pedley, S., Stuart, M. E., Tijani, M. N., and Wright, J.: Urban groundwater quality in sub-Saharan Africa: current status and implications for water security and public health, Hydrogeol. J., 25, 1093–1116, https://doi.org/10.1007/s10040-016-1516-6, 2017.

Lee, J., Jang, C., Wang, S., Liang, C., and Liu, C.: Delineation of spatial redox zones using discriminant analysis and geochemical modelling in arsenic-affected alluvial aquifers, Hydrol. Process., 22, 3029–3041, https://doi.org/10.1002/hyp.6884, 2008.

Mande, S. A.-S., Liu, M., Djaneye-Boundjou, G., Liu, F., Bawa, M. L., and Chen, H.: Nitrate in drinking water: A major polluting component of groundwater in gulf region aquifers, south of Togo, Int. J. Phys. Sci., 7, 144–152, https://doi.org/10.5897/ijps11.874, 2012.

MPT-Togo: Cadre de gestion environnemental et social du projet WARCIP-Togo, Rapport Final, SECDE & SOTED, 107 pp. + annexes, 2013.

Rodier, J., Legube, B., and Merlet, N.: L'analyse de l'eau, 9th Edn., Dunod, 2009.

Sylvain, J. P., Collart, J., Aregba, A., and Godonou, S.: Notice explicative de la carte géologique 1/500.0000 è du Togo, Direction Générale des Mines de la Géologie et du Bureau National de Recherches Minières Mémoire N°6, 1986.

Proc. IAHS, 384, 81–84, 2021
https://doi.org/10.5194/piahs-384-81-2021

Open Access PIAHS

Hydrology of Large River Basins of Africa

Assessment of Drinking Waters Quality Collected from Boreholes in Afe Babalola University Ado-Ekiti (ABUAD)

Adeshina S. Adebanjo

Department of Civil Engineering, Afe Babalola University, Ado Ekiti, Ekiti State, Nigeria

Correspondence: Adeshina S. Adebanjo (adebanjo.as@abuad.edu.ng)

Published: 16 November 2021

Abstract. The quality of drinking water is dependent of its source(s) and means of collection. The water available for drinking in the study area is accessible through groundwater exploration. This research was carried out to determine the quality of water from the selected boreholes within the university and checked by the World Health Organization (WHO) drinking water standards for the various parameters considered. Samples were gotten from seven (7) of the sited boreholes and various physico-chemical parameter tests such as turbidity test, conductivity test, pH test, copper test, manganese test, chloride test, sulphate test, chromium test, nitrate test and cadmium test and bacteriological tests as total plate count, total coliform count and faecal coliform count were carried out following due procedure, precautions and the results analyzed. The results show that most samples were satisfactory for the tests carried out while only two (samples from Behind Admin and Engineering) were unsatisfactory in turbidity while one sample (sample from Back of College 2) was unsatisfactory in the cadmium test. WQI showed that the water samples had excellent qualities except for that from Back of College 2 which had poor quality and is unsuitable consumption.

1 Introduction

Groundwater is one of the most important sources for drinking water and domestic activities in Africa. It is likely to be less affected than surface resources by climate variability, higher temperatures, evaporation and contamination. While the people of Ado Ekiti have access to various water sources for their domestic usage, the most sought and reliable source of drinking water is the groundwater accessed through borehole (Oyebode et al., 2019). It is the cheapest of the reliable drinking water sources. Studies carried out by Odeyemi et al. (2013) and Oyebode et al. (2019) showed that other water sources in Ekiti are exposed too much more contamination including physical, chemical and bacteriological. The quality of water tells a lot about the wellbeing of the consumer and also indirectly influences economic productivity. Afe Babalola University Ado-Ekiti (ABUAD) is the most growing Nigerian university in recent times, having a population of over 9000 people. Majority of students and academic staff members prefer to drink water from packaged water in sachets and bottles because of lack of confidence in the purity level of the borehole waters in the institutions. Water supplied through boreholes in ABUAD for so long has been perceived as unhealthy for drinking by staff and students and this perception has led to the limited use and has created an inconvenience financially as students of the university have to spend money on the purchase of drinking water daily. This has also affected the environment in a lot of ways because plastic sachets and bottles are found in excess as a part of waste generated in the university. It is for this reason that this research is being carried out. This study was embarked on to begin the needed periodic assessment of the quality of groundwater in this area.

2 Materials and Methods

2.1 Study area and Sampling sites

The study area is ABUAD, a young private university having a population of about 11 000 people. It is located in eastern part of Ado-Ekiti along Ijan road, Ado Ekiti, Ekiti State, Nigeria, lying on area of about 130 ha within latitude 7°35′59.16″ and 7°36′31.32″ N and longitude 5°18′6.61″ and 5°18′37.56″ E. This research was carried out on water

Figure 1. Study area with boreholes location (© Google Maps 2018).

Table 1. Method for testing parameters.

Parameter	Test Method
Turbidity (NTU)	Turbidimeter
pH	pH meter
Electrical Conductivity (µS/cm)	Conductivity meter
Nitrate (mg/L)	Spectrometric method
Sulphate (mg/L)	Gravimetric method
Chloride (mg/L)	Photometric method
Parameter	Test Method
Cadmium (mg/L)	Atomic Absorption Spectroscopy
Manganese (mg/L)	Atomic Absorption Spectroscopy
Chromium (mg/L)	Atomic Absorption Spectroscopy
Copper (mg/L)	Atomic Absorption Spectroscopy
Total Plate Count (cfu)	Culture Media
Total Coliform Count (MPN/100 mL)	Most Probable Number Method

samples taken from seven of the boreholes sited in ABUAD. ABUAD has a geology that is of basement complex, igneous rock of South-Western, Nigeria (Ogungbemi et al., 2013; Oyegoke et al., 2015; Oladimeji and Ogungbemi, 2013; Ogundana and Talabi, 2014) comprising of crystalline basement rocks which include coarse grained charnokite (the most abundant in Ado Ekiti), fine grained granite, medium grained granite and porphyritic biotite, medium grained granite and quartzite as its lithological rock units (Oyegoke et al., 2020). Its climate is tropical with distinct wet and dry seasons that are associated with the prevalence of maritime south westerly monsoon winds from the Atlantic Ocean and the dry continental north easterly harmattan winds from the Sahara Desert having abundant rainwater between May and October with substantial rainwater deficit between November and April (Oyegoke et al., 2020).

Physico-chemical parameters tests as well as bacteriological tests were carried out on the collected water samples from the boreholes according to the procedures outlined by the World Health Organisation standards for drinking water (WHO, 2004, 2006, 2008). The physico-chemical tests include turbidity test, electrical conductivity test, pH test, copper test, manganese test, chloride test, sulphate test, chromium test, nitrate test and cadmium test and the bacteriological tests include total plate count, total coliform count and faecal coliform count. Water quality Index (WQI) was used to analyse the parameters tested in the groundwater samples collected.

2.2 Water Quality Index

The weighted arithmetic index method was used to determine the WQI of groundwater from the selected locations. Thirteen physico-chemical parameters i.e. Turbidity, pH, EC, Cu^{2+}, NO_3^-, SO_4^{2-}, Cl^-, Cd^{2+}, Mn^{2+}, Cr^{2+}, TPC, TCC, and FCC were used to calculate WQI. The formula to calculate the WQI is given as (Yisa and Jimoh, 2010; Tyagi et al., 2014):

$$q_i = \frac{C_i}{S_i} \cdot 100 \quad (1)$$

Where, q_i, C_i, and S_i indicate quality rating scale, concentration of ith parameter, and standard value of ith parameter respectively.

$$\text{Relative weight, } w_i = \frac{1}{S_i}; \quad (2)$$

$$\text{and WQI} = \frac{\sum w_i q_i}{\sum w_i} \quad (3)$$

3 Result and Discussion

The results of the various parameter tests that were carried out on the collected samples are as presented in Table 2. Results in Table 2 when compared with the WHO standards for drinking water (WHO, 2004, 2006, 2008; Gray, 2008) show that, water samples taken from all boreholes except for those at Engineering (1.24 NTU) and Behind Admin (1.16 NTU) are of values greater than the set standard of 1 NTU in turbidity, samples from all the boreholes fall within the standards for pH of drinking water which between 6.5–8.5 and the electrical conductivity of 1000 µS/cm. Copper ions detected in the samples fall within acceptable limit of 2.00 mg/L by standard, the amount of nitrate contained in the samples falls within acceptable limit of 50.00 mg/L by standard, sulphate of the samples falls within acceptable limit of 100.00 mg/L by standard, the chloride ions in the samples fall within acceptable limit of 250.00 mg/L by standard, cadmium ion was only detected in the borehole at the Back of College 2 (0.004 mg/L) and was in an amount exceeding the acceptable limit of 0.003 mg/L by standard, the manganese ions in the samples fall within acceptable limit of 0.5 mg/L by standard, chromium ion was only detected in the samples from boreholes at WEMA Hostel, Main Gate, Amphi-Theatre, and Engineering, and were in amounts within the acceptable limit of 0.05 mg/L by standard. Total Plate Count results revealed

Table 2. Quality parameters for samples from boreholes.

Samples	WEMA Hostel	Main Gate	New Female Hostel	Back of College 2	Amphi-Theatre	Engineering	Behind Admin	WHO standards
Turbidity (NTU)	0.56	0.62	0.8	0.59	0.48	1.24	1.16	1
pH	7.56	7.29	7.7	8.36	7.22	7.74	8.24	6.5–8.5
Electrical Conductivity (µS/cm)	582.1	316.7	212.1	587.2	204.4	285.1	419.2	1000
Copper (mg/L)	< 0.01	0.04	< 0.01	< 0.01	< 0.01	0.03	< 0.01	2
Nitrate (mg/L)	0.02	0.01	0.01	0.02	0.04	0.22	0.01	50
Sulphate (mg/L)	10.57	0.62	0.83	0.5	0.14	0.26	0.44	100
Chloride (mg/L)	92	68.03	57.82	80.03	32.41	5.4	71.08	250
Cadmium (mg/L)	Not Detected	Not Detected	Not Detected	0.004	Not Detected	Not Detected	Not Detected	0.003
Manganese (mg/L)	0.06	0.09	0.04	0.07	0.06	0.06	0.08	0.5
Chromium (mg/L)	< 0.001	< 0.001	Not Detected	Not Detected	< 0.001	< 0.001	Not Detected	0.05
Total Plate Count (cfu)	0	0	0	0	0	0	0	Not Detectable
Total Coliform Count (MPN/100 mL)	0	0	0	0	0	0	0	Not Detectable
Faecal Coliform Count (cfu/mL)	0	0	0	0	0	0	0	Not Detectable

Table 3. Water Quality Index.

SN	Groundwater Location	WQI value	Water quality status
1	WEMA Hostel	0.2384	Excellent water
2	Main Gate	0.2852	Excellent water
3	New Female Hostel	0.2846	Excellent water
4	Back of College 2	124.77	Poor quality water
5	Amphi-Theatre	0.2079	Excellent water
6	Engineering	0.4342	Excellent water
7	Behind Admin	0.4423	Excellent water

that all the water samples contained no aerobic bacteria, Total Coliform Count results and analysis revealed that samples contained no coliform colonies, Faecal Coliform Count results and analysis revealed that all the borehole samples contained no coliform colonies.

This conforms to earlier study done by Adeosun and Omietimi (2020) on some parts of Ado-Ekiti where they found physico-chemical parameters of impurities to be insignificant. Although, other studies done in this region showed significant contaminants level, but they were on shallow well. Not much has been done on groundwater accessible through borehole.

4 Conclusions

The results and analysis of the physico-chemical quality parameters of the water samples from the seven boreholes taken for this study in ABUAD revealed that, all water samples when compared with the permissible limits presented by WHO standard are

satisfactory in pH, electrical conductivity, nitrate, sulphate and chloride; are satisfactory in examined heavy metal constituents except for that from borehole at Back of College 2 which was unsatisfactory in cadmium ion constituent. Turbidity level was within the limit for acceptability for samples from all boreholes except for those at Engineering and Behind Admin which will therefore need to be coagulated, filtered and disinfected before drinking. Investigation with Water Quality Index (WQI) calculation showed that all water samples were excellent quality except for samples collected at Back of College 2 which showed poor quality.

Results and analysis of the Total Plate Count, Total Coliform Count and Faecal Coliform Count revealed that there is no form of bacteriological contamination in all the borehole samples.

Water from the boreholes in ABUAD can generally be regarded as been good and fit for drinking and for domestic use while further investigation should be conducted on that at the Back of College 2.

Data availability. No data sets were used in this article.

Competing interests. The contact author has declared that there are no competing interests.

Disclaimer. Publisher's note: Copernicus Publications remains neutral with regard to jurisdictional claims in published maps and institutional affiliations.

Special issue statement. This article is part of the special issue "Hydrology of Large River Basins of Africa". It is a result of the 4th International Conference on the "Hydrology of the Great Rivers of Africa", Cotonou, Benin, 13–20 November 2021.

Acknowledgements. I appreciate the efforts and support from Paul Edoh of Afe Babalola University in the laboratory tests.

References

Adeosun O. A. and Omietimi E. J.: Groundwater quality in parts of Ado-Ekiti Metropolis, South-Western Nigeria, Int. J. Sci. Res. Publ., 10, 227–254, https://doi.org/10.29322/IJSRP.10.08.2020.p10431, 2020.

Gray, N. F.: Drinking Water Quality: Problems and Solutions, 2nd Edn., Cambridge University Press, UK, 520 pp., 2008.

Odeyemi, A. T., Adebayo, A. A., and Adeosun, O. M.: Bacteriological and physicochemical studies of three major dams in Ekiti State, Nigeria, J. Environ. Earth Sci., 3, 210–218, 2013.

Ogundana, A. K. and Talabi, A. O.: Groundwater Potential of College of Engineering, Afe Babalola University, Ado-Ekiti, Southwestern, Am. J. Water Res., 2, 25–30, https://doi.org/10.12691/ajwr-2-1-4, 2014.

Ogungbemi, O. S, Badmus, O. G., Ayeni, O. G., and Ologe, O.: Geoeletric Investigation of Aquifer Vulnerability within Afe Babalola University, Ado-Ekiti, Southwestern, Nigeria, IOSR J. Appl. Geol. Geophys., 1, 28–34, 2013.

Oladimeji, L. A. and Ogungbemi, O. S.: Evaluation of Aquifer Protective Capacity of Ground Water Resources within Afe Babalola University, Ado – Ekiti, Southwestern Nigeria. Transnational Journal of Science and Technology, 3, 1–16, 2013.

Oyebode, O. J., Adebanjo, A. S., ND-Ezuma, S. R.: Conditions of available sources of water for domestic use in selected communities in Ado-Ekiti, Afr. J. Environ. Sci. Technol., 13, 84–90, https://doi.org/10.5897/AJEST2018.2509, 2019.

Oyegoke, S. O., Oyebode, J. O., Olowe, K. O., and Ekom, E.: Exploitation of Groundwater in Fractured Basement of Ado-Ekiti, Nigeria, Am. J. Eng. Res., 4, 55–63, 2015.

Oyegoke, S. O., Ayeni, O. O., Olowe, K. O., Adebanjo, A. S., and Fayomi, O. O.: Effectiveness of Geophysical Assessment of Boreholes Drilled in Basement Complex Terrain at Afe Babalola University using Electromagnetic (E.M.) Method, Nigerian Journal of Technology, 39, 36–41, https://doi.org/10.4314/njt.v39i1.4, 2020.

Tyagi, S., Singh, P., Sharma, B., and Singh, R.: Assessment of water quality for drinking purpose in District Pauri of Uttarkhand India, Appl. Ecol. Environ. Sci., 2, 94–99, 2014.

WHO: Guidelines for Drinking-water Quality, 3rd Edn., 1, World Health Organization, China, 515 pp., 2004.

WHO: Guidelines for Drinking-water Quality, Incorporating 1st Addendum to the 3rd Edn., Volume 1-Recommendations, World Health Organization, Geneva, 515 pp., 2006.

WHO: Guidelines for drinking-water quality, 3rd Edn., 1, Recommendations, Incorporating first and second addenda, Vol. 1, World Health Organization, Geneva, 668 pp., 2008.

Yisa, J. and Jimoh, T.: Analytical studies on water quality index of river Landzu, Am. J. Appl. Sci., 7, 453–458, 2010.

Proc. IAHS, 384, 85–92, 2021
https://doi.org/10.5194/piahs-384-85-2021

Open Access
PIAHS

Hydrology of Large River Basins of Africa

Pollution des eaux à usages domestiques par les éléments traces métalliques des activités anthropiques : cas du sous bassin versant du fleuve Sassandra en amont du barrage de Buyo, Côte d'Ivoire

Kouakou Séraphin Konan[1], **Kouakou Benoit Kouamé**[1], **Félix Koffi Konan**[2,✝], **Koffi Charles Boussou**[2], **and Kouassi Lazare Kouakou**[1]

[1]Laboratoire des Sciences et Technologies de l'Environnement, Université Jean Lorougnon Guédé, Daloa, Côte d'Ivoire

[2]Laboratoire de Biodiversité et Gestion Durable des Ecosystèmes Tropicaux, Université Jean Lorougnon Guédé, Daloa, Côte d'Ivoire

[✝]deceased

Correspondence: Kouakou Séraphin Konan (konandks@yahoo.fr)

Published: 16 November 2021

Résumé. Pour résoudre les problèmes d'approvisionnement en eau potable des centres urbains, le gouvernement ivoirien a décidé d'utiliser les ressources pérennes telles que le fleuve Sassandra. Ainsi, le sous bassin du fleuve Sassandra en amont du barrage de Buyo est pressenti pour alimenter les villes de la région du Haut Sassandra et du département de Duekoué. Cependant, les intrants agricoles, les activités d'orpaillage et les déchets ménagers entrainent une forte contamination des eaux de ce fleuve en éléments traces métalliques (ETM). Ainsi, les tendances spatiales et temporelles de la pollution métallique ont été évaluées par le calcul des facteurs d'enrichissement, des indices de géo-accumulation, de l'indice de charge polluante, de l'indice de pollution sédimentaire et de l'indice de risque écologique individuel sur vingt-huit (28) échantillons provenant de sédiments de surface de ce fleuve. Les résultats montrent que des pollutions modérées à modérément sévères sont enregistrées au niveau du mercure (Hg) et du cuivre (Cu). Le Pb, l'As et le Cd connaissent un enrichissement modéré, avec des facteurs d'enrichissement respectifs de 1,77 ; 1,60 et 1,79 suggérant l'existence d'autres sources de métaux autres que la source naturelle terrigène. L'indice de pollution sédimentaire estimé à 5,95 (< 10) montre que les sédiments sont moyennement pollués. Le risque écologique moyen lié aux Cd, As, Pb, Cu et Zn est largement inférieur à 40 traduisant un risque bas en référence à ces métaux avec le risque écologique moyen lié au Hg (160) largement supérieur à 40. L'analyse temporelle couplée au test SEQ-Eau montre également que le stockage des éléments traces métalliques au niveau des sédiments n'est pas définitif. Les risques de relargage, de remobilisation, de biodisponibilité et donc de toxicité, sont à craindre et constituent ainsi un danger permanent pour toute la chaîne trophique. Des mesures de prévention doivent donc être prises pour anticiper sur les difficultés que pourrait rencontrer le traitement de ces eaux pour les usages domestiques.

1 Introduction

Les métaux lourds sont de nature persistants, toxiques et ont une très forte tendance à s'accumuler dans les organismes vivants et à s'adsorber sur les particules de sédiment (Wang et al., 2012). La pollution par les métaux des écosystèmes aquatiques augmente en raison des effets de l'urbanisation, de l'agriculture, des activités d'orpaillage et dans une moindre mesure de l'industrialisation dans notre zone d'étude. Le sous bassin versant du fleuve Sassandra (lac Guessabo) est susceptible d'être fortement contaminé par le Hg, le Pb, le Cd, le As, le Cu, le Zn, le Mn et le Fe en raison du rejet croissant de polluants dans le bassin par les eaux usées des stations-services, du ruissellement des pesticides et des engrais agricoles. Les sédiments sont connus pour être des réservoirs ultimes des métaux lourds rejetés dans l'environnement (Malferrari et al., 2009). Ainsi, les sédiments pourraient constituer une source potentielle de métaux lourds qui seront rejetés dans les eaux sus-jacentes par des moyens naturels et des processus anthropiques (Beutel et al., 2008), où ils pourraient avoir des effets néfastes sur la qualité de l'eau de boisson et la santé humaine. Certains métaux toxiques dans les sédiments peuvent être mobilisés par des procédés biochimiques tels que la prolifération de cyanobactéries, causant une pollution de l'eau (Olivares-Rieumont et al., 2009). D'autres métaux lourds sont nécessaires aux organismes mais peuvent être toxiques si leurs concentrations sont assez élevées. Selon plusieurs auteurs, plus de 90 % des teneurs en métaux lourds sont liées aux particules en suspension dans les hydrosystèmes soulignant ainsi l'importance de cette matrice dans l'étude de la qualité des systèmes aquatique (Wang et al., 2012). Ainsi, l'étude des sédiments permet de connaitre les changements environnementaux survenus dans les bassins fluviaux et fournissent des informations utiles sur l'accumulation de métaux lourds, reflétant les impacts naturels (géogéniques) et anthropiques. La surveillance de la concentration de métaux lourds et l'évaluation de leur niveau de contamination dans les sédiments du lac Guessabo sont des questions impératives et cruciales pour assurer la durabilité des écosystèmes. L'objectif de ce travail est de déterminer la distribution de certains métaux lourds (Hg, As, Cd, Pb, Cu, Zn, Mn et Fe) dans le sédiment et d'évaluer le degré de contamination afin d'aider à la prise de décision pour anticiper sur les difficultés que pourrait rencontrer le traitement de ces eaux pour les usages domestiques.

2 Matériel et Méthodes

2.1 Milieux d'étude

Le lac Guessabo est situé entre les latitudes 6°57′ et 7°2′ N et les longitudes 6°45′ et 6°46′ W (Fig. 1). Il se trouve en amont du barrage de Buyo construit sur le fleuve Sassandra. Il couvre une superficie de 27 km^2. L'agriculture constitue l'activité économique principale (Kouamé et al., 2019a). Le lac Guessabo est traversé par la nationale 6. Les activités économiques liées au milieu lacustre sont constituées essentiellement de la pêche.

2.2 Échantillonnage

L'échantillonnage a été fait à l'aide d'une benne Van Veen, qui a permis de prélever les 5 premiers centimètres de la colonne sédimentaire. Ces premiers centimètres correspondent à la zone la plus active (mobilité verticale et latérale, interface eau-sédiment). Les concentrations en ETMs sont parfois très faibles, c'est pour pourquoi nous avons redoublé de vigilance au moment des prélèvements et éviter toute source de pollutions externes. Il a été impératif d'utiliser des instruments en plastique décontaminés au préalable. Les échantillons recueillis ont été placés directement dans des sacs en polyéthylène scellés à l'air et conservés à 4 °C jusqu'à l'analyse. Au laboratoire, les échantillons de sédiments lacustres ont été soumis à un traitement mécanique pour obtenir une poudre dont les particules ont un diamètre inférieur à 63 µm.

2.3 Analyse

Avant l'analyse au spectrophotomètre d'absorption atomique, les échantillons prétraités de sédiments (0,1 g) sont minéralisés à chaud dans des tubes en téflon à l'aide de 3 mL d'acide hydrofluoridrique (HF) concentré pur (48 %) avec 1 mL d'eau régale ($HNO_3 + HCl$; $1/3v/v$). L'utilisation de HF est essentielle parce qu'il est le seul acide qui dissout complètement les treillis silicatés et libère tous les métaux. Après la digestion, les résidus sont récupérés avec une solution d'acide borique (2,7 g dans 20 mL d'eau bidistillée). Le volume final est ramené à 50 mL. Ce volume final est laissé au repos pendant un jour pour s'assurer de la digestion totale des échantillons. La méthode d'analyse correspond à celle décrit par Kouamé et al. (2019b).

2.4 Contamination assessment

Pour évaluer le niveau de contamination des sédiments par les métaux lourds, celui-ci a été évalué par le calcul de facteurs d'enrichissement, d'indices de géoaccumulation, de l'indice de charge polluante, de l'indice de pollution sédimentaire et de l'indices de risque écologique. La haute croûte continentale faisant référence à des formations sédimentaires (Wedepohl, 1995), est la référence utilisée dans ce présent travail. Aussi dans ce travail, le fer (Fe) a été choisi comme élément normalisant parce qu'il est associé à des surfaces solides fines, sa géochimie est similaire à celle de nombreux métaux traces ; et sa concentration dans le sédiment naturel tend à être uniforme (Daskalakis and O'Connor, 1995).

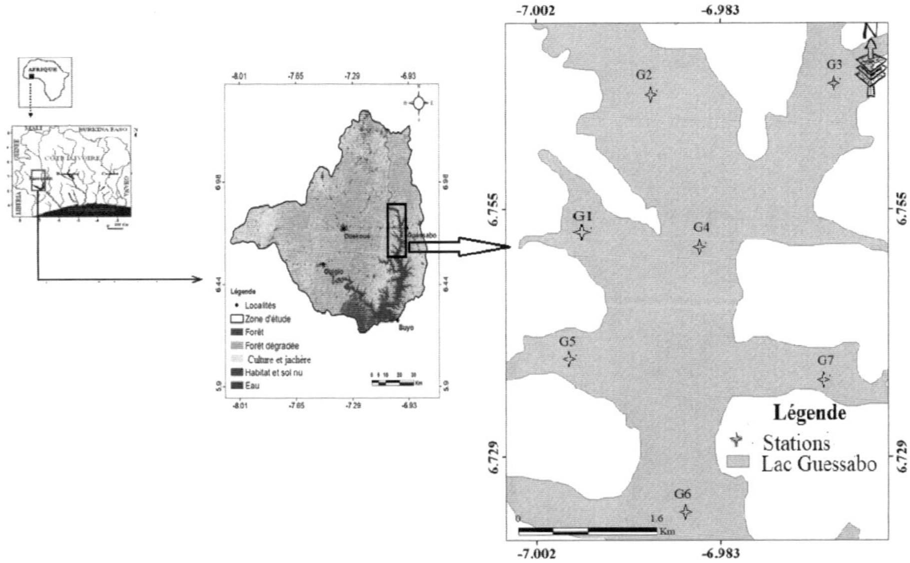

Figure 1. Stations d'échantillonnage et localisation de la zone d'étude.

2.5 Indice de géoaccumulation

Pour quantifier l'intensité de la pollution métallique l'index de géoaccumulation introduit par Müller (1981) est utilisé et définit par l'équation 1.

$$\text{Igeo} = \log_2 \left(\frac{C_n}{1,5 \times B_n} \right) \quad (1)$$

- Igéo : indice de géo accumulation ;
- Cn : concentration mesurée dans l'échantillon
- Bn : fond géochimique ;
- 1,5 : facteur d'exagération du fond

Igéo < 0 Non contaminé ; 0 < Igéo < 1 Non contaminé à modérément contaminé ; 1 < Igéo < 2 Modérément contaminé ; 2 < Igéo < 3 Modérément contaminé à sévèrement contaminé ; 3 < Igéo < 4 Sévèrement contaminé ; 4 < Igéo < 5 Sévèrement à extrêmement contaminé ; 5 < Igéo Extrêmement contaminé.

2.5.1 Facteur d'enrichissement

Le FE a été calculé pour déterminer l'origine anthropique ou naturelle des éléments dans les sédiments du lac Guessabo. Le facteur d'enrichissement fourni le nombre de fois qu'un élément est enrichi par rapport à l'abondance de cet élément dans le matériau de référence. FE est défini par l'équation 2

$$\text{FE} = \frac{([X]/[R])_{\text{échantillon}}}{([X]/[R])_{\text{matériel de reférence}}} \quad (2)$$

- R : concentration de l'élément normalisant dans le sédiment ($\mu g\, g^{-1}$)
- $[X]$ = concentration de l'élément trace métallique dans le sédiment ($\mu g\, g^{-1}$)
- FE ≤ 1 Aucun enrichissement ; 1 < FE ≤ 3 Enrichissement mineur ; 3 < FE ≤ 5 Enrichissement modéré
- 5 < FE ≤ 10 Enrichissement modérément sévère, 10 < FE ≤ 25 Enrichissement sévère ; 25 < FE ≤ 50 Enrichissement très sévère

2.5.2 Indice de pollution sédimentaire

Pour l'évaluation globale de la qualité des sédiments tenant en compte de la toxicité relative de chaque métal en lui affectant un facteur pondérateur, l'indice de pollution sédimentaire, définit comme une fonction linéaire de la somme de facteurs d'enrichissement et introduit par Müller (1969) a été calculé (équation 3).

$$\text{IPS} = \frac{\sum(\text{FE}_m \times W_m)}{\sum W_m} \quad (3)$$

- W_m : poids de toxicité ou facteur pondérateur du métal m.
- FE_m : facteur d'enrichissement du métal m ; 0 ≤ IPS < 2 Sédiment sain ; 2 ≤ IPS < 5 Sédiment faiblement pollué ; 5 ≤ IPS < 10 Sédiment moyennement pollué ; 10 ≤ IPS < 20 Sédiment très pollué ; IPS ≥ 20 Sédiment dangereux

2.5.3 Indice de charge Polluante

L'indice global (l'indice de charge métallique (PLI : Pollution Load Index en Anglais)) d'étude de la contamination métallique des sédiments à partir de la concentration totale de tous les métaux étudiés et qui permet d'estimer le degré de contamination global des sédiments d'une zone étudiée a été utilisé. L'équation (4) suivante a été développée par Tomlinson et al. (1980), pour le calcul de l'indice de charge polluante (PLI).

$$\mathrm{PLI} = \sqrt[n]{\mathrm{FC_1} \times \mathrm{FC_2} \times \mathrm{FC_3} \times \ldots \times \mathrm{FC_n}}$$

ou

$$\mathrm{FC}_i = \frac{C_D^i}{C_R^i} \quad (4)$$

FC_i : facteur de contamination du métal i. PLI = 0 pas de détérioration ; PLI = 1 présents aux concentrations de référence ; PLI > 1 détérioration progressive du milieu.

2.5.4 Indice de risque écologique potentiel (RI) et risque potentiel de chaque ETM pris individuellement

Cet indice est introduit pour évaluer le degré de risque écologique des éléments traces dans les sédiments. Les valeurs de RI se calculent à partir des formules de Häkanson (1980) suivantes (équation 5)

$$\mathrm{RI} = \sum\nolimits_i^n E_r^i ; E_r^i = T_f^i \times C_f^i ; C_f^i = \frac{C_D^i}{C_R^i} \text{ ou :} \quad (5)$$

C_f^i : coefficient de contamination ; C_D^i : concentration de l'ETM, C_R^i : concentration de l'ETM dans les sédiments non pollués ; Tif ; facteur de toxicité de l'ETM. Les poids respectifs assigné aux métaux utilisés pour les calculs de IPS et RI sont : Cd = 30 ; Hg = 40, As = 10 ; Pb = 5 ; Cu = 2 ; Zn = 1) ; RI > 150 risque bas : 150 ≤ RI < 300 risque modéré ; 300 ≤ RI < 600 risque considérable ; RI ≥ 600 risque très élevé.

3 Résultats et Discussion

3.1 Variabilité saisonnière et étude comparative

Les résultats des concentrations totales des ETMs dans les sédiments du lac Guessabo sont présentés dans le tableau 1.

Les concentrations moyennes des éléments traces métalliques lors de cette étude, comparées à celles des valeurs de pollution préindustrielle, mettent en évidence une contamination métallique des sédiments du lac.

Les valeurs des concentrations totales Mn et Zn (tableau 1) à toutes les stations sont inférieures aux valeurs préindustrielles, suggérant que les teneurs de ces éléments sont inférieures à ceux de la croute continentale. Cependant, les valeurs des concentrations totales de Hg, Pb, As et Cu (tableau 1) à toutes les stations sont supérieures aux valeurs préindustrielles excepté une station pour le plomb (G3) et l'arsenic (G4), ce qui suggère que ces stations sont enrichies par ces métaux. La forte croissance démographique et le développement économique ont contribué de manière significative à la détérioration actuelle de la qualité de l'eau, y compris l'accumulation de métaux lourds tels que le Hg, Cd, Pb, As et Cu dans le milieu aquatique (Rajeshkumar et al., 2017). Les résultats obtenus dans les sédiments du lac Guessabo du présent travail n'ont pu être comparés avec ceux d'un travail antérieur mené dans le même lac en raison d'absence d'étude de ce type dans cette ressource. En revanche, les sédiments du lac Guessabo ont fait l'objet d'une autre comparaison avec ceux du fleuve Comoé du même pays. Nous avons noté différents degrés de contamination métallique par les éléments Cu, Zn, Pb et Mn dans les sédiments de ce fleuve en raison de la proximité de l'estuaire fleuve Comoé avec la plus agglomération qui est Abidjan (Keumean et al., 2013).

En ce qui concerne la variabilité saisonnière de ces paramètres (Fig. 2), l'analyse a permis de discriminer la saison sèche de la saison des pluies ($P < 0,05$). Durant la saison sèche, les teneurs en Hg, en Pb, en Cd, en Cu, en Mn et en Fe étaient importantes tandis qu'en saison des pluies, c'était l'As, qui présentait des valeurs élevées. Au plan spatial, les stations G2 et G5 présentaient généralement des teneurs élevées pour tous les éléments traces métalliques. Un ordre général de disponibilité des métaux lourds dans les sédiments a pu être établi : Fe > Cu > Mn > Pb > As > Hg > Cd. Le faible taux d'accumulation des métaux (As et Cd) dans les sédiments serait dû aux formes carbonates qui ont un rôle de diluant dans le développement des éléments de transition (Li et al., 2018b). Les concentrations de Pb et de Cu étaient plus élevées dans les sédiments de surface par rapport aux Cd, probablement en raison de la plus grande stabilité de leurs complexes (les oxydes de fer amorphes et les minéraux argileux) de surface ou sa faible abondance dans la nature. Les concentrations de métaux (Hg, Pb, Cd, Cu, Mn et Fe) élevés dans les sédiments en saison sèche seraient dues au niveau bas de l'eau qui pourrait aider à accumuler les métaux lourds dans les sédiments (Mohammad Ali et al., 2016) ou aux apports d'eaux usées des sous- préfectures et villages en bordure du lac. De plus, la diminution de la concentration en saison pluvieuse pourrait être due aux précipitations, qui pourraient augmenter le processus de « lixiviation » et contribuer à la dilution des métaux lourds pendant la saison des pluies qui s'opèrent par des apports de sédiments moins ou non contaminés (Duman et Kar, 2012).

Tableau 1. Teneur moyenne des éléments traces et valeurs Préindustrielles.

Métaux louds	Valeurs préindustrielles	G1	G2	G3	G4	G5	G6	G7
Hg (µg/g)	0.056	0.47 ± 0.24	0.61 ± 0.11	0.68 ± 0.55	0.81 ± 0.52	0.41 ± 0.06	0.14 ± 0.19	0.35 ± 0.19
Cd (µg/g)	0.102	0.22 ± 0.29	0.26 ± 0.18	0.25 ± 0.18	0.03 ± 0.02	0.07 ± 0.04	0.04 ± 0.05	0.08 ± 0.14
Pb (µg/g)	17	47.91 ± 30.46	34.75 ± 17.95	14.88 ± 9.35	18.86 ± 20.96	30.20 ± 14.16	15.34 ± 8.53	11.28 ± 7.76
As (µg/g)	2	2.53 ± 1.71	2.67 ± 1.14	2.10 ± 0.77	1.41 ± 0.43	3.46 ± 0.36	2.55 ± 0.97	3.09 ± 0.61
Cu (µg/g)	14.3	39.29 ± 11.63	45.49 ± 15.80	163.20 ± 121.13	157.63 ± 137.59	278.42 ± 155.81	230.25 ± 195.55	178.60 ± 170.49
Mn (µg/g)	527	21.34 ± 2.85	43.83 ± 6.96	19.62 ± 8.73	77.09 ± 22.82	28.61 ± 4.56	51.13 ± 7.94	63.06 ± 13.62
Zn (µg/g)	52	7.96 ± 5.95	45.89 ± 27.74	14.97 ± 8.93	5.77 ± 5.27	14.16 ± 16.12	9.11 ± 3.72	10.07 ± 10.60
Fe (g/kg)	30. 890	19.36±11.31	33.81 ± 13.97	21.65 ± 10.96	32.86 ± 9.92	35.52 ± 19.35	30.09 ± 6.51	15.71 ± 5.39

3.2 Diagnostic des contaminations sédimentaires

3.2.1 Facteurs d'enrichissement (FE) et indices de pollution sédimentaires (IPS)

Le tableau 2 montre les valeurs moyennes du facteur d'enrichissement (FE) et de l'indice de pollution sédimentaire des éléments traces dans les sédiments des stations du lac Guessabo. Les résultats montrent une absence d'enrichissement des sédiments en Mn et en Zn à toutes les stations par ces métaux (FE < 1) dans le lac. Ils montrent que ces éléments sont principalement d'origine terrigène (Diop et al., 2015). Pour le Pb et le Cd l'enrichissement varie d'aucun enrichissement à enrichissement modéré pour les sédiments (1 > FE < 5). Toutes les stations présentent un enrichissement modéré à un enrichissement sévère pour le Hg (3 < FE < 20). Pour ce qui est du Cu, l'on a observé un enrichissement mineur à modérément sévère (1 < FE < 20) au niveau des sédiments. Toutes les stations présentent un enrichissement mineur à modérément sévère pour l'As (1 > FE < 10) au niveau des sédiments. Les stations G1, G3, G4 et G7 présentent les valeurs de facteurs d'enrichissement, les plus élevées (modérément sévères) pour le mercure indiquant qu'ils sont fortement enrichis et que l'origine de cet enrichissement est principalement anthropique (Diop et al., 2015). L'origine anthropique du cuivre aux stations G5, G6 et G7 serait dû aux additifs nutritionnels contenant du cuivre utilisés pour l'engraissement des porcs ou/et aux fertilisants dans l'agriculture. Pour l'évaluation de la qualité globale des sédiments, tenant en compte d'une approche multi-métaux, incluant de la toxicité relative de chaque métal, l'indice de pollution sédimentaire (IPS) a été calculé (tableau 2). Selon la classification de Singh et al. (2002), les sédiments ont été trouvés faiblement pollués (2 < IPS < 5) dans les stations G5 et G6, modérément pollués (5 < SPI < 10) aux autres stations. Dans le lac, les valeurs de IPS par points de prélèvement décroissent en suivant l'ordre : G3 > G1 > G7 > G4 > G2 > G5 > G6. Ce classement s'explique aisément par les caractéristiques des effluents : les sédiments des stations G1, G3 et G7 sont directement impactés par les eaux usées non traitées et le ruissellement agricole (engrais et pesticides).

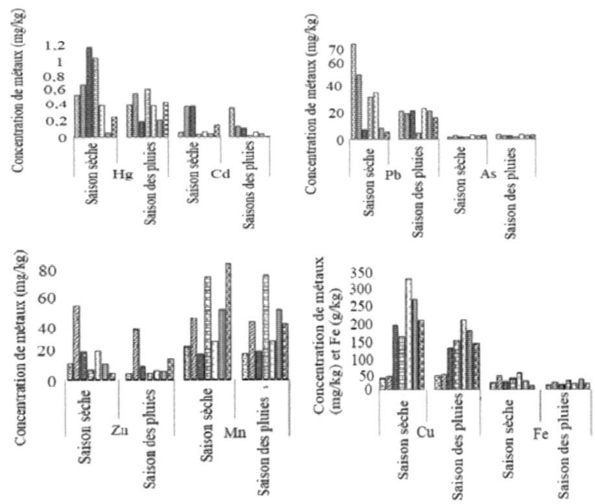

Figure 2. Variations des concentrations des éléments traces métalliques dans les sédiments du lac Guessabo.

3.2.2 Indice de géoaccumulation

Le tableau 3 indique les valeurs de l'indice de géoaccumulation des différents points d'échantillonnages et pour chaque métal. Les index de géoaccumulation du Fe, du Mn et du Zn sont tous < 0 à toutes les stations montrant que les sédiments du lac Guessabo ne sont pas pollués par ces ETMs. Les classes Igeo de Pb étaient respectivement " Modérément contaminé " et " Non contaminé à modérément contaminé " pour les sédiments des stations G7 et autres (G1, G2, G5), tandis que la classe Igeo de As et Cd était " Non contaminé à modérément contaminé " pour les sédiments des stations G1, G2, G3 pour Cd et G5, G7 pour As selon la classification de Müller (1981). Les classes Igeo de Hg étaient respectivement « Le Sévèrement contaminées » et « Modérément contaminées à sévèrement contaminées » pour les sédiments des stations G4 et autres (G1, G2, G3, G5, G7), tandis que les classes Igeo de Cu étaient respectivement « Sévèrement contaminées » et « Modérément contaminée à sévèrement contaminée » pour les sédiments des stations G5, G6, G7 et autres (G2, G3, G4) selon la classification de Muller (1981).

Tableau 2. Récapitulatif des valeurs moyennes du facteur d'enrichissement et de l'indice de pollution sédimentaire par station et par élément métallique dans les sédiments du lac Guessabo.

Millieu d'étude	Stations	FE							IPS
		Mn	Cu	Zn	Pb	As	Cd	Hg	
Guessabo Lake	G1	0.06	2.52	0.17	4.49	2.02	3.49	13.50	7.74
	G2	0.07	1.67	0.58	1.86	1.22	2.42	9.91	5.53
	G3	0.05	9.36	0.30	1.25	1.50	3.63	17.00	9.37
	G4	0.12	5.95	0.07	1.04	0.66	0.28	13.59	6.50
	G5	0.04	9.74	0.17	1.54	1.50	0.61	6.31	3.64
	G6	0.09	9.50	0.13	0.92	1.30	0.48	2.56	1.84
	G7	0.21	14.13	0.27	1.30	3.03	1.67	12.37	7.02
	Moyenne	0.09	7.55	0.24	1.77	1.60	1.79	10.75	5.95

Tableau 3. Récapitulatif des indices de géoaccumulation des éléments métalliques dans les sédiments du lac Guessabo.

Millieu d'étude	Stations	Indice de géoaccumulation							
		Fe	Mn	Cu	Zn	Pb	As	Cd	Hg
Guessabo Lake	G1	−1.26	−5.21	0.87	−3.29	0.91	−0.24	0.51	2.49
	G2	−0.45	−4.17	1.08	−0.76	0.45	−0.16	0.79	2.85
	G3	−1.09	−5.33	2.92	−2.38	−0.77	−0.51	0.73	2.99
	G4	−0.49	−3.35	2.87	−3.75	−0.43	−1.08	−2.35	3.26
	G5	−0.38	−4.78	3.69	−2.46	0.24	0.20	−1.13	2.27
	G6	−0.62	−3.95	3.42	−3.09	−0.73	−0.23	−1.76	0.74
	G7	−1.56	−3.68	3.05	−2.48	1.17	0.04	−0.84	2.07
	Moyenne	−0.84	−4.35	2.56	−2.60	−0.22	−0.28	−0.58	2.38

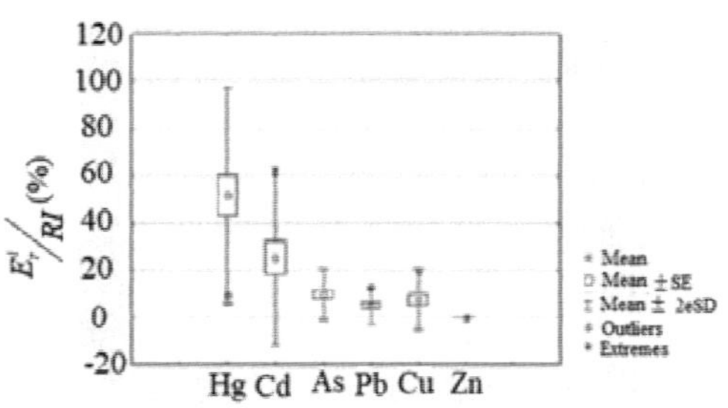

Figure 3. Contribution des métaux respective au RI.

Tableau 4. Valeurs moyennes du PLI des sédiments par station.

Millieu d'étude	Stations	PLI
Guessabo Lake	G1	0.96
	G2	1.45
	G3	1.1
	G4	0.94
	G5	1.23
	G6	0.87
	G7	0.97
	Moyenne	1.08

Ces résultats concordent assez bien avec ceux obtenus par l'évaluation au Facteur d'enrichissement.

3.2.3 Indice de charge polluante (PLI)

Le tableau 4 montre les valeurs de l'indice de charge polluante. Les stations G2, G3 et G5 ont des valeurs PLI supérieures à 1, indiquant une détérioration progressive du lac en ces stations. La raison probable de cette contamination serait les déchets agricoles (engrais et pesticides) et les eaux usées ainsi que les déchets des exploitations piscicoles trouvées dans la partie où se trouve la station G5 (Men et al., 2018).

3.2.4 Indices de risque écologique individuel (E_r^i) et de risque écologique potentiel (RI)

Le tableau 5 montre les valeurs des indices de risque écologique individuel et de risque écologique potentiel (RI) par station.

Le risque écologique moyen lié aux Cd, As, Pb, Cu et Zn est largement inférieur à 40 traduisant un risque bas en référence à ces métaux. En revanche, celui lié au mercure est de 86,6 (> 40), montrant un risque considérable. Sur l'ensemble des échantillons de sédiments du lac Guessabo analysés, le résultat révèle un indice du risque écologique potentiel égale à 153,19 indiquant une pollution modérée (tableau 5).

Tableau 5. Récapitulatif des Indices du risque écologique individuel (E_r^i) et du risque écologique potentiel (RI).

	Indices du risque écologique individuel (E_r^i)						Risque écologique potentiel (RI)
Stations	Hg	Cd	As	Pb	Cu	Zn	
G1	9.48	59.66	12.67	12.61	2.38	0.08	97
G2	121.5	72.27	13.38	9.15	2.75	0.48	219
G3	133.5	69.47	10.52	3.95	9.89	0.15	227
G4	162	8.18	7.06	4.96	9.55	0.06	191
G5	81.3	19.09	17.30	7.94	16.87	0.14	143
G6	28	12.27	12.75	4.04	13.95	0.09	71
G7	70.5	23.18	15.45	2.97	10.82	0.1	123

Les résultats de l'indice de risque écologique individuel indiquent que les concentrations de Hg sont susceptibles d'avoir des effets nocifs sur les organismes vivant dans les sédiments. Hg contribue en moyenne à 51,7 % au risque écologique potentiel suivit du Cd qui contribue à 25,57 % (Fig. 3). Ces métaux lourds sont susceptibles d'affecter négativement une diversité de communauté microbienne (Huang et al., 2016). Les concentrations élevées de mercure et de cuivre dans les sédiments du lac Guessabo constituent donc un danger potentiel et font de ces sédiments une source endogène de pollution. En effet, ces polluants métalliques peuvent passer dans la colonne d'eau lorsque les conditions sont favorables, engendrant ainsi des effets néfastes sur la qualité des eaux et sur la vie aquatique. L'enrichissement en Hg enregistré dans cette zone serait sans doute dû aux principales sources anthropogéniques qui sont l'agriculture (insecticides, fongicides, bactéricides et herbicides), la peinture, les appareils électriques et les produits pharmaceutiques. La mauvaise gestion des déchets dans la zone serait à l'origine de cette pollution. L'As, le Cd, le Pb et Hg sont des éléments non essentiels qui ont des effets toxiques avérés sur les humains, en particulier les enfants et les adolescents (Whitehead et Buchanan, 2019). Ces effets englobent les maladies immunitaires, la défaillance d'un organe interne, les lésions neurologiques et les maladies respiratoires (Chikkanna et al., 2019).

4 Conclusions

Le présent travail avait pour objectif principal de connaître l'état de contamination des sédiments par les éléments traces métalliques et leur toxicité, il constitue un premier état des lieux.

Les résultats de Igeo et FE calculés, ont montré que les sédiments devaient être considérées comme modérément contaminées à sévèrement contaminées en Hg, modérément contaminées en Cu et presque non contaminées par les autres éléments. Au regard de la pollution cumulative (PLI et IPS) et des risques écologiques potentiels, les sédiments du lac Guessabo se sont avérés être pollués par le Hg, qui a contribué à 51,7 % des risques écologiques potentiels et dans une moins mesure par Cd, As et le Pb.

Collaborateurs. KSK, corresponding author, he wrote the manuscript. KBK contributed in samplings and processing of data. KFK assisted in proofreading the manuscript KCB participated in the interpretation of the results KMK participated in sampling campaigns and data analysis KOUAKOU Kouassi Lazare helped in the making of maps.

Intérêts concurrents. Les auteurs déclarent qu'ils n'ont aucun conflit d'intérêts.

Déclaration du numéro spécial. This article is part of the special issue "Hydrology of Large River Basins of Africa". It is a result of the 4th International Conference on the "Hydrology of the Great Rivers of Africa", Cotonou, Benin, 13–20 November 2021.

Remerciements. Nous adressons nos sincères remerciements au directeur du Laboratoire National d'Appui au Développement Agricole. Cette institution nous a fourni un cadre et des facilités de travail remarquable pour l'analyse des échantillons. Les auteurs remercient les évaluateurs du Bulletin pour leurs corrections et commentaires.

Références

Beutel, M. W., Leonard, T. M., Dent, S. R., and Moor, B. C. : Effects of aerobic and anaerobic conditions on P, N, Fe, Mn and Hg accumulation in waters overlaying profundal sediments of an oligo-mesotrophic lake, Water Res., 42, 1953–1962, 2008.

Chikkanna, A., Mehan, L., Sarath, P. K., and Ghosh, D. : Arsenic Exposures, Poisoning, and Threat to Human Health : Arsenic Affecting Human Health. In Environmental Exposures and Human Health Challenges, IGI Global, 86–105, 2019.

Daskalakis, K. D. and O'Connor, T. P. : Normalization and elemental sediment contamination in the Coastal United States, Environ. Sci. Technol., 29, 470–477, 1995.

Diop, C., Dewaelé, D., Cazier, F., Diouf, A., and Ouddane, B. : Assessment of trace metals contamination level, bioavailability and toxicity in sediments from Dakar coast and Saint Louis estuary in Senegal, West Africa, Chemosphere, 138, 980–987, https://doi.org/10.1016/j.chemosphere.2014.12.041, 2015.

Duman, F. and Kar, M. : Temporal variation of metals in water, sediment and tissues of the European Chup (Squalius cephalus L.), Bull. Environ. Contam. Toxicol, 89, 428–433, 2012.

Häkanson, L. : An ecological risk index for aquatic pollution control. A sedimentological Approach, Water Res., 14, 975–1001, 1980.

Huang, Y., Zhang, D., Xu, Z., Yuan, S., Li, Y., and Wang, L. : Effect of overlying water pH, dissolved oxygen and temperature on heavy metal release from river sediments under laboratory conditions, Archives of Environmental Protection, 43, 28–36, 2017.

Keumean, K. N., Bamba, S. B., Soro, G., Soro, N., Metongo, S. B., et Biemi, J. : Concentration en métaux lourds des sédiments de l'estuaire du fleuve Comoé à Grand-Bassam (Sud-Est de la Côte d'Ivoire), J. Appl. Biosci, 61, 4530–4539, 2013.

Kouamé, K. B., Konan, K. S., Attoungbre, K. S., Konan, K. F., Boussou, K. C., and Kouamé, K. M. : Physicochemical and Bacteriological Parameters of Surface Water Quality in part of West Côte d'Ivoire : Potential Resources for Drinking Water Production (Guessabo Lake), J. Appl. Environ. Biol. Sci., 9, 17–29, 2019a.

Kouame, K. B., Konan, K. S., Konan, K. F., Boussou, K. C., and Kouassi, K. L. : Heavy metals contamination of an aquatic environment and health risks assessment in two fish species : case of the Guessabo lake, western Côte d'Ivoire, Orient. J. Chem., 35, 1742–1755, 2019b.

Li, Q., Na, G., Zhang, L., Lu, Z., Gao, H., and Li, R. : Effects of corresponding and non-corresponding contaminants on the fate of sulfonamide and quinolone resistance genes in the Laizhou Bay, China., Mar. Pollut. Bull., 128, 475–482, 2018b.

Malferrari, D., Brigatti, M. F., Laurora, A., and Pini, S. Heavy metals in sediments from canals for water supplying and drainage : mobilization and control strategies, J. Hazard. Mater. 161, 723–729, 2009.

Men, C., Liu, R. M., Xu, F., Wang, Q. R., Guo, L. J., and Shen, Z. Y. : Pollution characteristics, risk assessment, and source apportionment of heavy metals in road dust in Beijing, China., Sci. Total Environ., 612, 138–147, https://doi.org/10.1016/j.scitotenv.2017.08.123, 2018.

Mohammad, A. M., Mohammad, L. A., Saiful, I. M., and Zillur, R. M. : Preliminary assessment of heavy metals in water and sediment of Karnaphuli River, Bangladesh, Environ. Nanotec. Monit. Manage. 5, 27–35, 2016.

Müller, G. : Die Schwemetallbelastung der sedimente des Neckars und seiner Nebenflusse : eine Bestandsanfnahme, Chem. Zeitung, 105, 157–164, 1981.

Müller, G. : Index of geoaccumulation in sediments of the Rhine River, Geojournal, 2, 109–118, 1969.

Olivares-Rieumont, S., de la Rosa, D., Lima, L., Graham, D. W., Alessandro, K. D., Borroto, J., Martınez, F., and Sanchez, J. : Assessment of heavy metal levels in Almendares River sediments-Havana City, Cuba, Water Res., 39, 3945–3953, 2005.

Rajeshkumar, S., Liu, Y., Zhang, X., Ravikumar, B., Bai, G., and Li, X. : Studies on seasonal pollution of heavy metals in water, sediment, fish and oyster from the Meiliang Bay of Taihu Lake in China, Chemosphere, 191, 626–638, 2017.

Wang, C., Liu, S., Zhao, Q., Deng, L., and Dong, S. : Spatial variation and contamination assessment of heavy metals in sediments in the Manwan Reservoir, Lancang River., Ecotoxicology and Environmental Safety, 82, 32–39, 2012.

Whitehead, L. T. S. and Buchanan, S. D. : Childhood Lead Poisoning : A Perpetual Environmental Justice Issue, J. Public Health Manag Pract., 1 115–120, 2019.

Proc. IAHS, 384, 93–98, 2021
https://doi.org/10.5194/piahs-384-93-2021

Open Access
PIAHS

Modeling of Spacio-temporal Evolution of Salt Dispersion on Nokoue Lake

Esdras Babadjidé Josué Zandagba[1,2], Eric Adéchina Alamou[1,2], Ezechiel Obada[1,2], Amédée Chabi[1], Eliézer Iboukoun Biao[1,2], and Abel Afouda[1]

[1]Laboratory of Applied Hydrology, National Water Institute, University of Abomey-Calavi, Abomey Calavi, Benin

[2]Laboratory of Geosciences, Environment and Applications, National University of Sciences, Technology, Engineering and Mathematics, Abomey, Benin

Correspondence: Esdras Babadjidé Josué Zandagba (zjosuas@gmail.com)

Published: 16 November 2021

Abstract. The numerical modeling of spatio-temporal evolution of lagoon has an important role in predicting the behaviour of these systems. Knowing the concentration of the pollutant field distribution in time and space contributes significantly to the prediction of exceptional phenomena. The purpose of this paper is to simulate the transport and dispersion of salt at Nokoue Lake. To this end, the 2D hydrodynamic model SMS (Surface Water Modeling System) has been used. Results showed that in flood period the freshwater inflows produce a net seaward transport, while in low water period the tides lead to periodic seaward and landward transport. The developed numerical model is useful for predicting pollutants transport in this system, for water quality management of the Nokoue Lake, and therefore, fight against eutrophication.

1 Introduction

Ecological processes in coastal and estuarine areas strongly depend on dynamic physical processes, such as water surface inundation, residence time, circulation and salinity variation (Zhaoqing et al., 2010).

Estuaries are highly variable in physical, chemical, and biological properties. Tides and freshwater inflows are the two major external forcing mechanisms controlling estuarine processes. The interest and driving force in studying the ecology of the lagoon comes from increasing threats to the lagoon by siltation, choking of the mouth connecting the lagoon to the sea, eutrophication weed infestations, salinity changes and a decrease in fishery resources (Jayaraman et al., 2007). Due to their location at the junction of terrestrial, aquatic and marine ecosystems, lagoons constitute particularly important transitional ecosystems, whose hydrological and ecological functions are a source of ecosystem services (Tunin-Ley et al., 2014).

Faced to this, modern and efficient management of the water resources of these ecosystems is essential. This requires a detailed description of the various processes that take place, which necessarily involves in observation and modeling. Numerical modeling has been demonstrated as a useful and efficient tool to predict ecological change under varying physical processes and hydrological conditions and to provide guidance for estuarine and coastal restoration projects (Boumans et al., 2002; Roman et al., 1995; Silvestri et al., 2005; Yang et al., 2008).

Characterized by continental and marine interactions, various pollutions and the development of aquatic plants, Nokoue Lake has become the site of complex phenomena, which should be studied and mastered, in a context of climate change. Thus Zandagba et al. (2016) study the hydrodynamic functioning of Nokoue Lake and show that the dynamic of the lake is controlled by the tributaries flows in a flood period and by tide during a low water period. It is also characterized by the dephasing and the flattening of the tidal amplitude, which increase as we move away from the channel. In a flood period, contrary to the low-water period, incoming flows are higher than outflows, reinforced by the tidal amplitude. Adandedji et al. (2017) showed that winds, flow and

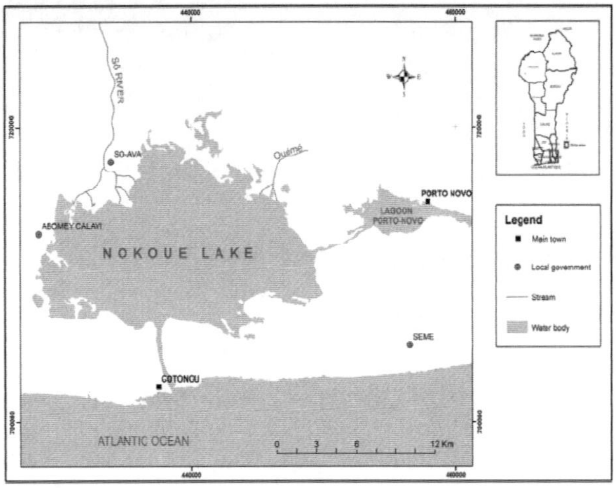

Figure 1. Location of Nokoue Lake (Benin). Source: Topographic bostom (IGN, 1992).

tides are key factors that governs the pathways and pollutants spreading in Nokoue Lake.

Pollutants dispersion processes are often encountered when a quantity of polluted water is discharged into the natural environment, through a controlled or uncontrolled regime. By introducing a pollutant in an aquatic environment, its polyphone constituents are dispersed through the whole volume of water. The mathematical modeling of spatio-temporal evolution of lagoon has an important role in predicting the behaviour of these systems. Knowing the concentration of the pollutant field distribution in time and space contributes significantly to the prediction of exceptional phenomena (Galina et al., 2015).

The purpose of this paper is to simulate the transport and dispersion of salt at Nokoue Lake.

2 Study area

Nokoue Lake is located in southern Benin, in West Africa, between 6°25′ N and 2°36′ E (Fig. 1). It is located in the sub-equatorial area, and underwent two rainy seasons and two dry seasons of unequal durations. The annual rainfall is 1309 mm, the average temperature of 27.7 °C with the maxima up to 33 °C and minimum at 23 °C. It covers an area of 150 km^2 and is the largest lake of the country. This lake is connected to Porto-Novo lagoon at the East by the Totche Canal, at south by the Atlantic Ocean through the channel of Cotonou, and at north by the Sô River and the Oueme River on which it depends (Mama, 2010). This communication, together with the impact of natural flood of Sô and Ouémé streams, causes very significant seasonal variations of salinity (Mama, 2010).

3 Methodology

3.1 Modeling

The numerical model used in this study is the Surface Water Modeling System (SMS) program, which is a software package for efficient management of the process of modeling surface water, temperature, salinity, sediment, and water quality constituents (SMS, 2009).

The hydrodynamic model is driven by tide, river flow, bathymetry and wind stress. It has been calibrated and validated by Zandagba et al. (2016). The development of salinity dispersion was conducted using the SMS Resource Management Associates module 4 (RMA4). To run RMA4, the hydrodynamic solution (velocity and water level simulations) obtained on the lake (Zandagba et al., 2016) were used as input to determine the field evolution of salinity concentrations. In fact RMA4 uses the flow solutions to compute the constituent concentration as it flows through the mesh (RMA4, 2011). The quality module (RMA4) is coupled to the hydrodynamic module, through the transport equation. It integrates the advection and the diffusion properties of the flow, as well as the basic processes occurring in the water column resulting from the interaction between the currents and at the surface and bottom interfaces as well as the major biochemical interactions (Panda et al., 2015).

$$\frac{\partial hC}{\partial t}+\frac{\partial hCu}{\partial x}+\frac{\partial hCv}{\partial y}=\frac{\partial}{\partial x}\left[hD\frac{\partial C}{\partial x}\right]+\frac{\partial}{\partial y}\left[hD\frac{\partial C}{\partial y}\right]-KhC \quad (1)$$

where h: water depth (m), C: concentration of pollutant (mg/L), t: time (s), u: velocity in the x direction (m/s), v: velocity in ydirection (m/s), and D: coefficients of turbulent diffusion in the x and y, K: decay constant (s^{-1}) (RMA4, 2011).

Because the water particles velocity depends on the flow and water level, the following boundary conditions were established. In low water period, the constant flowrate of 41 m^3/s was assigned to Sô river and 14 m^3/s to Oueme river. In flood period a constant flowrate of 135 and 271 m^3/s were respectively assigned to Sô and Oueme rivers. Bathymetric and flow data used was obtained from the General Direction of Water at 2017.

At the downstream, hourly tide level with average amplitude was affected.

A constant source of salinity of 12 ‰ during high water periods and 28 ‰ during low water periods in the Cotonou channel were considered.

During high and low water periods, the initial salinity throughout the lake is assumed to be respectively 20 ‰ and 9 ‰. The salinity is considered to be zero for the inflows of the rivers freshwater. The salinity data have been collected at Nokoue lake and cotonou channel, on October and February 2020, respectively for high and low water period.

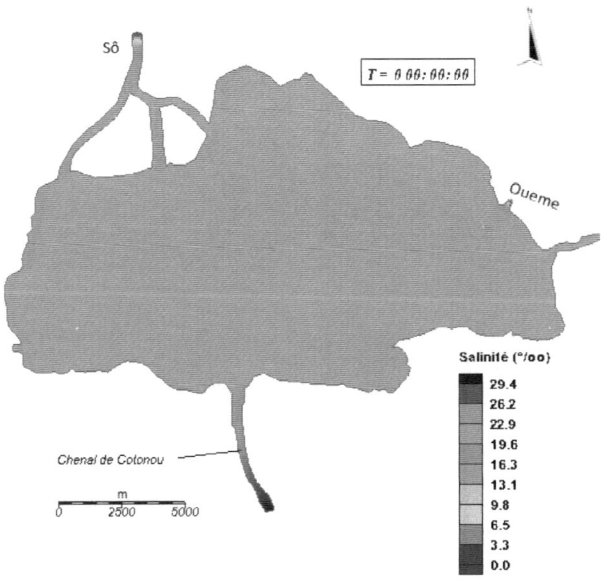

Figure 2. Spatio-temporal variation of salinity-fresh water during the low-flow period at the start of the simulation.

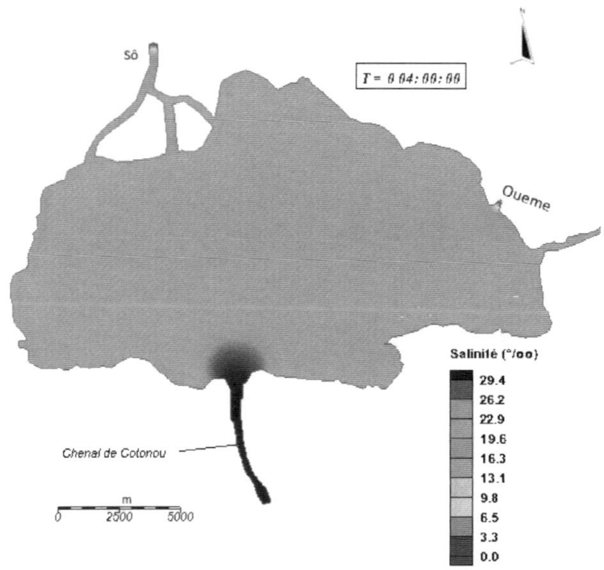

Figure 3. Spatio-temporal variation of salinity-fresh water during the low-flow period after 4 h of simulation.

To determine the turbulent diffusion coefficients, two methods are used: direct in which each element receives the respective values of these coefficients, or automatically using the Peclet number which is given by:

$$P_e = \frac{U \mathrm{d}x}{D} \tag{2}$$

where U: the mean velocity, dx: element length in the direction of flow, D: turbulent diffusion coefficient (SMS, 2009). The dispersion coefficient (D) is the calibration factor in salinity simulation. Scenarios were simulated with different values of D and optimized with value 1.1 m^2/s.

4 Results and discussion

4.1 Simulation of the dispersion of salinity during low water period

Figures 2 to 5 show the spatio-temporal evolution of salinity during the low-flow period. The Figs. 2 and 3 respectively show the spatial evolution of salinity at the start of the simulation and after 4 h of simulation.

The analysis of Fig. 3 indicates that after 4 h of simulation, the salt water of Ocean Atlantic migrated from Cotonou channel to entrance of the lake.

After 4 d 18 h of simulation (Fig. 4), the plume (salt water-freshwater) was observed beyond the center of the lake, about 7 km from the entrance of the channel into the lake and covers about 1/3 of the Lake.

From the 8th day of simulation (Fig. 5), salt water occupies the entire lake and we can also notice that marine waters goes up towards the rivers.

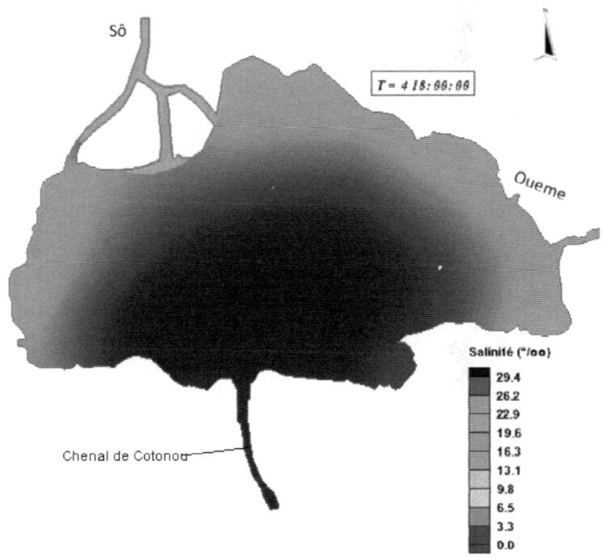

Figure 4. Spatio-temporal variation in salinity – fresh water during the low water period after 4 d 18 h.

4.2 Simulation of the dispersion of salinity during flood period

Figures 6, 7 and 8 show the spatio-temporal evolution of salinity during a flood period in the lake.

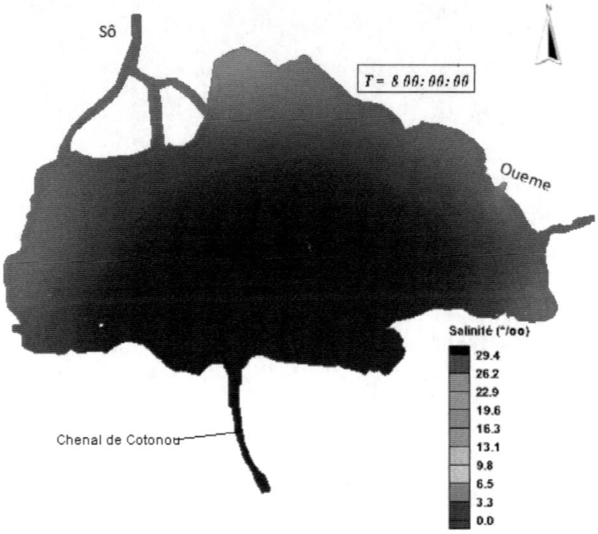

Figure 5. Spatio-temporal variation in salinity – fresh water during the low water period after 8 d of simulation.

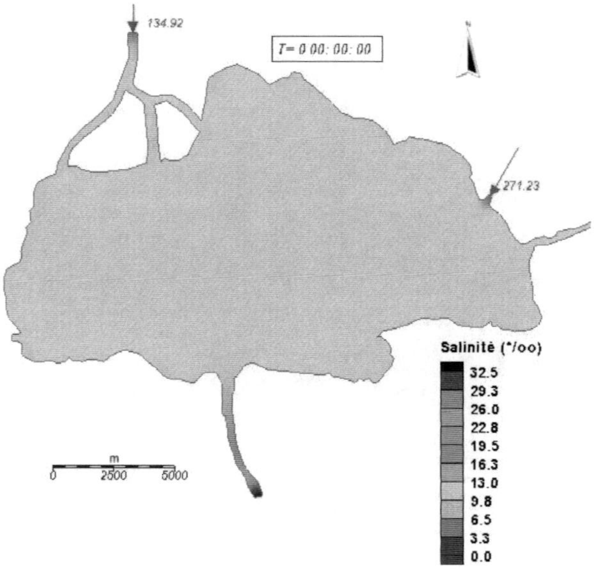

Figure 6. Spatio-temporal variation in salinity during a flood period at the start of the simulation.

Figures 6 and 7 respectively show the spatial evolution of salinity at the start of the simulation and after 8 h of simulation.

Their analysis indicate that from the start of the simulation, seawater rises through the channel to reach the entrance of the lake after 8 h simulation.

The Fig. 8 shows that from the 5th day of simulation the dispersion of salt water was completely stopped. Indeed, the three-quarters of the lake are controlled by the fresh water

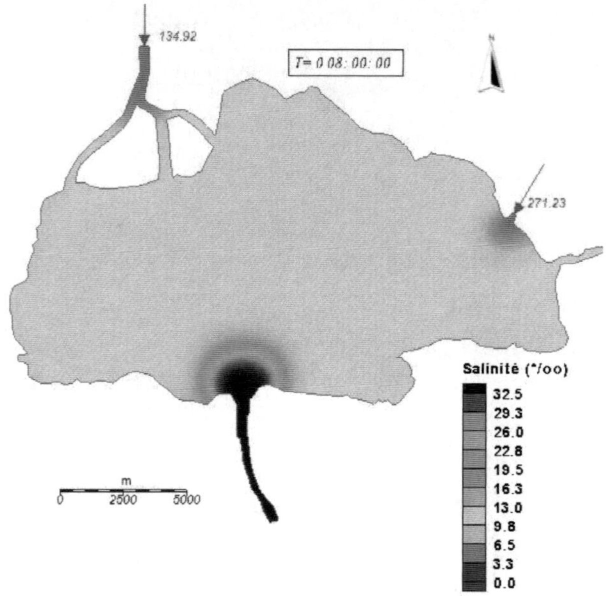

Figure 7. Spatio-temporal variation in salinity during a flood period after 8 h of simulation.

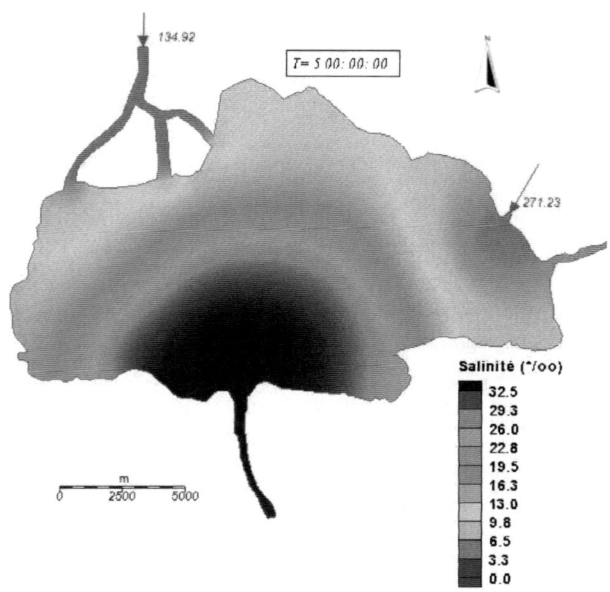

Figure 8. Spatio-temporal variation in salinity during a flood period after 5 d of simulation.

brought by the rivers whose quantity prevents the migration of salt water towards the bottom of the lake.

4.3 Discussion

Analysis of the dispersion and transport of salt reveals a clear increase in salinity from the ocean to the lake during low-flow periods. This evolution up rivers and was felt up to Bonou according to Mama (2010).

At the end of the simulation, we note that the migration of fresh water from Oueme and Sô rivers is limited to entering the lake during low-water periods, while in high-water periods it is limited to entering the lake reached 3 to 5 km in the lake. In fact, the large freshwater discharge makes it difficult for seawater to be transported into the lake from the open ocean. This is in accordance with the results of Jayaraman et al. (2007) according to which in this period the persistent large freshwater inflows flush the salt out to the ocean.

The contact zone between the salt water and the fresh water inflows characterized by a plume is located a little before the center of the lake during the flood period, while in the low water period it is close to the entrances of Oueme and Sô rivers. The intrusion of marine waters into the lake is therefore greater during low flow than during high water, which was in accordance with the results of Zandagba et al. (2016), according to whom the dynamic of the lake is controlled by the tide in period of low water. The freshwater inflows produce a net seaward transport, while the tides lead to periodic seaward and landward transport. The freshwater tends to float over the denser seawater, but tidal mixing reduces this stratification.

According to Dehotin et al. (2007), this behavior is common to all tropical ecosystems characterized by desalination in the flood period and high salinity in the dry season. The simulation confirmed the general pattern of salinity intrusion through inlet, low salinity of northern sector, medium salinity of central and southern sector (Panda et al., 2015; Adandedji et al., 2017).

The results also revealed that the temporal variability of the system is mainly due to the disparity of the seasonal changes in salinity like found by Wango (2019). The spatial variation in the salinity reveals the importance of tributary discharges.

Due to an array of physical processes taking place together, Nokoue Lake environment becomes more dynamic and there exists a greater degree of salinity variations. Lower salinity values in the northern sector could be related to severe dilution as a consequence of addition of massive freshwater into this sector by rivers.

5 Conclusions

The 2D hydrodynamic model SMS has been used to investigated the transport and dispersion of the salt on Nokoue Lake. To this end, the RMA4 program allowed to determine the evolution of salt dispersion in time and space on the studied area. The obtained results showed that freshwater inflow plays a major role in flood period while it is the tide that controls the dynamics during low water periods. It was also found that in low water period from 8th day of simulation, salt water occupies almost the entire lake. However in flood period, from the 4th day, the dispersion of salt water begins to be slowed down and the transport has become stationary from the 5th day. It is concluded that the entrance is a critical element affecting the exchange of water between the sea and lagoon.

The developed numerical model is useful for predicting pollutants transport in this system, for water quality management of the Nokoue Lake, and therefore, fight against eutrophication.

Data availability. All data used in this manuscript are available upon request from the corresponding author.

Author contributions. EBJZ, EAA, EO, AC, EIB and AA designed the study, developed the methodology and wrote the manuscript; EBJZ performed the field work, collected the data and conducted the computer analysis.

Competing interests. The contact author has declared that neither they nor their co-authors have any competing interests.

Disclaimer. Publisher's note: Copernicus Publications remains neutral with regard to jurisdictional claims in published maps and institutional affiliations.

Special issue statement. This article is part of the special issue "Hydrology of Large River Basins of Africa". It is a result of the 4th International Conference on the "Hydrology of the Great Rivers of Africa", Cotonou, Benin, 13–20 November 2021.

Acknowledgements. Thank to the International Foundation for Science (IFS) for supporting this work.

Financial support. This research has been supported by the International Foundation for Science (grant no. 1I2_W_5706).

References

Adandedji, F. M., Sintondji, L. O., Mama, D., and Vested, H. J.: Pollutants dispersion in tropical coastal lake based on numerical modeling approach: A case study of Nokoué Lake in the Republic of Benin, Int. J. Sci. Eng. Res., 8, 1369–1374, 2017.

Boumans, R. M. J., Burdick, D. M., and Dionne, M.: Modeling habitat change in salt marshes after tidal restoration, Restor. Ecol., 10, 543–555, 2002.

Dehotin, U. A., Laleye, P. A., Dauta, A., and Moreau, J.: Facteurs ecologiques et diversité piscicole d'une lagune Ouest Africaine: Le lac Nokoué au Bénin, Journal of Afrotropical Zoology, 49–55, 2007.

Galina, M., Ion, S., Violeta, V., Constantin, F., Nina, S., and Monica, A. C.: Modeling of Spacio-temporal Evolution of Fluoride Dispersion in "River-type" Systems, Rev. Chim. (Bucharest), 66, 503–506, 2015.

Institut Géographique National (IGN): Annal des données topographiques du Bénin, Vol. 2, 32 pp., 2007.

Jayaraman, G., Rao, A. D., Dube, A., and Mohanty P. K.: Numerical simulation of circulation and salinity structure in Chilika Lagoon, J. Coast. Res., 23, 861–877, 2007.

Mama, D.: Méthodologie et résultats du diagnostic de l'eutrophisation du lac Nokoué (Bénin), Mémoire de Thèse de l'Université de Lausanne, Lausanne, 157 pp., 2010.

Panda, U. S., Mahanty, M. M., Ranga Rao, V., Patra S., and Mishra, P.: Hydrodynamics and water quality in Chilika Lagoon – A modelling approach, 8th International Conference on Asian and Pacific Coasts (APAC 2015), Proc. Eng., 116, 639–646, 2015.

Roman, C. T., Garvine, R. W., and Portnoy, J. W.: Hydrologic modeling as a predictive basis for ecological restoration of salt marshes, Environ. Manag., 19, 559–566, 1995.

Silvestri, S., Defina, A., and Marani, M.: Tidal regime, salinity and salt marsh plant zonation, Estuar. Coast. Shelf Sci., 62, 119–130, 2005.

Surface Water Modeling System (RMA4): US Army Engineer Research and Development Center, 2011.

Surface Water Modelling System (SMS): Users Guide to RMA2, US Army Corps of Engineer-Waterways Experiment Station: Vicksburg, MS, USA, 296 pp., 2009.

Tunin-Ley, A., Malet, N., Fiandrino, A., Orsoni, V., and Baldi, Y.: Estimation des échanges entre la mer et la lagune de Biguglia: Utilisation de la méthode d'estimation des volumes d'eau échangés entre la mer et les lagunes par différence de hauteurs d'eau, IFREMER, 29 p., 2014.

Wango, T. E.: Modélisation de l'hydrodynamique, de la dispersion du sel et de l'eau douce dans le complexe lagunaire de la Côte d'Ivoire (Grand-Lahou, Ebrié et Aby), Thèse de Doctorat en Géologie Marine (océanologie), Université de Cocody, 188 pp., 2009.

Yang, Z., Khangaonkar, T., Calvi, M., and Nelson, K.: Evaluation of nonlinear cumulative effect of estuarine restoration projects using a three-dimensional hydrodynamic model, Ecol. Modell. 221, 969–977, https://doi.org/10.1016/j.ecolmodel.2008.12.006, 2008.

Zandagba, J., Moussa, M., Obada, E., and Abel Afouda: Hydrodynamic Modeling of Nokoué Lake in Benin, 2016, Hydrology, 3, 44, https://doi.org/10.3390/hydrology3040044, 2016.

Zhaoqing, Y., Kathryn, L. S., Danelle, H., Tarang, K., Ronald, T., and Roger, F.: Hydrodynamic and ecological assessment of nearshore restoration: A modeling study, Ecol. Modell., 221, 1043–1053, 2010.

Proc. IAHS, 384, 99–105, 2021
https://doi.org/10.5194/piahs-384-99-2021

Open Access PIAHS

Qualité et usages de l'eau de la retenue d'eau de Kogbétohoue, dans la commune d'Aplahoue (Sud-Ouest, Bénin)

Coffi Justin Noumon[1], Domiho Japhet Kodja[2], Ernest Amoussou[2], Luc O. Sintondji[1], Daouda Mama[3], and Euloge K. Agbossou[1]

[1]Laboratoire d'Hydraulique et de Maîtrise de l'Eau (LHME), Institut National de l'Eau, Université d'Abomey Calavi 01 BP 526 Cotonou, Bénin

[2]Laboratoire Pierre Pagney Climat Eau Ecosystèmes et Développement, Faculté des Lettres Arts et Sciences Humaines, Université d'Abomey Calavi 01 BP 526 Cotonou, Bénin

[3]Laboratoire d'Hydrologie Appliquée (LHA), Institut National de l'Eau (INE), Université d'Abomey-Calavi, 01 BP 526 Cotonou, Benin

Correspondence: Coffi Justin Noumon (justinoum2001@yahoo.fr)

Published: 16 November 2021

Résumé. Le comblement et l'eutrophisation des retenues d'eau constituent des contraintes majeures pour les usages qu'en font les populations riveraines. L'étude bathymétrique a révélé une diminution moyenne de 0,116 m/an de la profondeur sur sept (07) ans. Ce niveau élevé de comblement s'explique par l'érosion des terres due à la dégradation du couvert végétal et le surpâturage. Les mesures directes de la qualité physique : le pH, la température et l'oxygène dissous avec le multi paramètre PC HORIBA WATER QUALITY CHECKER U-10 ; les paramètres chimiques par la méthode chromatographique ICS 1000 et de dosage de la chlorophylle *a* avec le spectrophotomètre DR 5000 par la Norme AFNOR T90-117 ont permis d'évaluer le risque d'eutrophisation par la grille de diagnostic d'Ifremer (2000). Les eaux sont dans un état eutrophe. Le comblement et l'eutrophisation de la retenue constituent des menaces potentielles à court terme pour la pisciculture, l'irrigation et la consommation tandis que la disponibilité quantitative de la ressource est menacée à moyen terme. Des relevés bathymétriques périodiques et des actions d'aménagement du bassin versant de la retenue sont nécessaires à la gestion intégrée de la retenue d'eau.

1 Introduction

L'eau est une ressource nécessaire à la vie, au développement et à l'environnement. L'eau douce, principal objet d'enjeux, ne constitue que 3 % de la ressource mondiale dont les trois quarts sont stockés sous forme de glace (Perrin, 2000). La progression des dépôts solides engendre une perte de la capacité de stockage du réservoir, altère le rôle habituel des organes d'exploitation de l'ouvrage et raccourcit sa durée de vie (Bergahou et Ziadi, 1989).

Réalisée par le Projet d'Hydraulique Pastorale et Agricole (PHPA) en 2007, la retenue d'eau de Kogbétohouè vise essentiellement à : (i) réduire la vulnérabilité de la production agricole face aux aléas climatiques grâce à une meilleure maîtrise de l'eau ; (ii) assurer l'approvisionnement en eau pour le bétail et (iii) susciter l'émergence d'autres activités telles que la pêche, la pisciculture, le maraîchage et la production de plants forestiers.

Les observations de terrain ont montré que la retenue d'eau de Kogbétohouè connaît un développement des macrophytes, signe d'eutrophisation qui peut affecter les usages (pisciculture, agriculture, abreuvage) auxquels elle se prête et que la pression agricole sur les terres du bassin contribue également à la dégradation du couvert végétal, source d'érosion et de transport solide de sédiment dans la retenue. Entre autres, les deux forages existants dans le village souffrent des difficultés de fonctionnement et ont été abandonnés par les populations

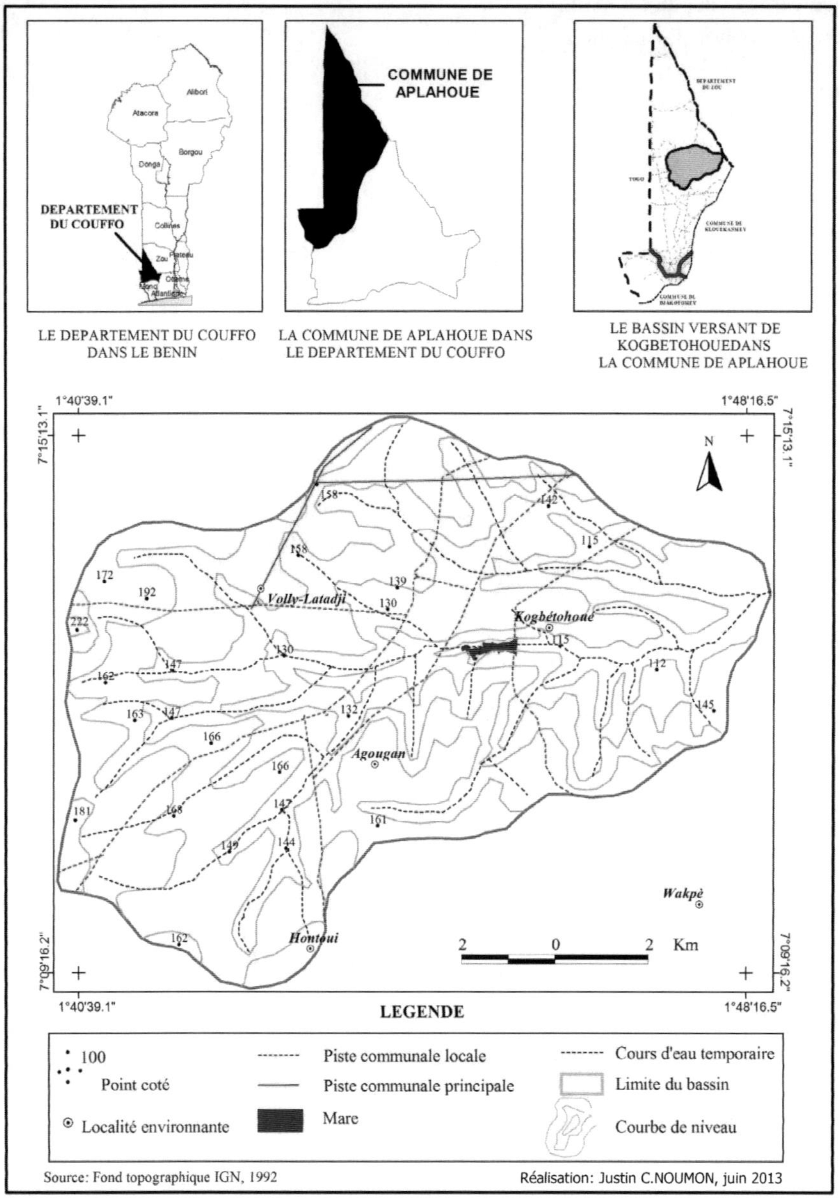

Figure 1. Carte de situation de la zone d'étude.

qui ont recours à l'eau de la retenue pour la consommation et les besoins domestiques.

Le comblement et l'eutrophisation des retenues d'eau ont été étudiés par plusieurs auteurs dont Mama (2010), Boukary (2012) et Ibouraïma et al. (2013) au Bénin. Ces travaux ont peu abordés les implications de ces deux phénomènes pour les usages de l'eau.

L'objectif de la présente étude est d'analyser les processus interdépendants du comblement et d'eutrophisation de la retenue d'eau de Kogbétohouè en relation avec les menaces à l'accès d'eau des populations riveraines.

2 Matériels et methodes

2.1 Matériels d'étude

La zone d'étude (Fig. 1), à l'instar de toute la commune d'Aplahoué (département du Couffo) qui abrite la retenue d'eau de Kogbétohouè, jouit d'un climat du type subéquatorial avec des précipitations annuelles moyennes de 1139 mm pour la période de 1983 à 2012 avec une moyenne des températures de l'ordre de 27 °C et des écarts de thermiques assez faibles (6 °C). L'évapotranspiration potentielle annuelle est en moyenne de 1482 mm pour la même période Ecrire

(Noumon, 2014). Le site de la retenue d'eau de Kogbétohoué est localisé au point de coordonnées géographique 07°12′44″ Latitude Nord et 01°46′54″ Longitude Est.

2.2 Matériels de collecte

Le matériel déployé pour la conduite de cette étude comporte essentiellement : la carte d'occupation du bassin versant de la retenue à partir des images satellitaires (Spot, Landsat de 1995, 2006 et 2012) pour estimer les superficies dégradées du couvert végétal à chacune de ces trois (03) dates ; le multi paramètre PC HORIBA WATER QUALITY CHECKER pour la mesure des paramètres physique de l'eau de la retenue ;

les matériels (bouteilles en plastiques, des glacières) de prélèvements d'eau pour les analyses biochimiques de laboratoire, une (01) fiche d'enquête pour collecter les données des effectifs des différentes catégories d'animaux d'élevage et apprécier les charges pastorales (de l'aire d'influence de la retenue d'eau) et d'abreuvement (à la retenue d'eau). Sept (07) points d'échantillonnage sont définis de manière à avoir une bonne distribution spatiale des paramètres indicateurs de l'état trophique de la retenue (Fig. 2, annexe 1).

2.3 Méthodes

2.3.1 Comblement de la cuvette

Le tableau 1 résume les paramètres d'estimation du comblement de la cuvette.

Le traitement des images satellitaires avec le logiciel « ArcView » a permis d'apprécier l'évolution des unités d'occupation du sol en 1995, 2006 et 2012.

2.3.2 Risque d'eutrophisation et état trophique

Deux campagnes de prélèvements ont été effectuées (juin et septembre). Pour chaque campagne, sept (07) mesures et prélèvements sont faits aux emplacements définis (Fig. A1) La détermination des paramètres biochimiques caractéristiques de l'état trophique et le risque d'eutrophisation de la retenue a été faite par la méthode chromatographique ICS 1000 et le dosage de la chlorophylle *a*, a été effectué avec le spectrophotomètre DR 5000 par la Norme AFNOR T90-117. La détermination de l'état trophique de la retenue d'eau a été faite suivant la grille d'évaluation du risque d'eutrophisation de la colonne d'eau selon la méthode Ifremer (2000) complétée par le diagramme de classification du niveau trophique suivant l'échelle du MDDEP (2004). Pour chacun des sept (07) prélèvements, le risque d'eutrophisation a été apprécié pour chaque paramètre indicateur de l'état trophique. Ainsi, l'état global découlant de l'appréciation des différents paramètres a été donné pour les différents points échantillonnés et ceci pour les deux campagnes de prélèvements : juin (hautes eaux, pleine campagne vivrière, forte utilisation d'engrais NPK), septembre (petite saison des pluies).

Afin de déterminer le caractère oxydant ou réducteur de l'eau (Rejset, 2002) de la retenue d'eau, le pouvoir oxydant (R_H) de l'eau a été évalué à partir de la formule suivante : pH = Potentiel en Hydrogène.

La valeur obtenue pour le pouvoir oxydant permet de situer l'eau dans les catégories suivantes selon Rejset (2002) :

- $R_H \geq 23$, le milieu est dit oxydant ;
- $15 < R_H < 23$, le milieu est qualifié d'anoxique
- $R_H < 15$, le milieu est dit réducteur

Les valeurs moyennes des différents paramètres ont permis de caractériser le pouvoir oxydant de l'eau de la retenue pour les deux campagnes de prélèvements.

2.3.3 Qualité de l'eau et usages

La qualité de l'eau a été appréciée en fonction des usages actuels de l'eau de la retenue (abreuvage, irrigation, aquaculture et consommation). Pour ce faire, quelques paramètres physico-chimiques indicateurs de l'aptitude de l'eau à ces usages selon le Système d'Evaluation de la Qualité de l'Eau des cours d'eau (SEQ-Eau, 2003) ont été utilisés.

Pour chaque paramètre, les résultats moyens des campagnes de prélèvements sont comparés aux valeurs limites des classes d'aptitude.

3 Résultats

3.1 Facteurs de comblement de la retenue d'eau

Le niveau annuel moyen de comblement de la retenue de 2007 à 2013 est estimé 0,116 m/an (tableau 3). La dégradation du couvert végétal et la charge animale en sont les principaux facteurs responsables.

3.1.1 Occupation du sol et état du couvert végétal

Les unités d'occupation du sol sont les types de formations végétales (forêt, savanes, jachères, cultures) et des agglomérations. Les superficies des unités d'occupation du bassin versant en 1995, 2006 et 2012 et leur regroupement selon le degré d'anthropisation est résumé dans le tableau B1 (annexe). Les Forêts-Savanes sont passées de 49 % en 1995 à 15 % en 2012 tandis que les Champs-Jachères ont augmenté, passant de 49 % à 83 % pour la même période.

3.1.2 Charge animale sur la retenue et les pâturage de son bassin versant

La charge animale sur la retenue d'eau est estimée à 952,5 UBT contre une capacité de charge évaluée à 221 UBT (Noumon, 2015). Ceci montre qu'il s'agit d'un surpâturage : facteur de dégradation du couvert végétal avec l'érosion comme corollaire.

Tableau 1. Paramètres d'estimation du comblement de la cuvette.

Profondeur maximale à la réalisation de la retenue en 2007	$H_{\text{max 0 (m)}}$
Profondeur maximale obtenue en 2013	$H_{\text{max (m)}}$
Age de la retenue	A (ans)
Niveau de comblement pour la période	$H_{\text{max 0(m)}} - H_{\text{max (m)}}$
Comblement moyen annuel	$H = (H_{\text{max 0 (m)}} - H_{\text{max (m)}})/A$ (m/an)
Durée de fonctionnement de la retenue d'eau	$H/H_{\text{max 0}}$

Tableau 2. Correspondance des valeurs des paramètres indicateurs de la qualité de l'eau.

Couleur	Très bon	Bon	Moyen	Médiocre	Mauvais

3.2 Qualité de l'eau

3.2.1 Pouvoir oxydant de l'eau

La valeur obtenue pour le pouvoir oxydant (tableau 4, annexe) est de 13 (inférieure à 15) aussi bien en juin qu'en septembre. Cette valeur montre que nous sommes en présence d'un milieu réducteur. En milieu réducteur, les nitrates et sulfates sont respectivement réduits en nitrites et sulfures. Ces deux dérivés sont de nature toxique à la vie aquatique notamment celle des poissons.

3.2.2 Etat trophique et risque d'eutrophisation de la retenue d'eau

Le niveau trophique de la retenue d'eau suivant le diagramme de classement du MDDEP (2004) est présenté dans les tableaux B3 et B4 (annexe). Le bilan général (tableau B5, annexe) donne des eaux hypereutrophes en Juin en période de pleine campagne de production agricole avec une utilisation des engrais NPK d'une part et eutrophes en septembre où l'intensification agricole est moindre.

3.2.3 Conséquences du comblement et de l'eutrophisation de la retenue

Conséquences physico-économiques relatives au comblement de la retenue

Le niveau annuel moyen de comblement de la retenue de estimé 0,116 m/an entraîne une perte d'eau estimée à 1660 m^3/an correspondant à une durée de fonctionnement de la retenue d'eau de 42 ans. Ceci constitue une menace réelle sur la disponibilité quantitative de l'eau de la retenue.

3.2.4 Conséquences économiques et sanitaires liées à l'eutrophisation de la retenue d'eau

Menaces sur la pisciculture

La concentration très élevée en phosphore (tableau 4) est la menace première sur la pisciculture en raison des effets secondaires notamment d'eutrophisation et son corollaire sur l'épuisement de l'oxygène.

Menaces sur l'irrigation

La concentration en microorganismes (tableau 5) suivie de la charge solide (environ 560 mg/L de matières sèches) constitue les menaces majeures sur l'usage de l'eau à des fins agricoles.

3.2.5 Menaces sur l'abreuvement des animaux

La qualité de l'eau ne présente pas de menace majeure à l'abreuvage (tableau 10).

Menaces sur la production d'eau potable

La forte teneur en coliformes totaux (tableau 9) et la turbidité (tableau 5) assez élevée constitue une contrainte pour l'aptitude de l'eau de la retenue à la production d'eau potable.

4 Discussion et conclusion

La dynamique de comblement (0,116 m/an) est plus élevée que celle obtenu par les études de Mama (2010) sur le lac Nokoué au Bénin où la diminution de la profondeur est de l'ordre de 0,03 m/an La durée de fonctionnement de la retenue de Kogbétohouè (42 ans) se retrouve bien dans la fourchette (12 à 114 ans) obtenues par Ibouraïma (2005) pour les retenues d'eau du département de l'Alibori au Bénin. L'évaluation du risque d'eutrophisation de la retenue d'eau de Kogbétohoué montre que les eaux sont dans un état général mauvais à médiocre vis-à-vis de leur état trophique (Noumon et al, 2015). Cette eutrophisation, caractéristique d'une pollution biologique du plan d'eau est marquée par la présence d'algues et le développement de plusieurs macrophytes. Elle

Tableau 3. Niveau de comblement et durée de fonctionnement de la retenue.

Paramètres d'estimation du comblement	2007	2013	Productivité de sédiment (m)	Apport annuel de sédiment (m)	Durée probable de la retenue
Profondeur max. (m)	5,7	5,4	0,3	0,05	96 ans
Niveau d'eau à l'échelle (m)	4,8	4,1	0,7	0,116	42 ans

Tableau 4. Aptitude de l'eau à la pisciculture en juin et septembre.

Paramètres	Aptitude en Juin	Aptitude en Septembre
Oxygène dissous (mg/L)		
PH		
Ammonium (mg/L)		
Nitrite (mg/L)		
Nitrates (mg/L)		
MES (mg/L)		
Phosphore Total (mg/L)		

Tableau 5. Aptitude de l'eau à l'irrigation en novembre.

Paramètres	Aptitude en Novembre
Résidus à sec à 105°C (mg/L)	
Chlorures (mg/L)	
Coliformes totaux (u/100ml)	

résulte de l'altération de plusieurs caractéristiques physico-chimiques de la qualité de l'eau notamment la transparence, l'oxygène dissous, la présence de l'azote, le phosphore et les sulfures à des concentrations élevées. L'eutrophisation constitue une menace sur la qualité des eaux pour les usages à des fins piscicoles (Chouti et al., 2010; Dèdjiho et al., 2013; Delhi et al., 2012; Francis-Lloyd, 2003). En effet, la prolifération des plantes aquatiques est à l'origine de la formation d'un écran en surface des eaux qui empêche l'oxygénation du milieu. De plus la décomposition de ces plantes, induit une augmentation de la quantité de matière organique dans le milieu dont la dégradation favorise l'asphyxie des espèces halieutiques (Mama et al., 2011). La production de sulfure et de nitrite (substances toxiques) inhibent le développement d'autres producteurs primaires et des poissons à grande sensibilité. Ceci peut provoquer un déséquilibre de la chaîne trophique (producteur/consommateur) d'un plan d'eau et entraîner des conséquences écologiques importantes (Dèdjiho et al, 2013). Mieux, Chambers et al, 2001 indiquaient que la présence des sulfures avec des concentrations de l'ordre de 10 mg/L est un signe précurseur de toxicité du milieu pour la vie aquatique, pour le bétail et la santé humaine.

Tableau 6. Aptitude de l'eau à l'abreuvage en juin et septembre.

Paramètres	Aptitude en Juin	Aptitude en Septembre
Nitrite (mg/L)		
Nitrates (mg/L)		
Sulfates		
Sulfures		

En conclusion, l'étude de la qualité de l'eau de retenue de Kogbétohoué en relation avec ses usages a permis d'évaluer les menaces sur la ressource. Il ressort que les processus interdépendants du comblement et d'eutrophisation constituent des menaces réelles pour la pisciculture, l'irrigation et la consommation. Les conséquences physiques, économiques et sociales ressorties appellent à l'instauration d'une approche intégrée de la gestion de la retenue d'eau et de son bassin versant pour garantir les différents les usages que procure la retenue aux populations riveraines.

Annexe A

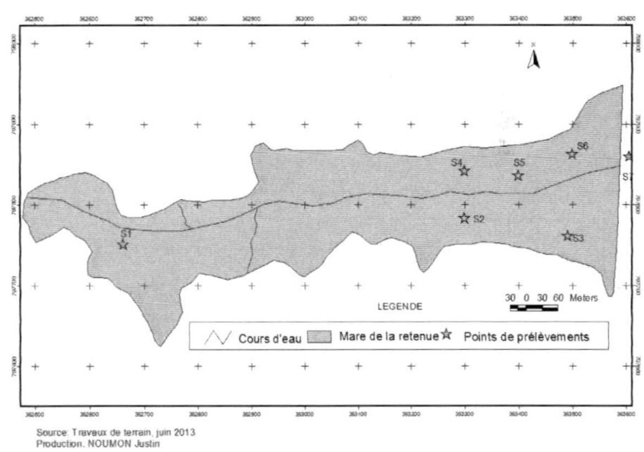

Figure A1. Carte de situation de points échantillons pour les analyses de qualité de l'eau.

Annexe B

Tableau B1. Evolution des unités d'occupation du bassin versant selon les degrés d'anthropisation.

Unité d'occupation du sol	Superficie en 1995 (ha)	Superficie en 2006 (ha)	Superficie en 2012 (ha)
Agglomération	63,6 (0,9 %)	84,7 (1,20 %)	112,13 (1,59 %)
Forêts et Savanes	3490 (49,39 %)	1919 (27,16 %)	1082 (15,31 %)
Champs et jachères	3511,7 (49,70)	5061,6 (71,64 %)	5871,2 (83,1 %)
Total	7065,3 (100 %)	7065,3 (100 %)	7065,3 (100 %)

Source : CENATEL : Images satellitaires du bassin versant de Kogbétohouè.

Tableau B2. Pouvoir oxydant de l'eau en juin et septembre.

Période	Eh	T ° absolu	pH	R_H	Nature de milieu
Juin 2013	−35,57	300,46	7,14	13,08	Réducteur
Septembre 2013	−54,14	300,48	7,42	13,02	Réducteur

Tableau B3. Niveau trophique (MDDEP, 2004) de la retenue d'eau au mois de juin.

Stations	1	2	3	4	5	S 6	7	Moy./Ecartype
Phosphore Total (µg/L)	1600	1300	1100	1900	2800	800	2700	1766/772
Chlorophylle a (µg/L)	81	15,3	3,6	58,5	36	15,3	93,6	43,3/35
Transparence (m)	0,4	0,43	0,5	0,6	0,55	0,48	0,55	0.50/0,07
Etat trophique	Hyper eutrophe	Eutrophe	Mésotrophe	Hyper eutrophe	Hyper eutrophe	Eutrophe	Hyper eutrophe	Hyper eutrophe

Tableau B4. Niveau trophique (MDDEP, 2004) de la retenue d'eau au mois de septembre.

Stations	1	2	3	4	5	6	7	Moy./Ecartype
Phosphore Total (µg/L)	300	150	270	180	210	290	120	217/71
Chlorophylle a (µg/L)	37,8	10,8	5,4	9	12,6	3,8	207	45,9/79
Transparence (m)	0,85	0,72	0,78	0,85	0,85	0,85	0,4	0,75/0,16
Etat trophique	Hyper eutrophe	Eutrophe	Méso eutrophe	Eutrophe	Eutrophe	Mésotrophe	Hypereutrophe	Eutrophe

Tableau B5. Evaluation du risque d'eutrophisation de la retenue d'eau (Ifremer, 2000).

Mois	P1	P2	P3	P4	P5	P6	P7	Bilan général
Juin	Mauvais	Médiocre	Moyen	Mauvais	Mauvais	Médiocre	Mauvais	Mauvais
Septembre	Mauvais	Médiocre	Moyen	Médiocre	Médiocre	Moyen	Mauvais	Médiocre

 https://doi.org/10.5194/piahs-384-99-2021

Disponibilité des données. Les données collectées et celles générées par la présente recherche sont stockées sur un disque dur externe privé et sont également conservées en version numérique au Laboratoire de l'Hydraulique et de La Maîtrise de l'Institut National de l'Eau de l'Université d'Abomey-Calavi (Bénin).

Collaborateurs. DJK a participé à l'analyse des données et à la relecture de l'article. La phase de collecte des données de terrain et celle de documentation sur la retenue d'eau ont été facilitées par EA et le LOS. Enfin DM et EKA ont suivi et encadré cette étude.

Intérêts concurrents. The contact author has declared that neither they nor their co-authors have any competing interests.

Déclaration du numéro spécial. This article is part of the special issue "Hydrology of Large River Basins of Africa". It is a result of the 4th International Conference on the "Hydrology of the Great Rivers of Africa", Cotonou, Benin, 13–20 November 2021.

Références

Bergahou, M. et Ziadi, K. : Quantification de l'érosion a l'échelle du bassin versant de l'Oued El Koudgi. Ecole Supérieure des Ingénieurs de l'Equipement Rural de Medjez El Bab (Tunisie), 9 pp. 1989.

Boukari, O. : Etude de l'eutrophisation de la retenue d'eau de l'Okpara à Kpassa (Parakou), Mémoire de Master Professionnel en Ecohydrologie, 85 pp, LHA/FAST/UAC (BENIN), 2012.

Chambers, P. A., Kent, R., Charlton, M. N., Guy, M., Gagnon, C., Roberts, E., Grove, E., et Foster, N. : Les éléments nutritifs et leurs effets sur l'environnement au Canada, Environnement Canada, 271 pp. 2001.

Chouti, W., Mama, D., et Alapini, F. : Études des variations spatiotemporelles de la pollution des eaux de la lagune de Porto-Novo (Sud-Bénin), Int. J. Biol. Chem. Sci., 4, 1017–1029, 2010.

Dèdjiho, C. A., Mama, D., Tomètin, L., Nougbodé, I., Chouti, W., Sohounhlouè, D. C. K., et Boukari, M. : Évaluation de la qualité physic-chimique de certains tributaires d'eaux uses du lac Ahémé au Bénin, J. Appl. Bioscie., 70, 5608–5616, 2013.

Delhi, R., Benzha, F., Hilali, A., Tahiri, M., Kaoukaya, A., Baidder, L., Rhinane, H., et Hangouet, J. P. : Caractérisation de la qualité des eaux de la retenue Sidi Mohamed Ben Abdellah sur l'oued Bouregreg, ScienceLib, Editions Mersenne, 4, 120401, 20 pp. 2012.

Francis-Floyd, R. : Dissolved Oxygen for fish production. Institute of Food and Agriculture Sciences, University of Florida, 3 pp., 2003.

Ibouraïma, S. : Comblement des retenues d'eau d'abreuvement en zone agropastorale soudano-sahélienne : dynamique, bilan et impact de la sédimentation intra-cuvette, Thèse de Doctorat (unique) en gestion de l'Environnement, Laboratoire de Sédimentologie, FAST/UAC, 221 pp., 2005.

Ibouraima, S., Oyédé, L. M., and Sinsin, B. A. : Contribution de la vase au comblement des retenues d'eau de Gogbèdè et Guéné dans le Département de l'Alibori au Nord-Est du Bénin, Bulletin de Recherche Agronomique du Bénin (BRAB) Numéro 73, Juin 2013, 1025–2355 et ISSN (online), 1840–7099, 2013.

Ifremer : Mise à jour d'indicateurs du niveau d'eutrophisation des milieux lagunaires méditerranéens – Rapport final-09-2000, Tome I – 236 pp., Tome II – 196 pp., 2000.

Mama, D., Aïna, M., Alassane, A., Chouti, W., Boukary, O. T., Deluchat V., Bowen, J., Afouda, A., et Baudu, M. : Caractérisation physico-chimique et évaluation du risque d'eutrophisation du lac Nokoué (Bénin), Int. J. Biol. Chem. Sci., 5, 2076–2093, 2011.

MDDEP-Ministère du Développement durable, de l'Environnement et des Parcs : Réseau de surveillance volontaire des lacs, Les méthodes Québec, ISBN 978-2-550-53364-1, 5 pp, 2004.

Noumon, C. J. : Analyse du comblement et de l'eutrophisation de la retenue d'eau de Kogbétohouè (commune d'Apahouè) et implications pour une exploitation durable, Mémoire de Master Professionnel en Ecohydrologie, 88 pp, LHA/FAST/UAC, 2014.

Noumon, C. J., Mama, D., Dedjiho, C. A., Agbossou, E., et Ibouraima, S. : Évaluation de la qualité physico-chimique et du risque d'eutrophisation de la retenue d'eau de Kogbétohouè (Sud-Bénin), J. Appl. Biosci., 85, 7848–7861, 2015.

Perrin, C. : Vers une amélioration d'un modèle global pluie-débit au travers d'une approche comparative, Thèse de doctorat de l'Institut National Polytechnique de Grenoble, 276 pp., 2000.

Rejset, F. : Analyse des eaux : aspects réglementaires et techniques, Centre Régional de documentation pédagogique d'Aquitaine, 89 pp., 2002.

SEQ Eau : Système d'évaluation de la qualité de l'eau des cours d'eau. Grilles évaluation, Version 2, 40 pp., 2003.

Proc. IAHS, 384, 107–112, 2021
https://doi.org/10.5194/piahs-384-107-2021

Open Access PIAHS

Quantification of water related soil erosion in the transboundary basin of the Bia (West Africa)

N'diaye Edwige Hermann Meledje[1], Kouakou Lazare Kouassi[1,2], and Yao Alexis N'Go[3]

[1]Marine Geology, Sedimentology and Environment Laboratory, Center of Ecology Research, University Nangui Abrogoua, Abidjan, 225, Ivory Coast
[2]Technical Sciences and Environment Laboratory, University Lorougnon Guédé, Daloa, 225, Ivory Coast
[3]Geosciences and Environment Laboratory, University Nangui Abrogoua, Abidjan, 225, Ivory Coast

Correspondence: N'diaye Edwige Hermann Meledje (meledjendiay@yahoo.fr)

Published: 16 November 2021

Abstract. In view of the complexity of the phenomenon of water related soil erosion in the Bia catchment area, linked to a large heterogeneity of soils, to a very scattered and in some places non-existent vegetation cover, and to a poorly distributed precipitation in both space and time, a mapping test of the "specific erosion" random variable is undertaken. The mapping of the intensity of the erosion hazard was carried out using the Universal Soil Loss Model (USLE). The map shows that the basin is generally characterized by relatively moderate erosion rates with an average erosion rate of 16 t/ha/year.

1 Introduction

Erosion is a complex process. In Côte d'Ivoire, few studies have focused on water erosion, particularly in the Bia basin. However, population growth, land clearing and intensification of agriculture would lead to degradation of plant cover, ploughing, deterioration of soil aggregates, loss of organic matter, soil compaction, loss of soil biota and subsequent increases in runoff and soil erosion (on-site damages). The transport of sediment resulting in nutrient inputs accelerates the process of silting, silting and eutrophication of lakes (off-site damages) (Kouassi et al., 2013). In light of these negative consequences, the phenomenon of erosion is of particular interest for the economic development of countries (Raissouni, 2011). However, quantification of sediment input into the fluvial system remains difficult due to the temporal and spatial variability of the hydrological regime coupled with a lack of available data related to it. Modelling erosion phenomena and solid transport will be better addressed. It should be contextualized by measurements. In this sense, erosion modelling and mapping has now become a necessity for decision makers in order to be able to target areas of major risk and seek appropriate solutions. The integration of existing models with methods such as remote sensing and geographic information systems has made it possible to minimize financial costs and save time for risk mapping and consequently, targeted and rapid response. In this context, this study aims to use a universal soil loss equation (USLE) based model integrated under a Geographical Information System to quantify and map the risk of water erosion over the Bia watershed

2 Presentation of study area

The Bia basin is a transboundary watershed between Côte d'Ivoire and Ghana (Fig. 1). This basin has an area of about 9500 km^2. the Bia River flowed entirely under dense forest with an average flow of 82 $m^3\,s^{-1}$. The region receives an average annual rainfall of about 1450 mm. The extreme south is occupied by very leached ferralitic soils. The digital elevation model (DEM) shows that about 20 % of the total area of the basin is between 76 and 111 m. The highest points in the center exceed 600 m. In the southern part of the sub-basin, the elevation is below 200 m. The average longitudinal slope is 3.5 %. Slopes greater than 15° represent 11.07 % of the Bia watershed area, slopes less than 15° represent 88.92 %. The average longitudinal slope is 3.5 %. The agriculture is varied and include palm trees, rubber trees, coffee trees, cocoa trees and banana trees. These crops are in the form of agro-industrial plantations

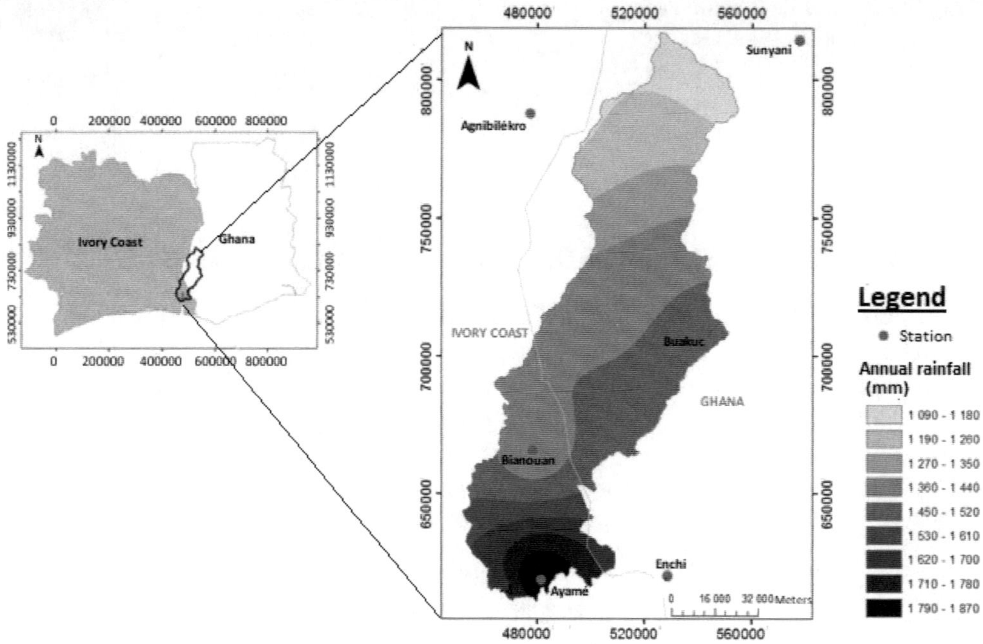

Figure 1. Situation map of the Bia basin.

3 Material and methods

3.1 Data Acquisition and Processing

The monthly rainfall data from six stations (Bianouan, Ayamé, Agnibilekro, Sunyani, Buakuc, Enchi) used in this study cover the period from 1990 to 2009. The data from the two agencies responsible for the rainfall network namely SODEXAM in Ivory Coast and the Department of Meteorological Services (Meteorological Services Department) in Ghana (Fig. 1).

Satellite images of the study area are of two types:

- The Shuttle Radar Topography Mission (SRTM) image of N05W004 coordinates, from 11 to 22 February 2000, from Space Shuttle Endeavour (STS-99), obtained from ftp://eosrp01u.ecs.nasa.gov/srtm/version1/Africa (last access: 1 January 2021) was used. It was used to produce a digital elevation model (DEM) with a resolution of 30 m and the slope map
- Landsat ETM+ images, scenes 195-056 and 195-055, 14 March 2003, provided by http://earthexplorer.usgs.gov/ (last access: 1 January 2021). Its operation led to the characterization of land use (Fig. 2a) and land cover (Fig. 2b).

3.2 Identification of Potential Erosion Areas

Modelling of the process of water related soil erosion was performed using the Universal Soil Loss Equation (USLE) model of Wischmeier and Smith (1978) and modified by Renard et al. (1997). Erosion (A), expressed in t/ha/year, is the product of 5 factors (Eq. 1 and Fig. 3):

$$A = RKLSCP, \tag{1}$$

Where

- Rainfall erosivity factor R (in MJ mm (ha h year)$^{-1}$): The factor R was calculated (Eq. 2) for six (6) stations in the region using the Arnoldus (1980) formula, adapted to tropical conditions. The formula takes into consideration only the average monthly (p) and annual (P) rainfall;

$$R = \sum_{n=1}^{12} \left(\frac{p^2}{P}\right), \tag{2}$$

- Soil erodibility factor K (in t h MJ^{-1} mm^{-1}): The K-factor was calculated using the texture triangle (Brown, 2003) and the matching table (Stone and Hilborn, 2000), thus allowing the soil erodibility values to be deduced from the texture.
- The topographic factor LS (dimensionless) represents the combination of "L" Slope length factor and "S" steepness factor. A depression less DEM is required to

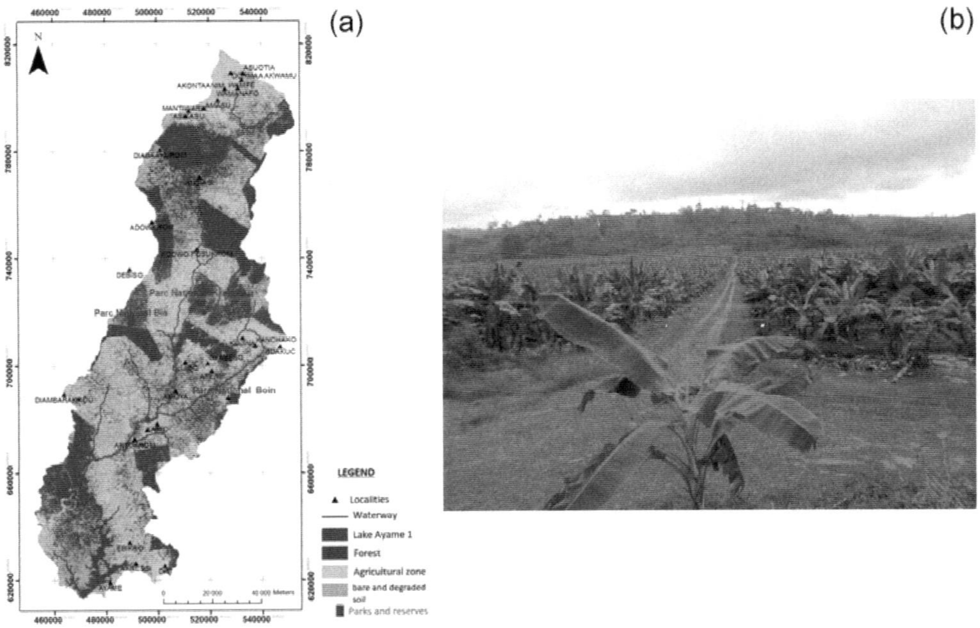

Figure 2. **(a)** Land cover and **(b)** land use of the Bia basin.

perform the necessary steps in finding the LS factor in which there are no sinks present. A 30 m spatial resolutions digital elevation model (DEM) was used to generate slope which was derived from the ASTER (GDEM). "Spatial Analyst Tool > Surface Slope" and "ArcHydro Tool" in ArcGIS 10.2 environment were applied for subsequent steps in calculating LS factor. Finally, spatially distributed value was determined using an equation proposed by Kumar et al. (2016) as in Eq. (3):

$$\mathrm{LS} = (\text{flow acc} \times [\text{cell size}]/22.1)^{0.4} \left((\mathrm{Sin}(\text{Slope in deg} \times 0.01745)/0.09)^{1.4}\right) \times 1.4 \tag{3}$$

The parameters "slope_degree" and "flow acc" represent respectively the slope map and the flow accumulation map of the Bia basin

- Cover and management factor C (dimensionless)

The value of the C-factor is estimated by the Normalized Difference Vegetation Index (NDVI). The NDVI calculation was based on the following formula (Gitas et al., 2009; Durigon et al., 2014) following Eq. (4):

$$\mathrm{NDVI} = (\mathrm{PIR} - R)/(\mathrm{PIR} + R) \tag{4}$$

PIR: the reflection of the near infrared portion of the electromagnetic spectrum and R: the reflection in the visible spectrum.

The C-factor map is generated using the regression line found using the Arcgis10 Spatial Analysis module following Eq. (5).

$$C = \exp(-a\,\mathrm{NDVI}/(\beta - \mathrm{NDVI})) \tag{5}$$

where α, β: parameters determining the shape of the NDVI-C curve.

Support and conservation practices factor P (dimensionless): Slope percentage map was prepared using "Spatial Analyst Tool > Surface Slope" in GIS environment using DEM which was derived from the ASTER GDEM. According to cultivation method and slope, the following P factor values were used as described by Shin (1999) cited by Kumar et al. (2016).

4 Results

4.1 Data Acquisition and Processing

The correspondence between the soil texture and the soil erosion revealed the most erodible soils in the catchment area (Fig. 3a). The K-factor ranges from 0 to $0.25\,\mathrm{t\,h\,MJ^{-1}\,mm^{-1}}$, for soils most susceptible to erosion. The most erosible soils correspond to lithosols in the Ayamé region in the northen part of the study area, around Lake Ayamé, in the extreme South and in the ferruginous soil strip between the Sayero, Abum and Buakuc regions in the North (Fig. 3a). The highest R values are located in the southern part of the basin, in Côte d'Ivoire with about $1160\,\mathrm{MJ\,mm\,(ha\,h\,year)^{-1}}$, while the low values are located in the northern part of the study area (Fig. 3b). The relatively flat topography of the Bia watershed naturally implies a distribution of LS values between 0 and 100. The lowest values are in

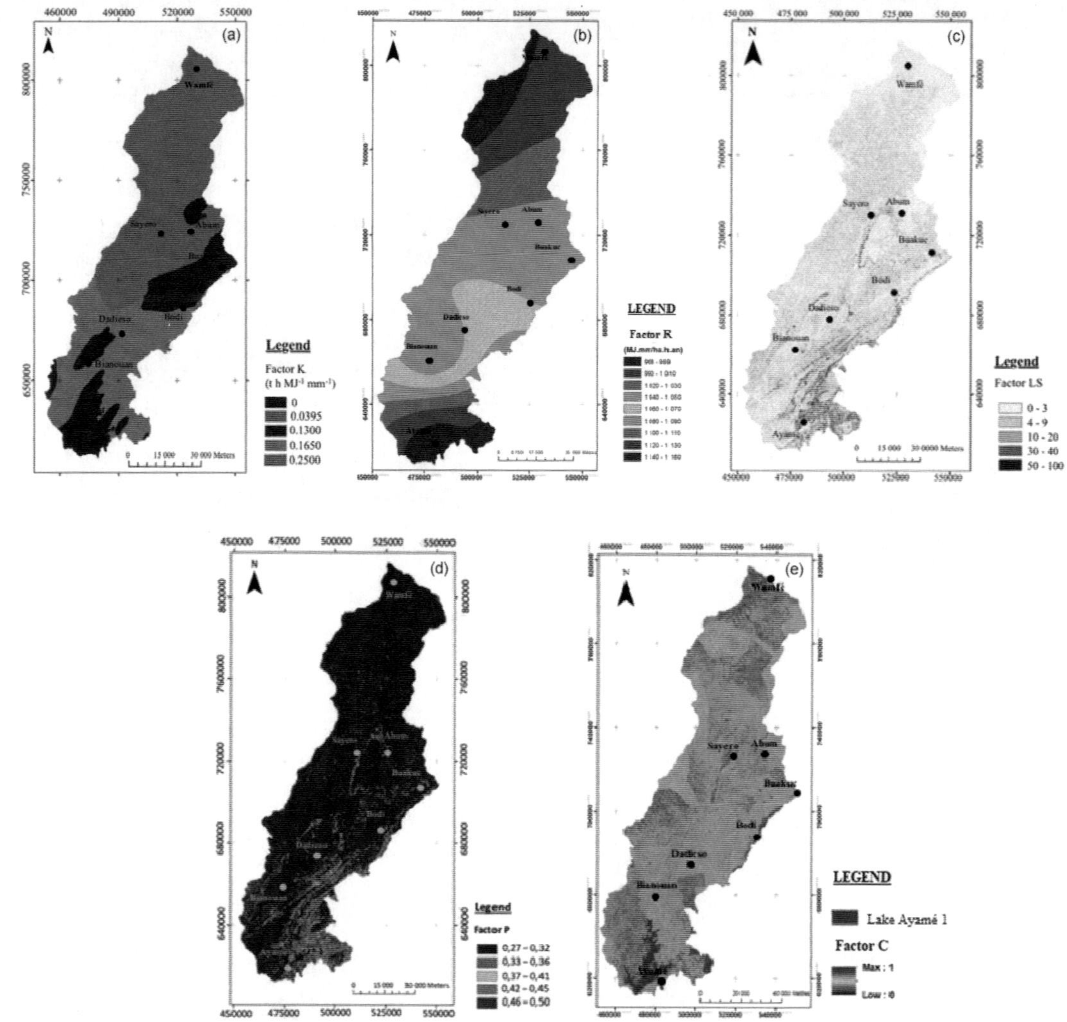

Figure 3. The different erosion factors of the Bia catchment area: **(a)** Soil erodibility factor K – **(b)** Rainfall erosivity factor R – **(c)** Topographical factor LS – **(d)** Support and conservation practices factor P – **(e)** Land cover factor C.

the north and the highest in the south. The average factor over the entire watershed is estimated at 3.3 (Fig. 3c). Most of the basin is characterized by C-factor values greater than 0.8. These areas are bare soil or fallow. The second largest class consists of values ranging from 0.6 to 0.7 in the South and South-West and are associated with large grain and forage crops. Areas with values of less than 0.6 are generally covered by forests (parks and reserves). Bare soils account for about 10.47 % of the study area, without taking into account areas of low vegetation cover (Fig. 3e). The values of anti-erosion practices are between 0.27 and 0.5 (Fig. 3d) reflecting the strip farming on the basin.

4.2 Soil Loss and Soil Erosion Sensitivity

Figure 4a presents a quantitative map with classes between 0 and > 1600 t/ha/year, with an average loss of 16 t/ha/year. The most sensitive areas of water related soil erosion, represent 189.41 km^2 (2.04 % of the catchment) and can deliver up to 1631 t/ha/year of materials. At the watershed level, 29.43 million t/year of soil can potentially be removed and are therefore likely to contribute to current sediment flow dynamics. The statistical distribution (Fig. 4b) shows that 84.92 % of the area of the Bia watershed reveals a low to moderate sensitivity to erosion while 2.04 % a very high sensitivity (greater than 150 t/ha/year). Through comparisons with maps of different factors, the sensitivity map shows low erosion in the Central-West (very low slope areas), North-East (forests) of the basin, and moderate erosion in the West and North. Very high erosion occupies only limited and con-

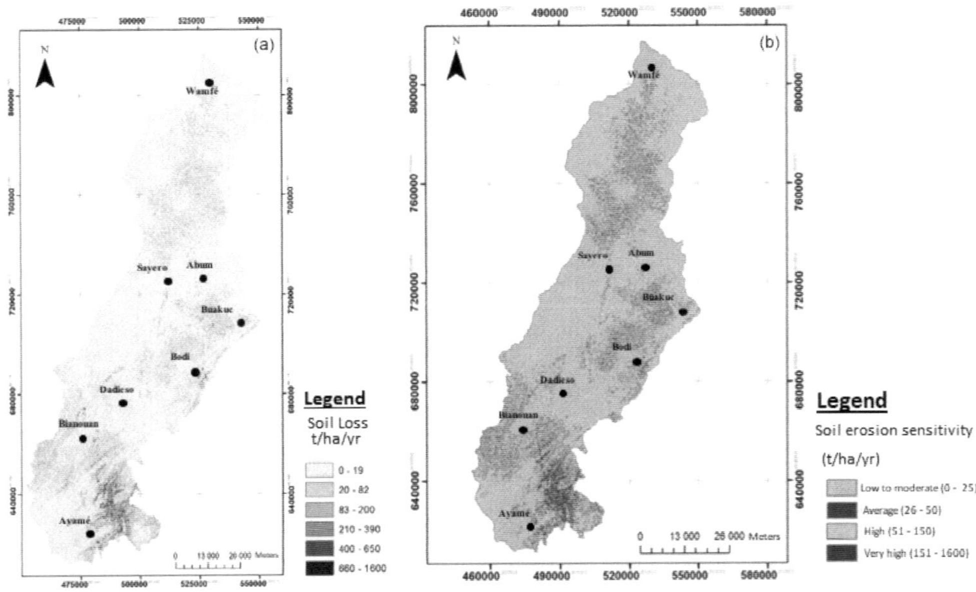

Figure 4. The different erosion factors of the Bia catchment area: (**a**) Soil loss – (**b**) Soil erosion sensitivity.

centrated areas, especially at the southern end of the basin where there is bare land on relatively steep slopes and high relief.

5 Discussion and conclusion

The Bia watershed is subject to strong climatic aggressiveness with an average of 1040 MJ mm (ha h year)$^{-1}$. The average value of weighted K for the entire basin is 0.13, which confirms the high susceptibility of the soils in the Bia watershed to water erosion. Vegetation plays a fairly good protective role over the majority of the Bia watershed since the value of C equals 0.35. The most sensitive areas of water related soil erosion, can deliver up to 1631 t/ha/year of materials. Our results are similar to those found by Payet et al. (2011) in southwestern Madagascar where they found values up to 1631 t/ha/year. At the watershed scale, 29.43 million tons of soil can potentially be removed and thus contribute to current sediment flow dynamics. This estimate therefore does not in any way predict the inflow to the lake from the Ayamé 1 hydroelectric dam, but provides information on the amount of material potentially supplied by the watershed and thus reveals its current alteration. The rate of erosion in the fragile Rif zones averages 17 t/ha/year according to Moukhchane et al. (1998). These different results are very close to those obtained by this study (16t/ha/year), which shows the relative reliability of the model applied. Hasbaia et al. (2012) explains the soil loss by the poor vegetation cover, by the vulnerability of the soil to erosion due to high temperatures and by the high intensity of rainfall. The most significant findings of this study are that soil loss in the area is well beyond the tolerance limit and that crop factor is more important in influencing soil loss. Soil loss tolerance limit varies from 1 to 12 t/ha/year depending on climate (Roose, 1994). The very fact that soil loss has crossed the tolerance limit is a matter of concern because this is an area under agriculture. If soil erosion is allowed to continue at the present rate, there may be eventual degradation of land, decline in soil fertility and reduced yield of crops. The results of the study clearly indicate that soil erosion in the area should be a cause for concern. Under such circumstances it becomes increasingly relevant to adopt necessary soil conservation measures so as to ensure the prolonged fertility of the land and prevention of degradation of land quality. Thus, it can be said that the USLE is a very useful tool for predicting soil loss under African conditions, its main advantage being its simplicity

Data availability. The precipitation data is not publicly available due insitutions policy. The Shuttle Radar Topography Mission (SRTM) image of N05W004 coordinates, from 11 to 22 February 2000, from Space Shuttle Endeavour (STS-99), obtained from ftp://eosrp01u.ecs.nasa.gov/srtm/version1/15Africackbox/ (last access: 1 January 2021). Landsat ETMC images, scenes 195-056 and 195-055, 14 March 2003, provided by http://earthexplorer.usgs.gov/ckbox/ (last access: 1 January 2021).

Author contributions. MNEH is responsable for data analysis and mansuscript writing. Others authors participate in data collection and analysis.

Competing interests. The contact author has declared that neither they nor their co-authors have any competing interests.

Disclaimer. Publisher's note: Copernicus Publications remains neutral with regard to jurisdictional claims in published maps and institutional affiliations.

Special issue statement. This article is part of the special issue "Hydrology of Large River Basins of Africa". It is a result of the 4th International Conference on the "Hydrology of the Great Rivers of Africa", Cotonou, Benin, 13–20 November 2021.

Acknowledgements. We would like to thank the IAHS SYSTA award for providing us with a financial support to attend the 4th International conference on the "Hydrology of the Great Rivers of Africa", Cotonou, Benin,13–20 November 2021.

References

Arnoldus, H. M. J.: An approximation of the rainfall factor in the USLE, Assessment of Erosion, Wiley, Chichester, England, 127–132, 1980.

Desmet, P. J. J. and Govers, G.: Comment on "Modelling topographic potential for erosion and deposition using GIS", International Journal of Geographical Information Science, 11, 603610, https://doi.org/10.1080/136588197242211, 1997.

Durigon, V. L., Carvalho, D. F., Antunes, M. A. H., Oliveira, P. T. S., and Fernandes, M. M.: NDVI time series for monitoring RUSLE cover management factor in a tropical watershed , Int. J. Remote Sens., 35, 441453, https://doi.org/10.1080/01431161.2013.871081, 2014.

Gitas, I., Douros, K., Minakou, C., Silleos, G., and Karydas, C.: Multi-temporal soil erosion risk assessment in N. Chalkidiki using a modified Usle raster model, EARSeL eProceedings, 8, 40–53, 2009.

Hasbaia, M., Hedjazi, A., and Benayada, L.: Variability of water erosion in the Hodna basin: case of the Elham wadi subwatershed, Rev. Mar. Sci. Agron. Vét., 1, 28–32, 2012.

Kouassi, K. L., Kouamé, K. I., Konan, K. S., Angoulo, M. S., Demé, M., and Meledje, N. E. H: Two-Dimensional Numerical Simulation of the Hydro-Sedimentary Phenomena in Lake Taabo, Côte d'Ivoire, Water Ressources Management, Water Resour. Manage., 27, 4379–4394, https://doi.org/10.1007/s11269-013-0417-x, 2013.

Kumar, B. J., Yu, I., and Jeong, S.: Estimation of Soil Erosion Using RUSLE Model and GIS Techniques for Conservation Planning from Kulekhani Reservoir Catchment, Nepal, Journal of Korean Society of Hazard Mitigation, 16, 323–330, https://doi.org/10.9798/KOSHAM.2016.16.3.323, 2016.

Moukhchane, M., Bouhlassa, S., and Chalouana, A : Approche cartographique et magnétique pour l'identification des sources de sédiments: cas du bassin versant Nakhla (Rif, Maroc), Sécheresse,3, 227–232, 1998.

Payet, E., Dumas, P., and Pennober, G.: Modélisation de l'érosion hydrique des sols sur un bassin versant du sud-ouest de Madagascar, le Fiherenana, VertigO, 11, https://doi.org/10.4000/vertigo.12591, 2012.

Raissouni, A.: Modeìlisation et cartographie de la sensibiliteì et de l'aleìa d'eìrosion des sols aÌ l'échelle régionale par USLE et SIG (Rif Nord occidental, Maroc), Theìse de Doctorat Es-Sciences, Fac. Sci. Tech. Tanger, Maroc, 208 pp., 2012.

Renard, K. G., Foster, G. R., Weesies, G. A., Mac Cool, D. A., and Yoder, D. C. (Eds): Predicting soil erosion by water: A guide to conservation planning with the Revised Universal Soil Loss Equation (RUSLE), Agric. Handb., number 703, U.S. Gov. Print. Office,Washington, DC., 383 pp., 1997.

Roose, E.: Introduction aÌ la gestion conservatoire de l'eau, de la biomasse et de la fertiliteì des sols (GCES), FAO Soils Bulletin, Rome, 70, 420 pp., 1994.

Shin, G. J.: The analysis of soil erosion analysis in watershed using GIS, Ph.D. thesis, Department of Civil Engineering, Gang-won National University, South Korea, 145 pp., 1999.

Stone, R. P. et Hillborn, D.: Universal Soil Loss Equation, Ontario, Canada, Ontario Ministry of Agriculture and Food (OMAFRA), [Enligne], http://www.giser.be/wpcontent/uploads/2012/05/USLE-infosCanada.pdf (last acces: 18 October 2021), 2000.

Wischmeier, W. H. and Smith, D. D.: Predicting Rainfall Erosion Losses, Agr. Han., 537, 285–291, https://doi.org/10.1029/TR039i002p00285, 1978.

Proc. IAHS, 384, 113–119, 2021
https://doi.org/10.5194/piahs-384-113-2021

Open Access PIAHS

Hydrology of Large River Basins of Africa

Erosion fluviale et mutations morphométriques du drainage dans le secteur aval du barrage de Lagdo (Bassin de la Bénoué, Nord-Cameroun)

Félix Mbele Abbo

Département de Géographie, Faculté des Arts, lettres et Sciences Humaines,
Université de Maroua, BP 644 Maroua, Cameroun

Correspondence: Félix Mbele Abbo (mbeleabbofelix2@gmail.com)

Published: 16 November 2021

Résumé. La présente étude s'intéresse à la dynamique du drainage de la Bénoué, par comparaison de deux situations à savoir, celles d'avant et après implantation d'un barrage à Lagdo (1975–2016). En effet, jusqu'en 1983, aucun barrage n'existait sur la Bénoué. La mise en place de cet ouvrage a non seulement modifié l'équilibre environnemental du bassin versant, mais actionné des dynamiques humaines et physiques importantes. On y note une modification des modules hydrologiques, tandis que les processus morphogéniques se sont accélérés dans le secteur aval, et se révèlent à travers des glissements fréquents de berges. Il est question dans ce travail de caractériser l'évolution du drainage par l'entremise des indicateurs morphométriques et d'en évaluer les impacts ou enjeux. Pour ce faire, des observations de terrain ont été faites, des enquêtes/entretiens ont été menés auprès des riverains, et une base de données géoreférencées a été constituée sous Qgis 2.6.1 afin de faciliter les analyses diachroniques du drainage. Il en ressort que les activités morphogéniques ayant cours dans ce secteur sont de plus en plus actives. Il y a également une revalorisation positive des caractéristiques actuelles des méandres, comparativement à la situation de 1975. Les longueurs d'onde et les amplitudes des méandres n'ont connus que de légères modifications, tandis que la largeur moyenne des sections est régulièrement passée de 121,77 m en 1975 à 169,08 m en 2016. En comparant les différentes sections définis dans le cadre de cette analyse, on se rend également compte que les largeurs érodées plein bord s'étendent en moyenne de 1,47 m an^{-1}, alors que les pertes en terre par section sont de l'ordre de 0,63 ha an^{-1}. Globalement, les impacts évalués ou ressentis s'avèrent non négligeables d'où les propositions faites afin de juguler l'aléa.

1 Introduction

L'érosion fluviale est de loin l'une des formes de météorisation les plus constantes et les plus dynamiques dans l'espace. Elle se développe notamment dans les hydrosystèmes sous les formes linéaire, régressive ou de masse, et dépendent des influences répétées de l'eau courante sur les berges. Celles-ci procèdent à l'enlèvement de la matière, à son transport ou à son accumulation, selon la force de l'écoulement, ou suivant les différents mouvements intrinsèques au fluide (Saint-Laurent et al., 2002 ; Lacoste, 2003). Vue sous cet angle, l'érosion fluviale est un processus délicat du fait de la sensibilité de ses variables, mais aussi de l'ampleur de ses conséquences, qui peuvent dans certains cas s'avérer importantes.

Les facteurs d'évolution des systèmes érosifs fluviaux sont multiples. Il est en effet admis depuis le Néolithique que nos paysages sont une co-construction régie à la fois par les dynamiques environnementales et sociétales (Carcaud, 2016). Même si ces deux familles de facteurs restent poignantes, l'acception la plus admise de nos jours par l'ensemble des scientifiques est que ces dynamiques environnementales et sociétales sont de plus en plus accélérées, notamment à cause d'une fréquente variabilité du climat et d'un accroissement démographique toujours constant. Ces dernières s'opèrent par une diversification des activités anthropiques, une densification des exploitations dans l'espace, ou parfois par une désorganisation des schémas d'occupation des terroirs, souvent liée à la mise en œuvre des pôles agroindustriels ou des mégastructures telles que les barrages hydroélectriques.

Le cadre de cette étude (bassin versant de la Bénoué) est une vaste cuvette dont l'altitude moyenne varie entre 300 et 400 m. Il couvre une superficie d'environ 340 000 km^2 et représente un peu plus de 8 % de la superficie totale du bassin du Niger (ABN, 2008). Dans l'espace camerounais, la Bénoué a une longueur totale estimée à 350 km, dont 220 km en amont du barrage et 130 en aval. Son lit majeur est une vaste zone d'inondation constituée de cuvettes latérales en amont de la ville de Garoua (Tchotsoua et al., 2007). On y distingue trois sous-bassins à savoir : le sous bassin de la haute Bénoué, celui du mayo Kébi, et le sous bassin du Faro (Fig. 1).

Le tronçon étudié dans le cadre de cet article va de la confluence mayo Kébi-Bénoué à celle du Faro-Bénoué. Il est long d'environ 100 km, et est compris entre les coordonnées 9°17′16.32″ N–13°32′48.74″ E et 9°19′59.99″N–12°55′0.16″ E. L'ensemble du bassin versant est soumis à un climat soudanien à deux saisons (sèches et de pluie), et reçoit des moyennes ombrothermiques de l'ordre 1000 mm par an de pluie et de 28 °C respectivement (Olivry, 1986).

La mise en eau du barrage sur la Bénoué a Lagdo a provoqué une reconfiguration de l'hydrographie de la Bénoué supérieure (mise en place du lac de retenu), et en même temps, une restructuration fonctionnelle de la section avale. En effet, avant la mise en place du barrage, les étiages étaient sévères à certaines périodes de l'année, tandis que de nos jours, on assiste à une valorisation du niveau des écoulements, assortie d'une accélération des processus morphogéniques, notamment le long du tronçon compris entre Lagdo et la frontière Cameroun-Nigéria. Dans ce contexte, on se pose alors la question de savoir comment s'opère la dynamique du drainage dans le secteur aval au barrage de Lagdo et quels en sont les enjeux et impacts ? Il est dès lors question de caractériser l'évolution du drainage par l'entremise des indicateurs morphométriques et d'en évaluer les enjeux. L'atteinte des objectifs ainsi énoncés passe cependant par la mise en œuvre d'une méthodologie qu'il convient de présenter.

2 Matériels et méthodes

La méthodologie développée dans le cadre de ce travail se fonde sur la collecte des données secondaires, les observations de terrain, les enquêtes menées auprès des riverains, les entretiens, et le montage d'une base de données géoréférencées sous Qgis 2.6.1 afin de procéder aux analyses diachroniques qui s'imposent.

2.1 La collecte des données secondaires

La collecte des données secondaires a été faite auprès des structures telles que la Mission d'Etudes pour l'Aménagement et le Développement de la région du Nord (MEADEN), dans les centres de documentation et les bibliothèques des Universités de Ngaoundéré, Maroua, et Yaoundé I, ou sur les moteurs de recherche d'Internet. Il était question de procéder à la recension des écrits afin d'asseoir les bases théoriques et épistémologiques de l'étude.

2.2 Les observations de terrain

Les observations de terrains ont été faites en suivant fidèlement le drainage d'amont en aval, dans l'optique d'apprécier les réalités morphohydrologiques via les formes de berges, la configuration des méandres, ou celles des bancs de sables s'organisant dans le lit mineur de la Bénoué. Il a été aussi question de faire des prises de vue afin d'illustrer le travail.

2.3 Les enquêtes menées auprès des riverains

Elles ont été menées auprès de 478 personnes réparties dans 15 villages. Il était question de collecter via l'administration des questionnaires préalablement apprêtés des informations sur la perception et l'évaluation locale du phénomène, de même que les solutions envisageables. Lors de cette activité, l'accent était particulièrement mis sur des personnes ayant passé au moins vingt ans dans le bassin, et ayant une certaine appréhension du phénomène étudié.

2.4 Le montage d'une base de données géoréferencées

Il s'est agi de compiler les résultats des classifications supervisées des images satellitaires de 1975 et 2016 sous Qgis 2.6.1 afin de procéder aux analyses diachroniques. Cette technique a permis à terme de quantifier les mutations morphométriques du drainage, ainsi que les impacts qui leur sont liés (longueur des flèches d'érosion, taux d'érosion des berges, etc.). L'ensemble des données collectées a été traité, puis analysé selon des procédés plus ou moins différents, conformément à leur nature.

3 Résultats

3.1 Une érosion fluviale fondée sur les glissements récursifs des berges

L'érosion est un des processus physiques dont les actions et formes varient selon les contextes bioclimatiques. La prise en compte de ce postulat permet d'identifier des modèles érosifs discrets ou lents, tandis d'autres s'avèrent brusques ou inattendus dans l'espace.

3.1.1 Les formes de glissement de berges observées en aval du barrage de Lagdo

L'érosion des berges s'appréhende comme un phénomène régressif contribuant à l'enlèvement progressif de matériaux suite à l'action des forces liées à un écoulement turbulent ou abondant. Cette forme d'érosion est inéluctablement associée au fonctionnement naturel d'un cours d'eau, et procède d'un

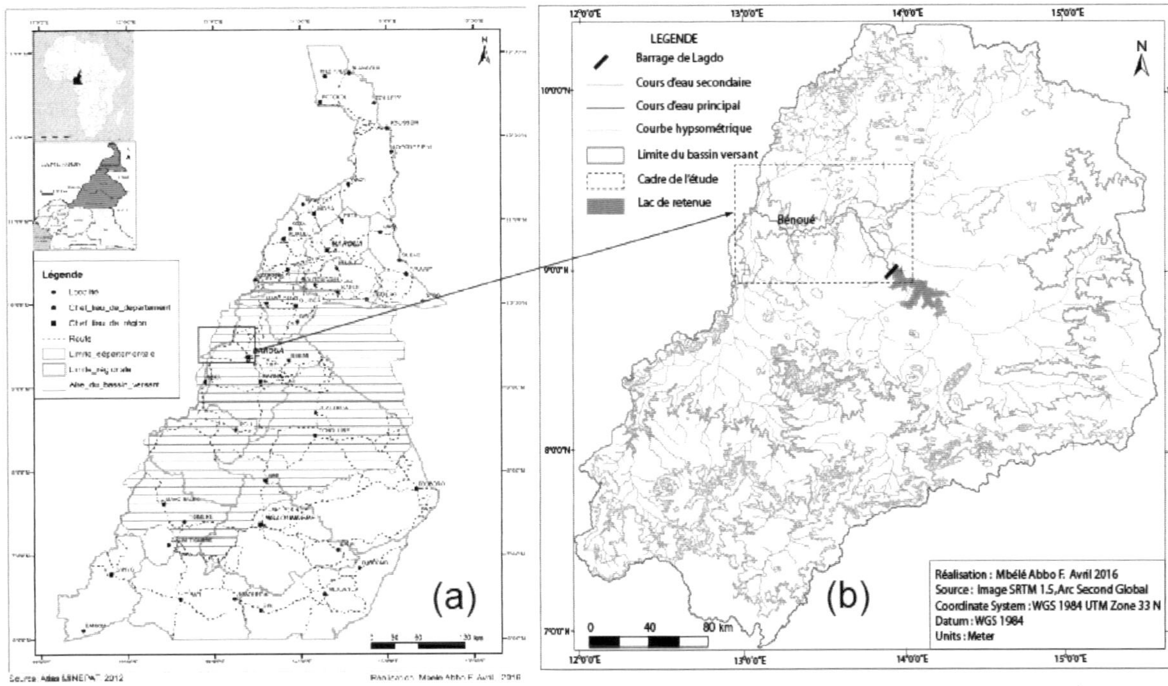

Figure 1. Localisation (**a**) et oro-hydrographie (**b**) du cadre de l'étude.

plan opératoire en fonction duquel les berges se déstabilisent progressivement par glissement. En se fondant sur la définition de l'érosion de berges, il s'en dégage que le glissement peut s'opérer soit par sapement ou par arrachement, suite à l'action opposée du fluide en mouvement. On distingue en fonction de l'un ou l'autre cas plusieurs formes de glissement de berges dans le cadre d'étude. Celles-ci se révèlent en surface à travers des fissures de rupture bien incisées dans les berges (Fig. 2) à partir desquelles le décollement s'amorce avant la rupture totale.

L'illustration ci-dessus révélé une entaille ouverte parallèlement au regard de la berge. Ce décollement en direction du cours d'eau démontre que la base de la masse sédimentaire est en train d'être sapé suite à l'action de l'érosion. Il s'agit d'un processus qui s'observe presque continuellement le long du cours d'eau, même si certains secteurs semblent plus exposés que d'autres du fait de l'organisation en méandres ou de la nature fortement sableuse des berges.

Figure 2. Amorce de déstabilisation d'une berge non loin de Kokoumi, 9° 18′22.76″ N–12° 59′43.99″ E. Cliché : Félix Mbélé Abbo.

3.1.2 Une prédisposition morphogénétique fondée sur la forme et la richesse sableuse du drainage

Le tronçon en étude est particulièrement structuré en méandres libres et semble très peu encaissé dans une importante masse sédimentaire, particulièrement sableuse. Il s'agit d'un ensemble constitué de sinuosités opposant alternativement des rives concaves aux rives convexes. De ce fait, les dynamiques hydro-sédimentaires s'opèrent de façons plus ou moins différentes selon que l'on est sur une ou l'autre forme de rive. La vulnérabilité des berges se fonde dès lors sur le fait que les sols dominants sont sableux à sablo-limoneux. Compte tenu de l'ampleur des écoulements, ces sols s'érodent facilement pour peu qu'une crue se manifeste.

3.2 La dynamique des caractéristiques morphométriques du tronçon d'étude

Le méandrage apparait de par sa définition comme une forme de drainage continuellement entretenue par l'équilibre

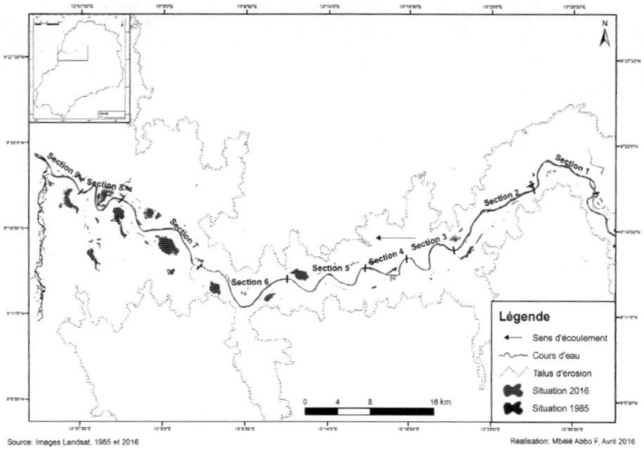

Figure 3. Visualisation des différentes sections d'analyse du tronçon d'étude.

hydrodynamique du cours d'eau. Même si l'appréhension de ce postulat semble évidente, Malavoi et Bravard (2010) proposent une méthodologie de caractérisation et d'évaluation qui se fondent sur un ensemble de dimensions précises et répertoriées.

3.2.1 Les dimensions du drainage avant la mise en eau du barrage

Les caractéristiques morphométriques mises en valeur dans cette étude sont de plusieurs types. Elles se rapportent globalement à la largeur du lit à plein bord (w), à la longueur d'onde (λ) du méandre, à son amplitude (A), au rayon de sa courbure du méandre (R_c), et à la longueur de son arc (L arc) entre autres.

Le tronçon d'étude fait environ 100 km de long et serpente un vaste lit sédimentaire s'étendant sur plusieurs centaines de km^2. Compte tenu donc de la diversité des formes développées par le cours d'eau, il semblait intéressant de structurer ledit tronçon en neuf (9) « sections morphologiquement homogènes ». Cela a permis de faire des mesures beaucoup plus précises, tant sur le terrain que via le système d'information géographique, d'opérer sur la base des réalités du terrain des analyses acceptables, et même de comparer les sections entre elles. Les différents morcellements du tronçon transparaissent sur la carte suivante (Fig. 3). La logique de distinction des sections tient principalement à leur forme, à l'homogénéité des manifestations érosives observées, et aux types d'occupation des sites en présence.

La délimitation des sections d'analyse telle que présentée par la figure précédente s'avère aléatoire, mais fait suite à une compilation des tronçons de deux époques distantes de quatre décennies. L'évaluation de la situation de 1975 vise à présenter l'état du bassin bien avant le début des travaux de mise en place du barrage hydroélectrique à Lagdo. Cela a permis d'appréhender « l'état initial » du drainage avant la restructuration des écoulements du fait du barrage. Les données obtenues se révèlent à travers le tableau suivant (Tableau 1).

Si l'on s'intéresse de façon détaillée à quelques-unes des variables contenues dans le tableau ci-dessus, on constate que la largeur moyenne des sections était de 121,77 m en 1975. La longueur d'onde et l'amplitude des méandres sont tout aussi impressionnantes. Leurs moyennes sont respectivement de 4,859 et de 2,101 km.

3.2.2 Les mutations morphométriques observables de nos jours

Contrairement à la situation de 1975, les caractéristiques de 2016 (Tableau 2) correspondent à une hydrodynamique plus ou moins régulée. En effet, depuis la mise en eau en 1983, le secteur aval du barrage est ancré dans une forme d'exacerbation de l'érosion fluviale du fait de la constance des écoulements et de l'effet des débits dérivants des phénomènes hydrologiques extraordinaires (inondations majeures).

Il se dégage de ce tableau, qu'il y a une revalorisation positive de l'ensemble des données obtenues, lorsqu'on compare les deux situations de 1975 et de 2016 (Tableau 3).

Au vu du tableau ci-dessus, on note une revalorisation de la largeur moyenne des sections qui est passée de 121,77 m (en 1975) à 169,08 m (en 2016). Contrairement à cette situation, les longueurs d'onde n'ont connues que de légères modifications (1,17 %), dans la mesure où on est respectivement passé d'une moyenne de 4,859 km (en 1975) à 4,917 km (2016). L'érosion fluviale s'exacerbe donc de plus en plus le long du cours de la Bénoué. Toutefois, les données obtenues semblent discutables.

3.3 Les formes de transmutation du lit et leurs conséquences

Ces transmutations du drainage se révèlent à travers des indicateurs précis et sont porteuses d'enjeux importants.

3.3.1 Les indicateurs d'évolutions morphométriques du drainage

En dehors des modifications mises en évidence plus haut (section par section), les transmutations observables dans le secteur aval de la Bénoué présentent plusieurs formes, mais obéissent à des principes physiques plus ou moins semblables. La compilation des tronçons d'époque différente dans un Système d'Information Géographique monté sous QGIS a permis d'approfondir les analyses sur la base du support suivant (Fig. 4).

L'observation de cette carte permet de se rendre compte que les méandres sont dynamiques dans l'espace. Les décalages par sections sont compris entre 60 et 200 cm soit un taux de recul des berges de 1,206 m an^{-1}.

Tableau 1. Caractéristiques morphométriques des méandres en 1975.

Sections	Variables (en km)				
	Largeurs plein bord (W)	Longueur d'onde du méandre (λ)	Rayon courbure du méandre (R_c)	Longueur de l'arc du méandre (L arc)	Amplitude du méandre (A)
Section 1	0,109	4,451	0,995	3,176	2,454
Section 2	0,132	2,226	0,761	2,523	1,601
Section 3	0,133	3,812	0,86	2,873	1,951
Section 4					
Section 5	0,081	5,126	1,028	3,165	1,806
Section 6	0,121	7,988	1,375	5,316	3,821
Section 7	0,146	7,975	1,002	4,304	2,052
Section 8	0,138	3,442	1,009	3,098	2,396
Section 9	0,111	3,858	0,44	2,026	0,727

Source : Base de données SIG, août 2016.

Tableau 2. Caractéristiques morphométriques des méandres en 2016.

Sections	Variables (en km)				
	Largeurs plein bord (W)	Longueur d'onde du méandre (λ)	Rayon courbure du méandre (R_c)	Longueur de l'arc du méandre (L arc)	Amplitude du méandre (A)
Section 1	0,185	2,317	1,508	2,091	2,660
Section 2	0,333	4,700	0,501	2,938	0,976
Section 3	0,152	3,976	1,203	3,950	2,067
Section 4	0,123	5,226	0,934	2,289	1,802
Section 5	0,119	5,075	0,847	2,828	1,880
Section 6	0,129	8,321	1,527	5,036	3,937
Section 7	0,150	7,130	0,818	2,948	2,306
Section 8	0,149	3,668	1,017	2,337	2,375
Section 9	0,176	3,838	0,483	1,719	0,932

Source : Base de données SIG, août 2016.

Tableau 3. Analyse comparée des largeurs et longueurs d'onde des méandres entre 1975 et 2016.

Sections	Variables (en km)			
	Situation en 1975		Situation en 1975	
	Largeurs plein bord (W)	Longueur d'onde du méandre (λ)	Largeurs plein bord (W)	Longueur d'onde du méandre (λ)
Section 1	0,109	4,451	0,185	2,317
Section 2	0,132	2,226	0,333	4,700
Section 3	0,133	3,812	0,152	3,976
Section 4			0,123	5,226
Section 5	0,081	5,126	0,119	5,075
Section 6	0,121	7,988	0,129	8,321
Section 7	0,146	7,975	0,150	7,130
Section 8	0,138	3,442	0,149	3,668
Section 9	0,111	3,858	0,176	3,838

Source : Base de données SIG, août 2016.

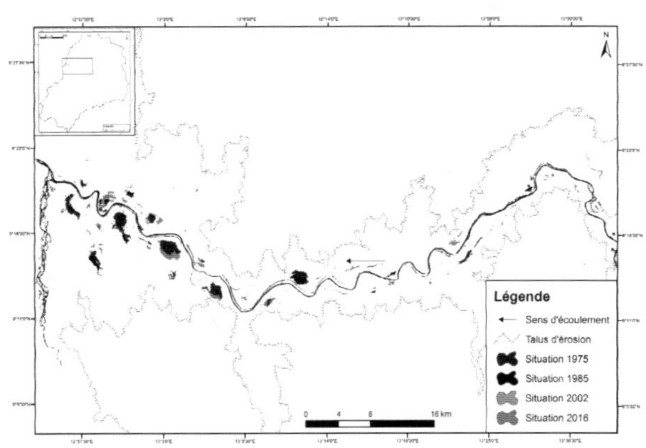

Figure 4. Dynamique des méandres en aval de Lagdo entre 1975 et 2016.

Il se dégage des analyses que le cours d'eau procède globalement à une translation qui s'effectue notamment dans le sens de l'écoulement. Mais à y voir de près, chaque section développe une évolution plus ou moins particulière. Cela s'explique par le fait que chaque maillon de ce tronçon se caractérise par un indice de maturité différent (compris entre 1,5 et 11,84) dérivant de la diversité des sous-contextes pédologique, géomorphologique et hydrologique observables le long du drainage. On assiste alors à une mise en évidence de deux anabranches (sections 1 et 8) et de plusieurs extensions de méandres, notamment au niveau des sections 2, 3 et 9, tandis que les types d'évolution les plus importants s'apparentent à la création des bras morts (entre les sections 6 et 7). Les largeurs érodées plein bord quant à elles s'étendent en moyenne de 1,47 m an^{-2}, avec des valeurs extrêmes comprises entre 0,78 m an^{-2} et 2,30 m an^{-2} respectivement.

3.3.2 Les enjeux et les solutions envisageables face à l'aléa

Les mutations du drainage de la Bénoué sont des réalités aux conséquences non négligeables. D'après les résultats d'enquête, le secteur d'étude concentre à lui seuls plus de la moitié (51 %) d'agriculteurs, 42 % d'éleveurs, et 65 % de pêcheurs du bassin. Il s'agit dans l'ensemble d'une population préoccupée par les conséquences de ces activités morphogéniques, qui se révèlent à travers des pertes en terre de l'ordre de 0,63 ha an^{-2}. Même si la population dans sa majorité (86 %) estime que l'érosion des berges est difficile à maitriser, les solutions envisagées se rapportent aux techniques telles que l'enrochement des berges ou la végétalisation des sites. Cependant, l'application de ces mesures gagnerait à passer par la vulgarisation de l'agroforesterie, à travers l'adoption des plantes locales, ayant des systèmes racinaires adéquats, et qui soient par ailleurs susceptibles d'apporter une plus-value (fruits, bois énergie, bois d'œuvre, haies-vives, etc.) aux populations déjà économiquement vulnérables.

4 Discussions

L'évolution morphométrique du drainage est une réalité courante à tous les cours d'eau, surtout en contexte de modification des paramètres d'écoulement. En effet, les ouvrages hydrauliques ont souvent eu pour effet de perturber les évolutions naturelles, soit en les limitant ou en déclenchant des processus parfois accélés (Dzana, 2000 ; Aubry, 2012 ; Amassou et al., 2012 ; Bahin et al., 2017 ; Mbélé Abbo, 2018 ; Nzango, 2018). Pour le cas d'espèce, l'analyse diachronique (1975 et 2016) du secteur aval de ce barrage révèle trois résultats importants. Il s'agit de la variabilité de l'élargissement du drainage (compris ici entre 121,77 et 169,08 m) qui peut se justifier par la hausse globale des débits journaliers, due à un régime d'exploitation mixte promouvant à la fois l'exploitation hydroélectrique du barrage et le transport fluvial d'une part, et les apports hydrologiques liés au libre fonctionnement du mayo kébi (sous affluent non barré et alimentant le secteur aval) d'autre part. Les deuxième et troisième résultats se rapportent respectivement à l'évolution des largeurs érodées plein bord (moyenne 1,47 m an^{-2}), et aux pertes en terre estimées en moyenne à 0,63 ha an^{-2}. L'évolution morphogénique de la Bénoué dans ce secteur est donc préoccupante, même si les manifestations érosives caractérisées n'ont ni la même fréquence encore moins la même occurrence partout. Ainsi, en se fondant sur les études menées sur le Logone ou le mayo kaliao par des auteurs tels que Kossoumna Liba'a et Djarsoumna Kainaramsou (2017), Kana (2017) ou Djarsoumna Kainaramsou et al. (2019), on note à quelques différences prêtes une convergence des taux de recul des berges (1,206 m an^{-1}) ou des pertes en terre, alors que les cours d'eau étudiés se situent dans des milieux bioclimatiques plus ou moins distincts. Il se dégage à contrario des travaux de Dzana (2000) et de Saint-Laurent et al. (2002) des taux de recul encore plus faibles (soit 0,50 m an^{-2}). Cela peut être lié à la méthodologie de quantification choisie les auteurs, à la faiblesse des écoulements, à l'effet limitant de la végétation rivulaire, ou simplement à la nature des berges en présence.

Quoi qu'il en soit, cette étude a en fin de compte le mérite d'avoir évalué les impacts morphométriques des écoulements plus ou moins contrôlés sur les berges de la Bénoué. Elle a permis de fournir plusieurs types de données caractéristiques d'une réalité longtemps identifiée, mais très peu quantifiée. La mise en œuvre d'une analyse diachronique associée aux observations de terrain aura permis d'obtenir des informations importantes sur les transmutations du drainage. Cependant, les données en présence auraient été plus significatives si elles avaient aussi subi des tests statistiques particuliers. Les enjeux et les impacts autour des glissements de berges ont été évalués. Toutefois, la quantification des masses sédimentaires évacuées par le cours d'eau, l'analyse de la dynamique des charges solides s'avèrent être des thématiques susceptibles de compléter la connaissance sur le sujet.

5 Conclusion

Il apparait au vu de tout ce qui précède que la dynamique morphométrique du tronçon aval au barrage de la Bénoué se fonde à la fois sur la modification des conditions naturelles de drainage et sur l'évolution des contextes d'exploitation du cours d'eau. Ces facteurs peuvent être associés aux conditions d'occupation des rives pour expliquer les écarts notés lorsqu'on compare les paramètres morphohydrologiques du tronçon d'étude entre 1975 et 2016. Les résultats évoqués plus haut ont démontré que les méandres sont en transmutation progressive selon le sens des écoulements, même si les impacts morphogéniques observés le long des berges diffèrent selon les sections. Toutefois, ces modifications ne comportent pas que des effets négatifs dans les mesures où elles procèdent aussi à des dépôts alluvionnaires qui enrichissent les berges et par ricochet favorisent l'émergence d'une agriculture de contre-saison. La stabilisation des berges s'avèrent donc être un impératif tant pour les acteurs que pour les décideurs, mais il convient d'en bien étudier les contours afin de les solutionner judicieusement.

Disponibilité des données. Les données utilisées dans cet article peuvent être demandées par l'auteur.

Intérêts concurrents. L'auteur déclare qu'il n'a aucun conflit d'intérêts.

Clause de non-responsabilité. Publisher's note : Copernicus Publications remains neutral with regard to jurisdictional claims in published maps and institutional affiliations.

Déclaration du numéro spécial. This article is part of the special issue "Hydrology of Large River Basins of Africa". It is a result of the 4th International Conference on the "Hydrology of the Great Rivers of Africa", Cotonou, Benin, 13–20 November 2021.

Remerciements. Ladite étude n'aurait été possible sans les apports multiformes de certaines personnes qu'il convient de remercier ici. Il s'agit notamment des riverains pour les informations fournies, des enquêteurs pour leur aide dans la collecte des données sur le terrains, et des relecteurs pour les critiques et suggestions ayant permis d'améliorer ce manuscrit. Qu'ils trouvent tous en ce travail l'expression de ma profonde gratitude.

Références

ABN : Étude pour la mise en place d'un réseau de recherche et développement à l'échelle du bassin du Niger, Rapport final provisoire, Programme de Lutte Contre l'Ensablement/Bassin du Niger (PLCE/BN), 109 pp., 2008.

Amassou, E., Camberlin, P., et Mahé, G. : Impact de la variabilité climatique et du barrage Nangbéto sur l'hydrologie du système mono-couffo (Afrique de l'ouest), Hydrol. Sci. J., 57, https://doi.org/10.1080/02626667.2011.643799, 2012.

Aubry, L. : Analyse des impacts du barrage de Rawdon Sur l'évolution hydromorphologique de la rivière Ouareau (Québec, Canada), Mémoire de Maîtrise en Sciences de l'environnement, Université du Québec, Trois-rivières, 55 pp., 2012.

Bahin, Y., Baudet, E., et Haida, S. : Analyse de la variabilité hydroclimatique et impacts des barrages sur le régime hydrologique d'une rivière de zone semi-aride : le Sebou au Maroc, ESJ, 13, 509–525, https://doi.org/10.19044/esj.2017.v13n5p509, 2017.

Carcaud, N. : L'hydrosystème et le changement : étude de cas dans le bassin de la Loire, BSGLg, 67, Dynamique fluviale, available at : https://popups.uliege.be/0770-7576/index.php?id=4454 (la date du dernier accès : 19 jouillet 2021), 2016.

Djarsoumna, K., Kossoumna Liba'a, N., et Boukar, A. : Cartographie des zones à risque d'érosion hydrique dans le bassin versant du Mayo Tsanaga : cas des berges du mayo Tsanaga et du Mayo Kaliao, ReMaGéom, 3, 75–92, 2019.

Dzana, J. G. : Réponses morphodynamiques d'un hydrosystème à faible énergie aux aménagements hydrauliques : l'exemple de la Seine supérieure (France), in : Géomorphologie : relief, processus, environnement, Juillet–septembre, Persées, France, 6, 161–176, 2000.

Kana, C. E. : Dynamique des berges du Logone entre la retenue de Maga et le lac Tchad et ses implications socio-économiques, Revue Science Eaux & Territoires, article hors-série, 7 pp., https://doi.org/10.14758/SET-REVUE.2017.HS.10, 2017.

Kossoumna Liba'a, N. et Djarsoumna, K. : Erosion des berges du mayo kaliao dans la ville de Maroua (extrême nord, Cameroun) : facteurs, types et conséquences, en : Risques et catastrophes en zone soudano-sahélienne du Cameroun : aléas, vulnérabilités et résiliences, collection Etude et recherche volume thématique no. 8, édité par : Kossoumna Liba'a, N., Djiangoué, B., et Wanie, C. M.,, éditions Cheikh Anta Diop, Douala, Cameroun, 23–45, 2017.

Lacoste, Y. : De la géopolitique aux paysages : dictionnaire de géographie, Armand colin/VUEF, Paris, France, 413 pp., 2003.

Malavoi, J. R. et Bravard, J.-P. : Éléments d'hydromorphologie fluviale, collection « comprendre pour agir », ONEMA, France, 224 pp., 2010.

Mbélé Abbo, F. : Analyse morphohydrologique du lit majeur de la Bénoué, de la confluence avec le Mayo-Kebi à la frontière Cameroun-Nigéria, Thèse de Doctorat PhD, Université de Ngaoundéré, Ngaoundéré, Cameroun, 387 pp., 2018.

Nzango, C. : Les barrages de l'Oubangui : de l'impact hydraulique actuel à la prospective environnementale, Thèse de Doctorat PhD, Ecole doctorale Sciences de l'homme et de la Société, Université d'Orléans, Orléans, France, 274 pp., 2018.

Olivry, J.-C. : Fleuves et rivières du Cameroun, MESRES-ORSTOM, Collection Monographies hydrologiques, MESRES-ORSTOM, Paris, France, 733 pp., 1986.

Saint-Laurent, D., Bérubé, M., Thériault, I., et Lemieux, M. : Phénomènes d'érosion fluviale et conditions hydrodynamiques de la rivière Saint-Maurice, Québec, Canada, Belgeo, https://doi.org/10.4000/belgeo.16018, 2002.

Tchotsoua, M., Fotsing, J.-M., et Moussa, A. : Évaluation des risques d'inondation dans la vallée de la Bénoué en aval du barrage de Lagdo (Cameroun), Actes des JSIRAUF, Hanoi 6–9 novembre 2007, 9 pp., 2007.

Proc. IAHS, 384, 121–126, 2021
https://doi.org/10.5194/piahs-384-121-2021

Open Access
PIAHS

Hydrology of Large River Basins of Africa

Soil loss vulnerability: the case study of Aghien lagoon watershed outskirts Abidjan city (Côte d'Ivoire)

Amidou Dao[1], Ehouman Serge Koffi[1], Dabissi Djibril Noufé[1], Bamory Kamagaté[1], Lanciné Droh Goné[1], Luc Séguis[2], and Jean Louis Perrin[2]

[1]Laboratoire de Géosciences et Environnement, Université Nangui Abrogoua, 02 BP 801 Abidjan 02, Côte d'Ivoire
[2]IRD, UMR Hydrosciences, Université de Montpellier 2, Montpellier, France

Correspondence: Amidou Dao (daoamidou@hotmail.fr)

Published: 16 November 2021

Abstract. Aghien lagoon is a source of fresh water outskirts of Abidjan city in the south of Côte d'Ivoire. For a better understanding of its functioning, we proposed to estimate its main tributaries (Bété and Djibi) soil loss during 2016 and 2017 as part of our research activities in the lagoon watershed in order to evaluate its vulnerability face to soil loss. The methodological approach is based on USLE (Universal Soil Loss Equation) incorporated into GIS (Geographic Information Systems). This equation takes into account five key factors: the erosivity of rainfall, the soil erodibility, the topographic factor integrating slope length and steepness, the cover-management factor and the support practice factor. The combination of these factors made it possible to obtain soil loss maps of the lagoon main tributaries. The analysis of them revealed that soil loss varying mostly between 0 and 250 $\mathrm{t\,ha^{-1}\,yr^{-1}}$ in 2016 and 2017. With regard to the two years, the vulnerability of the lagoon face to soil loss is "low" category. In fact, the soil loss class ranging from 0 to 20 $\mathrm{t\,ha^{-1}\,yr^{-1}}$ occupies more than 60 % of the two sub-basins area in 2016. This trend increased in 2017 with equivalent of 71 % of the area. On the over hand, the "very high" vulnerability ranging from 250 to 1050 $\mathrm{t\,ha^{-1}\,yr^{-1}}$, occupied in 2016, only 0.01 % of the area. In 2017, this category of vulnerability increased in intensity, occupying 0.05 % of it. Ultimately, the increasing observed in 2016 and 2017 seems to be related to annual rainfall of respectively 1553 and 2198 mm. The case study of Aghien lagoon, soil loss vulnerability can be improved by taking account a long time series of rainfall and land use data.

1 Introduction

Rain, characterized by its height, intensity and duration, defines potential erosion. Among the factors that modify the expression of climatic aggressiveness, the vegetation cover has by far the main role (variations 1 to 1000), then come the slope (1 to 50), the type of soil (1 to 10) and anti-erosion practices (1 to 10) (Roose and Lelong, 1976; Roose et al., 2014). However, human activities are catalysts that can profoundly alter the structure of soils. Thus, human activities, through agricultural practices, logging, grazing, transhumance or the construction of roads and buildings tend to modify erosion phenomena, often accelerating it considerably (Eblin et al., 2017). In Africa, the phenomenon of soil loss due to water erosion is very widespread, the majority of watersheds being characterized by severe specific degradation exceeding 2000 $\mathrm{t\,km^{-2}\,yr^{-1}}$, which leads to siltation of dam reservoirs at a rate of 125 million meter cubes (Zouagui et al., 2018).

In Côte d'Ivoire, more precisely studies on water erosion carried out in the commune of Attecoubé (Abidjan) by N'Dri et al. (2008) and in the region of Bonoua by Aké et al. (2012) have highlighted evidence of strong soil loss due to water erosion. This sad observation extends to the entire Abidjan city, which is facing strong population growth and rampant urbanization to Aghien lagoon in its outskirts. This source of fresh water has been identified by the State of Côte d'Ivoire to alleviate the drinking water problems that the megalopolis has been face for nearly two decades. Indeed, two main

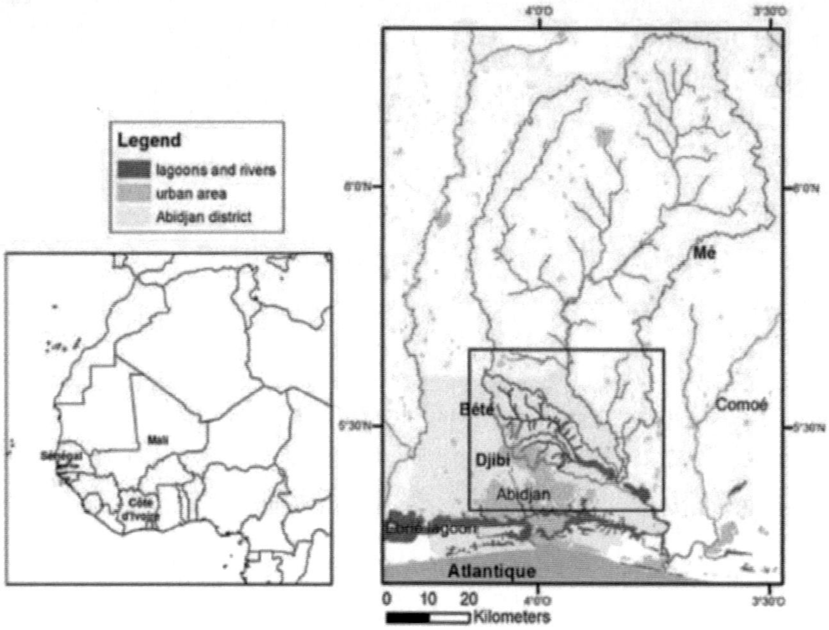

Figure 1. Study area (Ehouman et al., 2019).

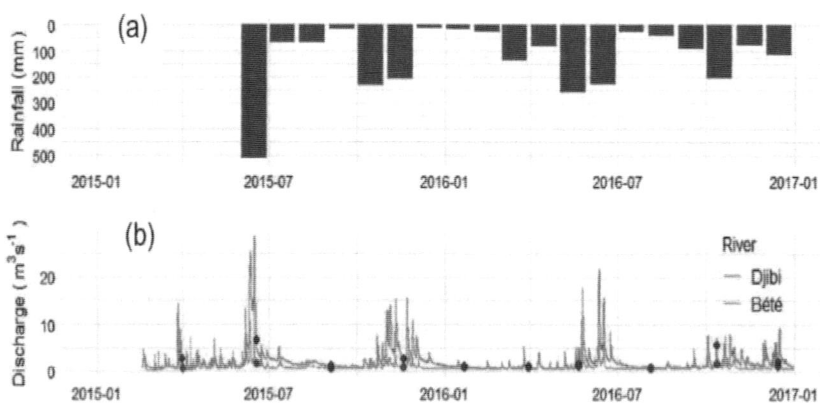

Figure 2. Rainfall **(a)** and Discharge **(b)** relationship in Aghien lagoon watershed (Ehouman et al., 2019).

tributaries (Bété and Djibi) feed this lagoon. Among them, Djibi's watershed is highly urbanized (Fig. 1).

Aghien lagoon watershed is influenced by a humid tropical climate characterized by 4 seasons: a large rainy season (March–July) and a short rainy season (October–November). The annual rainfall was 1553 mm in 2016 and 2198 mm in 2017. The other two periods correspond to more or less dry seasons. The hydrological regime is bimodal in close relation with the rainfall regime. Heavy rainfall occurs in June causing heavy floods, which define the potential water erosion (Fig. 2). According to Soro et al. (2004), Aghien lagoon watershed is located in a sedimentary environment formed mainly by tertiary and quaternary deposits made up of coarse sands, variegated clays, and ferruginous sandstones with iron ore. Since these formations are vulnerable to water erosion, Aghien lagoon watershed could be exposed to silting from soil loss. For a better understanding of how this lagoon works, we set out to estimate the soil loss that would contribute to its silting up due to water erosion from its main tributaries.

2 Material and methods

The material consists of a soil map of the sedimentary watershed of Côte d'Ivoire at a scale of 1 : 200000 from the archive of Vennetier (1973), a digital elevation model (DTM)

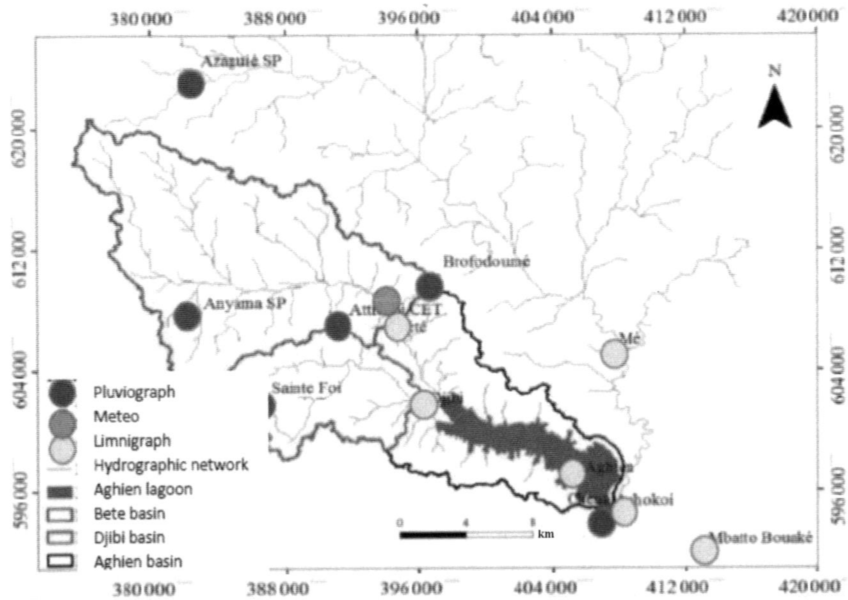

Figure 3. Hydroclimatic measurement network of Aghien lagoon watershed.

and Landsat 8 OLI images in resolution of 30 m for the years 2016 and 2017. The rainfall data come from six rainfallstations installed in the watershed as part of Aghien lagoon project (Fig. 3).

The method for estimating soil loss is based on USLE by taking account five key factors: the erosivity of rainfall, the soil erodibility, the topographic factor integrating slope length and steepness, the cover-management factor and the support practice factor. The USLE equation proposed by Wischmeier and Smith (1978) was the most widely used model in predicting the soil loss (Chadli, 2016). It is described by:

$$A = R \cdot K \cdot LS \cdot P \cdot C \tag{1}$$

where A is the annual soil loss ($\mathrm{t\,ha^{-1}\,yr^{-1}}$). R corresponds to the erosivity of rainfall ($\mathrm{MJ\,mm\,ha^{-1}\,h^{-1}\,yr^{-1}}$). This factor is expressed by the following equation:

$$R = \sum_{n=1}^{12} \left(P_i^2\right) / P. \tag{2}$$

P_i is the monthly average rainfall (mm), P is the annual rainfall (mm) and n is the number of month. K represents the soil erodibility factor ($\mathrm{t\,ha\,h\,ha^{-1}\,MJ^{-1}\,mm^{-1}}$). It is established using the correspondence table of Stone and Hilborn (2012).

LS is the topographic factor integrating slope length (L) and steepness (S) is expressed by:

$$LS = \left(\frac{\lambda}{22.1}\right)^m \left(65.4 \sin\theta^2 + 4.56 \sin\theta + 0.065\right). \tag{3}$$

λ is the slope length (m), θ is the inclination of slope (%) and m is the factor established as a function of the slope.

Table 1. Variation of m as a function of the slope (Payet et al., 2011).

Percentage of slope (%)	Factor m
≥ 5	0.5
$3.5 \leq \theta < 5$	0.4
$1 \leq \theta < 3.5$	0.3
< 1	0.2

Table 1 presents values of m as a function of the slope percentage (Payet et al., 2011).

C corresponds to the cover-management factor whose equation is:

$$C = e^{-\left(\frac{\alpha \mathrm{NDVI}}{\beta - \mathrm{NDVI}}\right)}. \tag{4}$$

α and β are determining the shape of the NDVI curve ($\alpha = 2$ and $\beta = 1$). When the vegetation cover is continuous, erosion and runoff remain very low despite the aggressiveness of tropical rains.

P corresponds to the support practice factor. The values of P are between 0 and 1, in which the value 0 represents a very good environment of resistance to erosion of human origin and the value 1 shows an absence of anti-erosion practice. See the soil conservation table recommended by Meliho et al. (2016).

All these factors cited are combined into GIS to develop soil loss maps. Spatial analysis, data combination, editing of thematic maps and setting up of GIS were carried out according to Fig. 4.

Table 2. Erosion values by vegetation cover in Côte d'Ivoire (Roose and Lelong, 1976).

Rainfall stations in Côte d'Ivoire	Erosion ($t\,ha^{-1}\,yr^{-1}$)		
	Natural environment	Bare ground	Culture
Adiopodoumé (forest)	0.03–1	60–570	0.1–90
Bouaké (dense savannah)	0.01–0.2	18–30	0.1–26
Korhogo (clear savannah)	0.1–0.2	8–9	–

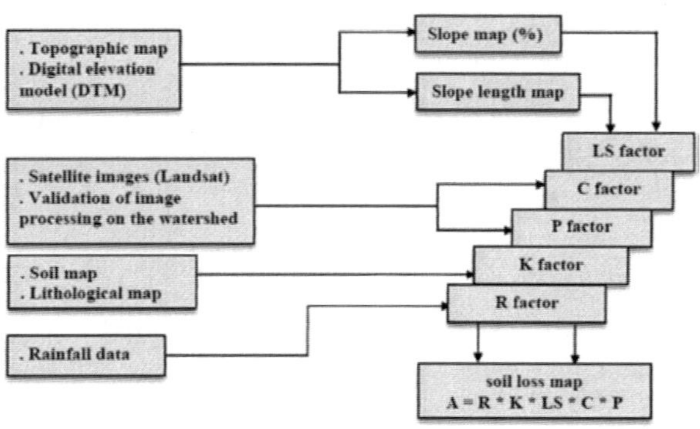

Figure 4. Methodology for estimating soil loss (Meliho et al., 2016).

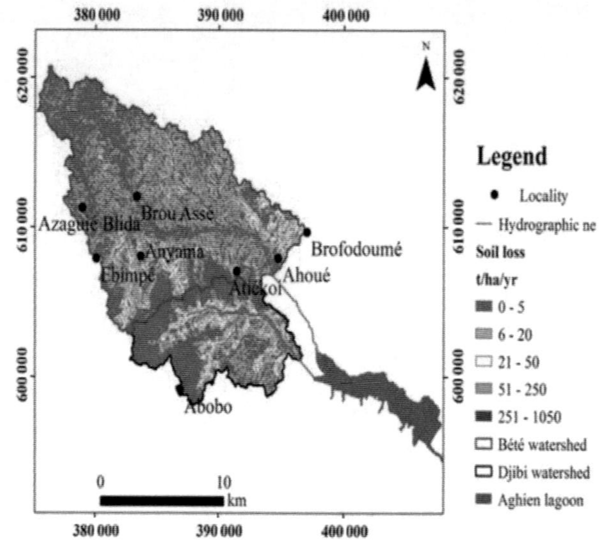

Figure 5. Soil loss in Aghien lagoon tributaries watershed (2016).

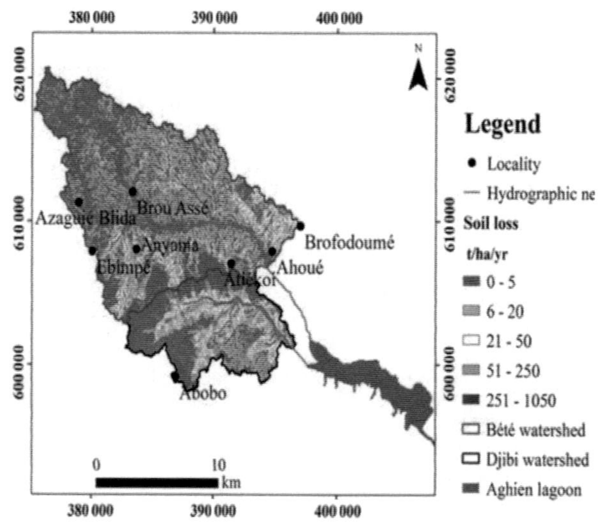

Figure 6. Soil loss in Aghien lagoon tributaries watershed (2017).

3 Results and discussion

3.1 Results

3.1.1 Estimation of Aghien lagoon tributaries' watershed soil loss

Figures 5 and 6 are the results of thematic maps of the main factors involved in soil loss. Indeed, these maps show the tributaries' watershed of Aghien lagoon (see Fig. 3). Soil loss is varying essentially between 0 and 250 $t\,ha^{-1}\,yr^{-1}$ in 2016. It can be seen that the low soil loss classes (0–20 $t\,ha^{-1}\,yr^{-1}$) are very widespread in the watershed. On the contrary, the classes of high soil loss (50–1050 $t\,ha^{-1}\,yr^{-1}$) are not widespread and seems to follow the hydrographic network. In 2017, this observed trend is clearly increasing for high soil loss classes. The very high one goes from 250 to 1050 $t\,ha^{-1}\,yr^{-1}$.

3.1.2 Vulnerability of Aghien lagoon face to its tributaries' watershed soil loss

Vulnerability of the lagoon face to its tributaries' watershed soil loss is estimated by the area percentages. Percentages of soil loss area make it possible to assess the soil loss vulnerability ranging from "very low" to "very high" (Tables 3 and 4). With regard to the two years (2016 and 2017), the dominant vulnerability of the lagoon in connection with water erosion is "low" category. In fact, in 2016, the soil loss class ranging from 6 to 20 $t\,ha^{-1}\,yr^{-1}$ occupies more than 60 % of the area of the two sub-basins. This trend increased in 2017 with soil loss area of 18 210 ha, equivalent of 71 % of the two tributaries area.

The "very low" vulnerability comes second with area percentages of 27 % and 24 % respectively in 2016 and 2017. The "very high" vulnerability ranging from 251 to 1050 $t\,ha^{-1}\,yr^{-1}$ in 2016 occupied only 0.01 % of the area.

Table 3. Vulnerability of Aghien lagoon to soil loss in 2016.

Class of soil loss ($t\,ha^{-1}\,yr^{-1}$)	Area (ha)	Area (%)	Vulnerability of the lagoon
0–5	7071	27.55	Very low
6–20	15 552	60.59	low
21–50	2650	10.32	medium
51–250	393	1.53	high
251–1050	3	0.01	Very high

Table 4. Vulnerability of Aghien lagoon to soil loss in 2017.

Class of soil loss ($t\,ha^{-1}\,yr^{-1}$)	Area (ha)	Area (%)	Vulnerability of the lagoon
0–5	6228	24.27	Very low
6–20	18 210	70.95	low
21–50	971	3.78	medium
51–250	244	0.95	high
251–1050	12	0.05	Very high

In 2017, this category of vulnerability increased in intensity, occupying 0.05 % of the sub-basins.

3.2 Discussion

The soil loss maps show that the areas where soil losses vulnerability are not dominant in the lagoon watershed (very low-to-low category), are very widespread and cover large areas (88 % in 2016 and 95 % in 2017). These percentages are corresponding respectively to soil loss of 0 to 20 $t\,ha^{-1}\,yr^{-1}$ during the two years. These values can be classified in the same category as those obtained by Roose and Lelong (1976) in the same climatic zone in Côte d'Ivoire (see Table 2, the case of Adiopodoumé). They can be explained by the presence of abundant vegetation and gentle slopes (Roose et al., 2014). However, highly vulnerable areas occupy an infinite part with soil loss going to 251 to 1050 $t\,ha^{-1}\,yr^{-1}$ covering respectively area percentages of 0.01 % and 0.05 % respectively in 2016 and 2017. These high values of soil loss and low percentages of area can be due to the topographic factor integrating slope length along the hydrographic network (Payet et al., 2011). Overall, the increase in the risk of erosion between 2016 and 2017 in the watershed could be explained by rampant urbanization (see Fig. 1) and overgrazing leading to the destruction of the vegetation in progress (Aké et al., 2012). In addition, this degradation exposes the watershed to very strong runoff and very little infiltration (El hafid et Akdim, 2018). Soil displaced by erosion carries nutrients, pesticides and other chemicals (Benkaci et al., 2018). In the case of Aghien lagoon watershed, most of these harmful products can be carried by the two rivers to the lagoon (Ehouman et al., 2019). The assessment of soil loss gave high soil loss from 50–250, these values assigned to bare ground and culture (Roose and Lelong, 1976), are between the tolerance threshold which is (1–12 $t\,ha^{-1}\,yr^{-1}$) set by Kouadio et al. (2007). The deposition of soil loss and substances in the lagoon could contribute in the long term to its pollution and siltation, thus causing a drastic reduction of surface water (Meliho et al., 2016; Zouagui et al., 2018).

4 Conclusion

The aim of this paper was to study the soil loss vulnerability in Aghien lagoon watershed. The combination of the factors of the universal soil loss equation made it possible to obtain soil loss maps of the lagoon tributaries' watershed (Bété and Djibi). The results showed that soil loss varying mostly between 0 and 250 $t\,ha^{-1}\,yr^{-1}$. With regard to the two years (2016 and 2017), the dominant vulnerability of the lagoon in connection face to water erosion is "low" category. In fact, the soil loss class ranging from 0 to 20 $t\,ha^{-1}\,yr^{-1}$ occupies more than 60 % of the two sub-basins area. This trend increased in 2017 with soil loss of 71 %. The "very low" vulnerability comes second with area percentages of 27 % and 24 % respectively in 2016 and 2017. The "very high" vulnerability ranging from 250 to 1050 $t\,ha^{-1}\,yr^{-1}$ in 2016 occupied only 0.01 % of the area. In 2017, this category of vulnerability increased in intensity, occupying 0.05 % of the sub-basins. Ultimately, this dynamic of water erosion correlated with strong anthropogenic pollution could compromise the use of Aghien lagoon as a freshwater reserve for Abidjan city. This work is a contribution to understand the functioning of Aghien lagoon. Applying the universal soil loss equation over a long series of rainfall and land use data could improve the results of this work.

Code availability. The GIS software used is Quantum GIS. The code is publicly accessible at https://www.qgis.org/fr/site/forusers/download.html (QGIS, 2021).

Data availability. Data is available via the link: https://dataverse.ird.fr/dataverse/root?q=Aghien (Dataverse, 2021).

Author contributions. AD rote the manuscript with the data collected and compilated under quantum GIS software by ESK. DDN, BK and LDG read and correct the document. LS and JLP are our research partners within the framework of the Aghien lagoon project.

Competing interests. The contact author has declared that neither they nor their co-authors have any competing interests.

Special issue statement. This article is part of the special issue "Hydrology of Large River Basins of Africa". It is a result of the 4th International Conference on the "Hydrology of the Great Rivers of Africa", Cotonou, Benin, 13–20 November 2021.

Acknowledgements. Our thanks go to the "Debt Reduction and Development Contract" (C2D) between France and Côte d'Ivoire which funded the research activities of "Aghien lagoon" project through the partnership PReSeD-CI. This partnership was a very good collaboration between researchers from Université Nangui Abrogoua (UNA), in particular those from the "Laboratoire de Géosciences et Environnement" (LGE) and researchers from the French Institute for Research and Development (IRD).

Financial support. This research has been supported by the Contrat Desendettement Developpement (C2D) (Lot no. 4 du projet AMRUGE-CI – Appui à la Modernisation et à la Réforme des Universités et Grandes Ecoles de Côte d'Ivoire).

References

Aké, G. E., Kouadio, B. H., Adja, M. G., Ettien, J. B., Effebi K. R., and Biémi, J.: Cartographie de la vulnérabilité multifactorielle à l'érosion hydrique des sols de la région de Bonoua (Sud-Est de la Côte d'Ivoire), Physio. Geo., 6, 1–42, https://doi.org/10.4000/physio-geo.2285, 2012.

Benkaci, S., Abir, D., Oumellal, A., and Boualem, R. B.: Modélisation de l'érosion du bassin haut et moyen Cheliff par l'application Model builder sur ArcGis, J. Mater. Eng. Struct., 5, 81–93, 2018.

Chadli, K.: Estimation of soil loss using RUSLE model for Sebou watershed (Morocco), Model. Earth Syst. Environ., 2, 2–10, https://doi.org/10.1007/s40808-016-0105-y, 2016.

Dataverse: Des données ouvertes pour une science durable au Sud, available at: https://dataverse.ird.fr/dataverse/root?q=Aghien, last access: 19 October 2021.

Eblin, S. G., Yao, A. B., Anoh, K. A., and Soro, N.: Cartographie de la vulnérabilité multifactorielle aux risques d'érosion hydrique des sols de la région d'Adiaké, Sud-Est côtier de la Côte d'Ivoire, Rev. Ivoir. Sci. Technol., 30, 197–216, 2017.

Ehouman, K. S., Koffi, K. T., Perrin, J. L., Séguis, L., Guilliod, M., Droh, L. G., and Kamagaté, B.: Hydrological and water quality assessment of the Aghien Lagoon hydrosystem (Abidjan, Côte d'Ivoire), Hydrolog. Sci. J., 64, 1893–1908, https://doi.org/10.1080/02626667.2019.1672875, 2019.

El hafid, D. and Akdim., B.: Quantification De L'érosion Hydrique En Utilisant Le Modèle Rusle Et Déposition Intégrée Dans Un Sig, Cas Du Bassin Versant De L'oued Isly (Maroc Oriental), Eur. Scient. J., 14, 1857–7881, https://doi.org/10.19044/esj.2018.v14n5p373, 2018.

Kouadio, B. H., Kouamé, K. F., Saley, M., Biemi, J., and Traoré, I.: Insécurité climatique et géorisques en Côte d'Ivoire: étude du risque d'érosion hydrique des sols dans la région semimontagneuse de Man (Ouest de la Côte d'Ivoire), Sécheresse, 18, 29–37, 2007.

Meliho, M., Khattabi, A., Mhammdi, N., and Zhang, H.: Cartographie Des Risques De L'erosion Hydrique Par L'equation Universelle Revisee Des Pertes En Sols, La Teledetection Et Les Sig Dans Le Bassin Versant De L'ourika (Haut Atlas, Maroc), Eur. Scient. J., 12, 32, https://doi.org/10.19044/esj.2016.v12n32p277, 2016.

N'Dri, B. E., N'Go, Y. A., Kadio, H. N., Ouattara, A., Touré, B., and Biémi, J.: Effect of soil slope and cover on runoff and rate soil loss from experimental plots in area of Attécoubé, Eur. J. Scient. Res., 21, 459–470, 2008.

Payet, E., Dumas, P., and Pennober, G.: Modélisation de l'érosion hydrique des sols sur un bassin versant du sud-ouest de Madagascar, le Fiherenana, Vertigo, 11, 3, https://doi.org/10.4000/vertigo.12591, 2011.

QGIS: Un logiciel open source utilisant la licence GNU General Public License ce qui signifie que son code source peut être téléchargé via des archives tar ou un dépôt git at: https://www.qgis.org/fr/site/forusers/download.html, last access: 20 October 2021.

Roose, E., Sabir, M., and Laouina, A.: Gestion durable des eaux et des sols au Maroc, OpenEdition Books, IRD Éditions, p. 343, https://doi.org/10.4000/books.irdeditions.294, 2014.

Roose, E. J. and Lelong, F.: Les facteurs de l'érosion hydrique en Afrique tropical, in: Etudes sur petites parcelles expérimentales de sol, 21, vol. xviii, fasc. 4, Revue de géographie physique et de géologie Dynamique, 18, 365–374, 1976.

Soro, N.; Lasm, T., Kouadio, B. H., Soro, G., and Ahoussi, K. E.: Variabilité du régime pluviométrique du Sud de la Côte d'Ivoire et son impact sur alimentation de la nappe d'Abidjan, Sud Sci. Technol., 14, 30–40, 2004.

Stone, R. P. and Hilborn, D,: Universal Soil Loss Equation (USLE), Agdex#: 572/751, 2012.

Vennetier, P.: Le milieu naturel de la Côte d'Ivoire, Les Cahiers d'Outre-Mer Année 1973 26-104, Les Cahiers d'Outre-Mer, p. 440, 1973.

Wischmeier, V. H. and Smith, D. D.: Predicting rainfall erosion losses- a guide to conservation planning, Agriculture Handbook No. 282, United States Department of Agriculture in cooperation with Purdue Agricultural Experiment Station, Washington, 1978.

Zouagui, A., Sabir, M., Naimi, M., and Benmansour, M.: Modélisation du risque d'érosion hydrique par l'équation universelle des pertes en terre dans le rif occidental: cas du bassin versant de Moulay Bouchta (Maroc), Eur. Scient. J., 14, 524–544, https://doi.org/10.19044/esj.2018.v14n3p524, 2018.

Proc. IAHS, 384, 127–131, 2021
https://doi.org/10.5194/piahs-384-127-2021

Open Access
PIAHS

Hydrology of Large River Basins of Africa

Water Erosion in the Moulouya Watershed and its Impact on Dams' Siltation (Eastern Morocco)

Abdelkader Sbai[1], Omar Mouadili[1], Mohamed Hlal[1], Khadija Benrbia[1], Fatima Zahra Mazari[2], Mostafa Bouabdallah[1], and Abdelouahid Saidi[3]

[1]Department of Geography, Mohammed Ist University, 60000, Oujda, Morocco
[2]English Department, Mohammed Ist University, 60000, Oujda, Morocco
[3]Department of Geography, Polydisciplinary Faculty of Taza, Taza, Morocco

Correspondence: Abdelkader Sbai (sbaiabdelkader59@gmail.com)

Published: 16 November 2021

Abstract. The Moulouya watershed presents high risks of water erosion. The dams built on this river lose each year, by siltation, a volume of water of about $10\,\mathrm{M\,m^3}$ (Lahlou, 1994). This worrying rate shows that water erosion can hinder the socio-economic development of the basin if soil protection and siltation measures are not taken. This study defines the areas most threatened by water erosion to protect them as a priority. Using a GIS and the Universal Earth Loss Equation (USLE), we mapped soil losses. The adopted model determines soil loss from erosivity of rainfall, soil erodibility, inclination and slope length, soil cover, and erosion control practices. The possible origins of the deposits alluviating the dams are related to the products of alteration of the Secondary's carbonate formations and the Tertiary and Quaternary formations. The modalities of the transport of deposits of banks of the network of Moulouya are mainly by a suspension. The specific degradation in the Moulouya watershed is estimated at $214\,\mathrm{t/km^2/year}$. Precipitation intensity, steep slopes and deforestation are responsible for these losses.

1 Introduction

The degradation of natural resources by water erosion is one of the worrying risks that threaten human stability and is difficult to reduce. Moreover, it has been greatly exacerbated in recent years by the threat of the current dynamics, which could directly threaten food security and the population's health.

Water erosion in the Moulouya watershed has led to profound changes in various economic, social and environmental fields. It has led to the emergence of problems that can only be alleviated by integrated measures ensuring decent life for the population and preserving natural resources and sustainability for future generations. It presents a high risk and causes siltation of dams and significant losses of water volumes.

Therefore, the main issue of this research centres on the risk of water erosion, characterized by sheet flow, concentrated flow and gullying. All slopes are marked by erosion, especially sheet erosion which is responsible for the loss of significant amounts of soil and their deposition in dams. This erosion is governed by the nature of the fragile geological formations, especially the superficial quaternary formations, the nature of the sandy loam soil, the steep degree of the slopes, the slopes' dry exposure, and the type of land use (cultivation and overgrazing).

The main objective of this study is to assess the degree of soil degradation and highlight the degradation processes and systems concerning Man, which is a factor in the current imbalance of the environment. The study also aims to highlight the priority of protecting the areas most threatened by water erosion.

2 The general geographical framework

The Moulouya wadi is 520 km long; it rises in the High Atlas at an altitude of 2000 m and drains the largest watershed in Morocco ($57\,500\,\mathrm{km^2}$). This basin located in the Orien-

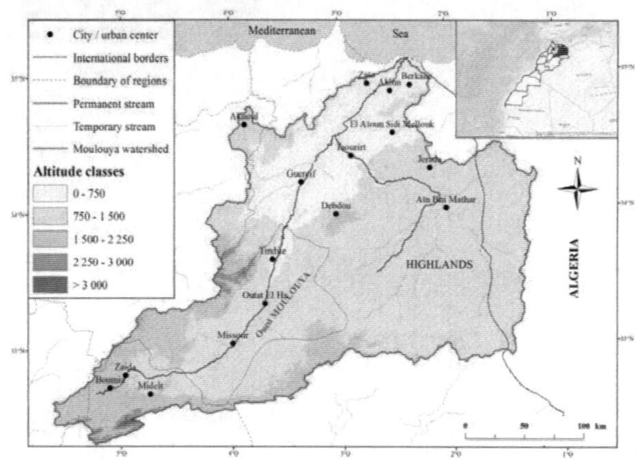

Figure 1. Location Map and the Moulouya Basin DEM (created by us based on layers of ASTER, GDEM, United States Geological Survey and https://www.openstreetmap.org/, last access: March 2021).

tal Region of Morocco (Fig. 1) presents a great diversity of morphological, climatic and socio-economic characteristics.

2.1 A massive and rugged relief

Young mountainous areas of high altitude (the eastern Rif, folded Middle Atlas, Beni Bou Yahi, Beni Snassen, High Atlas) and high plateaus (Highlands, Jerada chain) surround wide plains: Middle Moulouya, Guercif basin, Taourirt-El Aïoun corridor, Mediterranean plains. The mountain ranges inside the Moulouya basin are quite high and peak at 1839 m for the Beni Bou Yahi, 1535 m for the Beni Snassen, and 1726 m for the Jerada range. 18 % of the watershed exceeds 1500 m a.s.l., and 11 % is less than 500 m. The Highlands go west to Moulouya. These vast entablatures of sedimentary rocks rest on an old base (Oran Meseta). The plains of Moulouya are large basins filled with detrital deposits.

The Middle and High Atlas and the mountains located inside the basin are predominantly limestones (especially Lias and Upper Jurassic). Sedimentary rocks (limestones, marls, sandstones and conglomerates) occupy 97 % of the watershed's surface, while magmatic and metamorphic rocks cover only 3 %.

2.2 A Mediterranean climate with differentiated seasons

The Moulouya basin is a particularly arid region. It is located far from the humidifying and softening Atlantic influences and behind the mountainous screen formed by the Atlas and the Rif. Consequently, the climate becomes increasingly arid as one heads south. It is characterized by hot, dry summers and cool winters, with limited precipitations to winter periods. Total rainfall reaches more than 400 mm on average per year in the North and mountain areas, while in the South, it varies between 200 and 300 mm per year (Table 1).

2.3 Diverse vegetation

The Moulouya basin has a natural forest whose area was estimated in 2008 at 2.5 million hectares, or about 29 % of the total area of national forests (Royaume du Maroc, 2015). The alfa covers most of these forests: 2.2 million hectares. The aridity explains the generally poor nature of the plant cover, which comes in two main forms. On the one hand, natural forest formations, more or less degraded, are located on certain relatively rainy mountain slopes where the holm oak and thuja are largely dominated. On the other, steppe formations are extending mainly through the arid spaces of the center and the South of the region where alfa and sagebrush dominate.

Population growth and the high demand for cultivable land have led to the clearing of rangelands, which are unsuitable for agriculture.

2.4 An important hydrographic network with disturbed flow

Oued Moulouya is by far the main river in the Oriental region. Its average flow is estimated at 41 m^3/s at Mechra-Hamadi (85 m^3/s in April and 7.4 m^3/s in August, a month of low water). It has an even more irregular regime as it crosses more humid mountain regions and vast semi-desert plains and plateaus. The flows can exceed 6000 m^3/s in the Lower Moulouya River. Three large dams have been built in this basin to provide for the region's needs in terms of drinking water, irrigation and electricity production.

- *Machraâ Hammadi dam* built in 1956 on the Moulouya River with a height of 56 m; it has a holding volume of 6.6 million m^3 (42 million m^3 at the start) currently.
- *Mohammed V dam* commissioned in 1967 with a height of 64 m, is the largest water reservoir in the region, with currently a reservoir volume of 411 million m^3 (760 million m^3 at the start).
- *Hassan II dam* holding volume varies between 109 and million m^3 with a flow of 710 m^3/s during the rainy period and 450 m^3/s during summer. It was put in the water in 2000.

2.5 Population and land use

The population of the Moulouya basin exceeds 800 000 inhabitants according to the General Census of Population and Housing of 2014, while the total area of irrigated land is around 110 784 hectares, or 9 % of the total area of agricultural land (MI.DGCL, 2015). The northern part, composed mainly of irrigated land, constitutes the first agricultural area of the region (Trifa plain). The other irrigated lands are located in the other plains of the region and near-certain urban agglomerations (Guercif, Taourirt...). As for the agricultural land in Bour, it extends over large areas exceed-

Table 1. Monthly variations in rainfall: 1963–1999 (Météorologie nationale, Oujda, Maroc).

	J	F	M	A	M	J	J	A	S	O	N	D	Total
Taforalt	57	65	75	70	55	18	2	3	21	37	53	59	512
Midelt	17	18	27	33	29	12	4	4	18	19	23	16	220

ing 635 000 hectares. In the arid regions, livestock on natural rangelands constitutes the only possible source of agricultural income for the population of these regions. Cultivation techniques are an important anthropogenic factor that can either reduce or accelerate erosion. These various factors lead to intense erosion. The contributions of solid matter arriving in the dams were estimated at more than 10 M m^3/year (Lahlou, 1994).

3 Tools and working methods

First, the various studies related to the subject were consulted (Sadiki, 2005; Sbai and Ghzal, 2007). Subsequently, documents and data to describe the basin were used:

- The topographic maps at the 250 000-scale covering the basin, the geological maps (Oujda at 1/500 000), the vegetation map of the Waters and Forests (1/132 000), then the soil map of Morocco (1/1 000 000), in order to produce thematic maps;
- The numerical model of precipitation and temperature "WorldClim" was also used in order to calculate the precipitation aggressiveness index;
- "Landsat" and "Sentinel" satellite images, in order to study the degradation of plant cover and land use;
- The "Alaska" Digital Elevation Model, in order to study the topography of the basin;
- The precipitation data for several stations were used to know the rainfall and thermal regimes of the basin.

A quantitative approach was adopted to estimate soil loss using the Empirical Soil Loss Model to calculate the extent of soil degradation. Field measurements using the RAMP rain simulator have also been made (Roose, 1996). The universal soil loss equation (Wischmeier and Smith, 1978) is written as follows:

- $A = R \times K \times LS \times C \times P$
- A: The average annual soil loss (t/ha/year),
- R: the rainfall erosivity factor,
- K: the soil erodibility factor,
- LS: index combining the effect of the slope length factor and the slope steepness factor

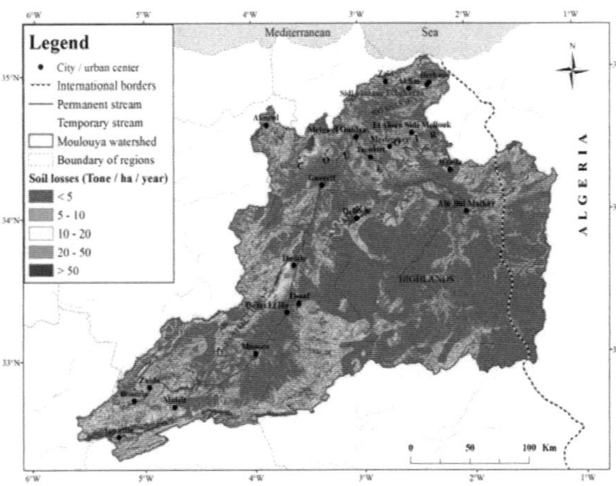

Figure 2. Soil erosion map (created by us based on layers of https://www.openstreetmap.org/, last access: March 2021).

- C: the cover management factor,
- P: the supporting practice factor.

Climate erosivity is represented by R and can be estimated by the rainfall intensity and amounts data. The soil erodibility monograph can be used to predict the K value. The L and S factors characterize the topography and hydrology effects on soil loss. For direct USLE applications, a combined LS factor was evaluated for each land cell. Land use and management are represented by P and can, with some difficulty, be inferred using remote sensing combined with ground-trusting (Sbai and Mouadili, 2021).

4 Results and discussion

The forms of current dynamics in the Moulouya basin are dominated by water erosion. This erosion is responsible for the loss of soil significant quantities and their deposit in dams. However, it is also governed by the nature of the geological formations and, especially, by the quaternary surface formations, the nature of the soil, the degree of slope, the orientation of the slopes, and the type of land use.

After calculating the various factors adopted in the universal soil loss equation, we calculated the areas of eroded land (Fig. 2).

In Fig. 2, we have tried to extract the amount of soil loss by dividing it into five categories according to the nature and morphological structure of the study area, and then we have

Table 2. Quantities and areas of soil losses (created from the Fig. 2).

Class	Soil loss (tones/ha/year)	Quantity in tones/year	%	Area in km^2	%
Very weak	< 5	751 086	8	39 398	65
Low	5–10	695 067	8	7968	13
Middle	10–20	934 785	10	5639	9
Strong	20–50	1 515 485	17	4279	7
Very strong	> 50	5 145 537	57	3119	5
Total		9 041 960	100	60 403	100

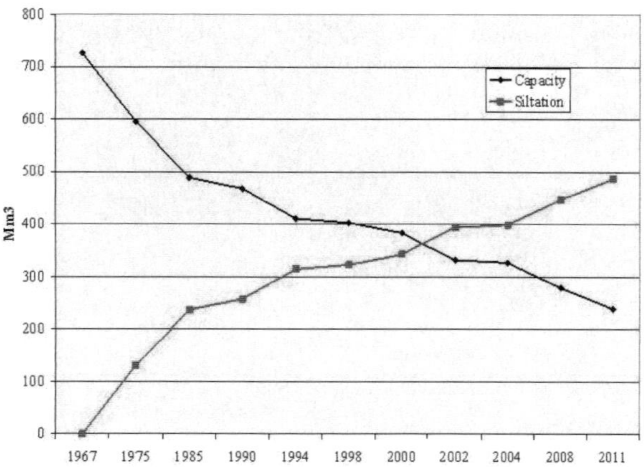

Figure 3. Photo 1: Extension of agricultural space and its removal by stone removal, which contributes to erosion (Evolution of Mohammed V dam's siltation (Agence du Bassin Hydraulique de la Moulouya, Oujda, Maroc)).

Figure 4. Evolution of Mohammed V dam's siltation (Extension of agricultural space and its removal by stone removal, which contributes to erosion (1 August 2017, national road no 19. Plain of Tafrata)).

drawn up a table that indicates the quantity of soil loss categories (Table 2).

From the diagnosis of the manifestations of erosion and its quantification by modeling and by direct measurements in the field, we discovered that the degradation had reached complex and advanced levels.

The average soil loss is 3.86 mm thick per year, or 37.64 tonnes/ha/year. Direct field measurements show that the material removed varies from 0.02 to 29.05 tons per hectare. By generalizing these values to the annual averages of the stations in the study area, we obtain an average soil loss of 39.18 tonnes/ha/year.

The areas with high erosion represent 7398 km^2 out of a total of 60 402 km^2; it is estimated at more than 6 660 000 tons per year. 74 % of the materials removed are distributed over only 12 % of the basin area. The average specific soil degradation in the basin is estimated at 13.5 tons/ha/year. Erosion values are relatively high in the Northwest (Msoun basin), in the Cretaceous marly limestones in the Southeast of Outat El Haj, in Hight Moulouya (Midelt region) and in some plains in the center. The quantities of eroded soil (less than 10 tons/ha/year) represent 16 % of the area of the basin. They are estimated at more than 1 446 153 tons per year spreading over 78 % of the basin.

The excessive and violent precipitations, the decline and degradation of the vegetation cover as well as the overexploitation of pastures and agriculture contribute to increase vulnerability and accelerate the development of the dynamics of water erosion. The soil is constantly deteriorating due to the rocky nature of the region and the natural conditions of drought, compounded to inadequate systems of human exploitation and degradation. Recent changes in land use have modified the spatial distribution of sediment feeding areas (Sbai and Ghzal, 2007).

Field observation clearly shows that agricultural work has extremely low production. Farmers are trying to gain more land at the expense of plant cover, which plays an important role in providing fodder for livestock and protecting the soil from water and wind erosion. This extension in inappropriate places causes erosion (Photo 1).

What is interesting is that the forms of degradation prevail in the absence of a global approach considering the needs and requirements of the rural population, whose absence in development projects and programs constitutes a major turning point for natural components and their importance for the preservation of natural resources and their sustainability.

The intense erosion processes in the Moulouya basin explain the high siltation rates of the Mechrâa Hammadi (1958) and Mohamed V (1967) dams and affect the irrigation of the Lower Moulouya river and the supply of drinking water for the main cities and urban centres of North-eastern Morocco. The evolution of siltation of the Mohamed V dam since 1967 shows that the reservoir has been subject to increasing sedimentation at variable speeds and an average rate of 10.8 M m^3/year, representing a capacity loss of 67 %. Figure 3 illustrates the temporal evolution of siltation.

5 Conclusions

The Moulouya basin experiences strong morphodynamical activity. Runoff develops on all slope classes and is intricately linked to the nature of the inappropriate exploitation of the fragile natural environment. Water erosion leads to the gullying and lateral extension of the valleys, which contributes to the removal of significant parts of the surface.

Thus, in this study, we have tried to evaluate the amount of soil loss by the universal Wischmeier equation and carry out direct measurements on experimental plots to show the magnitude of the current erosive dynamics. Field observations show that the transport of deposits by the various rivers is essentially due to suspension. This mode of transport can affect particles smaller than 400 µm. The remaining fraction is transported by saltation. These sediments are trapped in dams and explain the significant loss of water volumes.

It is necessary to seek this degradation's prospects and follow its evolution concerning local resources and the human factor, which is an essential element of the environment current imbalance.

Data availability. The data are available upon request (sbaiabdelkader59@gmail.com).

Author contributions. AS, OM, MB and AS carried out the rain simulation experiments in the field. OM, KB and MH took care of the data processing and the production of the maps. AS wrote the text with contributions from all the co-authors. FZM revised the text from a language point of view.

Competing interests. The contact author has declared that neither they nor their co-authors have any competing interests.

Disclaimer. Publisher's note: Copernicus Publications remains neutral with regard to jurisdictional claims in published maps and institutional affiliations.

Special issue statement. This article is part of the special issue "Hydrology of Large River Basins of Africa". It is a result of the 4th International Conference on the "Hydrology of the Great Rivers of Africa", Cotonou, Benin, 13–20 November 2021.

Acknowledgements. The authors would like to express their sincere and profound recognition to Mr. Mohamed DELLAL, Professor Emeritus of English at Mohammed First University in Oujda for his proofreading.

References

Lahlou, A.: Envasement des barrages au Maroc. Collection "Sciences et techniques" de la Faculté des Lettres et des Sciences Humaines: Série Thèses et Mémoires, Wallada, 1994.

Royaume du Maroc: Ministère de l'Intérieur, Direction Générale des Collectivités Locales (2015) – La Région de l'Oriental, Monographie générale, 66 pp., 2015.

Roose, E.: Land Husbandry: Components and Strategy, FAO Soils, Bulletin 70, Food and Agriculture organization of United Nations, Rome, 1996.

Sadiki, A.: Estimation des taux d'érosion et de l'état de dégradation des sols dans le bassin versant de Boussouab, Maroc Nord Oriental: Application du modèle empirique (USLE), de la technique du radio-isotope ^{137}Cs et de la susceptibilité magnétique " Thèse pour l'obtention du grade de Docteur d'Etat en géologie, l'Université Mohammed Premier, Faculté des Sciences, Oujda, 2005.

Sbai, A. et Ghzal, M.: Dégradation des parcours et érosion dans le couloir d'El Aioun - Taourirt (Maroc nord-est), Colloque National: Taourirt – Oued Za – la Kasbah. Histoire, espace et développement, 8-9 Mai 2007, Province de Taourirt, Travaux du Colloque, Collection Etudes & Séminaires, (en langue arabe), 169–182, 2007.

Sbai, A.: et Mouadili, O.: Le risque d'érosion hydrique entre fragilité des équilibres environnementaux et perspectives de durabilité: Cas du bassin versant d'oued El Abed, Maroc, Revue marocaine des Sciences Agronomiques et Vétérinaires, Vol. 9 No 4 (2021): (Décembre 2021), Soumis: avr 5, 2021, Publiée: déc 15, 2021. https://www.agrimaroc.org/index.php/Actes_IAVH2/issue/view/83, 2021.

Wischmeier, W. H. and Smith, D. D.: Rainfall energy and its relatioship to soil loss, Trans. AGU 39, 285–291, 1958.

Proc. IAHS, 384, 133–139, 2021
https://doi.org/10.5194/piahs-384-133-2021

Open Access PIAHS

Multi-criteria approaches to identify the shoreline retreat downstream of dams: the North African case

Oula Amrouni[1] **and Gil Mahé**[2]

[1]National Institute of Marine Sciences and Technologies, Laboratory of Marine Environment, University of Carthage, Tunis, Tunisia
[2]HSM, UMR 5151, CNRS, Université Montpellier, IRD, 300, Avenue du Professeur Emile Jeanbrau, 34090 Montpellier, France

Correspondence: Oula Amrouni (oula.amrouni@instm.rnrt.tn)

Published: 16 November 2021

Abstract. Northern African beaches are among the most vulnerable areas under the extreme climate change hazard. Mainly sedimentary low-lying platform, the coasts are supplied by terrestrial yields, which are increasingly interrupted by dams. Unfortunately, the sediment fluxes are rarely measured and monitored, so that it is quite impossible today to assess the contribution of continental sediments to the coast and its variability. The aim of our study is to determine the sampling protocol of delta sedimentation plain and nearshore seabed for better understanding of the anthropogenic driver in contrast to climate change. We adopt a multi-criteria analysis based both on the geomorphologic feature and the historic evolution from the River to the littoral plain. The shoreline evolution reveals an alarming retreat trend reaching $-20\,\mathrm{m} \pm 0.15\,\mathrm{m\,yr^{-1}}$ after the human-induced change where $\sim 50\,\%$ of sediment discharge has been trapped upstream the dam, including quite all the coarse material, like sand. The shoreline retreat and the decreasing sediment rate of fluvial flow are all due to the dam construction.

1 Introduction

Deltaic coastal plains are among the most productive environment. It hosts complex and rich ecosystems such as wetlands, evaporate flats-sabkha- salt marsh and are a natural inlet barrier against the flood and storminess events. Modern-River system is defined by geologic settings with associated environmental processes (Coleman and Roberts, 1989). Mostly composed by sedimentary fine-grained landforms, deltaic coast are shaped, reworked and destructed by generated current-wave (Anthony, 2015). Beach equilibrium sediment balance depends on interactions between waves and fluvial discharge at the river mouth. Sediment discharge responds both to the impact of human activities and to changes in climate. Since the mid-century, human influences on the sediment flux to the coast have a significant effect on floodplain (Syvitski and Kettner, 2011), a large coastal land loss and delta-plain retrogradation (Benmoussa et al., 2018, 2019). Dams construction and maritime infrastructures have both intercepted the sediment supply and disturbed the sediment movements along the longshore pathways. The Maghreb countries have launched since the 60s a vast program of dam's constructions to ensure the water supply for their developing populations and economies. Today the cumulated water storage capacity of the dams in Maghreb semi-arid region is of 28 billion cubic meters. But the dams also store large amounts of sediments. Unfortunately, the sediment fluxes are poorly monitored in most countries and thus it has been impossible up to now to quantify the amount of sediment that do not reach the sea anymore since the dams building. In the upcoming few decades, under the most plausible climatic and socio-economic scenario projection (SSP2-AVG), the North African countries will undergo water deficits ranging between 45 % to 90 % of their current water budget in 2050 (Mazzoni et al., 2018). Therefore, North African shoreline follows severe retreat trend since the last mid-century period and is perceived as vulnerable to the impacts of the natural hazard (Hzami et al., 2021). The main vulnerability drivers are rainfall fluctuation, sea level rise, and the increasing human impact. The combined climatic and human drivers will rapidly deteriorate the coastal aquifer, wetland ecosystem, agricultural, and tourism resources (Amrouni et al., 2019b).

This research aimed to identify the protocol and methods to determine how fluvial sediment discharge varies from source to sink, during the last modern-century under the increasing basin management constraint. This is helpful in understanding present trends in coastal regions, as well as attempting to predict future processes trends. It is worth saying that this type of research is developed in the 3 Maghreb countries, but is most advanced in Tunisia.

2 Physiographic setting

The study area is located in the North Africa coasts in the central Mediterranean – namely Maghreb – on the bay of the Gulf of Tunis, Tunisia. The littoral borders the low-lying upstream watershed of the Medjerda River plain. The Medjerda delta is located at 25 km northeast of Tunis capital city along the semi-arid central Mediterranean coast. The coast is about 35 km long and 20 to 200 m landward wide, microtidal (<2 m range), dissipative to intermediate and wave-dominated types.

The delta valley is characterized by the evolution of several geomorphological entities (Fig. 1a). The sandy spit barrier, lagoons, sabkhas, salt marshes, dune ridge, subtidal bars and river meanders and channels across the coastal plain. The low-lying coastal plain of the Medjerda downstream valleys are filled by Pliocene and marine Quaternary deposits (El Arrim, 1996). The morphology of the embayed beaches is the result of a terrestrial filling coming from Medjerda River since the end of the Holocene transgression (Pleuger et al., 2019). The bay is bordered by two promontories, Cape Gammarth and Cape Ferina, which range in age from Pliocene and the Neotyrrhenian coastal deposits (Paskoff and Sanlaville, 1983). Based on the 1889 bathymetric map (scale 1 : 334 650) for both the northern and central areas of the bay, nearshore setting reveals an irregular inner shelf bottom shape. In the northern ridge, the 10 m isobaths is 600 m far from the beach. In front of the old River mouth towards the North of the bay, it moves 3000 m from the coast. While in the central bay, in front of the new mouth of the Medjerda River, it gets closer by less than 1000 m. The nearshore slope is steeper both in the Ferina and Gammarth Capes, where the −6 m depth does not exceed 100 m from the shoreface (Fig. 1b).

The coast is dominated by the seasonal wind coming from the N to NE direction in Spring and Winter whereby storminess east to SE winds are responsible to the local incident-wave. Landward winds are blowing from the W to NW direction in Winter and sirocco winds from SW direction in summer season. The dominant longshore drift is from the NE to SE direction in Winter and Spring. A secondary SE/NW is generated during Summer from the eastern to southern winds. The average of significant wave height and period are ∼ 1.1 m and 8 s respectively (COADS, 1964). Maximum wave heights during high storms reached ∼ 5 m over the gulf of Tunis (INM, 2019). The relative sea level rise recorded in the Tunis Carthage tide-gauge station is about +1.3 mm yr^{-1} (Pirazzoli, 1986). The study area is under the temperate Mediterranean to semi-arid climate with rainfall in the north-eastern region averaging 400 to 600 mm yr^{-1} (INM, 2019). Rainfall pattern is highly variable, undergoing strong temporal and spatial fluctuations (Tramblay et al., 2013) with very high flood events, much reduced since the dams over the last 70 years with an average of 450 mm (Kotti et al., 2016).

The Medjerda River drains an area of 23 600 km^2. The average annual flow of 30 $m^3\,s^{-1}$, which can reach 3500 $m^3\,s^{-1}$ during exceptional extreme flood (Claude et al., 1977). Sediment load has been rarely studied, and a few recent data show a great variability from 50 $g\,L^{-1}$ in 1989 to 10 $g\,L^{-1}$ in 1990 (Soussi and Ben Mammou, 1989). The Medjerda River has formed the wetland system of the Gulf of Utica during the filling process since the last ∼ 1700 years (Delile et al., 2015). In the study area, the most urbanized area over the country, 9 dams were built upstream of the Medjerda River and tributaries since 1954 totalizing a storage capacity of more than 1.4 billion cubic meters, including the largest dam of Sidi Salem which covers 75 % of the surface of the basin (Fig. 1a) (Kotti et al., 2018a).

3 Methodology

The sediment yields into the Mediterranean Sea and their pattern are largely unknown. The scientific approach coupling continental and coastal investigation methods will help establish links between current attempts at quantifying and modeling them. They will be linked and calibrated with the sediment cores from floodplain areas, which will also help coastal environments researchers to better classify the vulnerability of coastal areas.

3.1 Sedimentological analysis

To determine the multi-decadal to pluri-annual sediment discharges from River downstream to coastal ridge we adopted a multi-criteria approach extended on the pre-estuarine areas of the Medjerda River new course artificially created since 1936. We propose first to analyze the sediment deposit fluctuations in regard of past events recorded on the inland basins lower Medjerda course in the floodplain, which is supposed to include effects of dam constructions, changes in agricultural practices and population density. Our second approach is based on the morphological evolution monitoring of the upstream basin including sandy and fine-grained beaches.

3.1.1 Sediment sampling

We collected 265 subsurface marine samples along the 20 km of the complex beaches covering the deltaic plain, lagoons,

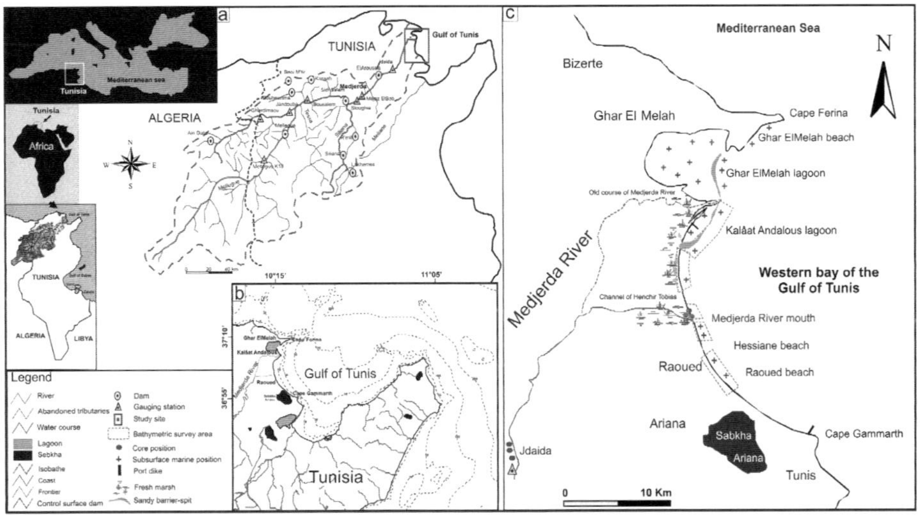

Figure 1. **(a)** Location map (from Kotti el al., 2018a) **(b)** Nearshore setting **(c)** Multi-criteria approaches methods in the Medjerda deltaic plain and coastal areas, Gulf of Tunis, Tunisia.

the sandy ridges, tidal inlets and the nearshore areas. For the medium to short temporal scales, field campaigns were organized seasonally and during the storm conditions. For the long-term monitoring, we collected 3 sediment cores between 0.42 to 1.68 m thickness in the pre-estuarine area, and 3 cores between 0.9 to 3 m thickness on terraces in the course of the river downstream of the last dams and far enough from the coast to avoid marine influence (Fig. 1c).

3.1.2 Sediment analysis

Collected sediments were undertaken particle-size distribution, by using classic indexes (average mean size Mz and/or median value Q2, Subsurface sorting index σ, asymmetry index SKI and kurtosis index KG), according to Folk and Ward (1957). The Malvern Mastersizer 2000 undertook the microgranulometric analysis for the measurement of particle size distributions within the range of 0.01–1000 microns. Modal statistical analysis are established from log-normal distribution and was determined to emphasize the percentage of the distribution of populations by grain-size (GSD) based on modal values in order to define the grain-size components and the sedimentary types to which they belong (Barusseau, 1973). Sediment Trend Analysis (STA) modeling tools according to the Le Roux approach (Le Roux, 1994) were applied to define the seasonal sediment transport pathways throughout the nearshore of the Medjerda delta.

3.2 Photogrammetry and remote sensing

In order to assess the long-term shoreline evolution, we perform time-series analysis of both photogrammetric Landsat, SPOT1, SPOT4 and sentinel 2A, 2B orbital scenes completed by the aerial photography, bathymetric and topographic maps dating from 1936–1948–1952–1974–1984–1988–1999 and 2016. The obtained datasets are analyzed using the ESRI ArcGIS 10.2.2 software to calculate shoreline evolution of change statistics from multiple coastline positions. Additionally, we use the Digital Shoreline Analysis System (DSAS) software to calculate the rate-of-change at the specified time interval (Thieler and Danforth, 1994). In our study area, transects were generated using DSAS every ~ 1 km length with ~ 200 m spacing to detect the high resolution changes (Hzami et al., 2018). DSAS tool provides a statistical calculation (Bush and Young, 2009) for each transect. The two statistical outputs calculated from the DSAS model are the Net Shoreline Movement (NSM), which provides the total reported separation between the oldest (1936) and current (2016) shoreline positions; and the End Point Rate (EPR), which is the annual rate of shoreline retreat.

4 Results and discussions

The present multi-criteria study reveals a complex sediment dynamic in the pre-estuarine and deltaic areas. We adopt multi-scale and regional approaches established by Cowell and Thom (1995) and Levoy (2000) to examine the beach sediment balance and the impact the anthropogenic factors on the littoral processes during the modern-decades.

4.1 Sediment discharge evolution from River to coasts before dams

In this study, sedimentological results reveal the dominance of two sediment types (ST): silt and clay type ST III (mean modal value Mo <0.063 mm) coming from the discontinuous supplies of rivers; and fine-grained sediment STI

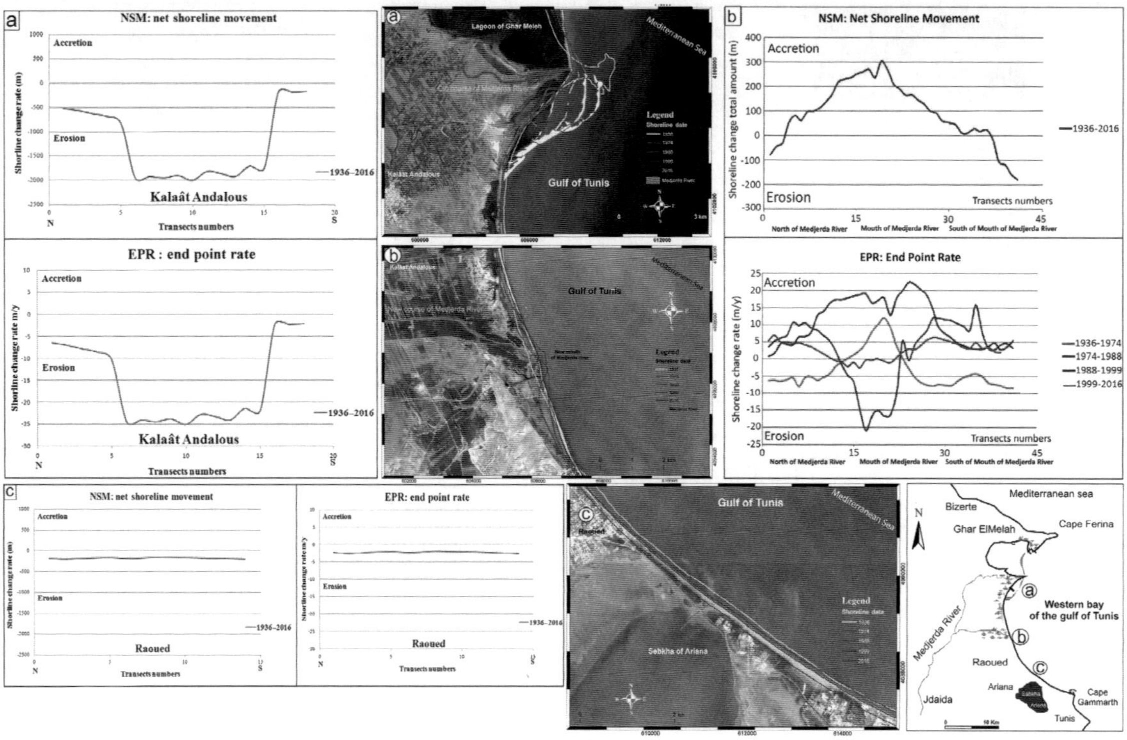

Figure 2. The diachronic shoreline evolution analysis at the deltaic and sandy beaches of the Gulf of Tunis, Tunisia. The Net Shoreline Movement (NSM) and The End-Point Rate (EPR) during the 1936–2016 period. The End-Point shoreline change Rates for the Gulf of Tunis beaches clearly show the erosion-dominated shoreline trend. Negative values indicate erosion. © Google Maps Software: Esri ArcGIS 10.2.

(Mo = 0.125 to 0.25 mm) originating from the continuous reworking of erosion of the cliffs and coastal drift throughout beach during storm events. Seasonal sediment nearshore dynamic show that grain size distribution (GSD) and Sediment Trend Analysis (STA) model pathways are determined by cross-shore geomorphology, location of the sediment-cell, seasonal incident wave and local terrestrial supply (Amrouni et al., 2019a).

Longshore and short-term deltaic plain assessment reveals sedimentological spatial variability in both the pre-estuarine and nearshore areas. In early century, the natural progression between 1936 to 1952 is marked by the deltaic plain progradation of the Medjerda River with a fluvial sedimentary unit mainly composed by silt and clay deposits rich in terrigenous components (D50 = 98 µm) (Benmoussa et al., 2018). The shoreline evolution shows a net accretion rate during this period with a positive rate of change NSM by $+300\,\mathrm{m} \pm 25\,\mathrm{m}$. The flood event occurred in 1973 is marked by the dissymmetric morphology of the mouth River clearly observed on aerial photography of 1974 dataset (Fig. 2a). The shoreline level shows accretion rate at the mouth of the new artificial channel, from 1952 to 1981 with an EPR average of $+13\,\mathrm{m} \pm 0.15\,\mathrm{m\,yr^{-1}}$. Geomorphological dissymmetry observed on the multi-temporal scenes of 1973 is the response of the reworking of the tidal-inlet by the wave action during storm surge.

4.2 Sediment discharge evolution from River to coasts after dams

After the dam construction, by the 1980th both the deltaic plain and the nearshore seabed are marked by decreasing of the grain size distribution with a clay deposit facies (D50 = 60 µm). The temporal evolution shows a rapid recession of the delta plain and shoreline level (Kotti et al., 2018b; Benmoussa et al., 2019). The dominance of clay deposits more than coarser fraction results from the impact of the construction of several large dams on the basin. Only the finer grain size fractions can feed the coast as revealed by the analysis of several cores in the lower alluvial plain of the Medjerda before the sea with Cesium, granulometry and geochemistry, coupled with the hydrological proxy (Kotti et al., 2018a).

The long term shoreline evolution between the periods 1974–1984 and 1988 shows a severe beach erosion trend extended in the entire delta shoreline with an EPR range of $-20 \pm 0.15\,\mathrm{m\,yr^{-1}}$ and a maximum NSM rate of $-219\,\mathrm{m} \pm 2\,\mathrm{m}$. Currently, the sediment balance of the gulf of Tunis coast shows an extended shoreline retreat under the

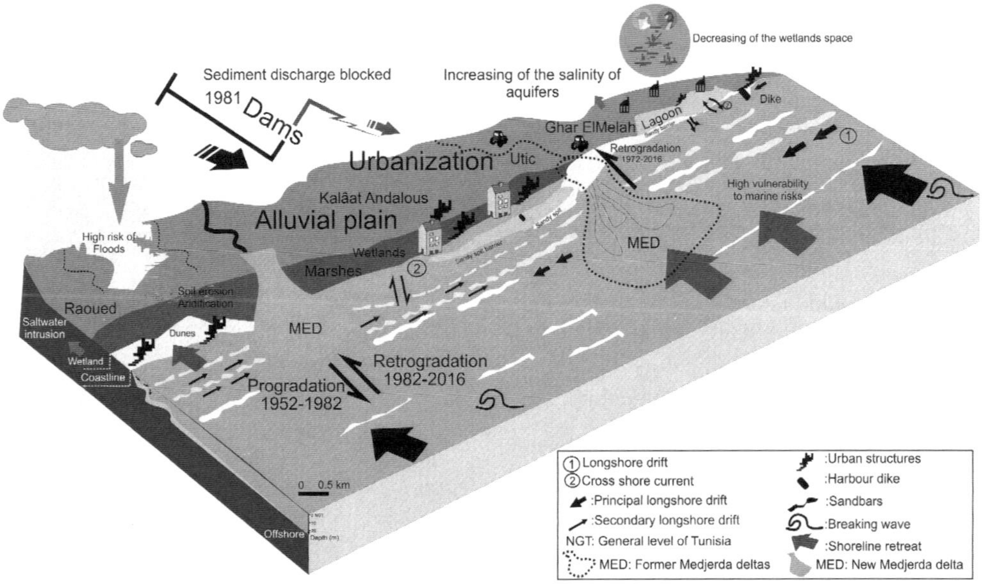

Figure 3. Environmental implication of the dam infrastructure on North African coasts ecosystems.

coarser sediment deficit, as at least 50 % of sediments no more reach the sea and are blocked by upstream dams. The old estuarine plain and delta mouth is experiencing a "hot spot" sediment loss with a Maximum Net Shoreline movement (NSM) of $-1570\,\mathrm{m} \pm 6\,\mathrm{m}$ inland between the period 1936 to 2016 (Fig. 2a). The progradation and retrogradation of the Medjerda Delta plain during the last century is strongly influenced by anthropogenic factor. Sediment deposit facies associated to geomorphological evolution of post-dam alluvial deposits confirms the shoreline retreat due to continental sediment reduction. Even though the southward beach shoreline seems not to be threatened by the deltaic regression (Fig. 2c), the beach stability is closely sensitive to rapid urban expansion, the dam construction and the overexploitation of natural resources over the entire river basin (i.e, coastal aquifer, wetland, sabkhas, etc.) (Fig. 3).

5 Conclusions

Our multi-criteria approach results confirm that the construction of dams greatly modify the quantity and the quality of the sediments downstream of dams to the sea. In the case of deltaic system of the Medjerda River, extended in semi-arid areas, deltaic plain is marked by the loss of sediments discharge that no more reach the sea, and the absence of coarse particle-size in recent decades cores sediment. Our study reveals an alarming shoreline retreat trend reaching $-20\,\mathrm{m} \pm 0.15\,\mathrm{m\,yr^{-1}}$ after the human-induced change where $\sim 50\,\%$ of sediment discharge has been trapped upstream the dam.

Thus, since the last decades, no more sand reaches the sea from the largest river in North African network into the Mediterranean Sea. Anthropogenic infrastructures have a great impact on coastal areas evolution such as the severe shoreline retreat associated to a large sandy beach-loss of $-1570\,\mathrm{m} \pm 6\,\mathrm{m}$ inland over the study period (1936 to 2016), alarming reduction of the deltaic plain and wetland areas, and changes in the ecohydrology balance, with impacts on the socio-economic activities. This coastal issue is superimposed to the beach vulnerability due to the marine hazard and climate changes disasters.

Code availability. The licence of the software code Esri ArcGIS 10.2 is provided by the Hydroscience of Montpellier Research website (HSM, Montpellier) (https://app.umontpellier.fr/securum/public/centralAuthentication, last access: 19 October 2021).

Data availability. Data set related to the physical parameters of the wave propagation are provided by the database of COADS (1964) (https://shen.sdsu.edu/researchgroup/coads.htm, last access: 20 October 2021).

Author contributions. OA and GM conceived the scientific project PHC-UTIQUE-RYSCMED 16G1005 (2016–2018) and the research monitoring and wrote the manuscript. OA coordinate and supervised the Tunisian-French workfield compaigns. GM supported the software acquirement and model development. Both OA and GM edited the paper and provided comments.

Competing interests. The contact author has declared that neither they nor their co-authors have any competing interests.

Disclaimer. Publisher's note: Copernicus Publications remains neutral with regard to jurisdictional claims in published maps and institutional affiliations.

Special issue statement. This article is part of the special issue "Hydrology of Large River Basins of Africa". It is a result of the 4th International Conference on the "Hydrology of the Great Rivers of Africa", Cotonou, Benin, 13–20 November 2021.

Acknowledgements. The authors are grateful to all the research members who contributed to the scientific and academic tasks during the project. The authors would also like to thank the Copernicus editorial support team for their editing that helped improve this manuscript.

Financial support. This research has been funded by the National Institute of Marine Sciences and Technologies in Tunisia, University of Carthage. This project was developed under the framework of the PHC-UTIQUE-RYSCMED (grant nos. 16G1005 and 34854QC) project and the institutional DYNAMISME & MOSBIO-CLIM research projects funded by the Tunisian High Ministry of Education and Scientific Research and Campus France.

References

Amrouni, O., Sánchez, A., Khélifi, N., Benmoussa, T., Chiarella D., Mahé G., Abdeljaouad S., and McLaren P.: Sensitivity assessment of the deltaic coast of Medjerda based on fine-grained sediment dynamics, Gulf of Tunis, Western Mediterranean, J. Coast. Conserv., 23, 571–587, https://doi.org/10.1007/s11852-019-00687-x, 2019a.

Amrouni, O., Hzami, A., and Heggy, E.: Photogrammetric assessment of shoreline retreat in North Africa: anthropogenic and natural drivers, ISPRS J. Photogramm., 157, 73–92, https://doi.org/10.1016/j.isprsjprs.2019.09.001, 2019b.

Anthony, E. J.: Wave influence in the construction, shaping and destruction of river deltas: A review, Mar. Geol., 361, 53–78, https://doi.org/10.1016/j.margeo.2014.12.004, 2015.

Barusseau, J. P.: Evolution du plateau continental rochelais (golfe de Gascogne) au cours du Pléistocène terminal et de l'Holocène. Les processus actuels de la sédimentation, Thèse Es-Science, Université de Bordeaux I, Bordeaux, France, 360 pp., 1973.

Benmoussa, T., Amrouni, O., Dezileau, L., Mahé, G., and Abdeljaouad, S.: The sedimentological changes caused by human impact at the artificial channel of Medjerda-River (Coastal zone of Medjerda, Tunisia), Proc. IAHS, 377, 77–81, https://doi.org/10.5194/piahs-377-77-2018, 2018.

Benmoussa, T., Amrouni, O., Hzami, A., Dezileau, L., Mahe, G., Condomines M., and Saadi A.: Progradation and retrogradation of the Medjerda delta during the 20th century (Tunisia, Western Mediterranean), Comptes Rendus Geosciences, 351, 340–350, 2019.

Bush, D. M., and Young, R.: Coastal features and processes, in: Geological Monitoring, edited by: Young, R. and Norby, L., Geological Society of America, Boulder Colorado, pp. 47–67, available at: https://www.nps.gov/articles/geological-monitoring-of-coastal-features-and-processes.htm (last access: 19 October 2021), 2009.

Claude, J., Francillon, G., and Loyer, J. Y.: Les alluvions déposées par l'Oued Medjerda lors de la crue exceptionnelle de mars 1973. Cahiers ORSTOM, Ser. Hydrol. 14, 37e109, available at: http.//horizon.documentation.ird.fr/exl-doc/pleins_textes/divers14-05/08532.pdf (last access: 19 October 2021), 1977.

COADS: Comprehensive Ocean-Atmosphere Data Set, Release 1, NOAA Environmental Research Laboratories [data set], Climate Research Program, Boulder, CO, 268 pp., available at: https://shen.sdsu.edu/researchgroup/coads.htm (last access: 19 October 2021), 1964.

Coleman, J. M. and Roberts, H. H.: Deltaic coastal wetlands, in: Coastal Lowlands, edited by: van der Linden, W. J. M., Cloetingh, S. A. P. L., Kaasschieter, J. P. K., van de Graaff, W. J. E., Vandenberghe, J., and van der Gun, J. A. M., Springer, Dordrecht, the Netherlands, 1989.

Cowell, P. and Thom, B.: Morphodynamics of coastal evolution, in: Coastal Evolution: Late Quaternary Shoreline Morphodynamics, edited by: Plassche, O., Carter, R., and Woodroffe, C., Cambridge University Press, Cambridge, UK, pp. 33–86, https://doi.org/10.1017/CBO9780511564420.004, 1995.

Delile, H., Abichou, A., Gadhoum, A., Goiran, J. P., Pleuger, E., Monchambert, J. Y., Wilson, A., Fentress, E., Quinn, J., Ben Jerbania, I., and Ghozzi, F.: The Geoarchaeology of Utica, Tunisia: the paleogeography of the Mejerda Delta and hypotheses concerning the location of the Ancient Harbor, Geoarchaeology, 30, 291–306, https://doi.org/10.1002/gea.21514, 2015.

El Arrim, A.: Étude d'impact de la dynamique sédimentaire sur la stabilité du littoral du Golfe de Tunis, Doctorat de spécialité, Unversité de Tunis El Manar, Tunis, Tunisia, 203 pp., 1996.

Folk, R. L. and Ward, W. C.: Brazos river bar: a study in the signification of grain size parameters, J. Sediment. Petrol., 27, 3–26, 1957.

Hzami, A., Amrouni, O., Romanescu, G., Constantin Stoleriu, C., Mihu-Pintilie, A., and Saâdi, A.: Satellite images survey for the identification of the coastal sedimentary system changes and associated vulnerability along the western bay of the Gulf of Tunis (northern Africa), Proc. IAHS, 377, 83–89, https://doi.org/10.5194/piahs-377-83-2018, 2018.

Hzami, A., Heggy, E., Amrouni, O., Mahe, G., Manaan, M., and Abdeljaouad, S.: Alarming coastal vulnerability of the deltaic and sandy beaches of North Africa, Sci. Rep., 11, 2320, https://doi.org/10.1038/s41598-020-77926-x, 2021.

INM: Tunisian National Institute of Meteorology, INM, Data-base, available at: http://www.meteo.tn/htmlfr/accueil.php (last access: 19 October 2021), 2019.

Kotti, F., Mahe, G., Habaieb, H., Dieulin, C., and Calvez, R.: Etude des pluies et des débits sur le bassin versant de la Medjerda, Tunisie, Bulletin de l'Institut Scientifique, 38, 19–28, 2016.

Kotti, F., Dezileau, L., Mahe, G., Habaieb, H., Benabdallah, S., Bentkaya, M., Calvez, R., and Dieulin, C.: The impact of dams and climate on the evolution of the sediment loads to the sea by the Medjerda River using a paleohydrological approach, J. Afr. Earth Sci., 142, 226–233, https://doi.org/10.1016/j.jafrearsci.2017.10.003, 2018a.

Kotti, F., Dezileau, L., Mahé, G., Habaieb, H., Bentkaya, M., Dieulin, C., and Amrouni, O.: Etude de l'impact des barrages sur la réduction des transports sédimentaires jusqu'à la mer par approche paléohydrologique dans la basse vallée de la Medjerda, Proc. IAHS, 377, 67–76, https://doi.org/10.5194/piahs-377-67-2018, 2018b.

Le Roux, J. P.: An alternative approach to the identification of the end sediment transport paths based on grain size trends, Sediment. Geol., 94, 97–107, 1994.

Levoy, F.: Morphodynamique des plages sableuses et trait de côte à différentes échelles spatio-temporelles, vol. I., Mémoire d'habilitation, UFR des Sciences de la Terre et de l'Aménagement Régional, Université de Caen Basse-Normandie, Caen, France, 150 pp., 2000.

Mazzoni, A, Heggy, E., and Scabbia, G.: Forecasting water budget deficits and groundwater depletion in the main fossil aquifer systems in North Africa and the Arabian Peninsula, Global Environmental Change, 53, 157–173, 2018.

Paskoff, R. and Sanlaville, P.: Les côtes de la Tunisie. Variations du niveau marin depuis le Tyrrhénien, Coll. Maison Orient Medit., 14, Sér. Géogr. et Préhist., 192 pp., available at: https://www.persee.fr/issue/mom_0243-2439_1983_mon_14_1 (last access: 19 October 2021), 1983.

Pirazzoli, P. A.: Secular trends of relative sea level (RSL) changes indicated by tide-gauge records, J. Coastal Res., 1, 1–26, available at: http://www.jstor.org/stable/44863318 (last access: 19 October 2021), 1986.

Pleuger, E., Goiran, J.-P., Mazzini, I., Delile, H. Abichou, A., Gadhoum, A., Djerbi, H., Piotrowska, N., Wilson, A., Fentress, E., Ben Jerbania, I., and Fagel, N.: Paleogeographical and paleoenvirenmental reconstitution of the Medjerda delta (Tunisia) during the Holocene, Quaternary Sci. Rev., 220, 263–178, 2019.

Soussi, N. and BenMammou, A.: Les apports solides en suspension de la Medjerda-Tunisie, 2ème congrès de sédimentologie, Paris, 1989.

Syvitski, P. M. and Kettner, A.: Sediment flux and the Anthropocene, Phil. Trans. R. Soc. A, 369, 957e975, available at: https://www.jstor.org/stable/41061709 (last access: 19 October 2021), 2011.

Thieler, E. R. and Danforth, W.: Historical Shoreline Mapping (II): Application of the Digital Shoreline Mapping and Analysis System (DSMS/DSAS) to Shoreline Change Mapping in Puerto Rico, J. Coastal Res., 10, 600–620, available at: https://pubs.er.usgs.gov/publication/70135638 (last access: 19 October 2021), 1994.

Tramblay, Y., El Adlouni, S., and Servat, E.: Trends and variability in extreme precipitation indices over Maghreb countries, Nat. Hazards Earth Syst. Sci., 13, 3235–3248, https://doi.org/10.5194/nhess-13-3235-2013, 2013.

Proc. IAHS, 384, 141–147, 2021
https://doi.org/10.5194/piahs-384-141-2021

Reflections on almost a century of hydrological studies on Africa's largest lake

Kevin Sene[1], **Helen Houghton Carr**[2], **and Wlodek Tych**[1]

[1]Lancaster Environment Centre, Lancaster, LA1 4YQ, UK
[2]UK Centre for Ecology & Hydrology, Wallingford, OX10 8BB, UK

Correspondence: Kevin Sene (k.sene1@lancaster.ac.uk)

Published: 16 November 2021

Abstract. Lake Victoria is the largest lake in Africa and its outflows strongly influence flows in the White Nile, including the availability of water for hydropower generation, irrigation and water supply. Understanding the water balance is a major challenge since the lake is large enough to influence the local climate and its catchment spans several countries. Hydrometeorological monitoring networks are also sparse in some parts of the basin. In this paper, we consider the history of water balance estimates for the lake and how the science has developed as new information and techniques have become available, including in areas such as seasonal flow forecasting and estimating the potential impacts of dam operations and climate change. These findings are placed into a wider context including the challenges arising from a changing climate and evolving ideas from international research programmes, which lead to some suggestions for future research priorities for Lake Victoria and other sub-Saharan/Rift Valley lakes.

1 Introduction

Lake Victoria is the largest lake in Africa and the main source of water for the White Nile. It has a mean surface area of approximately 68 000 km^2 and the surrounding catchment area is more than three times as large, at about 194 000 km^2 (e.g. WMO, 1974, 1982). Inflows to the lake arise from Rwanda, Burundi, Kenya, Tanzania, Uganda, and the Democratic Republic of the Congo.

Whilst the hydrology of the lake has been surveyed since the early 20th century, the first significant studies were associated with the design of Owen Falls (now Nalubaale) Dam which was commissioned in 1953. Typically, these focused on the long-term water balance of the lake (Sutcliffe and Parks, 1999) but, by the 1970s, several computer simulation studies had been conducted using raingauge averaging methods to estimate lake and catchment rainfall, and conceptual rainfall-runoff models to estimate ungauged inflows. Pioneering work had also been performed on the use of stochastic models to estimate future outflows. Piper et al. (1986) provide a useful overview of techniques in use at that time.

More recently, novel approaches have been developed to meet the challenges of new applications, such as for short-range weather forecasting, seasonal flow forecasting, and assessing the potential impacts of climate change. The hydropower generating capacity of the White Nile also continues to be exploited further, with several schemes now along the river. The techniques used have included regional and mesoscale atmospheric models, and stochastic approaches, as discussed later.

Following a brief literature review, this paper presents insights into how modelling techniques have evolved over the decades and some key lessons learned. This includes perspectives gained from new applications in areas such as climate change, seasonal flow forecasting and assessing the impacts of dam operations. These findings are then placed into a wider context including the challenges arising from a changing climate and evolving ideas from international research programmes.

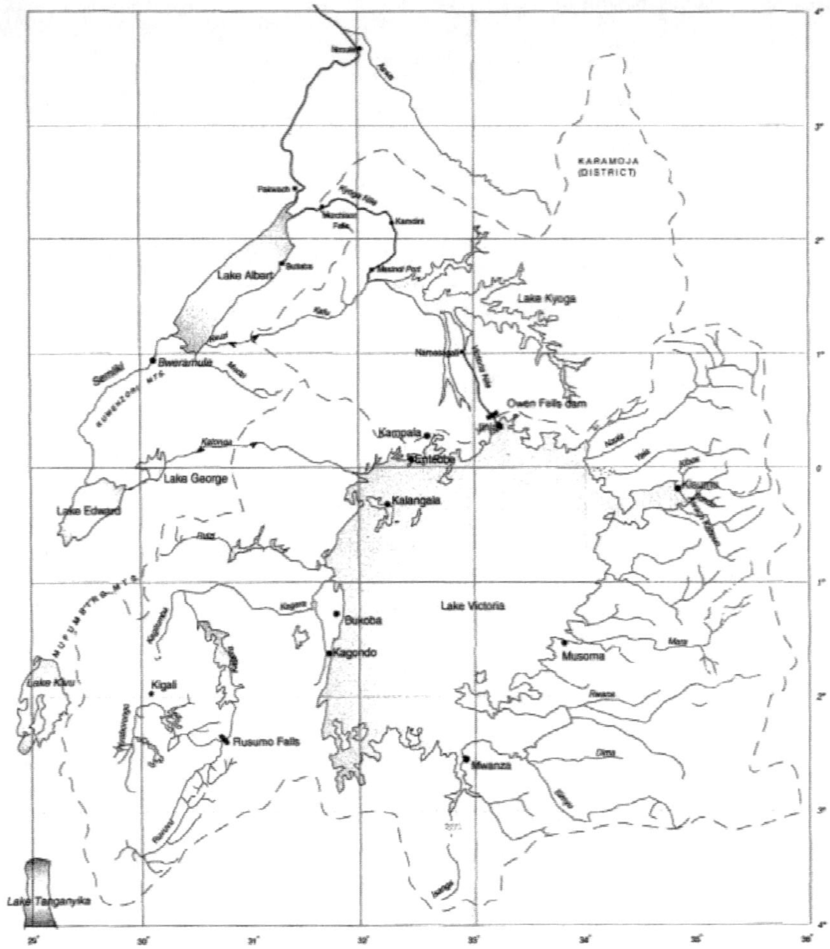

Figure 1. Map of the lake region (source: IAHS Press and Sutcliffe and Parks, 1999).

2 The study area

Figure 1 shows the location of Lake Victoria. The lake rainfall is dominated by the annual progression of the Intertropical Convergence Zone resulting in two rainfall seasons, although lake temperature influences are large enough to affect the local atmospheric circulation.

Regular lake level observations began in the 1890s. Some raingauge observations started in the following decade, whilst routine monitoring of outflows commenced in the 1940s. Monitoring reached a peak during the landmark World Meteorological Organisation (WMO) Hydrometeorological Survey of the 1960s, 1970s and 1980s. More recently, there been major improvements associated with regional bodies such as the Lake Victoria Basin Commission (LVBC) and the Nile Basin Initiative (NBI).

Figure 2 shows the lake level variations from 1896 to 2017. An interesting feature is how seasonal variations in levels are superimposed on much longer-term storage influences, meaning that long record lengths, over periods of decades, are required to understand the lake response.

The water balance for a lake can be expressed as Eq. (1):

$$A\Delta h = AN - Q_{\mathrm{o}} \tag{1}$$

where A is the surface area, Δh is the change in water level in a time interval, N is the net inflow and Q_{o} is the lake outflow (e.g. Piper et al., 1986). The term N is sometimes called the Net Basin Supply (NBS) and is given by Eq. (2):

$$N(t) = P(t) - E(t) + \frac{Q_{\mathrm{i}}(t)}{A(t)} + \epsilon(t) = \Delta h + \frac{Q_{\mathrm{o}}(t)}{A(t)} \tag{2}$$

where t is time, P is the lake rainfall, E is the lake evaporation and Q_{i} is the tributary inflow. The surface area is often assumed to be constant whilst the additional term ϵ represents relatively minor items which are usually neglected as they are difficult to estimate, such as water abstractions and groundwater seepage.

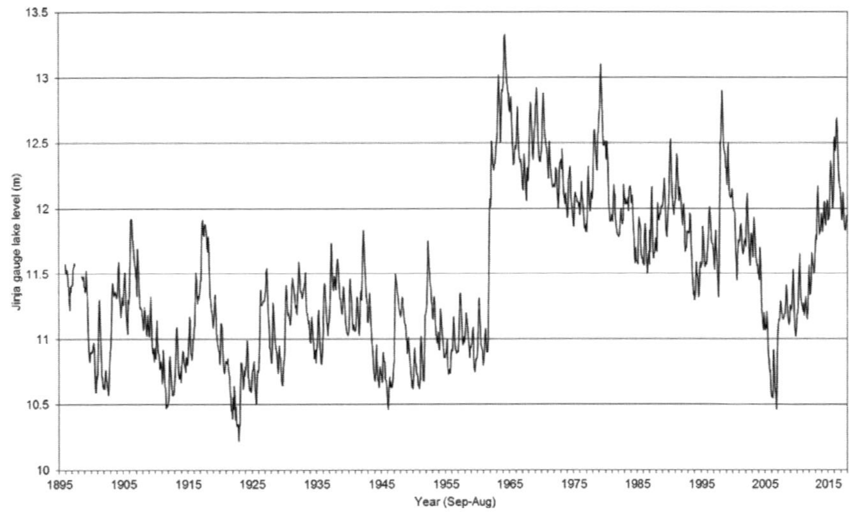

Figure 2. Lake Victoria water levels from 1896 to 2017 (source: NCSU et al., 2018).

In some studies, an empirical rainfall scaling term, c, is introduced so that the net basin supply can be expressed as Eqs. (3) or (4):

$$N = cP - E + Q_i/A \tag{3}$$

$$N = c(P - E) + Q_i/A. \tag{4}$$

This value is then estimated by specifying closure of the water balance over the chosen calibration period, with a further refinement being to consider individual factors for each of the monitoring stations considered.

3 Modelling challenges

3.1 Water balance approaches

Given the importance of Lake Victoria in the region, there have been numerous studies of its water balance; far too many to list here. Table 1 presents some landmark studies, selected both for the techniques considered and the large number of citations they include. Additional references appear later for specific applications.

Of the terms in the water balance, the direct rainfall on the lake surface provides the majority of the input but is the most difficult component to estimate. Typically, this is achieved by area-weighting raingauge observations although, as noted in the table, recent studies are increasingly using satellite precipitation estimates. However, one challenge with both approaches is that the lake is large enough to affect the local atmospheric circulation, causing rainfall to be higher over the water surface. This means that raingauge observations, even from the lakeshore, may not be representative, whilst the interpretation of satellite observations is made more challenging due to the impacts of enhanced cloud cover over the lake, affecting both rainfall and water temperatures.

The lake evaporation is another important component and is typically estimated from a Penman-type approach using observations from weather stations. As with raingauge observations, the local atmospheric circulation introduces additional uncertainty, with the added complication that there are relatively few weather stations. However, as might be expected, evaporation is much less variable than rainfall so often a long-term monthly average is assumed in simulations.

The tributary inflows, in comparison, are relatively small, but many studies have noted that much of the variability in lake levels arises from this term, due to the fine balance between lake rainfall and evaporation, which are of a similar magnitude. However, many of the tributaries are ungauged or have short record lengths, making rainfall-runoff model calibration difficult for infilling and extending records. Fortunately, though, the longest and best quality records are for the main rivers which flow into the lake, so the combined inflow from these rivers is often scaled to estimate the total inflow. For example, the catchments of the Kagera, which flows into the lake from Tanzania to the west, and the Nzoia, Yala, Sondu and Awash Kaboun from Kenya to the northeast, represent about 40 % of the total catchment area.

Given these challenges, one innovative aspect in many studies of Lake Victoria has been the recognition that much can be understood about the lake response by considering the NBS alone, estimated from lake levels and outflows. This approach has the attraction that both terms are monitored with greater precision than the inflow components so that, in principle, estimates should be more accurate. Also, since level records now span more than a century, there is the benefit of a dataset that represents a wide range of historical lake levels and their frequencies of occurrence. Linked to this, another innovation was the early adoption of stochastic models for the NBS to explore variations in future levels.

Table 1. Summary of key investigations of the water balance of Lake Victoria.

Publication date	Findings
1938	First published attempt to close lake water balance, based on minimal observations from around the lake (Hurst and Phillips, 1938)
1970	Technical note to Uganda Water Development Dept including a first lake rainfall series (De Baulny and Baker, 1970)
1974–1982	Initiated systematic measurements of tributary inflows and other components of the water balance (WMO, 1974, 1982)
1981	Demonstrated that the 1960s rise in lake levels was caused by natural variability and consecutive heavy rainfall seasons (Kite, 1981)
1984–1994	A series of reports to ODA[1] (now FCDO[2]) and associated papers, reviewing the hydrology of Lake Victoria, and presenting methodologies for estimating lake rainfall, evaporation and tributary inflow, including the use of stochastic techniques, and using a simple conceptual model to extend the monthly lake water balance back to 1925, and annual lake water balance back to 1900 (Institute of Hydrology, 1984, 1985, 1993; Piper et al., 1986; Sene and Plinston, 1994)
1996	Used 3 GCMs to drive a lake water balance model to investigate impacts of climate change on Nile flows (Conway and Hulme, 1996)
1998–2001	Reconstructed lake level trends back to 1780s using historical witness accounts and reports of meteorological conditions, and interpreted the trends in terms of components of water balance (Yin and Nicholson, 1998; Nicholson and Yin, 2001)
2002	Report to LVBC updating lake water balance (COWI Consulting Engineers and Planners AS, 2002)
2004	Updated water balance to 2000, and did sensitivity analysis of levels and outflows to climate change (Tate et al., 2004)
2008	Applied SWAT[3] model to Nzoia tributary using 5 GCMs with focus on climate change impacts on catchment high flows (Githui et al., 2008)
2011	Applied VHM[4] and NAM[5] models to Nyando tributary using 17 GCMs with focus on catchment hydrological extremes (Taye et al., 2011)
2017	AGU[6] conference presentation describing application of GR4J[7] model to Katonga tributary using CP4A[8] climate data (Stratton et al., 2018; Macdonald et al., 2017)
2018	Tested lake water balance for recent past climate then applied ensemble of 6 CORDEX[9] RCMs[10] to quantify future levels and outflows (Vanderkelen et al., 2018a, b)
2019	Weighted and post-processed remote sensing satellite products to quantify components of water balance (Khaki and Awange, 2019)

[1] ODA: Overseas Development Administration (UK); [2] FCDO: Foreign, Commonwealth and Development Office (UK); [3] SWAT: Soil and Water Assessment Tool; [4] VHM: Dutch abbreviation for "generalized lumped conceptual and parsimonious model structure identification and calibration"; [5] NAM: Nedbør–Afstrømnings-Model; [6] AGU: American Geophysical Union; [7] GR4J: "modèle du Génie Rural à 4 paramètres Journalier"; [8] CP4A: Pan-African Convection-Permitting Regional Climate Simulation with the Met Office Unified Model (UM); [9] CORDEX: Coordinated Regional Climate Downscaling Experiment; [10] RCM: Regional Climate Model.

3.2 Newer techniques from recent applications

3.2.1 Climate change studies

One key application of water balance models has been to help understand whether sudden changes in water levels were due to climatic or other causes.

For example, based on this approach, the sudden rise in levels in the period 1961-1964 has been widely attributed to increased rainfall in East Africa at that time, rather than to other factors such as dam operations (e.g. Piper et al., 1986; Yin and Nicholson, 1998; Sutcliffe and Parks, 1999). Water balance models have also suggested that the sudden fall in levels around 2005 was caused by a drier climatic period and increased releases at Nalubaale Dam (Sutcliffe and Petersen, 2007).

Water balance models can also be used to estimate future levels, with the potential impacts of climate change of particular interest. The earliest studies were typically based on simple scaling factors for rainfall from a single climate model, but nowadays a more sophisticated approach is used. For example, one of the most recent (NCSU et al., 2018) describes the application of change factors from an ensemble of 36 General Circulation Models (GCMs) under a high emissions scenario (Representative Concentration Pathway RCP08.5) to estimate future impacts on lake rainfall, lake evaporation and tributary inflows for input to a water balance model. Houghton-Carr et al. (2021a, b) also describe stochastic modelling approaches that generated many possible future sequences of climate-modified NBS.

An important point is that many of the current regional atmospheric models do not explicitly represent Lake Victoria (NCSU et al., 2018). For example, Thiery et al. (2016) have shown that temperature differences between the lake and surrounding land affect the strength of the lake-land breeze and, hence, the storms and precipitation that occur on and around the lake. As a result, one current research priority is to represent the rainfall due to convective storms over the lake in GCMs.

3.2.2 Lake operating rules

For hydropower operations, the outflows from Lake Victoria are governed by an internationally agreed relationship between lake levels and outflows. This is called the Agreed Curve and is designed to mimic the behaviour of the natural lake, before hydropower operations first began.

In recent years, proposed changes to the operating rules for Nalubaale Dam have led to further in-depth studies of the lake outflow characteristics (Houghton-Carr et al., 2021b). As part of these studies, one innovation has been to extend existing models to consider additional terms not usually included, such as water abstractions, allowing rapid assessment of the implications of increases in water use for domestic, agricultural and industrial water use. However, NCSU et al. (2018) demonstrate the potential effect of future changes in abstraction to be relatively minor compared to the possible impact of future changes in climate.

3.2.3 Seasonal flow forecasting

Another developing area is that of seasonal flow forecasting with operational or pre-operational systems now running in several countries at a regional or national scale.

For Lake Victoria, some forecasting challenges include the additional rainfall generated by the lake, its huge catchment area, and the availability of real-time data. In contrast, the large storage in the lake provides some persistence in outflows, which can be exploited in a seasonal forecasting model.

To explore some key drivers on lake response, Sene et al. (2018) used a data-driven approach to derive stochastic estimates for future lake levels and outflows, which demonstrated forecast skill several months ahead. Some innovative features included use of a data assimilation component, incorporating climate indices, and interpretation of the results using performance metrics better suited to the non-stationary response than more conventional measures. The climate indices chosen were linked to the El Niño Southern Oscillation and the Indian Ocean Dipole, which are the primary known drivers of East African inter-annual rainfall variability.

The potential for seasonal flow forecasting for the Nile has also been explored using indices representative of the Pacific and Indian Oceans (Siam and Eltahir, 2015). More generally, there have been many investigations into the links between regional rainfall and atmospheric and ocean conditions using both statistical and atmospheric modelling approaches (e.g. Nicholson, 2017). Looking to the future, the Lake Victoria basin is also a pilot project under the WMO Hydrological Status and Outlooks (HydroSOS) initiative (https://public.wmo.int/en, last access: 25 October 2021).

4 Conclusions

When considering the history of hydrological studies of Lake Victoria, there is much to celebrate, from innovations in estimating the water balance and the early adoption of stochastic techniques, to increasing recognition of the utility of remote sensing information across data-sparse regions and the embracing of basin-wide solutions through the auspices of the LVBC and others. The development and application of hydrological science in the basin has evolved as new data sources and more advanced computing techniques have become available.

One fundamental challenge is that, due to the lake dynamics, long-term records are required to understand its response, spanning many decades. This requires long, good quality observational records and a common recommendation in many of the studies cited here is the need to improve monitoring, whether ground-based and/or by satellite; and the need for further research to fill gaps in observations in both regional meteorology and hydrology, particularly for the lake rainfall and tributary inflows. This can also reduce uncertainties to improve decision-making going forward, helping to inform a risk-based approach combining probabilities and consequences.

Another priority is to improve forecasting skill at a range of timescales, from short-range weather forecasting to seasonal flow forecasting. Improved climate change impact projections are also required over timescales of years and decades ahead. Internationally, there is an increasing emphasis on a seamless approach, using the same or mutually consistent models and datasets at all timescales.

The challenges and issues faced for Lake Victoria are common to many regions and include the impacts of climate change and human intervention on water resources. In the Lake Victoria region these are compounded by rapid urbanisation, growing population and their associated water demands, and unsustainable exploitation of resources.

These themes are strongly endorsed by evolving ideas from international research programmes, such as the GEWEX (Global Energy and Water Exchanges programme) HyVic project (2012–present), the IAHS (International Association of Hydrological Sciences) Decade on Predictions in Ungauged Basins (PUB; 2003–2012) and the UNESCO (United Nations Educational, Scientific and Cultural Organization) Flow Regimes from International Experimental and Network Data programme (FRIEND-Water; 1984–present). In particular, these emphasise the need for a cooperative approach to the management of the water resources of international river basins, including the free exchange of data, information, expertise and ideas.

Some of the key challenges for practitioners are how future change will affect water resources, and what adaptation strategies are available to best equip key stakeholders. The recent and ongoing unprecedented changes in climate and population in the Lake Victoria basin, and the associated pressures on and demands for water and other natural resources, have driven the requirement for state-of-the-art scientifically-based assessments of the status and future impacts of various environmental and socio-economic change scenarios on water resources, to inform robust, coherent basin management plans. These issues apply to other sub Saharan/Rift Valley lakes too.

Data availability. No data sets were used in this article.

Author contributions. This paper draws on previous research by KS and HHC on the water balance of Lake Victoria, with contributions on approaches to seasonal flow forecasting from WT.

Competing interests. The contact author has declared that neither they nor their co-authors have any competing interests.

Disclaimer. Publisher's note: Copernicus Publications remains neutral with regard to jurisdictional claims in published maps and institutional affiliations.

Special issue statement. This article is part of the special issue "Hydrology of Large River Basins of Africa". It is a result of the 4th International Conference on the "Hydrology of the Great Rivers of Africa", Cotonou, Benin, 13–20 November 2021.

Acknowledgements. The UKCEH and NCSU studies cited were based on work funded by UK Aid through the Corridors for Growth Trust Fund administered by the World Bank, and on work funded by the Lake Victoria Basin Commission. The views expressed are the authors' own, and not necessarily those of the funding agencies or of any of the Nile basin countries.

Financial support. The support of the Lancaster Environment Centre and the UK Centre for Ecology & Hydrology in preparing this paper is gratefully acknowledged.

References

Conway, D. and Hulme, M.: The Impacts of Climate Variability and Future Climate Change in the Nile Basin on Water Resources in Egypt, Int. J. Water Res. Dev., 12, 277–296, https://doi.org/10.1080/07900629650178, 1996.

COWI Consulting Engineers and Planners AS, in association with DHI Water and Environment: Integrated Water Quality/Limnology Study for Lake Victoria: Final Report, COWI, Copenhagen, Denmark, 2002.

De Baulny, H. L. and Baker, D. L.: The water balance of Lake Victoria, Technical Note, Uganda Ministry of Water Development, Entebbe, 1970.

Githui, F., Gitau, W., Mutua, F., and Bauwens, W.: Climate change impact on SWAT simulated streamflow in western Kenya, Int. J. Climatol., 29, 1823–1834, https://doi.org/10.1002/joc.1828, 2008.

Houghton-Carr, H. A., Farquharson, F. A. K., Plinston, D. T., Matano, A. S., Canangire, C., Mwinjaka, O., Okurut, T., and Sutcliffe, J. V.: The development of a new water release and abstraction policy for the Lake Victoria basin, in preparation, 2021a.

Houghton-Carr, H. A., Smith, K., Plinston, D., Marsham, J., Farquharson, F., Finney, D., Ascott, M., Rowell, D., Semazzi, F., and Vanderkelen, I.: Plausible ranges of Lake Victoria level and outflow for decision-making under climate change, in preparation, 2021b.

Hurst, H. E. and Phillips, P.: The Nile Basin Vol. 5: The Hydrology of the Lake Plateau and Bahr El Jebel, Paper No. 35, Ministry of Public Works Physical Department, Egypt, 1938.

Institute of Hydrology: A review of the hydrology of Lake Victoria and the Victoria Nile, Institute of Hydrology, Wallingford, 1984.

Institute of Hydrology: Further review of the hydrology of Lake Victoria, Institute of Hydrology, Wallingford, 1985.

Institute of Hydrology: Review and update of the water balance of Lake Victoria in East Africa, Institute of Hydrology, Wallingford, 1993.

Khaki, M. and Awange, J.: Improved remotely sensed satellite products for studying Lake Victoria's water storage changes, Sci. Toal Environ., 652, 915–926, https://doi.org/10.1016/j.scitotenv.2018.10.279, 2019.

Kite, G. W.: Recent changes in level of Lake Victoria, Hydrol. Sci. Bull., 26, 233–243, https://doi.org/10.1080/02626668109490883, 1981.

Macdonald, D., Ascott, M., Lapworth, D., and Tindimugaya, C.: Application of scenario-neutral methods to quantify impacts of climate change on water resources in East Africa, in: AGU Fall Meeting 2017, New Orleans, USA, available at: http://nora.nerc.ac.uk/id/eprint/519528/ (last access: 25 October 2021), 2017.

NCSU – North Caroline State University, CEH – Centre for Ecology & Hydrology, and University of Leeds: Lake Victoria Levels under a Changing Climate, WP1 Report: Analysis of Historical Lake Levels, WP2 Report: Lake Levels under Climate Change, WP3 Report: Recommendations for Future Research and Monitoring to Support Decision-Making, World Bank, Washington, 2018.

Nicholson, S. E.: Climate and climatic variability of rainfall over eastern Africa, Rev. Geophys., 55, 590–635, https://doi.org/10.1002/2016RG000544, 2017.

Nicholson, S. E. and Yin, X.: Rainfall Conditions in Equatorial East Africa during the Nineteenth Century as Inferred from the Record of Lake Victoria, Climatic Change, 48, 387–398, https://doi.org/10.1023/A:1010736008362, 2001.

Piper, B. S., Plinston, D. T. and Sutcliffe, J. V.: The water balance of Lake Victoria, Hydrolog. Sci. J., 31, 25–37, https://doi.org/10.1080/02626668609491025, 1986.

Sene, K., Tych, W., and Beven, K.: Exploratory studies into seasonal flow forecasting potential for large lakes, Hydrol. Earth Syst. Sci., 22, 127–141, https://doi.org/10.5194/hess-22-127-2018, 2018.

Sene, K. J. and Plinston, D. T.: A review and update of the hydrology of Lake Victoria in East Africa, Hydrolog. Sci. J., 39, 47–63, https://doi.org/10.1080/02626669409492719, 1994.

Siam, M. S. and Eltahir, E. A. B.: Explaining and forecasting interannual variability in the flow of the Nile River, Hydrol. Earth Syst. Sci., 19, 1181–1192, https://doi.org/10.5194/hess-19-1181-2015, 2015.

Stratton, R. A., Senior, C. A., Vosper, S. B., Folwell, S. S., Boutle, I. A., Earnshaw, P. D., Kendon, E., Lock, A. P., Malcolm, A., Manners, J., Morcrette, C. J., Short, C., Stirling, A. J., Taylor, C. M., Tucker, S., Webster, S., and Wilkinson, J. M.: A Pan-African Convection-Permitting Regional Climate Simulation with the Met Office Unified Model: CP4-Africa, J. Climate, 31, 3485–3508, https://doi.org/10.1175/JCLI-D-17-0503.1, 2018.

Sutcliffe, J. V. and Parks, Y. P.: The Hydrology of the Nile, IAHS Special Publication no. 5, IAHS Press, Wallingford, 192 pp., ISBN 978-1-901502-75-6, 1999.

Sutcliffe, J. V. and Petersen, G.: Lake Victoria: derivation of a corrected natural water level series, Hydrolog. Sci. J., 52, 1316–1321, https://doi.org/10.1623/hysj.52.6.1316, 2007.

Tate, E., Sutcliffe, J., Conway, D., and Farquharson, F.: Water balance of Lake Victoria: update to 2000 and climate change modelling to 2100, Hydrolog. Sci. J., 49, 563–574, https://doi.org/10.1623/hysj.49.4.563.54422, 2004.

Taye, M. T., Ntegeka, V., Ogiramoi, N. P., and Willems, P.: Assessment of climate change impact on hydrological extremes in two source regions of the Nile River Basin, Hydrol. Earth Syst. Sci., 15, 209–222, https://doi.org/10.5194/hess-15-209-2011, 2011.

Thiery, W., Davin, E. L., Seneviratne, S. I., Bedka, K., Lhermitte, S., and van Lipzig, N. P. M.: Hazardous thunderstorm intensification over Lake Victoria, Nat. Commun., 7, 12786, https://doi.org/10.1038/ncomms12786, 2016.

Vanderkelen, I., van Lipzig, N. P. M., and Thiery, W.: Modelling the water balance of Lake Victoria (East Africa) – Part 1: Observational analysis, Hydrol. Earth Syst. Sci., 22, 5509–5525, https://doi.org/10.5194/hess-22-5509-2018, 2018a.

Vanderkelen, I., van Lipzig, N. P. M., and Thiery, W.: Modelling the water balance of Lake Victoria (East Africa) – Part 2: Future projections, Hydrol. Earth Syst. Sci., 22, 5527–5549, https://doi.org/10.5194/hess-22-5527-2018, 2018b.

WMO – World Meteorological Organisation: Hydrometeorological survey of the catchments of lakes Victoria, Kyoga and Albert: Burundi, Egypt, Kenya, Rwanda, Sudan, United Republic of Tanzania and Uganda: Vol. 1 Meteorology and hydrology of the basin, World Meteorological Organisation report, Geneva, Switzerland, 1974.

WMO – World Meteorological Organisation: Hydrometeorological survey of the catchments of lakes Victoria, Kyoga and Mobutu Sese Seko: project findings and recommendations, World Meteorological Organisation report, Geneva, Switzerland, 1982.

Yin, X. and Nicholson, S. E.: The water balance of Lake Victoria, Hydrolog. Sci. J., 43, 789–811, https://doi.org/10.1080/02626669809492173, 1998.

Proc. IAHS, 384, 149–153, 2021
https://doi.org/10.5194/piahs-384-149-2021

Open Access
PIAHS

Hydrology of Large River Basins of Africa

Changements historiques à long terme du débit écoulé dans le Haut bassin du fleuve Sénégal

Cheikh Faye

Département de Géographie, U.F.R. Sciences et Technologies, Université Assane Seck de Ziguinchor, Laboratoire de Géomatique et d'Environnement, BP 523 Ziguinchor, Sénégal

Correspondence: Cheikh Faye (cheikh.faye@univ-zig.sn)

Published: 16 November 2021

Résumé. In this study, the flow trends on either side of the Manantali dam, in the upper Senegal River basin, were assessed. The assessment was carried out for the period 1970 to 2019 using the Sen slope and the Mann-Kendall test. The Mann-Kendall trend test was used to indicate the direction of the trends, while the Sen slope was used to assess the magnitude of the change. The results of the study showed that there was an upward trend in the flow of the Senegal River both annually and for every month upstream (Bafing-Makana station) and downstream (Bakel station) from the barrage. The strongest upward trend is noted in October on the upstream station with 4.12 m^3/month and in September on the downstream station with 10.99 m^3/month. In return, the weakest upward trend is observed in May on the upstream station with 0.07 m^3/month and in August on the downstream station with 0.67 m^3/month. The annual flow also showed a significant increase upstream (with a rate of 0.35 m^3/year) and downstream (with a rate of 4.53 m^3/year) with a confidence level of 99 %. The results of the study also showed that an increase in precipitation causes an increase in runoff, despite the construction of hydraulic structures.

1 Introduction

Les fortes concentrations de gaz à effet de serre provoquent une augmentation de la température globale (Crawford et al., 2019). L'augmentation de la température joue un rôle défavorable dans la modification des composants du cycle hydrologique global (Ahmed et al., 2019). Les précipitations sont considérées comme la composante la plus vigoureuse du cycle hydrologique mondial, qui aurait changé dans plusieurs régions du monde (Asfaw et al., 2018). L'altération des caractéristiques des précipitations provoque de graves conséquences sur la société sous la forme d'inondations et de sécheresses, ce qui peut avoir un impact négatif sur la situation socioéconomique des populations (Ezzine et al., 2014). Les modifications du débit des cours d'eau ou des rivières sont souvent associées aux modifications des précipitations (Rawshan et al., 2019).

Les évaluations de tendance du débit des fleuves fournissent des informations de base relatives à l'eau (Gajbhiye et al., 2016). Le Sénégal, situé en Afrique tropicale, a également connu des changements dans les précipitations et la température au cours des cinquante dernières années (Faye et al., 2019). Devant la succession d'épisodes climatologiques (sécheresses et inondations) et hydrologiques (crues et étiages) extrêmes, les études faites sur le bassin versant du fleuve Sénégal ont donc procédé à une analyse des données pour caractériser le changement climatique dans ce bassin. Sur la période la plus récente, certaines études ont mis en exergue la hausse de la pluie et de l'écoulement dans la zone depuis les années 2000 et qui présage d'une amélioration du régime hydrologique (Ali et al., 2008) et une recrudescences des inondations. La connaissance de l'évolution du débit des cours d'eau au Sénégal est souvent limitée en raison de la disponibilité et de la qualité des données. La présente étude a examiné les méthodes non paramétriques pour caractériser les tendances et leur signification des cours d'eau à partir de données historiques. L'évaluation a été réalisée sur des données de cinquante années allant de 1970 à 2019.

Published by Copernicus Publications on behalf of the International Association of Hydrological Sciences.

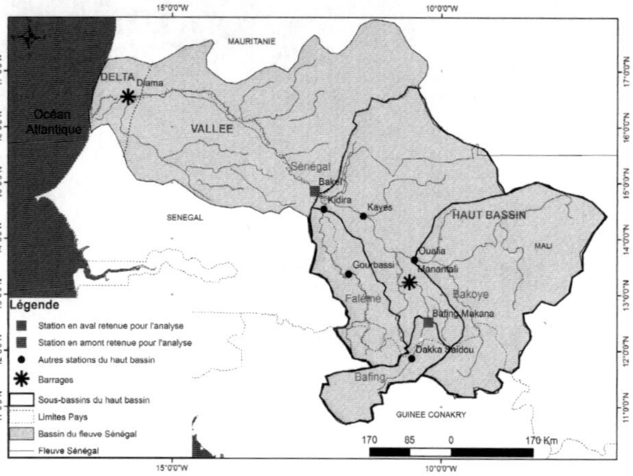

Figure 1. Localisation du bassin versant du fleuve Sénégal et de son haut bassin.

2 Zone d'étude

Le fleuve Sénégal, long de quelques 1700 km, draine un bassin de 300 000 km^2, à cheval sur quatre pays que sont, d'amont en aval, la Guinée, le Mali, le Sénégal et la Mauritanie (Fig. 1). Il va de 10°20′ à 17° N et de 7° à 12°20′ W et est formé de plusieurs affluents, dont les principaux sont le Bafing, le Bakoye et la Falémé qui prennent leurs sources en Guinée et forment le haut bassin (OMVS/Projet FEM/ BFS, 2008) (Fig. 1). Le fleuve Sénégal ainsi formé par la jonction entre le Bafing et le Bakoye, reçoit la Kolimbiné puis le Karokoro sur la droite et la Falémé sur la gauche, à 50 km en amont de Bakel. Dans la partie Sud du bassin, la densité du réseau hydrographique découle de la nature imperméable des sols et de la faiblesse de l'infiltration (Michel, 1973 ; Rochette, 1974).

3 Matériels et méthodes

3.1 Données

La base de données de la station amont (Bafing-Makana) et celle aval (Bakel) sont retenues pour ladite étude. Les données hydrométriques ont été mises à notre disposition par l'Organisation pour la Mise en Valeur du Fleuve Sénégal (OMVS). Ces données concernent les débits journaliers, (de 1970 à 2019), à partir desquels sont calculés les débits annuels et saisonniers sont utilisées dans l'analyse.

3.2 Méthodes

3.2.1 Analyse fréquentielle

Dans la présente étude, les distributions gamma, celle de la valeur extrême généralisée et la distribution normale, qui sont largement utilisées dans de nombreuses études climatiques, ont été appliquées aux jeux de données du bassin du fleuve Sénégal.

3.2.2 Estimation de l'ampleur du changement

Dans cette étude, la méthode de l'estimateur des pentes de Sen (Sen, 1968) a été appliquée pour calculer l'intensité du changement du débit de la rivière. La méthode est bien connue et largement utilisée pour le calcul des tendances (Gajbhiye et al., 2016).

3.2.3 Tests de tendance de Mann–Kendall

Le test de Mann Kendall (Mann, 1945) est un test non paramétrique fréquemment utilisé pour évaluer des tendances dans les séries chronologiques. Le test est également approuvé par l'Organisation météorologique mondiale (OMM) pour les évaluations des tendances hydrométéorologiques.

4 Résultats et discussion

4.1 Analyse fréquentielle

Les résultats obtenus sont indiqués dans le Tableau 1. Les valeurs en gras dans le Tableau indiquent que l'hypothèse nulle H0 : « absence de tendance dans la série » est rejetée au niveau de confiance de 95 %, tandis que les valeurs en italique indiquent le meilleur ajustement. A la station en amont (Bafing-Makana), il a été constaté que toutes les données suivaient la distribution gamma, tandis que l'hypothèse nulle avait été rejetée pour certains mois (juin à octobre), sur le module, sur le débit maximal et sur le débit des périodes de hautes et de basses eaux pour la distribution normale et GEV.

De même, pour la station en aval (Bakel), il a été constaté que toutes les données suivaient la distribution gamma. Dans la distribution généralisée des valeurs extrêmes, les mois de mars à septembre et de janvier à mars, le module, le débit maximal annuel et le débit des périodes de hautes et de basses eaux indiquaient une hypothèse nulle rejetée. De même, dans la distribution normale, l'hypothèse nulle rejetée sur les même variables (mois de mars à septembre et de janvier à mars, module, débit maximal annuel et débit des périodes de hautes et de basses eaux) en plus du mois d'octobre.

4.2 Distribution saisonnière du débit

Dans la partie amont barrage (Fig. 2), sur les deux périodes, on note une évolution similaire malgré des débits mensuels légèrement plus élevés sur la période 1988–2019. Les pics sont notés en septembre (avec 740 m^3/s de 1970 à 1987 et 867 m^3/s de 1988 à 2019) et les débits les plus faibles en avril en 1970–1987 (avec 3,8 m^3/s) et mai en 1988–2019 (avec 5,9 m^3/s). Les mois de la période de hautes eaux (juillet, août, septembre et octobre) affichaient les débits les

Tableau 1. Statistiques de test de Kolmogorov-Smirnov obtenues lors de la répartition par ajustement du débit du fleuve Sénégal aux stations en amont (Bafing-Makana) et en aval (Bakel) du barrage.

Paramètres	Station en amont			Station en aval		
	Gamma	GEV	Normale	Gamma	GEV	Normale
Mai	0,4287	0,3621	0,2297	0,4779	0,2466	0,1941
Juin	0,4183	0,0694	0,1631	0,4476	0,1100	0,1101
Juillet	0,4638	0,0737	0,1159	0,4717	0,0695	0,1270
Août	0,3907	0,1107	0,1143	0,5477	0,0869	0,1265
Septembre	0,4952	0,1125	0,1478	0,5078	0,1131	0,1052
Octobre	0,4166	0,0831	0,0933	0,5578	0,2996	0,1499
Novembre	0,5064	0,5165	0,2022	0,5144	0,7551	0,2028
Décembre	0,3534	0,2209	0,2052	0,4323	0,4774	0,2090
Janvier	0,3217	0,5292	0,2279	0,5338	0,1491	0,1676
Février	0,2990	0,5918	0,2444	0,4414	0,1144	0,1206
Mars	0,2883	0,4898	0,2055	0,4231	0,1654	0,1631
Avril	0,2907	0,5170	0,1645	0,4688	0,2225	0,2013
Module	0,3158	0,0997	0,1087	0,4808	0,0855	0,1256
Débit Maximum	0,4665	0,0618	0,0651	0,4897	0,0738	0,0758
Débit minimum	0,3711	0,6479	0,2338	0,4490	0,3125	0,2015
Hautes eaux	0,3884	0,1000	0,0938	0,5254	0,0650	0,1142
Basses eaux	0,3269	0,0695	0,1546	0,4474	0,1174	0,1099

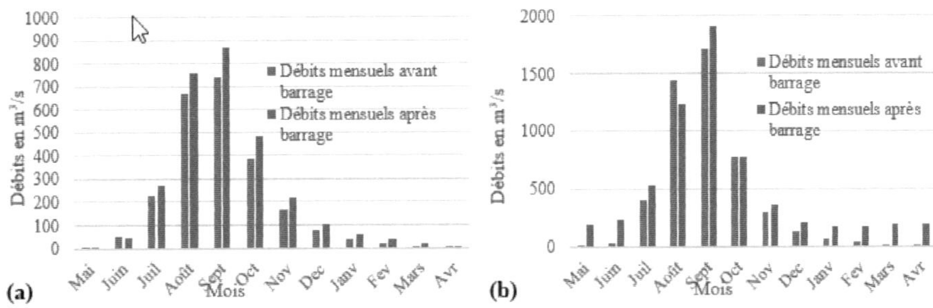

Figure 2. Distribution saisonnière du débit enregistré en amont (**a**) et en aval (**b**) du barrage de Manantali.

plus élevés sur les deux périodes. Sur la période de basses eaux (le reste de l'année), les débits étaient très faibles.

Dans la partie aval barrage (Fig. 2), l'évolution du débit montre une nette opposition entre la période 1970–1987 (dont l'évolution est assez semblable de celle notée à Bafing-Makana sur les deux périodes) et celle 1988–2019 qui a vu ses débits en période de basses eaux fortement augmentés (son débit minimal est plus de 90 fois plus important que celui de 1970–1987). Cette augmentation s'explique par la politique de soutien des étiages du barrage de Manantali. Toutefois, sur les deux périodes, les pics sont notés en septembre (avec 1714 m^3/s en 1970–1987 et 1902 m^3/s en 1988–2019) et les débits les plus faibles en mai en 1970–1987 (avec 2,07 m^3/s) et en 1988–2019 (avec 188 m^3/s). Contrairement à la période avant barrage, sur la période après barrage, on peut noter que le débit a commencé à augmenter à partir du mois de janvier. Cette augmentation s'est opérée après une diminution allant de septembre à janvier, ce qui est en contradiction avec le débit avant le barrage, où il a commencé à diminuer à partir du mois de septembre jusqu'en avril. Les augmentations de débit sont dues au surplus d'eau du réservoir et au maintien d'un débit minimal dans le fleuve (Faye, 2015). Il ressort aussi de la Fig. 2 que le débit était très faible de décembre à mai en amont, alors qu'il s'élève sur cette période au niveau de la partie aval.

4.3 Tendances des flux mensuels et annuels

Le Tableau 2 présente l'importance de la statistique Z (tau de Kendall) et la pente de Sen aux niveaux de confiance de 90 % à 95 % de chaque variable. Les résultats montrent des augmentations importantes des tendances des deux côtés du barrage (de toutes les variables, seul le mois de juin a connu une tendance à la baisse de son écoulement à la station de Bafing-Makana). En amont, les mois d'octobre à avril ont montré des tendances à la hausse significatives au niveau de confiance de 99 %. Les tendances de mai étaient significatives au niveau de confiance de 95 %, tandis que les ten-

Tableau 2. Tendances des débits mensuels et moyens annuels des rivières obtenues à l'aide de la pente de Sen et du test de tendance de Mann-Kendall situé en amont et en aval du barrage de Manantal.

Paramètres	Station en amont (Bafing Makana)			Station en aval (Bakel		
	Tau de Kendall	Significativité	Pente de Sen	Tau de Kendall	Significativité	Pente de Sen
Mai	0,2135	**	0,0712	0,6344	***	6,9206
Juin	−0,0034		−0,0077	0,6361	***	7,7636
Juillet	0,1650	*	1,8607	0,3844	***	5,6139
Août	0,1480		2,8723	0,0119		0,6723
Septembre	0,1361		2,3319	0,1378		10,9858
Octobre	0,3180	***	4,1147	0,0493		1,5702
Novembre	0,4286	***	2,6147	0,1259		1,5357
Décembre	0,3861	***	1,1942	0,2126	**	1,8892
Janvier	0,4082	***	0,7428	0,4337	***	3,2915
Février	0,4541	***	0,5293	0,6020	***	4,9282
Mars	0,4507	***	0,3203	0,6003	***	5,9834
Avril	0,4065	***	0,1739	0,6344	***	6,7668
Module	0,3452	***	1,821	0,2619	***	4,531
Débit Maximum	0,0255		2,1833	0,0952		10,8645
Débit minimum	0,5352	***	0,0709	0,6785	***	3,9356
Hautes eaux	0,2466	**	2,9862	0,0884		4,3142
Basses eaux	0,3384	***	0,8891	0,5748	***	5,3217

* Significative à 90 %, ** Significative à 95 % et *** Significative à 99 %

dances de juillet étaient significatives au niveau de confiance de 90 %. Il est important de mentionner ici que le module, le débit minimal et le débit en période de basses eaux avaient également des tendances à la hausse significatives au niveau de confiance de 99 %. La tendance significative la plus élevée avec 4,11 m^3/an a été observée pour le mois d'octobre, suivie du mois d'août avec 2,87 m^3/s/an. Par ailleurs, la tendance du module a montré une variation de 1,82 m^3/s/an avec un niveau de confiance de 99 %. On peut également noter qu'un seul débit mensuel a montré une légère tendance à la baisse (le mois de juin avec −0,0034 m^3/s/an. Une augmentation des tendances du débit des cours d'eau a également été observée dans d'autres parties du monde (Soro et al., 2014).

Le barrage indique une différence du nombre de mois de hautes eaux (c'est-à-dire mois à CMD > 1). Sur la période 1970–1987, le nombre de mois de hautes eaux est de 3 mois, alors que sur la période 1988–2019, il est de 4 mois. Ainsi, la mise en eau du barrage de Manantali change le régime du bassin à la station de Bakel qui passe d'un régime fluvial tropical pur (avant barrage) à un régime fluvial tropical de transition (après barrage). De plus, le débit de la période de hautes eaux est passé de 506 m^3/s (en 1970–1987) à 595 m^3/s (en 1988–2019) dans la partie amont du bassin (soit une hausse de 17,6 %) et de 1085 m^3/s (en 1970–1987) à 1109 m^3/s (en 1988–2019) dans la partie aval du bassin (soit une faible hausse de 2,21 %).

5 Conclusions

Cette étude avait comme objectif de caractériser les tendances du débit de part et d'autre du barrage de Manantali, dans le haut bassin du fleuve Sénégal. L'évaluation a été réalisée sur la période de 1970 à 2019 avec la méthode d'estimateur de la pente de Sen et le test de Mann-Kendall. Les résultats de l'analyse des tendances ont montré que le débit du fleuve avait augmenté presque tous les mois des deux côtés du barrage. Le mois d'octobre a enregistré le taux de changement le plus élevé en amont, tandis que le mois de septembre a enregistré le taux de changement le plus élevé en aval du barrage. Les mois de mai et d'août ont montré les changements les plus faibles respectivement en amont et en aval du barrage. Des changements importants ont été remarqués aux mois de mai, juin, janvier, février et mars. De même, les modifications du débit annuel du fleuve ont également été jugées importantes. Des changements significatifs positifs ont également été observés d'abord dans les débits maximum et minimum et ensuite dans les débits en périodes de hautes et de basses eaux des deux côtés du barrage. Il a été constaté que le débit maximal avait augmenté de manière relativement significative du côté amont du barrage, ce qui indique une augmentation des précipitations. De plus, la construction de structures hydrauliques peut être une cause majeure de la hausse du débit en période d'étiages. Les résultats de cette étude devraient fournir une dcumentation importante à la planification et à la gestion de l'agriculture et des ressources en eau du bassin du fleuve Sénégal.

Disponibilité des données. Aucun ensemble de données n'a été utilisé dans cet article.

Intérêts concurrents. The contact author has declared that there are no competing interests.

Clause de non-responsabilité. Publisher's note : Copernicus Publications remains neutral with regard to jurisdictional claims in published maps and institutional affiliations.

Déclaration du numéro spécial. This article is part of the special issue "Hydrology of Large River Basins of Africa". It is a result of the 4th International Conference on the "Hydrology of the Great Rivers of Africa", Cotonou, Benin, 13–20 November 2021.

Financement. This research has been supported by the Systematics Association.

Références

Ahmed, K., Shahid, S., Chung, E.-S., Wang, X.-J., and Harun S. B. : Climate Change Uncertainties in Seasonal Drought Severity-Area-Frequency Curves : Case of Arid Region of Pakistan, J. Hydrol., 570, 473–485, 2019.

Ali, A., Lebel, T. and Amami, A. : Signification et usage de l'indice pluviométrique au Sahel, Sécheresse, 19, 227–235, 2008.

Asfaw, A., Simane, B., Hassen, A., and Bantider, A. : Variability and time series trend analysis of rainfall and temperature in northcentral Ethiopia : A case study in Woleka sub-basin, Weather Clim. Extremes, 19, 29–41, 2018.

Crawford, J., Venkataraman, K., and Booth, J. : Developing climate model ensembles : A comparative case study, J. Hydrol., 568, 160–173, 2019.

Ezzine, H., Bouziane, A., and Ouazar, D. : Seasonal comparisons of meteorological and agricultural drought indices in Morocco using open short time-series data, Int. J. Appl. Earth Obs. Geoinf., 26, 36–48, 2014.

Faye, C. : Impact du changement climatique et du barrage de Manantali sur la dynamique du régime hydrologique du fleuve Sénégal à Bakel (1950–2014), Bulletin de la Société Géographique de Liège : BSGLg, 64, 69–82, 2015.

Faye, C., Grippa, M., and Wood, S. : Use of the Standardized Precipitation and Evapotranspiration Index (SPEI) from 1950 to 2018 to determine drought trends in the Senegalese territory, Climate Change, 5, 327–341, 2019.

Gajbhiye, S., Meshram, C., Singh, S. K., Srivastava, P. K., and Islam, T. : Precipitation trend analysis of Sindh River basin, India, from 102-year record (1901–2002), Atmos. Sci. Lett., 17, 71–77, 2016.

Michel, P. : Les bassins des fleuves Sénégal et Gambie : Etude géomorphologique, Mémoires ORSTOM n° 63-3tomes, 752 pp., 1973.

Mann, H. B. : Non-Parametric Test against Trend, Econometrica, 13, 245–259, https://doi.org/10.2307/1907187, 1945.

OMVS, Projet FEM/Bassin du fleuve senegal : Plan d'Action Stratégique de Gestion des Problèmes Environnementaux Prioritaires du Bassin du Fleuve Sénégal, Version finale, 133 pp., 2008.

Rawshan, A., Arez, I., Arien, H., and Nadeem, N. : Long Term Historic Changes in the Flow of Lesser Zab River, Iraq, Hydrology, 6, 1–12, https://doi.org/10.3390/hydrology6010022, 2019.

Rochette, C. : Monographie hydrologique du fleuve Sénégal, Coll. Mém. ORSTOM, 1442 pp., 1974.

Sen, P. K. : Estimates of the regression coefficient based on Kendall's tau, J. Am. Stat. Assoc., 63, 1379–1389, 1968.

Soro, G. E., Anouman, D. G. L., Goula, B. I. T. A., Srohorou, B., and Savane, I. : Caractérisation des séquences de sécheresse météorologique a diverses échelles de temps en climat de type soudanais : cas de l'extrême Nord-ouest de la cote d'ivoire, Larhyss/Journal, 18, 107–124, 2014.

Proc. IAHS, 384, 155–161, 2021
https://doi.org/10.5194/piahs-384-155-2021

Bilan de masse et érosion dans le bassin de l'Oued Mina Nord-ouest algérien

Faiza Hallouz[1,2], Abdelhadi Ammari[2], Mohamed Meddi[2], Ouarda Yasmine Nait Yahia[2], and Ali Rahmani Salah Eddine[3]

[1]University of Khemis Miliana, 44225 Ain Defla, Algeria
[2]LGEE, École Nationale Supérieur d'Hydraulique, BP 31, 09000 Blida, Algeria
[3]Geo-Environment Laboratory, Faculty of Earth Sciences, Geography and Spatial Planning, University of Sciences and Technology Houari Boumediene, 16111 Bab Ezzouar Algiers, Algeria

Correspondence: Faiza Hallouz (f.hallouz@univ-dbkm.dz)

Published: 16 November 2021

Résumé. Le taux de sédimentation des barrages du bassin hydrographique du Cheliff reste le plus élevé de toutes les régions de l'Algérie. L'oued Mina est un affluent majeur de l'oued Cheliff, le plus grand oued de l'Algérie qui transporte un taux de sédiment important et le barrage de Sidi M'Hamed Ben Aouda (Relizane) qui barre cet oued a reçu un volume de vase égal à 6,7 millions de m^3 par année. L'objectif de cette étude est de proposer un modèle d'estimation de l'érosion spécifique permettant de construire un ensemble d'indicateurs afin d'identifier les zones prioritaires nécessitant le renforcement des moyens de lutte contre l'érosion et une exploitation meilleure des barrages qui sont la seule ressource pour l'irrigation et l'activité agricole de la région en prolongeant leur durée de vie. Nous avons procédé à l'analyse des données journalières de pluie et de débits des stations hydro-pluviométriques d'oued El Abtal (de 1985–2010) du sous bassin versant dont les apports s'accumulent dans le barrage SMBA. Ce travail nous a permis d'élaborer une carte de sensibilisation qui délimite les zones vulnérables selon le taux d'érosion spécifique (Es oued Haddad = 18.3, Es oued El Abd = 7.3, Es oued Mina = 10.47 ($t\,ha^{-1}\,an^{-1}$) … calculés à partir du modèle trouvé dans cette étude. Enfin, l'état érosif des versants a été divisé en quatre classes principales de vulnérabilité des sols, à savoir : les niveaux « négligeable », « modéré », « élevé » et « très élevé ». Cette carte permet de faciliter la prise de décisions relatives à la priorisation des zones à préserver ou à restaurer.

1 Introduction

Depuis des siècles, les géographes ont considéré la zone des bassins méditerranéens comme très sensible à l'érosion sous toutes ses formes (Roose et al., 2012). L'Algérie, faisant partie de cette zone, n'est pas à l'abri de ce phénomène, il est parmi les pays les plus menacés par l'érosion hydrique (Achite et al., 2006).

En effet, en Algérie, le taux d'érosion spécifique atteint les valeurs les plus importantes d'Afrique du Nord, dépassant les 2000 $t\,km^{-2}\,an^{-1}$ et concerne la plupart des bassins versants de l'Atlas tellien (Rhiou, Sly, Fodda, Mazafran, Isser, Soummam…), il atteint 4000 $t\,km^{-2}\,an^{-1}$ sur la chaîne des fleuves côtiers Dahra et 5000 $t\,km^{-2}\,an^{-1}$ sur le bassin d'Ighil-Emda (Demmak, 1982). L'Algérie est de ce fait l'un des pays les plus menacés dans le monde par l'érosion.

L'envasement est la conséquence directe de l'érosion et la sédimentation, qui sont aussi la cause de la dégradation des sols agricoles ; 45 % des sols agricoles du Nord Algérien sont directement menacés par l'érosion, soit 12 Millions d'hectares (Chebbani et al., 1999).

Le taux de sédimentation des barrages du bassin hydrographique du Cheliff reste le plus élevé de toutes les régions de l'Algérie (Remini, 2006). L'oued Mina est un affluent majeur de l'oued Cheliff, le plus grand oued de l'Algérie qui transporte un taux de sédiment important. A titre d'exemple, le barrage de Sidi M'Hamed Ben Aouda (Relizane) qui barre cet oued a reçu un volume de vase égal à 6,7 millions de m^3

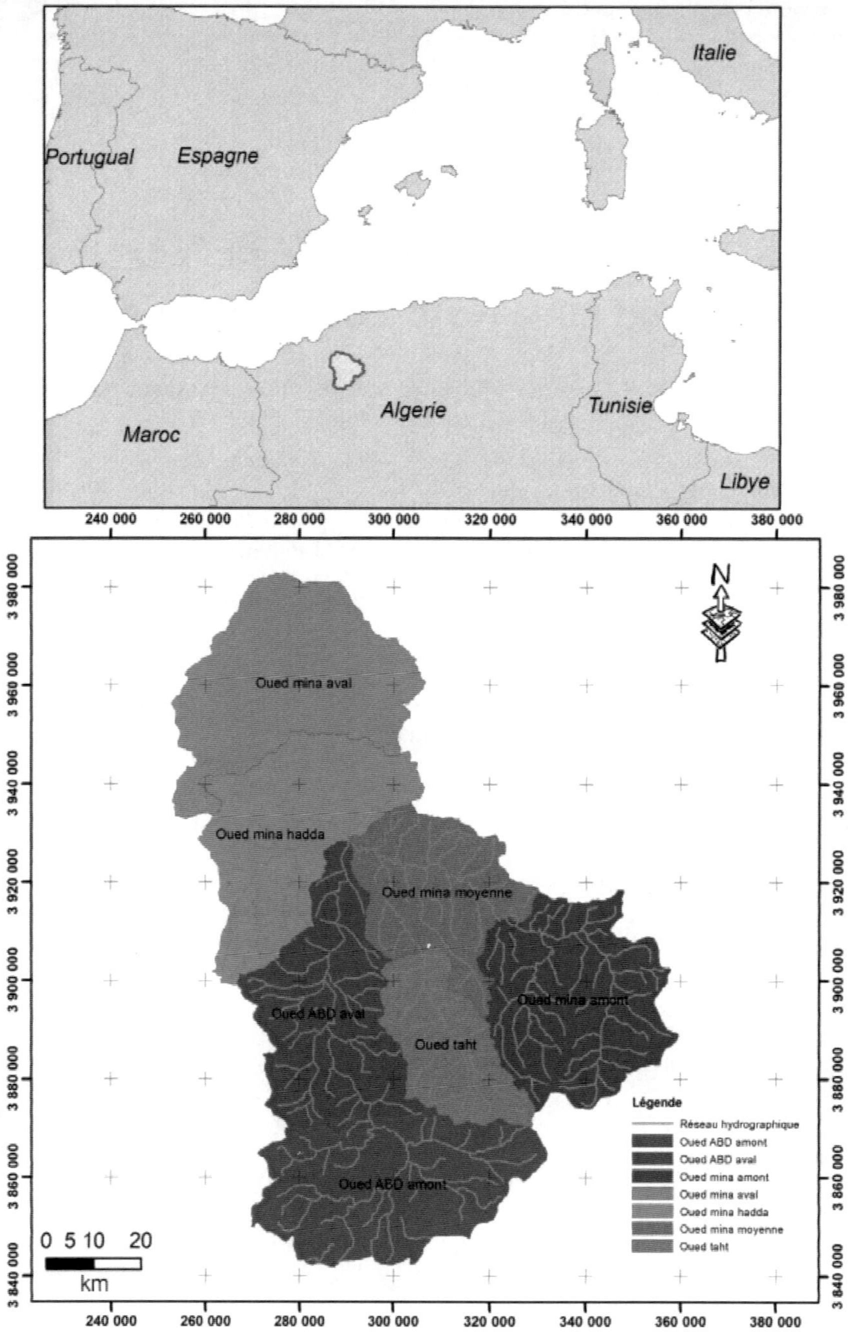

Figure 1. Bassin versant de l'oued Mina.

par année durant la période 1995–2003 (Remini et Bensafia, 2012 ; Meddi, 1992 ; Kouri, 1993 ; Touaïbia, 2000 ; Achite, 2002 ; Hallouz, 2013 ; Toumi, 2013).

La maîtrise de ce phénomène ne peut passer qu'aÌ travers des modèles qui nous permettent de décrire et de quantifier dans la mesure du possible les taux de dégradation des sols et ainsi les conséquences qui se suivent. La validitéì de ces modèles est tributaire de la qualité et la quantité des données disponibles. Le manque de données conduit souvent l'ingénieur à utiliser les méthodes empiriques disponibles afin d'évaluer les quantités de sédiments transportés annuellement par les cours d'eau (Meddi et al., 1998), sans que nous nous ayons des dispositifs de contrôle de ces résultats.

Le principal objectif de ce travail est de proposer un modèle d'estimation de l'érosion spécifique permettant de construire un ou un ensemble d'indicateurs permettant d'identifier les zones prioritaires nécessitant le renforcement des moyens de lutte contre l'érosion et déterminant des scénarios pour une gestion de l'occupation et l'utilisation du sol à long terme dans le but de minimiser l'érosion hydrique et une exploitation meilleure des retenues et des barrages qui sont la seule ressource pour l'irrigation et l'activité agricole de la région en prolongeant leur durée de vie. Nous tenterons de mettre en évidence les risques directs d'envasement des barrages situés dans le bassin de l'oued Mina, des risques liés directement aÌ la cadence accélérée de l'érosion hydrique, avec la dégradation continue du couvert végétal comme facteur principal.

Vue le changement continue du cycle hydrologique et du climat de la région, le modèle le plus représentatif est certainement celui baseì sur les données les plus récentes, malheureusement, la décennie noire qu'a connue l'Algérie a eu son impact sur la collecte des données, pour cela, nous nous sommes intéressés à la mesure hydrologique, en proposant des méthodes simples et précises, qui permettrons sans doute de pallier au manque de données hydrométriques et du transport solide, en augmentant facilement la fréquence des mesures avec une meilleur qualité de la donnée.

Nous proposons à la fin de cette étude une approche cartographique de la variation de l'érosion hydrique afin de prévoir le risque érosif et prendre les précautions nécessaires de lutte contre les dégâts causés au niveau des barrages et des surfaces du sol.

2 Zone d'étude, données et méthodes

2.1 Présentation de la zone d'étude

Situé en zone semi-aride, le bassin versant de l'Oued Mina répond à notre objectif en raison de sa forte érodabilité et de la disponibilité des données pluviométriques et hydrométriques. Situé à environ 300 km à l'ouest d'Alger, entre 0°20′ et 1°10′ E et entre 34°40′ et 35°40′ N, le bassin versant de l'Oued Mina draine, au niveau du Barrage de Sidi M'Hamed Benaouda (SMBA), une superficie de 6000 km^2 (Fig. 1). L'Oued Mina est parmi les principaux affluents de l'Oued Cheliff. Il parcourt une distance de 90 km environ entre les barrages de Bakhada et de Sidi M'Hamed Benaouda avec une orientation Sud Est–Nord-Ouest.

Le climat de la région est de type semi-aride méditerranéen avec une précipitation moyenne annuelle de 305 mm marquée par une irrégularité aussi bien saisonnière qu'interannuelle (Meddi, 1992). Par ailleurs, l'analyse de la répartition géographique et de la diversité des formations végétales sur le bassin a montré deux parties nettement distinctes (Mahieddine, 1997 ; Kouri, 1993) : la zone Nord purement marneuse, fortement érodée et dépourvue de végétation à l'exception de quelques îlots de reboisement et de plantations d'arbres fruitiers dans la vallée ; et la zone sud moins érodée et dont environ 50 % de la surface sont recouverts par une végétation de densité très variable et très parsemée, allant de la forêt (Pin d'Alep) localisée au maquis très éparse. Il y a lieu de noter qu'une agriculture d'autosubsistance y règne avec une surexploitation des sols, un défrichement permanent et un surpâturage intensif.

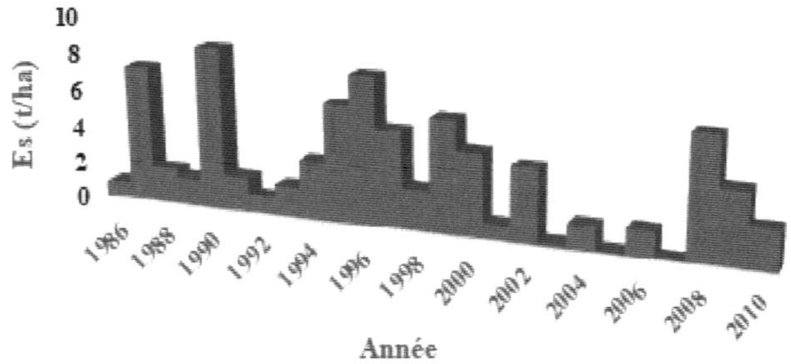

Figure 2. Dégradation spécifique dans le bassin de l'Oued Mina.

2.2 Acquisition et préparation des données

Dans cette étude nous avons utilisé les données journalières de pluies et de débits de la station d'Oued El Abtal sur une période allant de 1985 à 2010 (25 années). Ces données proviennent de la base de données de l'Agence Nationale des Ressources Hydrauliques (ANRH, 2010). Le débit liquide moyen interannuel dans ce bassin est égal à 2,44 m^3/s avec un coefficient de variation de 0.44 et une pluie moyenne annuelle de l'ordre de 430 mm (ANRH, 2010).

2.3 Estimation de la dégradation spécifique

La dégradation spécifique ou l'érosion spécifique (ES) donne une idée sur l'érodabilité de la région, et donc sur la vulnérabilité à l'envasement des ouvrages de stockage existants. L'érosion spécifique moyenne interannuelle est le rapport entre l'apport solide moyen interannuel et la superficie du bassin drainé (Touainia et Achite, 2003).

$$\mathrm{Es} = \frac{\mathrm{As}}{S} \text{ en } [\mathrm{t\,km^{-2}\,an^{-1}}] \text{ ou } [\mathrm{t\,ha^{-1}\,an^{-1}}] \quad (1)$$

Avec : As : Apport solide moyen interannuel [t km^{-2}], S : Superficie du bassin drainé [km^2].

2.4 Modélisation du transport solide

Afin de générer un modèle représentatif de l'érosion dans notre zone d'étude qui est d'une superficie de plus de 6000 km^2 en utilisant des indicateurs en rapport avec les conditions sur site, nous avons procédé à une analyse multivariée, avec des régressions multiples, de l'érosion spécifique en fonction de divers paramètres (climatiques, hydro

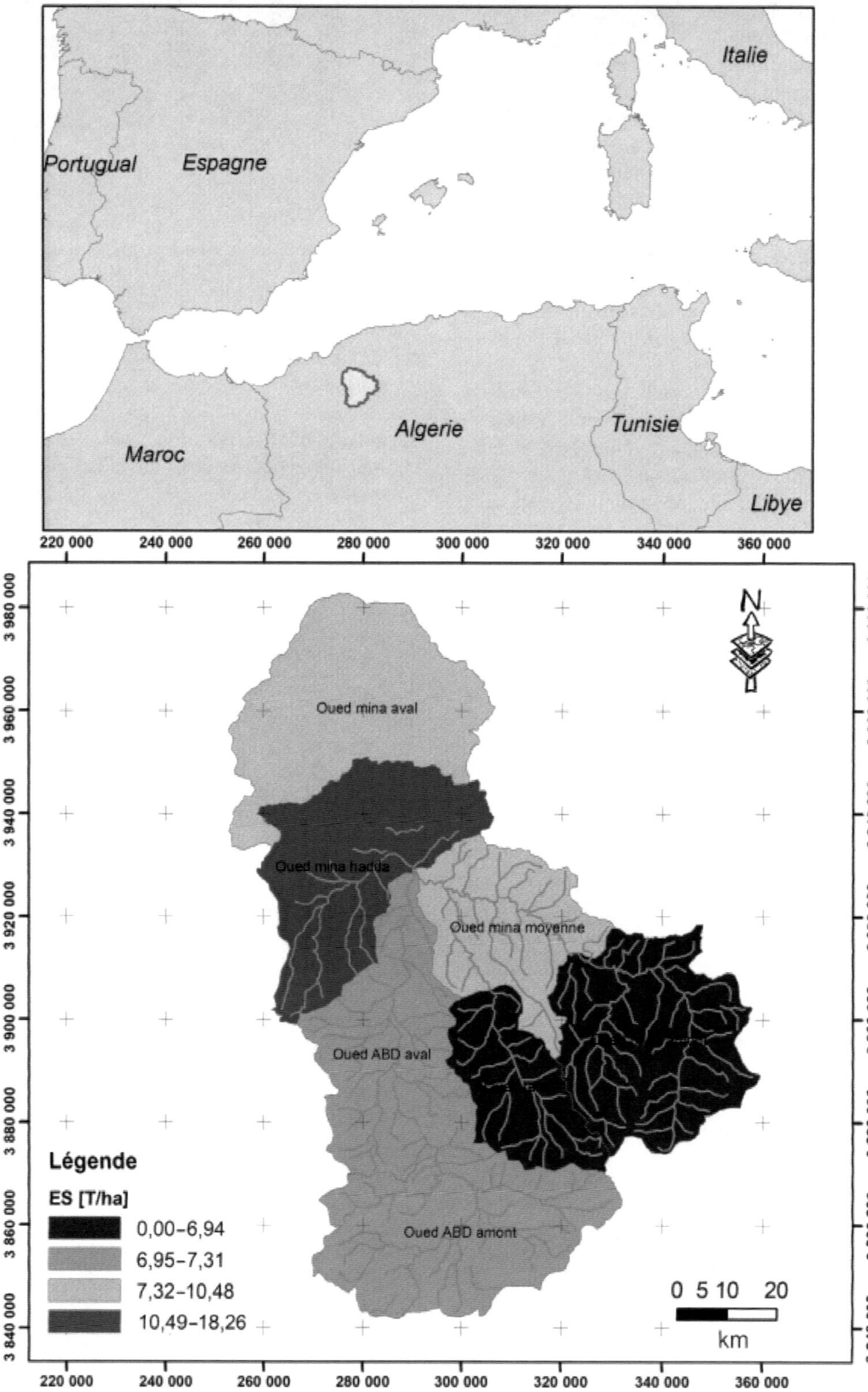

Figure 3. Évolution de l'érosion spécifique dans le bassin versant de l'Oued Mina.

Tableau 1. Récapitulatif des bassins versants.

Sous bassins	S	P	Ic	L	l	L_{TP}	D	Ds	Ig	Le	Dd	Ct	Tc
Oued Haddad	470	100	1.29	36.95	12.58	47	935	542	1.84	23.71	0.53	0.05	10.1
Oued EL ABTAL	5365	256	1.45	108.3	40.54	120	660	312	0.61	23.12	0.45	0.05	17.4
Oued Taht	852	153	1.46	62.9	19.78	67.5	660	307	1.05	16.70	0.46	0.03	10.4
Oued Mina	2190	275	1.64	119.1	41.3	130	870	342	0.73	14.77	0.4	0.04	17.5
Au droit du barrage SMBA	6132	352	1.26	128.5	47.73	135	1194	728	0.6	14.52	0.96	0.08	23.5

On a rassemblé tous les facteurs susceptibles d'influencer le phénomène à savoir (Ammari, 2012) : La superficie des sous bassins (S) (km^2). La précipitation moyenne interannuelle (P) (mm). L'indice de compacité (Ic). L : longueur du rectangle équivalent (km). l : largeur du rectangle équivalent (km). Longueur du talweg principal (L_{TP}) (km). Altitude moyenne (D) (m). Dénivelée spécifique (Ds) (m). L'indice de pente globale (Ig). La lame écoulée moyenne interannuelle (Le) (cm). La densité de drainage (Dd) (km/km^2). Le coefficient de torrentialité (Ct). Temps de concentration (Tc) (h).

Tableau 2. Matrice de corrélation.

Variable	Es ($t\,ha^{-1}\,an^{-1}$)	S (km^2)	Ic	Dd ($km\,km^{-2}$)	Ig	Ct	Le (cm)	P (cm)	Ce (%)
Es	1,00								
S	0,31	1,00							
Ic	−0,77	−0,01	1,00						
Dd	0,49	0,58	0,16	1,00					
Ig	−0,03	−0,90	0,00	−0,17	1,00				
Ct	0,83	0,70	−0,66	0,49	−0,52	1,00			
Le	0,01	0,58	0,42	0,47	−0,44	0,04	1,00		
P	−0,22	0,38	−0,13	−0,51	−0,70	0,09	0,28	1,00	
Ce	0,42	0,68	0,03	0,57	−0,46	0,40	0,91	0,24	1,00

Es : Érosion spécifique moyenne interannuelle ($t\,ha^{-1}\,an^{-1}$). S : La superficie des sous bassins (km^2). Ic : L'indice de compacité. Dd : La densité de drainage (km/km^2). Ig : L'indice de pente globale. Ct : Le coefficient de torrentialité. Le : La lame écoulée moyenne interannuelle (cm). P : La précipitation moyenne interannuelle (cm). Ce : Le coefficient d'écoulement.

morpho-métriques ...) contenus dans le Tableau 1. Ces regressions ont e´te´ calcule´es en utilisant le logiciel statistique XL-STAT (2013).

3 Résultats et discussion

3.1 Estimation de la dégradation spécifique

Dans le bassin versant de l'oued Mina, on estime une dégradation spécifique moyenne interannuelle de l'ordre de $4{,}4\,t\,ha^{-1}\,an^{-1}$ (Fig. 2).

Les valeurs de l'érosion spécifique calculées sont expliquées par la nature du relief, la géologie ainsi que la nature des cours d'eau dans le bassin versant; les dépôts alluvionnaires et les lâchées du barrage Bakhadda favorisent également le phénomène. Le couvert végétal contribue énormément à la réduction de l'action érosive aussi.

Toutefois, on constate que ces valeurs concordent avec un certain nombre de travaux d'aménagement de la région d'étude sur les différents niveaux (Meddi, 1992; Touaibia et Achite, 2003; Hallouz et al., 2012, 2018).

3.2 Modélisation du transport solide

La matrice des coefficients de corrélation est illustrée dans le Tableau 2.

La sélection finale des paramètres est basée sur l'étude de régression multiple.Nous utilisons le logarithme du paramètre pour linéariser le modèle.

Ainsi, le modèle le plus représentatif est :

$$\mathrm{Es} = e^{8.93} L_e{}^{0.83} C_t{}^{0.87} D_d{}^{0.23} S^{-0.87} \quad (2)$$

Avec des coefficients de régression multiple et partielle proche de 1.

Le test de Student a démontré la significativité du coefficient le plus petit au seuil de 5 % comme le suggère Laborde (2003) car il a moins de 5 % de chance qu'il est dû au hasard.

Sur la table de Fisher – Snedecor, pour un seuil de confiance de 5 % et pour des degrés de liberté de $(p-1)$ et $(n-p+1)$: $F_{0,95} = 9{,}55 < F_{obs}$. D'après le résultat obtenu, on peut dire que le modèle est significatif.

Après avoir développé un modèle de calcul de l'érosion spécifique avec analyse multivariée, une cartographie de la vulnérabilité à l'érosion hydrique a été réalisée (Fig. 3). La méthode est basée sur des données saisies sur SIG (QGis), en représentant l'érosion des sols de l'ensemble de la zone d'étude.

Ce support cartographique nous permet de visualiser les différentes zones à risque érosif sur l'ensemble du bassin de la Mina. Ce dernier partagé en quatre sous bassin versant et ainsi partagé en quatre régions suivant le degré d'érosion dans chacune d'elle.

On remarque un faible risque dans la région sud-ouest (oued Abd) et encore moins en est (oued Mina amont et oued Taht) car la région est caractérisée par des formations géologiques et lithologique denses et consolidées. Tandis que le risque augmente en allant vers le haut (vers le Nord) appelée Oued Haddad situé en amont du barrage SMBA ce qui explique clairement le risque d'envasement du réservoir de cet ouvrage hydraulique.

Enfin, cette étude nous a permis d'aboutir à des indicateurs permettant d'identifier les zones prioritaires (Fig. 3) nécessitant le renforcement des moyens de lutte contre l'érosion et déterminant des scénarios pour une gestion de l'occupation et l'utilisation du sol à long terme dans le but de minimiser l'érosion hydrique et une exploitation meilleure des retenues et des barrages qui sont la seule ressource pour l'irrigation et l'activité agricole de la région en prolongeant leur durée de vie.

4 Conclusions

En Algérie, l'érosion hydrique est un problème majeur. Elle devient dangereuse dès qu'elle dépasse le seuil tolérable et son intensitéì varie d'une zone aÌ l'autre.

L'objectif de ce travail est la mise au point d'une méthode d'estimation de l'érosion spécifique qui permettra aux gestionnaires des ouvrages et au projeteurs de projet de mieux prédire ce phénomène et ainsi prédéterminer l'importance et la nature des ouvrages et les pertes de capacité possible suite à l'envasement ; les résultats estimés devraient être confrontés aux observations et au mesures lors de l'études pour un meilleur choix.

Par ailleurs, nous avons analysé des données journalières de pluie et de débits des stations hydro-pluviométriques d'oued El Abtal (1985–2010) du sous bassin versant dont les apports s'accumulent dans le barrage SMBA.

Un modèle basé sur une analyse multivariée de divers facteurs climatiques, hydro morpho- métriques nous a permis d'aboutir à des indicateurs influençant les scénarios du risque érosif sur le bassin versant. Ce travail a permis d'élaborer une carte de sensibilisation qui délimite les zones vulnérables selon le taux d'érosion spécifique (Es oued Haddad = 18.3, Es oued El Abd = 7.3, Es oued Mina = 10.47 (t ha an^{-1}) …calculés à partir du modèle trouvé.

Enfin, le risque d'érosion augmente en allant vers le haut (vers le Nord) sous bassin de l'Oued Haddad situé en amont du barrage SMBA ce qui explique clairement le risque d'envasement du réservoir de cet ouvrage hydraulique.

Disponibilité du code. Le code du logiciel XL-STAT qui a été utilisé dans cette étude peut être trouvé à l'adresse suivante : XL-STAT : id de la license est 0117590, version du 04.02.2013 (Addinsoft, 2014).

Collaborateurs. FH wrote the main paper, and FH, AA, MM, OYNY, ARSE discussed the results and implications and commented on the manuscript at all stages. FH, AA, MM, OYNY, ARSE contributed extensively to the work presented in this paper.

Intérêts concurrents. Les auteurs déclarent qu'ils n'ont aucun conflit d'intérêts.

Clause de non-responsabilité. Publisher's note : Copernicus Publications remains neutral with regard to jurisdictional claims in published maps and institutional affiliations.

Déclaration du numéro spécial. This article is part of the special issue "Hydrology of Large River Basins of Africa". It is a result of the 4th International Conference on the "Hydrology of the Great Rivers of Africa", Cotonou, Benin, 13–20 November 2021.

Remerciements. Merci a` l'Agence Nationale des Ressources Hydrauliques (ANRH) pour les donne´es hydrome´triques et pluviome´triques. Ce travail a e´te´ re´alise´ dans le cadre du projet SICMED Indicateurs, composante de MISTRALS SICMED, INSU CNRS.

Références

Achite, M. : Approche statistique d'e´valuation du transport solide dans le bassin versant de l'oued Mina (nord ouest Alge´rien), Colloque international sur l'eau dans le bassin me´diterrane´en : Ressources et de´veloppement durable, 10–13 octobre 2002, Monastir (Tunisie), 894–899, 2002.

Achite, M., Touaibia, B., and Ouillon, S. : "Érosion hydrique en Algérie du Nord : Ampleur, Conséquences and Perspectives « Haddad (Nord-Ouest algérien) » 14th International Soil Conservation Organization Conference, Water Management and Soil Conservation in Semi-Arid Environments. Marrakech, Morocco, 14–19 May 2006, 2006.

Addinsoft : XLSTAT, Analyse de données et statistique avec MS Excel, Addinsoft, NY, USA, 2014.

ANRH (Agence Nationale des Ressources Hydrauliques) : Les changements climatiques and leur impact sur les ressources en eau en Algérie, in : Proceedings of the Assises Nationales sur L'Eau, Fez, Morocco, 25–29 October 2010, ANRH, Alger, Algeria, 41 pp., 2010.

Chebbani, R., Djilli, K., et Roose, E. : Étude des risques d'eìrosion dans le bassin versant de l'Isser, Algérie, Bulletin Réseau Érosion, 19, 85–95, 1999.

Demmak, A. : Contribution à l'étude de l'érosion et des Transports Solides en Algérie Septentrionale, Ph.D. Thesis, Université de Pierre et Marie Curie, Paris, France, 1982.

Hallouz, F. : Transport Solide Dans le Bassin D'El Oued Mina et Sédimentation du Barrage SMBA, Ph.D. Thesis, Es-Sciences, École Nationale Supérieure d'Hydraulique (ENSH), Soumaâ, Algeria, 219 pp., 2013.

Hallouz, F., Meddi, M., et Mahé, G. : Relation débit liquide-transport de matières en suspension dans le bassin de l'Oued Mina, en amont du barrage Sidi M'Hamed Ben Aouda (Oued Cheliff), Nord-Ouest d'Algérie, 1*ère* Conférence internationale IS-RIVERS, Recherches et actions au service des fleuves et grandes rivières, Lyon, France, 2012.

Hallouz, F., Meddi, M., Mahé, G., Toumi, S., and Rahmani, S. E. A. : Erosion, Suspended Sediment Transport and Sedimentation on the Wadi Mina at the Sidi M'Hamed Ben Aouda Dam, Algeria, Water, 2018, 10, 895, 31 pp., https://doi.org/10.3390/w10070895, 2018.

Kouri, L. : L'e´rosion hydrique des sols dans le bassin versant de l'Oued Mina. Etude des processus et type de fonctionnement des ravins dans la zone des marnes tertiaires, The`se de doctorat, Universite´ Louis Pasteur, Strasbourg, France, 238 pp., 1993.

Mahieddine, M. : Quantification and Variabilité Parcellaire Sous Simulation de Pluie Dans le Bassin Versant de L'Oued Mina, PhD. Thesis, École Nationale Supérieure des Sciences Agronomiques, Alger, Algeria, 120 pp., 1997.

Meddi, M. : Hydro-pluviométrie et transport solide dans le bassin versant de l'oued Mina (Algérie). Thèse de doctorat, Université, Louis Pasteur, Strasbourg, France, 285 pp., 1992.

Remini, B. : Valorisation de la vase des barrages – quelques exemples algériens, Larhyss Journal, No. 05, 75–89, Juin 2006.

Remini, B. et Bensafia, D. : Envasement du Barrage de SMBA Algérie. Comité Scientifique Projet SIGMED Rabat Maroc, 2012.

Roose, É., Sabir, M., Arabi, M., Morsli, B., et Mazour, M. : Soixante années de recherches en coopération sur l'érosion hydrique et la lutte antiérosive au Maghreb, Physio-Géo [En ligne], Vol. 6, mis en ligne le 3 mai 2012, 43–69, https://doi.org/10.4000/physio-geo.2319, 2012.

Touaïbia, B. et Achite, M. : Contribution à la cartographie de l'érosion spécifique du bassin versant de l'Oued Mina en zone semi-aride de l'Algérie septentrionale, Hydrol. Sci. J., 48, 235–242, 2003.

Toumi, S. : Application des Techniques Nucléaires et de la Télédétection à L'étude de L'érosion Hydrique Dans le Bassin Versant de L'oued Mina, Ph.D. Thesis, Es-Sciences, École Nationale Supérieure d'Hydraulique (ENSH), Soumaâ, Algeria, 189 pp., 2013.

Proc. IAHS, 384, 163–168, 2021
https://doi.org/10.5194/piahs-384-163-2021

Open Access PIAHS

Hydrology of Large River Basins of Africa

Évolution du régime hydrologique dans le bassin versant du Bouregreg, Maroc

Rajae El Aoula[1], Gil Mahé[2], Nadia Mhammdi[1], Abdellatif Ezzahouani[3], Ilias Kacimi[3], and Kenza Khomsi[4]

[1]Institut Scientifique, centre de recherche GEOPAC, laboratoire de Géophysque et Risques Naturels, Université Mohammed V de Rabat, Rabat, Maroc
[2]UMR HydroSciences Montpellier / IRD, Montpellier, France
[3]Déepartement des Sciences de la Terre, Faculté des sciences, Université Mohammed V de rabat, Rabat 10999, Maroc
[4]Service de la qualité de l'air, Centre national du climat, Direction générale de la météorologie, Boulevard Mohamed 4 Tayeb Naciri, Hay Hassani B.P., 8106 Oasis-Casablanca, Maroc

Correspondence: Rajae El Aoula (elaoula.rajae@gmail.com)

Published: 16 November 2021

Résumé. Le Bouregreg est le deuxième fleuve du Maroc. Il draine les eaux du centre du Maroc, du Moyen Atlas à l'Atlantique, qui sont stockées dans la retenue du barrage Sidi Mohamed Ben Abdellah créé en 1974 et situé à proximité de Rabat. Sa fonction exclusive est l'alimentation en eau potable pour les besoins individuels et industriels de 6 millions d'habitants de la bande littorale allant de Kénitra à Casablanca. Toutefois, l'exploitation de cette retenue présente aujourd'hui quelques incertitudes surtout en considérant l'augmentation de la population, la baisse des volumes précipités depuis 1979 et le comblement de la retenue par le transport de sédiments dus à l'érosion des sols. L'objectif de cette étude est de d'écrire l'évolution des pluies et des débits du bassin versant du Bouregreg à partir des séries de pluie et débit sur une période de 36 ans, de 1977 à 2013. Nous avons utiliséì les indices pluviométrique et hydrométrique, dans le but de caractériser l'hétérogénéité temporelle de la variabilité climatologique et dégager les grandes tendances des séries de données hydroclimatologiques. L'évolution des indices pluviométriques et hydrométriques met en relief la période déficitaire qu'a connu le Maroc au début des années 80. En effet, une baisse générale des pluies est remarquée dans le bassin versant du Bouregreg depuis 1979, ainsi il existe des relations entre la variabilité des pluies de certaines régions et l'Oscillation Nord Atlantique en 1996. Les méthodes statistiques de détection de ruptures montrent que la période de rupture majeure se trouve durant la deuxième moitié des années 70 et le début des années 80. Depuis cette date la baisse de la pluviométrie s'est installée de manière significative, et une perturbation du régime hydrologique est observée sur la presque totalité du bassin versant du Bouregreg, même si quelques années de meilleures pluviométries ont été observées à la fin des années 2000.

1 Introduction

Les précipitations représentent le facteur le plus important du climat tant pour les populations que pour les écosystèmes (Kouassi et al., 2010). Ainsi de nombreuses études portent sur la détection des tendances et des variations dans les longues séries hydroclimatologiques. Cette connaissance est pertinente pour l'élaboration des stratégies d'atténuation et d'adaptation au changement climatique dans les régions affectées. En effet, le Maroc fait partie des pays qui ont souffert économiquement, socialement et écologiquement de la sécheresse au cours des dernières décennies (80 et 90) (Tsiourtis, 2001 ; Singla, 2009). Les systèmes naturels et les activités humaines sont les plus à risque en raison du manque d'approvisionnement régulier en eau au cours de l'année. Ceci est principalement dû à la forte variabilité interannuelle et à l'importante saisonnalité des précipitations. Le bassin versant du Bouregreg il est l'un des cinq grands fleuves marocains par son débit et sa taille et il présente la particularité d'être limité à 18 km de son embouchure par le très grand barrage Sidi Mohamed Ben Abdellah d'un volume supérieur à 1 milliard de m^3. Son régime hydroclimatique est influencé par la NAO (North Atlantic Oscillation) et la MO (Mediterranean Oscillation) (Khomsi et al., 2013), avec la baisse des pluies de 1979 (Singla et al., 2010), et suite à une augmentation des évènements extrêmes des températures et des vagues de chaleur (Khomsi et al., 2012, 2016). Les conditions de l'érosion et du transport solide ont été modifiées sur ce bassin. Afin d'approfondir la compréhension du fonctionnement hydroclimatique du bassin versant du Bouregreg, il est judicieux de comparaitre l'évolution des pluies et des débits pour mieux décrire les évolutions climatiques dans la zone d'étude. Les stations utilisées pour ce travail présentent un jeu de données intéressant pour tester l'impact du changement climatique sur le régime hydrologique et connaître la variabilité de la ressource en eau qui serait disponible pour les aménagements en aval de ces stations, suivant une méthode qui a été appliquée partiellement au Maroc par Singla et al. (2010).

2 Matériels et méthodes

2.1 Zone d'étude

Le bassin versant du Bouregreg (situé entre 5.4–6.8° W et 32.8–34° N), occupe la presque totalité du plateau central marocain et s'étend partiellement sur la Wilaya de Rabat-Salé (Fig. 1), les provinces de Khénifra et de Khémisset. Son réseau hydrographique s'articule autour de deux principales rivières, l'oued Bouregreg et l'oued Grou, qui drainent respectivement une superficie de 4000 km et de 3600 km^2 (Lahlou, 1995). Les apports en eau du bassin sont étroitement liés aux précipitations. Le potentiel moyen en eau du bassin du Bouregreg est évalué à 680 mm^3 par an (Khomsi, 2014).

L'oued Bouregreg correspond à la partie nord-est du bassin Bouregreg, il prend naissance dans le massif du Moyen Atlas à l'altitude de 1627 m au niveau du Jebel Mtourzgane et de Grou (Beaudet, 1969), et se jette dans l'océan Atlantique à l'embouchure séparant les villes de Salé au nord et Rabat au sud, après avoir cheminé vers le littoral atlantique à travers la Meseta. Son chenal suit une direction globale rectiligne orientée SE–NW et s'étend sur une longueur de 240 km. Les 23 derniers kilomètres du cours inférieur du Bouregreg se terminent par un estuaire partiellement aménagé (Lahlou, 1995).

2.2 Jeu de données

Les données pluviométriques proviennent de la base SIEREM (Dieulin et al., 2006) et des données récentes sont issues de l'agence hydraulique du Bouregreg et de la Chaouia, sur la période 1977–2013. Les données de débit ont été recueillies auprès de 8 stations de mesure de débit par l'agence hydraulique du Bouregreg et de la Chaouia et couvrent une période allant de 1977 à 2007. Les stations étudiées sont situées dans la moitié amont du bassin, et sont peu influencées par la construction du barrage SMBA.

2.3 Indices pluviométriques et hydrométriques

Les indices pluviométriques et hydrométriques permettent de caractériser l'hétérogénéité temporelle de la variabilité climatologique et de dégager les grandes tendances dans les séries hydroclimatologiques. Pour mieux visualiser les périodes de déficit et d'excédent à l'échelle annuelle, ces indices ont été centrés et réduits à partir de la formule suivante :

$$\mathrm{Vcr} = (P - \mathrm{MML})/\sigma$$

Où, Vcr est la Variable centrée réduite (Indice pluviométrique (SPI) ou hydrométrique (SDI) selon la variable étudiée), P est le cumul de la variable sur une période t donnée, MML est la moyenne mobile lissée de la série chronologique sur la période d'étude et σ est l'écart type de la série sur la période d'étude (Hallouz et al., 2013).

2.4 Tests statistiques de détection de ruptures

Ces tests ont pour objectif de détecter une tendance (ou une rupture en moyenne dans les séries. Les tests utilisés dans ce travail sont regroupés dans le logiciel Khronostat (1998) réalisé par l'IRD et disponible sur le site d'HydroSciences Montpellier (http://www.hydrosciences.fr, la date du dernier accès : 2012).

Le test de Pettitt (Pettitt, 1979) a été utilisé sur tous les postes retenus en vue d'identifier une « rupture » moyenne dans les séries chronologiques de pluviométries et hydrométries annuelle et mensuelle étudiées. Ce test réputé pour sa robustesse, est non paramétrique et dérive de la formulation du test de Mann-Whitney. L'absence d'une rupture dans la série chronologique X constitue l'hypothèse nulle $H0$.

Ensuite, la méthode bayesienne de Lee et Heghinian (1977) permet de déterminer la date et l'amplitude de la rupture. Cette méthode fournit la probabilité qu'un changement se produise à un moment t, dans une série chronologique où on suppose à priori qu'il y a effectivement un changement à un moment indéterminé. De même, cette procédure donne une estimation de la probabilité que l'amplitude du changement ait la valeur du changement éventuel de la moyenne.

La segmentation de Hubert proposée par Hubert et al. (1989), consiste à découper la série en m segments ($m > 1$) de telle sorte que la moyenne calculée sur tout segment soit significativement différente de la moyenne du (ou des) segment(s) voisin(s). Une telle méthode est appropriée à la recherche de multiples changements de la moyenne.

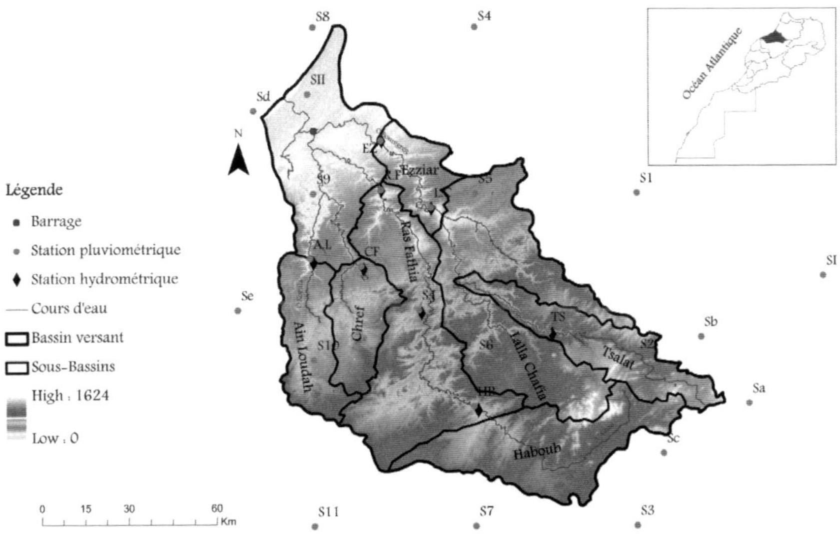

Figure 1. Le bassin versant du Bouregreg et les principaux sous-bassins.

3 Résultats et discussion

3.1 Évolution des relations pluie-débit pour l'ensemble du bassin du Bouregreg

3.1.1 Indices pluviométriques et hydrométriques

Sur la base des indices exprimés par les données pluviométriques et hydrométriques pour la période 1977–2007, on remarque que le graphique des précipitations suit le même sens que celui des débits mais avec des valeurs parfois sensiblement différentes. Cette période est marquée par deux années de déficit pluviométrique 1994 et 2006 (Fig. 2). Après la fin des années 70 on observe une augmentation non seulement de la fréquence des années sèches mais aussi de l'amplitude de la sécheresse par rapport aux deux années précédentes (1978 et 1979). Cette diminution des précipitations est déjà signalée par de nombreux chercheurs dont (Singla et al., 2010) qui ont évoqué la rupture des pluies au Maroc en 1979. Les décennies 80 et 90 sont particulièrement sèches avec le maximum de déficit en 1994. En dehors des années 1995 et 1996 caractéristiques de forte anomalie NAO favorable à la pluie au Maroc (Khomsi et al., 2012), les décennies 80 et 90 sont particulièrement sèches avec le maximum de déficit en 1994.

Sur la Fig. 3 on trace la différence entre l'indice pluviométrique et l'indice hydrométrique. On en déduit une différence positive depuis le début de la sécheresse (1979) jusqu'à l'événement NAO (1995/1996), puis à partir de 1996 la différence devient négative, avec un indice pluviométrique plus fort que l'indice hydrométrique. La diminution brusque enregistrée en 2006 est principalement due à la courte période de disponibilité de données, environ 8 stations régionales pour chaque année. Ce résultat semble être expliqué par le fait que

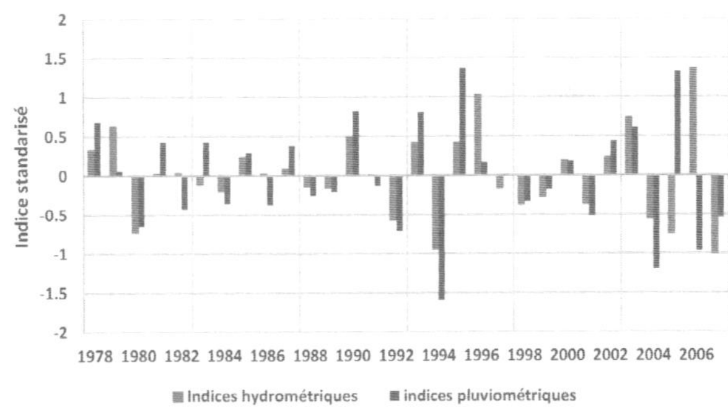

Figure 2. Indices pluviométriques et hydrométriques de 1978 à 2007 pour l'ensemble du bassin versant du Bouregreg (9970 km^2).

la variabilité du cumul annuel est plus liée à celle du nombre d'événements plutôt qu'à la quantité de pluie par événement pluvieux (Lawin et al., 2011). Au niveau du bassin versant du Bouregreg, le nombre de jours de pluie par an varie de 60 à 75 entre les montagnes et l'océan et de 75 à 100 sur les régions montagneuses, et les changements dans la variabilité du climat local sont influencés par les changements de la circulation atmosphérique à large échelle.

Khomsi et al. (2013) ont montré l'existence d'un lien statistique significatif entre l'évolution des épisodes chauds d'été et la MO sur les zones internes du Maroc. Une corrélation significative a été également trouvée entre la NAO et les épisodes chauds de Rabat-Salé. En effet, au cours de la saison froide, la NAO est dans sa phase négative, les perturbations s'engouffrent vers le Nord de l'Afrique, apportant ainsi des pluies sur la région. Les types de temps résultants peuvent adoucir le climat et diminuer la fréquence des épisodes chauds au Nord du Maroc. Alors qu'au cours de la

Tableau 1. Ruptures détectées par les tests de KhronoStat pour les séries de pluies et débits moyennes sur le bassin versant du Bouregreg, période 1977–2007. (NR = Pas de rupture).

Variable	Période	Pettitt	Lee and Heghinian	Hubert	Buishand	Probabilité de rupture		
						Faible	Moyen	Fort
Pluies	1977–2007	1997	1997	1994 1998 2006	NR	–	1997	–
Débits	1977–2007	NR	1979	1979	1979	–	–	1979

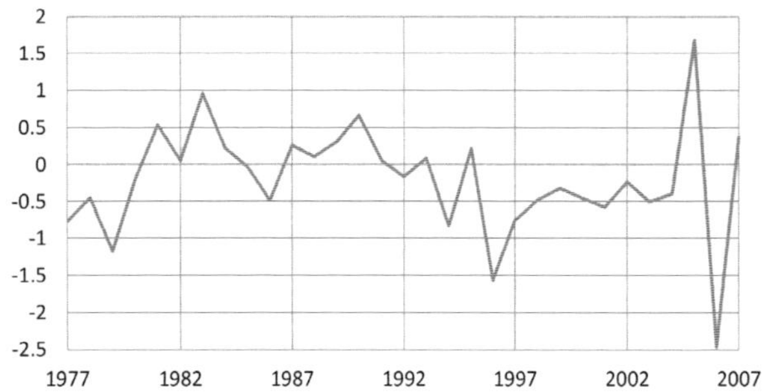

Figure 3. Différence indice pluviométrique moins indice hydrométrique.

saison chaude, la NAO est généralement en phase positive, une augmentation de cet indice renforce les vents d'Ouest sur l'Afrique du Nord. Ces vents devenant plus chauds suite à l'augmentation de la température de la surface de la mer affectent le Nord du pays dont fait partie le Bouregreg. Aussi, l'augmentation de l'indice MO provoque l'infiltration sur le pays, de masses d'air chaudes depuis la Méditerranée. Ces infiltrations peuvent occasionner des épisodes chauds.

3.1.2 Rupture de stationnarité

L'analyse et l'interprétation des tests statistiques a permis de constater que pour les pluies annuelles (Tableau 1) la rupture sur l'ensemble du bassin a été détectée au niveau des années 1994 et 1997. Celles-ci correspondent à une diminution des précipitations due à la variabilité climatique. Néanmoins, quelques ruptures et tendances peuvent être observées à la fin de la décennie 70 ou au début de la décennie 80, traduisant peut-être un changement qui se révélerait plus net à l'échelle de l'événement pluvieux. Ces résultats réaffirment les constatations des différentes études réalisées auparavant sur l'évolution des pluies en Afrique comme celles de Klaus (2002), les périodes où le réchauffement est le plus rapide en Afrique au cours du XXe siècle, sont observées à partir des années 1910 jusqu'aux années 1930 et après 1979. Ce qui confirme aussi les études démontrant une diminution des précipitations au Maroc, depuis les années 1960 (Driouech, 2006 ; DMN, 2007) et plus significativement depuis la fin des années 70/début des années 80 (Agoumi et Debbarh, 2006 ; Knippertz et al., 2003). Cette rupture est aussi confirmée à l'échelle régionale, par l'étude de la pluviométrie régionale du bassin versant du Bouregreg (Khomsi, 2014).

Il apparait ainsi qu'un changement du régime pluviométrique et hydrométrique a affecté une grande partie du bassin versant du Bouregreg à partir de 1979.

3.1.3 Régime mensuel avant et après la rupture (1979)

Le régime des pluies mensuelles sur les deux périodes avant et après la rupture (Fig. 4) montre une diminution sensible surtout entre décembre et février, estimée à 33 % sur l'ensemble du bassin et dû au changement de régime de pluie interannuel en 1979. On constate que la pluviométrie a baissé au cours des années 80, sauf en mars. Les totaux pluviométriques ont évolué au cours de la « grande » saison des pluies novembre–décembre, tandis que la « petite » saison des pluies de mars à avril a presque disparu, et qu'un maximum relatif apparaît en septembre–octobre. L'examen de la Fig. 3 indique que avant et après la période de la rupture, il constate qu'une diminution significative de 14,3 % de hauteur des précipitations, soit d'un total pluviométrique de 534 à 458 mm, alors qu'au mois de février on note une augmentation de 49 % de lame d'eau tombée. L'impact de cette baisse des pluies est plus important sur les débits qui présentent un changement de régime mensuel très significatif, traduits par une baisse des débits mensuels de printemps, et des débits plus élevés en février et mars sur la période courte 1977–1979. La différence mensuelle de débit entre les deux périodes est de près de 41 %.

4 Conclusion

L'analyse de l'étude de l'évolution hydroclimatique montrent que le régime pluviométrique général, sur le bassin du Bouregreg, est caractérisé par une baisse des pluies encore accentuée aux alentours de l'année 1979.

Les méthodes statistiques de détection de ruptures ont montré que la période de rupture majeure se trouve durant la deuxième moitié des années 70 et le début des années 80.

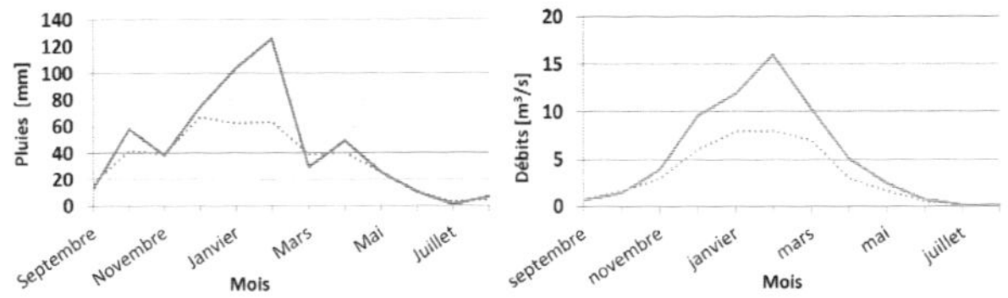

Figure 4. Évolution des pluies et débits mensuels sur le bassin du Bouregreg entre 1977 et 1997 (trait plein) et 1980 à 2013 (tireté) de septembre à août.

Après cette période l'hydraulicité du bassin versant du Bouregreg n'a cessé de se dégrader avec une alternance de périodes humides et sèches au regard de l'évolution des pluies et débits. Cette tendance persistante à la baisse doit être soulignée car elle montre une dégradation durable du système hydrologique malgré un retour assez sensible à de meilleures conditions de précipitation. Ainsi, le régime hydrologique du bassin est directement influencé par le régime de précipitations. Nous supposons donc que la diminution des débits est due principalement à la diminution des précipitations sur le bassin versant du Bouregreg, et non à une baisse majeure des ressources en eau souterraines.

Disponibilité du code et des données. Les données sont disponibles sur demande.

Collaborateurs. REA contributed to the conception and development of this research, data analysis and their interpretation, and drafting text. GM is the supervisor and leadership responsible for this research activity planning and execution. NM participated in the process of preparing this research. AE participated in the data processing and their interpretation. IK participated in the collection of data, and discussions. KK contributed to the preparation and creation of the published work, specifically critical review, commentary and revision- including pre and post-publication stages.

Intérêts concurrents. Les auteurs déclarent qu'ils n'ont aucun conflit d'intérêts.

Clause de non-responsabilité. Publisher's note : Copernicus Publications remains neutral with regard to jurisdictional claims in published maps and institutional affiliations.

Déclaration du numéro spécial. This article is part of the special issue "Hydrology of Large River Basins of Africa". It is a result of the 4th International Conference on the "Hydrology of the Great Rivers of Africa", Cotonou, Benin, 13–20 November 2021.

Remerciements. Ce travail s'inscrit dans le cadre du projet Toubkal No. TBK/15/CASBA (Caractères Anthropiques et environnementaux de l'érosion des versants du barrage Sidi Mohammed Ben Abdellah) entre le laboratoire Géophysique et Risques Naturels de l'Institut Scientifique, l'agence de bassin hydraulique du Bouregreg et de la Chaouia (ABHBC) et le laboratoire HydroSciences Montpellier.

Références

Agoumi, A. and Debbarh, A. : Ressource en eau et bassins versants du Maroc : 50 ans de développement (1955–2000), 62 pp., available at : http://www.rdh50.ma/fr/pdf/contributions/GT8-I.pdf (la date du dernier accès : 2018), 2006.

Beaudet, G. : Le plateau central marocain et ses bordures : étude géomorphologique, Thèse, Imprimeries françaises et marocaines, Rabat, p. 478, 1969.

Dieulin, C., Boyer, J. F., Ardoin-Bardin, S., Ardoin-Bardin, S., and Dezetter, A. : The contribution of GIS to hydrological modeling. Climate variability and change : hydrological impacts, FRIEND 2006, edited by : Demuth, S., Gustard, A., Pianos, E., Scatena, F., and Servat, E., Vth Int. Conf., La Havane, Cuba, IAHS Pub., 30868–30874, IAHS Press, Wallingford, UK, 2006.

Direction de la Météorologie Nationale (DMN) : Les changements climatiques au Maroc : Observations et projections, DMN, Secrétariat d'Etat auprès du Ministère de l'Energie, des Mines, de l'Eau et de l'Environnement, Chargé de l'Eau et de l'Environnement, Royaume du Maroc, Casablanca, 2007.

Driouech, F. : Etude des indices de changements climatiques sur le Maroc : températures et précipitations, DMN "INFOMET", Secrétariat d'Etat auprès du Ministère de l'Energie, des Mines, de l'Eau et de l' Environnement, Charge de l'Eau et de l'Environnement, Royaume du Maroc, Casablanca, 2006.

Hallouz, F., Meddi, M., and Mahe, G. : Modification du régime hydroclimatique dans le bassin de l'Oued Mina (nord-ouest d'Algérie), Journal of Water Science, 26, 33–38 2013.

Hubert, P., Carbonnel, J. P., and Chaouche, A. : Segmentation des séries hydrométéorologiques. Application à des séries de précipitations et de débits de l'Afrique de l'Ouest, J. Hydrol., 110, 349–367, 1989.

Khomsi, K. : Variabilité hydroclimatique dans les bassins versants du Bouregreg et du Tensift au Maroc : moyennes, extrêmes et projections climatiques, Thèse de Doctorat, spécialité, Climatologie, Université Mohamed V, Rabat, Maroc, 2014.

Khomsi, K., Mahe, G., Sinan, M., and Snousi, M. : Evolution des évènements chauds rares et très rares dans les bassins versants du Tensift et du Bouregreg (Maroc) et identification des types de 70 temps synoptiques associés, in : From prediction to prevention of hydrological risk in Mediterranean countries, edited by : Ferrari, E. and Versace, P., 4th international Workshop on Hydrological extremes MEDFRIEND group, University of Calabria, EdiBios, Cosenza, Italia, 169–182, 2012.

Khomsi, K., Mahe, G., Sinan, M., and Snousi, M. : Hydroclimatic variability in two Moroccan watersheds : A comparative analysis of temperature, rain and flow regimes. Climate and land surface changes in hydrology, edited by : Boegh, E., Blyth, E., Hannah, D. M., Hisdal, H., Kunstmann, H., Su, B., and Yilmaz, K. K., IAHS Publ. 359, 183–190, 2013.

Khomsi, K., Mahe, G., Tramblay, Y., Sinan, M., and Snoussi, M. : Regional impacts of global change : seasonal trends in extreme rainfall, run-off and temperature in two contrasting regions of Morocco, Nat. Hazards Earth Syst. Sci., 16, 1079–1090, https://doi.org/10.5194/nhess-16-1079-2016, 2016.

Khronostat : Logiciel de tests statistiques de ruptures dans des séries chronologiques, IRD, France, accessible à l'adresse suivante : http://www.hydrosciences.fr (la date du dernier accès : 2012), 1998.

Klaus, T. : Afrique changements climatiques. Tendance des variations de températures mondiales au cours du XXème siècle, Vital Graphics, p. 4, 2002.

Knippertz,, P., Christophn, M., and Speth, P. : Long term precipitation variability in Morocco and the link to the large-scale circulation in recent and future climates, Met. Atmos. Phys., 83, 67–88, 2003.

Kouassi, A. M., Kouamé, F. K., Koffi, Y. B., Dje B. K., Paturel, E. J., et Oulare, S. : Analyse de la variabilité climatique et de ses influences sur les régimes pluviométriques saisonniers en Afrique de l'Ouest : cas du bassin versant du N'zi (Bandama) en Côte d'Ivoire, CyberGeo, July 2010, 2010.

Lahlou, A. : Etude actualisée de L'envasement des barrages au Maroc, Revue des sciences de l'eau, 6, 337–356, 1995.

Lawin, A. E., Afouda, A., et Lebel, T. : Analyse de la variabilité du régime pluviométrique dans la région agricole d'Ina au Bénin, European Journal of Scientific Research, 50, 425–439, 2011.

Lee, A. F. S. and Heghinian, S. M. : A shift of the mean level in a sequence of independent normal random variables, a Bayesian approach, Technometrics, 19, 503–506, 1977.

Pettitt, A. N. : A non-parametric approach to the change point problem, Appl. Statist., 28, 126–135, 1979.

Singla, S. : Impact du changement climatique global sur les régimes hydroclimatiques au Maroc : tendances, ruptures et effets anthropiques sur les écoulements, Mémoire de Master, M2R2E, France, 2009.

Singla, S., Mahé, G., Dieulin, C. Driouech, F., Milano, M., El Guelai, F. Z., et Ardoin-Bardin, S. : Evolution des relations pluie-débit sur des bassins versants du Maroc, Global Change : Facing Risks and Threats to Water Resources, Proc. of the Sixth World FRIEND Conference, Fez, Morocco, October 2010, IAHS Publ. 340, 679–687, 2010.

Tsiourtis, S. : Drought management plans for the Mediterranean region, Report of the Water Engineer Water Development Department, Nicosia, Cyprus, 2001.

Proc. IAHS, 384, 169–173, 2021
https://doi.org/10.5194/piahs-384-169-2021

Open Access PIAHS

Hydrology of Large River Basins of Africa

Estimation des réserves en eau souterraine régulatrices dans la partie moyen-atlasique du fleuve Sebou (Maroc)

Sébastien Lebaut[1], Abdelghani Qadem[2], Brahim Akdim[3], Emmanuel Gille[1], and Mohamed Laaouane[3]

[1]Unité de recherche LOTERR UR 7304, Université de Lorraine, Metz, France
[2]Laboratoire Dynamique des Paysages, Risques, Patrimoine, Université Sultan Moulay Slimane, Béni Mellal, Maroc
[3]Laboratoire d'Analyses Géo-Environnementales et Aménagement-Développement Durable, Université Sidi Mohamed Ben Abdellah, Fès, Maroc

Correspondence: Sébastien Lebaut (sebastien.lebaut@univ-lorraine.fr)

Published: 16 November 2021

Résumé. L'estimation de la ressource en eau souterraine dans le Moyen-Atlas est investiguée à partir des débits mesurés dans l'oued Sebou à la station d'Azzaba, sur la période 1959–2015. Pour cela presqu'une centaine de phases de tarissement a été individualisé pour construire des courbes maîtresses de tarissement à partir desquelles le calcul des volumes des nappes est possible. Il est calculé mensuellement à partir du débit modal dont nous posons l'hypothèse qu'il représente la limite entre ruissellement et alimentation uniquement par les nappes. Les calculs donnent un volume de la réserve régulatrice moyen de 54 hm^3, mais extrêmement variable à l'échelle interannuelle et intra-annuelle. Ces résultats démontrent la faible inertie des aquifères du Moyen Atlas et soulignent la vulnérabilité du secteur agricole vis-à-vis de cette ressource même lors de courte période de sécheresse.

The estimate of the groundwater resource in the Middle Atlas is investigated from the runoff measured in the Sebou wadi at the Azzaba station, over the period 1959–2015. Almost a hundred recession curves have been individualized to build the master recession curves from which the calculation of the volumes of groundwater is possible. It was calculated at a monthly scale from the modal flow, which we assume is the limit between runoff and flow supply only by the aquifers. The results indicated a volume of the average regulatory reserve of 54 hm^3, but extremely variable on an inter-annual and intra-annual scale. These results demonstrate the low inertia of the Middle Atlas aquifers and underline the vulnerability of the agricultural sector to this resource even during short periods of drought.

1 Introduction

La notion de ressource en eau souterraine est née de la prise de conscience que des limites s'imposent aux possibilités d'exploitation globales d'une nappe. Dès lors se pose le problème de l'évaluation quantifiée de cette ressource. D'un point de vue théorique les systèmes d'eau souterraine peuvent être classés en trois catégories en fonction de leur profondeur et en fonction de la proximité du réseau drainant (Ophori et Toth, 1990) : peu profond, intermédiaire et régional. Dans le système peu profond, l'eau circule rapidement et répond rapidement à la recharge et décharge, a un temps de trajet court et constitue le débit de base dans les cours d'eau. Ce débit de base est la part de l'écoulement total naturel qui est différé sous l'effet régulateur de divers réservoirs naturels, notamment celui des aquifères à exutoire compris dans le bassin versant et compose tout l'écoulement en période de tarissement (Castany et Margat, 1977). La courbe de tarissement, qui caractérise la décroissance des débits durant cette période, renferme des informations concernant les caractéristiques des aquifères et le stockage (Tallaksen, 1995) et il existe dans la littérature plusieurs modèles conceptualisant les échanges nappe/rivière durant cette période non directement influencées par les pluies (Zécharias et Brutsaer, 1988).

Published by Copernicus Publications on behalf of the International Association of Hydrological Sciences.

Il est donc possible de calculer un volume dynamique V_d de la nappe dans un bassin versant (Sanz et Recio, 2011) qui correspond à la notion de réserve régulatrice proposée par Castany (1968). Tout l'enjeu de cette quantification à partir des hydrogrammes journaliers réside dans la bonne formulation du tarissement et dans la séparation de celui-ci en débit d'écoulement et débit de base. Dans cette communication nous présentons une démarche méthodologique pour apprécier la réserve régulatrice dans le bassin du haut Sebou à partir de l'hydrogramme de la station hydrométrique d'Azzaba verrouillant le bassin versant du Sebou (4677 km^2) à sa sortie du Moyen-Atlas. A notre connaissance, s'il existe de nombreuses estimations de volumes de nappe à partir de sources dans les basaltes ou karstiques, aucune évaluation sur l'ensemble du bassin n'est disponible.

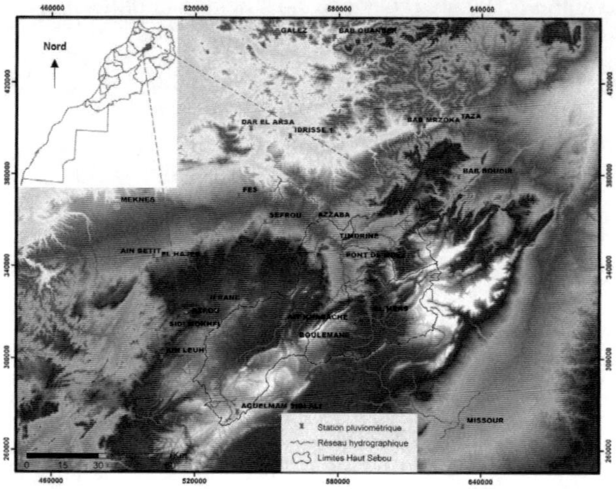

Figure 1. Localisation du bassin du Haut-Sebou.

2 Secteur d'étude

Le bassin versant du Haut Sebou s'inscrit exclusivement dans le Moyen-Atlas (Fig. 1). Du point de vue géologique le Moyen-Atlas appartient au système intracontinental atlasique qui traverse l'Afrique du Nord d'ouest en est, de l'Atlantique jusqu'à la Méditerranée centrale selon le parallèle 34° de latitude nord. Il en constitue une branche qui traverse en oblique le domaine des mesetas avec des altitudes qui dépasse les 3000 m (Michard et al., 2008). Le Moyen-Atlas central s'inscrit dans deux unités morphostructurales. A l'ouest, surplombant la meseta ouest c'est le Moyen-Atlas tabulaire : causse établit dans des séquences sédimentaires carbonatés du Jurassique inférieur (Trias, Lias) et jalonné d'appareils volcaniques et de coulées basaltiques datant du Cénozoïque (Frizon de Lamotte et al., 2008) dont les altitudes sont de l'ordre de 1200 m. La partie est correspond au Moyen-Atlas plissé : bordé par deux failles majeures de direction NE-SO, il correspond à une succession de cuvettes synclinales dans lesquelles des marnes affleurent et de lignes de crêtes anticlinales dans des dolomies correspondant à un relief de montagne dont le point culminant est de 3081 m (Jebel Bou Iblane).

La principale caractéristique du Moyen-Atlas central est l'empreinte du calcaire et des phénomènes karstiques (De Waele et Melis, 2009). Le fonctionnement hydrologique de l'hydrosystème découle bien évidemment de cette donne géologique (Fig. 2, Table 1). Le trait dominant de l'hydrosystème est l'infiltration généralisée et la circulation souterraine de l'eau dans le substratum très perméable qui alimente les résurgences en pied de versants au contact de formations imperméables (Amraoui et al., 2003).

Le contexte climatique favorable aux forts abats pluviométriques de la région, 523 mm en moyenne annuelle sur la période 1974/2010 (Qadem, 2015), couplé à une lithologie générale favorable à la formation d'aquifère induit de fortes potentialités de ressource en eau, néanmoins extrêmement variables.

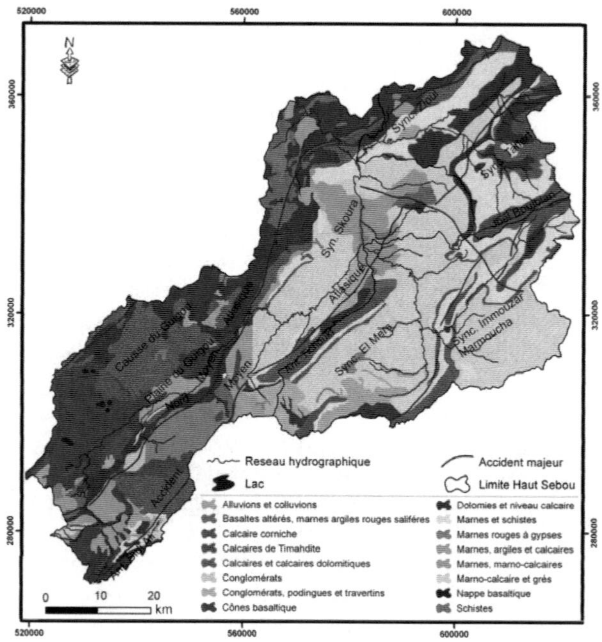

Figure 2. Lithologie du bassin du Haut-Sebou.

3 Données et méthode

3.1 Données

Les données hydrologiques à la station d'Azzaba (4677 km^2) (Fig. 1) sont produites par l'Agence du Bassin Hydraulique du Sebou (ABHS). Disponibles de 1959 à 2015, elles sont réputées de bonnes qualités.

Tableau 1. Synthèse lithologie/perméabilité dans le bassin versant du Haut Sebou.

Roches dominantes	Perméabilité	Surface (km^2)	%
Alluvions et colluvions	Perméable	238	5
Basaltes, poudingues, conglomérats, calcaires et dolomies	Perméable	2099	45
Alternances de marnes, calcaires et grés	Semi-perméable	1591	34
Marnes, argiles et schistes	Peu perméable	752	16

3.2 Méthode

La réserve régulatrice correspond au volume d'eau renouvelable dans les conditions climatiques de la période considérée. Ce dernier peut être apprécié en s'appuyant sur les phases de tarissement des cours d'eau, étant entendu qu'elles représentent la vidange de la réserve régulatrice par les sources et le drainage des cours d'eau. Dans ce cas $V = Q0/\alpha$ avec $Q0$, le débit au début du tarissement et α le coefficient de tarissement.

D'application relativement simple sur les sources, son application sur les cours d'eau est beaucoup plus délicate à mettre en œuvre : choix du débit $Q0$ à partir duquel calculer le volume, calcul de α. Nous proposons d'estimer les volumes mensuels d'eau souterraine constituant la réserve régulatrice des aquifères du bassin du Haut-Sebou sur la base du débit modal pour $Q0$ et le calcul d'un coefficient de tarissement.

3.2.1 Détermination du débit modal

A la suite de plusieurs auteurs, Gamez et Zumstein (1993), font l'hypothèse que dans une série de Qmj classés, la limite entre débits causés par le ruissellement et débits induits uniquement par la vidange des nappes est le débit modal. Compte tenu d'une distribution dissymétrique des Qmj les auteurs calculent le mode à partir d'une loi de Fréchet, loi de Gumbel à laquelle on substitut les Q à Log Q, suivant :

$$\mathrm{Log}(Qj) = \log(Qj_0) - \frac{1}{K}\mathrm{Log}\left[-\log(F)\right] \quad (1)$$

Avec :

$$\mathrm{Log}(Qj_0) = \text{mode} = \overline{\mathrm{Log}(Qj)} - 0,45S\left[\mathrm{Log}(Qj)\right] \quad (2)$$

$$\frac{1}{K} = \text{gradex de Fréchet} = 0,78S\left[\mathrm{Log}(Qj)\right] \quad (3)$$

$S\left[\mathrm{Log}(Qj\right]$ =écart type de la distribution des logarithmes naturels des débits. $\overline{\mathrm{Log}(Qj)}$ = moyenne des logarithmes naturels des débits.

3.2.2 Calcul du coefficient de tarissement

Les phases de tarissement sont des événements courts qui varient l'un par rapport à l'autre en fonction des variations de facteurs physiques (Moore, 1997) tel que l'évapotranspiration, l'inégale distribution des pluies, la présence de plusieurs aquifères.... Ainsi l'hétérogénéité lithologique du milieu d'application de cette expression linéaire a comme effet une variabilité temporelle du tarissement. Afin de pallier celle-ci et d'aboutir à un coefficient de tarissement moyen nous avons construit une « courbe maîtresse de tarissement » pour laquelle l'ensemble des phases de tarissement définies à partir des chroniques de débits moyens journaliers est utilisé simultanément. Deux méthodes sont utilisées pour l'évaluation : corrélation et matching strip.

- Méthode des corrélations : déterminer la pente moyenne d'un épisode de tarissement particulier, soit pour calculer un rapport αt en corrélant les débits $Q0$ et Q à t jours (Q_t), pour chaque phase observée (Tallaksen et Van Lanen, 2004). Le principe est d'organiser le nuage de points (Q_t, $Q_t + \mathrm{d}t$) de manière à dessiner une droite de régressions, dont le coefficient de tarissement α est exprimé par la relation suivante :

$$k = \exp(-\alpha) = \left(\frac{Q}{Q0}\right)^{1/t} \quad (4)$$

- Méthode « matching strip » (MSM) : à partir des débits de la phase de tarissement exprimés en ln de manière à former une droite, la méthode consiste ensuite à lier les différents épisodes de tarissement afin qu'ils forment une seule droite (Toebes and Strang, 1964).

4 Résultats et discussion

Sur la période considérée, 98 phases de tarissements supérieures à 5 jours ont été individualisées pour une durée moyenne de 14.6 jours. La méthode des corrélations renvoie un coefficient α de $1,2 \times 10^{-2}$ (Fig. 3), ce qui est en accord avec des formations aquifères à « perméabilité de fissures ». La méthode MSM (Fig. 4) permet de nuancer cette valeur en fonction des saisons et donc en fonction du remplissage des aquifères. Les valeurs α sont comprises entre $1,7 \times 10^{-2}$ au printemps avec un remplissage maximum et 1×10^{-2} en hiver (Fig. 4). En outre, le calcul du coefficient sur 4 longues phases de tarissement de 80 à 100 jours pour des débits $Q0$ compris entre 6,8 et 9,1 m^3/s a fourni des résultats compris entre $1,1 \times 10^{-2}$ et $1,3 \times 10^{-2}$. Afin de tenir compte de cette variabilité, le volume dynamique V_d a été calculé selon 3

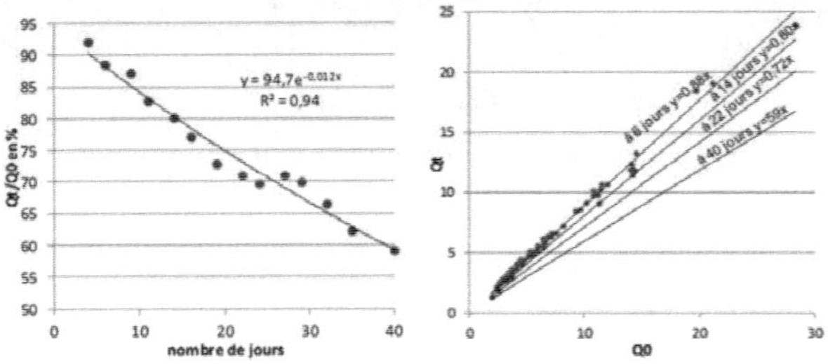

Figure 3. Méthode des correlations.

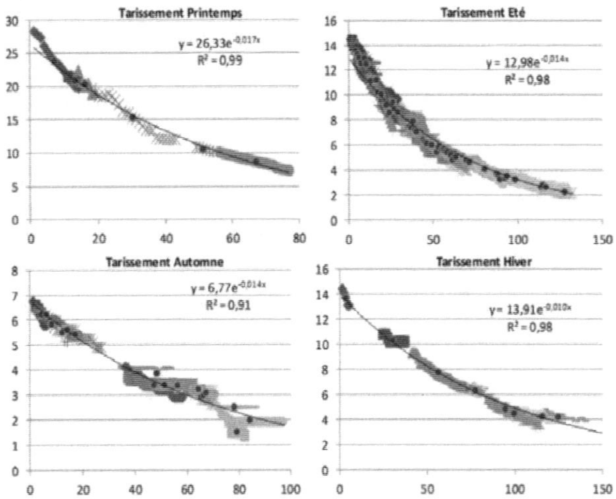

Figure 4. Méthode MSM.

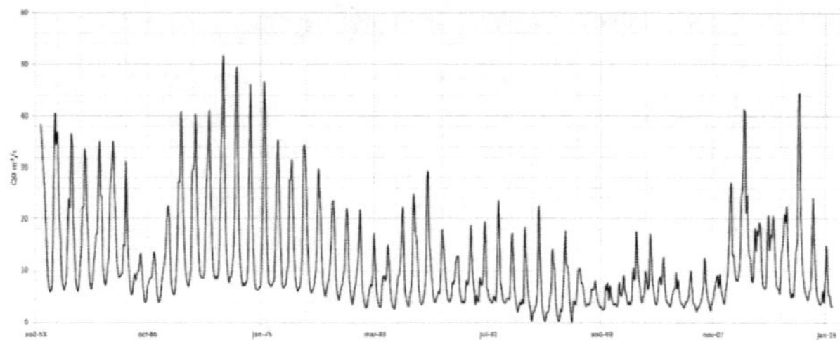

Figure 5. Chronique des débits modaux ($Q0$) moyens glissant sur 3 années à la station d'Azzaba, 1960–2015.

Tableau 2. Synthèse sur la ressource en eau dans le bassin du Haut Sebou.

		1959–2015	1959–1980	1981–2008	2009–2015
Moyenne		80	108	54	97
Max	Moyenne	160	212	113	170
	Min	59	95	59	106
	Max	298	298	183	261
	Ecart-type	53	38	28	42
	Coeff. Var.	33 %	18 %	25 %	24 %
Min	Moyenne	35	49	22	42
	Min	1	31	1	24
	Max	66	66	33	64
	Ecart-type	13	8	6	11
	Coeff. Var.	38 %	16 %	27 %	27 %

intervalles de débit $Q0$: supérieur à $20\,m^3/s$ avec un α de 1710^{-2}, compris entre la valeur précédente et $5\,m^3/s$ avec un α de $1,3 \times 10^{-2}$ et inférieur à la valeur précédente un α de 1×10^{-2}.

Les débits modaux mensuels ont été calculés sur trois années glissantes afin d'avoir un nombre de valeurs suffisant pour appliquer la méthode de détermination du débit modal (Fig. 5).

Les méthodes de segmentation des longues séries appliquées aux précipitations caractéristiques du secteurs (station de Sefrou, représentative des pluies dans le Haut Sebou) mettent en évidence 3 périodes homogènes : 1959/1980, 1981/2008, 2009/2015. En moyenne la ressource régulatrice en eau souterraine est de $80\,hm^3$, mais elle est extrêmement variable (Table 2). Dans la période sèche 1981–2008 elle n'était que de $54\,hm^3$, en moyenne, soit deux fois inférieures à celle de la période humide 1959–1980. Les contrastes sont davantage accusés pour les débits minimums, généralement en juillet, avec pour les années particulières de la moitié des années 1990 une ressource de l'ordre de $1\,hm^3$ seulement. Les chiffres pour la période récente sont relativement proches de ceux de la période humide. Notons également que les nappes contenues dans le système aquifère du haut Sebou sont très peu inertielles puisque le rapport entre le volume maximum et minimum annuel est de 1 à 4 en moyenne mais peut être supérieur à 10 sur des années particulières, comme en 1995 par exemple (Fig. 6).

La principale limite à nos résultats est liée à la nature des données utilisées. En effet, les débits sont mesurés à la station d'Azzaba qui est située à l'aval d'un réseau hydrographique principal de 262 km, le long duquel des prélèvements sont réalisés pour irriguer des cultures. Les débits soustraits minorent donc les débits à la station hydrométrique et donc les débits modaux à partir desquels sont calculés les volumes dynamiques V_d. La nature karstique des aquifères, impliquant des transferts d'eau du bassin du haut Sebou vers les bassins contigus (Akdim et al., 2012) et notamment vers l'Oum Er Rbia situé plus au sud, ne permet d'établir un bilan des nappes à partir de nos valeurs.

5 Conclusions

La démarche méthodologique mise en œuvre sur le bassin amont du Haut Sebou permet de quantifier les réserves en eau souterraine d'une fraction du Moyen-Atlas. Dans les limites du bassin versant fonctionnel, de $80\,hm^3$ en moyenne sur la période 1959/2015, les valeurs comprises entre 298 et $0.5\,hm^3$ montrent une variabilité extrême de cette ressource en eau en accord avec la nature calcaire et basaltique des aquifères. Ceci démontre que les nappes du « château d'eau du Maroc » sont très peu inertielles pouvant entraîner conjoncturellement de très faibles volumes d'eau dispo-

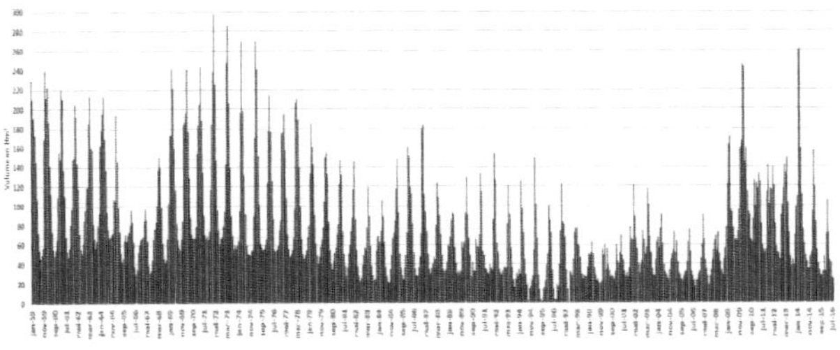

Figure 6. Volumes mensuels de la réserve régulatrice dans le bassin du Haut Sebou (1959–2015).

nibles en surface pour l'irrigation des parcelles agricoles via le système d'eau gravitaire traditionnel. A cet égard, les années 1980/90, marquées par une sécheresse pluviométrique sévère, ont vu cette ressource chuter de moitié par rapport à la période 1960/70 (54 hm^3 contre 108 hm^3) et ont été marquées par de profond changement hydro-agricole. Il sera intéressant d'étendre cette quantification des ressources en eau souterraine à l'ensemble du Moyen-Atlas en s'appuyant sur les débits des autres fleuves y prenant leur source comme l'Oum Er Rbia ou la Moulouya.

Disponibilité des données. Les données de débits ont été fournies par l'agence du bassin hydraulique du Sebou (ABHS) de Fès.

Collaborateurs. Cet article présente une partie des résultats d'un projet visant à quantifier et à spatialiser la ressource en eau dans le bassin du Haut Sebou. SL, BA, EG et ML ont participé à la conceptualisation du projet et à la recherche de son financement. Plus spécifiquement pour cet article SL et AQ ont contribué à élaborer la méthodologie, à analyser les données et à la rédaction.

Intérêts concurrents. Les auteurs déclarent qu'ils n'ont aucun conflit d'intérêts.

Clause de non-responsabilité. Publisher's note : Copernicus Publications remains neutral with regard to jurisdictional claims in published maps and institutional affiliations.

Déclaration du numéro spécial. This article is part of the special issue "Hydrology of Large River Basins of Africa". It is a result of the 4th International Conference on the "Hydrology of the Great Rivers of Africa", Cotonou, Benin, 13–20 November 2021.

Remerciements. Ce travail s'inscrit dans l'Action Intégrée franco-marocaine no. MA 10-244 d'un Partenariat Hubert Curien.

Références

Akdim, B., Sabaoui, A., Amyay, Mh., Laaouane, M., Gille, E., and Obda, K. : Influences hydro karstiques du système sourcier Aïn Sebou-Timedrine-Ouamender Sur l'hydrologie de l'oued Sebou (Moyen Atlas, Maroc), Z. Geomorph., 56, 165–181, 2012.

Amraoui, F., Razack, M., and Bouchaou, L. : Turbidity dynamics in karstic systems. Example of Ribaa and Bittit springs in the Middle Atlas (Morocco), Hydrolog. Sci. J., 48, 971–984, 2003.

Castany, G. : Prospection et exploitation des eaux souterraines. Paris, Dunod édition, 718 pp., 1968.

Castany, G. et Margat, J. : Dictionnaire français d'hydrogéologie, BRGM Service géologique national, Orléans, 249 pp., 1977.

De Waele, J. and Melis, M. T. : Geomorphology and geomorphological heritage of the Ifrane–Azrou region (Middle Atlas, Morocco), Environ. Geol., 58, 587–599, 2009.

Frizon de Lamotte, D., Zizi, M., Missenard, Y., Hafid, M., El Azzouri, M., Maury, R. C., Charrière, A., Taki, Z., Benammi, M., and Michard, A. : The Atlas system – Continental Evolution : The geology of Morocco, Lecture notes in earth sciences 116, Springer-Verlag Berlin Heidelberg, 2008.

Gamez, P. and Zumstein, J. F. : Méthode de détermination des débits de base et de ruissellement d'un cours d'eau, L'eau, la terre et les hommes – Hommage à René Frécaut, 113–120, Presse Universitaire de Nancy, 1993.

Michard, A., Frizon de Lamotte, D., Saddiqi, O., and Chalouan, A. : An outline of the geology of Morocco – Continental Evolution : The geology of Morocco, Lecture notes in earth sciences 116, Springer-Verlag Berlin Heidelberg, 2008.

Moore, R. D. : Storage-outflow modelling of streamflow recessions, with application to a shallow-soil forested cathment, J. Hydrol., 198, 260–270, 1997.

Ophori, D. and Toth, J. : Relationships in regional ground-water discharge to stream : an analysis by numerical simulation, J. Hydrol., 119, 215–244, 1990.

Qadem, A. : Quantification, modélisation et gestion de la ressource en eau dans le bassin versant du Haut Sebou (Maroc), Thèse de doctorat, Université de Lorraine et Fès, 272 pp., 2015.

Sanz, E. et Recio, B. : A method to assess annual average renewable groundwater reserves for large regions in Spain, Hydrolog. Sci. J., 56, 99–107, 2011.

Tallaksen, L. M. : A review of base flow recession analysis, J. Hydrol., 165, 359–370, 1995.

Tallaksen, L. M. et Van Lanen, H. A. J. : Hydrological Drought : Processes and Estimation Methods for Streamflow and Groundwater, Elsevier Edn., 579 pp., 2004.

Toebes, C. and Strang, D. D. : On recession curves – Recession equations, J. Hydrol. NZ, 3, 2–15, 1964.

Zécharias, Y. B. et Brutsaer, W. : Recession characteristics of groundwater outflow and base flow from montainous watersheds, Water Resour. Res., 24, 1651–1658, 1988.

Proc. IAHS, 384, 175–179, 2021
https://doi.org/10.5194/piahs-384-175-2021

Open Access PIAHS

Hydrology of Large River Basins of Africa

Evolution des apports en eau dans la plaine rizicole de Marovoay par le fleuve Betsiboka (Madagascar) : reconstitution et projections jusqu'en 2100

Dominique Dumas[1], **Sylvain Bigot**[2], **Yao Télesphore Brou**[3], **Rivo Ramboarison**[4], **Samuel Razanaka**[5], and **Laurent Roger Robison**[5]

[1]Université Lyon 3 – Environnement Ville Société (EVS, UMR 5600), 69362 Lyon, France
[2]UGA – Institut des Géosciences de l'Environnement (IGE, UMR 5001), 38000 Grenoble, France
[3]Université de La Réunion – ESPACE-DEV (UMR 228), 97744 Saint-Denis, France
[4]Cabinet d'études BEST-IDEES, Antananarivo, Madagascar
[5]Centre National de Recherche en Environnement (CNRE), Antananarivo, Madagascar

Correspondence: Dominique Dumas (dominique.dumas@univ-lyon3.fr)

Published: 16 November 2021

Résumé. Aujourd'hui, le fleuve Betsiboka alimente une vaste plaine agricole située au NW de Madagascar. Ce fleuve est pourtant très mal connu sur un plan hydrologique. Sur la base de mesures des écoulements entre 1958 et 1971, cette étude cherche dans un premier temps à reconstituer les débits contemporains de ce fleuve, puis dans un second temps à cerner comment les écoulements peuvent évoluer au cours du XXIème siècle. Les résultats de cette étude montrent qu'une baisse des débits semble se dessiner à la fin de ce siècle. Elle diminuerait la ressource en eau apportée par le fleuve, selon les scénarii climatiques retenus, de 2.4 à 2.7 km^3 d'eau par an. Cette diminution de la ressource serait par ailleurs plus marquée sur la période de décembre à février, durant justement la période de repiquage du riz qui implique un apport en eau élevé.

1 Introduction

La plaine rizicole de Marovoay, avec une superficie de presque 18 000 ha de cultures, constitue le deuxième grenier à riz de Madagascar, après la région du lac d'Alaotra. L'alimentation en eau de cette plaine est largement liée au fleuve Betsiboka (Bigot et al., 2018), ainsi qu'à ses affluents, dont les bassins s'inscrivent largement au sein du Parc National d'Ankarafantsika et de son espace forestier (l'une des dernières grandes forêts sèches malgaches). Certains secteurs de la plaine sont d'avantage tributaires directement du fleuve Betsiboka, et d'un apport en eau par pompage de ses eaux. La diminution des écoulements du fleuve et la possibilité de mobiliser des quantités d'eau rendent aujourd'hui l'irrigation plus délicate, voire aléatoire.

A l'inverse, il semblerait que la durée d'inondation, par débordement du fleuve, se soit allongée (Clark, 2012 ; Lawrence et Vandecar, 2015). Les agriculteurs doivent alors retarder la mise en culture. La période de la campagne culturale est donc non seulement décalée, mais aussi, sur la fin de campagne, raccourcie par le manque d'eau. Les agriculteurs ne peuvent plus compter totalement sur les apports du fleuve. La riziculture pluviale, encore marginale sur l'ensemble de la plaine de Marovoay il y a quelques années, s'étend de plus en plus. Sur ces secteurs, pour des raisons économiques (prix du carburant, vétusté du matériel de pompage), mais aussi probablement hydrologiques (étiages plus longs, excès d'eau au mauvais moment...), les parcelles sont aujourd'hui de plus en plus exploitées à travers une riziculture pluviale, où la maîtrise de l'eau, et donc du calendrier rizicole, n'est dès lors plus possible.

Quelle est la part des raisons consécutives à une irrigation défaillante ou mal maitrisée, et celles liées à des modifications à venir des écoulements du fleuve ? Une gestion efficace de cette plaine rizicole implique nécessairement de pouvoir répondre à cette interrogation et celles qui en découlent. Dans cet objectif, une connaissance des écoulements du fleuve reste indispensable. Or, le réseau hydro-

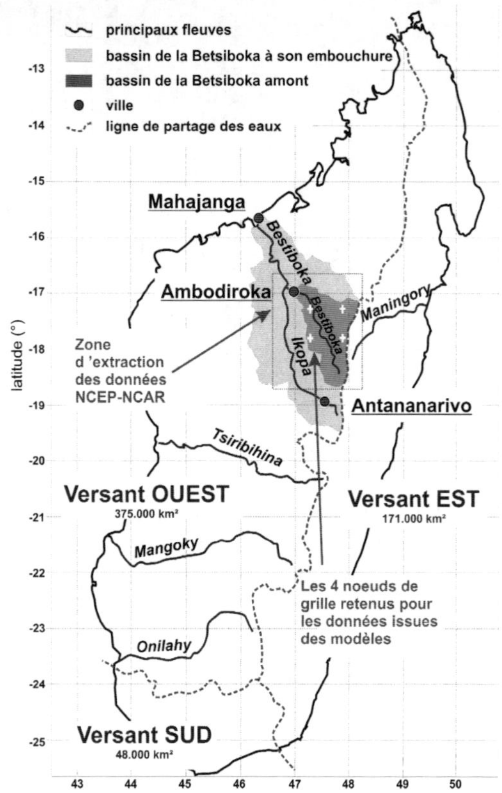

Figure 1. Localisation du bassin de la Betsiboka à Madagascar, et position des extractions des données NCEP-NCAR et des modèles climatiques globaux.

logique malgache, souvent à l'abandon, ne permet plus de fournir des informations suffisantes pour évaluer et cerner les évolutions des écoulements. Par ailleurs, les mesures et observations collectées dans les années 1970–1990 (Chaperon et al., 1993) ne sont plus toujours accessibles au sein des services hydrologiques malgaches, ou dans des bases de données. Certaines données acquises ont été visiblement perdues, malgré les moyens considérables mis en place pour les acquérir dans les années 70 et 80. Ainsi, sur la Betsiboka, au niveau de son embouchure, où se place la plaine de Marovoay, la connaissance des débits passés et actuels est devenue difficile. Pourtant, après la Mangoky (55 700 km^2) et la Tsiribihina (49 800 km^2), la Betsiboka est le troisième fleuve malgache par la taille de son bassin qui collecte des affluents sur plus de 49 000 km^2.

Au niveau de la plaine de Marovoay, l'apport en eau plus réduit est déjà une réalité ces dernières années par le contexte environnemental modifié, par la déforestation et par la variabilité climatique (Maina et al., 2012 ; Bigot et al., 2020). Ce travail se propose justement de mieux cerner les écoulements actuels de ce grand fleuve malgache, puis d'évaluer pour l'ensemble du XXIème siècle leurs fluctuations attendues. D'abord, cette étude cherche à reconstruire mensuellement les données hydrologiques sur une période récente, pour mieux saisir le comportement hydrologique du fleuve et ses fluctuations actuelles. Puis, à partir de cette modélisation des écoulements, et de données complémentaires issues de modèles climatiques, cette étude montre les évolutions possibles des écoulements de la Betsiboka pour l'ensemble du XXIème siècle.

2 Données et méthodologie

2.1 Reconstitution des écoulements de la Betsiboka à Ambodiroka

Avec des variations saisonnières marquées, son débit mensuel moyen passe de 120 à 4000 m^3/s, son module est à l'embouchure vraisemblablement proche de 800 m^3/s, dont 460 m^3/s sont apportés par son affluence principal de l'Ikopa (à partir des données *in* Chaperon et al., 1993). Notons que ces valeurs semblent éloignées des descriptions observées dans la littérature plaçant son module à 270 m^3/s (Raharimahefa et Kusky, 2010 ; Duvail et al., 2017) qui est en fait le module de la Betsiboka avant sa confluence avec l'Ikopa (approché à partir de la station hydrologique d'Ambodiroka). La raison tient probablement à la confusion liée au fait qu'il n'existe pas d'autres stations plus à l'aval, et qu'elle décrit la Betsiboka dans l'ouvrage monographique de référence de Chaperon et al. (1993). Le module y est d'ailleurs évalué plus exactement à 297 m^3/s sur la période 1958–1973.

Les écoulements de la Betsiboka ont été mesurés de 1957 à 1983, et de 1988 à 1993, à la station hydrologique d'Ambodiroka. Son bassin est alors de 11 800 km^2. Les débits ont été calculés avec une grande précision puisque plus de 150 jaugeages ont été effectués dans une large gamme de débits allant de 40 à 7800 m^3/s (Chaperon et al., 1993). De 1958 à 1971, les données hydrologiques présentent peu de lacunes, contrairement à celles inscrites après 1971, et plus encore entre 1988 et 1993. Les données après l'année 1971 sont ainsi écartées de cette étude par leur mauvaise qualité.

Si dans les années 60, on recensait plus de 250 stations météorologiques, aujourd'hui les stations sont peu nombreuses, et par ailleurs se placent pour l'essentiel sur le littoral ou à proximité de la capitale (Bigot et al., 2018). A partir des produits du National Centers for Environmental Prediction – National Center for Atmospheric Research (NCEP–NCAR), il est possible d'obtenir les données journalières de réanalyses atmosphériques de 1958 à 2020 (Fig. 1) avec une résolution spatiale de 1.875° (Kalnay et al., 1996 ; Grist et Nicholson, 2001). Sur la période 1950 à 2020, il est dès lors possible d'examiner les variations quotidiennes moyennes des pluies et des températures sur une zone intégrant le bassin de la Betsiboka à Ambodiroka. Dans ce travail, nous utiliserons cet ensemble de données NCEP-NCAR pour calculer l'évaporation potentielle puis calculer les écoulements à partir d'un modèle hydrologique conceptuel global (GR2M).

L'évaporation potentielle sur le bassin est appréciée à partir de la température journalière des réanalyses NCEP-NCAR. L'introduction du rayonnement extraterrestre journalier (Re_j), calculé ici pour la latitude moyenne du bassin de -17.5 °S permet d'affiner l'estimation de l'évaporation potentielle (Oudin, 2004 ; Oudin et al., 2005) (Eq. 1).

$$\mathrm{ETP}_j = \frac{Re_j}{\gamma\rho}\frac{T_j+5}{100} \tag{1}$$

avec γ, la chaleur latente de vaporisation de l'eau ($2{,}25\,\mathrm{MJ\,kg^{-1}}$) ; ρ, la masse volumique de l'eau ($1000\,\mathrm{kg\,m^{-3}}$) ; Re_j, le rayonnement extraterrestre ($\mathrm{MJ\,m^{-2}\,j^{-1}}$) calculé à la latitude $-17.5°$ et du jour julien j ; T_j, la température moyenne du jour julien j (°C).

L'estimation des écoulements mensuels utilise le modèle conceptuel global GR2M (Génie Rural Mensuel à deux paramètres), dont la description est détaillée par de nombreux travaux (Mouelhi et al., 2006, 2013). Il est constitué de deux réservoirs, que l'on alimente avec des données mensuelles de pluie et d'évaporation potentielle, et par deux coefficients qui caractérisent le dimensionnement et le fonctionnement de ces réservoirs. Ces deux paramètres caractérisent dès lors le fonctionnement hydrologique de l'ensemble du bassin versant. Ils sont définis par comparaison avec les débits mesurés – pour la Betsiboka de 1958 à 1971 – par itération avec le critère d'optimisation de Nash–Sutcliffe (Fig. 2, Éq. 2). Cet indicateur juge de la qualité de l'estimation des débits modélisés par rapport à des débits observés (Nash et Sutcliffe, 1970 ; Perrin et al., 2003). La valeur de cet indice varie entre $-\infty$ et 100 %, et se rapproche de 100 % lorsque les débits calculés se calent sur les débits observés. L'équation suivante le définit comme :

$$\mathrm{NS}(\%) = 100\left[1 - \frac{\sum(Q_{\mathrm{obs}} - Q_{\mathrm{cal}})^2}{\sum(Q_{\mathrm{obs}} - \bar{Q}_{\mathrm{obs}})^2}\right] \tag{2}$$

L'indice de Nash est de 71 % dans cette reconstitution. Au regard de la sévérité de cet indicateur, une performance supérieure à 60 % est généralement jugée satisfaisante dans les modèles hydrologiques (Perrin et al., 2003).

2.2 Estimations des pluies et des températures mensuelles jusqu'en 2100

Les projections climatiques de 7 modèles climatiques globaux, calculées mensuellement par vingtaines jusqu'en 2100, et sur 4 points de grille (Fig. 1, Tableau 1), ont été utilisées. Pour les données de projections climatiques futures, cette étude a utilisé deux scénarii d'émission RCP (Representative Concentration Pathways) ; le scénario moyen RCP4.5, et celui, plus pessimiste, RCP8.5 supposant un forçage élevé dans l'émission des gaz à effet de serre. Le choix d'utiliser des moyennes calculées sur des périodes de 20 ans a permis de s'affranchir de la variabilité interannuelle pour ne conserver que les principales modifications tendancielles. Cette approche multi-modèle a limité également le poids et l'incertitude liés aux résultats issus d'un unique modèle.

Le Tableau 1 synthétise les moyennes annuelles des sorties des 7 modèles climatiques pour les deux scénarii retenus. On observe naturellement des différences entre les modèles pour les deux paramètres climatiques, même si les variabilités sont plus marquées pour les précipitations. La moyenne des 7 modèles de projection montre une hausse des températures sur l'ensemble du XXIème siècle serait de 1.7 °C dans le cas du scénario RCP4.5, et de 4.4 °C pour le scénario RCP8.5. En revanche, et quel que soit le scénario, les pluies n'indiquent pas d'évolution simple au cours du XXIème siècle.

3 Résultats et discussion

3.1 Reconstitution des écoulements mensuels de 1950 à 2019

Sur la période 1958 à 1971, et à partir des températures et précipitations journalières issues des réanalyses NCEP–NCAR, il est dès lors possible de caler le modèle GR2M. La performance obtenue, de 71.1 %, montre qu'il est possible de reconstituer les débits à partir de ces données à une relative robustesse (Fig. 2). Dès lors, il est possible de calculer les débits sur l'ensemble de la période 1958–2019 pour laquelle les températures et précipitations sont connues mensuellement à partir des réanalyses NCEP-NCAR. Les débits mensuels moyens par vingtaine peuvent, par exemple, être précisés (Tableau 2).

3.2 Diminution des écoulements sur l'ensemble du XXIème siècle

Le Tableau 3 montre les modules de la Betsiboka par vingtaines allant de 2020 à 2099. Les valeurs avant 2019 sont obtenues à partir des données de réanalyses, après, les valeurs des écoulements sont calculées à partir des modèles climatiques globaux selon les deux scénarii RCP retenus.

L'utilisation de deux types de données climatiques, et sur des espaces différents, produit un biais dans la reconstitution des écoulements entre les vingtaines 2000–2019 et 2020–2039 (Tableau 3). Il implique de prendre les valeurs absolues estimées avec prudence. En revanche, l'évolution générale des débits est plus signifiante, et plus encore si on ne regarde que les projections hydro-climatiques jusqu'en 2100.

Les écoulements calculés varient naturellement fortement avec les modèles utilisés. Par exemple, sur la période 2060 à 2079, avec le scénario RCP4.5, les modules reconstitués de la Betsiboka peuvent varier de 187 à $359\,\mathrm{m^3\,s^{-1}}$ selon le modèle retenu. Curieusement, l'analyse des modules par vingtaine montre des valeurs très proches quel que soit le scénario RCP retenu, les moyennes des débits des 9 modèles sont quasiment identiques sur toutes les vingtaines.

Ces évolutions illustrent sans doute l'aspect systémique du contexte climatique, toute modification d'un paramètre a des

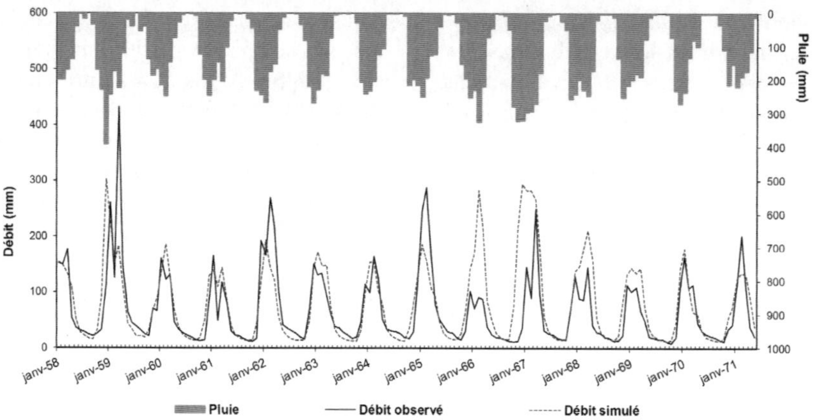

Figure 2. Modélisation des débits mensuels de la Betsiboka à Ambodiroka à partir du modèle hydrologique CR2M, et débits mesurés de 1958 à 1971.

conséquences sur d'autres. Ici, l'augmentation plus forte des températures (dans le cas du scénario RCP8.5), et donc de l'évaporation, est compensée par des pluies accrues. Mais à l'échelle de ce petit bassin, cette évolution aussi équilibrée n'était pas forcément imaginable au départ. Les tendances des évolutions des écoulements restent les mêmes, une baisse moyenne de $140\,m^3\,s^{-1}\,100\,ans^{-1}$ dans le cas du scénario RCP4.5, contre $125\,m^3\,s^{-1}\,100\,ans^{-1}$ pour le scénario RCP8.5. Le scénario le plus pessimiste montre une baisse légèrement plus réduite des écoulements que celui plus optimiste. La diminution moyenne des débits entre 2020 et 2100 serait respectivement de 86 et $77\,m^3\,s^{-1}$ selon les scénarii, soit une perte annuelle de la ressource en eau de 2.4 à $2.7\,km^3$, ce qui est considérable.

Pour les prochaines décennies (2020–2100), à l'échelle mensuelle (Tableau 4), les modifications de l'écoulement de la Betsiboka apparaissent plus contrastées, parfois imperceptibles, parfois très marquées. A l'échelle mensuelle, les modifications restent aussi relativement similaires quel que soit les scénarii RCP retenus. La diminution de l'écoulement mensuel, abordée à partir de la moyenne des 7 modèles climatiques globaux, montre clairement une intensité plus marquée en décembre, janvier et février. La diminution moyenne tendancielle sur l'ensemble de la période étudiée serait même en janvier de $957\,m^3\,s^{-1}\,100\,ans^{-1}$ à $914\,m^3\,s^{-1}\,100\,ans^{-1}$ selon les scénarii RCP. En période de basses eaux, de juin à novembre, les écoulements ne sont que légèrement modifiés.

4 Conclusions

Les reconstitutions des débits mensuels avec des données climatiques NCEP-NCAR produisent des résultats satisfaisants (Nash = 71 %). Avec l'introduction des projections climatiques issues des modèles globaux, la modélisation hydroclimatique offre alors la possibilité de déceler les grandes tendances hydrologiques pour les prochaines décennies. Sur la base de plusieurs projections climatiques issues de différents modèles numériques, l'étude des débits de la Betsiboka indique que les écoulements risquent de diminuer ces prochaines décennies. Curieusement, quel que soit l'hypothèse d'émission envisagée, la baisse des ressources en eau semble relativement identique, l'accroissement du prélèvement par évaporation, dans le cas du scénario RCP8.5, est compensé par une majoration des précipitations. A l'échelle saisonnière, la baisse des ressources en eau est fortement contrastée selon les mois. Elle est très marquée en janvier, et forte en décembre et février.

Au niveau de la plaine de Marovoay, l'apport en eau plus réduit est déjà une réalité ces dernières années par le contexte environnemental modifié, par la déforestation et par la variabilité climatique (Maina et al., 2012 ; Bigot et al., 2020). Ces projections hydro-climatiques sont néanmoins inquiétantes, et montrent que sur la fin de ce siècle, la diminution de l'écoulement sera probablement guidée en premier lieu par le changement climatique. Cette baisse de la ressource en eau ne sera pas sans conséquence sur les pratiques agricoles de la région. La première campagne de riz notamment risque d'être fortement menacée ces prochaines années. Pour cette campagne, la période de repiquage se place de novembre à janvier, et implique un apport en eau élevé, or elle coïncidera tout particulièrement à une baisse très forte des écoulements. En revanche, la seconde campagne de riz, de contre-saison, avec un repiquage d'avril à juin, sera a priori moins affectée par cette évolution hydro-climatique. Dans des perspectives de projets d'aménagement hydro-agricole, les gestionnaires du fleuve, ou ceux de la plaine de Marovoay, doivent pouvoir s'appuyer sur les dynamiques hydrologiques actuelles, mais également sur des projections au cours du XXIème siècle.

Disponibilité du code. Le modèle GR2M peut être utilisé à partir de feuilles Excel : https://webgr.inrae.fr/modeles/mensuel-gr2m/ (Perrin et Andréassian, 2021).

Disponibilité des données. Données hydrologiques : https://www.documentation.ird.fr/hor/fdi:37307 Chaperon et al. (1993). Les données NCEP-NCAR : https://psl.noaa.gov/data/gridded/data.ncep.reanalysis.html (National Centers for Environmental Prediction et al., 1994).

Collaborateurs. DD a réalisé la conceptualisation de la méthode, le traitement et l'écriture du manuscrit. Les co-auteurs ont contribué comme suit : SB et TYB ont examiné les données NCEP-NCAR, examiné le processus de rédaction et contribué à la version finale du manuscrit ; RR, SR et LRR ont participé à l'acquisition des données, et examiné le processus de rédaction.

Intérêts concurrents. Les auteurs déclarent qu'ils n'ont aucun conflit d'intérêts.

Clause de non-responsabilité. Publisher's note : Copernicus Publications remains neutral with regard to jurisdictional claims in published maps and institutional affiliations.

Déclaration du numéro spécial. This article is part of the special issue "Hydrology of Large River Basins of Africa". It is a result of the 4th International Conference on the "Hydrology of the Great Rivers of Africa", Cotonou, Benin, 13–20 November 2021.

Remerciements. Les auteurs remercient les relecteurs anonymes pour leurs remarques qui ont contribué à améliorer ce manuscrit.

Références

Bigot, S., Dumas, D., Brou, Y. T., Ramboarison, R., Razanaka, S., Descroix, L., Philippon, N., and Gwiazdzinski, L. : La difficulté du monitoring climatique à une échelle fine dans une région fortement impactée par la déforestation tropicale : la forêt sèche du parc national d'Ankarafantsika (Madagascar), Publications de l'Association Internationale de Climatologie, 31, 356–361, 2018.

Bigot, S., Dumas, D., Brou, T., Ramboarison, R., Razanaka, S., and Philippon, N. : Etude hydroclimatique dans le nord-ouest de Madagascar : apport des données satellitaires GRACE, Publications de l'Association Internationale de Climatologie, 33, 121–126, 2020.

Chaperon, P., Danloux, J., and Ferry, L. : Fleuves et rivières de Madagascar, Monographies hydrologiques, 10, ORSTOM, Paris, France, 874 pp., 1993.

Clark, M. : Deforestation in Madagascar : Consequences of population growth and unsustainable agricultural processes, Global Majority E-Journal, 3, 61–71, 2012.

Duvail, S., Hamerlynck, O., Paron, P., Hervé, D., Nyingi, W. D., and Leone, M. : The changing hydro-ecological dynamics of rivers and deltas of the Western Indian Ocean : Anthropogenic and environmental drivers, local adaptation and policy response, C. R. Geosci., 349, 269–279, https://doi.org/10.1016/j.crte.2017.09, 2017.

Grist, J. P. and Nicholson, S. E. : A study of the dynamic factors influencing the rainfall variability in the West African Sahel, J. Climate, 14, 1337–1359, https://doi.org/10.1175/1520-0442(2001)014<1337 :ASOTDF>2.0.CO2, 2001.

Kalnay, E., Kanamitsu, M., Kistler, R., Collins, W., Deaven, D., Gandin, L., and Zhu, Y. : The NCEP/NCAR 40-year reanalysis project, B. Am. Meteorol. Soc., 77, 437–472, https://doi.org/10.1175/1520-0477(1996)077<0437 :TNYRP>2.0.CO2, 1996.

Lawrence, D. and Vandecar, K. : Effects of tropical deforestation on climate and agriculture, Nat. Clim. Change, 5, 27–36, https://doi.org/10.1007/s00382-014-2203-8, 2015.

Maina, J., de Moel, H., Vermaat, J. E., Bruggemann, J. H., Guillaume, M. M., Grove, C. A., Madin, J. S., Mertz-Kraus, R., and Zinke, J. : Linking coral river runoff proxies with climate variability, hydrology and land-use in Madagascar catchments, Mar. Pollut. Bull., 64, 2047–2059, 2012.

Mouelhi, S., Michel, C., Perrin, C., and Andréassian, V. : Stepwise development of a two-parameter monthly water balance model, J. Hydrol., 318, 200–214, https://doi.org/10.1016/j.jhydrol.2005.12.022, 2006.

Mouelhi, S., Madani, K., and Lebdi, F. : A structural overview through GR (s) models characteristics for better yearly runoff simulation, Open Journal of Modern Hydrology, 318, 200–214, https://doi.org/10.1016/j.jhydrol.2005.12.022, 2013.

Nash, J. E. and Sutcliffe, J. V. : River flow forecasting through conceptual models Part one : A discussion of principles, J. Hydrol., 27, 282–290, https://doi.org/10.1016/0022-1694(70)90255-6, 1970.

National Centers for Environmental Prediction/National Weather Service/NOAA/U.S. Department of Commerce : updated monthly, NCEP/NCAR Global Reanalysis Products, 1948–continuing, Research Data Archive at NOAA/PSL, accessible à l'adresse suivante : https://psl.noaa.gov/data/gridded/data.ncep.reanalysis.html (la date du dernier accès : juin 2021), 1994.

Oudin, L. : Recherche d'un modèle d'évapotranspiration potentielle pertinent comme entrée d'un modèle pluie-débit global, Sciences of the Universe, ENGREF (AgroParisTech), 2004.

Oudin, L., Hervieu, F., Michel, C., Perrin, C., Andréassian, V., Anctil, F., and Loumagne, C. : Which potential evapotranspiration input for a lumped rainfall–runoff model ? : Part 2 – Towards a simple and efficient potential evapotranspiration model for rainfall–runoff modelling, J. Hydrol., 303, 290–306, https://doi.org/10.1016/j.jhydrol.2004.08.026, 2005.

Perrin, C. et Andréassian, V. : Á partir du modèle mise au point par Mouelhi (2003), Application du modèle pluie-débit mensuel GR2M sous Excel, accessible à l'adresse suivante : https://webgr.inrae.fr/modeles/mensuel-gr2m/, la date du dernier accès : octobre 2021.

Perrin, C., Michel, C., and Andréassian, V. : Improvement of a parsimonious model for streamflow simulation, J. Hydrol., 279, 275–289, https://doi.org/10.1016/j.jhydrol.2004.08.026, 2003.

Raharimahefa, T. and Kusky, T. M. : Environmental monitoring of Bombetoka Bay and the Betsiboka Estuary, Madagascar, using multi-temporal satellite data, J. Earth Sci., 21, 210–226, https://doi.org/10.1007/s12583-010-0019-y, 2010.

Proc. IAHS, 384, 181–186, 2021
https://doi.org/10.5194/piahs-384-181-2021

Open Access PIAHS
Hydrology of Large River Basins of Africa

Investigating links between rainfall variations in the Ogooué River basin and ENSO in the Pacific Ocean over the period 1940–1999

Sakaros Bogning[1,6], Frédéric Frappart[2], Gil Mahé[3], Adrien Paris[4], Raphael Onguene[1], Fabien Blarel[2], Fernando Niño[2], Jacques Etame[5], and Jean-Jacques Braun[6]

[1]Laboratory of Technology and Applied Sciences, University Institute of Technology, P.O. Box 8698 Douala, Cameroon
[2]LEGOS, Université de Toulouse, France, OMP, 14 Av. E. Belin, 31400 Toulouse, France
[3]HydroSciences Montpellier, Université de Montpellier, 300 Av. Pr E. Jeanbrau, Montpellier, France
[4]Collecte Localisation Satellite (CLS), Ramonville-Saint-Agne, 31520 France
[5]Département de Sciences de la Terre, Université de Douala, BP 24 157 Douala, Cameroon
[6]LMI DYCOFAC, IRD, BP 1857 Yaoundé, Cameroon

Correspondence: Sakaros Bogning (sakarosb@gmail.com)

Published: 16 November 2021

Abstract. This paper investigates links between rainfall variability in the Ogooué River Basin (ORB) and El Niño Southern Oscillation (ENSO) in the Pacific Ocean. Recent hydroclimatology studies of the ORB and surrounding areas resulting in contrasting conclusions about links between rainfall variability and ENSO. Thus, to make the issue clearer, this study investigates the links between ENSO and rainfall in the ORB over the period 1940–1999. The principal component analysis of monthly rainfall in the ORB was done. The temporal mode of the first component corresponds to the interannual variations of rainfall on the ORB. Also, the pattern of the spatial mode of the first component shows that the ORB is a homogeneous hydroclimatic zone. However, no leading mode is significantly correlated to the ENSO index. A cross-wavelet analysis of the time series of basin-scale rainfall and the ENSO index was therefore carried out. The result is a set of periodogram structures corresponding to some ENSO episodes recorded over the study period. And wavelet coherence analysis of both time series confirms that there are significant links between ENSO and rainfall in the ORB.

1 Introduction

The Ogooué River is about a 1200 km long stream-flow (Kittel et al., 2018), and the principal river of Gabon in west-central Africa. The river rises in the Mount Ntalé, situated northwest of the Plateaus Batéké in The Republic of Congo. Its watershed drains more than 70 % of the country of Gabon, with some tributaries reaching neighboring countries of Cameroon, Equatorial Guinea, and The Republic of Congo. This area annually receives large amounts of rainfall (the mean annual precipitation was 1425 mm between 1940–1999; Bogning et al., 2021) which leads to about 4750 m^3/s average annual discharge at Lambaréné, making this river the second-largest of the region by volume discharge, trailing only the Congo (Lambert et al., 2015; Mignard et al., 2017).

In recent decades, the Ogooué River Basin (ORB) has been subject to regional climate change in Western Equatorial Africa characterized by a heterogeneous reduction in rainfall and consequently a reduction in runoff (Hua et al., 2016; Nicholson et al., 2018). Among factors of rainfall variability, some authors showed that central Pacific SSTs governed by El Niño Southern Oscillation (ENSO) are highly correlated to rainfall in central Africa (Camberlin et al., 2001; Farnsworth et al., 2011). However, despite the identification of ENSO as one of the leading factors of regional changes in rainfall, very weak correlations were found between rainfall in the ORB and El Niño index (Bogning et al.,

2021). Since these latter correlations suggest poorly known nonlinear ENSO-rainfall relationships in the ORB, the aim of this work is to investigate links between ENSO and rainfall in the ORB between 1940 and 1999, using cross-wavelet analysis.

2 Data and methods

2.1 The Ogooué River

The ORB is approximately 224 000 km^2, 90 % of which lies within Gabon (Kittel et al., 2018). This basin is located between 9/15° E, and 4° S/2.5° N (Bogning et al., 2018). The ORB is one of the world's best-preserved ecosystems and probably the last remaining witness of the great natural diversity of Central Africa's rainforests. For example, the rural penetration rate in Gabon is about 1 inhabitant per km^2 (Braun et al., 2015; Sannier et al., 2016) and many studies report a deforestation rate close to 0 %, with even full natural regeneration (Fichet et al., 2014; Sannier et al., 2016). It is also drained by a dense hydrographic network with about one river every 600 m. The main tributaries of the Ogooué River in order of discharge are the Ivindo which flows into the Ogooué from the right bank close to the equator at Kankan and the Ngounié which flows into the Ogooué from the left bank a few dozen kilometers upstream from Lambaréné (see Fig. 1).

The landscape of the Ogooué is mainly mountainous rainforest, although the elevations are not very high with an average height of about 450 m (see Fig. 1) (Kittel et al., 2018). The coastal part is plain that extends from the river mouth to the Atlantic Ocean for about 100 km into the mainland and covers the lake area. The second relief is made up of plateaus very often carved by rivers, which covers a very large part of the basin extending northwards over Ogooué-Ivindo, southeast over Haut-Ogooué (the Batéké plateau), southwards over the Ogooué-Lolo and Ngounié regions. The third structure observed on the ORB is the mountainous massifs consisting of mountain ranges, some of which are 1000 m high, such as the Crystal Mountains, the Chailu massif (Bogning et al., 2021).

2.2 Datasets

The rainfall data used in this study are 0.5° × 0.5° gridded monthly precipitation generated from the reference series inherited from observations of the ORSTOM (Office de la Recherche Scientifique et Technique de l'Outre-Mer, now Institut de Recherche pour le Développement – IRD) and now on free access at the SIEREM (Système d'Informations Environnementales sur les Ressources en Eau et leur Modélisation) website of Hydrosciences Laboratory Montpellier (Rouché et al., 2010; Dieulin et al., 2019). These data were made available for the entire African continent over a 60-year period (1940–1999) and

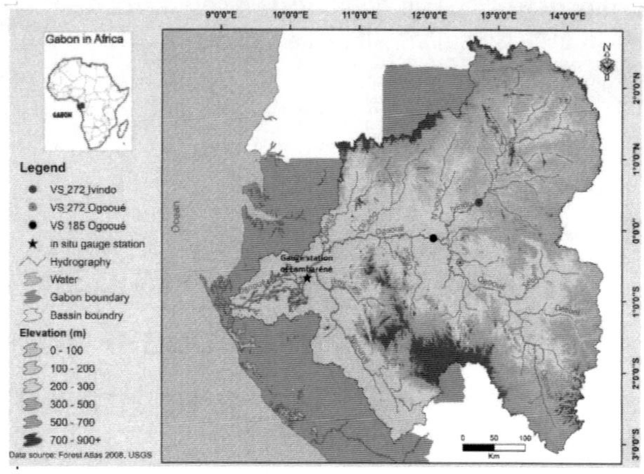

Figure 1. Topography and main rivers of the ORB.

are available at http://www.hydrosciences.fr/sierem/produits/index.asp?frame=grille (last access: 29 July 2021).

The ENSO index used in this study is Niño 3.4. Monitoring of ENSO conditions primarily focuses on sea surface temperature (SST) anomalies in 4 geographic regions of the equatorial Pacific. The Niño 3.4 region is comprising portions of Niño regions 3 and 4, from 170 to 120° W longitude. In the Niño 3.4 region, SST anomalies equal to or greater than 0.5 °C in are indicative of ENSO warm phase (El Niño) conditions, while anomalies less than or equal to −0.5 °C are associated with cool phase (La Niña) conditions (Bunge and Clarke, 2009; Tippett and L'Heureux, 2020).

2.3 Methods

The Empirical Orthogonal Functions (EOF) analysis was carried out to understand the spatial and temporal dynamics of rainfall in the ORB between 1940 and 1999. The correlations between the temporal principal components associated with the first EOFs analysis and the time series of the El Niño index were computed to determine the effects of ENSO on rainfall variations in the ORB. In addition, given the time lags between changes of SSTs in the central Pacific and the appearance of divers ENSO-induced effects in certain regions of the earth's surface (Plisnier et al., 2000; Giannini et al., 2001; Rotstayn et al., 2010; Anyamba et al., 2018; Kolusu et al., 2019), cross-wavelet analysis was applied to the ORB basin-scale time series of rainfall (computed based on Eq. 1; Ramillien et al., 2006) and the El Niño index time series to identify the temporal scales presenting similar variations in these two-variables. It should be noted that the seasonal variability is firstly removed from precipitation data.

$$\delta h(t) = \frac{R^2}{S}\sum_{j \in S} h(\lambda_j, \theta_j, t)\sin(\theta_j)\delta\lambda\delta\theta \quad (1)$$

where $h\left(\lambda_j, \theta_j, t\right)$ is monthly rainfall expressed in mm per month, with $j = 1, 2, 3, \ldots, N$, S is the area of the basin, λ_j

and θ_j are longitude and co-latitude, and are the grid steps in longitude and latitude (generally $\delta\lambda = \delta\theta$), R is the mean radius of the Earth ~ 6371 km.

Wavelet cross-correlation analysis is used to analyze the link between two signals from a common power spectrum. The wavelet transformation technique was introduced into geophysics in the early 1980s by Morlet (1983) for use in seismic research. Since then, the technique has been widely popularized in geosciences (Kumar and Foufoula-Georgiou, 1997; Torrence and Compo, 1998; Torrence and Webster, 1999; Labat et al., 2000; Grinsted et al., 2004; Pozo-Vázquez et al., 2001; Vu et al., 2019; Bogning et al., 2021). Wavelet Cross-correlation and associated coherence and phase, are used to identify scales and temporal locations where two signals co-vary significantly. The cross-wavelet analysis of both time series $X = x(t_i)$ and $Y = y(t_i)$ is formulated as follow:

$$W_j^{XY} = \langle W_j^X(s) W_j^Y(s)* \rangle \qquad (2)$$

where $W_i^{XY}(s)$ is the cross-wavelet transform at time t_i on a scale s of the time series X and Y, $W_i^X(s)$ and $W_j^Y(s)$ are wavelet transform of time series X and Y respectively.

3 Results and discussion

The first two spatial components of the EOF analysis are shown in Fig. 2. The first component explains more about 55.5 % of the total variance and shows the first pattern of visibly homogeneous variation across the basin. The correlation between the temporal component associated with this first mode and the computed Standardized Anomaly Index (SAI) of rainfall in the ORB (not shown because of brevity) gives $R = -0.97$, p value $= 4.72 \times 10^{-39}$. This high correlation denotes a strong link between the SAI and this leading mode, reflecting the abundance of rainfall in the total ORB and suggests homogeneous interannual variability of rainfall in all the ORB. This hydroclimatic homogeneity at the basin-scale was not expected because of the strong climatic heterogeneity usually reported in this region (Balas et al., 2007; Nicholson et al., 2018; Nicholson, 2018).

The pattern observed in the second mode of variation shows a dipolar dynamic between the north and south of the basin. Calculation of the correlation between the temporal component associated with this second mode and the SAI applied to precipitation in the boreal summer (JJA) results in $R = -0.44$, p value $= 3.79 \times 10^{-4}$. This phase opposition behavior is singular in the ORB which is a coastal basin with a pure equatorial climate (Lienou et al., 2008). Indeed, this behavior is due to the important occurrences of rains in the northern part of the basin, while in the south one observes the great dry season. The climate of the northern part of the ORB changes from a pure equatorial regime to a northern equatorial regime.

However, the correlations between the temporal components of these two modes and the ENSO result in $R = 0.05$,

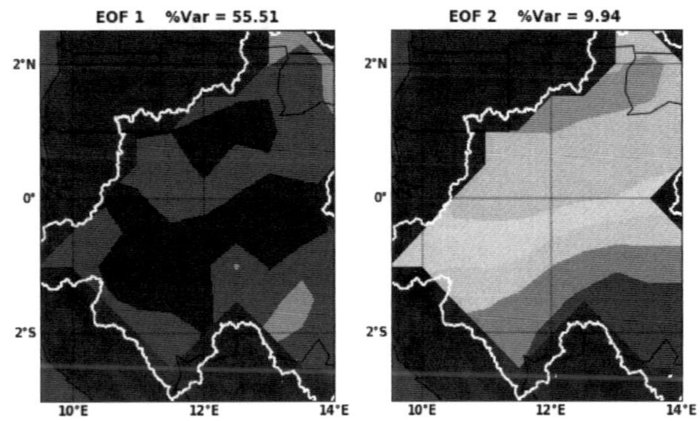

Figure 2. First two components of EOF analysis of rainfall in the ORB from 1940 to 1999.

p value $= 0.13$ and $R = 0.05$, p value $= 0.10$ for the first and second modes respectively. These correlation values suggest that there are no significant linear relationships between the leading modes of rainfall variabilities in the ORB and ENSO. For further investigation, taking into account the non-linear dynamics of the ENSO phenomenon, the cross-wavelet transform of the El Niño time series and that of rainfall estimated based on Eq. (1) has been computed and the result is presented in Fig. 3a. In this figure, many patterns with high power spectra are observed, but those with a significance level of 5 % are surrounded by the black solid lines. They are mainly observed around the 6-month frequency, a little less at one year and a broad and almost restricted structure in the 1980s. These structures of large cross-wavelet power with a 5 % significance level denote either the high wavelet power spectrum of both series or the high wavelet power spectrum of a single time series. It is therefore difficult to know whether there is a link between the two phenomena or not.

The obvious way to provide additional information to determine if both time series are linked or not was the computation of their wavelet coherence. Figure 3b shows the wavelet coherence of ENSO and rainfall in the ORB. In reality, only the simultaneously significant patterns of the cross-wavelet transform and the wavelet coherence of both time series denote links between these phenomena. Around the 6-month frequency, 14 observed patterns are significant in both cross-wavelet transform and wavelet coherence represented contours. These patterns are distributed along the study period such that from 1950 to 1990, there were three patterns per decade, and the 1940s and 1990s each had only one. At the one-year frequency, two significant patterns were also recorded, one in 1959 and the other in 1965. Also at the 5-year frequency, the high power spectrum of the 1990s observed in the cross-wavelet transform of both time series is partially confirmed by wavelet coherence. These observations confirm links between El Niño in the central Pacific and precipitation in the ORB as reported by some authors in regional studies (Balas et al., 2007; Nicholson and Dezfuli,

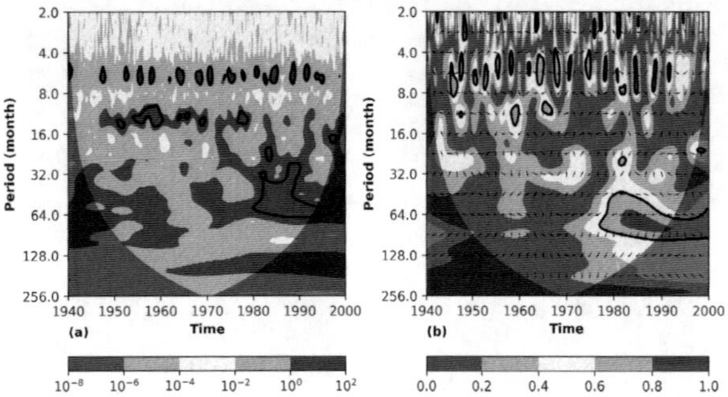

Figure 3. Contours of cross-wavelet transform **(a)** and wavelet coherence **(b)** between ENSO and rainfall in the ORB from 1940 to 1999.

2013; Dezfuli and Nicholson, 2013; Maloba Makanga, 2015; Nicholson et al., 2018).

This study shows that instead of the close links generally described between the rainfall in West-central Africa and ENSO (Camberlin et al., 2001; Farnsworth et al., 2011; Nicholson and Dezfuli, 2013; Dezfuli and Nicholson, 2013), the ORB rains are linked to ENSO only during these extreme phases (i.e. El Niño and La Niña). In addition, black arrows indicating the phase difference between both time series in areas of significant common power are not horizontal (see Fig. 3), suggesting time delays between local optima in wavelet transforms of ENSO and rainfall in the Ogooué basin. It means that links found between ENSO and rainfall in the ORB are not cause-and-effect relationships. The observed variation of this phase difference indicates that the time lag between the onset of El Niño (La Niña) in the Pacific Ocean and the decrease (increase) in ORB rainfall is not constant, making it difficult to describe the impacts of ENSO on rainfall in the ORB at seasonal and annual scales like stated in Bogning et al. (2021).

4 Conclusion

This study aims to improve the understanding of the link between rainfall in the ORB and the ENSO. It was, therefore, noted that although analysis of rainfall variability in West Equatorial Africa usually pointed out ENSO as a leading factor, its links to rainfall in the ORB are very ambiguous. The linear correlation generally used to highlight the links between rainfall and associated near and remote forcings has proved ineffective to find the ENSO-induced rainfall variability in the ORB. This is because only El Niño-like and La Niña-like effects influence significantly on rainfall of the ORB. This study was able to identify and localize the links between both phenomena in time and frequency thanks to the cross-wavelet analysis.

Code availability. The code is not available due to ongoing scientific research involving this code and additional climate indices.

Data availability. Datasets used in this study are available online. Related links are specified in the text.

Author contributions. SB, FF and GM conceived and designed the study; SB performed data processing with the help of AP, RO, FB, FF, FN and GM; all the authors analyzed the results and contributed to the writing of the paper.

Competing interests. The contact author has declared that neither they nor their co-authors have any competing interests.

Disclaimer. Publisher's note: Copernicus Publications remains neutral with regard to jurisdictional claims in published maps and institutional affiliations.

Special issue statement. This article is part of the special issue "Hydrology of Large River Basins of Africa". It is a result of the 4th International Conference on the "Hydrology of the Great Rivers of Africa", Cotonou, Benin, 13–20 November 2021.

Acknowledgements. The authors want to thank the "Laboratoire Mixte International (LMI) Dynamique des écosystèmes continentaux d'Afrique Centrale en contexte de changements globaux (DYCOFAC)" for its support in the accomplishment of this work and the anonymous reviewers for their helpful comments.

References

Anyamba, A., Glennie, E., and Small, J.: Teleconnections and Interannual Transitions as Observed in African Vegetation: 2015–2017, Remote Sensing, 10, 1–16, 2018.

Balas, N., Nicholson, S. E., and Klotter, D.: The relationship of rainfall variability in West Central Africa to sea-surface temperature fluctuations, Int. J. Climatol., 27, 1335–1349, 2007.

Bogning, S., Frappart, F., Blarel, F., Niño, F., Mahé, G., Bricquet, J.-P., Seyler, F., Onguéné, R., Etamé, J., Paiz, M.-C., and Braun, J.-J.: Monitoring Water Levels and Discharges Using Radar Altimetry in an Ungauged River Basin: The Case of the Ogooué, Remote Sensing, 10, 1–18, 2018.

Bogning, S., Frappart, F., Paris, A., Blarel, F., Niño, F., Saux Picart, S., Lanet, P., Seyler, F., Mahé, G., Onguene, R., Bricquet, J.-P., Etame, J., Paiz, M.-C., and Braun, J.-J.: Hydro-climatology study of the Ogooué River basin using hydrological modeling and satellite altimetry, Adv. Space Res., 68, 672–690, 2021.

Bogning, S., Frappart, F., Mahé, G., Niño, F., Paris, A., Sihon, J., Ghomsi, F., Blarel, F., Bricquet, J.-P., Onguene, R., Etame, J., Seyler, F., Paiz, M.-C., and Braun, J.-J.: Long-term Hydrological Variations of the Ogooue River Basin, edited by: Alsdorf Douglas, Tshimanga Raphael, Moukandi, Guy, Congo Basin Hydrology, Climate, and Biogeochemistry, John Wiley & Sons, in press, 2021.

Braun, J. J., Jeffery, K., and Koumba Pambo, A. F.: CZO perspective in Central Africa: The Lopé watershed, Lopé National Park, Ogooué River basin, Gabon, 2015.

Bunge, L. and Clarke, A. J.: A Verified Estimation of the El Niño Index Niño-3.4 since 1877, J. Climate, 22, 3979–3992, 2009.

Camberlin, P., Janicot, S., and Poccard, I.: Seasonality and atmospheric dynamics of the teleconnection between African rainfall and tropical sea-surface temperature: Atlantic vs. ENSO, Int. J. Climatol., 21, 8, 973–1005, 2001.

Dezfuli, A. K. and Nicholson, S. E.: The Relationship of Rainfall Variability in Western Equatorial Africa to the Tropical Oceans and Atmospheric Circulation. Part II: The Boreal Autumn, J. Climate, 26, 66–84, 2013.

Dieulin, C., Mahé, G., Paturel, J.-E., Ejjiyar, S., Tramblay, Y., Rouché, N., and EL Mansouri, B.: A New 60-Year 1940/1999 Monthly-Gridded Rainfall Data Set for Africa, Water, 11, 1–17, 2019.

Farnsworth, A., White, E., Williams, C. J. R., Black, E., and Kniveton, D. R.: Understanding the Large Scale Driving Mechanisms of Rainfall Variability over Central Africa, in: African Climate and Climate Change: Physical, Social and Political Perspectives, edited by: Williams, C. J. R. and Kniveton, D. R., Springer Netherlands, Dordrecht, 101–122, 2011.

Fichet, L. V., Sannier, C., Massard, K., Makaga, E., and Seyler, F.: Assessing the Accuracy of Forest Cover Map for 1990, 2000 and 2010 at National Scale in Gabon, IEEE J. Selected Topics in Applied Earth Observation and Remote Sensing, 7, 1346–1356, 2014.

Giannini, A., Chiang, J. C. H., Cane, M. A., Kushnir, Y., and Seager, R.: The ENSO Teleconnection to the Tropical Atlantic Ocean: Contributions of the Remote and Local SSTs to Rainfall Variability in the Tropical Americas, J. Climate, 14, 4530–4544, 2001.

Grinsted, A., Moore, J. C., and Jevrejeva, S.: Application of the cross wavelet transform and wavelet coherence to geophysical time series, Nonlin. Processes Geophys., 11, 561–566, https://doi.org/10.5194/npg-11-561-2004, 2004.

Hua, W., Zhou, L., Chen, H., Nicholson, S. E., Raghavendra, A., and Jiang, Y.: Possible causes of the Central Equatorial African long-term drought, Environ. Res. Lett., 11, 1–13, https://doi.org/10.1088/1748-9326/11/12/124002, 2016.

Kittel, C. M. M., Nielsen, K., Tøttrup, C., and Bauer-Gottwein, P.: Informing a hydrological model of the Ogooué with multi-mission remote sensing data, Hydrol. Earth Syst. Sci., 22, 1453–1472, https://doi.org/10.5194/hess-22-1453-2018, 2018.

Kolusu, S. R., Shamsudduha, M., Todd, M. C., Taylor, R. G., Seddon, D., Kashaigili, J. J., Ebrahim, G. Y., Cuthbert, M. O., Sorensen, J. P. R., Villholth, K. G., MacDonald, A. M., and MacLeod, D. A.: The El Niño event of 2015–2016: climate anomalies and their impact on groundwater resources in East and Southern Africa, Hydrol. Earth Syst. Sci., 23, 1751–1762, https://doi.org/10.5194/hess-23-1751-2019, 2019.

Kumar, P. and Foufoula-Georgiou, E.: Wavelet analysis for geophysical applications, Rev. Geophys., 35, 385–412, 1997.

Labat, D., Ababou, R., and Mangin, A.: Rainfall–runoff relations for karstic springs. Part II: continuous wavelet and discrete orthogonal multiresolution analyses, J. Hydrol., 238, 149–178, 2000.

Lambert, T., Darchambeau, F., Bouillon, S., Alhou, B., Mbega, J.-D., Teodoru, C. R., Nyoni, F. C., Massicotte, P., and Borges, A. V.: Landscape control on the spatial and temporal variability of chromophoric dissolved organic matter and dissolved organic carbon in large African rivers, Ecosystems, 18, 1224–1239, 2015.

Lienou, G., Mahe, G., Paturel, J. E., Servat, E., Sighomnou, D., Ekodeck, G. E., Dezetter, A., and Dieulin, C.: Evolution des régimes hydrologiques en région équatoriale camerounaise: un impact de la variabilité climatique en Afrique équatoriale?, Hydrol. Sci. J., 53, 789–801, 2008.

Maloba Makanga, J. D.: variabilité pluviométrique de la petite saison sèche au gabon, in: XXVIIIe Colloque de l'Association Internationale de Climatologie, Liège 2015, 6, 2015.

Mignard, S. L.-A., Mulder, T., Martinez, P., Charlier, K., Rossignol, L., and Garlan, T.: Deep-sea terrigenous organic carbon transfer and accumulation: Impact of sea-level variations and sedimentation processes off the Ogooue River (Gabon), Mar. Pet. Geol., 85, 35–53, 2017.

Morlet, J.: Sampling Theory and Wave Propagation, https://doi.org/10.1007/978-3-642-82002-1_12, 1983.

Nicholson, S. E.: The ITCZ and the Seasonal Cycle over Equatorial Africa, B. Am. Meteorol. Soc., 99, 337–348, 2018.

Nicholson, S. E. and Dezfuli, A. K.: The Relationship of Rainfall Variability in Western Equatorial Africa to the Tropical Oceans and Atmospheric Circulation. Part I: The Boreal Spring, J. Climate, 26, 45–65, 2013.

Nicholson, S. E., Funk, C., and Fink, A. H.: Rainfall over the African continent from the 19th through the 21st century, Glob. Planet. Change, 165, 114–127, 2018.

Plisnier, P. D., Serneels, S., and Lambin, E. F.: Impact of ENSO on East African ecosystems: a multivariate analysis based on climate and remote sensing data, Glob. Ecol. Biogeogr., 9, 481–497, 2000.

Pozo-Vázquez, D., Esteban-Parra, M. J., Rodrigo, F. S., and Castro-Díez, Y.: A study of NAO variability and its possible non-linear influences on European surface temperature, Clim. Dynam., 17, 701–715, 2001.

Ramillien, G., Frappart, F., Güntner, A., Ngo-Duc, T., Cazenave, A., and Laval, K.: Time variations of the regional evapotranspiration rate from Gravity Recovery and Climate Experiment (GRACE) satellite gravimetry, Water Resour. Res., 42, 1–8, 2006.

Rotstayn, L. D., Collier, M. A., Mitchell, R. M., Qin, Y., Campbell, S. K., and Dravitzki, S. M.: Simulated enhancement of ENSO-related rainfall variability due to Australian dust, Atmos. Chem. Phys., 11, 6575–6592, https://doi.org/10.5194/acp-11-6575-2011, 2011.

Rouché, N., Mahé, G., Ardoin-Bardin, S., Brissaud, B., Boyer, J.-F., Crès, A., Dieulin, C., Bardin, G., Commelard, G., Dezetter, A., Paturel, J.-E., and Servat, É.: Constitution d'une grille de pluies mensuelles pour l'Afrique (période 1900–2000), https://doi.org/10.1684/sec.2010.0269, 2010.

Sannier, C., McRoberts, R. E., and Fichet, L.-V.: Suitability of Global Forest Change data to report forest cover estimates at national level in Gabon, Remote Sens. Environ., 173, 326–338, 2016.

Tippett, M. K. and L'Heureux, M. L.: Low-dimensional representations of Niño 3.4 evolution and the spring persistence barrier, Climate and Atmospheric Science, 3, 1–11, 2020.

Torrence, C. and Compo, G. P.: A Practical Guide to Wavelet Analysis, B. Am. Meteorol. Soc., 79, 61–78, 1998.

Torrence, C. and Webster, P. J.: Interdecadal Changes in the ENSO–Monsoon System, J. Climate, 12, 2679–2690, 1999.

Vu, P. L., Ha, M. C., Frappart, F., Darrozes, J., and Ramillien, G.: Identifying 2010 Xynthia Storm Signature in GNSS-R-Based Tide Records, Remote Sens., 11, 1–16, 2019.

Proc. IAHS, 384, 187–194, 2021
https://doi.org/10.5194/piahs-384-187-2021

Open Access PIAHS

Hydrology of Large River Basins of Africa

Heavy rainfall frequency analysis in the Benin section of the Niger and Volta Rivers basins: is the Gumbel's distribution a one-size-fits-all model?

Djigbo Félicien Badou[1,2], Audrey Adango[1], Jean Hounkpè[2], Aymar Bossa[3], Yacouba Yira[4], Eliezer Iboukoun Biao[5], Julien Adounkpè[2], Eric Alamou[5], Luc Ollivier C. Sintondji[3], and Abel Akambi Afouda[2]

[1]Laboratoire d'Hydrologie Appliquée, Institut National de l'Eau, Université d'Abomey-Calavi, Abomey-Calavi, 01 BP526 Cotonou, Benin
[2]Ecole d'Horticulture et d'Aménagement des espaces Verts, Université Nationale d'Agriculture, Kétou, BP43 Kétou, Benin
[3]Laboratoire d'Hydraulique et de Maitrise de l'Eau, Institut National de l'Eau, Université d'Abomey-Calavi, Abomey-Calavi, 01 BP526 Cotonou, Benin
[4]Applied Science and Technology Research Institute – IRSAT/CNRST, P.O. Box 7047, Ouagadougou, Burkina Faso
[5]Laboratoire de Géoscience, de l'Environnement et Applications (LaGEA), Université Nationale des Sciences, Technologies, Ingénierie et Mathématiques, Abomey, Benin

Correspondence: Djigbo Félicien Badou (fdbadou@gmail.com)

Published: 16 November 2021

Abstract. West African populations are increasingly exposed to heavy rainfall events which cause devastating floods. For the design of rainwater drainage facilities (to protect populations), practitioners systematically use the Gumbel distribution regardless of rainfall statistical behaviour. The objective of this study is twofold. The first is to update existing knowledge on heavy rainfall frequency analysis in West Africa to check whether the systematic preference for Gumbel's distribution is not misleading, and subsequently to quantify biases induced by the use of the Gumbel distribution on stations fitting other distributions. Annual maximum daily rainfall of 12 stations located in the Benin sections of the Niger and Volta Rivers' basins covering a period of 96 years (1921–2016) were used. Five statistical distributions (Gumbel, GEV, Lognormal, Pearson type III, and Log-Pearson type III) were used for the frequency analysis and the most appropriate distribution was selected based on the Akaike (AIC) and Bayesian (BIC) criteria. The study shows that the Gumbel's distribution best represents the data of 2/3 of the stations studied, while the remaining 1/3 of the stations fit better GEV, Lognormal, and Pearson type III distributions. The systematic application of Gumbel's distribution for the frequency analysis of extreme rainfall is therefore misleading. For stations whose data best fit the other distributions, annual daily rainfall maxima were estimated both using these distributions and the Gumbel's distribution for different return periods. Depending on the return period, results demonstrate that the use of the Gumbel distribution instead of these distributions leads to an overestimation (of up to +6.1 %) and an underestimation (of up to −45.9 %) of the annual daily rainfall maxima and therefore to an uncertain design of flood protection facilities. For better validity, the findings presented here should be tested on larger datasets.

1 Introduction

Climate change is one of the greatest challenges of the 21st century. In West Africa, extreme hydroclimatic events such as heavy rainfall, floods and droughts severely impact populations and socio-economic development (Amoussou et al., 2014). Since the late 2000s, most West African countries are experiencing unprecedented flooding events (Badou et al., 2019). Ouagadougou, the largest city of Burkina Faso, was struck by extreme rainfall on the 1 September 2009 causing the death of nine persons, and losses estimated to be more than USD 14 million. That very year 2019, Dakar in Senegal, and Lomé in Togo recorded heavy rainfall events with comparable damages. The following year, in 2010, Benin, Cote d'Ivoire and Niger recorded extreme devastating floods. Benin, for example, was severely struck with 680 000 people affected and 46 deaths (World Bank, 2011). While the construction of rainwater drainage facilities is a common measure to control rainfall-induced floods, the design of such facilities is rainfall characteristic-dependent and requires a good knowledge of associated statistical distribution. In this perspective, frequency analysis of heavy rainfall must guide the selection of the best statistical distribution (e.g. Agué and Afouda, 2015; Goubanova and Li, 2007; Habibi et al., 2013; Hamzaoui, 2016; Ousmane, 1992; St-hilaire, 2007). Notwithstanding, based on expert elicitation, most civil engineers in West Africa systematically use the Gumbel distribution to design rainwater drainage systems regardless of rainfall statistical behaviour. Previous works on frequency analysis of heavy rainfall in West Africa have demonstrated the "suitability" of the Gumbel distribution for regions of high rainfall regime (Goula Bi Tie et al., 2007) and the majority of investigated stations (Agué and Afouda, 2015) seem to have supported that practice of civil engineers. However, given the high spatial and temporal rainfall variability of the West African region, are these previous findings still valid? For example, the analysis of Goula Bi Tie et al. (2007) covered the period 1947–1995 and that of Agué and Afouda (2015) the period 1921–2001; both periods not including the decades the 2000s and 2010s characterised by the increase in the frequency and magnitude of heavy rainfall (Hounkpè et al., 2016). Recent heavy rainfall frequency analyses in the Ouémé River Basin (Kodja, 2018) and the Oti River Basin (Koungbanane et al., 2020) considered the 2000s and 2010s but were limited to the GEV distribution in the case of the Oti River Basin and Gumbel, GEV, and Pearson type III distribution in the case of the Ouémé River Basin. To guide the design of drainage water facilities and provide an alternative to civil engineers who systematically use Gumbel distribution, this study takes stock from the previous ones to: (i) update existing knowledge on heavy rainfall frequency analysis in West Africa, (ii) quantify biases induced by the systematic use of the Gumbel distribution. Northern Benin, a region of transition between the Guinean and Sahelian rainfall regimes is chosen as a case study.

2 Material and methods

2.1 Research area

The study area is located in the northern part of Benin between latitudes 10° and 12°30′ N and longitudes 1° and 3°40′ E. With a total area of 57 093 km^2, it consists of the Beninese portion of the Niger River Basin (39 726 km^2) and that of the Volta River Basin (17 367 km^2) (Fig. 1). The rainfall regime is uni-modal of the Sudanian type with a single rainy season from April to October. Average annual rainfall varies between 700 and 1100 mm (Badou et al., 2015). On a seasonal scale, rainfall variability is characterized by a high concentration of rainfall in August sometimes causing devastating flooding (MEHU, 2011). The latest census in Benin amounted the population of the study area to about 2 031 534 (INSAE, 2016) whose mainly rely on rainfed agriculture and ruminant breeding for their livelihoods.

2.2 Data

The daily rainfall time series used in this study are from Benin Meteorological Agency. For the choice of rain gauges to consider, priority was given to those located in the most affected municipalities by rainfall-induced floods such as Banikoara and Kandi. Other stations located in the vicinity of the first ones and those with data available were also considered. Subsequently, the percentage of missing data was determined for each station (Table 1). In line with Sect. 2.1, the search for missing data was limited to the rainy season which starts in April and is mainly from June to October. Thus, for each station and each year, the ratio of the number of missing data within June to October to the total number of rainy days was computed, and the mean percentage of missing data (for all years and for each rainfall station). Stations with more than 25 % missing data were discarded in order not to bias the results. For stations with less than 25 % missing data, gaps were filled using the inverse distance method. Thereupon, for each gauge and each year, daily rainfall maxima were extracted. The time series used in this study consist of the annual daily rainfall maxima of 12 rainfall gauges listed in Table 1.

2.3 Methods

In this study, frequency analysis was applied to extreme rainfall time series. The statistical adjustment is performed using of the HYFRAN (HYdrological FRequency ANalysis) software with the prior objective of testing whether the Gumbel distribution is a one-size-fits-all model. The steps of the statistical adjustment fitting are briefly presented below.

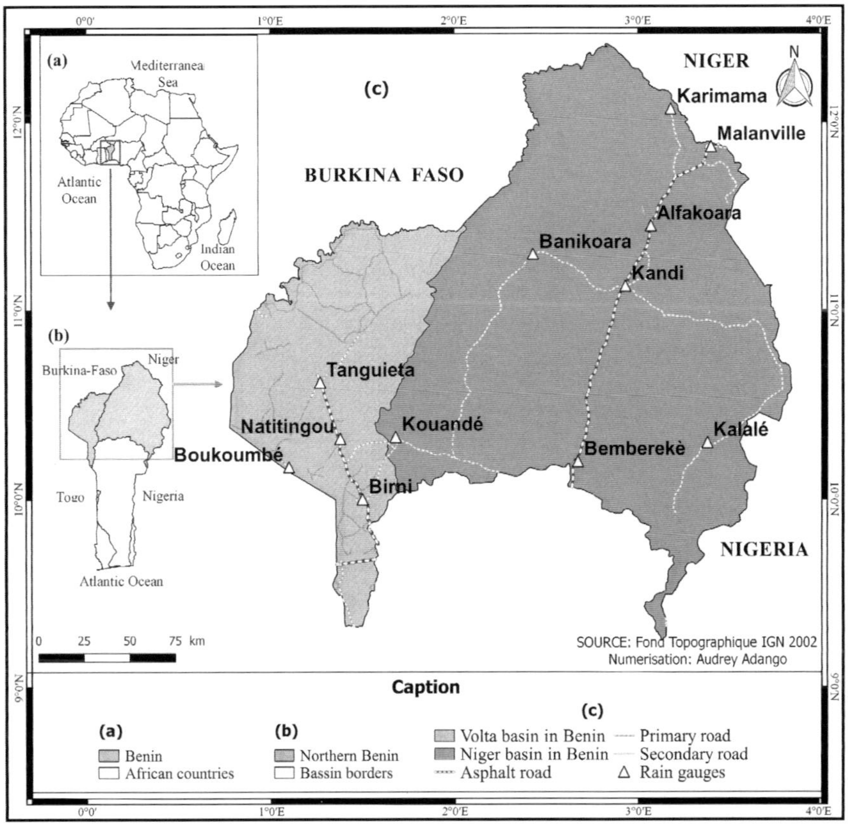

Figure 1. Location of the research area displaying the rain gauges used in this study represented by triangles.

Table 1. Rainfall series used in this study.

Station	Long. (°)	Lat. (°)	Period	Data length (year)	Missing data (%)
Alfakoara	3.07	11.45	1969–2010	42	4
Banikoara	2.43	11.30	1954–2010	57	1
Bemberekè	2.67	10.20	1921–2014	94	0
Birni	1.50	10.00	1953–2012	60	18
Boukoumbé	1.10	10.17	1923–2014	92	17
Kalalé	3.38	10.30	1957–2012	56	3
Kandi	2.93	11.13	1921–2016	96	0
Karimama	3.18	12.07	1976–2010	35	10
Kouandé	1.68	10.33	1931–2014	84	5
Malanville	3.40	11.87	1942–2010	69	4
Natitingou	1.38	10.32	1921–2016	96	0
Tanguiéta	1.27	10.62	1939–2010	72	4

2.3.1 Identification of the “best” statistical distributions fitting rainfall extremes

This section deals with the determination of the statistical distribution which best fits the data. Before the statistical fitting, hypotheses testing were carried to check the stationary, independence and homogeneity of the data using the tests of Kendall, Wald-Wolfowitz and Wilcoxon as in El Adlouni et Bobée (2014). The five distributions (Gumbel, Lognormal, GEV, Pearson type III, and Log-Pearson type III) embedded in HYFRAN software were considered.

The Akaike Information Criterion (AIC) and the Bayesian Information Criterion (BIC) were used to compare the five statistical distributions with the best score given to the distribution associated with the lowest values of AIC and BIC (Avahounlin et al., 2013).

2.3.2 Estimating the parameters of selected statistical distributions and rainfall quantiles

The estimation of the parameters of the selected statistical distributions was based on two methods. The maximum likelihood method for Gumbel, Lognormal, GEV and the Pearson type III distributions and the L-moments method for the log Pearson type III distribution. The method of moments consists in equaling the theoretical moments of the distribution (which are a function of the parameters) and their estimates obtained from the sample. Solving the resulting system of equations makes it possible to obtain the parameters of the distribution. The maximum likelihood method involves a likelihood function L defined by Eq. (1(as in Avahounlin et al. (2013).

$$L = \prod_{l}^{N} f(x_i; \theta_1, \ldots\ldots \theta_k) \tag{1}$$

$$f(x_i)\mathrm{d}x_i = P\left[x_i \subseteq X \subseteq x_i + \mathrm{d}x_i\right]$$

This latter method consists in finding the parameters θ which maximize the likelihood function, i.e. the probability of observing the sample $(x_i, \ldots\ldots x_N)$ which represents the best information available on the population.

Afterwards, the so-determined parameters were used to estimate rainfall quantiles for different return periods. According to Perreault et al. (1994), for the estimation of a quantile X_T of return period T referring to a random variable X from a sample which follows a given probability distribution F having k parameter, the formula below is used:

$$X_T = F^{-1}\left(\theta_1, \theta_2 \ldots \theta_k; -\frac{1}{T}\right) \tag{2}$$

where F^{-1} is the inverse function of the distribution function F and $\frac{1}{T}$ is the probability of exceeding this event.

Furthermore, spatial interpolation of the estimated quantiles was carried out using the kriging method embedded in the Surfer 11 software.

2.3.3 Quantifying biases induced by the systematic use of the Gumbel distribution

To quantify biases induced by the systematic use of the Gumbel distribution on stations fitting other distributions, quantiles estimated using these distributions were compared to quantiles estimated using the Gumbel distribution.

3 Results and discussion

3.1 Best statistical distributions fitting rainfall extremes

The hypotheses testing shows time series are stationary, homogeneous and independent except the time series of Boukoumbé (non-stationary and not independent) and Natitingou (non-stationary). The 5 distributions were fitted to the 12 times series (including Boukoumbé and Natitingou added for consistency and comparison purposes) then compared using the AIC and BIC criteria. Figure 2 displays the spatial distribution of stations according to the statistical distributions which best represent their time series. The statistical distributions which best represent the series vary from one station to the other indicating that the time series within the same basin (or climatic zone) is not necessarily fitted with the same statistical distribution. This applies especially to the Beninese section of the Volta Basin where "best" statistical distributions are systematically distinct.

Figure 2. Spatial distribution of selected statistical distributions.

This result is consistent with that of Agué and Afouda (2015) who demonstrated that the probability distributions of annual maximum daily rainfall are not rainfall regime-dependent. However, for some specific stations, our results are slightly different from those of Agué and Afouda (2015) who found that for the station at Bemberekè the Pearson type III instead of the Gumbel distribution fit better extreme rainfall data and for the station at Boukoumbé the Gumbel is the best statistical distribution fitting the data instead of the Lognormal distribution. The difference might stem from the length of the time series as the time series considered in this study ranges from 1921–2016 (and includes the wet decade of 2001–2010) against the period of 1921–2001 for the study of Agué and Afouda (2015) confirming the influence of data length and the need for updates of heavy rainfall frequency analysis is suggested here.

Also, Fig. 2 shows the predominance of the Gumbel distribution which fits the data of $2/3$ of stations contradicting the results of Goula Bi Tie et al. (2007) who compared the Gumbel and Lognormal distributions for 34 stations located in Côte d'Ivoire found instead the predominance of the Lognormal distribution (56 % of stations) over the Gumbel distribution (44 % of stations). The divergence of both results further shows that there is no rationale for taking the Gumbel distribution as one-size-fits-all-models.

3.2 Parameters of selected statistical distributions and rainfall quantiles

Applying the maximum likelihood and L-moments methods, the parameters of the selected statistical distributions were

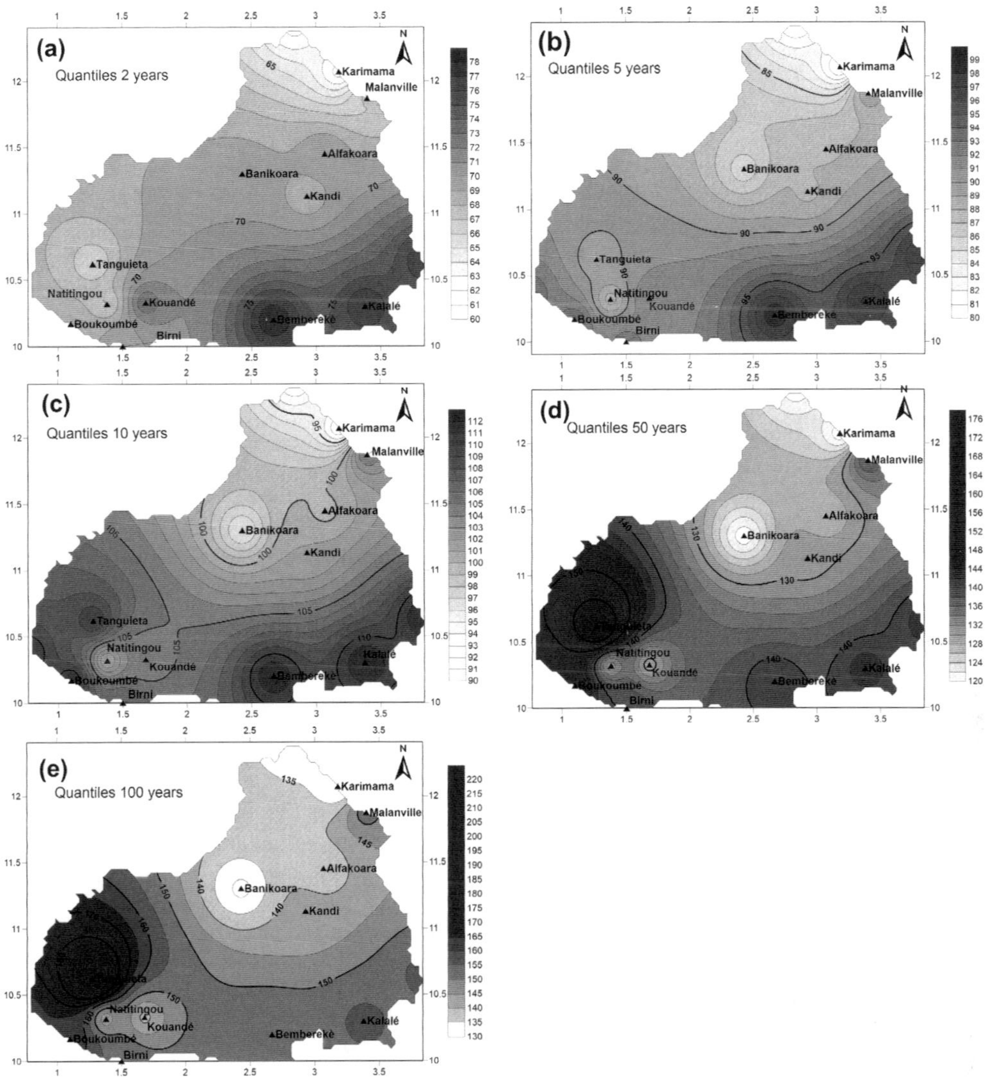

Figure 3. Extreme rainfall quantiles for 2 years **(a)**, 5 years **(b)**, 10 years **(c)**, 50 years **(d)** and 100 years **(e)** return periods.

determined then used to estimate the quantiles of maximum daily rainfall for five return periods (Table 2).

For short return periods (2, 5 and 10 years), minimums are observed at the station at Karimama but maximums at Bembereké and Kalalé. For longer return periods (50 and 100 years), the highest quantiles are recorded at Tanguiéta then Boukoumbé and Kalalé. Minimums are observed at the stations at Banikoara and Karimama.

Spatial interpolation of the quantiles is displayed in Fig. 3. The highest daily rainfall quantiles are generally recorded in the southern part of the research area (Boukoumbé and Kalalé). Isohyets decrease northward. For longer return periods (50 and 100 years), the maxima are observed at the Tanguiéta station and its surroundings. Besides, it is observed in the mountainous area of Atacora (between latitudes 10° and 11°30′), a very high rainfall variability characterized by the parallelism and the approximation of isohyets testifying to the important role of orography in local precipitation (Agué and Afouda, 2015; Goula Bi Tie et al., 2007).

3.3 Biases induced by the systematic use of the Gumbel distribution

For the stations at Boukoumbé, Natitingou and Tanguiéta, the Gumbel distribution slightly overestimates rainfall quantiles of 2 years return periods (Table 3). This also applies to Kouandé stations for the return periods for 5, 10, 50 and 100 years return periods. On the contrary for the other stations, the longer the return period the greater the underestimation of quantiles estimated using the Gumbel distribution. This is particularly true for the GEV distribution which best fits Tanguieta's data. Apart from the station at Kouandé,

Table 2. Quantiles in mm for different return periods $T = 2, 5, 10, 50$, and 100 years; in brackets are the 95 % confidence intervals.

Stations	Distribution	T_2	T_5	T_{10}	T_{50}	T_{100}
Alfakoara	Gumbel	70 (64.4–75.6)	87.9 (79.2–96.5)	99.7 (88.6–111)	126 (109–143)	137 (117–156)
Baniko	Gumbel	69.6 (65.3–73.8)	85.5 (78.9–92.1)	96 (87.6–105)	119 (106–132)	129 (114–144)
Bembereke	Gumbel	78 (73.6–82.3)	98.6 (92–105)	112 (104–121)	142 (129–156)	155 (140–170)
Birni	Gumbel	69.9 (64–75.7)	92.4 (83.3–101)	107 (95.6–119)	140 (122–158)	154 (134–175)
Boukoumbe	Log normale	68.2 (62.9–73.5)	93.7 (85.3–102)	111 (99.1–122)	148 (128–168)	164 (140–189)
Kalalé	Gumbel	76.8 (71.1–82.6)	98.1 (89.2–107)	112 (101–124)	143 (126–161)	156 (136–177)
Kandi	Gumbel	68.4 (64.4–72.4)	87.8 (81.6–93.9)	101 (92.7–108)	129 (117–141)	141 (127–155)
Karimama	Gumbel	61.7 (55.4–68.0)	80 (70.3–89.8)	92.2 (79.7–105)	119 (99.8–138)	130 (108–152)
Kouandé	Log normale	72.8 (68.4–77.1)	92 (85.6–98.4)	104 (95.6–112)	129 (115–143)	139 (123–155)
Malanville	Gumbel	67.1 (61.5–72.6)	89.8 (81.2–98.3)	105 (93.8–116)	138 (121–155)	152 (133–171)
Natitingou	Pearson type III	67.1 (63.1–71.1)	87.6 (80.9–94.3)	101 (92.1–101)	131 (115–147)	143 (125–162)
Tanguiéta	GEV	66 (61.1–71)	89.3 (79.1–99.6)	110 (91.7–128)	176 (115–236)	216 (121–311)

Table 3. Comparison of rainfall quantiles estimated with both Gumbel distribution and selected distributions; in bracket are the per cent of bias. A sign + (−−) signifies an overestimation (underestimation) of rainfall quantiles. Values are in mm.

Stations	Distribution	T_2	T_5	T_{10}	T_{50}	T_{100}
Boukoumbé	Gumbel	68.5 (+0.4)	93.2 (−0.5)	110 (−0.9)	146 (−1.4)	161 (−1.9)
	Log normal	68.2	93.7	111	148	164
Kouandé	Gumbel	72.4 (−0.6)	92.7 (+0.8)	106 (+1.9)	136 (+5.1)	148 (+6.1)
	Log normal	72.8	92	104	129	139
Natitingou	Gumbel	68.2 (+1.6)	86.2 (−1.6)	98.1 (−3.0)	124 (−5.6)	135 (−5.9)
	Pearson type III	67.1	87.6	101	131	143
Tanguiéta	Gumbel	69.6 (+5.2)	90.6 (+1.4)	104 (−5.8)	135 (−30.4)	148 (−45.9)
	GEV	66	89.3	110	176	216

for the other stations, the per cent of underestimation (up to −45.9 %) is by far greater than that of overestimation (up to +6.1 %). This underestimation of rainfall quantiles was reported by Koutsoyiannis (2004) who studied 169 stations (of more than 100 years of data length) located in the US, UK and southern France and found that the Gumbel distribution underestimates quantiles particularly for return periods ranging from 50 to 100 years.

4 Conclusions

In this study, the statistical distributions that best describe the statistical behaviour of heavy rainfall in the Benin section of the Niger and Volta Rivers basins were determined and daily rainfall quantiles associated with different return periods were estimated. This was done with the primary objective of testing whether the widely applied Gumbel distribution is a one size fit all model for the research area. Therefore, a frequency analysis of the rainfall extremes of 12 stations was carried out on time series ranging from 35 to 96 years in length. The HYFRAN (HYdrological FRequency ANalysis) software was used for the assignment. Before the frequency analysis, hypothesis testing was performed to check the stationarity, independence and homogeneity of the data. The investigations indicate that the time series of annual daily maxima rainfall are stationary, homogeneous and independent except the time series of Boukoumbé (which are non-stationary and not independent) and of Natitingou (which are non-stationary) and therefore verifying the hypotheses for applying the frequency analysis. However, for comparison purposes, the time series of Boukoumbé and Natitingou were also considered for the frequency analysis.

The results reveal that though predominant and fitting the best two-third of the time-series, Gumbel distribution is not a one-size-fits-all-time series. Three of the four other statistical distributions (GEV, Lognormal, and Pearson type III) fits the best at least one time series with the Lognormal distribution fitting the best two times series, and the GEV and Pearson type III one time series. Corresponding quantiles computed using selected statistical distributions vary between 67 and 216 mm respectively for the return periods of 2 and 100 years with the highest rainfall values obtained in mountainous areas (between latitudes 10° and 11°30′) testifying to the important role of orography in local rainfall. This work provides useful science-based evidence of the systematic mistake done when civil engineers apply Gumbel distribution as a one-size-fits-all-time series for the design of rainwater drainage systems for flood control. It is expected that it will guide the design of flood control infrastructures in the future to protect the livelihood of millions of people exposed to extreme rainfall-induced damages.

Data availability. Daily rainfall data is not publicly available due to the policies of Benin Meteorological Agency (http://meteobenin.bj/services/demande-de-donnees/, last access: 15 October 2021) (METEO BENIN, 2021).

Author contributions. AAA and DFB conceived the presented research. AA and JH collected the data and performed the computations. AB, YY and EA encouraged AA to quantify the biases induced by the use of the Gumbel distribution instead of appropriate distributions. EIB, JA, and LOCS reviewed the very first draft of the manuscript. All authors discussed the results and contributed to the final manuscript.

Competing interests. The contact author has declared that neither they nor their co-authors have any competing interests.

Special issue statement. This article is part of the special issue "Hydrology of Large River Basins of Africa". It is a result of the 4th International Conference on the "Hydrology of the Great Rivers of Africa", Cotonou, Benin, 13–20 November 2021.

Acknowledgements. The authors are grateful to the two anonymous reviewers whose comments helped in improving the initial manuscript.

References

Agué, A. I. and Afouda, A.: Analyse fréquentielle et nouvelle cartographie des maxima annuels de pluies journalières au Bénin, Int. J. Biol. Chem. Sci, 9, 121–133, https://doi.org/10.4314/ijbcs.v9i1.12, 2015.

Amoussou, E., Camberlin, P., Tramblay, Y., and Mahe, G.: Évolution des précipitations extrêmes dans le bassin versant du Mono (Bénin- Togo) en contexte de variabilité/changement climatique, in: XXVIIe Colloque de l'Association Internationale de Climatologie, Dijon, France, 2014.

Avahounlin, F. R., Lawin, A. E., Alamou, E., Chabi, A., and Afouda, A.: Analyse Fréquentielle des Séries de Pluies et Débits Maximaux de L'ouémé et Estimation des Débits de Pointe, Eur. J. Sci. Res., 107, 355–369, 2013.

Badou, D. F., Afouda, A. A., Diekkrüger, B. and Kapangaziwiri, E.: Investigation on the 1970s and 1980s droughts in four tributaries of the Niger River basin (West Africa), in: E-proceedings of the 36th IAHR World Congress 28, the Hague, 1–5, 2015.

Badou, D. F., Hounkpè, J., Yira, Y., Ibrahim, M., and Bossa, A. Y.: Increasing Devastating Flood Events in West Africa: Who is to Blame?, in: Regional Climate Change Series, WASCAL Publishing, Ouagadougou, Burkina Faso, 84–90, 2019.

El Adlouni, S. and Bobée, B.: Analyse Fréquentielle Hydrologique: Logiciel HYFRAN-PLUS (Version-V2.1), available at: https://www.wrpllc.com/books/HyfranPlus/hyfranplusdescrip.html (last access: 17 June 2021), 2014.

Goubanova, K. and Li, L.: Extremes in temperature and precipitation around the Mediterranean basin in an ensemble of future climate scenario simulations, Global Planet. Change, 57, 27–42, https://doi.org/10.1016/j.gloplacha.2006.11.012, 2007.

Goula Bi Tie, A., Konan, B., Brou, Y. T., Issiaka, S., Fadika, V., and Srohourou, B.: Estimation des pluies exceptionnelles journalières en zone tropicale: cas de la Côte d'Ivoire par comparaison des lois Lognormale et de Gumbel, Hydrolog. Sci. J., 52, 49–67, https://doi.org/10.1623/hysj.52.1.49, 2007.

Habibi, B., Meddi, M., and Boucefiane, A.: Analyse fréquentielle des pluies journalières maximales Cas du Bassin Chott-Chergui, Nat. Technol., 8, 41–48, 2013.

Hamzaoui, S.: Modélisation des courbes IDF à travers des stations pluviométriques de la région de l'Ouest du pays, Mémoire de Master, Ecole Nornale Supérieur d'Hydraulique, Blida, Algérie, available at: https://library.ensh.dz/images/site_lamine/pdf/these_master/2016/6-0030-16.pdf (last access: 16 July 2020), 2016.

Hounkpè, J., Diekkrüger, B., Badou, D. F., and Afouda, A.: Change in Heavy Rainfall Characteristics over the Ouémé River Basin, Benin Republic, West Africa, Climate, 4, 1–23, https://doi.org/10.3390/cli4010015, 2016.

INSAE: Effectifs de la population des villages et quartiers de ville du Bénin (RGPH-4, 2013), Cotonou, Bénin, available at: https://insae.bj/images/docs/insae-statistiques/demographiques/population/Effectifsdelapopulationdesvillagesetquartiersdevilledubenin/CahierVillageRGPH42013.pdf (last access: 5 June 2020), 2016.

Kodja, D. J.: Indicateurs des évènements hydroclimatiques extrêmes dans le bassin versant de l'Ouémé à l'exutoire de Bonou en Afrique de l'Ouest. Thèse de doctorat, Université Montpellier, Université d'Abomey-Calavi, Bénin, available at: https://tel.archives-ouvertes.fr/tel-01869842/document (last access: 15 June 2021), 2018.

Koungbanane, D., Zahiri, P. E., Sourou, H., Totin Vodounon, H. S., Amoussou, E., Lare, L. Y., and Koubodana, H. D.: Analyse fréquentielle et détermination des seuils pluvio-hydrologiques de risques d'inondation dans le bassin-versant de l'Oti au Togo, Afrique Sci., 17, 73–88, 2020.

Koutsoyiannis, D.: Statistics of extremes and estimation of extreme rainfall: II. Empirical investigation of long rainfall records, Hydrolog. Sci. J., 49, 591–610, https://doi.org/10.1623/hysj.49.4.591.54424, 2004.

MEHU: Deuxieme communication nationale de la République du Bénin sur les changements climatiques, Cotonou, Benin, available at: https://unfccc.int/resource/docs/natc/bennc2f.pdf (last access: 5 June 2020), 2011.

METEO BENIN: Demande de données, available at: http://meteobenin.bj/services/demande-de-donnees/, last access: 15 October 2021.

Ousmane, D.: Ajustement statistique à des échantillons de pluies et débits: Application aux données du bassin versant de la Faleme, Mémoire de DEA, Université Cheikh Anta Di, Dakar, Sénégal, Dakar, Hann., 1992.

Perreault, L., Bobée, B., and Perron, H.: Ajuste II: hydrological frequency analysis software. PART 1: theoretical aspects, Trans. Ecol. Environ., 8, 243–250, https://doi.org/10.2495/HY940292, 1994.

St-hilaire, A.: Analyse fréquentielle des précipitations de la station Assomption, Rapport de recherche, Québec, Canada, available at: http://espace.inrs.ca/id/eprint/1277/1/R000937.pdf (last access: 5 June 2020), 2007.

World Bank: Inondation au Benin: Rapport d'evaluation des Besoins Post Catastrophe, Cotonou, Benin, p. 84, 2011.

Proc. IAHS, 384, 195–201, 2021
https://doi.org/10.5194/piahs-384-195-2021

Open Access PIAHS

Hydrology of Large River Basins of Africa

Estimation of dry events duration in Northern Tunisia – Analysis of extremes trends

Majid Mathlouthi[1] **and Fethi Lebdi**[2]

[1]Research Laboratory in Sciences and Technology of Water at INAT, Tunis, 1082, Tunisia
[2]Rural Engineering, Water and Forestry Department, National Agronomic Institute of Tunisia, University of Carthage, Tunis, 1082, Tunisia

Correspondence: Majid Mathlouthi (majid_mathlouthi@yahoo.fr)

Published: 16 November 2021

Abstract. Modeling of extremes dry spells in Northern Tunisia, in order to detect the severity of the phenomenon, is carried out. Dry events are considered as a sequence of dry days (below a threshold) separated by rainfall events from each other. The maximum dry event duration follows the Generalized Extreme Value distribution. The data series adherence to the probability distribution was verified by the Anderson-Darling test. The positive trend and non-stationarity of dry spells was verified respectively by the Mann–Kendall test and Dickey–Fuller and augmented Dickey–Fuller tests. The irregular distribution of rainfall in the growing season for Sidi Abdelbasset station has increased the number of dry spells. The increase of rainy days in Ghézala dam and Sidi Salem gauge stations resulted in a decrease of dry spells in this area. Regarding the return period of one year (wet season), dry events occurred from 14 to 27 d in this region constitute an agricultural potential risk. The Southern region was the most vulnerable.

1 Introduction

Although northern Tunisia is subhumid, the rainfall is not uniformly distributed in all seasons. The phenomenon of drought during the rainy season is recurring. Worldwide, scientific research into rainfall modeling is active. The results are useful in planning and management of water resources. Several authors have used different statistical distributions to model the dry spell occurrence. Lana et al. (2006) used the extreme value distribution. The Gamma distribution in Dan'azumi and Shamsudin (2011); the Geometric distribution in Mahamud et al. (2011), Log-Normal distribution in Dan'azumi and Shamsudin (2011), the Pareto distribution in Lana et al. (2006) and the Negative Binomial distribution in Mathlouthi and Lebdi (2009).

The objective of this study was to model the occurrence of extreme dry events in Ichkeul Lake basin of Tunisia and to estimate the return period and the severity of the phenomenon.

2 Data

The data used in this analysis are the daily rainfall data that spans for 37 to 60 years of five rain gauges (Table 1) located in the Ichkeul lake basin Northern Tunisia (Fig. 1) collected from the Directorate General of Water Resources in Tunisia. The rainy season starting at September and lasting until April. The average annual rainfall is 600 mm. The dry season lasts from May to August. Daily values of precipitation are quite variable. There is also considerable variation from year to year.

3 Method

A rainfall event is an *uninterrupted sequence* of rainy days including at least one rainy day with rainfall greater than 3.6 mm. This rainfall threshold value (3.6 mm) equal the expected daily evapotranspiration rate, marking the lowest physical limit for considering rainfall that may produce utilizable surface water resources (Mathlouthi and Lebdi, 2020). The rainfall event m in a given rainy season n will

Table 1. Mann–Kendall test and Sen's slope for the maximum yearly dry spell length for Ichkeul lake basin.

Rain gauge	Time period	Mann–Kendall test	p-value	Sen's slope
Fretissa SM	1982–2019	−0.14	>0.1	0
Ghézala Dam	1968–2019	−0.01	>0.1	0
Wadi Joumine Antra	1962–2019	0.97	>0.1	0.066
Sidi Abdelbasset	1968–2019	−2.38	0.05	−0.5
Sidi Salem	1959–2019	−0.22	>0.1	0

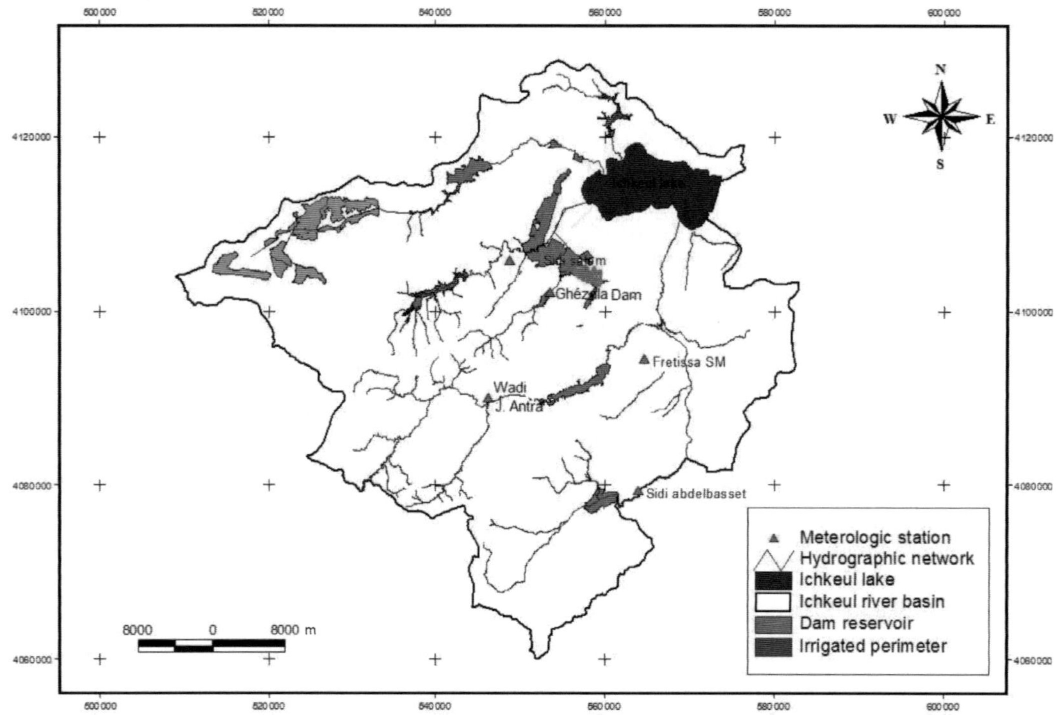

Figure 1. The Ichkeul Lake Basin and the location of gauge stations (the names and the time periods are listed in Table 1).

be characterized by its duration $D_{n,m}$, the temporal position within the rainy season, the dry event $Z_{n,m}$ and by the cumulative rainfall amounts of $H_{n,m}$ of $D_{n,m}$ rainy days in mm (Fig. 2). The rainy season (of variable duration, L_n) is defined as the time spans between the start of the first and the end of the last rainfall event (Fig. 2). We extract N_n rainfall events (variable depending on the year) per rainy season n, and then the number of dry events will be $N_n - 1$. For the set of dry spells calculated for each rainfall station, the maximum number of dry spell duration per year was used.

In this work, four models, generalized extreme value distributions (GEV), Lognormal, Pareto and Gamma distributions were fitted to the observed values – obtained.

To test if a sample of data came from a population with a specific distribution, the Anderson-Darling test (Stephens, 1974) is used, using SPC for Excel package (Buttrey, 2009). The Anderson-Darling test is defined as: H0: the data follow a specified distribution, Ha: the data do not follow the specified distribution (NIST/SEMATECH, 2013).

To verify the no systematic variations in the observed period, the nonparametric Mann–Kendall test proposed by Mann (1945) and improved by Hirsch and Slack (1984), is the most widely used.

To test the stationarity of the observed series, the unit root test was applied (Dickey and Fuller, 1981). The return period of the maximum annual dry spell duration was estimated by (Carvalho et al., 2013):

$$Z(F(x)) = \frac{1}{\frac{1}{\text{year}}(1 - F(x))} \tag{1}$$

Where $F(x)$ is the cumulative probability of occurrence of a given value of the maximum dry spell duration and 1/year denominator is the average sample frequency of the annual dry spell maximum duration.

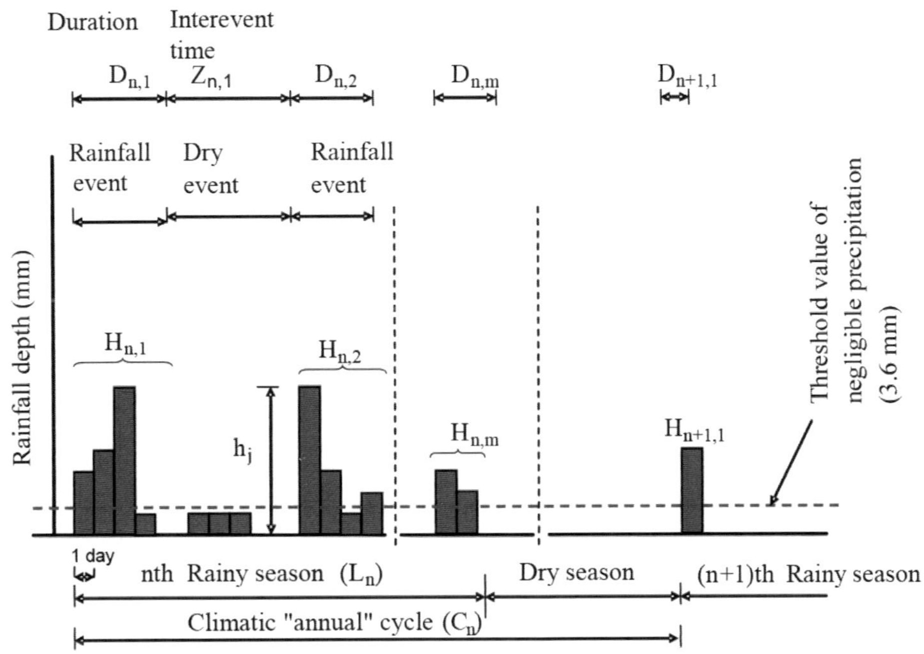

Figure 2. Dry and wet spell (year) representation.

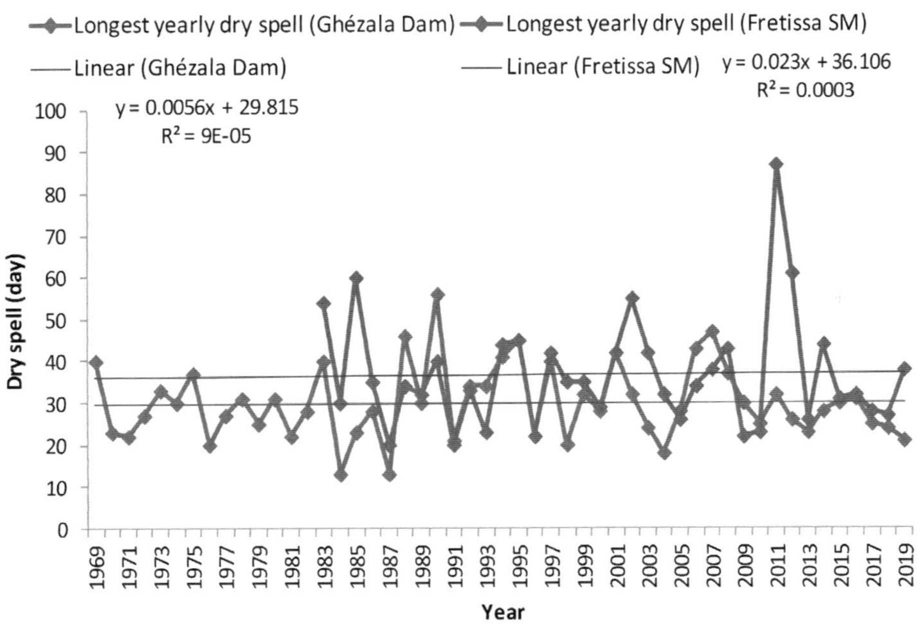

Figure 3. Maximum yearly dry event in days for Ghézala Dam and Fretissa SM rain gauges.

4 Results and discussion

Figure 3 shows the maximum annual dry spells observed for this basin during the rainy season (growing season). There is a slightly increased trend in the annual maximum dry spell of the Ghézala dam and Fretissa SM (Meteorological Station) stations. For medium and early crop cycles, the increasing trend increases the risks for productivity losses, particularly for corn.

The statistical trend obtained can be seen in Table 1 which shows the results of applying the nonparametric Mann–Kendall test. For four stations, the calculated p-value is more than the significance level ($\alpha = 0.05$), the null hypothesis (H0: there is no trend in the series) should not be rejected in

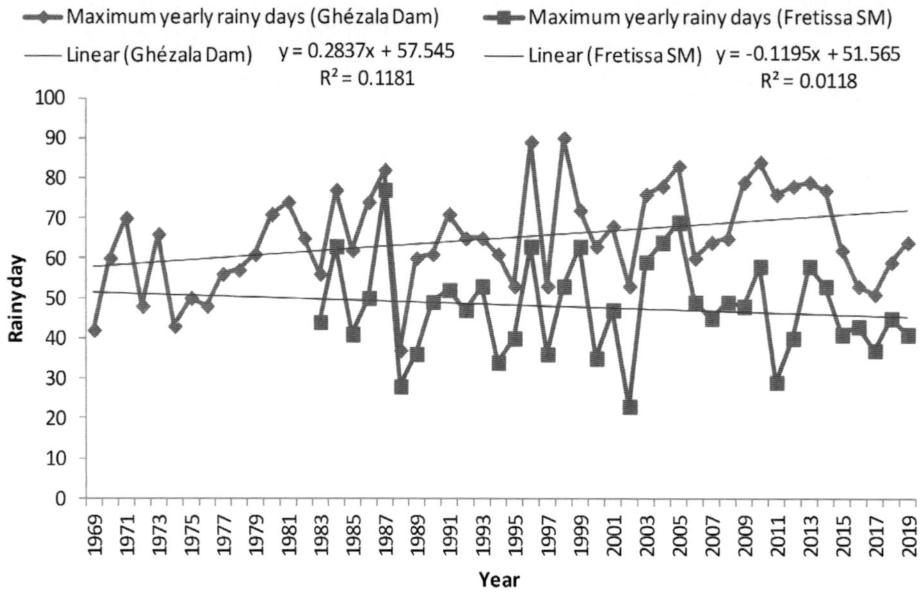

Figure 4. Maximum yearly rainy days for Ghézala Dam and Fretissa SM rain gauges.

favour of the alternative hypothesis (HA: there is a positive trend in the series). The risk of rejecting the null hypothesis when it is true is greater than 10 %. As the calculated p-value for Sidi Abdelbasset station is less than the significance level ($\alpha = 0.05$), the null hypothesis H0 should be rejected in favour of the alternative hypothesis HA. The risk of rejecting the null hypothesis when it is true is 0.05 %.

The rainfall event duration for each rain gauge was calculated and the maximum yearly non-consecutive rainy days for the growing season can be seen visually in Fig. 4. There is an increasing trend in the number of rainy days for Ghézala Dam station.

Table 2 shows the descriptive statistics for the maximum yearly rainy days for the region. The data series seen here presents similar mean values for the Fretissa SM and Sidi Abdelbasset in southern region and Wadi Joumine Antra and Sidi Salem in West and Est respectively. However the dispersion around the mean being greater for the Fretissa SM station. The average for the maximum yearly rainy days increases around 61 % for the Ghézala dam in relation to Sidi Abdelbasset. The dispersion around the mean can be considered low.

The statistical confirmation of the trend obtained visually in Fig. 4 can be seen in Table 3 where the results of applying the nonparametric Mann–Kendall test. As the p-value calculated for the Ghézala Dam and Sidi Salem rain gauge is less than the significance level ($\alpha = 0.05$), the null hypothesis (H0:) is rejected in favor of the alternative hypothesis (HA). The risk of rejecting the null hypothesis when it is true for Ghézala Dam and Sidi Salem northern region is respectively 0.05 % and 0.01 %. For the others stations (other areas), the p-value is greater than calculated, so the null hypothesis should not be rejected with a risk >10 %.

Table 2. Descriptive statistics for the maximum yearly rainy days at growing time.

Rain gauge	Mean	SD deviation	SD error mean
Fretissa SM	47.62	11.92	1.96
Ghézala Dam	64.92	12.27	1.71
Wadi Joumine Antra	58.21	11.98	1.58
Sidi Abdelbasset	40.16	9.43	1.55
Sidi Salem	56.87	13.07	1.68

Table 3. Mann–Kendall test and Sen's slope for the maximum number of rainy days at growing time.

Rain gauge	Mann–Kendall test	p-value	Sen's slope
Fretissa SM	−0.52	>0.1	−0.109
Ghézala Dam	2.44	0.05	0.286
Wadi Joumine Antra	−1.21	>0.1	−0.134
Sidi Abdelbasset	0.47	>0.1	0.061
Sidi Salem	2.65	0.01	0.272

As the value of maximum yearly rainy days for the growing period is increasing significantly in Ghézala dam and Sidi Salem rain gauges, there is no trend in the maximum yearly dry spell length in these stations. As there is no significant trend in the series of maximum yearly rainy days for Sidi Abdelbasset in southern region, the increasing maximum num-

Table 4. Maximum likelihood estimation of the Anderson-Darling statistic for Ichkeul lake basin.

Rain gauge	Anderson-Darling test critical values	GEV	Lognormal	Pareto	Gamma
Fretissa SM	Anderson-Darling statistic (AD)	0.431	0.384	5.526	0.523
	p-value	>0.25	0.372		0.206
	Akaike information criterion (AIC)	213.5	215.1		214.5
Ghézala Dam	Anderson-Darling statistic (AD)	0.269	0.209	10.736	0.207
	p-value	>0.25	0.855		>0.25
	Akaike information criterion (AIC)	311.5	312.0		311.9
Wadi Joumine Antra	Anderson-Darling statistic (AD)	0.228	0.231	9.669	0.215
	p-value	>0.25	0.792		>0.25
	Akaike information criterion (AIC)	353.6	355.3		355
Sidi Abdelbasset	Anderson-Darling statistic (AD)	0.825	0.762	3.172	0.761
	p-value	0.035	0.044		0.049
	Akaike information criterion (AIC)	337.4	338.6		338.4
Sidi Salem	Anderson-Darling statistic (AD)	0.217	0.201	9.684	0.224
	p-value	>0.25	0.876		>0.25
	Akaike information criterion (AIC)	384.6	386.3		385.6

ber of dry spell length is associated with the uneven distribution of rainfall in the growing period for this area.

Table 4 presents the result of Anderson-Darling statistic adherence test to check the degree of adjustment of the x variables series to the probability density function of the selected distributions, obtained by the maximum likelihood estimation (MLE) technique. Table 4 shows that the Generalized Extreme Value distribution that best describes the frequency of occurrence of the annual values of maximum periods of dry spell with the lowest values of AIC for $\alpha = 5\,\%$. AIC compares the relative "quality" of a model versus the other models. The distribution with the smallest AIC value is usually the preferred model (Mohammed et al., 2015).

The non-stationarity was rejected by the Dickey–Fuller test (DF) and Dickey–Fuller augmented test (ADF) (Dickey and Fuller, 1981), through the use of the autoregressive procedure of EViews 11 software (IHS Global Insight, 2017) using Unit Root Test (Table 5).

At 5 % significance level, the probabilities obtained are lower indicating therefore that the series with the maximum dry spell length values are apparently stationary, that is, the classical statistical assumptions remain valid. The results of the augmented Dickey–Fuller test reported in Table 5, show that the inclusion of parameters for the stations does not change at the 5 % significance level. The test confirms in this case there is no unit root.

The preceding analysis is a support for the use of the extreme value theory in describing the frequencies of occurrence of the maximum annual dry spells values for the studied region.

It can be seen in Table 6 that Sidi Abdelbasset station has a higher dry spells than the other stations. For a 60 d dry spell, the return periods for Sidi Abdelbasset, Fretissa SM and Sidi Salem stations at the agricultural growing season are respectively 20, 50 and 72 years. However, dry spells over as little as 21 d can cause reduced productivity, and for a return period of one year, it is possible to have an occurrence of dry spells between 14 and 27 d in Sidi Abdelbasset station. This situation, happening during grain filling season, increases the productivity losses for this region. The dry spell length for the return period of one year is higher in Sidi Abdelbasset and Fretissa SM stations respectively, both situated in southern region. Otherwise, this southern region is the most exposed to the risk of agricultural losses due to periods of drought for a return period of one year. If the dry spell duration is from 40 to 50 d, also the most vulnerable is this region. The observed maximum dry spell length for the return period of 10 years in the Southest was 49–51 d while in the Middle it was 43–45 d. So the gradient is North South.

5 Conclusions

The dry spell length increased in growing seasons for the Sidi Abdelbasset station and decreased in the other stations. Although the number of rainy days has increased in the growing season for the Ghézala dam and Sidi Salem stations, the distribution of rainfall in the Mideast is irregular at growing time. The result for time series with fewer years, indicate that the irregularity in the distribution of rainfall for this region has intensified in recent years. The average number of rainy days in the growing season for Ghézala Dam is 36 % to 61 % higher than for the South region.

For all distributions studied, the generalized extreme value distribution was the best fit to the data series of dry spell duration for the five stations.

Table 5. The Dickey–Fuller test and the Dickey–Fuller augmented test (ADF).

Rain gauge	DF	p-value	ADF	p-value
Fretissa SM	−4.8135	0	−5.7909	0
Ghézala Dam	−6.0991	0	−7.3252	0
Wadi Joumine Antra	−6.0029	0	−5.4492	0
Sidi Abdelbasset	−4.3247	0.0001	−5.1735	0
Sidi Salem	−7.5726	0	−7.5274	0

Table 6. Return period (years) for Ichkeul lake basin based on the maximum annual period of dry spell during the growing season.

	X (d)				
Return period (year)	Fretissa SM	Ghézala Dam	Wadi Joumine Antra	Sidi Abdelbasset	Sidi Salem
1	8	4	7	14	6
1	10	6	9	15	7
1	11	8	10	16	8
1	14	11	12	17	11
1	16	12	13	18	12
1	17	14	14	19	14
1	20	17	17	21	16
1.1	23	19	19	22	19
1.2	26	23	22	25	22
1.4	29	25	25	27	24
2	34	30	29	31	29
3	39	34	33	36	34
5	44	38	38	42	39
10	49	43	43	51	45
20	54	47	48	60	51
50	60	52	54	74	58
100	64	55	59	87	63
200	67	58	63	102	68
1000	75	63	72	144	79
>1000	78	65	76	167	84

It has been showed that 14–27 d dry spell durations occur in return period of one year for the series analyzed from 1968 to 2019. Previous works clearly show that 40–43 d dry spell durations were observed in return periods of 10 years (Mathlouthi and Lebdi, 2009). The dry spells duration become more frequent in recent years, throughout the region.

Data availability. The data are printed in paper documents stored (archived) in office of the General Directorate of Water Resources and the Division of Dam Operation of Extreme North and Ichkeul of the Ministry of Agriculture of Tunisia, (http://www.agriculture.tn/, Ministry of Agriculture of Tunisia, 2021). These data are the property of this organization, and are available in situ.

Author contributions. MM collected and analyzed the data, developed the main idea, performed the calculations, discussed the results and wrote the manuscript; FL contributed results discussion and revised manuscript.

Competing interests. The contact author has declared that neither they nor their co-author have any competing interests.

Disclaimer. Publisher's note: Copernicus Publications remains neutral with regard to jurisdictional claims in published maps and institutional affiliations.

Special issue statement. This article is part of the special issue "Hydrology of Large River Basins of Africa". It is a result of the 4th International Conference on the "Hydrology of the Great Rivers of Africa", Cotonou, Benin, 13–20 November 2021.

Acknowledgements. The authors acknowledge the help of General Directorate of Water Resources and Division of Dam Operation of Extreme North and Ichkeul of the Ministry of Agriculture of Tunisia. The authors thank the reviewers for their relevant remarks that contributed to the improvement of this article.

References

Buttrey, S. E.: An Excel Add-In for Statistical Process Control Charts, J. Stat. Softw., 30, 1–12, https://doi.org/10.18637/jss.v030.i13, 2009.

Carvalho, J. R. P., Assad, E. D., Evangelista, S. R. M., and Pinto, H. S.: Estimation of dry spells in three Brazilian regions – Analysis of extremes, Atmos. Res., 132–133, 12–21, https://doi.org/10.1016/j.atmosres.2013.04.003, 2013.

Dan'azumi, S. and Shamsudin, S.: Modeling the distribution of inter-event dry spell for peninsular Malaysia, J. Appl. Sci. Res., 7, 333–339, https://doi.org/10.1007/s00704-010-0254-2, 2011.

Dickey, D. A. and Fuller, W. A.: Likelihood ratio statistics for autoregressive time series with a unit root, Econometrica, 49, 1057–1072, https://doi.org/10.2307/1912517, 1981.

Hirsch, R. M. and Slack, J. R.: A nonparametric trend test for seasonal data with serial dependence, Water Resour. Res., 20, 727–732, https://doi.org/10.1029/WR020i006p00727, 1984.

IHS Global Insight: The EViews 10 Demo, 2nd edn., ISBN 978-1-880411-43-8 , available at: http://www.eviews.com (last access: 20 October 2021), 2017.

Lana, X., Martínez, M. D., Burgueno, A., Serra, C., Martínez-Vide, J., and Gómez, L.: Distributions of long dry spells in the Iberian Peninsula, years 1951–1990, Int. J. Climatol. 26, 1999–2021, https://doi.org/10.1002/joc.1354, 2006.

Mahamud, N., Deni, S. M., and Jemain, A. A.: Probability models for distribution of weekly dry and wet spells in Sabah and Sarawak, J. Stat. Model. Anal., 2, 34–44, 2011.

Mann, H. B.: Nonparametric tests against trend, Econometrica, 13, 245–259, https://doi.org/10.2307/1907187, 1945.

Mathlouthi, M. and Lebdi, F.: Analyse statistique des séquences sèches dans un bassin du nord de la Tunisie, Hydrol. Sci. J., 54, 442–455, https://doi.org/10.1623/hysj.54.3.442, 2009.

Mathlouthi, M. and Lebdi, F.: Estimating extreme dry spell risk in Ichkeul Lake Basin (Northern Tunisia): a comparative analysis of annual maxima series with a Gumbel distribution, Proc. IAHS, 383, 241–248, https://doi.org/10.5194/piahs-383-241-2020, 2020.

Ministry of Agriculture of Tunisia: Homepage, available at: http://www.agriculture.tn/, last access: 20 October 2021.

Mohammed, E. A., Naugler, C., Far, B. H.: Emerging business intelligence framework for a clinical laboratory through big data analytics, in: Emerging trends in computational biology, bioinformatics, and systems biology, edited by: Tran, Q. N. and Arabnia, H., Elsevier/Morgan Kaufmann, New York, USA, 577–602, 2015.

NIST/SEMATECH: e-Handbook of Statistical Methods, https://doi.org/10.18434/M32189, 2013.

Stephens, M. A.: EDF Statistics for Goodness of Fit and Some Comparisons, J. Am. Stat. Assoc., 69, 730–737, https://doi.org/10.2307/2286009, 1974.

Proc. IAHS, 384, 203–211, 2021
https://doi.org/10.5194/piahs-384-203-2021

Open Access
PIAHS

Hydrology of Large River Basins of Africa

Characterization of the variability in climate extremes in the Ferkessédougou sugar complexes (Northern Côte d'Ivoire)

Affoué Berthe Yao[1], Sampah Georges Eblin[1], Loukou Alexis Brou[1], Kouakou Lazare Kouassi[1], Gla Blaise Ouede[1], Ibrahim Salifou[1], Arona Diedhiou[2], and Bi Crépin Péné[3]

[1]Department of Earth Sciences, Jean Lorougnon Guédé University, Daloa, BP 150, Côte d'Ivoire
[2]IRD, CNRS, Grenoble INP, IGE, University Grenoble Alpes, Grenoble 3800, France
[3]Research and Development Department, SUCAF-CI, Abidjan, 01 BP 1967, Côte d'Ivoire

Correspondence: Affoué Berthe Yao (y_berth@yahoo.fr)

Published: 16 November 2021

Abstract. This study aims to analyse the frequency, intensity and duration of extreme climate events in order to optimise sugarcane production in the Ferkessédougou sugar complexes. The methodological approach is based on the calculation of extreme climate indices defined by the Expert Team on Climate Change Detection and Indices (ETCCDI) from daily rainfall and temperature data observed at the Ferké 2 station over the period 1999–2018. The results show that the rainfall indices are negative, except for the number of consecutive dry days (CDD); this shows a decreasing trend in rainfall with, however, insignificant trends. Over the period 1999–2006, the number of intense rainfall days (R10 mm) decreased from 40 to 28 d with an average decrease of 0.3 d yr^{-1} and the number of very intense rainfall days (R20 mm) fluctuated between 26 and 2 d, with a slope of 0.083. The extreme temperature indices show statistically significant positive trends for the warm sequences; this confirms the rising of temperatures on both a local and national scale. This study could enable the Ferkessédougou sugar complexes managers to develop strategies for adaptation to climate change.

1 Introduction

The consequences of climate change are clearly visible with the advent of extreme events such as floods, droughts, melting glaciers, scarcity of water resources, increase in infectious diseases, etc. (Handmer et al., 2012). Of these extreme events, drought is one of the extreme weather conditions that affects more people than any other form of natural disaster (Wilhite, 2000). The occurrence of drought in recent decades in West Africa highlights the importance of this phenomenon (Bhaga et al., 2020; Faye et al., 2017). Moreover, climate projections have shown that West Africa will experience a significant increase in temperature, with high variability in rainfall (Kouakou et al., 2014; Kouamé et al., 2014; Ly et al., 2013; Yapo et al., 2020). In Côte d'Ivoire, rainfall is the fundamental climatic element that conditions the development of various agricultural activities. Thus, their absence, scarcity, excess/deficit or poor spatial and temporal distribution generates crises (Péné and Assa, 2003). This is the case of the sugar complexes located in the departments of Ferkessédougou and Tafiré, in the north of Côte d'Ivoire. These sugar complexes exploit the water of the Bandama river for sugar cane irrigation, sugar cane production and domestic needs. Following the example of the large agri-food industry in Côte d'Ivoire, water needs for irrigation and for the populations of this region are progressively increasing and water resources are more and more in demand. However, studies of climate variability in the area have shown that the decline in rainfall that began in the 1970s is continuing (Soro et al., 2013). Annual and inter-annual irregularities in rainfall result in rainy seasons with poor harvests (Péné et al., 2010). Projections show that extreme weather events are expected to become more frequent as the climate warms (Yapo et al., 2020). These events have a negative impact on agriculture, livestock and natural resources, which are sectors on which the bulk of the Ivorian economy is based. It is therefore es-

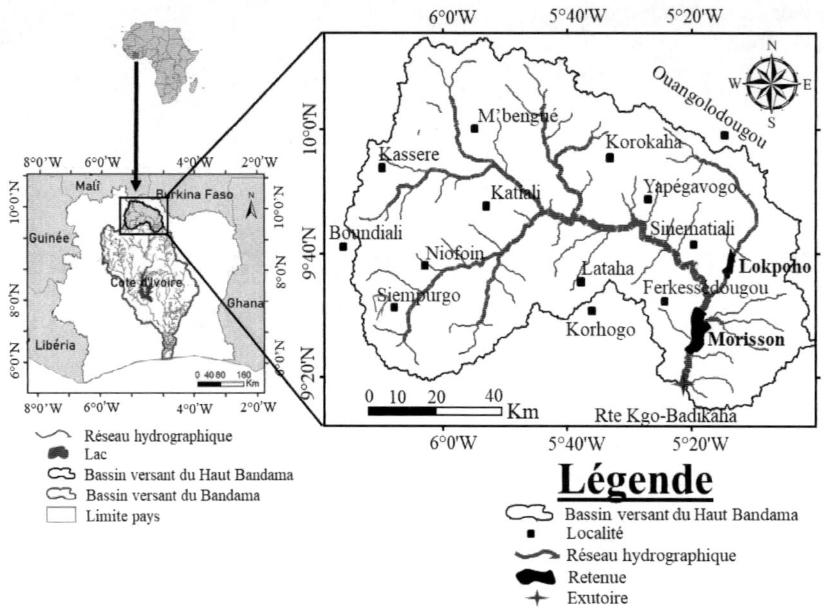

Figure 1. Location of the Upper Bandama watershed.

sential to analyse the variability of rainfall and temperature extremes observed in this region, which is an area of interest for Côte d'Ivoire because of its sugar production capacity. The aim of this study is to analyse the frequency, intensity and duration of extreme climate events with a view to optimising sugarcane production in the Ferkessédougou sugar complexes. The methodological approach is based on the calculation of climate indices defined by the Expert Team on Climate Change Detection and Indices (ETCCDI).

2 Presentation of the study area

The study area is located in the Upper Bandama watershed, at the Korhogo-Badikaha road station, between longitudes 5°00′ and 6°20′ W and latitudes 9°00′ and 10°20′ N (Fig. 1). It covers the square degrees of the Poro, Tchologo and part of the Bagoue regions with an estimated area of 9850 km^2 and represents about 10 % of the entire Bandama watershed. The two sugar complexes of Ferkessédougou (Ferké 1 & Ferké 2) cover an area of 42 721 ha. Sugar cane is grown under both irrigated and rain-fed conditions.

The study area is under the influence of the tropical transitional regime characterised by two main seasons (Soro et al., 2013) (Fig. 2).

- a long rainy season, from April to October, with average monthly rainfall ranging from 129.3 to 250.4 mm and
- a long dry season, from November to March, with average rainfall ranging from 1.9 to 82.5 mm.

Average monthly temperature ranges from 25 °C in August to 29 °C in March.

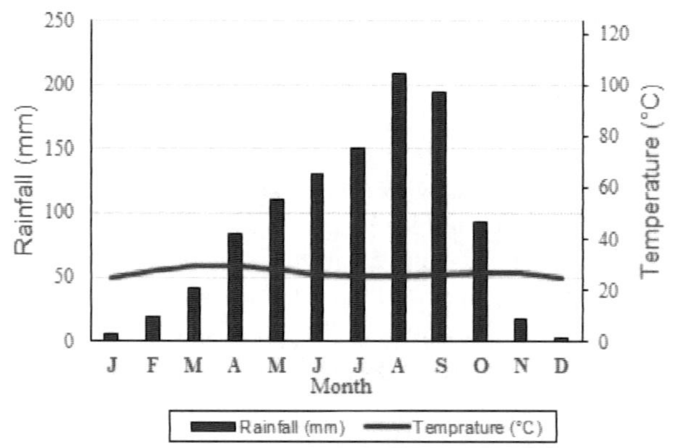

Figure 2. Umbrothermal diagram of Ferké 2 station for the period 1999–2019.

3 Material and methods

3.1 Material

The material consists of climatic data such as daily rainfall and temperature data for the period 1999–2018. These data come from the Ferkessédougou sugar complexes and were recorded at the Ferké 2 station. The synoptic overview of the variability of climate extremes in the study area is carried out with the RClimDex software. The software provides a data quality check to identify recording errors that may exist in the daily data (WMO, 2009). The quality control process is based on the following principles:

Table 1. List of ETCCDMI core Climate Indices used (Zhang and Yang, 2004).

Identification	Indicator name	Definitions	units
Precipitations			
PRCPTOT	Annual total wet-day precipitation	Annual total PRCP in wet days (RR $\geq$ 1 mm)	mm
RX1day	Max 1 d precipitation amount	Monthly maximum 1 d precipitation	mm
RX5day	Max 5 d precipitation amount	Monthly maximum consecutive 5 d precipitation	mm
R95p	Very wet days	Annual total PRCP when RR > 95th percentile	mm
CDD	Consecutive dry days	Maximum number of consecutive days with RR < 1 mm	Days
CWD	Consecutive wet days	Maximum number of consecutive days with RR $\geq$ 1 mm	Days
R10	Number of heavy precipitation days	Annual count of days when PRCP $\geq$ 10 mm	Days
R20	Number of very heavy precipitation days	Annual count of days when PRCP $\geq$ 20 mm	Days
Temperatures			
TXx	Max Tmax	Monthly maximum value of daily maximum temperatures	°C
TNx	Max T_{min}	Monthly maximum value of daily minimum temperatures	°C
TXn	Min T_{max}	Monthly minimum value of daily maximum temperatures	°C
TNn	Min T_{min}	Monthly minimum value of daily minimum temperatures	°C
WSDI	Warm spell duration indicator	Annual count of days with at least 6 consecutive days when TX > 90th percentile	Days
CSDI	Cold spell duration indicator	Annual count of days with at least 6 consecutive days when TN < 10th percentile	Days
DTR	Diurnal temperature range	Monthly mean difference between TX and TN	°C

- erroneous values of the maximum daily temperature must be replaced by −99.9 if it is lower than the minimum daily temperature;
- it must not have more than 365–366 daily observations per year;
- February should have no more than 28 observations in any given year;
- missing or negative data (for rainfall) are replaced by −99.9.

3.2 Methods

The analysis of rainfall and temperature extremes was possible through the calculation of climate indices. Although there are many climate indices, 27 have been defined by the Expert Team on Climate Change Detection and Indices (ETCCDI). This method of analysis is commonly used to study extreme events (Aguilar et al., 2009; Filahi et al., 2015; Hountondji et al., 2011; New et al., 2006; Santos et al., 2010).

Detailed descriptions of these indices and the RClimDex are available on the ETCCDI website (https://github.com/ECCC-CDAS/RClimDex, last access: 19 April 2021). In this study, 8 rainfall indices and 7 temperature indices were used (Table 1). The choice of these indices was made to highlight extreme rainfall events and their frequency on the one hand and to assess temperature variations and their duration on the other hand. The parameters selected are those that can have a positive or negative impact on sugarcane yield. Indeed, Saithanu et al. (2017) studied the effect of minimum temperature on the different phenological phases of sugarcane. According to Mali et al. (2014), high humidity leads to better sugarcane growth and increased sugar content. Linear trend analysis (expressed as % per year) was carried out from 1999 to 2018 by linear regression between the different rainfall indices and time (in years). The slopes thus estimated were grouped into two classes indicating significant and non-significant trends. The boundary of the two classes considered is defined from the Student's t-statistic used to test the hypothesis of a slope equal to 0. The trend is therefore qualified as significant if the probability p (p value) of the t-test applied to the regression slope is less than 0.05, whereas it is not significant when it exceeds the 0.05 threshold.

4 Results and Discussion

4.1 Results

4.1.1 Variability of extreme rainfall in Ferkessédougou

Figures 3 and 4 show the trends in the variability of the different extreme rainfall indices at the Ferké 2 station over the period 1999–2018. The linear trend line analysis (Fig. 3a) shows a regression of the annual rainfall amounts. The index trend is decreasing by 2.76 mm yr^{-1}. This regression is statistically insignificant with a p value of 0.699. The smoothing curve shows three phases in the evolution of the total annual rainfall. The period 1999–2006 is marked by a drop in annual totals. The period 2006–2014 is characterised by a slight increase in cumulative rainfall. This phase is followed by a further decline from 2015. Similarly, the R95p index

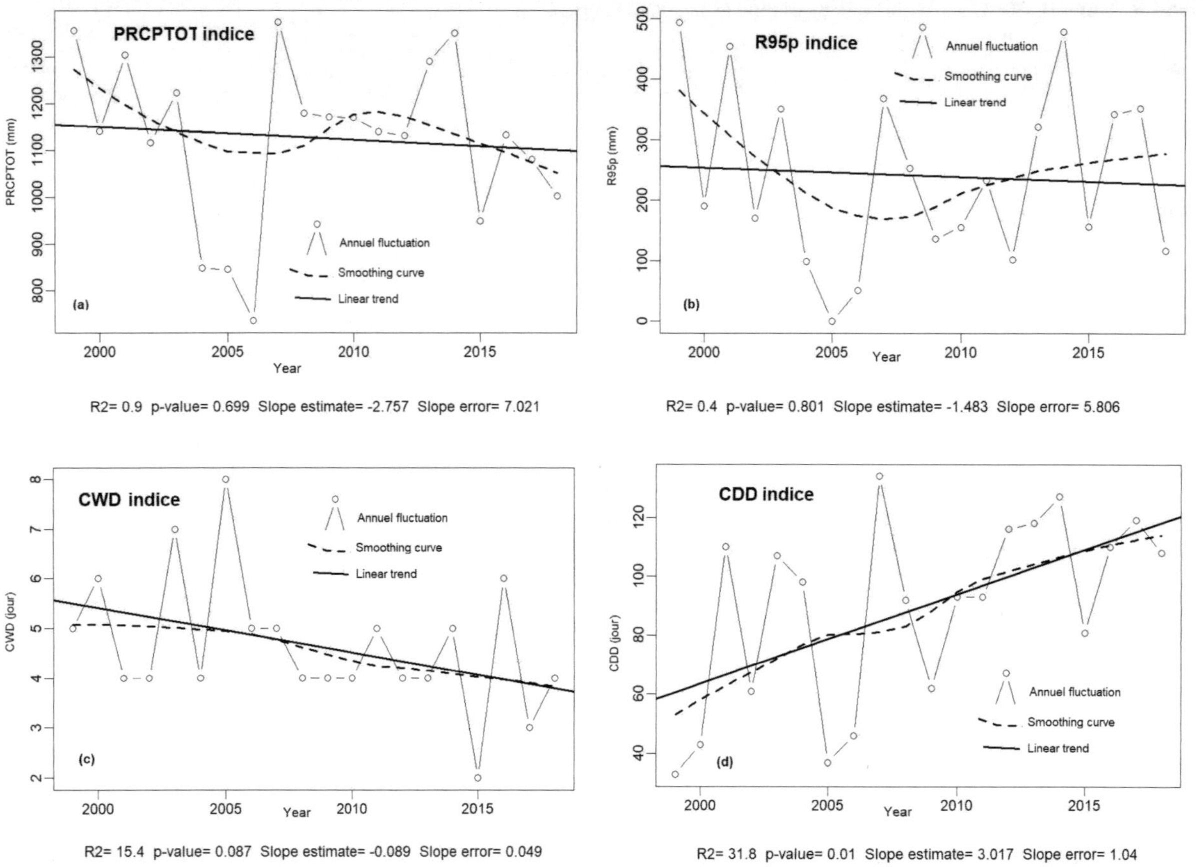

Figure 3. Trends in extreme rainfall indices calculated over the period 1999–2018 at the Ferké 2 station: **(a)** PRCPTOT indice, **(b)** R95p indice, **(c)** CWD indice and **(d)** CDD indice.

(Fig. 3b), which defines the cumulative rainfall above the 95th percentile of the rainfall time series, is marked by alternating downward and upward trends, with one downward trend that is not very significant (with a p value of 0.801).

Figure 3c and d show the evolution of consecutive wet days (CWD) and consecutive dry days (CDD) at the Ferké 2 station. It can be seen that the CDD (Fig. 3d) is increasing while that of CWD (Fig. 3c) is decreasing. In Ferkessédougou, the maximum number of CWD is 5 to 6 d while CDD reach 120 d (Fig. 3c). The decrease in CWD and CDD remains significant in the Ferkessédougou sugar complexes with p values of 0.087 and 0.01 respectively. In sum, these results confirm the decrease in average annual precipitation in the Ferkessédougou sugar complexes.

Indices Rx1day and Rx5day represent the monthly maximum 1 d precipitation and the monthly maximum consecutive 5 d precipitation respectively during the year (Fig. 4a and b). There was a slight upward trend in both indices. However, this increase is non-significant with p values of 0.475 and 0.175 respectively.

The evolution of intense (Fig. 4c) and very intense (Fig. 4d) rainfall days is similar to that of annual total wet-day precipitation (PRCPTOT). Over the period 1999–2006, the number of intense rainfall days (R10 mm) decreased from 40 to 28 d with an average decrease of 0.3 d yr^{-1} and the number of very intense rainfall days (R20 mm) fluctuated between 26 and 2 d, with a slope of 0.083. From 2006 to 2018, the overall trend for both indices is downward, with little significance. In sum, these analyses show that the decrease in precipitation appears to be due to a decrease in consecutive wet days and an increase in consecutive dry days.

4.1.2 Variability of extreme temperatures in Ferkessédougou

The analysis of Fig. 5 shows a generalized trend of increasing maximum (Fig. 5a) and minimum (Fig. 5b) temperatures, with a highly significant trend (p value equal to 0). The Min T_{max} (Fig. 5c) in Ferké 2 station remains almost stationary while the Min T_{min} indicates a regression (Fig. 5d).

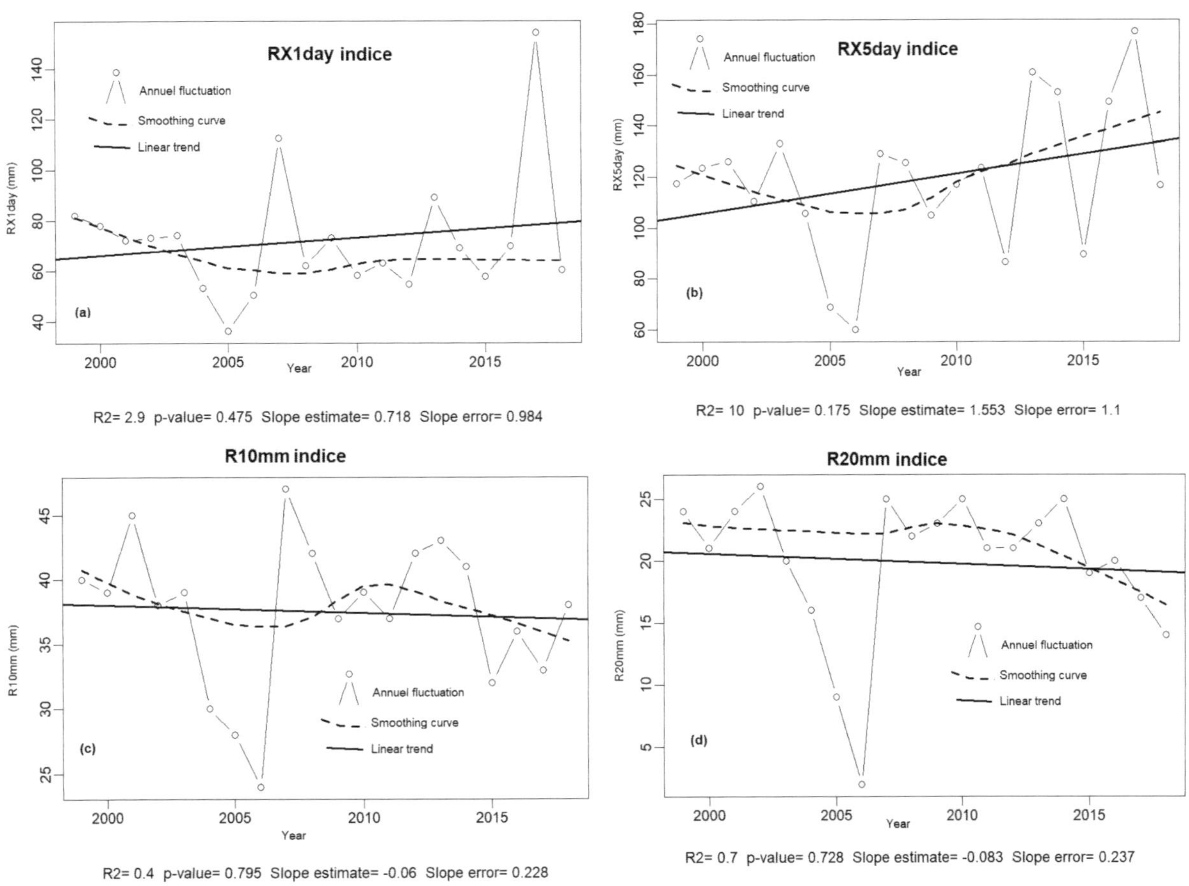

Figure 4. Trends in extreme rainfall indices calculated over the period 1999–2018 at the Ferké 2 station: **(a)** RX1day indice, **(b)** RX5day indice, **(c)** R10 mm indice and **(d)** R20 mm indice.

The trend in the indices of the cold and warm sequences evolves inversely at Ferké 2 station as shown in Fig. 6. According to this figure, the number of hot days (Fig. 6a) in the year increased by 1.15 % yr^{-1}. Conversely, the number of cold days (Fig. 6b) in the year decreased by 0.13 % yr^{-1} in Ferkessédougou. The evolution of the diurnal temperature range (DTR) (Fig. 6c), which is also increasing, confirms the global warming. Indeed, the p value of the duration indices of the warm sequences and the thermal amplitudes are lower than 0.05, which indicates that the climatic warming observed in the complexes is very significant.

4.2 Discussion

The application of extreme climate indices in the Ferkessédougou sugar complexes revealed a decrease in cumulative annual rainfall, the number of consecutive wet days and the number of heavy precipitation days at the Ferké 2 station over the period 1999–2018. The analysis reveals a significant trend in the number of consecutive wet days only. These results are similar to those obtained by N'Guessan et al. (2020) in the Lobo watershed. According to this study, all precipitation indices have decreased over the period 1984–2013. Indeed, the drought trend observed in the north of Côte d'Ivoire is confirmed by the index of consecutive dry days, which is on the rise in the Ferkessédougou area with a very significant regression trend. Thus, these results on the decreasing trend in the number of heavy precipitation days (R10 mm) and very heavy precipitation days (R20 mm) confirm this. Kouassi et al. (2010) reported that the decrease in the frequency of rainy days is synchronous with the decrease in annual rainfall. This decrease in the number of heavy (R10 mm) and very heavy (R20 mm) precipitation days observed in this study would be due to the decrease in the frequency of daily rainfall of between 10 and 50 mm. These different trends indicate that the drought that began in the 1960s and 1970s (Soro et al., 2013) is continuing in the Ferkessédougou sugar complexes, which could have considerable impacts on agricultural production. Indeed, according to studies by Atchérémi et al. (2018), the occurrence of a sequence of consecutive dry days within the rainy season can have a strong impact on agricultural production. The dry sequences create a lack of water in the plant.

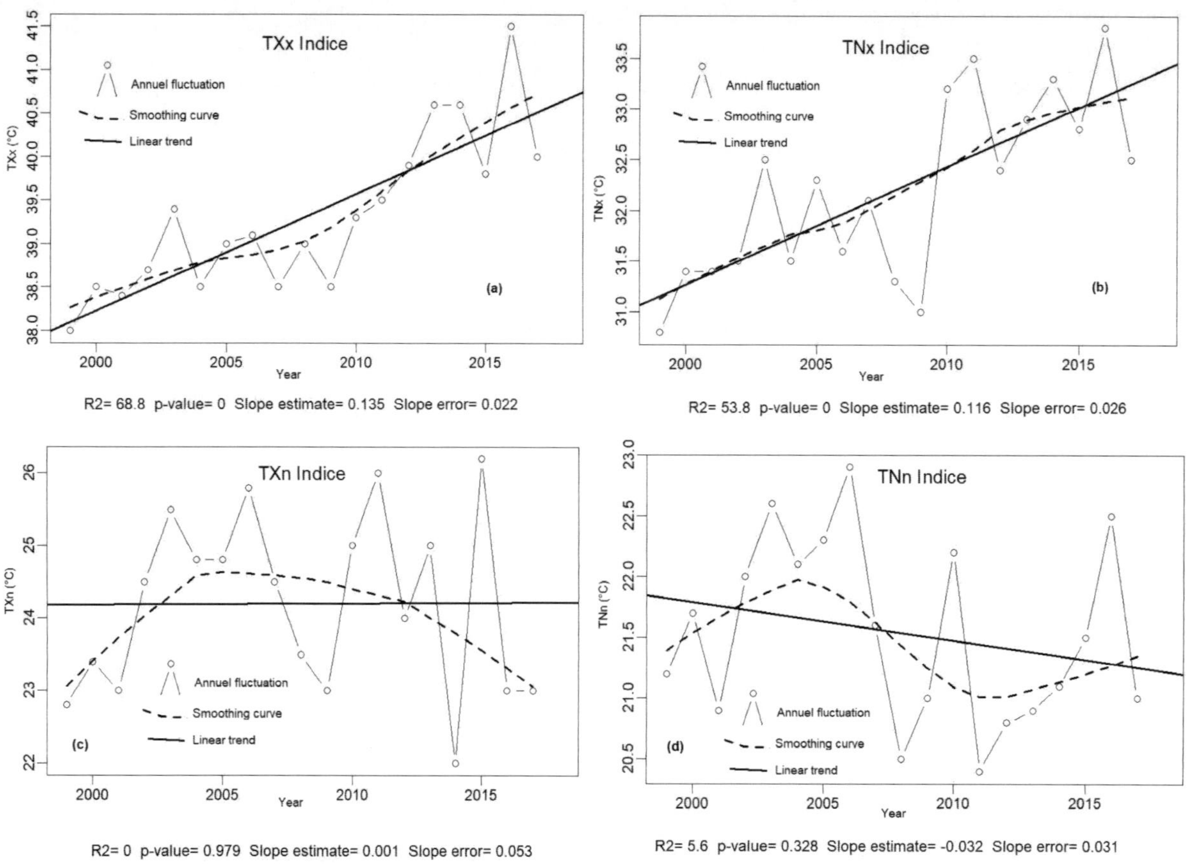

Figure 5. Trend in extreme temperature indices from 1999–2018 at the Ferké 2 station: **(a)** Max $T_{\max}$ indice, **(b)** Max $T_{\min}$ indice, **(c)** Min $T_{\max}$ indice, and **(d)** Min $T_{\min}$ indice.

When they occur at the beginning or in the middle of the growing season, they can cause a false start to the season or even mortgage an entire crop year. These trends highlighted in our work are in line with those of Djossou et al. (2020), on the Pendjari watershed at the Porga outlet in Benin.

The trend in maximum and minimum temperatures in Ferké 2 station is generally upwards. This increase seems to be accompanied by a significant decrease in the duration of cold sequences while that of warm sequences is increasing. These results are similar to those of Filahi et al. (2015), Karimou-Barke et al. (2015) and Djossou et al. (2020) in their studies on the evolution of peak temperatures in West Africa. These various studies indicate a more or less significant positive trend in the various indices of temperature extremes at local or regional level. These authors have shown a generalized trend of increasing hot days (TX90p index) and a significant increase in the duration of dry sequences (WSDI, TXx and TNn). The increase in temperature extremes can be attributed to heavy anthropisation and more particularly to the development of land use leading to the destruction of vegetation cover, which has increased over the last twenty years throughout the world (Abdalah, 2005). Thus, future projections also predict an upward trend in these extreme temperature indices (Diedhiou et al., 2018; Yapo et al., 2019) namely heat wave episodes, heat waves, extremely hot and more frequent nights. These extreme events are thought to be related to wind anomalies opposing the mean circulation in the upper and lower troposphere of West Africa (Fontaine et al., 2013).

5 Conclusion

This study examined the evolution of rainfall and extreme temperature indices in the Ferkessédougou sugar complexes over the period 1999–2018. Trend analysis of these indices showed that temperatures have increased significantly, but trends in indices of precipitation extremes are globally insignificant. The temperature extremes show statistically significant positive trends for the warm sequences and the maximum of the max and min temperatures. This confirms the warming of temperatures both locally and nationally.

The precipitation indices are negative, with the exception of the number of consecutive dry days; this shows a decreasing trend in rainfall with, however, insignificant trends. The

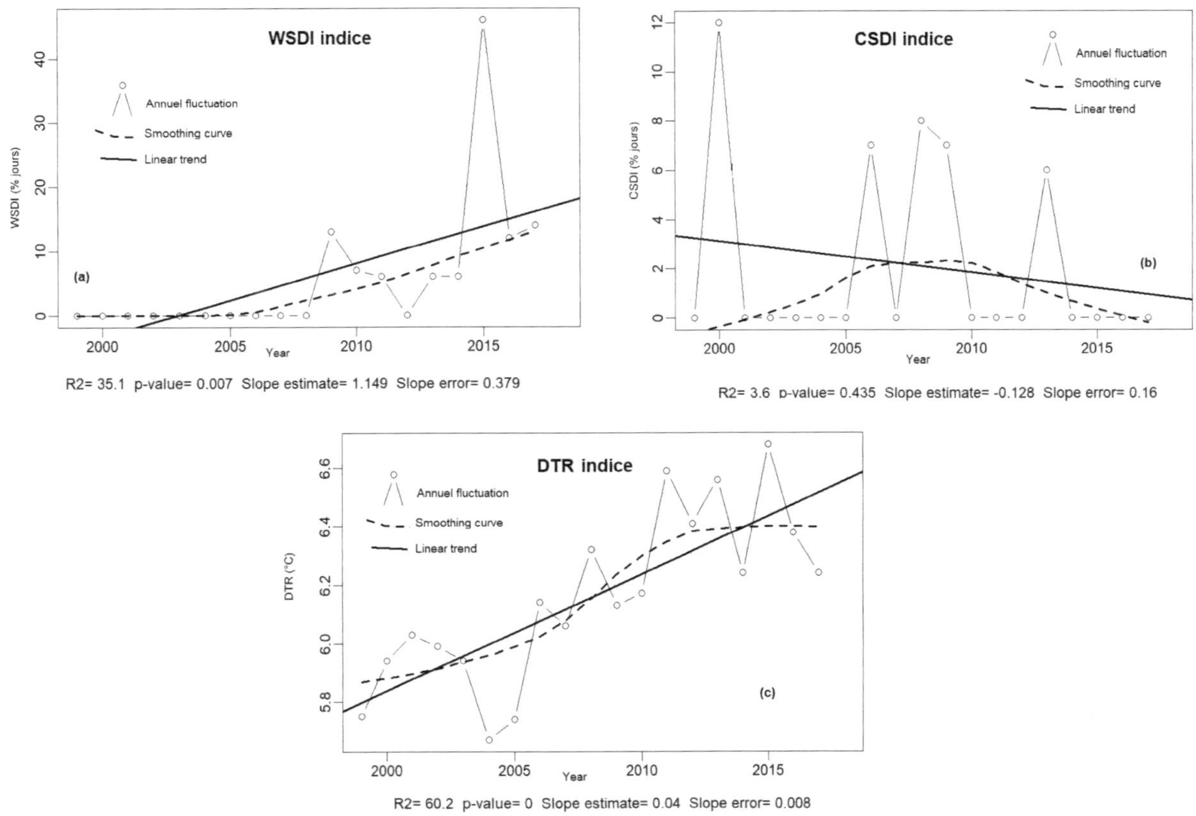

Figure 6. Trend of extreme temperature indices over the period 1999–2018 at Ferké 2 station: **(a)** WSDI indice, **(b)** CSDI indice and **(c)** DTR indice.

decrease in the number of heavy (R10 mm) and very heavy (R20 mm) precipitation days observed in this study would be due to the decrease in the frequency of daily rainfall of between 10 and 50 mm. Over the period 1999–2006, the number of heavy precipitation days (R10 mm) decreased from 40 to 28 d with an average decrease of 0.06 d yr^{-1} and the number of very heavy precipitation days (R20 mm) fluctuated between 26 and 2 d, with a slope of 0.083. From 2006 to 2018, the overall trend for both indices is downwards. Thus, dry seasons are getting longer while the rainy seasons are getting shorter. Further investigations are underway to analyse the impact of temperature and precipitation extreme variability on sugarcane yields in the Ferkessédougou sugar complexes.

Code availability. The Rclimdex software is freely available to the public. The manual and the software can be downloaded from the website: https://github.com/ECCC-CDAS/RClimDex, last access: 19 April 2021 (Chan, 2021).

Data availability. Precipitation and temperature data are not publicly available due to the structure's research policy. However, in the framework of a research cooperation as in this study, these data can be provided free of charge.

Author contributions. ABY analyzed the data and wrote the manuscript; SGE and LAB assisted in the preparation of the manuscript and reviewed the document; GBO and IS contributed to the data processing in the RClimDex software, KLK and AD supervised the study through the CLIMSUCAF project and BCP facilitated access to the SUCAF-CI data.

Competing interests. The contact author has declared that neither they nor their co-authors have any competing interests.

Special issue statement. This article is part of the special issue "Hydrology of Large River Basins of Africa". It is a result of the 4th International Conference on the "Hydrology of the Great Rivers of Africa", Cotonou, Benin, 13–20 November 2021.

Acknowledgements. This work was carried out as part of the AFD-CLIMSUCAF project (Convention IRS/ECO/No. 499-2019). The authors would like to thank the CLIMSUCAF project and SUCAF-CI company for providing funds and data for this study.

References

Abdalah, M.: Besoins d'observation pour les évènements extrêmes et pour l'adaptation au changement climatique. In: Rapport de l'atelier régional sur le système mondial d'observation du climat pour le bassin Méditerranéen, Marrakech, Maroc, du 22 au 24 novembre 2005, 61–64, 2005.

Aguilar, A., Aziz Barry, A., Brunet, M., Ekang, Fernandes, L. A., Massoukina, M., Mbah, J., Mhanda, A., Do Nascimento, D. J., Peterson, T. C., Thamba Umba, O., Tomou, M., and Zhang, X.: Changes in temperature and precipitation extremes in western central Africa, Guinea Conakry and Zimbabwe, 1955–2006, J. Geophys. Res., 114, 1–11, https://doi.org/10.1029/2008JD011010, 2009.

Atchérémi, K. N. D., Jourda, J. P. R., Saley, M. B., Kouamé, K. J., and Balliet, R.: Etude de l'évolution des extrêmes pluviométriques et de température dans le bassin versant de rivière Davo (Sud-Ouest de la Côte d'Ivoire) à partir de certains indices du logiciel RCLIMDEX, Larhyss Journal, 36, 99–117, 2018.

Bhaga, T. D., Dube, T., Shekede, M. D., and Shoko, C.: Impacts of climate variability and drought on surface water resources in Sub-saharan Africa using remote sensing: A review, Remote Sens., 12, 1–36, https://doi.org/10.3390/rs12244184, 2020.

Chan, R.: RClimDex, available at: https://github.com/ECCC-CDAS/RClimDex, Github [code], last access: 19 April 2021.

Diedhiou, A., Bichet, A., Wartenburger, R., Seneviratne, S. I., Rowell, D. P., Sylla, M. B., Diallo, I., Todzo, S., Toure, N. E., Camara, M., Ngatchah, B. N., Kane, N. A., Tall, L., and Affholder, F.: Changes in climate extremes over West and Central Africa at 1.5 °C and 2 °C global warming, Environ. Res. Lett., 13, 065020, https://doi.org/10.1088/1748-9326/aac3e5, 2018.

Djossou, M. L. D., Kodja, D. J., and Vissin, E. W.: Analyse de la Variabilité des Extrêmes Climatiques dans le Bassin Versant de la Pendjari à l'Exutoire de Porga au Bénin, Int. J. Prog. Sci. Technol., 22, 176–182, 2020.

Faye, C., Ndiaye, A., and Mbaye, I.: A comparative assessment of sequences of drought weather indices, by time scales and climate in Senegal areas, J. Water Environ. Sci., 1, 11–28, 2017.

Filahi, S., Mouhir, L., Tanarhte, M., and Tramblay, Y.: Tendance et variabilité des événements extrêmes au Maroc, XXVIIIe Colloque de l'Association Internationale de Climatologie, 1–4 Juillet 2015, Liège, Belgique, 463–468, 2015.

Fontaine, B., Janicot, S., and Monerie, P.: Recent changes in air temperature, heat waves occurrences and atmospheric circulation in Northern Africa, J. Geophys. Res., 118, 8536–8552, https://doi.org/10.1002/jgrd.50667, 2013.

Handmer, J., Honda, Y., Kundzewicz, Z. W., Arnell, N., Benito, G., Hatfield, J., Mohamed, I. F., Peduzzi, P., Wu, S. Sherstyukov, B., Takahashi, K., and Yan, Z.: Changes in impacts of climate extremes: human systems and ecosystems, in: Managing the Risks of Extreme Events and Disasters to Advance Climate Change Adaptation, edited by: Field, C. B., Barros, V., Stocker, T. F., Qin, D., Dokken, D. J., Ebi, K. L., Mastrandrea, M. D., Mach, K. J., Plattner, G.-K., Allen, S. K., Tignor, M., and Midgley, P. M., A Special Report of Working Groups I and II of the Intergovernmental Panel on Climate Change (IPCC), Cambridge University Press, Cambridge, UK, and New York, NY, USA, 231–290, 2012.

Hountondji, Y. C., De Longueville, F., and Ozer, P.: Trends in extreme rainfall events in Benin (West Africa), 1960–2000, Proceedings of the 1st International Conference on Energy, Environment and Climate Change, 26–27 August 2011, Ho Chi Minh City, Vietnam, available at: http://hdl.handle.net/2268/96112 (last access: 18 October 2021), 2011.

Karimou-Barke, M., Ambouta, K., Sarr, B., and Tychon, B.: Analyse des phénomènes climatiques extrêmes dans le Sud-Est du Niger, XXVIIIème Colloque de l'Association Internationale de Climatologie, 1–4 Juillet 2015, Liège, 537–542, 2015.

Kouakou, K. E., Kouadio, Z. A., Kouassi, F. W., Goula, B. T. A., and Savané, I.: Modélisation de la température et de la pluviométrie dans un contexte de changement climatique: cas de l'Afrique de l'Ouest, Afr. Sci., 10, 145–160, 2014.

Kouamé, Y. M., Soro, G. E., Kouakou, K. E., Kouadio, Z. A., and Meledje, N. E. H.: Scénarios des changements climatiques pour les précipitations et les températures en Afrique subsaharienne tropicale humide: Cas du bassin versant de Davo, Côte d'Ivoire, Larhyss Journal, 18, 197–213, 2014.

Kouassi, A. M., Kouamé, K. F., Koffi, Y. B., Djé, K. B., Paturel, J. E., and Oularé, S.: Analyse de la variabilité climatique et de ses influences sur les régimes pluviométriques saisonniers en Afrique de l'Ouest: cas du bassin versant du N'Zi (Bandama) en Côte d'Ivoire, Cybergéo: European Journal of Geography, Environnement, Nature, Paysage, 513, 1–29, https://doi.org/10.4000/cybergeo.23388, 2010.

Ly, M., Traoré, S. B., Alhassane, A., and Sarr, B.: Evolution of some observed climate extremes in the West African sahel, Weather Climate Extreme, 1, 19–25, https://doi.org/10.1016/j.wace.2013.07.005, 2013.

Mali, S. C., Shrivastava, P., and Thakare, H. S.: Impact of weather changes on sugarcane production, Res. Environ. Life Sci., 7, 243–246, 2014.

New, M., Hewiston, B., David, B., Tsiga, S. A., Kruger, A., Manhinsue, A., Gomez, B., Coelho, C. A. S. D., Masisi, N., Kalulanga, E., Bulane, E., Fortunata, L., Mdoka, M. L., and Lajoie, R.: Evidence of trends in daily climate extremes over southern and West Africa, J. Geophys. Res., 111, D14102, https://doi.org/10.1029/2005JD006289, 2006.

N'Guessan, B. V. H., Adjakpa, T. T., Allechy, F. B., Youan Ta, M., Assa, Y. F., and Affian, K.: Variabilité et prévision des extrêmes climatiques dans le bassin de la Lobo (Centre-Ouest de la Côte d'Ivoire): Apport des chaines de Markov et du générateur de temps mulgets, International Journal of Engineering Science Invention (IJESI), 9, 44–54, https://doi.org/10.35629/6734-0906024454, 2020.

Péné, C. B. and Assa, A. D.: Variations interannuelles de la pluviométrie et de l'alimentation hydrique de la canne à sucre en Côte d'Ivoire, Sécheresse, 14, 43–52, https://doi.org/10.1029/2005JD006289, 2003.

Péné, C. B., Ouattara, M. H., and Koulibaly, S. G.: Late season sugarcane performance as affected by soil water regime at the yield formation stage on commercial farms in northern Ivory Coast,

19th World Congress of Soil Science, Soil Solutions for a Changing, World, 1–5 August, Brisbane, Queensland, Australia, 2010.

Saithanu, K., Sittisom, P., and Mekparyup, J.: Estimation of Sugar Cane Yield in the Northeast of Thailand with MLR Model, Burapha Sci. J., 22, 197–201, 2017.

Santos, C. A. C. D., Neale, C. M. U., Rao, T. V. R., and Silva, B. B. D.: Trends in indices for extremes in daily temperature and precipitation over Utah, USA, Int. J. Climatol., 31, 813–1822, https://doi.org/10.1002/joc.2205, 2010.

Soro, T. D., Kouakou, B. D., Kouassi, E. A., Soro, G., Kouassi, A. M., Kouadio, K. E., Yéi, M.-S. O., and Soro, N.: Hydroclimatologie et dynamique de l'occupation du sol du bassin versant du Haut Bandama à Tortiya (Nord de la Côte d'Ivoire), VertigO, 13, 1–22, https://doi.org/10.4000/vertigo.14468, 2013.

Wilhite, D. A.: Drought as a Natural Hazard: Concepts and Definitions. Drought: A Global Assessment Routledge, London, UK, 2000.

WMO: Troisième conférence mondiale sur le climat, Bulletin de l'OMM, 58, 139–223, 2009.

Yapo, A. L. M., Diawara, A., Yoroba, F., Kouassi, K. B., Sylla, B. M., Kouadio, K., Odoulami, R. C., and Tiémoko, T. D.: Twenty-First Century Projected Changes in Extreme Temperature over Côte d'Ivoire (West Africa), Int. J. Geophys., 1–19, https://doi.org/10.1155/2019/5610328, 2019.

Yapo, A. L. M., Diawara, A., Kouassi, B. K., Yoroba, F., Sylla, M. B., Kouadio, K., Tiémoko, D. T., Koné, D. I., Akobé, E. Y., and Yao, K. P. A. T.: Projected changes in extreme precipitation intensity and dry spell length in Côte d'Ivoire under future climates, Theor. Appl. Climatol., 140, 871–889, https://doi.org/10.1007/s00704-020-03124-4, 2020.

Zhang, X. and Yang, F.: RClimDex (1.0): User Manual, Climate Research Branch Environment Canada, Downsview, Ontario, Canada, 23 pp., 2004.

Proc. IAHS, 384, 213–218, 2021
https://doi.org/10.5194/piahs-384-213-2021

Open Access PIAHS

Hydrology of Large River Basins of Africa

Evaluation de deux méthodes de correction de biais des sorties de modèles climatiques régionaux Cordex-Africa pour la prévision des pluies : cas du bassin côtier oranais

Sabrina Taïbi[1], **Ayoub Zeroual**[2], **and Naziha Melhani**[1]

[1]Département des Sciences de l'Eau et l'Environnement, L-PPRE, Université de Blida, Blida, Algérie
[2]École Nationale Supérieure d'Hydraulique, L-GEE, 09000, Blida, Algérie

Correspondence: Sabrina Taïbi (taibisabrina86@gmail.com)

Published: 16 November 2021

Résumé. Ce travail vise à évaluer les pluies simulées issues des sorties de modèles climatiques régionaux Cordex-Africa dans le bassin côtier oranais en Algérie. Pour cela les simulations du modèle RCA4 *(Rossby Centre Atmosphere model, version 4)* forcé par deux modèles de circulation globale *(MPI-ESM-LR et CNRM-CM5)* sous deux scenarios de forçages radiatifs « Representative Concentration Pathways » (RCPs) RCP 4.5 et RCP 8.5 sont comparées aux pluies observées au niveau de cinq stations pluviométriques, au cours de la période de contrôle 1981–2005 à l'échelle mensuelle. Les données futures simulées sont ensuite corrigées à l'aide de deux méthodes de correction de biais, à savoir, la méthode quantile-quantile et la méthode Delta, afin de mieux analyser leur évolution au cours de la période de projection 2075–2099. Les coefficients d'échange estimés au cours de la période 2075–2099 montrent que les simulations corrigées par la méthode Delta sont moins biaisées que les simulations corrigées par la méthode quantile-quantile. L'analyse de l'évolution future des pluies met en évidence une réduction de $-12\,\%$ à $-38\,\%$ d'ici la fin du 21*ème* siècle selon le RCP 4.5. Cette réduction qui est encore plus importante selon le scénario pessimiste RCP 8.5, risque d'affecter la disponibilité des ressources en eau dans la région qui a connu par le passé une période de sècheresse sévère et persistante. Enfin, cette étude peut être utilisée comme outil d'aide à la décision destiné aux parties prenantes de la gestion intégrée des ressources en eau et de l'agriculture. Néanmoins, pour une meilleure appréciation des impacts socio-économiques, une étude plus approfondie en considérant plusieurs modèles climatiques et d'autres paramètres climatiques, est recommandée.

1 Introduction

Du fait de sa position géographique, l'Algérie fait partie des pays les plus vulnérables aux changements climatiques. Des études ont déjà mis en évidence une augmentation des températures de 0.2–0.4 °C/décennie (Zeroual et al., 2017) ainsi qu'une une baisse de la pluviométrie particulièrement dans la partie Nord-Ouest de l'Algérie (Meddi et al., 2010; Bekkoussa et al., 2008) caractérisée par un déficit de 13 %–35 % (Taibi et al., 2013). Ainsi, les potentialités en eaux superficielles du Nord de l'Algérie sont passées de 12,5 milliards de m^3 à 10 milliards de m^3 (ANRH, 2007).

Les évènements extrêmes (sécheresse, pluviométrie déficitaire, augmentation des températures) ont eu des conséquences néfastes sur la satisfaction des besoins en eau des différents secteurs socio-économiques. Pour le GIEC (2014), cette situation risque de s'exacerber pour les années à venir.

En effet, selon les différents scénarios d'émissions (GIEC, 2014) les modèles climatiques prédisent une augmentation de la température de 1 °C à 4 °C d'ici la fin du 21*ème* siècle. Ceci risque encore d'accentuer les événements extrêmes particulièrement les sécheresses dans le bassin méditerranéen (GIEC, 2014) et affectera sensiblement la disponibilité des ressources en eau pour les différents usagers particulièrement l'agriculture, secteur le plus consommateur d'eau.

Face à cette situation, les études d'impact socio-économique et environnemental des changements climatiques et surtout pour la prévision, nécessitent l'utilisation des données sorties de modèles climatiques régionaux à haute résolution, tel que les modèles développés par le projet CORDEX qui offre des sorties de modèles climatiques pour différentes résolutions 0.44, 0.22 et 0.11° (https://cordex.org/, la date du dernier accès : 16 juin 2021). Cependant, l'utilisation directe des sorties des modèles régionaux dans les études d'impacts reste relativement limitée du fait qu'elles sont biaisées par rapport aux observations et nécessitent une correction afin d'affiner les simulations (Teutschbein and Seibert, 2012 ; Ly et al., 2019). Pour y remédier, des méthodes de correction de biais sont utilisées, telles que la méthode Delta et la méthode quantile-quantile (Mandez et al., 2020 ; Dobler et al., 2008 ; Fang et al., 2015).

C'est dans ce contexte que s'inscrit ce travail qui porte sur l'évaluation de deux méthodes de correction de biais des sorties de modèles climatiques pour la prévision des pluies dans le bassin côtier oranais en Algérie. Pour cela les simulations du modèle RCA4 *(Rossby Centre Atmosphere model, version 4)* forcé par deux modèles de circulation globale (MPI-ESM-LR et CNRM-CM5) sous deux scenarios de forçages radiatifs « Representative Concentration Pathways » (RCPs) RCP 4.5 et RCP 8.5 ont été utilisées. Les données futures des pluies simulées sont corrigées par les méthodes quantile-quantile et Delta avant d'être utilisées pour l'analyse de l'évolution des pluies du bassin côtier oranais au cours de la période de projection 2075–2099.

2 Présentation du secteur d'étude et données utilisées

Le bassin Côtier Oranais fait partie de la région hydrographique Oranie-Chott Chergui qui se trouve à l'Ouest de l'Algérie. Il est situé au Nord des bassins de la Tafna, la Macta et du Chott Chergui et limité au Nord par la mer Méditerranée.

L'étude a été menée sur cinq stations pluviométriques réparties sur l'ensemble de la zone d'étude (Fig. 1). Les données mensuelles ont été collectées auprès de l'Agence Nationale des Ressources Hydrauliques (ANRH) pour la période d'étude 1981–2010.

La pluie moyenne annuelle de la zone d'étude varie entre 249 mm à la station d'El braya à 353 mm à Oued Berkeche (Tableau 1). Effectivement, l'Ouest de l'Algérie est caractérisé par une pluviométrie inférieure à celle du centre et de l'Est du pays. C'est également la partie du pays qui a connu une longue période de sécheresse particulièrement au cours des années 80. Afin d'évaluer les sorties de modèles climatiques régionaux de Cordex-Africa, les modèles RCA4-MPI-ESM-LR et RCA4-CNRM-CM5 ont été utilisés. Ces modèles ont été jugés satisfaisants pour l'Algérie par rapport aux autres modèles disponibles (Zeroual et al., 2019) et offrent des simulations à une résolution de 50 km.

3 Méthodes

Pour évaluer la capacité des modèles climatiques à reproduire la variabilité pluviométrique passée du secteur d'étude, les données simulées par les deux MCR sont comparées aux pluies mensuelles observées au niveau de chacune des cinq stations au cours de la période de référence 1981–2005 par l'estimation du biais Eq. (1).

$$\text{Biais} = \frac{\text{Pluie simulée-pluie observée}}{\text{pluie observée}} \tag{1}$$

Il existe différentes méthodes de correction de biais qui permettent de corriger les simulations futures en se basant sur la différence entre les données observées et simulées au cours de la période de contrôle. Toutefois, les méthodes Delta et quantile-quantile sont le plus souvent utilisées (Tong et al., 2020 ; Mandez et al., 2020). Ces deux méthodes ont été appliquées dans le cadre de ce travail, afin de corriger les simulations futures issues des deux MCR (MPI et CNRM) au cours de la période de projection 2075–2099 et pour les deux scénarios d'émissions RCP 4.5 et RCP 8.5.

La méthode du Delta consiste à déterminer un changement (ou une anomalie) qui correspond à la différence des moyennes entre une simulation de climat futur moins la simulation de climat présent (Lenderink et al., 2007), en calculant simplement un changement moyen sur l'ensemble de la distribution des observations (Colin, 2010).

La méthode de correction quantile-quantile (*Q-Q*), ou quantile mapping consiste à estimer la correction en fonction des quantiles de la distribution de probabilité de la variable en question (Déqué, 2007). Un quantile de la distribution simulée est remplacé par le même quantile de la distribution observée.

Une fois estimée, la fonction de correction est appliquée à la variable issue des simulations de climat futur qui peut être finalement prise en compte dans des études d'impacts socio-économiques dans le contexte des changements globaux.

4 Résultats et discussions

4.1 Évaluation des modèles climatiques régionaux au cours de la période de contrôle 1981–2005

De manière générale, la représentation du cycle saisonnier des pluies observées et simulées au cours de la période de référence (1981–2005) montre que les deux modèles RCA4-MPI-ESM-LR et RCA4-CNRM-CM5 ont du mal à reproduire les pluies mensuelles observées à l'échelle du bassin du côtier oranais (Fig. 2). Le biais estimé entre les pluies observées et simulées présente en moyenne une variation de −6 % à −70 % (Tableau 2), il dépasse souvent −50 % pendant les mois d'hiver. Ainsi, les modèles CNRM et MPI sous-

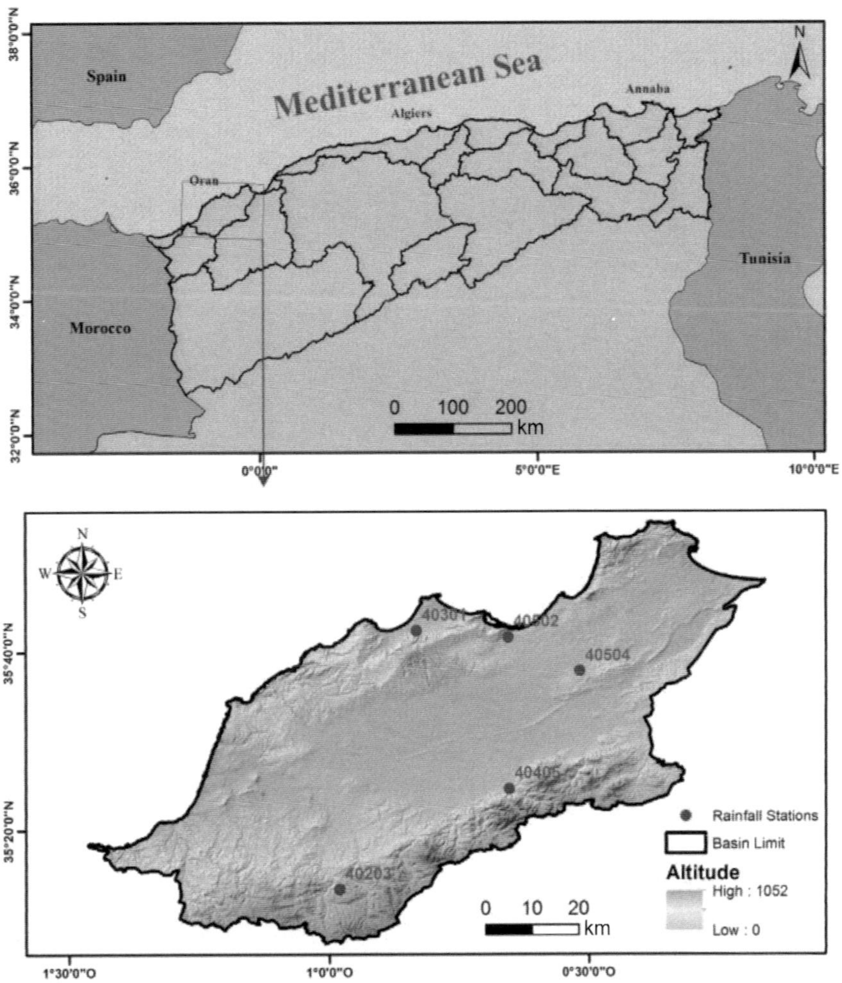

Figure 1. Localisation géographique des postes pluviométriques.

Tableau 1. Caractéristiques géographiques et statistiques des stations pluviométriques de la zone d'étude.

Code	Nom stations	X (km)	Y (km)	Z (m)	P. Moy (mm)
040203	Oued berkeche	165.1	220.2	450	352,69
040301	Bousfer	180.4	273.3	84	317,05
040405	Tamzourah	195.5	239.9	189	331,73
040502	Oran pépinière	196.4	271.3	82	344,51
040504	El braya	208.7	263.9	110	248,86

estiment fortement les pluies à l'échelle du bassin du Côtier oranais en période humide, alors que le biais estimé en période sèche montre une surestimation des pluies. Les faibles quantités de pluies enregistrées en cette période sèche font que les biais relatifs estimés sont plus élevés.

Compte tenu de ces résultats, une correction du biais est primordiale afin de corriger les données brutes simulées par les modèles climatiques régionaux et améliorer la qualité des simulations futures.

4.2 Correction de biais et projections futures

Afin d'analyser l'évolution future des pluies après application des méthodes de correction de biais, un coefficient d'échange est estimé entre les pluies simulées futures corrigées et les pluies observées au niveau de chaque station (Fig. 3). L'analyse de la variabilité future des pluies concerne uniquement la période humide de la zone d'étude (Octobre–Mars) d'autant plus que les quantités de pluies les plus im-

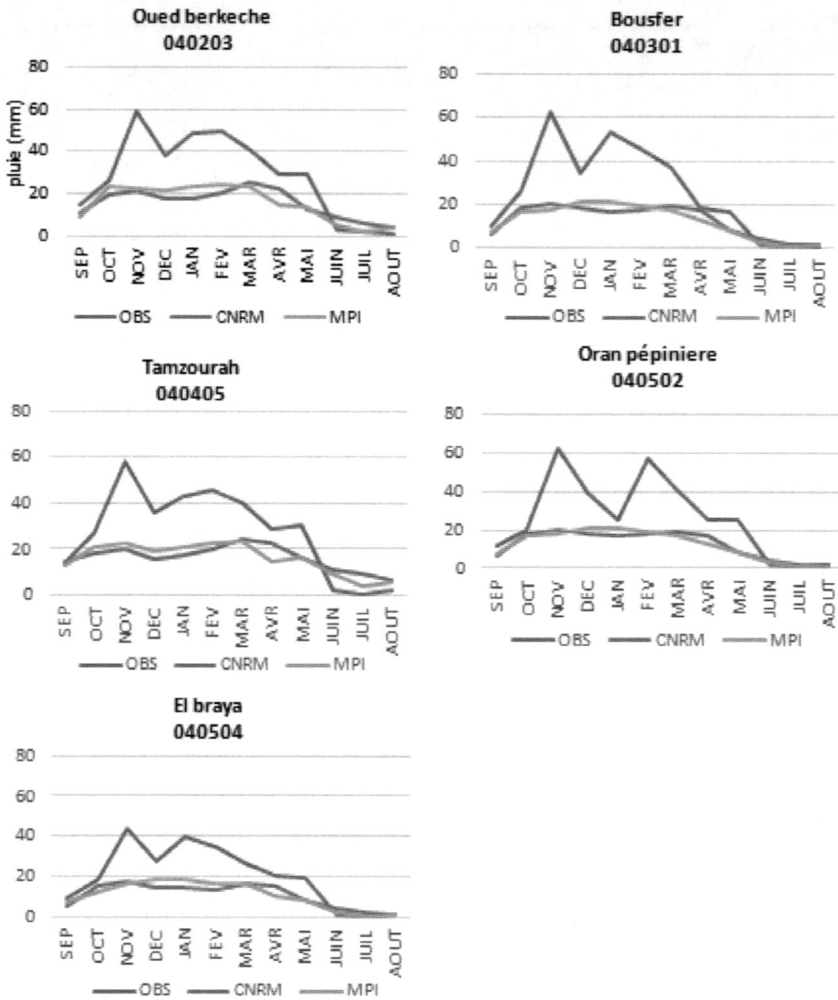

Figure 2. Représentation du cycle saisonnier par les MCR pour chaque station.

portantes qui participent aux écoulements sont enregistrées au cours de cette période (Fig. 2).

De manière générale, les variations des pluies simulées par le modèle MPI pour le scénario climatique RCP 4.5 au cours de la période 2075–2099 sont pratiquement similaires pour les deux méthodes de corrections et montrent une diminution de la pluviométrie comprise entre −19 % et −38 % à l'échelle du bassin oranais. Pour le RCP 8.5, les deux méthodes de correction montrent des changements différents des pluies pour l'ensemble des stations qui oscillent entre −27 % et −47 % pour la méthode Delta et entre −41 % et −65 % pour la méthode Q-Q.

Les simulations du modèle CNRM, sous le scénario RCP 4.5 mettent en évidence une diminution de −7 % à −61 % estimée par la méthode Delta à l'exception de la station 040502 (Oran pépinière) qui montre une augmentation de +34 %. Les simulations issues de la méthode Q-Q montrent une diminution des pluies de −25 % à −41 % exceptée pour les stations 040502 et 040504 qui indiquent respectivement une augmentation de +68 % et +48 % des pluies d'ici la fin du 21ème siècle. Pour le RCP 8.5, les projections futures corrigées par la méthode Delta révèlent une réduction des pluies de −31 % à −47 %, alors que la méthode Q-Q simule une réduction de −34 % à −59 % des pluies d'ici la fin du 21ème siècle.

L'ensemble de ces résultats témoignent de la vulnérabilité du bassin côtier oranais à la variabilité climatique. Ainsi, ils permettent d'appréhender, de manière globale, non seulement les conséquences sur la disponibilité des ressources en eau mais aussi sur l'ensemble des activités tributaires à l'eau et à enjeu socio-économique telles que : l'agriculture, l'industrie, la santé et le tourisme. Néanmoins, pour une meilleure appréciation des impacts socio-économiques, une analyse plus approfondie en considérant plusieurs modèles climatiques et paramètres climatiques, est recommandée.

Tableau 2. Biais estimé entre les pluies observées et simulées au cours de la période de référence 1981–2005.

MCR	Sept.	Oct.	Nov.	Déc.	Jan.	Fév.	Mars	Avril	Mai	Juin	Juil.	Aout
Oued berkeche (040203)												
CNRM	−27 %	−24 %	−64 %	−53 %	−63 %	−58 %	−39 %	−21 %	−55 %	219 %	204 %	171 %
MPI	−38 %	−12 %	−62 %	−42 %	−53 %	−52 %	−42 %	−50 %	−53 %	93 %	11 %	−38 %
Bousfer (040301)												
CNRM	28 %	39 %	−36 %	3 %	−37 %	−23 %	4 %	80 %	4 %	446 %	430 %	825 %
MPI	37 %	28 %	−44 %	21 %	−20 %	−17 %	−7 %	36 %	3 %	230 %	119 %	37 %
Tamzourah (040405)												
CNRM	11 %	−32 %	−65 %	−58 %	−59 %	−57 %	−41 %	−22 %	−46 %	332 %	857 %	245 %
MPI	−6 %	−24 %	−61 %	−46 %	−52 %	−52 %	−41 %	−50 %	−47 %	257 %	358 %	−6 %
Oran pépinière (040502)												
CNRM	−42 %	−8 %	−67 %	−53 %	−32 %	−69 %	−51 %	−33 %	−66 %	89 %	8 %	−25 %
MPI	−37 %	−16 %	−71 %	−46 %	−52 %	−66 %	−56 %	−49 %	−66 %	14 %	−56 %	−37 %
El braya (040504)												
CNRM	−38 %	−21 %	−60 %	−49 %	−64 %	−59 %	−37 %	−24 %	−58 %	252 %	1065 %	99 %
MPI	−23 %	−33 %	−63 %	−34 %	−53 %	−53 %	−40 %	−47 %	−57 %	137 %	282 %	−23 %

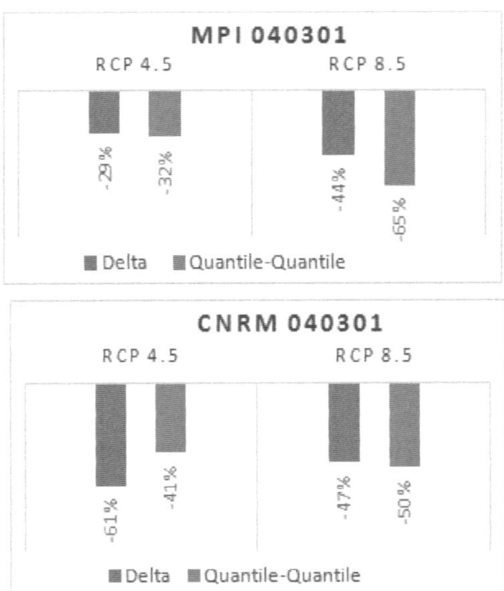

Figure 3. Variation des pluies futures (2075–2999) par rapport aux observations (1981–2005) après correction de biais par la méthode Delta et Q-Q (cas de la station Bousfer 040301).

5 Conclusions

Au terme de cette recherche, il convient de retenir que d'une manière générale, les résultats de simulations ont mis en évidence une réduction de −15 % à −40 % des pluies à l'exception de la station côtière d'Oran pépinière qui montre une augmentation d'environ 34 % des pluies à l'horizon 2099. Fort des résultats obtenus, cette recherche est une contribution pour la mise en place d'un outil d'aide à la planification et l'élaboration d'un plan d'actions opérationnel pour la gestion durable des ressources en eau dans le secteur d'étude. Toutefois, pour proposer les mesures d'adaptations les plus adéquates possibles, il est nécessaire de tenir compte des points suivants ; (i) des erreurs liées aux modèles climatiques régionaux et aux méthodes de corrections de biais utilisées ; (ii) des paramètres climatiques susceptibles d'affecter sensiblement l'hydrologie du bassin versant et avoir des répercussions sur les activités agricoles à savoir ; l'évapotranspiration ainsi que l'intensité et la fréquence des pluies.

Disponibilité du code. Le calcul est basé sur les packages du langage R, qmap, dont les informations supplémentaires sont disponibles dans Gudmundsson (2012) et https://cran.r-project.org/web/packages/qmap/index.html, la date du dernier accès : 19 avril 2020.

Disponibilité des données. Les données in-situ sont fournies par l'Agence Nationale des Ressources Hydraulique (ANRH). Les données projetées des modèles RCA4-MPI-ESM-LR et RCA4-CNRM-CM5 sont fournies par l'étude coordonnée de recherche sur le climat à échelle régionale sur l'Afrique (CORDEX-Africa) https://esg-dn1.nsc.liu.se/search/cordex/, la date du dernier accès : 4 avril 2020) (SMHI, 2020).

Collaborateurs. ST, AZ et NM ont conceptualisé le travail ; ST et AZ ont élaboré la méthodologie. AZ a fourni les simulations des modèles climatiques avec et sans correction des biais, et a supervisé le travail. NM a participé au traitement et à l'analyse des données. ST a fait l'analyse des données ; a rédigé le manuscrit préliminaire ; a fait la révision et l'édition de l'article. AZ a participé à la révision et à l'édition de l'article. la version finale de l'article a été approuvé par l'ensemble des auteurs.

Intérêts concurrents. Les auteurs déclarent qu'ils n'ont aucun conflit d'intérêts.

Clause de non-responsabilité. Publisher's note : Copernicus Publications remains neutral with regard to jurisdictional claims in published maps and institutional affiliations.

Déclaration du numéro spécial. This article is part of the special issue "Hydrology of Large River Basins of Africa". It is a result of the 4th International Conference on the "Hydrology of the Great Rivers of Africa", Cotonou, Benin, 13–20 November 2021.

Remerciements. Les auteurs souhaitent exprimer leur gratitude envers l'Agence Nationale des Ressources Hydrauliques (ANRH) pour avoir fourni les données pluviométriques sur lesquelles sont basées les analyses de cette étude. Nos remerciements sont adressés au Groupe de travail sur le climat régional du Programme mondial de recherche sur le climat, le CORDEX-Afrique (Coordinated Regional climate Downscaling Experiment), initiative du projet CMIP5 pour le Programme mondial de recherche sur le climat. Nous remercions également le Rossby Center pour la production et la mise à disposition de leur modèle atmosphérique régional, sortie RCA4. Enfin, nous tenons à remercions les reviewers pour leurs commentaires constructifs qui ont permis d'améliorer le manuscrit.

Références

ANRH : Impact des changements climatiques sur les ressources en eaux. Note de synthèse, 12p, 2007.

Bekkoussa, B., Meddi, M., and Jourde, H. : Forçage climatique et anthropique sur la ressource en eau souterraine d'une région semi-aride : cas de la plaine de Ghriss (Nord-Ouest algérien), Sécheresse 18, 173–184, 2008.

Colin, J., Déqué, M., Radu, R., and Somot, S. : Sensitivity study of heavy precipitation in Limited Area Model climate simulations : influence of the size of the domain and the use of the spectral nudging technique, Tellus A, 62, 591–604, 2010.

Déqué, M. : Frequency of precipitation and temperature extremes over France in an anthropogenic scenario : Model results and statistical correction according to observed values, Glob. Plane. Change, 57, 16–26, 2007.

Dobler, A. and Ahrens B. : Precipitation by a regional climate model and bias correction in Europe and South Asia, Meteorol. Z., 17, 499–509, https://doi.org/10.1127/0941-2948/2008/0306, 2008.

Fang, G. H., Yang, J., Chen, Y. N., and Zammit, C. : Comparing bias correction methods in downscaling meteorological variables for a hydrologic impact study in an arid area in China, Hydrol. Earth Syst. Sci., 19, 2547–2559, https://doi.org/10.5194/hess-19-2547-2015, 2015

GIEC : Changements climatiques 2014 : Rapport de synthèse, Contribution des Groupes de travail I, II et III au cinquième Rapport d'évaluation du Groupe d'experts intergouvernemental sur l'évolution du climat, Sous la direction de l'équipe de rédaction principale, édité par : Pachauri, R. K. et Meyer, L. A., GIEC, Genève, Suisse, p. 161, 2014.

Gudmundsson, L. : qmap : statistical transformations for postprocessing climate model output, R package version 1.0–3 [code], available at : http://cran.r-project.org/web/packages/qmap/ (last access : 19 April 2020), 2012.

Lenderink, G., Buishand, A., and van Deursen, W. : Estimates of future discharges of the river Rhine using two scenario methodologies : direct versus delta approach, Hydrol. Earth Syst. Sci., 11, 1145–1159, https://doi.org/10.5194/hess-11-1145-2007, 2007.

Ly, M., Segnon, A.C., D'haen, S., Totin, E., Noblet, M., Camara, I., and Pfleiderer, P. : Comprendre et interpréter les sorties de modèles climatiques pour la conduite des études de vulnérabilité : Guide à l'attention des praticiens, Climate Analytics gGmbH, Berlin, 21 pp., 2019.

Meddi, M., Assani, A. A., and Meddi, H. : Temporal variability of annual rainfall in the Macta and Tafna catchments, Northwestern Algeria, Water Resour. Manag., 24, 3817–3833, 2010.

Mendez, M., Maathuis, B., Hein-Griggs D., and Alvarado-Gamboa, L.F. : Performance Evaluation of Bias Correction Methods for Climate Change Monthly Precipitation Projections over Costa Rica, Water, 12, p. 482, https://doi.org/10.3390/w12020482, 2020.

SMHI (Swedish Meteorological and Hydrological Institute) : ESGF@LiU/CORDEX [data set], available at : https://esg-dn1.nsc.liu.se/search/cordex/, last access : 4 April 2020.

Taibi, S., Meddi, M., Souag, D., and Mahe, G. : Évolution et régionalisation des précipitationspluies au nord de l'Algérie (1936–2009), Climate and land surface changes in hydrology, IAHS Publ, 359, 191–197, 2013.

Teutschbein, C. and Seibert, J. : Bias correction of regional climate model simulations for hydrological climate-change impact studies : Review and evaluation of different methods, J. Hydrol., 456, 12–29, 2012.

Tong, Y., Gao, X., Han, Z., Xu, Y., Xu, Y., and Giorgi, F. : Bias correction of temperature and precipitation over China for RCM simulations using the QM and QDM methods, Clim. Dynam., 57, 1425–1443, https://doi.org/10.1007/s00382-020-05447-4, 2020.

Zeroual, A., Assani, A. A., and Meddi, M. : Combined analysis of temperature and rainfall variability as they relate to climate indices in Northern Algeria over the 1972–2013 period, Hydrol. Res., 48, 584–595, 2017.

Zeroual, A., Assani, A. A., Meddi, M., and Alkama, R. Assessment of climate change in Algeria from 1951 to 2098 using the Köppen–Geiger climate classification scheme, Clim. Dynam., 52, 227–243, 2019.

Proc. IAHS, 384, 219–224, 2021
https://doi.org/10.5194/piahs-384-219-2021

Open Access PIAHS

Hydrology of Large River Basins of Africa

Assessment of flood discharge sensitivity to climate indexes in West Africa

Jean Hounkpè[1,2], Djigbo F. Badou[3], Aymar Y. Bossa[1], Yacouba Yira[4], Julien Adounkpè[2], Eric A. Alamou[5], Emmanuel A. Lawin[1,2], Luc O. C. Sintondji[1], Abel A. Afouda[1], and Ernest Amoussou[6]

[1]National Water Institute, University of Abomey Calavi, Abomey Calavi, BP 2008 Abomey Calavi, Bénin
[2]West Africa Science Service Centre on Climate change and Adapted Land Use, Abomey Calavi, 03 BP 526 Cotonou, Bénin
[3]Ecole d'Horticulture et d'Aménagement des espaces Verts, Université Nationale d'Agriculture, BP 43 Kétou, Bénin
[4]Applied Science and Technology Research Institute – IRSAT/CNRST, Ouagadougou, Burkina Faso
[5]Laboratoire de Géoscience, de l'Environnement et Applications (LaGEA), Université Nationale des Sciences Technologies, Ingénierie et Mathématiques, Abomey, Bénin
[6]Département de Géographie et Aménagement du Territoire (DGAT), Université de Parakou (UP), BP 123 Parakou, Bénin

Correspondence: Jean Hounkpè (jeanhounkpe@gmail.com)

Published: 16 November 2021

Abstract. Floods are natural disasters that widely affect people and goods. Its frequency and magnitude are projected to substantially increase due to the ongoing environmental change. At regional and national levels, some efforts have been made in predicting floods at a short-term range. However, the usefulness of flood prediction increases as the time lead increases. The objective of this work is therefore to investigate flood sensitivity to climate indexes in West Africa as a basis for seasonal flood forecasting. The methodology consists of optimizing the relationship between Annual Maximal Discharge (AMD), a proxy for flood discharge and various climate indexes using correlation coefficient, linear regression and statistical modeling based on 56 river gauging stations across West Africa. The climate indexes considered are the Sea Surface Temperature (SST) of the Tropical Northern Atlantic (TNA), SST of the Tropical Southern Atlantic (TSA), the Sea Level Pressure (SLP) of the Southern Oscillation Indexes (SOI) and the detrended El-Nino Southern Oscillation indexes. It was found that SOI/SLP indexes are the most strongly related to the AMD for the investigated stations with generally high, positive, and statistically significant correlation. The TSA/SST indexes indicated both positive and negative statistically significant correlations with river discharge in the region. The percentage change in AMD per unit change in SOI/SLP for most of the statistically significant stations is within 10 % and 50 % indicating a strong relationship between these two variables. This relationship could serve as a basis for seasonal flood forecasting in the study area.

1 Introduction

Sea Surface Temperature (SST) plays a remarkable role in spatial and temporal rainfall variability world widely. The teleconnection between different phases of ENSO or El-Nina and rainfall variability over Africa have been found in many studies (Nicholson and Selato, 2000; Nicholson and Kim, 1997). A clear relationship between rainfall and SST has been demonstrated for Eastern Africa (Mutai and Ward, 2000), Sahelian Africa (Dyer et al., 2017; Giannini et al., 2004) and Western Guinean Africa (Balas et al., 2007). Warming in the Indian ocean induces an increase in the Sahelian rainfall while the opposite implies a decrease in the Sahelian rainfall (Dyer et al., 2017). In contrast to the Sahelian region, the relationships between rainfall and SST in

West Guinean Africa are not generally symmetric as the same SST pattern may enhance rainfall in one season, but reduce it in the following season (Balas et al., 2007). Factors inducing wet conditions are not necessarily the opposite or the reverse of the one leading dry condition.

Likewise, the links between extreme rainfall in West Africa and the large circulation pattern have received attention. Diatta et al. (2020) show that extreme precipitation in the Sahel and Guinean coast is strongly associated with El Niño 3.4 index (NINO.3.4) and other climate indexes. Extreme rainfall pattern over the Sahel is likely influenced by El Niño-Southern Oscillation (ENSO) and Mediterranean variability (Diakhaté et al., 2019) while extreme precipitation in Ghana is shown to be significantly and positively correlated with the Atlantic Ocean SST and negatively correlated with the Pacific and Indian basin SSTs (Atiah et al., 2020). The teleconnection between rainfall characteristics (including extremes) and climate indexes such as SST and SLP could serve as a basis for seasonal rainfall forecasting in West Africa during the regional climate outlook forum (http://acmad.net/rcc/presao.php, last access: 15 February 2020).

The knowledge of extreme rainfall is a good proxy for flood assessment. However, other flood-related characteristics such as high river discharge, flood magnitude, and flood loss can also be influenced by the global circulation pattern. At a global scale, flooding duration appears to be more sensitive to the ENSO index than for flood frequency (Ward et al., 2016) and ENSO exerts strong and widespread influences on both flood hazard and flood risk (Ward et al., 2014c). While the linkage between flood factors and climate indicators has been extensively investigated in Europe, Asia and America (See a review by Kundzewicz et al., 2019), information on this teleconnection is very limited in Africa and mainly in West Africa. Li et al. (2016) analyzed flood events for 55 countries in Africa relatively to ENSO (SOI) years and found a statistically significant relationship at a continental level. A strong linear correlation between annual discharges and ENSO was found in South Africa (Alemaw and Chaoka, 2006) and the Nile river basin (Siam and Eltahir, 2015). To the best of our knowledge, no study specifically investigated this topic in West Africa. There is therefore a need for filling this gap. The objective of this work is to assess the sensitivity of the annual maximal river flow of West Africa to different climate indexes and explore their use for seasonal flood forecasting.

2 Data and methodology

Data used are the discharge data obtained from the Global Runoff Data Center (GRDC) and some national hydrological services in West Africa. The climate indexes used are the tropical southern Atlantic indexes (TSA; NOAA, 2020), the tropical northern Atlantic indexes (TNA; NOAA, 2020), the El-Nino southern oscillation indexes (ENSO, JISAO, 2011) and the Southern oscillation indexes (SOI; Climatic Research Unit, 2019).

The block maxima approach (BMA) over the calendar year was used for discharge data sampling since high discharge is an excellent proxy for the river flood magnitude. However, to avoid a preponderant missing data effect on the BMA, the period of extraction of the annual maximal discharge was limited to the rainfall season which spans generally from June to mid-November in West Africa. In addition, the boxplot approach was used to identify possible outliers in the extracted data following the approach of Chambers et al. (2018). For any given station, after computing the interquartile range $\mathrm{IQR} = Q(0.75) - Q(0.25)$, the lower adjacent value of the boxplot defined as the smallest value greater of equal to $Q\,(0.25) + 1.5 \times \mathrm{IQR}$ is identified. The upper adjacent value of the boxplot is defined as the largest observation lesser than or equal to $Q\,(0.75) + 1.5 \times \mathrm{IQR}$. Values below the lower adjacent are considered possible outliers due to missing data and are particularly screened.

The sensitivity of the AMD to the climate indexes was first evaluated using the Pearson correlation. For stations indicating significant correlation at 10 % level, a linear relationship between the AMD and a weighted mean of a given climate index for different time steps was established. The ith weighted mean (WMCI_i) of four-monthly climate indexes (JFMA for $i = 1$, FMAM for $i = 2, \ldots,$ MJJA for $i = 5$) was computed as

$$\mathrm{WMCI}_i = \sum\nolimits_i^{i+3} \alpha_i \times \mathrm{CI}_i$$

where CI_i are the monthly climate indexes from January ($i = 1$) to August ($i + 3 = 8$). The coefficient α_i were obtained by optimizing the absolute correlation between the logarithm of the AMD and WMCI_i.

A significant correlation between the logarithm of the Annual maximal discharge of the jth station $\ln\left(Q_{\max j}\right)$ and the climate indexes (WMCI_i) imply that there is a linear relationship between these two variables. Therefore, when both correlation is statistically significant, the associated sensitivity β to the variation of climate indexes is determined following the approach of Ward et al. (2014) such as

$$\ln\left(Q_{\max j}\right) = \beta_{i,j} \times \mathrm{WMCI}_i + \alpha_{i,j}$$

with $\beta_{i,j}$ and $\alpha_{i,j}$ the regression coefficients. A unit change in WMCI_i is associated with an average change of $100 \times (\exp(\beta_{i,j}) - 1)$ in $Q_{\max j}$ and it can be referred to as "sensitivity" in percentage after Ward et al. (2014a).

3 Results and discussions

3.1 Selected River basins

In addition to the criteria indicated in the methodology, the river gauging stations were further selected based on a minimum data length of 30 years leading to the 56 stations (out of

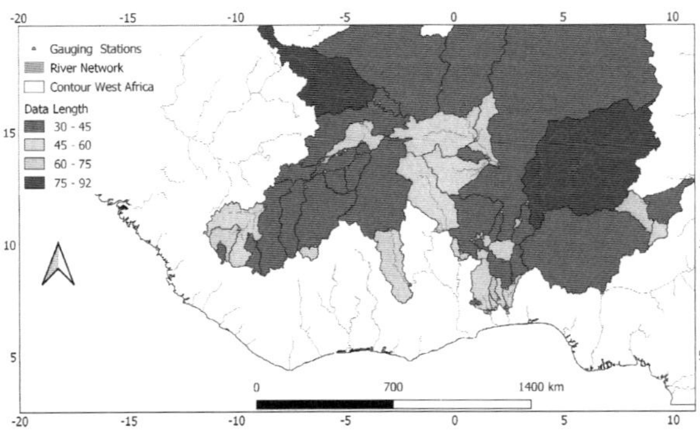

Figure 1. Selected West Africa basins, the associated data length, and gauging stations.

150 initial stations) spread throughout the region. The study area encompasses many transboundary basins such as the Niger basin, the volta basin, and the complex Ouémé-Mono-Couffo basin. Figure 1 presents the data length equivalent to the number of years of data available. Among these stations, 50 % have a data length between 30 and 45 years, while for 45 to 60 years, it is 32.1 % of the stations. In the same way, 14 % of the stations indicated a data length between 60 and 75 years. For the data with the highest length (between 75 and 92 years), their percentage is about 3.6 % and these stations are in the Niger river.

3.2 Relationship between annual maximal discharge and climate indexes

Figure 2 displays the correlation coefficient between the annual maximal discharge over the selected basins and the four-monthly weighted mean of four climate indexes (TSA/SST, TNA/SST, SOI/SLP and El-Nino) with information shown for statistically significant correlation. The strength of the relationship with the AMD depends on the climate indexes considered and the associated period. The strongest teleconnection with the river discharge in West Africa was found for the SOI/SLP while the weakest was obtained for TNA/SST indexes.

The sea level pressure of the southern oscillation indexes shows a very high and positive correlation with the AMD in West Africa compared to other climate indexes implying that estimated flood discharge based on these indexes could provide a substantial basis for seasonal flood forecasting. The SST indexes of the TSA indicated a good relationship with river discharge in the region with no general spatial pattern between the positive and negative correlation implying diverging effects. The maximal number of stations indicating a significant correlation with climate indexes are obtained in the period JFMA for TNA/SST (37.5 %), in MAMJ and AMJJ for the detrended El-Nino indexes (44.6 %), in AMJJ for TSA/SST (53.6 %) and, in MAMJ for SOI/SLP (83.9 %). The SOI/SLP outperformed the other climate indexes irrespective of the period considered. Its minimum number of stations indicating significant correlation is greater than the maximum number of significant stations indicated by the other climate indexes.

When considering the climate indexes of the first four months of the year (JFMA), obtaining a significant and relatively high correlation with a given station implies that there are possibilities to estimate AMD of that station several months ahead of flood season which generally is between July and October depending on the station location. The time lead for the forecast would be two and one months for the four-monthly climate indexes of respectively FMAM and MAMJ.

The statistically significant correlation of the SST indexes of the Tropical Southern Atlantic did not show any clear direction about the sign of the correlation in contrast to the TNA/SST and SOI/SLP which generally are positively correlated with the AMD. The El-Nino indexes indicated a mostly negative correlation with the AMD. The positive correlation implies that AMD increases with an increase in the climate indexes while the negative correlation implies that a decrease in AMD is associated with an increase in climate indexes. A combination of many climate indexes would help to improve the obtained correlation. Generally, all the AMDs are statistically correlated with at least one climate index.

3.3 Flood discharge sensitivity

Figure 3 presents the percentage change in annual maximal discharge per unit change in climate indexes. The flood discharge sensitivity varies substantially across basins and this sensitivity increases as the SOI period gets closer to the flood period. Generally, in the basins with significant correlation, AMD varies mostly between 10 % and 50 % per unit change in SOI/SLP implying a positive relationship. This positive relationship is consistent with the findings of Ward et al. (2014a) for the West Africa region. The highest sensitivities are spread throughout the region with an emphasis on the high latitude. The strength of the teleconnection between the SOI/SLP and AMD varies depending on the period considered for the SOI/SLP. For instance, the highest percentage (in term of the number of stations) for the sensitivity between 10 % and 50 % is 62.5 (Table 1) obtained while considering the SOI/SLP for March to June which correspond to the first rainy season in the West Africa Guinean Coast.

For the SOI of January to April, 39.3 % of the stations exhibited a sensitivity between 10 % and 50 % implying that substantial variation in the AMDs in the region could be explained ahead using the SOI of January to April. This opens the door for exploring seasonal annual maximal discharge forecasting, a strong proxy for seasonal flood forecasting. This kind of relationship has been established between rainfall amount in the Sahel and the SST anomaly in the Equatorial Atlantic upwelling, and the position of the Saint-Helena high-pressure center (Mahe and Citeau, 1993).

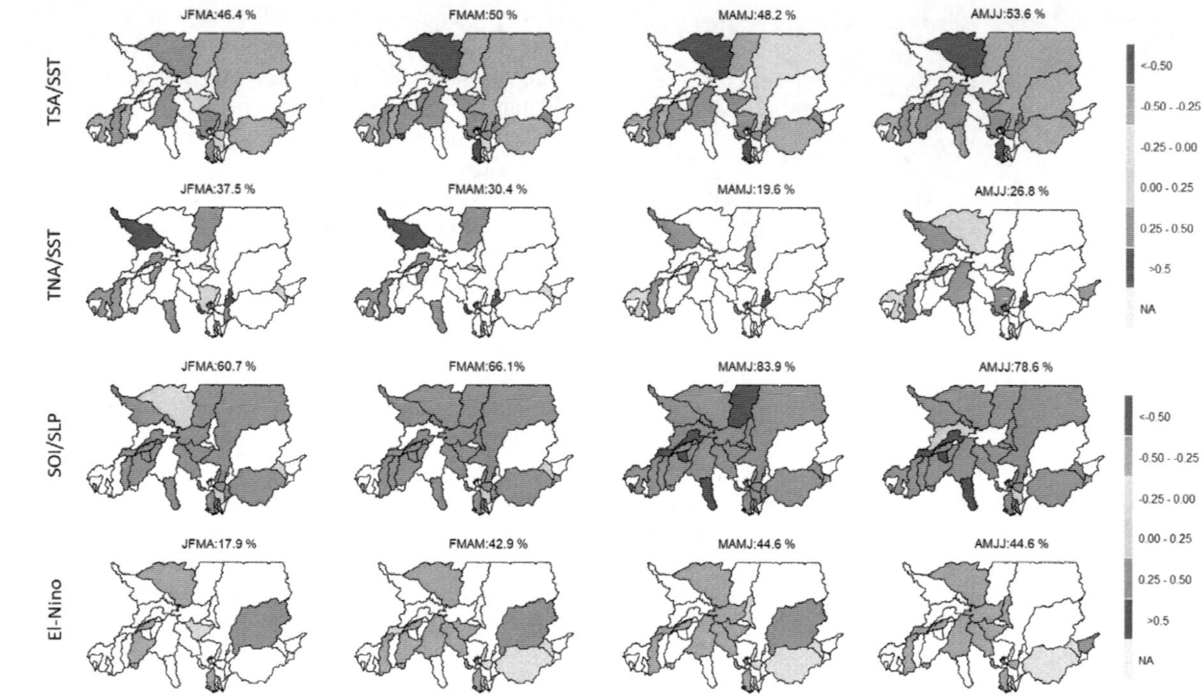

Figure 2. Correlation coefficient between the four monthly climate indexes and the logarithm of the annual maximal.

Table 1. Percentage of stations with the associated sensitivity level computed using the SOI/SLP (See Fig. 3).

Sensitivity (%)	< -50	$-50 : -10$	$-10 : 0$	$0 : 10$	$10 : 50$	> 50	Total
JFMA	0	0	0	21.4	39.3	0	60.7
FMAM	0	1.8	0	14.3	48.2	1.8	66.1
MAMJ	0	1.8	0	17.9	62.5	1.8	83.9
AMJJ	0	1.8	0	26.8	48.2	1.8	78.6

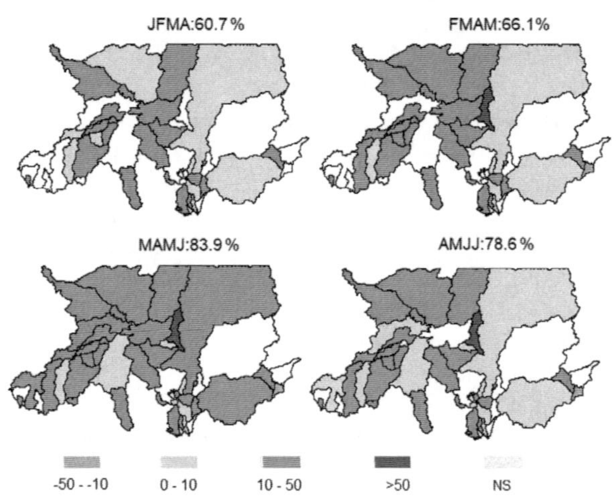

Figure 3. Percentage change of annual maximal discharge per unit change SOI/SLP for stations with significant correlation at 10 %.

Strong connections between precipitations in West Africa and the oceanic zones of the Atlantic frontage of Africa located on the course of the African monsoon were found by (Hamatan et al., 2004) and this relationship helps improve substantially the scores of the seasonal rainfall forecast in the region. However, estimating the discharge based on the climate indexes through a direct relationship is beyond the scope of this work.

4 Conclusions

This study evaluated the relationship between the annual maximal discharge (AMD) and four climate indexes using statistical methods. The results indicated a strong relationship between AMD and the climate indexes. The Sea Level Pressure (SLP) of the Southern Oscillation indexes (SOI) is the most correlated and statistically significant (10 % level) with flood discharges in the region irrespective of the period considered. Most of the sensitivity indexes meaning the change in AMD per unit change in climate indexes are within 10 % and 50 % confirming the strong relationship previously found. The established strong relation between climate indexes and AMD could serve as a basis for exploring the seasonal annual maximal discharge forecasting, a strong proxy for seasonal flood forecasting. A promising research path would be to use the potential predictability of climate indexes such as SOI to provide probabilistic estimates of flood hazard with lead times up to several months (Ward et al., 2014a, b).

Data availability. The discharge data is not publicly available due to research institution policy. The climate indexes used such as the tropical southern Atlantic indexes (https://psl.noaa.gov/data/correlation/tna.data, last access: 15 February 2020, NOAA, 2020), the tropical northern Atlantic indexes (https://psl.noaa.gov/data/correlation/tsa.data, last access: 15 February 2020, NOAA, 2020), the El-Nino southern oscillation indexes (http://research.jisao.washington.edu/data_sets/globalsstenso/, last access: 15 February 2020, JISAO, 2011) and the Southern oscillation indexes (SOI, https://crudata.uea.ac.uk/cru/data/soi/, last access: 15 February 2020, Climatic Research Unit, 2019) were freely downloaded from the indicated links.

Author contributions. The contribution of JH to the work consists in the collection and processing of hydrological data and climate indexes, design of graphic materials, presentation and discussion of the results. The contribution of DFB, AYB, YY and EAL is the formulation of the problem, preparation of the manuscript, analysis of literature sources and discussion of the results. JA, EAA, LOCS, AAA, EA reviewed the manuscript and reinforced its different sections.

Competing interests. The contact author has declared that neither they nor their co-authors have any competing interests.

Disclaimer. Publisher's note: Copernicus Publications remains neutral with regard to jurisdictional claims in published maps and institutional affiliations.

Special issue statement. This article is part of the special issue "Hydrology of Large River Basins of Africa". It is a result of the 4th International Conference on the "Hydrology of the Great Rivers of Africa", Cotonou, Benin, 13–20 November 2021.

Financial support. This research has been supported by the Bundesministerium für Bildung und Forschung. This research was supported by DAAD within the framework of the climapAfrica programme with funds of the Federal Ministry of Education and Research. The publisher is fully responsible for the content.

References

Alemaw, B. F. and Chaoka, T. R.: The 1950-1998 warm ENSO events and regional implications to river flow variability in Southern Africa, Water SA, 32, 459–464, 2006.

Atiah, W. A., Mengistu Tsidu, G., Amekudzi, L. K., and Yorke, C.: Trends and interannual variability of extreme rainfall indices over Ghana, West Africa, Theor. Appl. Climatol., 140, 1393–1407, https://doi.org/10.1007/s00704-020-03114-6, 2020.

Balas, N., Nicholson, S. E., and Klotter, D.: The relationship of rainfall variability in West Central Africa to sea-surface temperature fluctuation, Int. J. Climatol., 27, 1335–1349, https://doi.org/10.1002/joc, 2007.

Chambers, J. M., Cleveland, W. S., Kleiner, B., and A., T. P.: Graphical Methods for Data Analysis, Taylor & Francis Group., 2018.

Climatic Research Unit: Southern Oscillation Index (SOI), Data [online] Available from: https://crudata.uea.ac.uk/cru/data/soi/ (last access: 15 February 2020), 2019.

Diakhaté, M., Rodríguez-Fonseca, B., Gómara, I., Mohino, E., Dieng, A. L., and Gaye, A. T.: Oceanic forcing on interannual variability of Sahel heavy and moderate daily rainfall, J. Hydrometeorol., 20, 397–410, https://doi.org/10.1175/JHM-D-18-0035.1, 2019.

Diatta, S., Diedhiou, C. W., Dione, D. M., and Sambou, S.: Spatial variation and trend of extreme precipitation in west africa and teleconnections with remote indices, Atmosphere (Basel)., 11, 999, https://doi.org/10.3390/atmos11090999, 2020.

Dyer, E. L. E., Jones, D. B. A., Li, R., Sawaoka, H., and Mudryk, L.: Sahel precipitation and regional teleconnections with the Indian Ocean, J. Geophys. Res., 122, 5654–5676, https://doi.org/10.1002/2016JD026014, 2017.

Giannini, A., Saravanan, R., and Chang, P.: Oceanic Forcing of Sahel Rainfall on Interannual to Interdecadal Time Scales, Science, 19, 113–128, https://doi.org/10.1016/S1364-8152(03)00114-2, 2004.

Hamatan, M., Mahe, G., Paturel, J. E., and Amani, A.: Synthèse et évaluation des prévisions saisonnières en Afrique de l'Ouest, Sécheresse, 15, 279–286, 2004.

JISAO: Global-SST ENSO index, available at: http://research.jisao.washington.edu/data_sets/globalsstenso/ (last access: 15 February 2020), 2011.

Kundzewicz, Z. W., Szwed, M., and Pińskwar, I.: Climate variability and floods-A global review, Water (Switzerland), 11, 1399, https://doi.org/10.3390/w11071399, 2019.

Li, C., Chai, Y., and Yang, L.: Spatio-temporal distribution of flood disasters and analysis of influencing factors in Africa, Nat. Hazards, 82, 721–731, https://doi.org/10.1007/s11069-016-2181-8, 2016.

Mahe, G. and Citeau, J.: Interactions between the Ocean, Atmosphere and Continent in Africa, Related to the Atlantic Monsoon Flow: General Pattern and the 1984 Case Study, Veill. Clim. Satellitaire, 44, 34–54, 1993.

Mutai, C. C. and Ward, M. N.: East African rainfall and the tropical circulation/convection on intraseasonal to interannual timescales, J. Climate, 13, 3915–3939, https://doi.org/10.1175/1520-0442(2000)013<3915:EARATT>2.0.CO;2, 2000.

Nicholson, S. and Selato, J.: The influence of La Nina on the rainfall of the Western Pacific, Int. J. Climatol., 20, 1761–1776, 2000.

Nicholson, S. E. and Kim, J.: The relationship of the El Nino–southern oscillation to african rainfall, Int. J. Climatol., 17, 117–135, 1997.

NOAA: Linear Correlations in Atmospheric Seasonal/Monthly Averages, [online] available from: https://psl.noaa.gov/data/correlation/tna.data, https://psl.noaa.gov/data/correlation/tsa.data, last access: 15 February 2020.

Siam, M. S. and Eltahir, E. A. B.: Explaining and forecasting interannual variability in the flow of the Nile River, Hydrol. Earth Syst. Sci., 19, 1181–1192, https://doi.org/10.5194/hess-19-1181-2015, 2015.

Ward, P. J., Eisner, S., Flörke, M., Dettinger, M. D., and Kummu, M.: Annual flood sensitivities to El Niño–Southern Oscillation at the global scale, Hydrol. Earth Syst. Sci., 18, 47–66, https://doi.org/10.5194/hess-18-47-2014, 2014a.

Ward, P. J., Pelt, S. C. Van, Keizer, O. De, Aerts, J. C. J. H., Beersma, J. J., and Hurk, B. J. J. M. Van Den: Including climate change projections in probabilistic flood risk assessment, J. Flood Risk Manage., 7, 141–151, https://doi.org/10.1111/jfr3.12029, 2014b.

Ward, P. J., Jongman, B., Kummu, M., Dettinger, M. D., Weiland, F. C. S., and Winsemius, H. C.: Strong influence of El Niño Southern Oscillation on flood risk around the world, P. Natl. Acad. Sci. USA, 111, 15659–15664, https://doi.org/10.1073/pnas.1409822111, 2014c.

Ward, P. J., Kummu, M., and Lall, U.: Flood frequencies and durations and their response to El Niño Southern Oscillation: Global analysis, J. Hydrol., 539, 358–378, https://doi.org/10.1016/j.jhydrol.2016.05.045, 2016.

Proc. IAHS, 384, 225–231, 2021
https://doi.org/10.5194/piahs-384-225-2021

Open Access
PIAHS

Hydrology of Large River Basins of Africa

Caractérisation hydro-climatique, analyse comparative des termes du bilan hydrologique du bassin versant d'Ambouli (République de Djibouti)

Golab Moussa Omar[1,2], **Jean-Emmanuel Paturel**[1], **Christian Salles**[1], **Gil Mahé**[1], **and Mohamed Jalludin**[2]

[1]HSM, Univ. Montpellier, CNRS, IRD, Montpellier, France
[2]Centre d'Etudes et de Recherches de Djibouti, Institut des Sciences de la Terre, Route de l'aeroport, Djibouti-ville, Djibouti

Correspondence: Golab Moussa Omar (golabmoussa@gmail.com)

Published: 16 November 2021

Résumé. Le bassin versant exoréique de l'Oued Ambouli situé au sud-est de la République de Djibouti, s'étend sur plus de 794 km^2 soit 3.5 % de la surface du pays. Il est le plus grand bassin versant du district de Djibouti. Cet oued est soumis à des crues rares et brèves qui causent de lourds dégâts en vies humaines et économiques, accentués par l'expansion urbaine de la ville de Djibouti. Cette étude vise à calculer les différentes composantes du bilan hydrologique. Sur la période 2008–2013, ce bassin a reçu une pluie moyenne annuelle de 93.6 mm ; l'évapotranspiration réelle (ETR) sur le bassin versant est estimée à 79.6 mm/an soit 85 % des précipitations moyennes annuelles ; l'écoulement total approche un volume de 6.5×10^6 m^3/an, soit 3.7 mm/an ; par déduction, selon la formule du bilan hydrologique, l'infiltration totale est estimée à 10.3 mm/an. Ces résultats comparés aux résultats obtenus antérieurement, apparaissent comme très différents (principalement une baisse des valeurs). Les données disponibles sont très fragmentées dans le temps et ne couvrent que des périodes trop courtes pour pouvoir prétendre fournir des valeurs de référence pour un bilan hydrologique et permettre un suivi évolutif de celui-ci.

1 Introduction

La ville de Djibouti, capitale de la République de Djibouti, s'est développée sur le bassin versant de l'Oued Ambouli. Cet oued est à l'origine de crues rares et brèves avec de lourds dégâts en vie humaine et économiques qui sont accentués par l'expansion urbaine de la ville. Afin de contribuer à la prévention contre les risques d'inondations, qui surviennent de plus en plus fréquemment, et à la protection de la ressource en eau dans le bassin de l'Oued Ambouli, nous avons collecté toutes les données et tous les travaux disponibles qui permettent, dans un premier temps, d'établir un bilan hydrologique. Bien que d'une importance stratégique pour la ville de Djibouti, cet oued n'a fait l'objet que de rares suivis hydrologiques, sur des périodes très courtes et à des moments très différents de son développement sans qu'un suivi de son évolution ne puisse être établi. Cette étude permet de rassembler l'état (maigre) des connaissances sur les composantes du bilan hydrologique du bassin afin d'en faire une synthèse et d'en déduire des pistes de recherche scientifiques et/ou de données.

Nous avons rassemblé et analysé les données climatiques et hydrologiques disponibles auprès du Centre d'Etude et de Recherche de Djibouti sur la période 2008–2013 (CERD). Ces données ont permis d'établir un "instantané" du bilan hydrologique du bassin qui a, alors été, comparé aux estimations de ce bilan des travaux de la Coopération Hydrologique Allemande (CHA, 1982) et de Gamal-Eldin (1988). In fine, ces résultats seront les premiers éléments de guide pour la mise en place d'une stratégie holistique de compréhension du cycle de l'eau, depuis les eaux de surface jusqu'aux eaux souterraines. Ces premiers résultats nous permettront d'orienter la suite de nos travaux.

2 Presentation de la zone d'etude

La République de Djibouti, d'une superficie de 23 200 km^2, présente un climat tropical aride qui se caractérise par l'irrégularité et la faiblesse des précipitations annuelles, avec des températures élevées durant toute l'année, l'absence de cours d'eau pérenne et une évaporation très intense. La saison fraiche, d'octobre à mars, présente des températures modérées (entre 22 et 30 °C) et une humidité relativement élevée. La saison chaude de juin à septembre est marquée par des températures entre 30 et 40 °C et un vent sableux, sec et chaud (le Khamsin). La période intermédiaire (avril–mai) fait la transition entre saison fraiche et saison chaude.

Le bassin versant exoréique de l'Oued Ambouli, situé au sud-est de la République de Djibouti, s'étend sur plus de 794 km^2 soit 3.5 % de la surface du pays (Fig. 1). C'est le plus grand bassin versant du district de Djibouti qui accueille les deux tiers de la population de la République de Djibouti (570 000 habitants). Ce bassin versant, de par sa localisation géographique et ses ressources en eau souterraine, est d'une importance majeure pour le développement de la ville de Djibouti. En effet d'une part, le delta de l'oued dans l'Océan Indien constitue un vaste cône aplati prolongé par une presqu'île corallienne sur lequel est implantée la ville de Djibouti (SAFEGE, Consortium STDE, 2007), d'autre part, le sous-sol de la partie aval du bassin versant est constitué d'un épandage basaltique daté fin tertiaire / début quaternaire de plusieurs dizaines de mètres d'épaisseur qui contient une nappe aquifère abondante et de bonne qualité. Depuis plusieurs décennies, la quasi-totalité de l'eau nécessaire à l'alimentation de la ville de Djibouti est captée dans cet aquifère. A l'heure actuelle, le prélèvement annuel dépasse 13 millions de m^3 (Houmed Gaba, 2009) L'augmentation de la population s'accompagne d'une croissance rapide des besoins qui entraîne une surexploitation et une dégradation de cette ressource.

3 Données et methodes

3.1 Presentation des données

Les données hydrométéorologiques (principalement pluviometrique) disponibles sur le bassin d'Ambouli sont très disparates dans le temps et dans l'espace (Figs. 1 et 2). L'équipement des capteurs pluviometriques du bassin s'est fait progressivement avec une interruption des observations sur la période 1990–2008. La plus longue série disponible est la pluviométrie enregistrée à la station météorologique de Djibouti-Aérodrome en continu depuis 1953. C'est à partir de cette station que l'ETP et L'ETR ont été calculées.

Les résultats produits par la CHA ont été obtenus sur la base de données mensuelles des stations limnimétriques d'Ambouli et Oueah et des stations pluviométriques d'Ambouli, Oueah et Arta collectées sur la période 1980 à 1982 (Fig. 2). Les mesures de débit antérieures à 1980 n'ont pas pu être utilisées car elles sont très incomplètes. Les stations sont installées de manière à ne mesurer que les hauteurs d'eau supérieures à 1 m dans le cours d'eau.

Les travaux de Gamal-Eldin (1988) se sont appuyés sur le même réseau d'observation que la CHA, complété par les séries pluviometriques observées sur la période 1983 à 1986.

C'est à partir de 2008, suite aux inondations dévastatrices de 1994 et 2004, que le CERD a renforcé le suivi hydrométéorologique du bassin d'Ambouli. Sur la période 2008–2013, le bassin était équipé de :

- 8 stations pluviométriques avec pour la plupart des données infra journalières (données numérisées).
- 1 station de mesure de la température qui est la station Aérodrome (Djibouti-aérodrome 1962–2018).
- 3 stations limnimétriques (Boulleh, Kalaloho et Oueah) avec des données infra journalières (données numérisées).

Les données disponibles pour chacune des études passées (CHA, 1982 ; Gamal-Eldin, 1988) et présentes, sont trop fragmentées dans le temps et ne couvrent que des périodes trop courtes pour pouvoir prétendre fournir des valeurs de référence.

3.2 Méthodes de calcul des termes du bilan hydrologique

Simple opération comptable, le bilan hydrologique vise à établir le bilan hydrique d'une unité hydrologique définie (le bassin versant) pendant une période de temps donné. Il s'écrit :

$$P = Q + \text{ETR} + \Delta R(u+h)$$

La pluie (P) qui tombe sur le bassin versant dans un laps de temps donné soit s'écoule (Q), soit repart dans l'atmosphère par évapotranspiration (ETR), soit participe à la recharge des réserves en eau du sol (Ru) ou du sous-sol (Rh). Les variations de réserve peuvent être négatives et contribuer aux écoulements et/ou à l'évapotranspiration.

Pour déterminer la précipitation moyenne P du bassin versant, on utilisé les trois méthodes d'interpolation les plus courantes sont :

- La méthode arithmétique qui est une méthode simple qui consiste à déterminer les moyennes arithmétiques entre stations.
- La méthode des isohyètes (L'Hote, 1993), méthode graphique qui se base sur l'interpolation des moyennes de précipitation,
- La méthode des polygones de Thiessen (Schumann, 1998), méthode qui consiste à effectuer la moyenne pondérée des précipitations observées dans les différentes stations pluviométriques du bassin versant en fonction de leur aire d'influence géométrique.

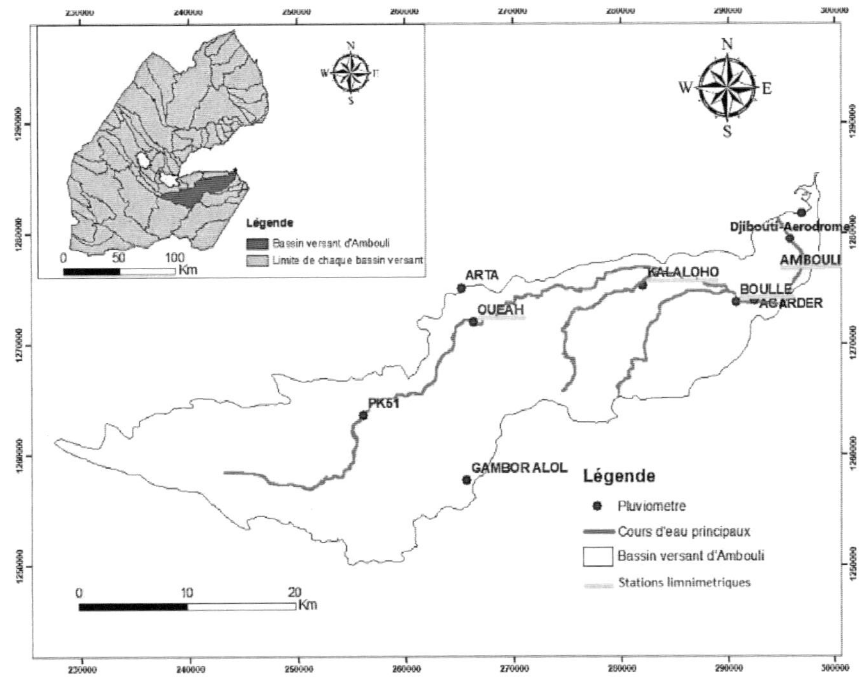

Figure 1. Bassin de l'Oued Ambouli et localisation des stations pluviométriques et limnimétriques (© CERD).

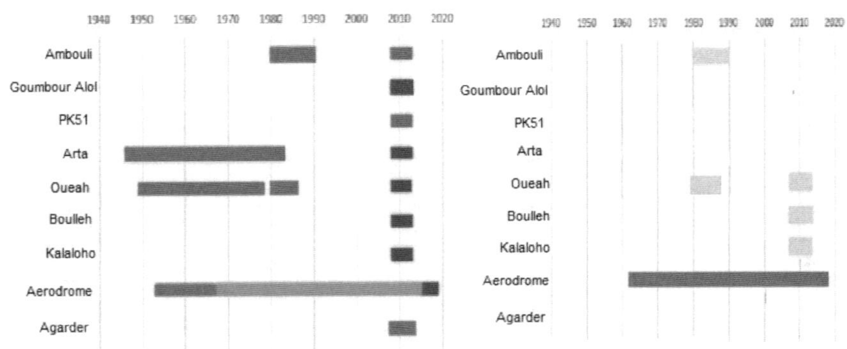

Figure 2. Données disponibles sur le bassin versant d'Ambouli. La Pluviométrie (Rouge : infra-journalier, Orange : journalier, Bleu : mensuel) ; la Limnimétrie (Jaune : infra-journalier) ; la Temperature (Vert : Journalier).

Le concept de l'évapotranspiration potentielle (ETP) a été introduit par Thornthwaite et Penman car l'évapotranspiration réelle est difficile à mesurer à l'échelle d'une parcelle ou d'une région. Pour déterminer l'évapotranspiration potentielle moyenne, les trois méthodes les plus courantes sont :

- La formule de Thornthwaite (Thornthwaite et Mather, 1955) qui n'utilise que la température et un indice thermique comme donnée climatique.
- La formule de Turc (Turc, 1961) qui fait intervenir plusieurs paramètres climatologiques (température, rayonnement et humidité).
- La formule de Penman simplifiée (Penman, 1963).

L'évapotranspiration réelle est en règle générale obtenue grâce à une réduction de l'évapotranspiration potentielle :

- La méthode du bilan de Thornthwaite (Thornthwaite et Mather, 1955) qui consiste à calculer l'ETR mensuelle à partir des précipitations moyennes mensuelles, l'ETP et la réserve facilement utilisable contenue dans le sol (RFU),
- La méthode de Turc (Turc, 1961) qui n'utilise que deux paramètres climatologiques, la température et la précipitation.

L'écoulement Q est mesuré aux stations hydrologiques d'Ambouli et Oueah (CHA, 1982 et Gamal-Eldin, 1988)

Tableau 1. Tableau comparatif des composantes du bilan hydrologique.

	Méthode	CHA (1980–1982)	Gamal-Eldin (1982–1986)	CERD (2008–2013)
Précipitation (mm/an)	Arithmétique Thiessen Isohyètes	176.5 ND* 200.0	104 ND ND	91.7 94.7 94.4
Evapotranspiration (mm/an)	Thornthwaite Turc	157 ND	85 ND	80 98
Ecoulement (mm/an et %)	Coeff. de ruissellement	11.3 mm/an 6 %	5.2 mm/an 5 %	3.7 mm/an 4 %
Infiltration (mm/an et %)	Déduite du bilan hydrologique	21.6 mm/an 11.5 %	11.4 mm/an 11 %	10.3 mm/an 11 %
Station		Ambouli Oueah	Ambouli Oueah	7 stations

* ND : non disponible.

et Boulleh, Kalaloho et Oueah pour les données du CERD (2008–2013).

L'infiltration ΔR est déduite à partir de la formule du bilan hydrologique et de la connaissance des 3 termes : P, Q et ETR pour la CHA (1982) et Gamal-Eldin (1988). Les travaux de Jalludin et Razack (2008) basés sur la modélisation numérique de l'aquifère confirment les évaluations de l'infiltration obtenues par la CHA (1982) et Gamal-Eldin (1988).

4 Analyse et comparaison des resultats issus des differentes methodes de calcul de chaque composante du bilan hydrologique

4.1 La pluie

La CHA (1982) a utilisé deux méthodes afin de déterminer la précipitation moyenne annuelle, la méthode arithmétique (176.5 mm/an) et la méthode des isohyètes (200 mm/an).

Gamal-Eldin (1988), à partir des stations d'Ambouli, Oueah et Djibouti Aérodrome, a obtenu une précipitation moyenne annuelle de 104 mm/an (Tableau 1) en utilisant la méthode arithmétique. L'écart avec le calcul précédent est important.

Sur la période 2008–2013, nous disposons d'observations plus fines et plus complètes sur la pluie (Fig. 3). Cependant, cette courte période de 5 années d'observation ne permet pas une représentativité fiable des moyennes et des extrêmes saisonniers. La pluie moyenne calculée selon les trois méthodes avoisine 94 mm (min–max). Elle est légèrement plus faible que la valeur calculée par Gamal-Eldin (1988). Nous constatons que les precipitations moyennes mensuelles des mois de juin et juillet sont très faibles, moins de 5 mm par mois (Fig. 3), contrairement aux mois de mars, avril et mai qui représentent plus de 55 % de la pluviométrie annuelle. Nous remarquons également que pour cinq des huit stations analysées, les valeurs maximales des précipitations moyennes mensuelles sont enregistrées au mois de mars et les valeurs minimales au mois de juin pour les sept stations. Nous remarquons enfin une forte irregularité spatiale des cumuls mensuels sur le bassin versant (Fig. 3).

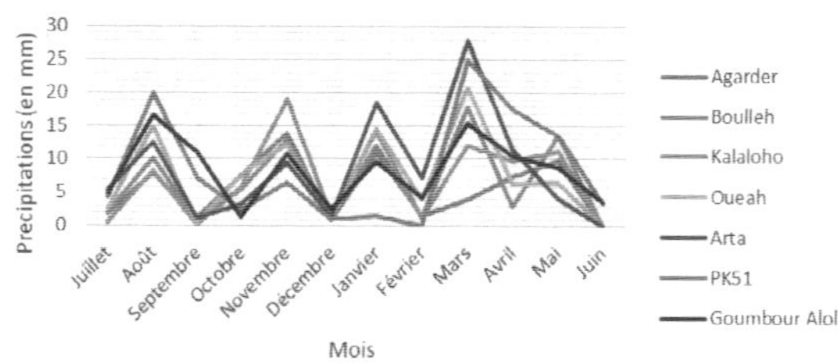

Figure 3. Variation des précipitations moyennes mensuelles dans le bassin versant d'Ambouli (2008–2013).

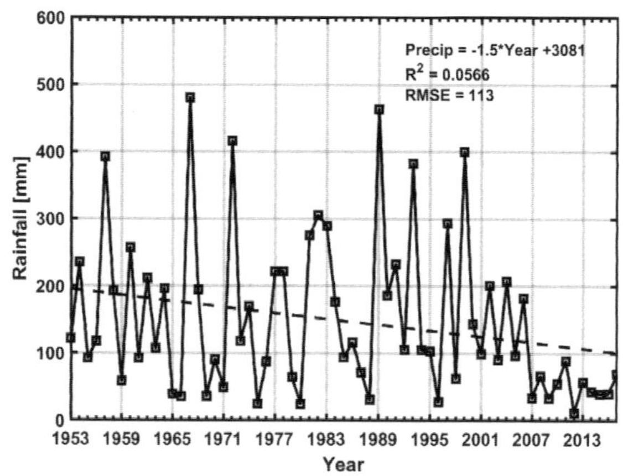

Figure 4. Evolution de la pluie annuelle de la station Djibouti-Aérodrome (1953–2018).

Concernant les précipitations annuelles, le maximum est observé à la station de Goumbour Alol (avec 103 mm/an en

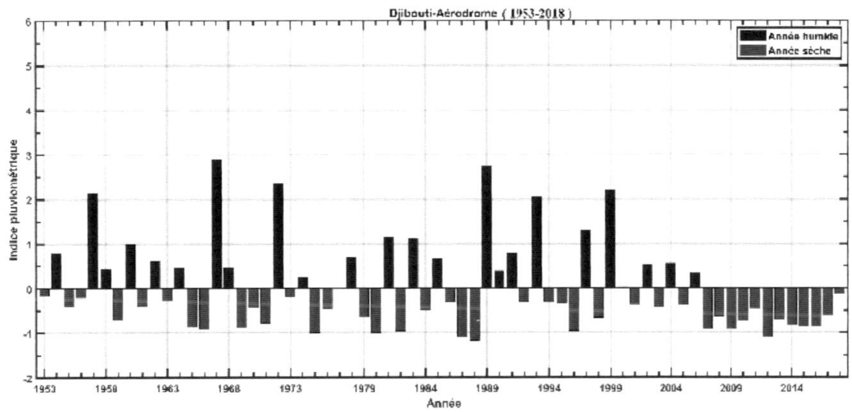

Figure 5. Anomalie pluviometrique (Indice de Lamb) de la station Djibouti aérodrome (1953–2018).

moyenne sur la période 2008–2013) et le minimum au niveau de la station Djibouti-Aérodrome (avec 55 mm). Cette variation des précipitations dépend fortement de la localisation de la station climatologique : on constate que les précipitations augmentent avec l'altitude au niveau de la zone d'etude.

Concernant l'évolution de la pluviométrie annuelle au niveau de la station Djibouti-Aérodrome (1953–2018), l'analyse de la Fig. 4 montre une forte variabilité temporelle des données pluviométriques avec un maximum de 500 mm (en 1967) et un minimum de 11 mm (en 1987).

Cette variabilité interannuelle semble être moindre depuis une dizaine d'années avec une valeur de pluie annuelle inférieure à la moyenne annuelle sur la période 1953–2018 (Djama and Gascon, 1997). Nous constatons également une tendance à la baisse (Fig. 4) et une période sèche qui commence à partir de 2007 (Fig. 5). Cet indice pluviometrique de Lamb (Fig. 5) est la difference entre la precipitation annuelle et sa moyenne puis divisé par l'ecart de type de ce dernier.

4.2 L'évapotranspiration

La CHA a mesuré l'ETP aux stations de Djibouti Aérodrome et Arta. À la station Arta, l'ETP a été mesurée à l'aide de l'évaporomètre de Piche, alors qu'à la station Djibouti-Aérodrome, elle a été mesurée sur un bac évaporatoire Classe A. Afin de rendre comparable ces données d'ETP, les ingénieurs de la CHA ont établi la relation suivante Eq. (1) :

$$E_{\text{pot,BC}} = 0.947 E_{\text{pot,piche}} - 0.096 \qquad (1)$$

$E_{\text{pot,BC}}$: Evaporation du bac évaporatoire Classe A (mm/j). $E_{\text{pot,piche}}$: Evaporation mesurée par l'evaporimetre Piche (mm/j).

Après cette conversion des données de l'ETP, la CHA a utilisé la méthode du bilan de Thornthwaite (Thornthwaite et Mather, 1955) sur les stations Djibouti Aérodrome et Arta pour estimer une ETR moyenne annuelle à 157 mm/an.

Gamal-Eldin (1988) en utilisant la même méthode que la CHA sur la période 1980–1986 a obtenu une ETR moyenne annuelle de 85 mm/an (Tableau 1). L'écart avec le calcul précédent est important en valeur absolue tout comme ce qui avait été observé sur les pluies.

L'ETP et l'ETR sont estimées au niveau des stations Djibouti Aérodrome et Arta qui toutes deux se trouvent légèrement en bordure de l'oued Ambouli. Les pluies etant très variables sur le bassin (Fig. 3), on peut s'interroger sur la représentativité de ces données utilisées par ces deux etudes pour l'élaboration des bilans hydrologiques.

Les données de 2008–2013 donnent une ETR moyenne annuelle de 80 mm/an (Tableau 1) en utilisant la méthode de bilan de Thornthwaite (Thornthwaite et Mather, 1955). La méthode de Turc (Turc, 1961) donne une ETR moyenne de 98.5 mm/an. On remarquera que cette valeur moyenne de l'ETR est supérieure à la précipitation moyenne annuelle (93.6 mm). De ce fait cette formule s'avère être inadéquate, mais en revanche ce résultat permet de constater que la valeur de l'ETR se rapproche beaucoup de celle de la précipitation annuelle moyenne.

Il ressort de cette analyse une grande variabilité de la valeur de l'ETR dans le temps : l'ETP obtenue à Djibouti-Aérodrome avec la formule de Penman simplifiée est de 1800 mm/an. Lorsque nous comparons les valeurs calculées avec cette formule à l'evaporation mesurée nous constatons que ces valeurs sont proches contrairement aux autres methodes de calculs. Cette variabilité est principalement due au pas de temps mensuel (dans un climat aride) car il n'y a pas d'adéquation entre le régime des pluies (sur quelques jours de pluie seulement) et l'évapotranspiration car dans un pas de temps mensuel, on sous-estime en général les pluies efficaces.

4.3 Ecoulement

Pour évaluer le ruissellement produit par ce bassin versant, la CHA (1982) et Gamal-Eldin (1988) ont analysé les crues événement par événement et en ont déduit un coefficient d'écoulement moyen aux stations limnimétriques Ambouli et Oueah. En s'appuyant également sur une étude antérieure (Hauquin, 1978), la CHA (1982) a obtenu une valeur de coefficient d'ecoulement moyen de 6 %. Gamal-Eldin (1988), prenant en considération les études d'Hauquin (1978) et celle de CHA (1982) et les mesures de débits effectuées entre 1980 et 1986, parvient à une valeur de coefficient d'ecoulement de 5 %. Mais ce coefficient est très influencé par certains facteurs tels que l'imperméabilisation de la surface du sol et la précision des mesures de débit. Pour notre part, nous n'avons tenu compte que des coefficients d'ecoulement calculés par la CHA de 1980 à 1982 car ces calculs ne présentaient aucune lacune et avons obtenu une valeur de coefficient de ruissellement de 4 %.

Malgré une variabilité temporelle de ces données, ces valeurs de coefficient d'ecoulement sur une moyenne de valeurs évènementielles sont du même ordre de grandeur.

4.4 Infiltration

La CHA (1982) et Gamal-Eldin (1988) ont déterminé deux types d'infiltration, l'infiltration des eaux pluviales sur les surfaces des versants et l'infiltration dans les lits de l'oued. Ne disposant que de peu de données, l'association des deux types d'infiltration en infiltration totale a été nécessaire afin de déterminer le bilan hydrologique. La CHA (1982) et Gamal-Eldin (1988) ont effectué des mesures directes d'infiltration sur le bassin (Ambouli et Oueah) et noté la variabilité de l'ETR d'un mois à un autre. Le calcul de l'infiltration totale s'est fait par déduction de la formule du bilan hydrologique. Ainsi la CHA (1982) a obtenu une valeur de 21 mm/an et Gamal-Eldin (1988) une valeur de 11.4 mm/an.

Nous avons pris en considération les études antérieures ainsi qu'une estimation de la recharge. Jalludin and Razack (2008) ont également tenu compte d'une forte incertitude à cause de l'effet du colmatage de la fissuration et de la porosité par les sédiments fins aux cours des périodes où l'eau stagne (Houmed-Gaba, 2009). L'infiltration a alors été estimée à 10.3 mm/an.

Il ressort de l'ensemble des travaux visant à estimer l'infiltration que c'est l'hétérogénéité du sol et la fluctuation mensuelle de l'ETR qui est à l'origine des écarts d'estimatiosn de l'infiltration mais relativement ces valeurs restent proches.

5 Conclusion et perspectives

La présente étude essaie d'approcher les termes du bilan hydrologique du bassin versant de l'Oued Ambouli à Djibouti et les compare à des travaux antérieurs (CHA et Gamal-Eldin). Les données disponibles pour chacune des études passées et présentes sont trop fragmentées dans le temps et ne couvrent que des périodes trop courtes pour pouvoir prétendre servir de référence d'un bilan hydrologique. De plus, les caractéristiques statistiques entre les périodes des données montrent des differences importantes dont il va falloir tenir compte en développant une analyse de l'évolution du contexte climatique. Les données disponibles apparaissent insuffisantes en l'état actuel pour avoir une évaluation précise des termes du bilan et de leur évolution.

Cependant cette étude fournit quelques éléments guide de mise en place d'une stratégie holistique de compréhension du cycle de l'eau, depuis les eaux de surface jusqu'aux eaux souterraines. Elle fournit des éléments de connaissance nécessaires pour la protection de la ressource en eau et à la prévention contre les risques d'inondations, qui surviennent de plus en plus fréquemment dans le bassin de l'Oued Ambouli à Djibouti.

Suite à cette première étude, plusieurs travaux vont être menés afin de mieux approfondir les connaissances concernant le bilan hydrologique de ce bassin versant :

- Recherche de données climatiques alternatives, issues de grilles de données observées ou de produits satellitaires afin de mieux prendre en compte la variabilité spatiale et temporelle et l'évolution dans le temps.

- Analyse statistique de l'évolution de la pluviométrie (analyses des tendances ...) et de la temperature et mise en relation avec la diminution de l'écoulement au sein du bassin versant et l'augmentation constatée des phénomènes d'inondation sur le bassin d'Ambouli.

- Critique, voire reprise, des évaluations antérieures, notamment en travaillant à des pas de temps plus fins (quotidien au lieu de mensuel) et à l'échelle de l'événement : saisie et/ou numérisation de données anciennes.

- Analyse de l'évolution de l'emprise urbaine dans la vallée et le delta de l'Oued Ambouli, pour suivre l'évolution de la vulnérabilité aux inondations.

- Mise en place d'un modèle hydrologique (afin de mieux comprendre le processus d'écoulement et les aléas associés) et, si possible, d'un modèle de prévention de crue.

- Mise en place d'un suivi continu et fiable des écoulements à une ou plusieurs stations.

Disponibilité des données. Les données utilisées sont issues des bases de données du Centre d'Etudes et de Recherches de Djibouti (CERD) et de l'Agence Météorologique Nationale (ANM) sont disponibles sur demande aupres de ces institutions.

Collaborateurs. GMO a fait la méthodologie, collecte de données, l'analyse de données, la rédaction-revue et l'édition. JEP a fait la méthodologie, l'analyse de données, la rédaction-revue et l'édition. CS a fait la méthodologie, l'analyse de données, la rédaction-revue et l'édition. GM a fait la méthodologie, l'analyse de données, la rédaction-revue et l'édition. MJ a fait la méthodologie, l'analyse de données, la rédaction-revue et l'édition.

Intérêts concurrents. Les auteurs déclarent qu'ils n'ont aucun conflit d'intérêts.

Déclaration du numéro spécial. This article is part of the special issue "Hydrology of Large River Basins of Africa". It is a result of the 4th International Conference on the "Hydrology of the Great Rivers of Africa", Cotonou, Benin, 13–20 November 2021.

Financement. This research has been supported by the Campus France (Bourses doctorale du gouvernement francais).

Références

Coopération Hydrogéologique Allemande (CHA) : Inventaire et mises en valeur des ressources en eau de la République de Djibouti, 4 vol., République de Djibouti, 1982.

Djama, M. and Gascon, A. : "La Corne dans tous ses États", Cahiers d'Études africaines, 37, 146, 277–500, 1997.

Gamal-Eldin, H. A. : Etude de l'aquifère basaltique de la zone de captage de la ville de Djibouti, République de Djibouti, Thèse de doctorat, Univ. Albert-Ludwig de Fribourg, Fribourg-en-Brisgau, Allemagne Fédérale, 159 pp., 1988.

Hauquin, J. P. : La nappe côtière de la région de Djibouti. Géologie et Hydrogéologie, Thèse de Doctorat, Université de Bordeaux III, Bordeaux, France, 86 pp., 1978.

Houmed-Gaba, A. :. Hydrogéologie des milieux volcaniques sous climat aride. Caractérisation sur site expérimental et modélisation numérique de l'aquifère basaltique de Djibouti (Corne de l'Afrique), PhD thesis, University of Poitiers, Poitiers, France, 220 pp., 2009.

Jalludin, M. and Razack, M. : Recharge simulation in stream bed alluviums overlying volcanic aquifer under semi-arid climate, Republic of Djibouti (Horn of Africa), in : Proceedings International Symposium SIHD, Djibouti, 14–17 December 2008, pp. 123–128, 2008.

L'Hote, Y. : Mesure et étude des précipitations en Hydrologie, Deust – Genie hydro sanitaire et aquacole, Univ. Montpellier, Montpellier, France, 38 pp., 1993.

Penman, H. L. : Vegetation and hydrology, Technical Communication, no. 53, Commonwealth Bureau of Soils, Harpenden, Commonwealth Agricultural Bureaux, Farham Royal, 1963.

SAFEGE, Consortium STDE : Projet d'Aménagement Intégré de L'Oued Ambouli. Etude d'identification des ouvrages de rétention des ruissellements et de recharge des nappes sur l'ensemble du bassin versant, Volume I et II, Republique de Djibouti, 127 pp., 2007.

Schumann, A. H. : Thiessen polygon, in : Encyclopedia of Hydrology and Lakes, Encyclopedia of Earth Science, Springer, Dordrecht, 1998.

Thornthwaite, C. W. and Mather, J. R. : The water balance, Climatology, 8, 5–86, 1955.

Turc, L. : Évaluation des besoins en eau d'irrigation, évapotranspiration potentielle, formule climatique simplifiée et mise à jour, Ann. Agron., 12, 13–49, 1961.

Proc. IAHS, 384, 233–239, 2021
https://doi.org/10.5194/piahs-384-233-2021

Open Access PIAHS

Hydrology of Large River Basins of Africa

Etude de la variabilité saisonnière des précipitations sur le plateau de Mbé au Pool-Nord (Congo-Brazzaville) : impacts sur les eaux de surface et souterraines

Harmel Obami-Ondon[1], **Médard Ngouala Mabonzo**[2], **Urbain Gampio Mbilou**[3], **and Bernard Mabiala**[1]

[1]Laboratoire Mécanique, Energétique et Ingénierie, Ecole Nationale Supérieure Polytechnique (Université Marien Ngouabi), B.P. 69 – Brazzaville, République du Congo
[2]Département de Géographie, Faculté des Sciences Sociales et des Arts, B.P. 69 – Brazzaville, République du Congo
[3]Département de Géologie, Faculté des Sciences et Techniques (Université Marien Ngouabi, B.P. 69 – Brazzaville, République du Congo

Correspondence: Harmel Obami-Ondon (harmelobami@gmail.com)

Published: 16 November 2021

Résumé. Cette étude porte sur l'impact de la variabilité saisonnière des précipitations sur les eaux du Plateau de Mbé. 120 échantillons d'eau ont été prélevés dans huit (08) points d'eau (1 puits, 4 forages et 3 rivières) entre novembre 2017 et août 2018. L'analysés au laboratoire de ces échantillons, permettent de caractériser les éléments physico-chimiques des eaux. Ces résultats sont traités à partir d'une méthode hydrochimie, en utilisant les diagrammes : de Piper, de Schoeller Berkaloff, de Stiff, et de Wilcox. Les données traitées montrent que les valeurs obtenues après les analyses des eaux des saisons ne présentent pas une variation considérable et sont toutes conformes aux normes prescrites par l'OMS pour l'eau de boisson, en dehors de la température qui est légèrement supérieure à la normale avec une moyenne de 27 °C. Le pH a une moyenne de 5.0 ; ce qui confirme le caractère acide des eaux dans toute cette zone d'étude.

Abstract. This study examines the impact of seasonal rainfall variability on the Mbé Plateau waters. 120 water samples were taken from eight (08) water points (1 well, 4 boreholes and 3 rivers) between November 2017 and August 2018, analyzed in the laboratory, to characterize the physicochemical elements of the water. These results are processed from a hydrochemical method, using the diagrams : of Piper, Schoeller Berkaloff, Stiff, Wilcox. They show that the water values analyzed between the 4 seasons do not vary considerably and all comply with the standards prescribed by the WHO for drinking water, except for the abnormal temperature with an average of 27 °C and average pH of 5.0 which confirms the acidity of the waters throughout this study area.

1 Introduction

Depuis une vingtaine d'années, les différents États accordent de plus en plus un intérêt croissant à la notion de changement climatique qui demeure une notion très vaste (réduction de la pluviométrie, augmentation de la température, effet de serre, etc.). En Afrique particulièrement, la notion de changement climatique se résume en grande partie à une baisse drastique de la pluviométrie, principal facteur climatique comme identifié dans plusieurs études (Hubert et al., 1989 ; Mahé et Olivry, 1995 ; Bricquet et al., 1997 ; Servat et al., 1999). Cette baisse de la pluviométrie a eu plusieurs conséquences, le plus souvent dramatiques (sécheresse des années 1970 et 1980, 1984, 2006 dans plusieurs pays de l'Afrique de l'Ouest).

Au Congo-Brazzaville, sur le plateau de Mbé (Fig. 1), la question de la qualité et de la quantité de l'eau s'est posée avec acute étant donné le manque de points d'eau en surface, la mauvaise qualité des eaux dans et dans bien des cas leur assèchement en saison sèche (Le Marechal, 1966).

Figure 1. Localisation de la zone d'étude et les points de prélèvements.

Plusieurs rapports font le point de cette situation : Sautter, 1952, 1953 ; Hudeley, 1952 ; Mermollod, 1961 ; Archambault, 1960 ; Palausi, 1960 et les rapports de quelques projets. Mais, il n'existe aucune étude détaillée sur l'impact de la variabilité saisonnière des pluies sur les eaux du plateau de Mbé. Il convient donc de caractériser ces eaux en fonction des saisons, afin d'étudier son impact sur les ressources en eau qui peut différer d'une saison à une autre.

2 Problématique

La rareté des eaux de surface et la non pérennité de cette ressource en surface sur le plateau de Mbé est due aux conditions géologiques constituées essentiellement des sables, qui ont favorisé l'utilisation de la nappe profonde comme palliatif pour satisfaire les besoins des populations (Le Marechal, 1966 ; Moukolo, 1992). Cette nappe est alimentée essentiellement par les infiltrations des eaux de pluie dans le sol (Le Marechal, 1966 ; Moukolo, 1992). Actuellement, beaucoup d'opérateurs économiques ont entrepris différentes activités (l'élevage, l'agriculture, industrie agroalimentaire, etc.) sur ce plateau (Obami-Ondon et al., 2018). Cette étude se propose donc de voir l'impact de ces activités humaines sur les eaux de la zone d'étude.

3 Méthodologie

Les huit (08) points d'eau (1 puits, 4 forages, 3 cours d'eau) échantillonnés en 4 saisons ont été localisés par leurs coordonnées géographiques à l'aide du GPS et 120 échantillons ont été recueillis pour les analyses physico-chimiques entre novembre 2017 et mai 2018 (Fig. 1). Les prélèvements

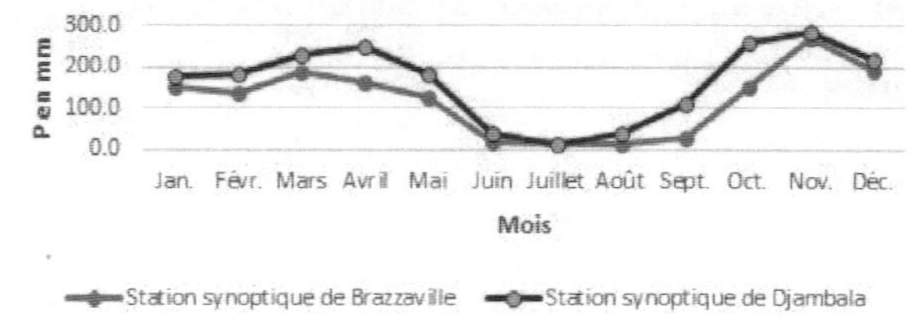

Figure 2. Variation des précipitations moyennes mensuelles sur les stations de Brazzaville et Djambala pour la période (1987–2018), source ANAC de Brazzaville.

ont été effectués et conditionnés dans des flacons de 1.5 L en polyéthylène spécialement préparés à cet effet. Les prélèvements pour les forages munis de robinets ont été effectués après avoir fait couler l'eau dans le vide pendant plusieurs minutes. Les flacons d'échantillonnage avant remplissage ont été rincés plusieurs fois avec l'eau à prélever et immédiatement stockés dans une glacière à 4 °C. Les analyses ont été effectuées rapidement dans les 24 h qui suivaient. L'analyse a concerné les éléments physiques, majeurs, quelques métaux lourds et bien d'autres au laboratoire de l'IRSEN à l'aide d'un spectrophotomètre en utilisant les méthodes classiques préconisées par les normes françaises (AFNOR). Une comparaison des teneurs en éléments physiques et chimiques des eaux en fonction des saisons et des normes de l'Organisation Mondiale de la Santé (OMS, Genève, 2004 et 2017) a été effectuée. Ensuite pour connaitre les faciès des eaux, une étude géochimique a été réalisée à l'aide des digrammes de Piper, de Schoeller-Berkalof et de Wilcox pour la potabilité. La qualité des analyses a été contrôlée grâce à la balance ionique pour la fiabilité des résultats.

4 Résultats

4.1 Fluctuation des moyennes mensuelles interannuelles de la pluviométrie entre 1987–2018

La variation des indices pluviométriques de la zone d'étude est représentée pour les stations synoptiques de Brazzaville et Djambala (Fig. 2). L'analyse de l'évolution des indices pluviométriques montre une période pluvieuse (sept-mai) de 9 mois avec une diminution des pluies (mi-janvier-février) et une période sèche (juin-aout) de 3 mois.

La Fig. 2 montre la variation des précipitations moyennes mensuelles sur les stations de Brazzaville et Djambala. Cette figure montre clairement une diminution des pluies entre le mois de mai et septembre, ce qui marque la saison sèche sur la zone d'étude.

La variation des paramètres physiques (pH, T, °C, CE et TDS) des 8 points prélevés est représentée sur la Fig. 3 et cette analyse de l'évolution des paramètres physiques (Fig. 3) est faite en fonction de la variabilité saisonnière. Sur la Fig. 3 nous constatons que les températures diminuent sur tous les points échantillonnés au mois d'aout et cela correspond à la

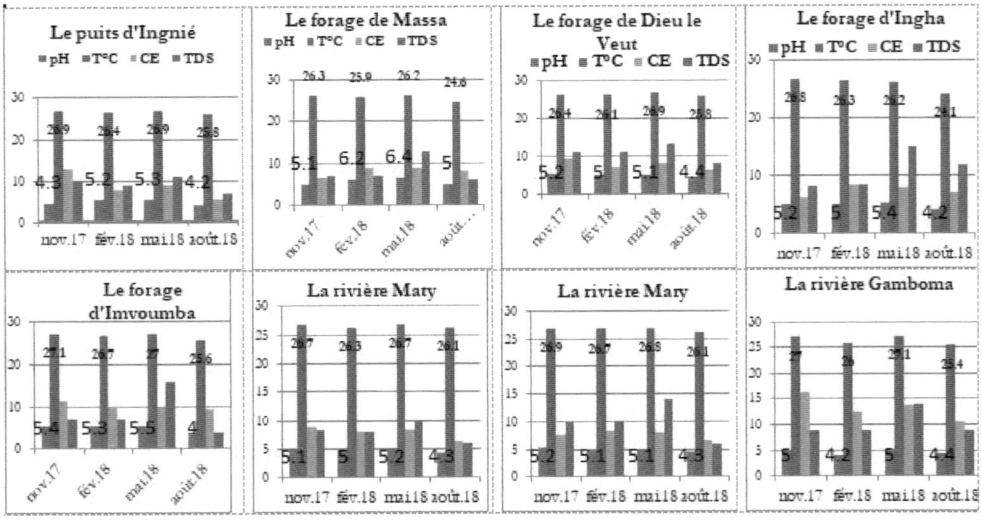

Figure 3. Variation saisonnière du pH, T, CE et TDS au cours de l'année 2017–2018.

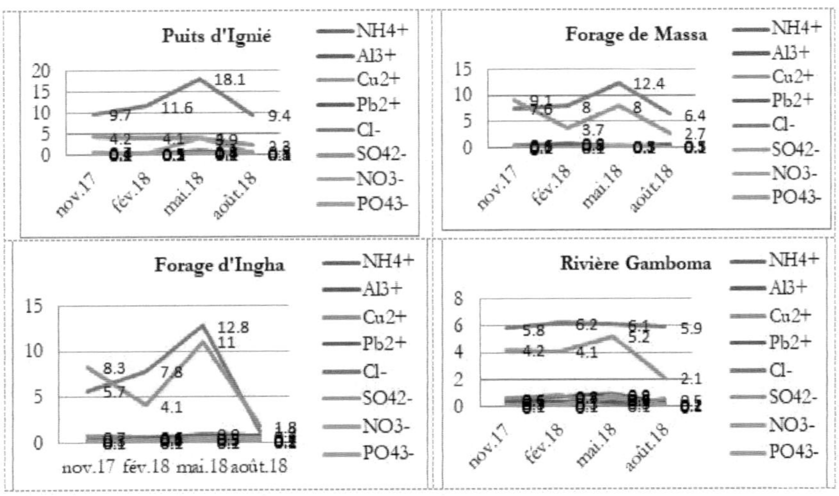

Figure 4. Variation saisonnière des métaux lourds et les polluants mesurés au cours de l'année 2017–2018.

grande saison sèche (Fig. 2). Cependant, les TDS présentent une augmentation dans tous points pendant la campagne de Mai. Ceci peut s'expliquer par le fait qu'au mois d'avril on observe le deuxième pic de la pluviosité dans la zone et le mois de Mai étant le dernier mois de la saison de pluie enregistre des conséquences dues aux apports des précipitations du mois d'avril ; mais toutefois, les valeurs restent inférieures aux normes prescrites par l'OMS pour l'eau de boisson (Directive, 2004). On peut aussi noter que, le pH connait une dynamique (de mai jusqu'en aout, il décroit) peu significative dans l'ensemble de la zone d'étude. On peut donc dire, que la variabilité saisonnière n'influe pas de façon directe sur les paramètres physiques des eaux de notre région.

Parmi les métaux et polluants étudiés (Fig. 4), on note que tous les autres métaux sont quasiment inexistants. Par contre les Cl^- et les SO_4^{-2}, bien que les valeurs soient inférieures à celles prescrites par l'OMES (Directive, 2017), ils présentent une dynamique exceptionnelle dans tous les points pendant la campagne de mai. Cette variation, serait due au temps de transfert des eaux de pluie dans la nappe. Ce même mois, les TDS (Fig. 3) ont connus aussi une évolution dans tous les points d'eau de la zone d'étude.

Il découle de la projection des résultats des campagnes, qu'un (01) seul faciès du diagramme de Piper est représenté (Fig. 5) : le faciès chloruré calcique. Mais, la campagne d'août 2018 (saison sèche) s'est avérée différente en présentant deux faciès, notamment le faciès chloruré calcique et le Faciès bicarbonaté calcique. Ce deuxième faciès est marqué par le forage de Massa qui s'écarte des autres. On constate que, dans le triangle des cations, le même forage du village

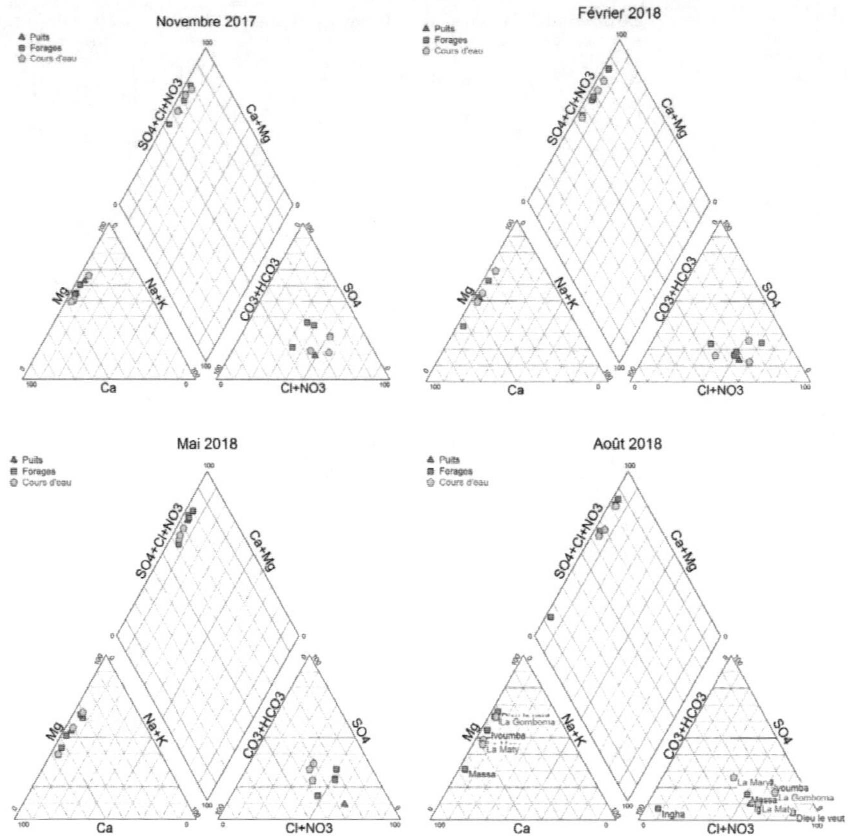

Figure 5. Diagramme de piper au cours de l'année 2017–2018.

Massa s'éloigne des autres. Ceci, serait certainement dû aux activités domestiques qui déversent les eaux usées des ménages et la présence des cochons aux alentours du puits.

En général, les eaux du plateau de Mbé sont chlorurées calcique toute l'année et ne présentent pas une dynamique importante.

Les cations sont représentés à gauche et les anions à droite. Selon le diagramme de Schöeller-Berkaloff (Fig. 6), les ions dominants sont les chlorures pour les anions et le magnésium pour les cations. Sur la Figure de la saison sèche, le forage de « Dieu Le Veut » présente la concentration des chlorures plus élevée que celle obtenue par Obami-Ondon et al. (2018) (saison des pluies). Le même constat est fait dans le forage d'Ingha par la concentration de $HCO_3 + CO_3$. On pourrait donc souligner que, pendant la saison sèche le profil des forages de Dieu le Veut, Ingha et le cours d'eau La Mary, présente un contraste très différent de la saison des pluies. Ceci peut s'expliquer par la présence des activités humaines autour des forages. Car les éléments qui présentent plus de variation sont les chlorures.

Dans le cadre de la potabilité des eaux de la région, les 120 échantillons représentés sur le diagramme de Wilcox (Fig. 7) se répartissent en 1 seule groupe (Excellent) en saison des pluies tout comme en saison sèche.

5 Discussion

Sur le Plateau de Mbé, la température moyenne des eaux est de 27 °C. Cette température correspond aux variations saisonnières des températures atmosphériques ambiantes définies par les données météorologiques (30 °C). Le pH des eaux varie de 4.02 à 6.4 ; avec une moyenne de 5.0. Cela indique que les eaux sont acides, donc corrosifs. En effet, l'acidité des eaux a été observée dans plusieurs régions du monde. Il s'agit notamment du plateau Mbé (Essouli et al., 2019), de Brazzaville (Matini et al., 2009), des régions de Tiassalé (Oga et al., 2009), de Katiola (Oga et al., 2010) et région de San-Pédro (Lasm et al., 2011). L'acidité des eaux en zone tropicale humide est selon Matini et al. (2009) principalement liée à la décomposition de la matière organique végétale, avec la production de CO_2 dans les premières couches du sol.

Les teneurs en métaux lourds observées dans les eaux de la zone sont faibles. Cela serait dû à la géologie et au manque d'industries dans la zone d'étude. Ces résultats sont en re-

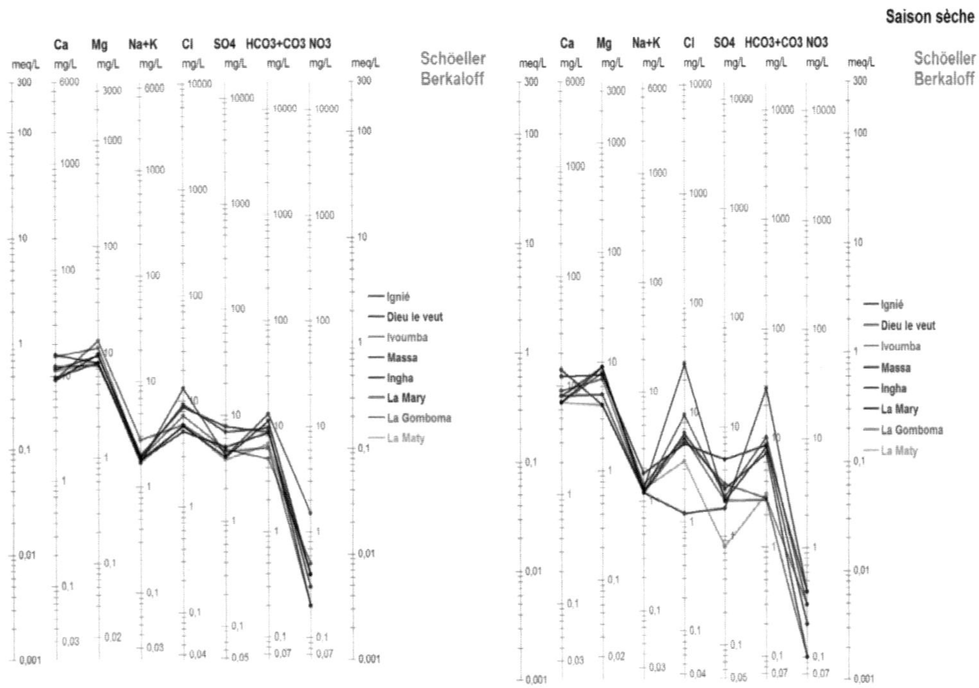

Figure 6. Diagramme de SchöellerBerkaloff entre les 2 grandes saisons.

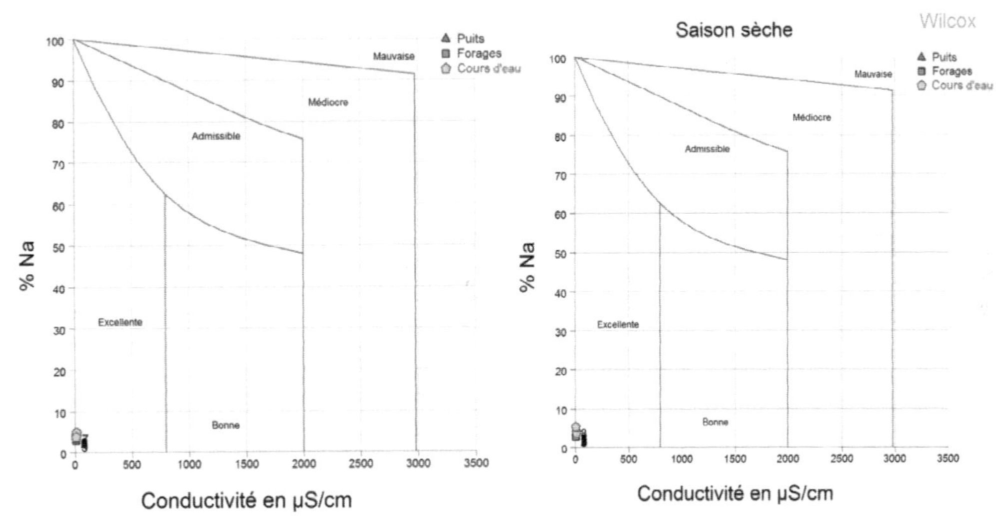

Figure 7. Diagramme de Wilcox en saison des pluies et sèche.

joignant ceux des travaux les plus récents Obami-Ondon et al. (2018) et d'Essouli et al. (2019) dans la zone d'étude.

Du point de vue hydrofacies, les eaux de la zone d'étude appartiennent au faciès chloruré calcique. Mais, la campagne d'août 2018 (saison sèche) s'est avérée différente en présentant deux faciès, notamment : le faciès chloruré calcique et le Faciès bicarbonaté calcique. Ces résultats sont en conformité avec celui obtenu par Matini et al., 2009; Obami-Ondon et al. (2018); Essouli et al. (2019) dans la zone d'étude. Selon ces auteurs; la quasi-homogénéité des hydrofaciès de la zone d'étude est liée à sa faible hétérogénéité lithologique.

Les valeurs de la conductivité électrique obtenues montrent que les eaux de cette zone sont faiblement minéralisées et cela a été montré dans les travaux de Matini et al., 2009; Essouli et al. (2019). Ces résultats en majorité coïncident avec ceux Obami-Ondon et al. (2018).

6 Conclusions

Cette étude a permis de caractériser les paramètres physico-chimiques en suivant la variabilité saisonnière de la région. Elle a révélé que tous les paramètres mesurés pendant les quatre (04) saisons sont tous conformes aux normes de l'OMS pour les eaux de boisson et ne présentent pas une variation correspondante à la variation des saisons. Cependant, les températures des eaux de surface et souterraines sont comprises entre 24.1 et 27.1 °C, et ne présente pas une dynamique particulière en fonction des saisons ; ces températures sont proches de celles recommandées par l'OMS. Elles sont acides avec des pH qui varient entre 4.02 et 6.4 ; elles sont donc agressives. Cette variation ne suit pas celle des saisons. Quant aux valeurs de conductivité électrique, elles varient de 5.14 à 16.3 µS/cm.

En somme, nous retenons que les indices locaux de contaminations par les activités humaines sont faibles et que ces eaux peuvent être consommées sans danger majeur par les populations qui n'ont pas d'autres sources de ravitaillement en eaux.

Collaborateurs. HOO : il a fait la conception de la recherche, l'élaboration, la rédaction et discussion des résultats, la révision et l'acceptation de la version finale. MNM : Il a participé à la conception et élaboration de la recherche. UGM : il a participé à la discussion des résultats et la révision de la version finale avant soumission. BM : il a fait la relecture et la mise en forme de la version finale avant soumission.

Intérêts concurrents. The contact author has declared that neither they nor their co-authors have any competing interests.

Déclaration du numéro spécial. This article is part of the special issue "Hydrology of Large River Basins of Africa". It is a result of the 4th International Conference on the "Hydrology of the Great Rivers of Africa", Cotonou, Benin, 13–20 November 2021.

Remerciements. Nous tenons à remercier la bonne collaboration de l'Ambassade de France en République du Congo (SCAC) avec l'Université Marien Ngouabi et le laboratoire HydroSciences de Montpellier (France).

Références

Archambault, J. : Note sur le Plateau Koukouya, BUBGEAPR., 282, Rapport ronéo 4 pp., 1960.

Bricquet, J. P., Bamba, F., Mahe, G., Toure, M., et Olivry, J. C. : Variabilité des ressources en eau de l'Afrique Atlantique, Journal des sciences hydrologiques, 16, 83–95, 1997.

Directives de qualité pour l'eau de boisson : 3e Ed., volume 1, recommandations, Organisation mondiale de la Santé, Genève, 2004.

Directives de qualité pour l'eau de boisson : 4e Ed., intégrant le premier additif [Guidelines for drinking-water quality : 4th Ed. incorporating first addendum], Organisation mondiale de la Santé, Genève, 2017.

Essouli, O. F., Matini, L., Miyouna, T., Fouroux, M. A., Boudzoumou, et F., Faye, S. : Prospection Hydrogéologique et Géophysique du Système Aquifère du Plateau de Mbé, District de Ngabé, Congo, European Scientific Journal October 2019 edition Vol. 15, No. 30 ISSN : 1857-7881 (Print) ISSN 1857-7431, 2019.

Frédéric, L. : Fonctionnement hydrodynamique de l'aquifère du Miocène du bassin de Carpentras, Thèse Doctorat Hydrogéologie, LHA, Univ. Avignon, 2006.

Hubert, P., Servat, E., Paturel, J. E., Kouame, B., Bendjoudi, H., Carbonnel, J. P., et Lubes-Niel, H. : La procédure de segmentation, dix ans après, Publication IAHS, 250, 267–273, 1998.

Hudeley, H. : Etude hydrogéologique du Plateau Koukouya et de la région de Djambala, Rapport ORSTOM, 13 pp. + annexes, 1952.

Lasm, T., De Lasme, O., Oga, Y.M.-S., Youan, T. M., Baka, D., Kouamé, F., et Yao, K. T. : Caractérisation hydrochimique des aquifères fissurés de la région de SanPedro (Sud-Ouest de la Côte d'Ivoire), International Journal of Biological and Chemical Sciences, 5, 642–662, 2011.

Le Marechal, A. : Contribution à l'élude des plateaux batéké ; géologie, géomorphologie hydrogéologie, Rapp.ORSrOM-Brazzaville Cote M.C., 137, 43 pp., 1966.

Mahé, G. et Olivry, J. C. : Variations des précipitations et des écoulements en Afrique de l'Ouest et centrale de 1951 à 1989, Bulletin sécheresse, 6, 109–170, 1995.

Matini, L., Moutou, J. M., et Kongo, M. S. : Evaluation hydrochimique des eaux souterraines en milieu urbain au sud-ouest de Brazzaville, Congo, Afrique Science, 5, 82–98, 2009.

Mermollod, M. : Compte rendu de mission sur le Plateauau Koukouya, (8-12 Juin1960) rapport ronéo IRGM Brazzaville. Réf. AH-T., 20, 1961.

Obami-Ondon, H., Mbilou, U. G., Tomodiatounga, D. N., Mabonzo, M. N., Elenga, R. G., et Mabiala, B. : Physicochemical Characterization of Water of the Plateau of Mbe in Pool-North in Republic of Congo Brazzaville, American Journal of Environmental Protection, 7, 40–54, https://doi.org/10.11648/j.ajep.20180703.11, 2018.

Oga, M. S., Lasm, T., Yao, T. K., Soro, N., Saley, M. B., Kouassi, D., et Gnamba, F. : Caractérisation chimique des eaux des aquifères de Fracture : Cas de La Région de Tiassalé en Côte d'Ivoire, European Journal of Scientific Research, 31, 72–87, 2009.

Oga, M. S., Gnamba, F. M., Lasm, T., Gnangne, T., Baka, D., Soro, N., et Biemi, J. : Caractérisation physico-chimique des eaux souterraines de Katiola (Centre-Nord de la Côte d'Ivoire), BIOTERRE, Revue Internationale des Sciences de la Terre, 10, 7–25, 2010.

Palausi, G. : Note préliminaire sur le forage du Plateauau Koukouya, BRGM Brazzaville, 1960.

Sautter, G. : L'eau sur les Plateaux Batékés, Rapport ronéo, ORSTOM géographie Brazzaville, 1952.

Sautter, G. : Les villages des Plateaux Batékés et le problème de l'eau, Rapport ronéo ORSTOM Géographie, Brazzaville, D35, 17 pp., 1953.

Servat, E., Paturel, J.-E., Lubes-Niel, H., Kouame, B., Masson, J. M., Travaglio, M., et Marieu, B. : De différents aspects de la variabilité de la pluviométrie en Afrique de l'Ouest et centrale non sahélienne, Revue des sciences de l'eau, 12, 363–387, 1999.

Proc. IAHS, 384, 241–246, 2021
https://doi.org/10.5194/piahs-384-241-2021

Open Access PIAHS

Hydrology of Large River Basins of Africa

Dynamique des extrêmes hydrologiques du système Chari-Logone et risques naturels dans la région de l'extrême-nord du Cameroun

Fréderic Saha[1], Mesmin Tchindjang[1], Jean-Guy Dzana[1], and Djasrabé Nguemadjita[2]

[1]Département de Géographie de l'Université de Yaoundé 1, Yaoundé, Cameroun
[2]Direction des ressources en eau et de la météorologie, N'Djamena, Tchad

Correspondence: Fréderic Saha (fredericsaha@yahoo.fr)

Published: 16 November 2021

Résumé. Le système Chari-Logone est principal cours d'eau drainant l'Extrême-Nord du Cameroun. À l'image des autres parties du bassin versant de ce cours d'eau, cette région a connu plusieurs inondations catastrophiques. Cet article examine la dynamique des extrêmes hydrologiques du Logone et du Chari à travers les stations de Bongor et N'Djamena en prenant en compte le contexte des changements climatiques et l'occurrence des risques hydro-climatiques. Les analyses portent essentiellement sur les données journalières maximum (Qmax) et minimum (Qmin) sur la période 1960–2015. Les logiciels XLStat et ChronoStat ont permis de calculer un ensemble d'indices. Il en ressort que les Qmax présentent une rupture en 1971 aux deux stations marquant la fin de la période humide. Les Qmin de Bongor ont connu deux ruptures successives en 1994 et en 2007. Les Qmin de N'Djamena par contre ont connu une rupture en 1986. Le coefficient de tarissement est globalement faible (0,020/jour en moyenne). L'occurrence des inondations catastrophiques est très contrastée, car seulement 30 % s'inscrivent dans les années excédentaires. Cela témoigne du rôle important que jouent les autres facteurs de construction du risque comme la pédologie, la pluviométrie et la vulnérabilité socioéconomique.

1 Introduction

Pendant le XXème siècle et le début du siècle en cours, l'Afrique a connu plusieurs séquences hydro-climatiques (Lévêque, 2006). Les années de la période de faibles écoulements entre 1970–1990 par exemple ont connu les étiages les plus sévères (Olivry, 1997). De nombreux pays (Mali, Niger, Tchad, etc.) avaient alors eu des difficultés supplémentaires dans l'approvisionnement en eaux pour différents besoins (boisson humaine et animale, irrigation, etc.). Pour les périodes d'excédents, les crues sont parfois à l'origine d'inondations catastrophiques. Il faut remarquer que ces périodes ne sont pas toujours homogènes. En période de déficits, se mêlent parfois les années de forts écoulements et inversement. En République Centrafricaine, Nguimalet (2017) met en exergue une importante dynamique des débits extrêmes de l'Oubangui entre 1911–2014. L'évolution de ce cours d'eau suit globalement les tendances hydro-climatiques qu'a connues l'Afrique centrale et de l'Ouest. Les débits extrêmes Qmin et Qmax ont eu de graves conséquences dans la ville de Bangui. En zone tropicale sahélienne, les extrêmes de nombreux cours d'eau furent encore plus marqués avec occurrence de sècheresses et inondations catastrophiques dans de nombreux pays (Olivry, 1997).

Le bassin endoréique du Lac Tchad a durement subi la sècheresse qui a débuté en 1970. On parle d'une diminution de près de 90 % de la superficie du lac. Cela étant dû au déficit d'alimentation ; notamment le Logone-Chari qui est la principale source d'alimentation de ce lac (Olivry et al., 1996). Dans le même bassin, quelques crues exceptionnelles ont été à l'origine de graves inondations au Tchad et au Cameroun (OCHA, 2012). Cet article s'intéresse au sous-bassin du Logone et au bassin du Chari. L'objectif est d'évaluer la dynamique de ce système hydrographique sous l'angle des débits extrêmes (Qmin et Qmax). Les résultats de cette dynamique sont mis en corrélation avec l'occurrence et l'ampleur des dégâts des catastrophes naturelles (inondations et sècheresses) que la région de l'Extrême-Nord du Cameroun a connues pendant la période d'observation (1977–2017).

2 Données et méthodes

Près de 90 % de la région de l'Extrême Nord du Cameroun fait partie du grand bassin hydrologique du lac Tchad en zone tropicale (Olivry, 1986) (Fig. 1). On peut y distinguer trois nuances climatiques : le climat tropical d'altitude qui règne essentiellement sur l'Adamaoua au Cameroun où le Logone prend sa source. Les précipitations moyennes se situent entre 1300–1500 mm. La seconde nuance climatique est soudanienne avec 1000 mm de pluies en moyenne par an. La partie au nord est le domaine du climat soudano sahélien avec des précipitations qui diminuent selon un gradient sud-nord.

Les données de débits de base utilisées concernent la période 1955–2016 (61 ans). Pour chacune des stations, quelques années sont lacunaires. Ces données proviennent d'une part de l'ouvrage « fleuves et rivières du Cameroun » de Olivry (1986), d'autre part de la Commission du Bassin du Lac Tchad (CBLT, 2010) et enfin de la Direction des Ressources en Eau et de la Météorologie (DREM) au Tchad.

Les valeurs annuelles ont été calculées du 1er avril au 31 mars, ce qui correspond à l'année hydrologique. Les traitements et analyses concernent spécifiquement les modules annuels et les débits journaliers extrêmes (maximum et minimum). La variabilité interannuelle du régime hydrologique est appréciée à travers les variations des différents indices. Le coefficient de tarissement (Lang et Gille, 2006) permet de juger de l'intensité des sècheresses hydrologiques, lorsque le débit n'est plus soutenu que par la vidange de la nappe phréatique. La puissance des écoulements de crue est traduite par le coefficient A de Pardé (Loup, 1965). L'indice d'irrégularité (Q_{max} annuel/Q_{min} annuel) mesure l'écart entre les débits les minima et les maxima annuels (Nguimalet, 2017).

Dans l'optique de détecter les tendances dans la distribution des données annuelles, nous avons appliqué différents tests d'homogénéité sur KhronoStat 1.01 et XLStat. Il s'agit des tests de Pettitt, d'Hubert et de la méthode Bayésienne de Lee-Heghinian. Les données sur les risques naturels (sècheresses et inondations) dans la région de l'Extrême-Nord du Cameroun sont issues des archives du service du gouverneur. Les variables retenues sont les dates d'occurrence des catastrophes et les dégâts enregistrés (pertes en vies humaines, personnes déplacées, pertes matérielles).

3 Résultats et analyses

3.1 Dynamique des extrêmes hydrologique du Logone et du Chari à travers quelques indices

3.1.1 Indice d'irrégularité

La moyenne interannuelle de l'indice d'irrégularité est de $R = 44,5$ à Bongor (1955–2016). Il présente une légère tendance générale à la baisse. Pour une moyenne de 48,6 entre 1955–1988, on est passé à 38,5 pour la période 1989–2016 soit une baisse de 20 %. Globalement, cela traduit une réduction de l'écart entre Q_{min} et Q_{max}.

Pour ce qui est de N'Djamena, le coefficient R moyen est de 67,4 (1955–2016). Le maximum de la série est de 668 pour l'année 1988–1989 et le minimum est de 13 correspondants à l'année 1963–1964. Avant la rupture de 1971–1972 le coefficient R était de 29 ; il est passé à 86,5 pour la période 1972–2016 ; soit une augmentation de 246 %.

3.1.2 Coefficient de tarissement

Le temps moyen de tarissement est de 190 jours sur le Logone à Bongor. Cela se passe entre le 23 septembre et le 20 mars. Cette moyenne cache une forte variabilité interannuelle. Le coefficient moyen de tarissement interannuel se situe alors à 0,019/jour. Le coefficient le plus élevé est 0,031/jour correspondant à l'année 1955–1956. Le plus faible est celui de l'année 1982–1983 (0,012/jour). Le coefficient d'asymétrie est de 0,56 ce qui traduit un étalement de la série de données à droite de la moyenne. Les données de N'Djamena sur le Chari présentent un coefficient de tarissement moyen plus élevé qu'a Bongor (0,018/jour).

3.2 Risques hydro-climatiques dans la région de l'Extrême-Nord du Cameroun

3.2.1 Différents aspects de la sècheresse hydrologique dans l'Extrême-Nord du Cameroun

Les données de débits journaliers du système Logone-Chari (Bongor et N'Djamena) mettent en évidence quelques aspects de la sècheresse hydrologique affectant la région de l'Extrême-Nord du Cameroun :

- les sècheresses s'étendant sur plusieurs années consécutives ;
- les déficits de l'écoulement total annuel ;
- les déficits d'étiage et
- le niveau des hautes eaux par rapport à la moyenne.

Pour la station de N'Djamena, le module (1955–2016) est de 903 $\mathrm{m^3\,s^{-1}}$. La période 1955–1971 (1304 $\mathrm{m^3\,s^{-1}}$ en moyenne) est de 44 % plus élevée que cette moyenne générale. À cette période succède la sècheresse qui se poursuit

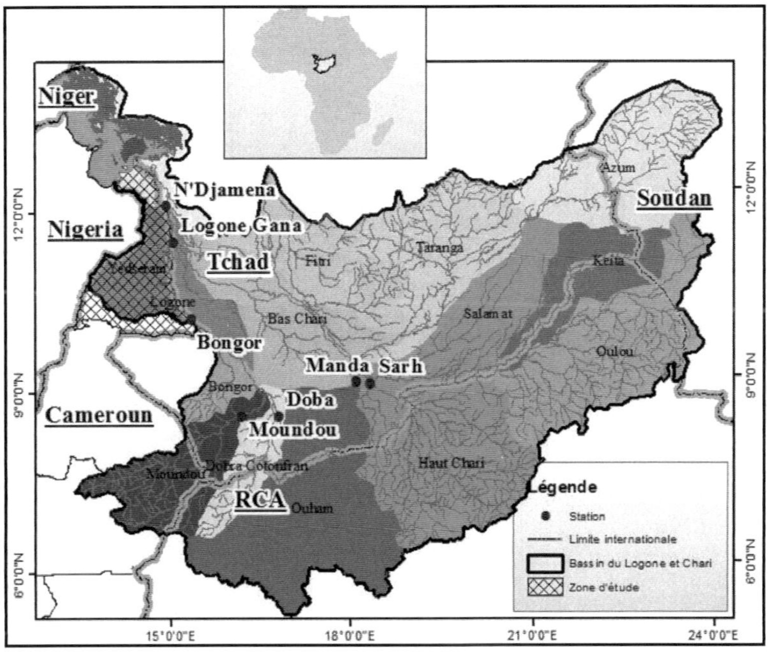

Figure 1. La région de l'Extrême-Nord du Cameroun dans le bassin du Logone-Chari (© CBLT, 2010).

jusqu'à la fin de la série avec des niveaux de sévérité différents. De 1972–1998, le déficit d'écoulement est le plus élevé : 27 %. La station de Bongor présente un module de 457 $m^3 s^{-1}$ (1955–2016). Une sècheresse persistante commence en 1977–1978. Elle se maintient jusqu'en 1994–1995 lorsque la sévérité baisse.

Pour les données de N'Djamena, les excédents de Q_{max} pendant les années déficitaires étaient récurrentes (1975–1976, 1976–1977, 1989–1990, etc.). La Fig. 2 présente un indice synthétique de sècheresse résultant de l'addition des variables centrées réduites des modules, Q_{min} et leurs Q_{max}. Les années de sècheresse se caractérisent par un résultat négatif.

Le contexte socio-économique de l'Extrême-Nord en fait une région très vulnérable aux risques naturels en général et à la sècheresse en particulier. En effet, l'insécurité alimentaire affecte 18 % de la population (DSCN, 2002). Il s'agit aussi de la région la plus pauvre du Cameroun avec 74,3 % de la population vivant en dessous du seuil de pauvreté (INS, 2015). Près de 60 % de la population dépend des sources non aménagées pour l'eau de boisson. L'agriculture (pluviale, irriguée et de contre saison) de subsistance représente 65 % de l'activité économique de la région (DSCN, 2002). Il en est de même de l'activité pastorale non mécanisée. C'est dans ces conditions que les déficits hydrologiques des décennies 1970 et 1980 ont occasionné de véritables catastrophes dans la région de l'extrême nord. Pour les deux décennies les populations l'aggravation : des pénuries en denrées alimentaires, des épidémies de méningite, des pénuries

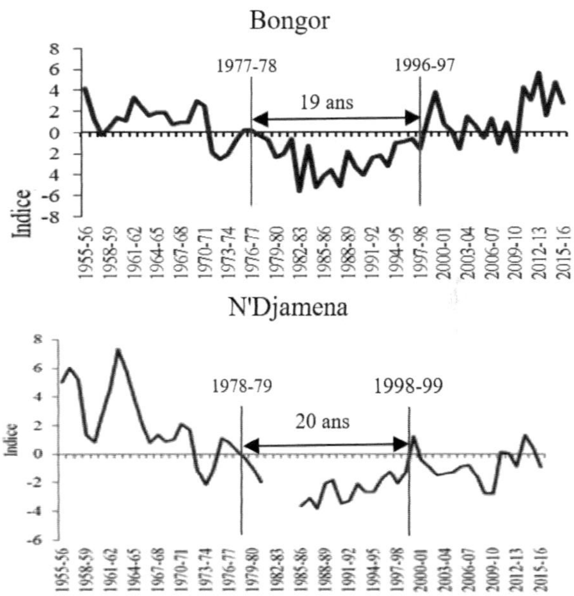

Figure 2. Valeurs de l'indice synthétique de sècheresse hydrologique du Logone à Bongor et du Chari à N'Djamena.

d'eau potable, des invasions de criquets, du tarissement des mares d'eau, de la dégradation des pâturages, de la désertification des terres. Par l'indice de rendement agricole espéré (IRESP), L'Hôte (2000) situe à 27,2 % en moyenne la baisse de rendement sur les 10 cultivars les plus semés dans

Tableau 1. Facteurs hydrologiques et inondations majeures dans la région de l'Extrême-Nord du Cameroun.

Année	Variable centrée réduite des indices								Arrondissement les plus affectés	Principaux dégâts
	Q_{moyen}		Q_{max}		DCC		Coefficient A			
	B	N	B	N	B	N	B	N		
2015–2016	0,16	0,1	1,7	−1, 1	0,9	−1, 1	1,6	−1, 1	Maroua, Gazawa et Ndoukoula	5 morts, 6200 sans abris
2014–2015	2,37	0,01		0,3	0,7	0,4		0,4	Zina et Kaikai	168 sans abris
2012–2013	2,76	−0, 9	2,3	0,8	2,1	0,9	2,3	1	Mayo Danay et Logone et Chari	20 morts; 20000 sans abris
2010–2011	0,49	−1, 07		0,5		0,59		0,4	Pouss	12 morts; 3000 sans abris
2007–2008	−0, 86	−0, 64	−0, 6	−0, 2	0,3	−0, 1	−0, 6	−0, 2	Mayo Danay	8 morts; 4421 sans abris
2006–2007	0,37	0,43	0,3	−1	−0, 9	−1, 0	−0, 3	−1	Maroua	2 morts; 1042 sans abris
2005–2006	−0, 41	−0, 88	−0, 1	−0, 1	−0, 2	−0, 1	−0, 1	−0, 1	Maroua	14 morts; 453 sans abris
2000–2001	−0, 16	0,2	−0, 6	−0, 7	1,6	−0, 7	−0, 6	−0, 8	Maroua	17 morts; 206 sans abris
1997–1998	−0, 54	−0, 26	−0, 6	−1, 3	−0, 7	−1, 2	−0, 6	−1, 3	Maroua	7 morts; 248 sans abris
1994–1995	−0, 49	−1, 12	0,1	−0, 4	0,2	−0, 3	0,1	−0, 4	Maroua, Mokolo, Kousseri	41 morts; 1004 sans abris
1991–1992	−0, 98	−1, 4	−0, 5	−0, 8	−0, 6	−0, 7	−0, 5	−0, 8	Maga et Kousseri	2 morts; 200 sans abris
1988–1989	−0, 28	−1, 6	−0, 2	0,7	−0, 2	0,9	−0, 2	−0, 8	Maroua et Kousseri	10 morts; 165 sans abris
1977–1978	−0, 33	0,07	0,4	0,1	-0,1	0,2	0,5	0,1	Maroua	7 morts

B = Bongor. N = N'Djamena.

la zone de Kousseri-N'Djamena en comparant les périodes 1950–1967 et 1968–1985. Ce pourcentage est de 2,2 % dans la zone de Maroua. Il faut remarquer que dans son approche, L'Hôte (2000) se fonde essentiellement sur les données pluviométriques et donc sur la diminution des précipitations, laquelle a affecté plus sévèrement la zone la plus septentrionale. En réponse, le gouvernement du Cameroun avait mis en place un ensemble de mesures pour assister les populations.

3.2.2 Paramètres hydrologiques et inondation dans l'Extrême-Nord du Cameroun

Les inondations dans la région de l'Extrême-Nord du Cameroun relève d'un ensemble de facteurs physiques et humains (relief plat, pluviométrie concentrée sur trois mois, sols argileux peu perméables, mauvaise qualité du bâti, précarité générale des modes de vie, etc.). Ainsi, l'eau qui inonde provient directement des précipitations ou alors du débordement des cours d'eau. En abordant l'aspect hydrologique de ce phénomène, les excédents sur les modules et de débits de pointes donnent des perspectives d'analyse. En plus de ces deux variables, il est tout aussi adéquat de s'intéresser aux débits égalés ou dépassés pendant 10 jours par an ou débits caractéristiques de crues (DCC). La moyenne des DCC est de 1714 $m^3 s^{-1}$ sur le Logone (Bongor) et 2604 $m^3 s^{-1}$ sur le Chari (N'Djamena). À l'image des autres paramètres de ces cours d'eau, les DCC sont très variables à l'échelle interannuelle. Les valeurs les plus élevées se situent au début de la série d'observation (avant 1971). Toutefois, à Bongor, c'est l'année 2012–2013 qui présente le maximum (2710 $m^3 s^{-1}$) ; soit un excédent de 58 %. Pour les deux stations, on remarque un retour d'années humides avec un DCC excédentaire à partir des années 2000.

Les années d'inondations catastrophiques ne se caractérisent pas toujours par des débits moyens au-dessus du module, et/ou des DCC et débits de pointe supérieurs aux moyennes interannuelles. La distribution de la puissance de crue (coefficient de Pardé) ne permet pas non plus de tirer des conclusions claires. Le Tableau 1 présente quelques exemples d'inondations catastrophiques dans la région de l'Extrême-Nord du Cameroun en relation avec les variables hydrologiques du Logone et du Chari pour les stations de Bongor et de N'Djamena.

Il se dégage du Tableau 1 que seulement 23 % (3/13) des inondations dans la région de l'Extrême-Nord du Cameroun surviennent pendant les années humides lorsqu'on considère les Q moyens, les débits de pointe et les DCC du Logone et du Chari. Ces années sont les plus récentes (2010–2011, 2012–2013 et 2014–2015).

4 Discussion

Nguimalet (2017) analysant les données de 1911–2014 sur l'Oubangui, met en exergue une tendance à la baisse des Q_{min} et Q_{max}. La rupture de 1971 ressort aussi des analyses de cet auteur. La même tendance a été observée sur le Sénégal et l'ensemble des bassins Ouest africains (Olivry, 1997). Il faut remarquer que sur les grands fleuves africains en général, le comportement des Q_{max} est identique aux tendances observées sur les débits moyens annuels (Olivry et al., 1998). La variabilité des Q_{min} se conforme très peu à cette tendance générale. Les cours d'eau d'Afrique de l'Ouest ont subi plus sévèrement la péjoration des écoulements par rapport à l'Afrique centrale (Olivry et al., 1998). En effet, sur le Sénégal, le déficit est de 57 % entre 1971–1993 avec un regain d'humidité à partir de 1994 (Cisse et al., 2014).

Olivry et al. (1997) relève une faible puissance des crues sur les fleuves d'Afrique tropicale (coefficients A de PARDÉ de 6 à 28 pour le Niger, de 2 à 20 pour le Sénégal). On peut évoquer l'immensité des bassins de ces cours d'eau comme facteur explicatif. L'année 1984, qui a connu la plus faible puissance de crue sur le Logone et le Chari, est identifiée comme la plus déficitaire de la période sèche sur plusieurs cours d'eau d'Afrique (Olivry et al., 1996). Les tarissements du Logone et du Chari sont nettement plus lents (190 jours) que ceux des fleuves d'Afrique de l'Ouest où deux mois seulement séparent Q_{max} de Q_{min} (Olivry et al., 1994). La situation du Logone et du Chari s'apparente plus à la tendance des cours d'eau d'Afrique équatoriale, comme l'Oubangui (affluent du Congo) où Nguimalet (2017) trouve pratiquement la même valeur moyenne du coefficient de tarissement (0,020/jour).

5 Conclusion

Il ressort de cet article que du point de vue temporel, la tendance majeure est celle de la baisse des débits de pointe à partir de 1971–1972 évaluée à 25,5 % à Bongor et 36 % à N'Djamena. La tendance des minimums d'écoulement est plus contrastée. La baisse des débits de pointe se répercute sur la puissance des crues. La régularité ou l'irrégularité du rapport entre Q_{max} et Q_{min} se situe entre 13 et 668 ce qui traduit une forte variabilité.

En considérant les variables centrées réduites, de 1977–1978 à 1996–1997 pour Bongor et de 1978–1979 à 1998–1999 pour N'Djamena, tous les indices (Q_{min}, Q_{max}, coefficient de tarissement et puissance de crues) étaient négatifs. Cette période correspond à la réalité locale de la sécheresse hydrologique ayant affecté l'Afrique. Il faut y ajouter d'autres années et des périodes plus courtes qui ont aussi connu des déficits d'écoulement. Du point de vue des crues, il ressort des analyses que ce ne sont pas toujours pendant les années excédentaires que les inondations catastrophiques sont enregistrées.

Disponibilité des données. Les données utilisées sont accessibles uniquement auprès des auteurs.

Collaborateurs. Cet article est redigé par FS sous la supervision de MT et JGD. DN a collecté les données hysrologiques et participé a leurs traitements.

Intérêts concurrents. Les auteurs déclarent qu'ils n'ont aucun conflit d'intérêts.

Clause de non-responsabilité. Publisher's note : Copernicus Publications remains neutral with regard to jurisdictional claims in published maps and institutional affiliations.

Déclaration du numéro spécial. This article is part of the special issue "Hydrology of Large River Basins of Africa". It is a result of the 4th International Conference on the "Hydrology of the Great Rivers of Africa", Cotonou, Benin, 13–20 November 2021.

Remerciements. Nous exprimons notre gratitude a l'endroit de IAHS SYSTA award qui a financé la présentation et la discussion des resultats de cette étude à la 4ème Conférence de Friend Water sur l'hydrologie des grands fleuves d'Afrique qui s'est tenue du 13 au 20 novembre 2021 à Cotonou.

Références

Bureau de la coordination des affaires humanitations des Nations Unies (OCHA) : Note de synthèse : impact des inondations. Afrique de l'Ouest et du Centre, septembre 2012, Rapport du Bureau de la Coordination des Affaires Humanitaires des Nations Unies, 5 pp., en ligne, 2012.

Cisse, M. T., Sambou, S., Dieme, Y., Diatta, C., et Bop, M. : Analyse des écoulements dans le bassin du fleuve Sénégal de 1960 à 2008, Revue des Sciences de l'Eau, vol. 27, no. 2, 167–187, https://doi.org/10.7202/1025566ar, 2014.

Commission du bassin du lac Tchad (CBLT) : Création et vulgarisation d'une charte de l'eau du lac Tchad. Phase 1 – Diagnostic A – Les défis de gestion de l'eau et des écosystèmes à relever en commun, CBLT, N'Djamena, Tchad, 177 pp., 2010.

Direction de la Statistique et de la Comptabilité Nationale (DSCN) : Conditions de vie des ménages et profil de pauvreté à l'Extrême-Nord Cameroun en 2001, Édit. Direction de la Statistique et de la Comptabilité Nationale, étude réalisée dans la cadre du PREPAFEN, 131 pp., accessible à l'adresse suivante : http://catalog.ihsn.org/catalog/3229/download/47794/ECAM2-ProfilExtrOme-Nord.pdf (la date du dernier accès : 25 janvier 2020), 2002.

Institut National de la Statistique du Cameroun (INS) : Rapport national sur les objectifs du millénaire pour le développement en 2015, Institut National de la Statistique, Yaoundé, Cameroun, 48 pp., 2015.

Lang, C. et Gille, E. : Une méthode d'analyse du tarissement des cours d'eau pour la prévision des débits d'étiage, Norois, 201, 31–43, https://doi.org/10.4000/norois.1743, 2006.

Lévêque, C. : Variabilité du climat et des régimes hydrologiques, in : Les poissons des eaux continentales africaines. Diversité, écologie, utilisation par l'homme, édité par : Lévêque, C. et Paugy, D., IRD, Paris, accessible à l'adresse suivante : https://horizon.documentation.ird.fr/exl-doc/pleins_textes/ed-06-08/010037336.pdf (la date du dernier accès : 14 mars 2019), 2006.

L'Hôte, Y. : Climatologie, in : Atlas de la province de L'Extrême-Nord du Cameroun, IRD, Paris, 17–20, accessible à l'adresse suivante : https://books.openedition.org/irdeditions/11550?lang=fr (la date du dernier accès : 25 novembre 2020), 2000.

Loup, J. : Pardé Maurice. – Sur la puissance des crues en diverses parties du monde, Rev. Géogr. Alp., 53, 157–159, 1965.

Nguimalet, C. R. : Changements enregistrés sur les extrêmes hydrologiques de l'Oubangui à Bangui (République centrafricaine) :

analyse des tendances, Revue des Sciences de l'Eau, 30, 183–196, 2017.

Olivry, J. C. : Fleuves et Rivières du Cameroun, collection "Monographies hydrologiques ORSTOM", vol. 9, MESRES/ORSTOM, Paris, 733 p. + cartes hors texte, accessible à l'adresse suivante : https://horizon.documentation.ird.fr/exl-doc/pleins_textes/pleins_textes_6/Mon_hydr/25393.pdf (la date du dernier accès : 14 mars 2019), 1986.

Olivry, J. C. : Long term effects of rain shortage : the ill rivers of Western and Central Africa, in : FRIEND General Report/UNESCO IHP V, UNESCO, Paris, 158–162, 1997.

Olivry, J. C., Bricquet, J. P., Bamba, F., et Diarra, M. : Le régime hydrologique du Niger supérieur et le déficit des deux dernières décennies, in : Quelques données préliminaires sur l'environnement et la qualité des apports du Niger au Sahel, édité par : OLIVRY, J. C., Diallo Iam, M., et Bricquet, J. P., ORSTON-CNRST, Bamako, 9–25, 1994.

Olivry, J. C., Chouret, A., Vuillaume, G., Lemoalle, J., et Bricquet, J. P. : Hydrologie du lac Tchad, ORSTOM, Paris, 302, accessible à l'adresse suivante : https://core.ac.uk/download/pdf/39853684.pdf (la date du dernier accès : 6 novembre 2020), 1996.

Olivry, J. C., Bricquet, J. P., et Mahé, G. : Variabilité de la puissance des crues des grands cours d'eau d'Afrique intertropicale et incidence de la baisse des écoulements de base au cours des deux dernières décennies, in : Water resources variability in Africa during the XXth Century, actes de colloque (Abidjan, Côte d'Ivoire), édité par : Servat, É., Hughes, D., Fritsch, J. M., et Hulme, M., IAHS, 252, 189–197, 1998.

Proc. IAHS, 384, 247–253, 2021
https://doi.org/10.5194/piahs-384-247-2021

Open Access PIAHS

Hydrology of Large River Basins of Africa

Changement climatique dans le bassin versant de l'Ogooué : évolution récente et impact sur les écoulements

Valentin Brice Ebodé[1], **Gil Mahé**[2], **and Ernest Amoussou**[3]

[1]Département de Géographie, Université de Yaoundé 1, Yaoundé, Cameroun
[2]Institut de Recherche pour le Développement, Marseille, France
[3]Département de Géographie et Aménagement du Territoire, Université de Parakou, Parakou, Bénin

Correspondence: Valentin Brice Ebodé (ebodebriso@gmail.com)

Published: 16 November 2021

Résumé. L'objectif de cet article est de faire une évaluation des tendances récentes des grandeurs hydroclimatiques dans le bassin de l'Ogooué en contexte de changement climatique. Pour cela, les données de pluies et de débits de ce bassin ont été analysées au moyen du test de Pettitt. Les résultats de cette étude révèlent une diminution statistiquement significative des modules annuels que le test de Pettitt situe en 1972–1973, mais rien de tel pour les pluies à ce même pas de temps. Les écarts des moyennes décennales à la moyenne interannuelle montrent cependant des liens plus nets entre ces variables, se traduisant essentiellement par une diminution synchrone au cours des décennies 1970 et 1980. Cette diminution est suivie d'une reprise au cours des deux décennies d'après (1990 et 2000), laquelle s'estompe à nouveau au cours de la décennie 2010. Les écoulements des saisons pluvieuses (printemps et automne) ont enregistré les modifications les plus importantes, suite aux variations des régimes pluviométriques des saisons sèches (hiver et été) qui les précèdent. La saison sèche d'hiver a connu entre les décennies 1970–1990 une diminution importante des précipitations qui a provoquée une chute des écoulements du printemps. Le rehaussement et la flexion respectivement notés au cours des décennies 2000 et 2010 s'accompagnent des mêmes tendances dans les écoulements du printemps. A l'inverse, entre les décennies 1980–1990, il a été noté une hausse conjointe des pluies d'été et des écoulements d'automne. La flexion des pluies d'été notée depuis la décennie 2000 est également perceptible dans les écoulements d'automne. Ces résultats pourraient servir à renforcer les capacités de gestion des ressources en eau dans le bassin versant concerné et dans la région. Ils apportent également de nouveaux éléments pour étudier et comprendre la variation saisonnière et la disponibilité de l'eau douce en aval, dans les estuaires et les zones côtières des rivières régionales.

The objective of this article is to assess recent trends of hydroclimatic quantities in Ogooue basin in the context of climate change. For this, the rainfall and discharges data of this basin were analyzed using the Pettitt test. The results of this study reveal a statistically significant decrease in runoff that the Pettitt test situates in 1972–1973, but nothing like that for rainfall at this same time scale. The decadal deviations from the interannual average, however show links between these variables essentially marked by a synchronous decrease in the 1970s and 1980s. This decrease is followed by a recovery in the two decades following (1990 and 2000), which stops again during the 2010s. The flow of the rainy seasons (Spring and Autumn) recorded the most important modifications, following variations in the rainfall regimes of the dry seasons (Winter and Summer) which precede them. The dry Winter season experienced a significant decrease in precipitation between the 1970s and 1990s, which caused a decrease in Spring flows. The increase and flexion noted respectively during the 2000 and 2010 decades are accompanied by the same trends in Spring flows. Conversely, between the 1980s and the 1990s, there was a joint increase in Summer rainfall and Autumn flows. The decrease of Summer rainfall noted since the 2000s is also noticeable in the Autumn flows. These results could be used to strengthen water resources management capacities in the watershed concerned and the region. They also provide new insights to study and understand seasonal variation and availability of freshwater downstream, in estuaries and coastal areas of regional rivers.

1 Introduction

L'intérêt désormais consacré à la compréhension de la variabilité climatique en Afrique centrale provient de son impact sur les écoulements, et ainsi sur les économies des pays en développement de cette région. L'Ogooué, l'un des plus grands fleuves d'Afrique centrale équatoriale par sa longueur (environ 1000 km), constitue un énorme espoir de vitalité pour les populations des pays qu'il traverse. Prenant sa source au Congo-Brazzaville, dans la partie occidentale des plateaux Batékés, sur les monts Massa, il parcourt essentiellement le territoire gabonais avant de se jeter dans l'océan Atlantique. Ce fleuve a fait l'objet d'un nombre relativement modeste de pertinentes études tant hydrologiques que climatologiques depuis le début de ce vingtième siècle ; et pour cause, le manque de données d'observation sur cette période. Le peu d'études existantes a porté aussi bien sur l'évaluation de la ressource en eau que sur des projets d'aménagements potentiels dans les domaines hydro-agricole et hydroélectrique ou encore sur la propagation des crues (E. D. F., 1961 ; Carré, 1978 ; Buisson, 1985a et b ; Mahé et al., 1990, 2013 ; Bogning et al., 2020 ; Ebodé, 2020). La station de Lambaréné est parmi les plus anciennes et l'une des mieux étudiées sur ce cours d'eau. Les études sur cette station (Mahé et al., 1990, 2013 ; Bogning et al., 2020 ; Ebodé, 2020) ont déjà montré des changements climatiques avec des modifications du régime hydrologique du fleuve à partir de la décennie 1970. Bien qu'ayant déjà une certaine connaissance des modifications du régime hydrologique de l'Ogooué à cette station, il convient néanmoins de l'actualiser en intégrant les données des deux récentes décennies, compte tenu de ce que la quasi-totalité d'études sus-évoquées se fondent sur des données qui n'excèdent pas la décennie 1990. L'objectif principal de cet article est donc de documenter le type et l'ampleur des changements de régime hydrologique de l'Ogooué dus aux changements climatiques sur la période récente (1929–1930 à 2017–2018).

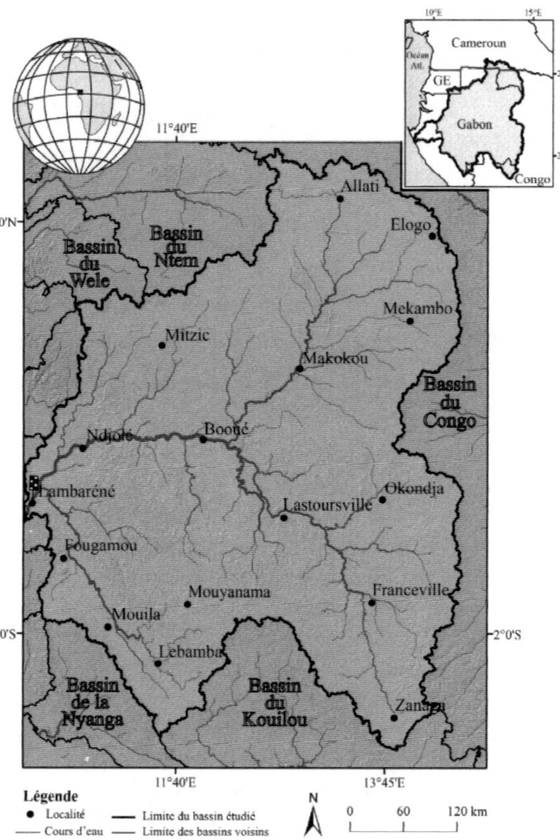

Figure 1. Localisation du bassin versant de l'Ogooué.

2 Cadre, données et méthodes

L'étude porte sur le bassin versant de l'Ogooué à Lambaréné (215 000 km^2). Ce bassin est situé en Afrique centrale, approximativement entre les latitudes 2°42′ N et 3°10′ S et les longitudes 10°9′ E et 14°30′ E (Fig. 1). Il appartient au domaine équatorial avec des précipitations annuelles abondantes, oscillant autour de 1800 mm, réparties sur quatre saisons d'inégale importance, dont deux sèches et deux pluvieuses. Les saisons de pluies (printemps et automne) sont généralement très humides avec des totaux records avoisinant 800 mm et marquées par de nombreux orages qui occasionnent parfois des crues importantes. La saison sèche d'été est la seule véritable saison sèche de la région, avec des précipitations moyennes oscillant autour de 200 mm ; la saison sèche d'hiver étant considérablement humide dans la région, avec des pluies moyennes sensiblement deux fois plus importantes que celles d'été.

Les séries de précipitations et de débits ont été utilisées dans ce travail. Les débits (journaliers) de l'Ogooué proviennent du SIEREM (Système d'Informations Environnementales sur les Ressources en Eau et leur Modélisation). Elles contiennent des lacunes sur les intervalles 1950–1953 et 1989–2000. Les débits de base ont été extraits du débit total à partir du filtre récursif Aquapak, en téléchargement libre sur : https://www.jacobs.com/natural-resource-management (la date du dernier accès : 5 mai 2019).

Les séries pluviométriques proviennent de deux sources. Sur la période 1940–1999, les grilles de pluies mensuelles SIEREM ont été utilisées. Leur unité d'espace est une maille de surface 0,5° × 0,5°, soit approximativement 2750 km^2 dans la zone considérée (Dieulin et al., 2019). Ces grilles sont en accès libre aux formats ASCII et NetCDF sur le site http://www.hydrosciences.org/spip.php\unskip\penalty\@M\hskip.5\fontdimen2\font\relax?article1387 (la date du dernier accès : 7 mai 2019). Avant toute utilisation, il a fallu les évaluer d'abord au moyen des comparaisons avec les pluies issues des stations. Entre 2000 et 2017, le choix a été porté sur les estimations de pluies journalières TRMM 3B42 V7, remplacées elles-mêmes par les pluies GPM (Global Precipitation Measurement) depuis 2015. L'ensemble de ces estimations de pluies sont téléchargeables gratuitement sur le site https://giovanni.gsfc.nasa.gov (la date du dernier accès : 11 juin 2019), au format ASCII et à une résolution spatiale de 0,25° × 0,25°. Les cumuls mensuels des pluies TRMM et GPM calculés ont d'abord été validés au moyen des comparaisons avec les pluies issues des stations, puis avec les pluies SIEREM sur les années de disponibilité commune (1998 et 1999). Les coefficients de correction (essentiellement compris entre 1,34 et 1,65) ont par la suite été établis dans le cas des mois pour lesquels les écarts entre les deux sources (SIEREM et TRMM) étaient importants (juin et août).

L'analyse des pluies, des débits et des coefficients d'écoulement a été réalisée à l'aide de tests statistiques de détection de rupture (corrélation sur le rang et test de Pettitt ; Lubès et al., 1994) au seuil de signification 95 %. Le test de Pettitt semble être le plus approprié pour l'analyse des séries lacunaires comme celles utilisées dans cette étude, parce qu'il sépare la série uniquement en deux périodes ayant un comportement globalement distinct, ce qui évite la détection des fausses ruptures comme on peut parfois l'observer avec d'autres tests comme la segmentation d'Hubert. Son choix au détriment des tests de tendance (test de Mann–Kendall par exemple) se justifie par le fait qu'il indique une date à partir de laquelle le changement devient significatif statistiquement parlant (rupture), ce qui permet une meilleure appréciation de la variabilité.

3 Résultats et discussion dans le contexte régional

3.1 Evolution des pluies annuelles et saisonnières de l'Ogooué

Les précipitations annuelles du bassin versant de l'Ogooué évoluent de manière non significative à la diminution entre 1940–1941 et 2017–2018 (Fig. 2). Le test de Pettit ne détecte d'ailleurs aucune rupture dans leur série chronologique. L'analyse des moyennes décennales révèle cependant dans leur évolution, une alternance des séquences sèches et humides plus ou moins longues. Ainsi, il a été observé au cours des décennies 1990 et 2000 une légère reprise consécutive à la sécheresse des décennies 1970 et 1980, qui semble avoir été interrompue à nouveau au cours de la décennie 2010 (Tableau 1). Ce retour à des conditions plus humides au cours de la décennie 1990 a également été suggéré partout en Afrique de l'ouest comme au Sénégal (Alhassane et al., 2013), au Burkina Faso (Lodoun et al., 2013) et au Niger (Ozer et al., 2009). Cependant, le rehaussement signalé dans cette étude s'estompe au cours de la décennie 2010, tandis que celui relevé en Afrique de l'ouest persiste jusqu'à présent (Descroix et al., 2020).

Les pluies des deux saisons sèches évoluent en tendances croisées sur le bassin versant de l'Ogooué (Fig. 2). Les cumuls de la saison sèche d'été évoluent statistiquement à la hausse entre 1940–1941 et 2017–2018. Le test de Pettit met en évidence dans leur série chronologique une rupture en 1979–1980 (Fig. 2). Le surplus par rapport à la moyenne enregistré suite à cette dernière est estimé à $+12,5\,\%$. Pourtant excédentaire dans son ensemble, il est observé pour les précipitations de cette saison un relâchement progressif après la forte augmentation des décennies 1980 et 1990. Ce relâchement devient davantage important au cours de la décennie 2010, pour laquelle il est noté un écart négatif par rapport à la moyenne interannuelle de $-3,2\,\%$ (Tableau 1). Les pluies de la saison sèche d'hiver évoluent de manière non significative à la baisse. Cependant, on observe dans leur évolution une chute progressive entre les décennies 1970 et 1990, puis un relèvement au cours des deux décennies d'après, qui semble quand-même beaucoup plus important pour la décennie 2000 (Tableau 1).

Les cumuls des saisons pluvieuses (automne et printemps) n'ont pas connu de diminution significative d'après le test de Pettit. Les écarts des moyennes décennales à la moyenne interannuelle sont certes faibles dans l'ensemble, mais ils laissent transparaître comme dans le cas des totaux annuels, plusieurs séquences aux tendances évolutives différentes (Tableau 1). De l'analyse de ces écarts, il ressort que les pluies du printemps ont diminué de 2 % et 4,5 % au cours des décennies 1970 et 1980. Mais depuis le début de la décennie 1990, on note en général dans leur évolution un très léger rehaussement (Tableau 1). Dans le cas des pluies d'automne, une hausse de $+1,5\,\%$ et $+4,3\,\%$ a été enregistrée durant les décennies 1980 et 1990, suivie d'une faible diminution depuis la décennie 2000.

3.2 Evolution des modules annuels et saisonniers de l'Ogooué

Les modules de l'Ogooué (annuels et saisonniers) évoluent à la baisse sur la période d'étude (Fig. 2). Cette diminution n'est pas significative statistiquement parlant dans le seul cas de l'automne (Fig. 2). Après la rupture observée en 1972–1973 dans les différents cas, les déficits par rapport à la moyenne interannuelle calculés pour les modules annuels, du printemps, de l'été et de l'hiver sont respectivement de $-6,4\,\%$; $-10,3\,\%$; $-6,9\,\%$ et $-6,4\,\%$ (Fig. 2). En dehors de l'automne, les écarts décennaux montrent une chute géné-

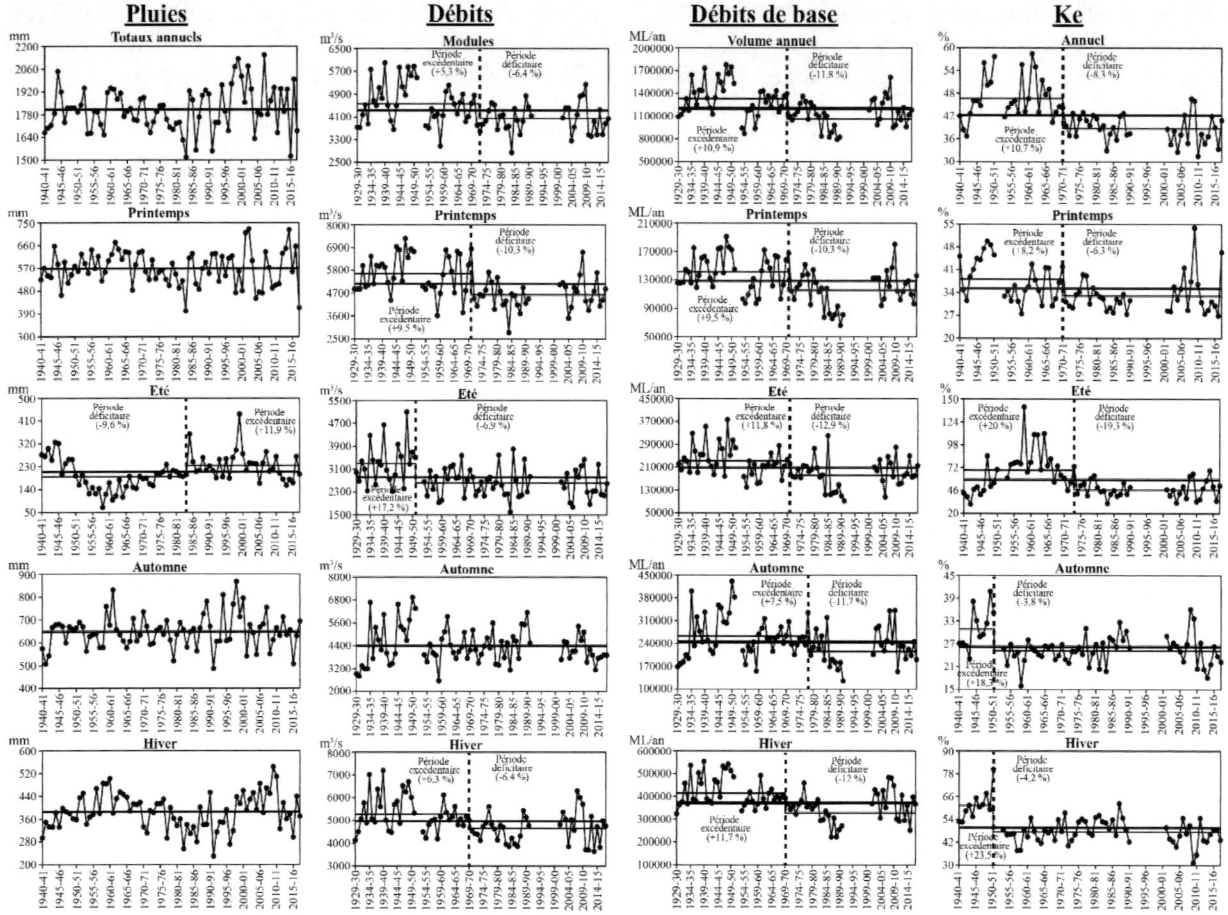

Figure 2. Evolution des pluies, débits, débits de base et coefficients d'écoulement annuels et saisonniers du bassin du l'Ogooué sur leurs périodes d'études respectives.

ralisée ayant débuté au cours de la décennie 1970, qui s'est considérablement amplifiée durant la décennie 2010. Cette décennie est la plus sèche de l'ensemble de la période d'observation (Tableau 1).

Les résultats de cette étude corroborent les conclusions des premières études menées dans la région. Cette étude confirme une rupture dans la série des modules annuels de l'Ogooué durant le décennie 1970 (en 1972–1973), ce qui est en accord avec les travaux de Liénou et al. (2008) et Conway et al. (2009) qui mettent en évidence des ruptures respectivement sur le Ntem en 1970 et l'Ogooué en 1977. Le présent travail relève également une diminution plus importante de la crue du printemps comparée à celle d'automne. Un constat identique a été fait par Mahé et al. (2013) sur ce même bassin.

3.3 Sensibilité des écoulements aux variations interannuelles des précipitations

Il est certes vrai que la stationnarité des pluies et des écoulements ne montre pas de liens apparents sur le bassin versant de l'Ogooué aux pas de temps étudiés (annuel et saisonnier), mais l'analyse des écarts décennaux à la moyenne interannuelle montre des liens relativement importants entre ces variables. Au pas de temps annuel, les tendances observées pour les précipitations entre les décennies 1970–1980 (baisse), 1990–2000 (rehaussement) et au cours de la décennie 2010 (rechute) sont identiques à celles enregistrées pour les écoulements (Tableau 1). On remarque cependant une diminution de débits plus forte que celle des pluies. Ceci ne pourrait s'expliquer que par une diminution des apports souterrains, ce que semble corroborer l'étude des débits de base. En effet, les débits de base diminuent considérablement, ce qui indique une vidange des nappes. Après les séries de sécheresse des années 1970 et 1980, on a assisté à un appauvrissement sévère des nappes phréatiques si bien que l'amé-

Tableau 1. Ecarts (%) des moyennes décennales annuelles et saisonnières des pluies, débits, débits de base et coefficients d'écoulement par rapport à leurs moyennes interannuelles sur le bassin étudié.

Variables	Périodes	Ecarts à la moyenne interannuelle (%)				
		Annuel	Printemps	Eté	Automne	Hiver
Précipitations	1940–1949	−0,5	−2,8	29,6	−3,5	−8,1
	1950–1959	−1,7	2,1	−31,1	−1,1	7,5
	1960–1969	1,7	7,5	−28,8	0,9	11,1
	1970–1979	−2,7	−1,9	−7,6	−1,4	−3,4
	1980–1989	−3,1	−5,9	10	1,5	−14,7
	1990–1999	2,3	0,9	20,3	4,3	−8,6
	2000–2009	4,2	−1,3	12,1	−0,1	15,5
	2010–2017	−1	1,9	−3,1	−1	0,2
Débits	1930–1939	8,8	8,2	17,4	−1,5	13,6
	1940–1949	14,9	17,7	17,3	12,7	12,7
	1950–1959	−5,1	−2,4	−8,8	−4,5	−6,3
	1960–1969	5,9	11,3	4,3	1,9	4,9
	1970–1979	−4,6	−3,4	−6,3	−2,6	−6,8
	1980–1988	−8,6	−17,8	−10,5	6,1	−10,9
	1990	–	–	–	–	–
	2001–2009	0,1	−3,2	−4	1,5	4,6
	2010–2017	−13,2	−11,6	−15,1	−14,2	−13,1
Débits de base	1930–1939	14,1	10	21	5,3	18,2
	1940–1949	23,4	22,5	22,5	22,5	23,1
	1950–1959	−6,6	−11,9	−7,1	−3,3	−4,8
	1960–1969	10,6	12,5	8,2	8,3	9,3
	1970–1979	0,6	0,3	−1,2	2,5	−1,3
	1980–1988	−33,5	−26	−22,3	−11,6	−22,8
	1990	–	–	–	–	–
	2001–2009	6,5	2,5	−1,4	9,6	11
	2010–2017	−10,4	−10,1	−10,1	−14,4	−10
Ke	1940–1949	7,5	20,9	−15,9	17,1	19,8
	1950–1959	11,6	−2,7	39,2	−6,3	−1,6
	1960–1969	13,6	3,1	42,7	−3,4	−4,7
	1970–1979	−3,7	−2,3	−5,8	−5,2	−2
	1980–1988	−8,5	−12,7	−24,5	2,4	6,8
	1990	–	–	–	–	–
	2001–2009	−7,6	1,3	−18,6	7,3	−9,5
	2010–2017	−12,4	−8,4	−17,8	−14,2	−8,9

lioration de la pluviométrie des décennies 1990 et 2000 sur le bassin n'a pas pu entraîner leur reconstitution. C'est ce qui a provoqué des disproportions parfois importantes entre les diminutions de la pluviométrie et de l'écoulement à partir de la décennie 1970.

Durant le printemps, les précipitations ont évolué modérément, de sorte qu'aucune rupture n'a été détectée dans leur série. A l'inverse, les débits de base, les débits moyens et les coefficients d'écoulement de cette saison diminuent de manière statistiquement significative depuis la décennie 1970. De même, l'examen des écarts à la moyenne de ces variables montre généralement des diminutions importantes pour ces mêmes variables, pendant que les pluies ne diminuent pas ou diminuent peu (Tableau 1). Le fléchissement des pluies d'hiver depuis la décennie 1970 semble être en rapport étroit avec celui des écoulements du printemps. Le rythme d'évolutions de ces deux variables au fil des décennies est similaire. Cela est clairement visible au cours de la décennie 2000 pour laquelle, une hausse des pluies d'hiver a entraîné un relèvement considérable des débits de base, des coefficients d'écoulement et des débits du printemps (Tableau 1). Il en est de même au cours de la décennie d'après où un fléchissement des pluies d'hiver est synchrone avec celui des variables d'écoulement étudiées, pendant que les pluies du printemps ont diminué elles-mêmes d'environ 2%. Cette influence des précipitations d'hiver a déjà été suggérée par Liénou et al. (2008). Ces auteurs soutiennent qu'à partir de la décennie 1970, la baisse des pluies d'hiver crée un plus important déficit hydrique (évaporation et réserve en eau du sol) du bassin versant au début de la première saison des pluies du printemps. Cette aridification de l'hiver fait qu'une partie plus importante des précipitations reçues pendant le printemps participe d'abord au comblement de ce déficit hydrique, et donc la fraction qui génère effectivement les écoulements se trouve réduite. Ce qui se traduit alors par des lames écoulées plus faibles pour les mêmes hauteurs moyennes de précipitations durant le printemps, c'est ce qui explique la baisse du coefficient d'écoulement.

En automne, les pluies et les écoulements diminuent très légèrement sur l'ensemble de la période d'étude. Mais cependant, au cours des décennies 1980 et 2000, les variables d'écoulement augmentent, alors que les pluies ne varient pas beaucoup dans l'ensemble. Cette augmentation résulterait de la hausse des pluies d'été. L'impact des précipitations d'été sur les écoulements d'automne est davantage perceptible durant les décennies 2010 pour laquelle il est noté une diminution conjointe des pluies d'été, des débits de base, des coefficients d'écoulement et des débits d'automne, malgré une variation négligeable des pluies d'automne (Tableau 1). Au regard de ces évolutions, il ressort que, l'augmentation des pluies au cours de la saison sèche d'été durant les décennies allant de 1970–1990, réduit considérablement le déficit d'évaporation et des réserves en eau du sol au début de la saison de pluies d'automne, favorisant le ruissellement (Liénou et al., 2008 ; Ebodé et al., 2020). La partie des précipitations qui participe effectivement à l'écoulement s'accroît, d'où une augmentation du coefficient d'écoulement d'automne au cours de ces décennies (Tableau 1). La diminution des pluies d'été durant les décennies 2000 et 2010 entraîne l'effet inverse, c'est pourquoi il est constaté au cours de ces dernières une réduction des coefficients d'écoulement, responsable d'une baisse des débits.

4 Conclusion

Jusqu'à présent, peu d'études d'hydroclimatologie ont porté sur le bassin de l'Ogooué à Lambaréné, du fait des lacunes d'observation dans les séries chronologiques des débits et de l'inexistence des données de pluies dans certaines parties du bassin après les années 2000. Pour effectuer cette étude, nous avons réalisé un travail de validation de deux sources de données pertinentes disponibles pour la région (SIEREM et TRMM). Un réajustement (au moyen des coefficients correcteurs) des données TRMM a par la suite été opéré pour les mois présentant des écarts importants entre les deux sources. La série complète des pluies obtenue explique assez bien la variabilité des écoulements sur ce bassin. L'analyse de cette série des précipitations et celle des écoulements montre en effet une évolution synchrone entre les deux variables comparées. En plus de la concordance notée entre elles à l'échelle annuelle au fil du temps, on constate également que les décennies 1970 et 1980, marquent le début d'un changement important dans l'évolution des précipitations des saisons sèches dans le bassin, caractérisé par une augmentation des volumes précipités en été et une diminution en hiver. Ces modifications observées jusqu'à la fin de la décennie 1990 ont provoqué une modification des écoulements des saisons de pluies qui les suivent directement. Il a ainsi été relevé une augmentation des écoulements d'automne et une baisse de ceux d'hiver sur cette même période. Le début de la décennie 2000 est marqué par une inversion des tendances. On note désormais une baisse des pluies d'été accompagnée d'une diminution des écoulements d'automne, et inversement une hausse de celles d'hiver synchrone avec un relèvement des écoulements du printemps.

Disponibilité du code. Les logiciels utilisés pour l'analyse et le traitement des données dans cette étude (Khronostat 1.01 et QGIS 3.20) sont gratuits. Les liens donnant accès aux sites sur lesquels ils peuvent être téléchargés sont les suivants : http://www.hydrosciences.org/index.php/2020/09/04/khronostat/ (pour le premier) (HydroSciences Montpellier, 2021) et https://www.qgis.org/fr/site/forusers/download.html (pour le second) (Open Source Geospatial, 2021).

Disponibilité des données. Les données de débits utilisées dans cette étude proviennent du SIEREM. Pour entrer en possession de ces données, il faudrait nécessairement entrer en contact avec la responsable actuelle de la banque des données à l'adresse : nathalie.rouche@umontpellier.fr. Les données de pluies (SIEREM et TRMM) par contre, sont disponibles gratuitement en ligne sur les sites : http://www.hydrosciences.fr/sierem/ (HydroSciences Montpellier, 2019) et https://giovanni.gsfc.nasa.gov (NASA, 2019).

Collaborateurs. Ce travail a été conceptualisé par VBE. La collecte, le traitement et l'analyse des données ont été effectués par VBE. Le projet a été dirigé par GM. La rédaction du manuscrit et la relecture ont été faites par VBE, EA et GM.

Intérêts concurrents. Les auteurs déclarent qu'ils n'ont aucun conflit d'intérêts.

Déclaration du numéro spécial. This article is part of the special issue "Hydrology of Large River Basins of Africa". It is a result of the 4th International Conference on the "Hydrology of the Great Rivers of Africa", Cotonou, Benin, 13–20 November 2021.

Remerciements. Les auteurs remercient chaleureusement les agents de la Représentation de l'IRD (Institut de Recherche pour le Développement) au Cameroun et la direction du LMI DYCOFAC (notamment Jean-Jacques Braun), à Yaoundé, pour leur soutien administratif.

Références

Alhassane, A., Salack, S., Ly, M., Lona, I., Traore, S. B., et Sarr, B. : Evolution des risques agroclimatiques associés aux tendances récentes du régime pluviométrique en Afrique de l'Ouest soudano-sahélienne, Sécheresse, 24, 282–93, https://doi.org/10.1684/sec.2013.0400, 2013.

Bogning, S., Frappart, F., Paris, A., Blarel, F., Ninõ, F., Picart, S. S., Lanet, P., Seyler, F., Mahé, G., Onguene, R., Bricquet, J. P., Etame, J., Paiz, M. C., and Braun, J. J. : Hydroclimatology study of the Ogooué River basin using hydrological modeling and satellite altimetry, Adv. Space Res., 68, 672–690, https://doi.org/10.1016/j.asr.2020.03.045, 2020.

Buisson, A. : La grande saison sèche 1985 au Gabon : Situation climatique en Afrique intertropicale, La Météorologie, 8 , 5–13, 1985a.

Buisson, A. : Une anomalie climatique au Gabon en 1984, La Météorologie, 7, 36–47, 1985b.

Carré, P. : Fleuves et rivières du Gabon. Diversité des régimes hydrologiques, République gabonaise, Centre national de la recherche scientifique et technologique, Section d'études et recherches en hydrologie, ORSTOM, Libreville, Gabon, 22 pp., 1978.

Conway, D. P., Persechino, A., Ardoin-Bardin, S., Hamandawana, H., Dieulin, C., and Mahé, G. : Rainfall and river flow variability in sub-saharan Africa during the 20th century, J. Hydrometeorol., 10, 41–59, 2009.

Descroix, L., Sané, Y., Mamadou, T., Manga, S-P., Boubacar, D., Mingou, J., Mendy, V., Coly, S., Dièye, A., Badiane, A., Senghor, M.-J., Diedhiou, A.-B., Djiby, S., Bouaita, Y., Soumaré, S., Diop, A., Bakary, F., Bamol, A., Machu, E., Montoroi, J.-P., Andrieu, J., and Vandervaere, J.-P. : Inverse estuaries in West Africa : Evidence of the rainfall recovery ?, Water, 12, 647, https://doi.org/10.3390/w12030647, 2020.

Dieulin, C., Mahé, G., Paturel, J. E., Ejjiyar, S., Tramblay, Y., Rouché, N., and Mansouri, B. E. : A new 60-year monthly-gridded rainfall data set for Africa, Water, 11, 387, https://doi.org/10.3390/w11020387, 2019.

Ebodé, V. B. : Variabilité hydroclimatique en Afrique centrale occidentale forestière : entre analyse des fluctuations observées, recherche des facteurs explicatifs et modélisation prédictive, Generis Publishing, Miron Costin, République de Moldavie, 382 pp., 2020.

Ebodé, V. B., Mahé, G., Dzana, J. G., and Amougou, J. A. : Anthropization and climate change : impact on the discharges of forest watersheds in Central Africa, Water, 12, 2718, https://doi.org/10.3390/w12102718, 2020.

Electricité de France (E. D. F.) : Prospection hydro-électrique générale des bassins de l'Ogooué et de la Nyanga, Rapport intérimaire de la première campagne d'étude, Ministère des travaux publics, Libreville, République Gabonaise, 103 pp., 1961.

HydroSciences Montpellier : Grilles de pluies, accessible à l'adresse suivante : http://www.hydrosciences.fr/sierem/, la date du dernier accès : 7 mai 2019.

HydroSciences Montpellier : Khronostat, accessible à l'adresse suivante : http://www.hydrosciences.org/index.php/2020/09/04/khronostat/, la date du dernier accès : 20 octobre 2021.

Liénou, G., Mahé, G., Paturel, J.-E., Servat, E., Sighomnou, D., Ekodeck, G. E., Dezetter, A., et Dieulin, C. : Evolution des régimes hydrologiques en région équatoriale camerounaise : un impact de la variabilité climatique en zone équatoriale, Journal des Sciences Hydrologiques, 53, 789–800, 2008.

Lodoun, T., Giannini, A., Traore, P. S., Some, L., Moussa, S., Vaksmann, M., and Rasolodimby, J. M. : Changes in seasonal descriptors of precipitation in Burkina Faso associated with late 20th century drought and recovery in West Africa, Environmental Development, 5, 96–108, https://doi.org/10.1016/j.envdev.2012.11.010, 2013.

Lubès, H., Masson, J. M., Servat, E., Paturel, J. E., Kouame, B., et Boyer, J. F. : Caractérisation des fluctuations dans une série chronologique par applications de tests statistiques, Etudes bibliographiques UR2-CEC, Orstom, Montpellier, France, 1994.

Mahé, G., Lerique, J., et Olivry J. C. : L'Ogooué au Gabon : Reconstitution des débits manquants et mise en évidence de variations climatiques à l'équateur, Hydrologie Continentale, 5, 105–124, 1990.

Mahé, G., Liénou, G., Descroix, L., Bamba, F., Paturel, J.-E., Laraque, A., Meddi, M., Habaieb, H., Adegea, O., Dieulin, C., Chahnez, K. F., and Khomsi, K. : The rivers of Africa : witness of climate change and the human impact on the environment, Hydrol. Process., 27, 2105–2114, 2013.

NASA : Earth Data, NASA, accessible à l'adresse suivante : https://giovanni.gsfc.nasa.gov, la date du dernier accès : 11 mai 2019.

Open source Geospatial : QGIS, accessible à l'adresse suivante : https://www.qgis.org/fr/site/forusers/download.html, la date du dernier accès : 22 octobre 2021.

Ozer, P., Hountondji, Y.-C., et Laminou, M. O. : Evolution des caractéristiques pluviométriques dans l'est du Niger de 1940 à 2007, Geo-Eco-Trop., 33, 11–30, https://orbi.uliege.be/handle/2268/78267 (la date du dernier accès : 22 octobre 2021), 2009.

Proc. IAHS, 384, 255–260, 2021
https://doi.org/10.5194/piahs-384-255-2021

Open Access PIAHS

Hydrology of Large River Basins of Africa

Impact of climate change on water availability in the Oueme catchment at the outlet of the Save's bridge (Benin, West Africa)

Amedée Chabi[1,2], **Esdras Babadjidé Josué Zandagba**[1,2], **Ezekiel Obada**[1,2], **Eliezer Iboukoun Biao**[1,2], **Eric Adéchina Alamou**[1,2], and **Abel Afouda**[1]

[1]Laboratory of Applied Hydrology, National Water Institut, Abomey-Calavi, Benin
[2]Laboratoire de Géosciences de l'Environnement et Applications/Université Nationale des Sciences, Technologies, Ingénierie et Mathématiques (UNSTIM), Abomey, Benin

Correspondence: Amedée Chabi (amedeechabi1@gmail.com)

Published: 16 November 2021

Abstract. One of the major threats to water resources today remains climate change. The objective of this study is to assess the impact of climate change on water availability in Oueme catchment at Savè. Precipitation provided by three regional climate models (RCMs) was analyzed. Bias in these data was first corrected using the Empirical Quantile Mapping (EQM) method be for etheir use as input to hydrological models. To achieve the objective, six hydrological models were used (AWBM, ModHyPMA, HBV, GR4J, SimHyd and Hymod). In projection, the results showed that the AWBM model appears to be the best. The multi-model approach further improves model performance, with the best obtained with combinations of the models AWBM-ModHyPMA-HBV. The AWBM model showed a fairly good capability for simulating flows in the basin with only HIRHAM5 climate model data as input. Therefore, the simulation with the HIRHAM5 data as inputs to the five (05) hydrological models, showed flows that vary at the horizons (2025, 2055 and 2085) under the scenarios (RCP4.5 and RCP8.5). Indeed, this variation is largely due to anthropogenic greenhouse gas (GHG) emissions.

1 Introduction

The successive reports of the Intergovernmental Panel on Climate Change (IPCC) have confirmed that major climate changes are under way and their effects are now proven. The first volume of the fifth report, published in 2013, reveals that, projections for 2100 depend on human activities and model uncertainties (Quae, 2013). Indeed, the planet is warming up because of man-made greenhouse gas emissions (IPCC, 2014). Due to its geographical location, West Africa will suffer more from the consequences of climate change. Indeed, it is marked by the repeated drought which has worsened since the 1970s and 1980s. This vulnerability to climate disturbances is due to the diversity of impacts, multiple stresses, greater magnitude and low adaptive capacity (Obada, 2017). A decrease in average annual rainfall of 20 % to 40 % was also observed between 1931–1960 and 1968–1990 (IPCC, 2007), which led to a greater drop in the flow of rivers from 40 % to 60 %, and was followed by increased vulnerability to natural disasters (Afouda et al 2007). The Oueme catchment at Savè is not spared from these climatic variations. There is a considerable decrease in the amount of water available, and this decrease is aggravated with the current context of climate change. It is therefore necessary to understand the relationships between hydrological functions and all interactions with climate. Thus, the study of the impacts of climate change on river flows in the Oueme catchment at Savè is of paramount importance in order to better develop adaptation policies. This study aims to assess the impacts of climate change on water availability in the Oueme catchment at Save.

2 Study area and methodology

2.1 Study area and data

The Oueme catchment at Savè covers an area of approximately 23 600 km^2. it is located between 8°00′40″ and 10°03′20 Lat. North, 1°30′20″ and 2°32′25″ Long. East (Le Barbé and Borel, 1987). The climate in this basin is tropical, but with a strong tendency towards the Sudanese type. It is characterized by two main seasons: a dry season (November to March) and a rainy season (April to October) (Akognongbe, 2007; Zandagba, 2012) (Fig. 1). The data used in this study are of two types: The observed daily hydro climatic data (precipitation, potential evapotranspiration (PET) and flow rates). They have been extracted from the database of Meteo-Benin and General Direction of Water of Benin, and covering the period of 1960 to 2014. Simulated data from regional climate models (RCM) (precipitation and PET). The period 1970 to 2005 has been chosen as reference period and 2006 to 2100 was chosen for future projection. All these data come from the CORDEX AFRICA https://cordex.org/data-access/regional-climate-change.simulation (last access: 18 June 2020) project database and have a spatial resolution of 0.5° × 0.5°.

2.2 Methods

2.2.1 Bias correction

To extract data from a grid of 0.5 × 0.5° spatial resolution, the EQM method has been used. This downscaling method preserves the long-term absolute (relative) trend of the simulated temperature or precipitation data and modify the daily variability of the simulated data about their monthly means to match the observed daily variability.This method uses empirical distribution functions (Sennikovs and Bethers, 2009) and is constructed by calculating empirical Probability Distribution Functions (PDF) and then using Cumulative Distribution Functions (CDF) for correction (Eq. 1):

$$Y = F_{\text{obs}}^{-1}(F_{\text{MCR}}(x)) \tag{1}$$

where Y is the bias corrected value of the meteorological parameter and x its value simulated by the model; F_{MCR} is the CDF (Cumulative Distribution Function) of the data simulated by the MCRs and F_{obs}^{-1} is the inverse of the CDF of the observed data

2.2.2 Hydrological modeling

In order to characterize the impacts of climate change on the availability of water resources in the Oueme catchment at Savè, the rainfall and PET resulting from the correction of the biases of the 03 investigated RCMs are used as input data to 06 hydrological models (Table 1).

The selected hydrological models have been calibrated from1974 to 1978 and validated from 2003 to 2007. Indeed, these periods has been chosen because of the availability of continuous data and necessity to considered different type of period (wet and dry period) which must be distinct from each other, in order to appreciate the robustness of the used model.

2.2.3 Performance criteria

The performance of both bias correction methods and models is evaluated using the Nash-Sutcliffe Efficiency Coefficient (NSE) (Nashand Sutcliffe, 1970), percent bias (PBIAS) (Guptaetal., 1999), root mean square deviation or square base error (RMSE), which in corporate both the variance of the estimate and its bias (Lehmannand Casella, 1998) and the error (RSR).

2.2.4 Analysis of the rate of variation of river discharge

Future changes relative to the reference period (1970–2005) are evaluated from Eq. (2). For this purpose, we have chosen to study the impacts of climate change over three (03) projection sub-periods: 2014–2040 (horizon, 2025), 2041–2070 (horizon, 2055) and 2071–2100 (horizon, 2085) to assess the changes in the investigated hydroclimatic series according to the RCP4.5 and RCP8.5 scenarios.

$$\text{change rate} = \frac{\overline{X}_{\text{P}} - \overline{X}_{\text{r}}}{\overline{X}_{\text{r}}} \times 100, \tag{2}$$

Where $\overline{X}_{\text{P}}$ the average of the hydro climatic parameter is over the projection period considered and $\overline{X}_{\text{r}}$ is the average of the series over the reference period.

3 Results and discussions

3.1 Rainfall bias correction performance

Figure 2 shows an example of performance of the EQM method in the correction of the biases of precipitations simulated by the models HIRHAM5, REMO, RCA4 and their ensemble. From the analysis of this figure, it can be seen that the corrected monthly precipitation data are close to those of the observations.

3.2 Performance of hydrological models and multi-models

Table 2 and Fig. 3 show the performance of models and multi-models in calibration and validation. From the analysis of Fig. 3, it was found that AWBM, ModHyPMA, HBV, GR4J models give good performance in calibration and validation with Nash performance criteria of over 60 %. In view of the results obtained and the number of parameters which characterize each of these hydrological models considered in this study, AWBM model appears superior to the others in terms of performance. From the analysis of Fig. 3, it also emerges that the multi-models AWBM-ModHyPMA-HBV

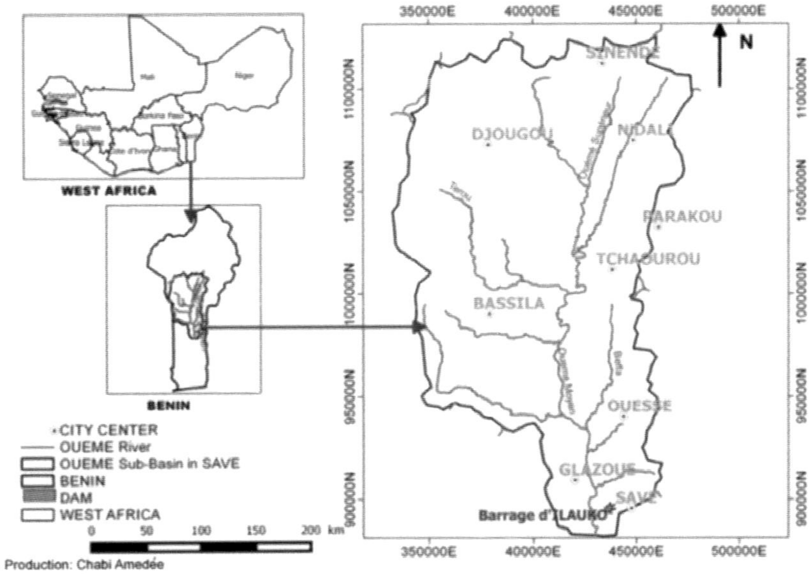

Figure 1. Geographic location in the Oueme at Save catchment.

Table 1. Description of selected models.

Modèle	Formal	Par	Res	Forcage	References
ModPMA	Phys	2	–	P, ETP + Q	Afouda et al. (2010)
GR4J	Conp	4	3	P, ETP + Q	Perrin (2000)
HBV	Conp	9	3	P, ETP + Q	Bergström and Forsman (1973)
SimHyd	Conp	7	2	P, ETP + Q	Chiew and Siriwardena (2005)
AWBM	Conp	7	3	P, ETP + Q	Boughton (1993)
Hymod	Conp	5	4	P, ETP + Q	Boyle (2001)

Formal = Formalisme, Par = Parameters, Res = Number of returns; Comp = Conceptuel, ETP + Q (calage).

Table 2. Performance of hydrological models in calibration and validation.

Modèles	ModHyPMA	GR4J	HBV	SimHyd	AWBM	Hymod
Calage (01-01-1974 au 31-12-1978)						
NSE	0.82	0.83	0.85	0.86	0.89	0.56
R^2	0.85	0.86	0.86	0.86	0.88	0.65
RMSE	0.25	0.30	0.23	0.23	0.19	0.38
MAE	0.05	0.13	0.02	0.04	0.01	0.12
Validation (01-01-2003 au 31-01-2007)						
NSE	0.70	0.75	0.81	0.65	0.82	0.54
R^2	0.73	0.79	0.81	0.83	0.85	0.74
RMSE	0.5	0.42	0.37	0.51	0.36	0.57
MAE	0.00	0.14	0.03	0.11	0.01	0.15

and the ensemble of the five models show good performance with the Nash value of 0.90 in calibration and 0.81 in validation.

Table 3. Comparison of RCMs data as in put to reproduce observed flow.

	Calibration				Validation			
Mod data	NSE	R^2	RMSE	MAE	NSE	R^2	RMSE	MAE
HIRH5	0.51	0.51	0.42	0.01	0.50	0.50	0.4	0.10
REMO	0.31	0.30	0.48	0.00	0.30	0.4	0.69	0.18
RCA4	0.25	0.25	0.51	0.01	0.22	0.35	0.74	0.15
Ensemble	0.34	0.34	0.47	0.00	0.40	0.50	0.65	0.13

3.3 Comparison of performance of AWBM models for different input data of RCMs

Figure 4, shows the performance of the AWBM model in calibration and validation with different climatic model data. The Analysis of this figure shows that HIRHAM5 models data using allow to get the best flow estimation on the basin has showed by NSE and R^2 value of 0.51 in calibration against 0.50 in validation (Table 3). The values of RMSE and MAE obtained are relatively low for each data item of the RCM models considered. In view of all the above, it emerges

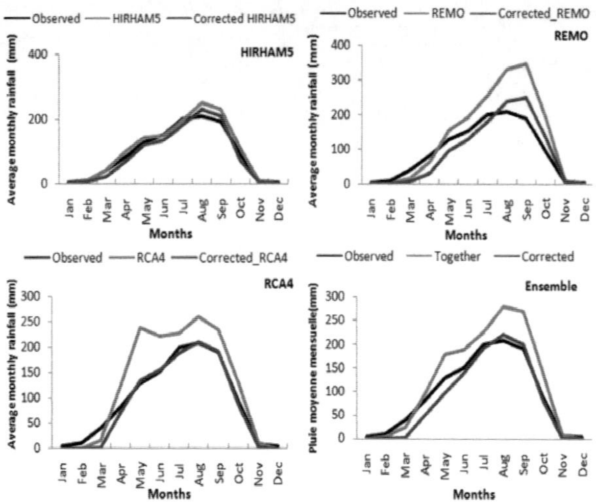

Figure 2. Performance of the EQM method in correcting biases in simulated precipitation by (HIRHAM5, REMO, and RCA4) and the ensemble of RCMs over the period 1970–2005.

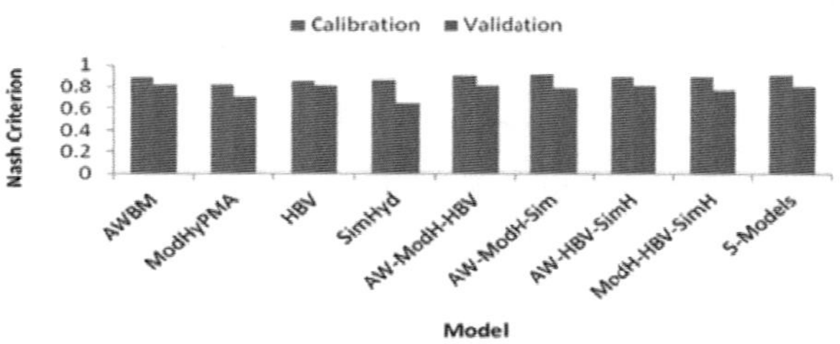

Figure 3. Performance of M-M to 3 and the ensemble of 05 models.

from this analysis that, AWBM model, showed a fairly good capacity for the simulation of the flows in the basin with only the data of HIRHAM5 as inputs.

3.4 Evolution of annual average flows

Figure 5 shows the evolution of the mean annual flows in the basin from the HIRHAM5 data as inputs to the 05-hydrological models considered together. Under the RCP4.5 scenario, the simulation with the five (05) models predicts by 2025 an increase of the flows, with an average value of 0.37 mm/day compare to observations period. By 2055, a slight decrease in flows compared to observations with an average of 0.35 mm/day, lower than observations mean can be observed. By 2085, we will observe a strong increase of the flows with an average of 0.45 mm/day, which is significantly higher compare to observations period with an equivalent change rate of 19.07 %. Under the RCP8.5 scenario, the simulation with the 05-models predicts a strong increase in the flows at the horizons 2055 and 2085 respectively with the mean of 0.4 and 0.51 mm/day, and is higher than those of the observation period. During this horizon, we note a rate of change of 19.56 % by 2085. By 2025 the models predict a

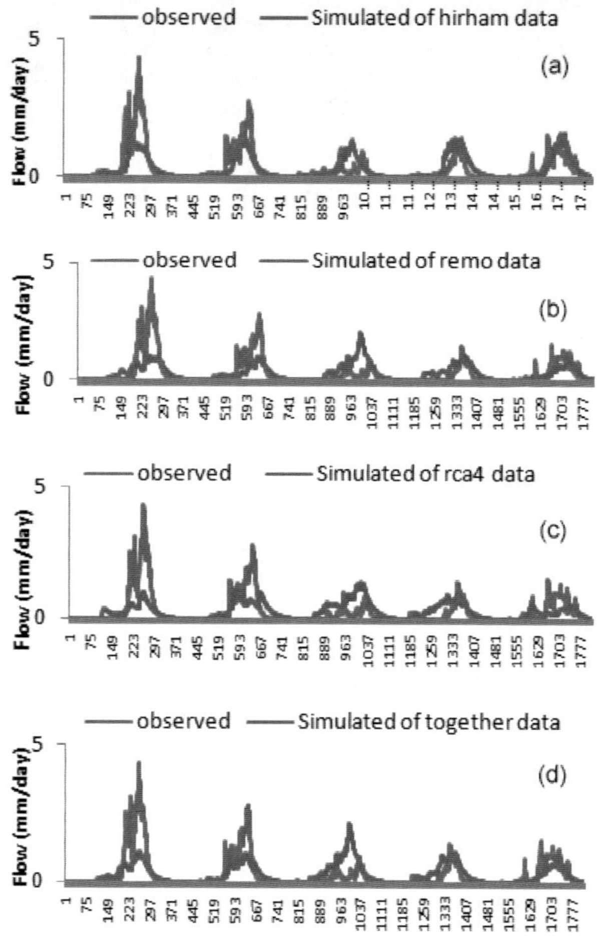

Figure 4. Comparison of observed and A flows for different input data.

slight decrease of the flows with an average of 0.30 mm/day. Overall, the horizons2055 and 2085 will be more affected by a strong increase in flows under the RCP8.5 scenario than under the RCP4.5 scenario.

3.5 Discussion

The applications of the models (AWBM, ModHyPMA, HBV, GR4J and SimHyd) give convincing results in terms of performance with quality criteria of more than 60 % from Nash. These results are consistent with those of (Paturel et al., 2009). In fact, these authors have shown that a performance criterion greater than 60 % indicates a satisfactory performance of the hydrological model in the reproduction of the basin flows. It can also be noticed that the combination in the case of a multi-model can gradually improve the performance of the models in calibration by an average of 1 % and 2 % in validation. Consequently, it would be appropriate to use a combination of hydrological models for the prediction of flows in the Ouémé to Savè basin instead of a single

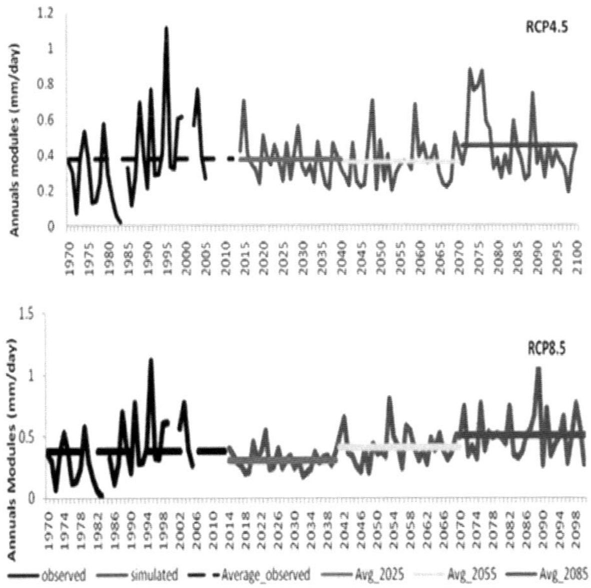

Figure 5. Evolution of the annual flow modules in the basin under the (RCP4.5 and 8.5) scenarios of the HIRHAM5 models as inputs to the 05 hydrological models.

model. Then, the different variations of the flows obtained in an alternative way with the set of 05-models used following the different scenarios show that, the future changes could be due to the anthropogenic activities of the man due in large part to the emissions of greenhouse gases and climate vulnerabilities.

4 Conclusions

This article focuses on the impacts of climate change on water availability in the oueme basin at Save. To this end AWBM model, the multi model AWBM-ModHyPMA-HBV and the set of five models have shown a fairly good capacity for simulating flows in the Oueme basin at Save. In addition, the simulation of flows with HIRHAM5 data as input to the hydrological models showed flow that vary with different horizon (2025, 2055 and 2085) under RCP4.5 and RCP8.5 scenarios.

Data availability. All the data used in this manuscript can be obtained by requesting from the corresponding author.

Author contributions. AC, EB, JZ, EO, EAA, EIB, and AA designed the study, developed the methodology and wrote the manuscript; Amedée Chabi performed the field work, collected the data and conducted the computer analysis.

Competing interests. The contact author has declared that neither they nor their co-authors have any competing interests.

Disclaimer. Publisher's note: Copernicus Publications remains neutral with regard to jurisdictional claims in published maps and institutional affiliations.

Special issue statement. This article is part of the special issue "Hydrology of Large River Basins of Africa". It is a result of the 4th International Conference on the "Hydrology of the Great Rivers of Africa", Cotonou, Benin, 13–20 November 2021.

References

Afouda, A. and Alamou, E.: Hydrologique basé sur le principe de moindre action (MODHYPMA), Ann. Sci. Agron. Bénin, 13, 23–45, 2010.

Afouda, A., Ould Baba Sy, M., Gaye, A. T., Cabral, A., Nazoumou, Y., Compaore, A. J., and Sanoussi, R.: Impact of Climate Change and Variability on Water Resources in West African Watersheds, What Are the Prospects?, Synthesis Report-Writeshop: Dakar, Senegal, 2007.

Akognongbe, A.: Contribution à l'étude de la caractérisation hydropluviométrique avec le modèle GR4J sur le bassin de l'Ouémé à Savè et Atchérigbé, Mémoire de maîtrise en géographie, UAC., 79 pp., 2007.

Bergström, S. and Forsman, A.: Development of a conceptual deterministic rainfall runoff model, Nordic Hydrology, 14, 147–170, 1973.

Boughton, W. C.: A hydrograph-based model for estimating the water yield of ungauged catchments, in: Proceedings of the 1993 Hydrology and Water Resources Conference, Institution of Engineers, Australia, National Conference Publication no., 93/14, 317–324, 1993.

Boyle, D.: Multicriteria calibration of hydrological models, PhD thesis, Dep. of Hydrol. and Water Resour., Univ. Of Ariz., Tucson, 2001.

Chiew, F. H. S. and Siriwardena, L.: Estimation of SIMHYD parameter values for application in ungauged catchments, in: MODSIM 2005 International Congress on Modelling and Simulation, Modelling and Simulation Society of Australia and New Zealand, Melbourne, Australia, 2883–2889, 2005.

IPCC: Impact, adaptation and vulnerability. Contribution of working Group II to the fourth assessment report of the intergovnmental Panel on Climate Change, 2007.

IPCC: Impacts, Adaptation, and vulnerability, Contribution of Working Group II to the Fifth Assessment Report of the Intergovernmental Panel on Climate Change, Cambridge University Press, Cambridge, United Kingdom and New York, NY, USA, 2014.

Nash, J. E. and Sutcliffe, J. V.: River flow forecasting through conceptual models part I – A discussion of principles, J. Hydrol., 10, 282–290, 1970.

Obada, E., Alamou, A. E., Chabi, A., Zandagba, J., and Afouda, A.: Trends and Changes in Recent and Future Penman-Monteith

Potential Evapotranspiration in Benin, Article published in Hydrology (MDPI), 18 pp, 2017.

Paturel, J. E. and Mahe, G.: 1896–2006 Sahelian annual rainfall variability and runoff increase of Sahelian RiversVariabilité des pluies annuelles au Sahel entre 1896 et 2006 et augmentation des écoulements des rivières sahéliennes, Comptes Rendus, Geosciences, 341, 538–546, https://doi.org/10.1016/j.crte.2009.05.002, 2009.

Perrin, C.: Vers une amélioration d'un modèle global pluie-débit au travers d'une approche comparative, Thèse de Doctorat, INPG: Grenoble, Cemagref, Antony, 530 pp., 2000.

Quae, J.-F. S.: S'adapter au changement climatique, Agriculture écosystèmes et territoires, France, EdiTionS QuAE, 282 pp., 2013.

Sennikovs, J. and Bethers, U.: Statistical downscaling method of regional climate model results for hydrological modelling, in: 18 th World IMACS/MODSIM Congress, Australia, 2009.

Zandagba, J.: Impact des changements climatiques sur les ressources en eau du bassin de l'oueme à Savè, Master of Science en Hydrologie et Gestion Intégrée des Ressources à la CIPMA-Chaire UNESCO-Bénin/ Université d'Abomey-Calavi (UAC), République du Bénin, 2012.

Proc. IAHS, 384, 261–267, 2021
https://doi.org/10.5194/piahs-384-261-2021

Open Access PIAHS

Impact de la variabilité climatique et de l'anthropisation sur les écoulements de la Bénoué (nord Cameroun)

Valentin Brice Ebodé[1], **Gil Mahé**[2], **and Ernest Amoussou**[3]

[1]Département de Géographie, Université de Yaoundé 1, Cameroun
[2]Institut de la Recherche pour le Développement, France
[3]Département de Géographie et Aménagement du Territoire, Université de Parakou, Bénin

Correspondence: Valentin Brice Ebodé (ebodebriso@gmail.com)

Published: 16 November 2021

Résumé. La sécheresse observée en Afrique tropicale vers la fin des années 1960, a également affecté le bassin de la Bénoué en Afrique centrale, avec une persistance remarquable qui s'est répercutée sur les écoulements. Les ruptures à la baisse ont été mises en évidence dans les séries hydropluviométriques de ce bassin au pas de temps annuel en 1970–1971 (pluies) et 1971–1972 (débits). Les déficits associés à cette rupture sont de $-2,9\,\%$ pour les pluies et $-14,2\,\%$ pour les débits, par rapport à la moyenne climatologique (1950–1951 à 2014–2015). La saison humide a connu des évolutions pratiquement identiques. Cependant, depuis la décennie 1990, il est observé un relèvement significatif dans les écoulements moyens annuels de ce cours d'eau, et cela coïncide avec le retour des pluies. Le maintien de cette hausse au cours des récentes décennies pourrait être envisagé en raison de l'accroissement des espaces imperméabilisés dans le bassin, qui compenseraient le déficit engendré par la rechute des pluies après la décennie 1990 via une accentuation du ruissèlement. De plus, depuis la mise en eau du barrage de Lagdo en 1983, il est observé non seulement une augmentation de l'ensemble des gammes de débits minima suivant des taux allant de $+57,8\,\%$ (minimum sur 1 jour) à $+70,1\,\%$ (minimum sur 90 jours), mais aussi un accroissement général de la variabilité de l'ensemble des débits extrêmes (minima et maxima). L'augmentation progressive du stockage en saison de pluies à la retenue de Lagdo, pour fournir l'électricité et l'eau d'irrigation durant la saison sèche à une population grandissante, rendent imperceptibles les effets de l'accroissement des espaces imperméabilisés non seulement sur les écoulements de la saison humide, mais aussi sur les écoulements maxima. Au demeurant, l'opérationnalisation du barrage de Lagdo a eu un impact significatif sur le régime de la Bénoué, se traduisant essentiellement par le caractère désormais permanent de l'écoulement tout au long de l'année hydrologique et un important fléchissement de la crue d'août à octobre.

Abstract. A tendency to drought conditions appeared in tropical Africa from the end of 1960s, also affect the Benoue watershed in central Africa, with a remarkable persistence in streamflow. Negative breakpoints are found in the annual hydroclimatic time series of the Benoue catchment at annual time step in 1970–1971 for rainfall, and in 1971–1972 for discharges. Compared to the climatological mean (1950–1951 to 2014–2015), the deficits associated with these negative breakpoints are approximating $-2.9\,\%$ for rainfall, and $-14.2\,\%$ for discharges. The wet season shows similar developments. However, from the 1990s, there has been a significant increase in the mean annual discharges of the Benoue River, which coincides with that of the rainfall during the same decade. The maintenance of this increase over the recent decades could also be expected in response to an increase in impervious surface areas (ISA) in the catchment area, which could compensate the deficit generated by the post-1990s rainfall deficit through increasing in runoff. From the watering of the Lagdo dam in 1983, an increase in all ranges of minimum flow, ranging from $+57.8\,\%$ (1 d minimum) to $+70.1\,\%$ (90 d minimum), as well as an increase in the variability of extreme flows (minima and maxima), has been detected. During the wet season, the increase in water storage capacity at the Lagdo reservoir, which is used to provide electricity and irrigation water to a growing population, mitigates the increase in runoff, especially in maximum runoff. Moreover, the operationalization of the Lagdo dam has a significant impact on the Benoue hydrological regime, resulting mainly in a new perennial nature of the flow, and a significant decrease in flood from August to October.

1 Introduction

L'effet du déficit pluviométrique sur les ressources en eau de surface a été mis en évidence par de nombreuses études en milieu tropical africain (Paturel et al., 2003 ; Liénou et al., 2008 ; Sidibé et al., 2018, 2020 ; Tramblay et al., 2020 ; Ebodé, 2020 ; Ebodé et al., 2020a). Les changements climatiques observés au cours des cinquante dernières années dans cette région, notamment au nord Cameroun, induisent des impacts visibles dans le cycle hydrologique du bassin versant de la Bénoué (Fita, 2020). Outre cette variabilité climatique, qui détermine la dynamique hydrologique de cet hydrosystème, le récent phénomène d'anthropisation (aménagement hydraulique et changements d'occupation du sol) a certainement provoqué un changement des relations pluie/débit et du régime hydrologique, qui nécessite une nouvelle étude de la variabilité hydrologique. Les aménagements hydrauliques ont en effet connu un essor très important durant la décennie 1980. Cependant, nous avons à présent suffisamment de recul pour évaluer leurs performances réelles, ainsi que leurs impacts socio-environnementaux. Par exemple, la construction du barrage de Lagdo entre 1977 et 1982, et sa mise en service depuis 1983, pourrait avoir entraîné une modification du régime hydrologique de la Bénoué. Il en est de même de l'accroissement de la population observé dans le bassin (BUCREP, 2011), qui a pour corollaire l'accroissement des surfaces imperméabilisées dont le rôle sur les écoulements a largement été prouvé ailleurs (Aulenbach et al., 2017 ; Diem et al., 2017 ; Ebodé et al., 2020a).

Cette étude est consacrée à l'analyse des relations pluie/débit sur la période récente (1950–1951 à 2014–2015) dans le bassin de la Bénoué, au moyen des séries hydroclimatiques actualisées et des classifications obtenues à partir du traitement des images satellitaires. Cette étude apparaît ainsi fondamentale pour la communauté, en raison du rôle majeur de ce bassin sur la production hydroélectrique, et des disfonctionnements observés lors des dernières décennies, et souvent. En outre, les données et les nouveaux résultats de cette étude aideraient à la planification à long terme de la demande et de l'utilisation de l'eau, ainsi qu'à l'amélioration des simulations futures du débit de la Bénoué. L'objectif principal de cet article est donc de documenter le type et l'ampleur des changements de régime hydrologique de la Bénoué dus aux changements climatiques et à l'anthropisation.

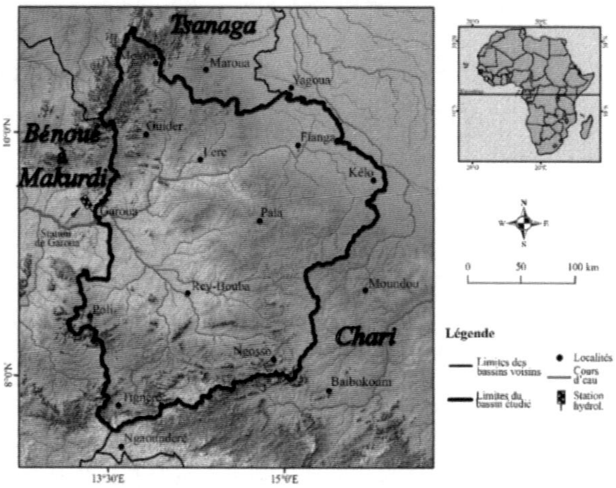

Figure 1. Localisation du bassin versant de la Bénoué.

2 Cadre, données et méthodes

Couvrant une aire de 61 810 km^2 à Garoua, le bassin de la Bénoué est situé en zone tropicale nord entre le Cameroun et le Tchad (7°32′–10°53′ N et 13°9′–15°52′ E), la plus grande partie se situant sur le territoire camerounais (Fig. 1). Il est soumis à un climat tropical soudanien enregistrant en moyenne 830 mm de précipitations par an. Ce climat est caractérisé par une saison sèche et chaude s'étendant en moyenne de novembre à avril, et une saison humide et moins chaude de mai à octobre, au cours de laquelle se produit l'essentiel des précipitations. La saison humide enregistre à peu près 887 mm an^{-1} de précipitations, tandis que la saison sèche n'enregistre qu'environ 43 mm an^{-1}.

Plusieurs jeux de données (hydropluviométriques et d'occupation du sol) ont été utilisés dans ce travail. Disponible sur la période 1950–2015, la série des débits de la Bénoué provient de l'Autorité du Bassin du Niger (ABN). Elle comporte d'importantes lacunes entre 1998 et 2009.

Les données de pluies utilisées dans cette étude sont celles du CRU (Climatic Research Unit) de l'université d'East Anglia au Royaume Uni. Ces données sont disponibles depuis 1901 via le site https://climexp.knmi.nl/selectfield_obs2.cgi\unskip\penalty\@M\hskip.5\fontdimen2\font\relax?id=_2833fad3fef1bedc6761d5cba64775f0/ (la date du dernier accès : 10 mai 2021) au format NetCDF, sur un pas temps mensuel et à une résolution spatiale de 0,25° × 0,25°.

Les données d'occupation du sol dans le bassin de la Bénoué sont issues des images satellitaires Landsat 8 de janvier à mars 2015 et Landsat TM de janvier à mars 1987. L'ensemble de ces images sont mises à disposition du grand public gratuitement par la National Aeronautics and Space Administration (NASA), via le site de l'US Geological Survey (https://earthexplorer.usgs.gov/, la date du dernier accès : 13 janvier 2020), au format GeoTIFF.

L'analyse des séries de pluies, des débits moyens et des coefficients d'écoulement a été réalisée à l'aide des tests statistiques de détection de rupture (corrélation sur le rang et test de Pettitt ; Lubès et al., 1994). Dans l'optique d'apprécier le comportement des débits extrêmes, l'outil Indicators of Hydrologic Alteration (IHA), version 7.1, développé par The Nature Conservancy a été utilisé. La classification supervisée par maximum de vraisemblance des images satellitaires Landsat téléchargées, et ce, à l'aide du logiciel SNAP (https://www.theia-land.fr, la date du dernier accès : 20 février 2020), nous a permis d'effectuer une analyse diachronique de l'évolution de l'occupation du sol dans le bassin étudié.

3 Résultats et discussion

3.1 Evolution des pluies annuelles et saisonnières

Les pluies annuelles diminuent de manière statistiquement significative sur le bassin de la Bénoué. Le test de Pettitt situe dans leur série une rupture majeure intervenue en 1970–1971 (Fig. 2). Cette dernière marque le passage d'une période excédentaire, dont le surplus par rapport à la moyenne climatologique est de $+6,1\,\%$, à une période déficitaire, ayant un déficit par rapport à cette même moyenne de $-2,9\,\%$ (Fig. 2). L'analyse des écarts décennaux révèle une chute depuis la décennie 1970, qui n'a été périodiquement interrompue qu'au cours de la décennie 1990 (Tableau 1). Les volumes précipités durant la saison humide suivent sensiblement les mêmes tendances (Fig. 2). Notamment, ces pluies présentent également une rupture à la baisse en 1970–1971, selon le test de Pettitt. L'excédent et le déficit enregistrés de part et d'autre de cette rupture sont, respectivement, de $+5,2\,\%$ et $-2,3\,\%$ (Fig. 2). La décennie 1990 apparaît aussi plus excédentaire dans les pluies de cette saison par rapport aux décennies 1970 et 1980 (Tableau 1). Comme en saison humide, les volumes précipités en saison sèche diminuent significativement, selon le test de Pettitt. Une rupture a été identifiée par le test de Pettitt dans la série des pluies de la saison sèche en 1970–1971 (Fig. 2). Les excédents et déficits par rapport à la moyenne climatologique de chacune des périodes situées avant et après ladite rupture sont, respectivement, de $+21,3\,\%$ et $-11,5\,\%$. La chute des pluies de la saison sèche observée au cours de la décennie 1970 est continue jusqu'à la décennie 2010 (Tableau 1). Un retour à des conditions plus humides au cours de la décennie 1990 a été suggéré en Afrique de l'ouest (Alhassane et al., 2013 ; Descroix et al., 2020). Cela concorde avec les résultats de notre étude, qui mettent en évidence un rehaussement des précipitations annuelles au cours de la décennie 1990, consécutif à la sécheresse des décennies 1970 et 1980. Cependant le rehaussement signalé dans cette étude concerne uniquement cette décennie, tandis que celui relevé en Afrique de l'ouest persiste jusqu'à présent (Descroix et al., 2020).

3.2 Dynamique de l'occupation du sol dans le bassin de la Bénoué

Les classifications effectuées à partir des images Landsat des années 1987 et 2015 montrent d'importants changements dans les principaux modes d'occupation du sol ayant un lien direct avec les écoulements dans le bassin versant de le Bénoué (Fig. 3). En 28 ans, il a été enregistré une diminution des surfaces forestières (îlots de forêt sous forme de petits massifs ou de forêts galeries et savanes boisées) de 24 % (Tableau 2). Inversement, les espaces imperméabilisés (bâti, voirie, sols nus et zones cultivées) et les plans d'eau (lacs, étangs et grandes rivières) ont augmenté, respectivement, de 133 % et de 16,4 % (Tableau 2).

3.3 Les répercussions des forçages environnementaux sur les écoulements

3.3.1 Le rôle de la pluviométrie

Les débits moyens annuels de la Bénoué ont enregistré une rupture à la baisse en 1971–1972. Le déficit par rapport à la moyenne de la période étudiée observé suite à cette rupture est de $-14,2\,\%$ (Fig. 2). L'impact de la pluviométrie dans l'évolution des écoulements de la Bénoué semble assez net à ce pas de temps. On note pour les deux variables comparées des diminutions statistiquement significatives sur la période d'étude (Fig. 2). A ce pas de temps, la stationnarité du régime hydrologique, et celui des précipitations, obéit à la logique générale qui admet que les ruptures hydrologiques interviennent à la suite des ruptures pluviométriques, puisqu'une rupture à la baisse est intervenue dans la série des précipitations annuelles en 1970–1971 et dans celle des débits moyens annuels en 1971–1972 (Fig. 2). Comme au pas de temps annuel, l'influence de la pluviométrie de la saison humide est également perceptible dans les écoulements de la même saison (Fig. 2). On note alors, dans les séries respectives de pluies et débits, des ruptures à la baisse, avec une légère précocité pour les pluies (1970–1971), permettant d'envisager un éventuel rôle de la diminution des pluies sur celle des écoulements.

3.3.2 L'impact du barrage

Sur le bassin de la Bénoué, depuis la mise en eau du barrage de Lagdo (1983), on note une augmentation importante de l'ensemble des gammes de débits minima sur des jours consécutifs (Tableau 3). Ces augmentations sont comprises

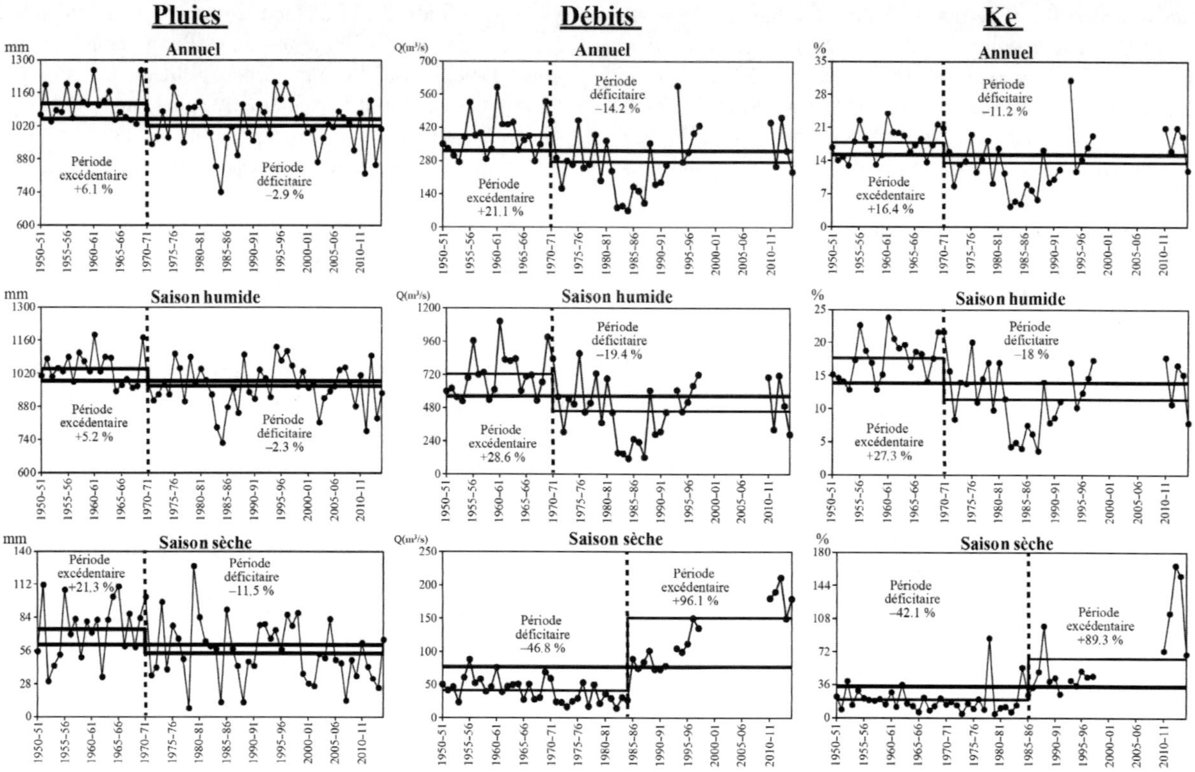

Figure 2. Evolution des pluies, débits et coefficients d'écoulement (Ke) annuels et saisonniers de la Bénoué entre 1950–1951 et 2014–2015. Traits horizontaux fins (épais) : moyenne des périodes homogènes (moyenne climatologique), tireté vertical : année de rupture.

Tableau 1. Ecarts (%) des moyennes décennales annuelles et saisonnières des pluies, débits et coefficients d'écoulement par rapport à leurs moyennes climatologiques sur le bassin de la Bénoué.

	Pluies			Débits			Ke		
Décennies	Annuel	Saison humide	Saison sèche	Annuel	Saison humide	Saison sèche	Annuel	Saison humide	Saison sèche
1950	6	5,7	11,9	11,3	16,6	−34, 2	7	15,3	−39, 2
1960	6,3	5,1	26,2	29,2	38,5	−39, 3	23	36,5	−48, 5
1970	0	−0, 4	5,2	−6, 5	0,6	−58, 5	−4, 8	4,7	−41, 5
1980	−7, 3	−6, 9	−13, 4	−43, 7	−46	−27, 3	−40, 5	−42, 2	2,3
1990	4	3,5	11,9	9,9	−6, 4	128,8	8	−6, 5	20,5
2000	−5, 2	−3, 7	−29, 7	–	–	–	–	–	–
2010	−6, 4	−5, 1	−26, 6	7,1	−10, 6	136,7	16,7	−2, 4	242,7

entre 57,8 % (débit minimum sur 1 jour) et 70,1 % (débit minimum sur 90 jours). Au-delà du retard moyen de cinq jours noté dans l'apparition des minima (Tableau 3), il est également noté un accroissement de la variabilité interannuelle pour les deux catégories de débits extrêmes évaluées (maxima et minima), qui semble beaucoup plus important pour les maxima (Tableau 3). L'ensemble de ces évolutions semblent liées au fonctionnement du barrage. Cet ouvrage permet en effet de stocker de l'eau en saison humide et de la relâcher en saison sèche, dans l'optique de continuer la production hydroélectrique et l'irrigation en aval durant cette saison, modifiant ainsi le comportement naturel de ce cours d'eau. La hausse des volumes lâchées (également perceptible au travers des minimums) sans cesse croissante au fil du temps (Fig. 4), traduit un accroissement de plus en plus important de la population et de ses besoins grandissants en énergie auxquels il faut subvenir en permanence, y compris en saison sèche. Le barrage de Lagdo a considérablement affecté le régime de la Bénoué. Depuis sa mise en place, on note le développement d'un caractère permanent de l'écou-

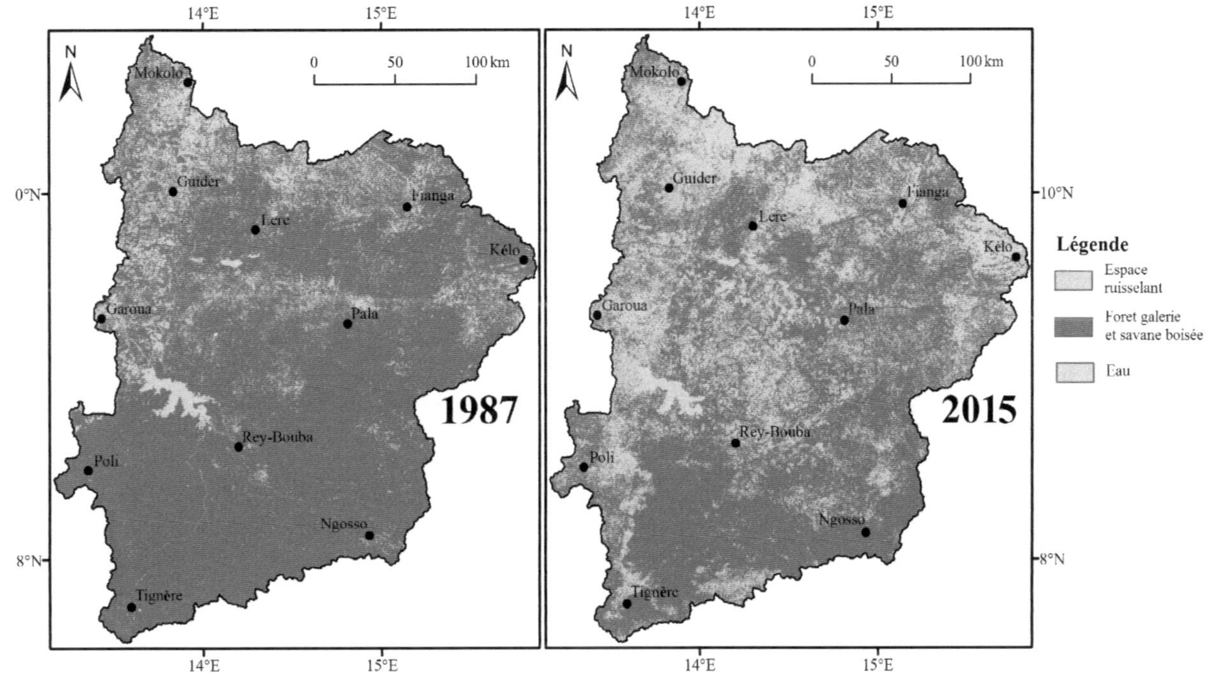

Figure 3. Principaux modes d'occupation du sol dans le bassin de la Bénoué en 1987 et 2015.

Tableau 2. Evolution des principaux modes d'occupation du sol dans le bassin de la Bénoué entre 1987 et 2015.

Modes d'occupation du sol	Superficie dans le bassin (km^2)		Evolution entre 1987 et 2015	
	1987	2015	km^2	%
Forêt galerie et savane boisée	51 977	39 480	−12 497	−24
Plan d'eau	505	588	83	16,4
Espace ruisselant	9328	21 742	12 414	133

lement tout au long de l'année hydrologique, ainsi que le fléchissement de la crue d'août à octobre (Fig. 4). La hausse des débits d'étiage est identifiée comme conséquence majeure des aménagements hydrauliques (barrages), comme proposé précédemment par Amoussou et al. (2012).

3.3.3 La contribution des changements des modes d'occupation du sol

Il est certes vrai que les écoulements de la Bénoué diminuent statistiquement depuis le début de la décennie 1970, en rapport avec la chute des précipitations entamée durant cette même décennie. Cependant, l'analyse des écarts décennaux confirme un relèvement important des écoulements depuis la décennie 1990. Les écarts relevés sont de $+9,9\,\%$ pour la décennie 1990, et $+7,1\,\%$ pour la décennie 2010 (Tableau 1). Il est tout à fait possible d'associer la hausse des écoulements de la décennie 1990 à celle de la pluviométrie. Par contre, rien n'explique le maintien de cette hausse sur les récentes décennies, puisque les pluies diminuent durant cette période. Les déficits calculés sont de $-5,2\,\%$ pour la décennie 2000 et $-6,4\,\%$ pour la décennie 2010 (Tableau 1). Nous pouvons suggérer que la hausse des écoulements, associée à l'augmentation des pluies dans la décennie 1990, a été immédiatement relayée par l'accroissement des espaces imperméabilisés dans le bassin. Ces formations imperméables agissent dans le sens d'un accroissement du ruissellement et, par conséquent, des écoulements. Des études effectuées ailleurs sont parvenus à des résultats similaires (Aulenbach et al., 2017 ; Ebodé et al., 2020a). Il est aussi tout à fait possible de s'interroger sur les causes de l'augmentation récente des écoulements observée sur le bassin de la Bénoué, et en particulier sur son absence durant la saison humide et dans les maxima. L'augmentation de plus en plus importante des quantités d'eau stockées en saison de pluies, et de celle relâchées en saison sèche (Fig. 4), pourraient être en rapport avec cette situation. Cette augmentation du stockage en saison de pluies à Lagdo implique que le surplus d'eau, occasionné par le ruissellement, ne soit pas perceptible dans les écoulements de la saison de pluies et les maxima à Garoua.

Tableau 3. Statistiques des débits maxima et minima de la Bénoué avant et après le barrage.

Statistiques IHA	Moyennes ($m^3 s^{-1}$)		Cv (%)		Variation	
	Avant	Après	Avant	Après	$m^3 s^{-1}$	%
Débits minimum						
Minimum sur 1 jour	0,6	58,4	1,13	0,96	57,8	9633
Minimum sur 3 jours	0,7	59,7	0,9	0,96	59	8428
Minimum sur 7 jours	0,8	61,6	0,85	0,95	60,8	7600
Minimum sur 30 jours	1,5	66,6	0,84	0,92	65,1	4340
Minimum sur 90 jours	6,1	76,2	0,6	0,89	70,1	1149
Débits maximum						
Maximum sur 1 jour	2957	1589	0,34	0,58	−1368	−46, 2
Maximum sur 3 jours	2851	1540	0,33	0,58	−1311	−46
Maximum sur 7 jours	2608	1436	0,32	0,57	−1172	−45
Maximum sur 30 jours	1979	1143	0,32	0,57	−836	−42, 2
Maximum sur 90 jours	1274	763,6	0,32	0,54	−510, 4	−40
Dates juliennes moyennes						
Du minimum	113	118				
Du maximum	252	245				

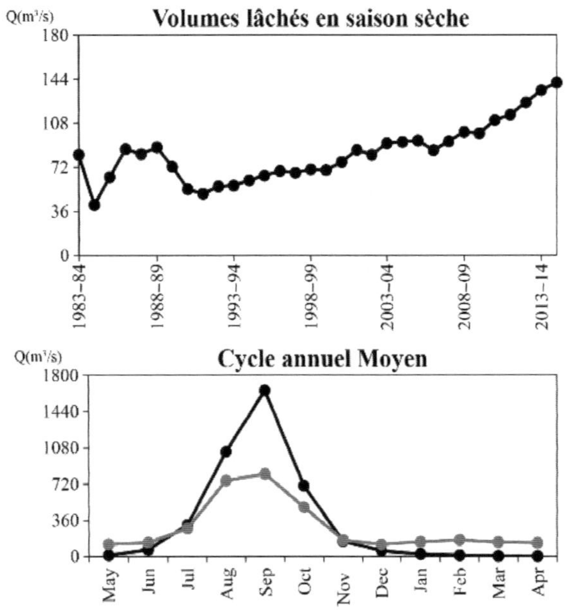

Figure 4. Evolution interannuelle des volumes lâchés à Lagdo depuis 1983 et cycle annuel moyen de la Bénoué avant (en noir) et après (en gris) le barrage.

4 Conclusion

A partir de la décennie 1970, le bassin de la Bénoué a connu une diminution générale des précipitations qui n'a été interrompue qu'au cours de la décennie 1990. Ce changement intervenu dans l'évolution des pluies a entraîné une modification des écoulements de son collecteur principal. Ainsi, il a été relevé, depuis 1971–1972, une diminution des écoulements moyens de la saison de pluies et annuels, dont les déficits respectifs par rapport à leurs moyennes climatologiques atteignent −19, 4 % et −14, 2 %. Certes en diminution depuis la décennie 1970, les écoulements moyens annuels augmentent depuis la décennie 1990. La hausse des écoulements de la décennie 1990 est imputable à son abondante pluviométrie, mais son maintien au cours des récentes décennies semble être la conséquence des changements de modes d'occupation du sol, puisque, durant cette période, la pluviométrie baisse. La hausse des minima observés sur ce cours d'eau, ainsi que l'accentuation de la variabilité des écoulements extrêmes semblent être associées à la mise en service du barrage de Lagdo depuis 1983. Cet ouvrage permet en effet de stocker de l'eau en saison de pluies et de la relâcher en saison sèche, dans l'optique de continuer de produire l'hydroélectricité et d'irriguer les champs situés en aval. L'augmentation dudit stockage pour le comblement des besoins d'une population grandissante pourrait d'ailleurs être à l'origine d'une diminution des crues à Garoua.

Disponibilité du code. Les logiciels utilisés dans cette étude sont gratuits. Les différents liens proposés plus haut dans cet article donnent accès aux sites sur lesquels ils peuvent être téléchargés.

Disponibilité des données. Certaines données (débits) utilisées dans cette étude proviennent d'une institution (Autorité du Bassin du Niger). Leur collecte nécessite des moyens financiers conséquents. Cette institution a la responsabilité entière sur ces données. Pour entrer en possession de ces données, il faudrait entrer en contact avec cette institution.

D'autres données (précipitations et images satellitaires) par contre, sont disponibles gratuitement en ligne sur des sites proposés plus haut dans cet article.

Collaborateurs. Ce travail a été conceptualisé par VBE. La collecte, le traitement et l'analyse des données ont été effectués par VBE. Le projet a été dirigé par GM. La rédaction du manuscrit et la relecture ont été faites par l'ensemble des auteurs.

Intérêts concurrents. Les auteurs déclarent qu'ils n'ont aucun conflit d'intérêts.

Clause de non-responsabilité. Publisher's note : Copernicus Publications remains neutral with regard to jurisdictional claims in published maps and institutional affiliations.

Déclaration du numéro spécial. This article is part of the special issue "Hydrology of Large River Basins of Africa". It is a result of the 4th International Conference on the "Hydrology of the Great Rivers of Africa", Cotonou, Benin, 13–20 November 2021.

Références

Alhassane, A., Salack, S., Ly, M., Lona, I., Traore, S. B., et Sarr, B. : Evolution des risques agroclimatiques associés aux tendances récentes du régime pluviométrique en Afrique de l'Ouest soudano-sahélienne, Sécheresse, 24, 282–293, https://doi.org/10.1684/sec.2013.0400, 2013.

Amoussou, E., Camberlin, P., et Mahé, G. : Impact de la variabilité climatique et du barrage Nangbéto sur l'hydrologie du système Mono-Couffo (Afrique de l'ouest), J. Sci. Hydrol., 57, 805–817, https://doi.org/10.1080/02626667.2011.643799, 2012.

Aulenbach, B. T., Landers, M. N., Musser, J. W., and Painter, J. A. : Effects of impervious area and BMP implementation and design on storm runoff and water quality on eight small watersheds, J. Amer. Wat. Ress. Ass., 53, 382–399, https://doi.org/10.1111/1752-1688.12501, 2017.

BUCREP : Rapport de présentation des résultats définitifs du recensement de la population en 2005, Yaoundé, Cameroun, 66 pp., 2011.

Descroix, L., Sané, Y., Mamadou, T., Manga, S.-P., Boubacar, D., Mingou, J., Mendy, V., Coly, S., Dièye, A., Badiane, A., Senghor, M.-J., Diedhiou, A.-B., Djiby, S., Bouaita, Y., Soumaré, S., Diop, A., Bakary, F., Bamol, A., Machu, E., Montoroi, J.-P., Andrieu, J., and Vandervaere, J.-P. : Inverse estuaries in West Africa : Evidence of the rainfall recovery ?, Water, 12, 647, https://doi.org/10.3390/w12030647, 2020.

Diem, J. E., Hill, T. C., and Milligan, R. A. : Diverse multi-decadal changes in streamflow within a rapidly urbanizing region, J. Hydrol., 556, 61–71, https://doi.org/10.1016/j.jhydrol.2017.10.026, 2017.

Ebodé, V. B. : Variabilité hydroclimatique en Afrique centrale occidentale forestière : entre analyse des fluctuations observées, recherche des facteurs explicatifs et modélisation prédictive, Generis Publishing, 382 p., 2020.

Ebodé, V. B., Mahé, G., Dzana, J. G., and Amougou, J. A. : Anthropization and climate change : impact on the discharges of forest watersheds in Central Africa, Water, 12, 2718, https://doi.org/10.3390/w12102718, 2020a.

Ebodé, V. B., Mahé, G., Dzana, J. G., Amougou, J. A., et Batha, R. A. : Changement climatique dans le bassin versant de l'Ogooué : évolution récente et impact sur les écoulements, Afr. Sci., 17, 110–125, 2020b.

Fita, D. E. : Impact de la variabilité climatique sur la réponse hydrologique du bassin versant supérieur de la Bénoué (nord-Cameroun), Thèse de doctorat, Université de Maroua, Cameroun, 2020.

Liénou, G., Mahé, G., Paturel, J. E., Servat, E., Sighomnou, D., Ekodeck, G. E., Dezetter, A., et Dieulin, C. : Evolution des régimes hydrologiques en région équatoriale camerounaise : un impact de la variabilité climatique en zone équatoriale ?, J. Sci. Hydrol., 53, 789–800, https://doi.org/10.1623/hysj.53.4.789, 2008.

Lubès, H., Masson, J. M., Servat, E., Paturel, J. E., Kouame, B., et Boyer, J. F. : Caractérisation des fluctuations dans une série chronologique par applications de tests statistiques, Etudes bibliographiques ; Programme ICCARE, Rapport no. 3, ICCARE, ORSTOM, Montpellier, France, 1994.

Paturel, J. E., Ouédraogo, M., Mahé, G., Servat, E., Dezetter, A., and Ardoin, S. : Influence of the spatialization of data on the modelling of monthly river regimes in West Africa, Hydrol. Sci. J., 48, 881–890, https://doi.org/10.1623/hysj.48.6.881.51422, 2003.

Sidibé, M., Dieppois, B., Mahé, G., Paturel, J. E., Amoussou, E., Anifowose, B., and Lawler, D. : Trend and variability in a new, reconstructed streamflow dataset for West and Central Africa, and climatic interactions, 1950–2005, J. Hydrol., 561, 578–493, https://doi.org/10.1016/j.jhydrol.2018.04.024, 2018.

Sidibé, M., Dieppois, B., Eden, J., Mahé, G., Paturel, J. E., Amoussou, E., Van de Wiel, M., Anifowose, B., and Lawler, D. : Near-term impacts of climate variability and change on hydrological systems in West and Central Africa, Clim. Dynam., 54, 2041–2070, https://doi.org/10.1007/s00382-019-05102-7, 2020.

Tramblay, Y., Villarini, G., and Zhang, W. : Observed changes in flood hazard in Africa, Environ. Res. Lett., 15, 1040b5, https://doi.org/10.1088/1748-9326/abb90b, 2020.

Proc. IAHS, 384, 269–273, 2021
https://doi.org/10.5194/piahs-384-269-2021

Open Access
PIAHS

Impacts des changements climatique et environnemental sur la réponse hydrologique du bassin du Nakanbé à Wayen (Burkina Faso) à travers le cadre de budyko

Yetchékpo Patrick Gbohoui[1,2], **Jean-Emmanuel Paturel**[2], **Tazen Fowe**[1], **Harouna Karambiri**[1], **and Hamma Yacouba**[1]

[1]Laboratoire Eaux, HydroSystèmes et Agriculture (LEHSA), Institut International d'Ingénierie de l'Eau et de l'Environnement (2iE), Rue de la Science, Ouagadougou 01 01-BP-594, Burkina Faso
[2]HydroSciences Montpellier, Univ Montpellier, CNRS, IRD, 163 rue Auguste Broussonnet, 34090, Montpellier, France

Correspondence: Y. Patrick Gbohoui (patrickgbohoui@gmail.com)

Published: 16 November 2021

Résumé. Les changements climatique et environnemental ont entraîné l'augmentation exacerbée des coefficients d'écoulement dans certains hydrosystèmes du Sahel ouest africain. Ce « paradoxe hydrologique sahélien » a fait l'objet de plusieurs études, mais très peu d'entre elles ont abordé les contributions de chacun des changements climatique et environnemental. L'objectif de cette étude est de quantifier les contributions de chacun des facteurs (climat et environnement) au changement des écoulements dans la partie sahélienne du bassin du Nakanbé sur la période 1965–1994. Les tests statistiques de stationnarité ont permis de découper la période d'étude en sous-périodes de base et d'impact. La quantification des impacts a été effectuée à travers l'application des techniques d'élasticité et de décomposition aux modèles Budyko de Fu et de Choudhury. L'analyse des chroniques de données hydro-climatiques du bassin a confirmé le paradoxe hydrologique avec une rupture dans les écoulements en 1977 (test de Pettitt, p value = 0,021). Ainsi, la période 1978–1994 comparativement à 1965–1977 a été caractérisée par une diminution des précipitations de 8,2 %, une augmentation de l'évapotranspiration potentielle de 1,3 % et une augmentation des écoulements de 91,5 %. Il ressort de cette étude que les contributions à l'augmentation des écoulements de 91,5 % ont été de -29 ± 2 % pour le changement climatique, $+90 \pm 1$ % pour la dégradation environnementale et $+33$ % pour les interactions couplées climat-environnement. Ces résultats montrent que l'impact de l'environnement est plus prépondérant et représente environ 3 fois celui du climat en intensité. L'adoption de meilleures pratiques de gestion de la dynamique environnementale pourrait donc permettre de réguler les changements dans les écoulements du bassin.

Abstract. Climate and environmental changes have caused the increasing runoff coefficients of some sahelian catchments in West Africa. Many previous studies have focused on this "sahelian hydrological paradox", but few have addressed the individual contribution of climate and the environment. This study aims to quantify the contributions of each factor (climate and environment) to the change of runoff in the sahelian part of the Nakanbé catchment over the period : 1965–1994. Based on time-series stationarity statistical tests, the study period was divided into baseline and impacted periods. The quantification of impacts was carried out by applying elasticity and decomposition techniques to the Budyko type models of Fu and Choudhury. Statistical analysis of the annual hydro-climatic data detected 1977 as the year of break in the runoff (Pettitt test, p value = 0.021) and confirmed the sahelian hydrological paradox. Thus, the period 1978–1994 compared to 1965–1977 was characterized by a decrease of 8.2 % in precipitation while increases of 1.3 % and 91.5 % have been observed respectively for potential evapotranspiration and annual runoff. The results indicated that the contributions to the runoff increase of 91.5 % were -29 ± 2 % for climate change, $+90 \pm 1$ % for environmental degradation and +33 % for interactions. It appears that, environmental change is the main cause of the increase in the runoff coefficient in the sahelian catchment of Nakanbé and the intensity of its impact is approximately 3 times more than the one induced by climate change. Therefore, good management of environmental dynamics at the catchment scale could help to regulate runoff changes.

1 Introduction

Le changement climatique et la dégradation de l'environnement constituent les causes de changements dans les écoulements. Dans certains hydrosystèmes du Sahel ouest africain, ils se traduisent par une augmentation exacerbée des coefficients d'écoulement. Cet accroissement du coefficient d'écoulement (communément appelé « paradoxe hydrologique sahélien »), constaté par plusieurs auteurs dont une revue a été faite par Descroix et al. (2018), résulterait d'un profond changement du comportement hydrologique de ces hydrosystèmes. La quantification de l'impact de chacune des causes s'avère donc indispensable pour expliquer le paradoxe et en tenir compte dans l'élaboration et/ou l'exécution des politiques d'aménagement, de gestion et de protection des ressources hydriques.

Dans la littérature, les méthodes utilisées pour estimer les impacts peuvent être regroupées en des approches descendante et ascendante (Blöschl et al., 2007). L'approche ascendante (encore appelée approche Newtonienne) englobe l'utilisation des modèles physiques distribués/semi distribués qui permettent les contrôles de causalité, mais les résultats des simulations sont en grande partie le reflet des nombreux paramètres. L'approche descendante (encore appelée approche Darwinienne) concernent les études statistiques/bayésiennes sur de longues séries de données hydro-climatiques et environnementales, les méthodes conceptuelles (cadre de Budyko, cadre de Tomer-Schiling) et les méthodes analytiques (élasticité du climat, sensibilité hydrologique). L'approche ascendante nécessite beaucoup de données difficilement disponibles en quantité et en qualité dont celles de la dynamique environnementale. Les méthodes analytiques et conceptuelles basées sur le cadre conceptuel de Budyko ne nécessitent pas de données environnementales, mais produisent des résultats pratiques utiles et utilisables par les décideurs (Hasan et al., 2018).

Le manque de données fiables de suivi de la dynamique environnementale en Afrique a limité la plupart des études d'impacts uniquement à la quantification des impacts du changement climatique ou à des études qualitatives (Boubacar, 2012 ; Grijsen et al., 2013 ; Fowe et al., 2012). Les approches descendantes constituent donc une alternative fiable pour quantifier les impacts réels des changements climatique et environnemental sur la ressource en eau de l'Afrique et surtout du Sahel dont le comportement hydrologique est très fragile.

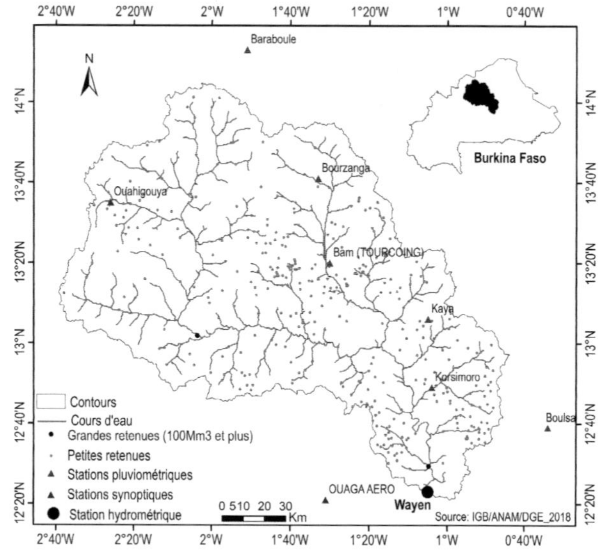

Figure 1. Zone d'étude : Bassin du Nakanbé à Wayen.

L'objectif principale de cette étude est de quantifier les contributions de chacun des changements climatique et environnemental au changement des écoulements dans la partie sahélienne du bassin du Nakanbé au Burkina Faso. Pour ce faire, des tests statistiques de détection de tendance et de rupture ont été utilisés pour choisir les périodes de référence et d'impact. Les impacts ont été quantifiés à travers deux approches et deux modèles, soit de quatre manières différentes.

2 Données et méthodes

2.1 Zone et données de l'étude

La zone d'étude est la partie sahélienne du bassin du Nakanbé, le bassin du Nakanbé à Wayen qui draine une superficie de 21 178 km^2 (Fig. 1). Ce bassin est très anthropisé et abrite 371 retenues d'eau (Fig. 1) dont les plus importantes (Ziga et Dourou) ont été mises en eau entre 1995 et 2000. La Fig. 1 montre la position des stations hydro-climatiques utilisées. Les données d'écoulements (R), de précipitations (P) et d'évapotranspirations potentielles (PE) utilisées proviennent des services météorologique et hydrologique du Burkina Faso et de la base SIEREM (Boyer et al., 2006 ; http://www.hydrosciences.fr/SIEREM/, la date du dernier accès : 20 October 2021). Elles concernent huit stations de pluie, deux synoptiques et une hydrométrique (Fig. 1). L'évapotranspiration réelle (AE) est déduite du bilan hydrique à long terme du bassin en négligeant les échanges avec les nappes souterraines. La pluie moyenne du bassin versant a été calculée en appliquant la méthode du polygone de Thiessen aux huits stations pluviométriques. L'évapotranspiration potentielle du bassin versant a été obtenue en faisant une moyenne pondérée des données aux deux stations synoptiques. L'analyse des données disponibles à chaque station a conduit au choix de la période 1965–1994 pour l'étude. Le choix de cette période se justifie par le fait qu'elle couvre la période de sécheresse (1970–1989) au cours de laquelle le paradoxe hydrologique sahélien a été mis en évidence et qu'elle n'intègre pas la période de mise en eau des grands aménagements sur le bassin (la capacité des deux grands aménagements faisant plus de 80 % du volume de toutes les retenues). Les autres retenues sont supposées intégrées au relief du bassin, car elles sont de petites capacités et beaucoup ont été réalisées autour des années 1960. Aussi, la longueur de cette période est suffisante pour une étude d'impacts (Dey et Mishra, 2017).

2.2 Cadre conceptuel de Budyko et modèles Budyko adaptés à la zone d'étude

Le cadre de Budyko (Eq. 1) est une approche basée sur le bilan hydrique et est utilisé par plusieurs auteurs à travers le monde pour quantifier les contributions du climat et de l'environnement au changement des écoulements. Une synthèse de ces travaux a été faite par Dey et Mishra (2017).

$$\frac{\mathrm{AE}}{P} = f(\varphi, e) = f \tag{1}$$

où AE est l'évapotranspiration réelle ; P la précipitation ; φ est le rapport entre l'évapotranspiration potentielle (ETP) et P ; e est un paramètre qui prend en compte l'état de l'environnement du bassin et f est la fonction Budyko.

L'analyse de sensibilité des modèles de type Budyko avait montré que les modèles de Fu (1981) et de Choudhury (1999) sont les plus adaptés au contexte sahélien où les indices d'aridité sont supérieurs à 2 (Gbohoui et al., 2021). Ces deux modèles sont utilisés dans cette étude car les indices d'aridités du bassin ont varié entre 2 et 5 sur la période 1965–1994.

2.3 Méthodologie de quantification d'impacts

2.3.1 Approche d'élasticité améliorée

En s'inspirant de Gbohoui et al. (2021), les expressions des contributions du changement climatique (Eq. 2), du changement environnemental (Eq. 3) et de l'interaction climat-environnement (Eq. 4) au changement d'écoulement par rapport à l'écoulement sur la période de référence ont été obtenues.

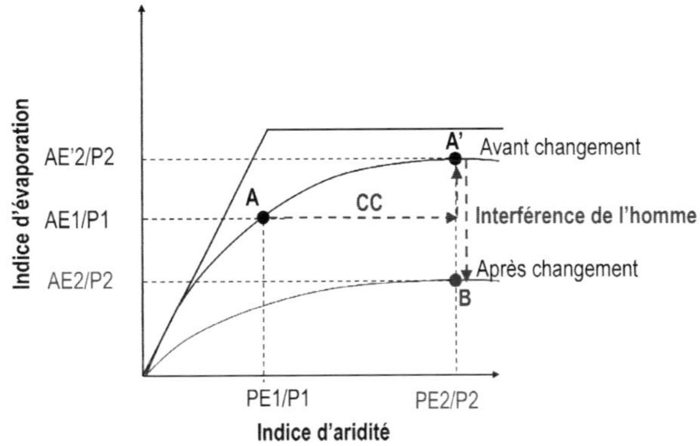

Figure 2. Décomposition de la courbe de Budyko. CC : Changement climatique ; AE : Evapotranspiration réelle ; PE : Evapotranspiration potentielle ; P : Précipitation.

$$\frac{\Delta R_\mathrm{C}}{R} = (1+\beta)\frac{\Delta P}{P} - \beta\frac{\Delta \mathrm{PE}}{\mathrm{PE}}; \beta = \frac{\varphi}{1-f(\varphi,e)}\frac{\partial f}{\partial \varphi} \tag{2}$$

$$\frac{\Delta R_\mathrm{e}}{R} = \left[\frac{-e}{1-f(\varphi,e)}\frac{\partial f}{\partial e}\right]\frac{\Delta e}{e} \tag{3}$$

$$\frac{\Delta R_\mathrm{ice}}{R} = \frac{\Delta R}{R} - \frac{\Delta R_\mathrm{C}}{R} - \frac{\Delta R_\mathrm{e}}{R} \tag{4}$$

où $\frac{\Delta P}{P}$, $\frac{\Delta \mathrm{PE}}{\mathrm{PE}}$, $\frac{\Delta R}{R}$ sont les taux de variations de la précipitation, de l'évapotranspiration potentielle et de l'écoulement de la période d'impact par rapport à la période de référence. $\frac{\Delta R_\mathrm{C}}{R}$, $\frac{\Delta R_\mathrm{e}}{R}$, $\frac{\Delta R_\mathrm{ice}}{R}$ sont les contributions respectives du climat, de l'environnement et de l'interaction climat-environnement à $\frac{\Delta R}{R}$.

2.3.2 Approche de décomposition

La décomposition de la courbe de Budyko est une approche conceptuelle qui se base sur le fait qu'à l'échelle d'un bassin versant, tous les termes (AE, P et PE) du bilan hydrique sont influencés par le climat alors que la dynamique environnementale n'impacte que l'évapotranspiration réelle (AE). La courbe de Budyko étant une relation entre l'indice d'évaporation (AE / P) et l'indice d'aridité (PE / P), si les changements intervenus sur le bassin ne sont dus qu'au climat, sa position bougera mais restera sur la même courbe (ex : A à A′ sur la Fig. 2). Tout changement de courbe indique qu'il y a eu d'autres modifications que le changement climatique, donc qu'il y a eu un impact environnemental. L'équation (5) permet de quantifier les variations d'écoulement dues à l'environnement et au climat sans dissocier l'interaction climat-environnement.

$$\Delta R_\mathrm{E} = P_2\left(\frac{\mathrm{AE}_2'}{P_2} - \frac{\mathrm{AE}_2}{P_2}\right) \tag{5}$$

Tableau 1. Impacts du climat et de l'environnement sur les écoulements du Nakanbé à Wayen.

Période d'impact	$\frac{\Delta P}{P}$	$\frac{\Delta PE}{PE}$	$\frac{\Delta R}{R}$	Approches/Modèles	Elasticité			Décomposition	
					$\frac{\Delta R_c}{R}$	$\frac{\Delta R_e}{R}$	$\frac{\Delta R_{ice}}{R}$	$\frac{\Delta R_c}{R}$	$\frac{\Delta RE}{R}$
1978–1994	−8,2 %	1,3 %	91,5 %	Fu	−30,6 %	89,3 %	32,7 %	−27,2 %	118,7 %
				Choudhury	−32,0 %	90,4 %	33,0 %	−28,3 %	119,8 %

$\frac{\Delta P}{P}$, $\frac{\Delta PE}{PE}$, $\frac{\Delta R}{R}$ sont les taux de variations de chacune des variables par rapport à la période de référence : 1965–1977. $\frac{\Delta R_c}{R}$, $\frac{\Delta R_e}{R}$, $\frac{\Delta R_{ice}}{R}$ sont les contributions respectives du climat, de l'environnement et de l'interaction climat-environnement à $\frac{\Delta R}{R}$. Sur la période de référence, les moyennes annuelles des variables hydro-climatiques sont de 636,30/2189,76/9,89 mm respectivement pour la précipitation, l'évapotranspiration potentielle et la lame d'eau écoulée. De 1965 à 1994, les paramètres environnementaux sont passés de 3,38 (w de Fu) et 2,58 (n de Choudhury) sur la période de référence à respectivement 2,79 et 2,01 sur la période d'impact 1978–1994.

et

$$\Delta R_C = P_2\left(1 - \frac{AE_2'}{P_2}\right) - R_1$$

R_1 désigne la lame d'eau moyenne annuelle écoulée sur la période de base, ΔR_C et ΔR_E sont les variations d'écoulement dues respectivement aux changements climatique et environnemental.

3 Résultats et discussion

3.1 Confirmation du paradoxe hydrologique sahélien dans le bassin du Nakanbé à Wayen

Les tests de détection de tendance de Mann-Kendall et de rupture de Pettitt ont été utilisés pour analyser les séries de données de précipitation, d'évapotranspiration potentielle et d'écoulement sur la période 1965–1994. Les résultats de l'étude de tendance ont montré une augmentation des écoulements annuels (Pente de Sen = 0,547 ; p value = 0,005) et de l'évapotranspiration potentielle (Pente de Sen = 1,44 ; p value = 0,014) alors que les précipitations annuelles du bassin sont en décroissance (Pente de Sen = −4,17 ; p value = 0,256). Ce résultat confirme le premier paradoxe hydrologique sahélien (Descroix et al., 2018) caractérisé par une augmentation exacerbée du ruissellement malgré la sécheresse de 1970–1989 et le réchauffement climatique. Le test de Pettitt a indiqué une rupture significative dans les écoulements à l'exutoire du bassin à partir de 1977 (p value = 0,021). Ainsi, la période 1965–1977 a été considérée comme la période de référence pour les études d'impacts.

3.2 Impacts des changements climatique et environnemental sur les écoulements du Nakanbé à Wayen

Nous avons quantifié les impacts des changements climatique et environnemental subis par le bassin sur la période 1978–1994 par rapport à 1965–1977 en utilisant les modèles Budyko de Fu et de Choudhury. Les résultats (Tableau 1) montrent que :

- L'approche d'élasticité permet de déduire l'impact des interactions environnement-climat ;
- Les contributions des différents facteurs à l'accroissement des écoulements de 91,5 % sur la période 1978–1994 par rapport à la période 1965–1977 ont été de -29 ± 2 % pour le climat, $+90 \pm 1$ % pour l'environnement et +33 % pour les interactions climat-environnement ;
- L'impact de l'environnement quantifié selon l'approche de décomposition ($\Delta RE / R$) inclus les interactions environnement-climat et a été considéré comme l'impact brut de l'environnement ;
- Les deux modèles donnent des résultats similaires : le modèle de Fu surestime les impacts de 1 à 5 % par rapport au modèle de Choudhury ;
- L'approche d'élasticité surestime les impacts de 2 à 13 % par rapport à celle de décomposition.

Ces résultats confirment que le premier paradoxe hydrologique mis en évidence est principalement dû à la dégradation environnementale (Descroix et al., 2018) et ont montré une dégradation des paramètres environnementaux de 20 % sur la période 1978–1994 par rapport à 1965–1977. L'adoption de meilleures pratiques de gestion environnementale permettra de réduire les coefficients de ruissellement car il existe une relation consistante et significative entre le taux de couverture végétale et le paramètre des modèles Budyko (Gbohoui et al., 2021 ; Li et al., 2013). Les résultats obtenus sont similaires à ceux de Gbohoui et al. (2021) à l'échelle du Nakanbé à Wayen, malgré le peu de stations utilisées dans la présente étude. Cela montre la robustesse des approches utilisées.

4 Conclusion

Dans cette étude, nous avons quantifié les contributions du climat et de l'environnement au premier paradoxe hydrologique sahélien dans le bassin du Nakanbé à Wayen au Burkina Faso sur la base de huit stations de pluie, deux d'évapotranspiration potentiel et d'une station hydrométrique. Nos résultats ont montré que l'intensité de l'impact de l'environnement fait plus du triple de celui du climat. L'adoption de meilleures pratiques de gestion de la dynamique environnementale du bassin pourrait donc permettre de réguler les écoulements.

Disponibilité du code. Le code utilisé dans cet article est disponible à https://doi.org/10.5281/zenodo.5590923 (Gbohoui, 2021).

Disponibilité des données. Les données utilisées sont la propriété des organismes nationaux du Burkina Faso : Direction Générale des Ressources en Eau (DGRE), Agence Nationale de la Météorologie (ANAM).

Collaborateurs. YPG a fait la conceptualisation, le traitement et l'analyse des données ; élaboré la méthodologie et le code informatique de calcul ; rédigé et révisé l'article.

JEP, TF, HK et HY ont participé à la conceptualisation, à l'élaboration de la méthodologie, à l'analyse des résultats et à la rédaction de l'article, et ont supervisé les activités de recherche.

Intérêts concurrents. Les auteurs déclarent qu'ils n'ont aucun conflit d'intérêts.

Clause de non-responsabilité. Publisher's note : Copernicus Publications remains neutral with regard to jurisdictional claims in published maps and institutional affiliations.

Déclaration du numéro spécial. This article is part of the special issue "Hydrology of Large River Basins of Africa". It is a result of the 4th International Conference on the "Hydrology of the Great Rivers of Africa", Cotonou, Benin, 13–20 November 2021.

Remerciements. Les auteurs remercient toutes les personnes qui ont contribué à cet article en lisant ou en donnant de judicieux conseils.

Financement. This research has been supported by the African Development Bank Group (Project no. P-Z1-IA0– 013, Grant no. 2100155032824).

Références

Blöschl, G., Ardoin-Bardin, S., Bonell, M., Dorninger, M., Goodrich, D., Gutknecht, D., Matamoros, D., Merz, B., Shand, P., and Szolgay, J. : At what scales do climate variability and land cover change impact on flooding and low flows ?, Hydrol. Process., 21, 1241–1247, https://doi.org/10.1002/hyp.6669, 2007.

Boubacar, I. : Caractérisation des saisons de pluies au Burkina Faso dans un contexte de changement climatique et évaluation des impacts hydrologiques sur le bassin du Nakanbé, Thèse de doctorat, 2iE et Université Pierre et Marie Curie, France, 2012.

Boyer, J.-F., Dieulin, C., Rouche, N., Cres, A., Servat, E., Paturel, J.-E., and Mahe, G. : SIEREM : an environmental information system for water resources, IAHS publication, 308, 19–25, 2006.

Choudhury, B. J. : Evaluation of an empirical equation for annual evaporation using field observations and results from a biophysical model, J. Hydrol., 216, 99–110, 1999.

Descroix, L., Guichard, F., Grippa, M., Lambert, L., Panthou, G., Mahé, G., Gal, L., Dardel, C., Quantin, G., Kergoat, L., Bouaïta, Y., Hiernaux, P., Vischel, T., Pellarin, T., Faty, B., Wilcox, C., Malam Abdou, M., Mamadou, I., Vandervaere, J.-P., Diongue-Niang, A., Ndiaye, O., Sané, Y., Dacosta, H., Gosset, M., Cassé, C., Sultan, B., Barry, A., Amogu, O., Nka Nnomo, B., Barry, A., and Paturel, J.-E. : Evolution of Surface Hydrology in the Sahelo-Sudanian Strip : An Updated Review, Water, 10, 748, https://doi.org/10.3390/w10060748, 2018.

Dey, P. and Mishra, A. : Separating the impacts of climate change and human activities on streamflow : A review of methodologies and critical assumptions, J. Hydrol., 548, 278–290, https://doi.org/10.1016/j.jhydrol.2017.03.014, 2017.

Fowe, T., Yacouba, H., Karambiri, H., Barbier, B., Paturel, J.-E., et Boubacar, I. : Impacts des scenarii climatique et de l'occupation des sols sur les ressources en eau du bassin versant du Nakambé (Burkina Faso), Climat et Développement, 14, 123–137, 2012.

Fu, B. : On the calculation of the evaporation from land surface, Sci. Atmos., 5, 23–31, 1981.

Gbohoui, Y. P. : R script for impacts quantification based on Budyko framework (Version V0). Hydrology of Large River Basins of Africa, Cotonou, Bénin, Zenodo [code], https://doi.org/10.5281/zenodo.5590923, 2021.

Gbohoui, Y. P., Paturel, J.-E., Tazen, F., Mounirou, L. A., Yonaba, R., Karambiri, H., and Yacouba, H. Impacts of climate and environmental changes on water resources : A multi-scale study based on Nakanbé nested watersheds in West African Sahel, J. Hydrol. Reg. Stud., 35, 100828, https://doi.org/10.1016/j.ejrh.2021.100828, 2021.

Grijsen, J. G., Brown, C., Tarhule, A., Ghile, Y. B., Taner, Ü., Talbi-Jordan, A., Doffou, H. N., Guero, A., Dessouassi, R. Y., Kone, S., Coulibaly, B., and Harshadeep, N. : Climate Risk Assessment for Water Resources Development in theNiger River Basin PartI : Context and Climate Projections, Climate Variability : Regional and Thematic Patterns, 37, BoD–Books on Demand, https://doi.org/10.5772/56707, 2013.

Hasan, E., Tarhule, A., Kirstetter, P.-E., Clark III, R., and Hong, Y. : Runoff sensitivity to climate change in the Nile River Basin, J. Hydrol., 561, 312–321, https://doi.org/10.1016/j.jhydrol.2018.04.004, 2018.

Li, D., Pan, M., Cong, Z., Zhang, L., and Wood, E. : Vegetation control on water and energy balance within the Budyko framework, Water Resour. Res., 49, 969–976, https://doi.org/10.1002/wrcr.20107, 2013.

Proc. IAHS, 384, 275–281, 2021
https://doi.org/10.5194/piahs-384-275-2021

Assessing the effect of soil parameterization in land use change impact modeling

Yacouba Yira[1,2], Aymar Yaovi Bossa[2,3], Ernest Amoussou[4], Djigbo Félicien Badou[5,2], Jean Hounkpè[3,2], and Kpade Ozias Laurentin Hounkpatin[6,2]

[1]Applied Science and Technology Research Institute – IRSAT/CNRST, P.O. Box 7047, Ouagadougou, Burkina Faso
[2]Hydro-Climate Services, P.O. Box 374, Ouagadougou 14, Burkina Faso
[3]Department of Water for Agriculture and the Society, National Institute of Water, University of Abomey–Calavi, P.O. Box 526, Cotonou 01, Benin
[4]Département de Géographie et Aménagement du Territoire de l'Université de Parakou, BP 123 Parakou, Benin
[5]Ecole d'Horticulture et d'Aménagement des espaces Verts, Université Nationale d'Agriculture, BP 43 Kétou, Benin
[6]Department of Soil and Environment, Swedish University of Agricultural Sciences, P.O. Box 7014, 75007, Uppsala, Sweden

Correspondence: Yacouba Yira (yira_y@uni-bonn.de)

Published: 16 November 2021

Abstract. This study addresses the importance of integrating the effect of land use on soil infiltration rate into land use change impact modeling. Based on a validated version 9.05.04 of the Water balance Simulation Model-WaSiM (statistical quality measures > 0.7), and field measurement of the infiltration rate under cropland and fallow, sixteen model simulations were performed. The impact of land use change is computed comparing LULC status of years 1990 and 2013. The effect of soil parameterization is computed using a refined soil map integrating land use change impact of soil infiltration rate and a classic soil map not considering this interaction. The results show differences in model results as an effect of soil parameterization approaches, indicating that the model is sensitive to the integration of LULC related effects on soil hydraulic conductivity. These differences are more pronounced with increasing modeling time steps (24 and 28 h). The signal-to-noise-ratio indicates that, results achieved in LULC impact assessment with a classic and a refined soil parameterization are very comparable except for interflow.

1 Introduction

Land use change impact is a major research topic in hydrology (Wagner et al., 2017). By affecting the partitioning of precipitation and influencing many hydrological process, land use and land cover (LULC) changes strongly affect the quality and quantity of water resources (Aghsaei et al., 2020). However, quantifying the net effect of LULC change impact on water resources remains challenging, as this effect is multifactorial (Yira et al., 2016), and includes a modeling approach component (Beven, 2012). Besides the modeling objective, a modeling approach often heavily relies on data availability.

Field evidences, by Yira and Bossa (2019), of land use impacts on soil infiltration rate provide the opportunity to assess the importance of considering land use dependent soil properties in the hydrological impact assessment of LULC change on water resources in the Dano catchment. A modeling exercise incorporating knowledge gained from field investigations with regards to land use impacts on soil infiltration rate may require additional parameterization effort, but the method has proven to improve hydrological processes representation (Hölzel and Diekkrüger, 2010).

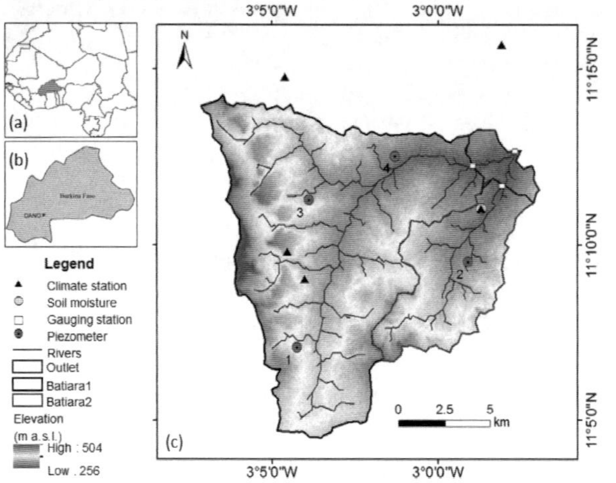

Figure 1. Location and instrumentation map of the Dano catchment.

The current study addresses the importance of integrating the influence of land use on soil infiltration rate into land use change impact assessment. It aims to analyze how soil parameterization and modeling time step could affect the results in land use change impact assessment. It has the following specific objectives: (i) assess the sensitivity of the hydrological simulation model to land use effect on soil infiltration rate, (ii) assess the consequences of integrating land use effect on soil infiltration rate in LULC impact assessment, and (iii) to evaluate whether the importance of integrating land use related effect on soil infiltration rate in the modeling exercise depend on the modeling time step.

2 Materials and methods

2.1 Study area

The study was carried out in the Dano catchment (Fig. 1). The catchment is located is in the Sudanese climate zone of Burkina Faso and covers an area of 195 km^2. It is characterized by an alternation of two seasons including a dry season of 6 to 7 months (November to April) and a rainy season of 5 months spanning from May to October (Yira et al., 2016). The natural vegetation of the catchment is of Sudanian region. Average monthly temperatures and annual rainfall range between 24 and 32 °C, and 800 to 1200 $\mathrm{mm\,yr^{-1}}$, respectively (Yira et al., 2016).

2.2 Land use change

The land use and land cover maps after Forkour (2014) and Landmann et al. (2007) corresponding to LULC status for the years 2013 and 1990 respectively, were used for LULC change impact assessment (Table 1). The reclassified version of both maps (Yira et al., 2016) was used, and the year 1990 was considered as baseline.

Table 1. Percentage of the catchment area per land use class in 1990 and 2013.

Land use class	Years	
	1990	2013
Cropland (%)	11.6	38
Savannah (%)	87	58
Water (%)	0.05	0.05
Urban area (%)	1	3

2.3 Infiltration measurement

A hood infiltrometer (Schwärzel and Punzel, 2007) was used to perform infiltration (K_s) tests on adjacent croplands and fallow areas. Measurements showed on average (1.16-fold) higher K_s under fallow compared to cropland (Yira and Bossa, 2019).

2.4 Soil map refinement

Soil infiltration rate is parameterized in WaSiM through the soil map and soil classes' characteristics, while no infiltration rate dependent parameter is associated with this parameterization. Following the infiltration test, the Eq. (1) by Hölzel and Diekkrüger (2010) was adopted to integrate land use influence on K_s into the soil parameterisation.

$$K_{s_{x_i}}^{\mathrm{ref}} = K_{\mathrm{sat}}^{\mathrm{map}} \cdot \frac{\hat{K}_{s_{x_i}}^{\mathrm{meas}}}{\hat{K}_{s_x}^{\mathrm{meas}}} \tag{1}$$

where $\hat{K}_s$ is the mean of the infiltration rate, x refers to all samples, x_i is the samples of the specific area (cropland or savannah), "ref" refers the refined values, "meas" refers to measured values, and "map" indicates the initial values given by the classic soil map.

The reference basis for K_s was provided by the K_s values from the classic soil map (CS), which were derived from laboratory measurements for each soil type and different soil horizons following Saxton and Rawls (2006).

2.5 Modeling approach and simulation experiment

Version 9.05.04 of WaSiM, after Yira et al. (2016) and Schulla (2015), calibrated and validated (Fig. 2) at a daily time step using a classic soil parameterization and the LULC map of 2013 was the starting point of the study.

The impact of land use change is computed comparing LULC status of years 1990 and 2013. Two additional factors are considered; i.e. (i) the interaction between soil and land use parameterization (with a soil parameterization integrating LULC dependent soil infiltration rate (refined soil-RS), and a classic soil parameterization ignoring this dependance (classic soil-CS), and (ii) the influence of the modeling time

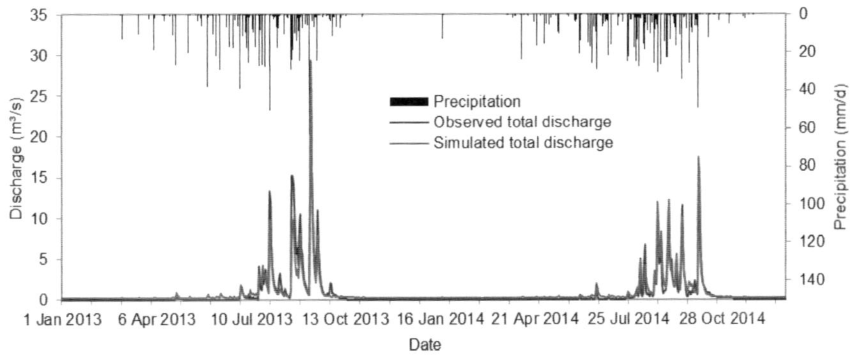

Figure 2. Daily observed and simulated discharges for the calibration (2013) and validation (2014) years at the outlet of the Dano catchment.

Table 2. Combined land use scenarios, soil parameterizations and time scales experiments.

Classic soil (CS)				Refined soil (RS)			
Land use status 2013 (LU2013)							
CS_LU2013_8h	CS_LU2013_12h	CS_LU2013_24h	CS_LU2013_48h	RS_LU2013_8h	RS_LU2013_12h	RS_LU2013_24h	RS_LU2013_48h
Land use status 1990 (LU1990)							
CS_LU1990_8h	CS_LU1990_12h	CS_LU1990_24h	CS_LU1990_48h	RS_LU1990_8h	RS_LU1990_12h	RS_LU1990_24h	RS_LU1990_48h

step on this interaction (8, 12, 24, and 48 h time steps). Overall, 16 model configurations were set up combining these two factors and LULC status (Table 2). The fitted calibrated parameters of the initial model (Yira et al., 2016) were kept constant for all the simulations experiments.

2.6 Comparison of land use and soil parameterization effects

The "signal-to-noise-ratio" (SNR) (Bormann et al., 2009) was adapted to compare LULC change impacts with soil parameterizations effects (Eq. 2). In this study, the "noise" is represented by soil parameterization effect, and "signal" is represented by land use change impact. *Positive* SNR values indicate that land use change impacts are larger than soil parameterization effects and *Negative* SNR values imply the opposite.

$$\mathrm{SNR} = \left[\frac{|X_{\mathrm{CS\_LU1990}} - X_{\mathrm{CS\_LU2013}}|}{|X_{\mathrm{CS\_LU1990}} - X_{\mathrm{RS\_LU1990}}|}\right] - 1 \qquad (2)$$

where SNR is the "signal-to-noise-ratio", X is the simulated variable (discharge, $\mathrm{ET_a}$, etc.).

3 Results and discussion

3.1 Modeling time step and model performance

The initial model, calibrated and validated at a daily time step using a classic soil parameterization and the LULC of 2013 (CS_LULC2013_24h), and the other seven model set ups using the same LULC map but different soil parameterizations and time steps (see Table 2) are compared to the observed discharge in order to relate change in model performance to soil parameterization and time step (Fig. 3). The annual variation of discharge appears well reproduced by each model set up, although difference in simulated peak discharges can be noticed.

Soil refinement appears beneficial to all simulations performed at 8, 12 and 48 h time steps irrespective of the considered model statistical quality measure; this may suggest that refined K_s is more effective than classic K_s. However, at the 24 h time step, classic soil (CS) outscores refined soil (RS) for Pearson product-moment-correlation-coefficient.

3.2 Modeling time step and water balance

Time step effects on water balance components are shown in Fig. 4. Change in water balance components with increasing modeling time step comparing land use status of 2013 and 1990 are very similar: (i) an increase of actual evapotranspiration, (ii) a decrease in total discharge, (iii) a decrease in surface runoff, (iv) an increase in baseflow, (v) a decrease in interception, and (vi) an increase followed by a decrease for interflow.

The comparison of RS and CS histograms for the same time step shows the effect of soil parameterization on dis-

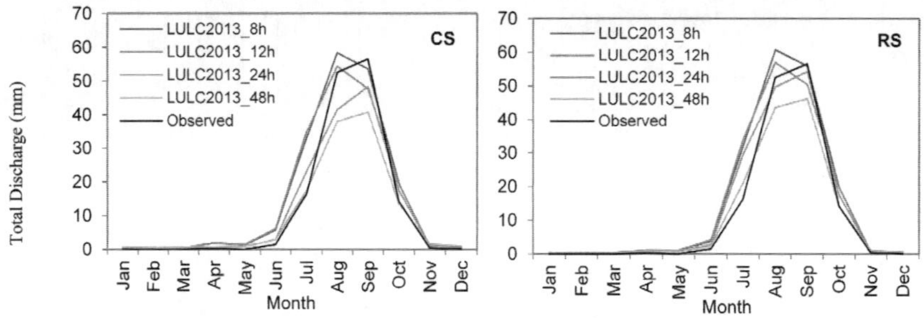

	CS				RS			
	8h	12h	24h	48h	8h	12h	24h	48h
R^2	0.90	0.83	0.94	0.98	0.92	0.87	0.93	0.97
NSE	0.92	0.90	0.95	0.90	0.93	0.91	0.96	0.96
KGE	0.69	0.69	0.82	0.76	0.71	0.72	0.82	0.86

Figure 3. Model performance using different modeling time steps and soil parameterization approaches. Period of January 2013 to December 2014. Performance is calculated using mean monthly discharges.

charge components. For all simulation time steps, total discharge and discharge components clearly appear sensitive to soil parameterization.

Consequences of LULC conversion from year 1990 to 2013 are the reduction of LAI, vegetation cover fraction, root depth etc. which leads to the increase of throughfall and the reduction of water that is evaporated by interception. Therefore, more water can infiltrate into the soil or flow as infiltration excess runoff. Interflow and baseflow increase with throughfall.

Change in the water balance between classic soil and refined soil parameterization has to be interpreted solely by the difference in hydraulic conductivity at the soil surface, as this is the only difference between CS and RS. Change in hydraulic conductivity at the uppermost numerical soil layer due to soil refinement leads CS and RS to partition throughfall differently which affects surface runoff generation and water balance

3.3 Soil refinement and land use change impact

Figure 5 presents the change in the water balance components due to LULC change (LULC of 1990 compared to LULC of 2013) following a classic soil parameterization (Fig. 5a) and a refined soil parameterization (Fig. 5b). A very comparable decrease in actual evapotranspiration and an increase in discharge (all components) characterize LULC change impacts for both soil parameterization approaches. However, differences in the water balance components between soil parameterization approaches can be noticed as shown by Fig. 5c. These differences vary with the modeling time step and water balance components. A LULC change impact assessment performed with a classic soil leads to a higher decrease in ET_a compared to an assessment performed with a refined soil; this difference appears to further increase with the modeling time step. Increase in surface runoff following LULC change is higher with the RS than the CS. Furthermore, the difference in surface runoff between RS and CS decreases with the modeling time step. Simulated change in total discharge, baseflow and interflow following LULC change are lower with the refined soil compared to the classic soil.

It can be noted from Fig. 5c that irrespective of modeling time step, a LULC change impact assessments performed with a RS and CS parameterization do not yield the same water balance. Differences range from 0.2 to 18 mm yr^{-1} depending on water balance components. Figure 5 further indicates that simulated changes in water balance due to LULC change are larger than the change due to soil parameterization, suggesting that soil refinement might have a marginal impact in LULC impact assessment.

The SNR is calculated to test whether the LULC change effect (Table 3) is significantly larger than the soil refinement effect. Table 3 shows both positive and negative SNR. Except interflow, all water balance components show positive SNR, indicating that LULC change impact is larger than soil refinement, meaning the noise created by soil refinement does not affect LULC change signal. In other words, results achieved in LULC impact assessment with both soil parameterizations (CS and RS) are very comparable except for interflow. Noise is larger than signal only for simulations performed at 24 and 48 h, even though this noise is low. Therefore, for a LULC impact assessment focusing on discharge components it is necessary to use a refined soil as omitting this information produced an effect higher than LULC change effect. However, LULC change impact appears larger than soil refinement effect for the water balance, as confirmed by positive SNR indices. It is noteworthy that for interflow, soil refinement effect was larger than the LULC change impact for some modeling time steps, implying that a consistent assessment of LULC change impact on interflow (and consequently, on discharge components) should integrate LULC related effects into soil properties.

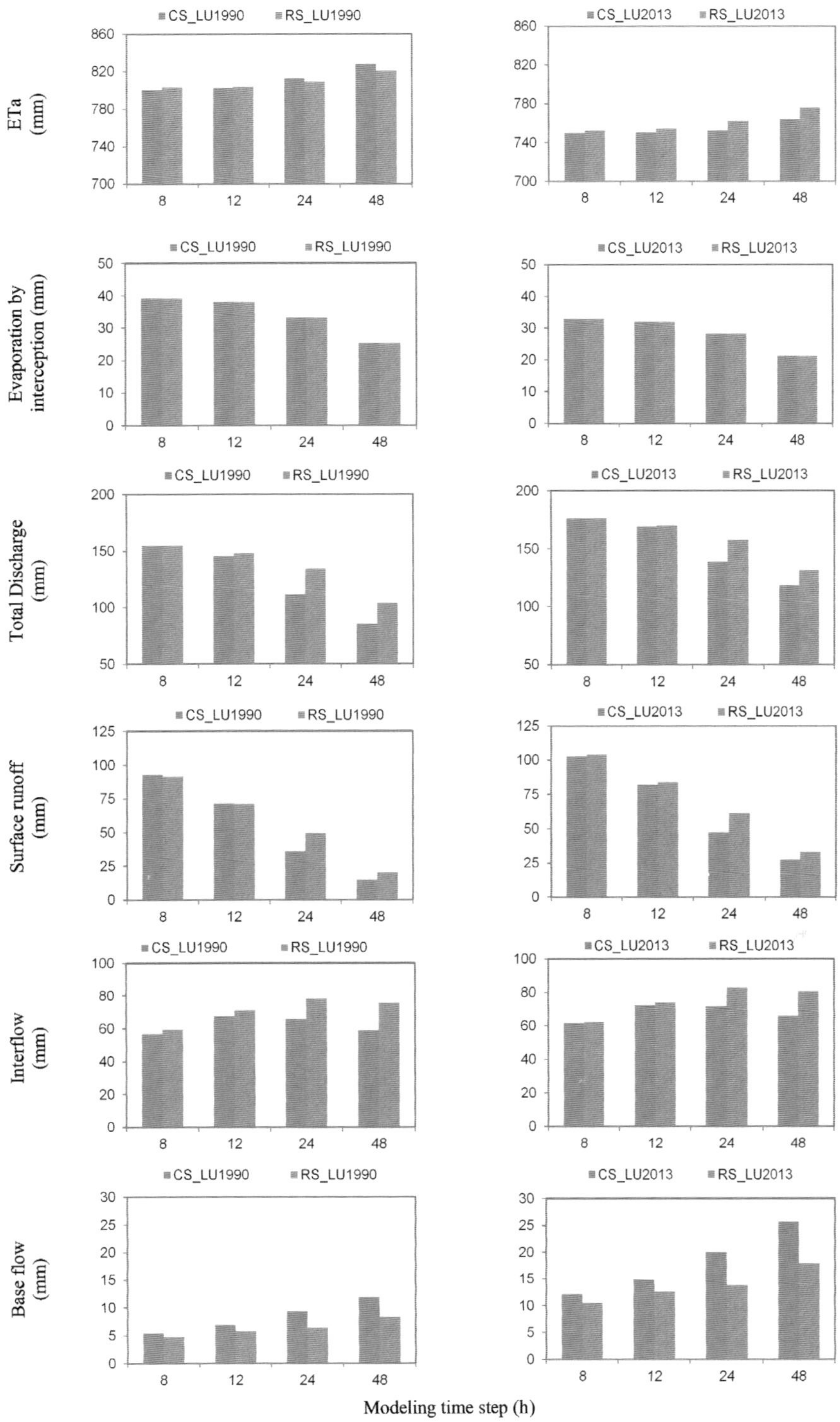

Figure 4. Annual water balance components for different modeling time steps and soil parameterization approaches for the period of January 2013 to December 2014.

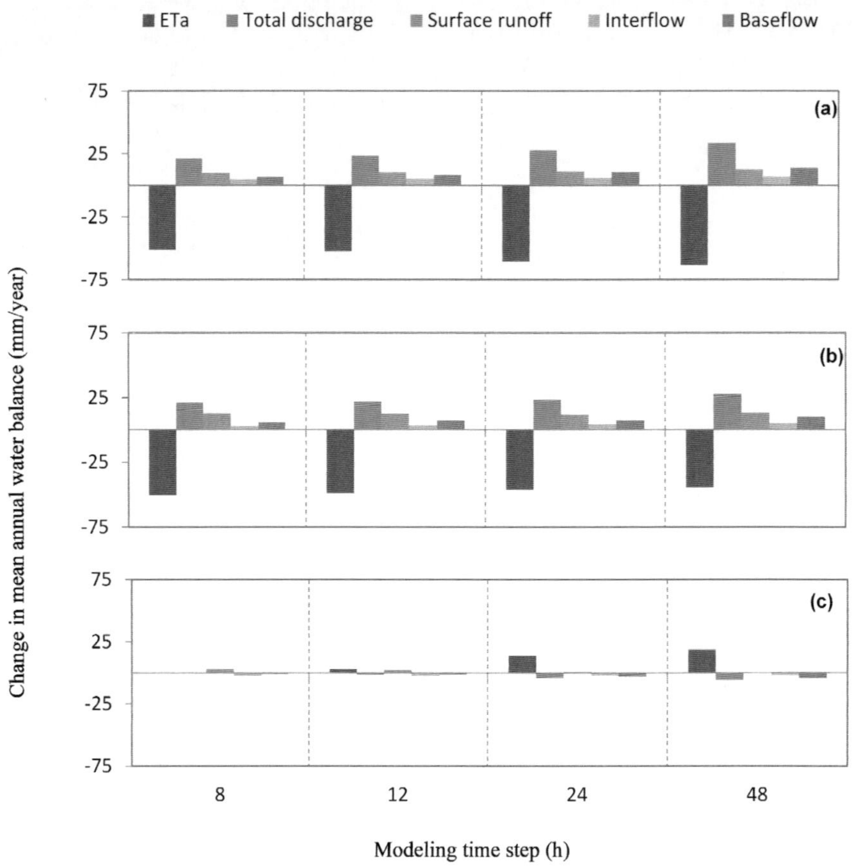

Figure 5. Change in water balance as a result of LULC change, modeling time step and soil parameterization. Part **(a)** compares LULC1990 to LULC2013 using a classic soil-CS, part **(b)** compares LULC1990 to LULC2013 using a refined soil-RS, and **(c)** shows the difference between soil parametrization approaches **(a, b)**.

Table 3. Signal-to-noise ratio for different modeling time steps. Signal refers to LULC change effect, and noise is soil refinement effect.

	Signal to noise ratio				
Modeling time step	Actual evapotranspiration	Total discharge	Surface runoff	Interflow	Base flow
8	22.91	39.83	5.94	0.82	9.61
12	69.76	9.35	27.51	0.42	6.20
24	17.18	0.20	0.07	−0.52	2.63
48	8.13	0.81	1.34	−0.58	2.86

4 Conclusion

In this study, land use change related effects on soil infiltration rate was integrated in the modeling of LULC change impact on the Dano catchment hydrology. Previous field investigation showed that soil surface hydraulic conductivity was significantly affected by land use. This additional information was integrated in the modeling as refined soil, and was compared to a simulation ignoring this information. Modeling time step was further considered as a potential factor affecting the model outputs. Several models were then set up combining soil parameterization, modeling time step, and LULC change.

The results show that trends in water balance and discharge components following LULC change was similar for classic and refined soil parameterization irrespective of the modeling time step. However, differences in model results could be observed between soil parameterization approaches, indicating that the model is sensitive to the integration of LULC related effects on soil hydraulic conductivity. This supports the idea that integrated land use related effects on soil properties renders LULC change scenarios more plausible.

Code availability. WaSIM code is not publicly accessible. The applied Richards version of the software is available for download at http://www.wasim.ch/en/products/wasim_richards.htm (last access: 7 October 2021) (WaSiM-ETH, 2021).

Data availability. The applied data are available on the WASCAL Geoportal (https://wascal-dataportal.org/geonetwork/apps/search/?east_collapsed=true&s_search=&s_timeType=true&s_scaleOn=false&s_E_hitsperpage=20, last access: 7 October 2021) (WASCAL Geoportal, 2021).

Author contributions. YY, AYB and EA developed the concept. All authors contributed to implementing the concept and writing the manuscript.

Competing interests. The contact author has declared that neither they nor their co-authors have any competing interests.

Disclaimer. Publisher's note: Copernicus Publications remains neutral with regard to jurisdictional claims in published maps and institutional affiliations.

Special issue statement. This article is part of the special issue "Hydrology of Large River Basins of Africa". It is a result of the 4th International Conference on the "Hydrology of the Great Rivers of Africa", Cotonou, Benin, 13–20 November 2021.

Acknowledgements. The authors are grateful to Bernd Diekkrüger for supervising this work.

References

Aghsaei, H., Mobarghaee D. N., Moridi, A., Asadolahi, Z., Delavar, M., Fohrer, N., and Wagner, P. D.: Effects of dynamic land use/land cover change on water resources and sediment yield in the Anzali wetland catchment, Gilan, Iran, Sci. Total Environ., 712, 136449, https://doi.org/10.1016/j.scitotenv.2019.136449, 2020.

Beven, K.J.: Rainfall-Runoff Modelling: The Primer, 2nd Edition, 2nd ed. John Wiley & Sons, Ltd, Lancaster, England, 2012.

Bormann, H., Breuer, L., Gräff, T., Huisman, J. A., and Croke, B.: Assessing the impact of land use change on hydrology by ensemble modelling (LUCHEM) IV: Model sensitivity to data aggregation and spatial (re-) distribution, Adv. Water Resour., 32, 171–192, https://doi.org/10.1016/j.advwatres.2008.01.002, 2009.

Forkour, G.: Agricultural Land Use Mapping in West Africa Using Multi-sensor Satellite Imagery, PhD Thesis, Julius-Maximilians-Universität Würzburg, Germany, available at: http://opus.uni-wuerzburg.de/files/10868/Thesis_Gerald_Forkuor_2014.pdf (last access: 6 June 2019), 2014.

Hölzel, H. and Diekkrüger, B.: Hydrological Analyses as a Prerequisite for Soil Erosion Modeling – Landscape Related Studies in a Mesoscale Hydrological Catchment, in: Landform – Structure, Evolution, Process Control, Lecture Notes in Earth Sciences, edited by: Otto, J.-C. and Dikau, R., Springer, Berlin, Heidelberg, 127–149, 2010.

Landmann, T., Herty, C., Dech, S., and Schmidt, M.: Land cover change analysis within the GLOWA Volta basin in West Africa using 30-meter Landsat data snapshots, in: Geoscience and Remote Sensing Symposium, 23–27 July 2007, Barcelona, Spain, https://doi.org/10.1109/IGARSS.2007.4424058, 2007.

Saxton, K. E. and Rawls, W. J.: Soil water characteristic estimates by texture and organic matter for hydrologic solutions, Soil Sci. Soc. Am. J., 70, 1569–1578, 2006.

Schwärzel, K. and Punzel, J.: Hood Infiltrometer – A New Type of Tension Infiltrometer, Soil Sci. Soc. Am. J., 71, 1438, https://doi.org/10.2136/sssaj2006.0104, 2007.

Schulla, J.: Model DescriptionWaSiM, available at: http://wasim.ch/downloads/doku/wasim/wasim_2015_en.pdf (last access: 17 June 2016), 2015.

Wagner, P. D., Bhallamudi, S. M., Narasimhan, B., Kumar, S., Fohrer, N., and Fiener, P.: Comparing the effects of dynamic versus static representations of land use change in hydrologic impact assessments, Environ. Model. Softw., 122, 103987, https://doi.org/10.1016/j.envsoft.2017.06.023, 2017.

WASCAL Geoportal: https://wascal-dataportal.org/geonetwork/apps/search/?east_collapsed=true&s_search=&s_timeType=true&s_scaleOn=false&s_E_hitsperpage=20, last access: 7 October 2021.

WaSiM-ETH: WaSiM (Richards) [code], available at: http://www.wasim.ch/en/products/wasim_richards.htm, last access: 7 October 2021.

Yira, Y. and Bossa, A. Y.: Agricultural Expansion-Induced Infiltration Rate Change in a West African Tropical Catchment, Appl. Environ. Soil Sci., 2019, 2434512, https://doi.org/10.1155/2019/2434512, 2019.

Yira, Y., Diekkrüger, B., Steup, G., and Bossa, A. Y.: Modeling land use change impacts on water resources in a tropical West African catchment (Dano, Burkina Faso), J. Hydrol., 537, 187–199, https://doi.org/10.1016/j.jhydrol.2016.03.052, 2016.

Proc. IAHS, 384, 283–288, 2021
https://doi.org/10.5194/piahs-384-283-2021

Open Access PIAHS

Hydrology of Large River Basins of Africa

Future changes in precipitation, evapotranspiration and streamflows in the Mono Basin of West Africa

Lamboni Batablinlè[1], **Lawin E. Agnidé**[2], **Kodja Domiho Japhet**[3], **Amoussou Ernest**[3], **and Vissin Expédit**[3]

[1]Institute of Mathematics and Physical Sciences, University of Abomey-Calavi, BP 613 Porto-Novo, Benin
[2]Applied Hydrology Laboratory, University of Abomey-Calavi, BP 4521 Cotonou, Benin
[3]Laboratory Pierre Pagney, Climate, Water, Ecosystems and Development, University of Abomey-Calavi, BP 4521 Cotonou, Benin

Correspondence: Lamboni Batablinlè (lbata1982@yahoo.fr)

Published: 16 November 2021

Abstract. The impact of climate change on precipitation and water availability is of major concern for policy makers in the Mono Basin of West Africa, whose economy mainly depends on rainfed agriculture and hydropower generation. The objective of this study is to project rainfall, flows and evapotranspiration (ET) in the future period and understand their changes across Mono River Basin. Observed data were considered for the historical period 1980–2010, and a Multi-model ensemble for future projections data of eight selected Regional Climate Models under RCP 4.5 and RCP 8.5 over the periods 2011–2100 was used. The GR4J model was used to simulate daily flows of the Mono watershed. The ensemble mean shows a decrease and increase streamflows between −54 % and 42 %, −58 % and 31 % under the RCP4.5, RCP8.5 scenario, respectively. The greatest decreases of high flows is projected to occur in the near term under RCP8.5, whereas the greatest decrease of low flows is projected to occur in the long term under the same RCP. For the rainfall and ET, the both scenarios (RCP4.5 and RCP8.5) predict an increase of ET while the rainfall will decrease. The results of this study of would be very useful in the choice of management and adaptation policies for water resources management.

1 Introduction

The future projection of climate change reported by the 5th Intergovernmental Panel on Climate Change (IPCC) assessment indicated that the global mean temperature will continue to rise for the rest of 21th century, and the amplitude for 2081–2100, relative to 1986–2005 ranges from 0.3 to 4.8 °C, under four Representative Concentration Pathways (RCP) scenarios (IPCC, 2007). An increase in temperature results in a higher evapotranspiration (ET) demand and will, in combination with a decrease in precipitation, severely stress the water resources in the region. In particular, all of Africa is highly likely to experience warming during this century, with the warming expected to exceed the global average (IPCC, 2007). Climate change and increasing climate variability might add to these trends and further degrade the availability and quality of water and arable land (Giertz and Diekkrüger, 2003). However, the spatiotemporal behaviour of precipitation, river discharge, ET, minimum and maximum temperature depend on the regional and local forcing.

The impact of climate change on precipitation and water availability is of major concern for policy makers in the Mono Basin of West Africa. In order to proffer solutions to the problems associated with water resources management within the Mono, there is a need to understand the relationship between these variables at the basin's scale.

The aim of this work is to project the rainfall, flows and ET in the future period and understand their changes.

2 Study area location, Data used and Methodological Approach

2.1 Study area location

This study area has already been described by certain authors (Batablinlè et al., 2018; Ernest et al., 2014). The Mono river basin is in West Africa and extends over 560 km from

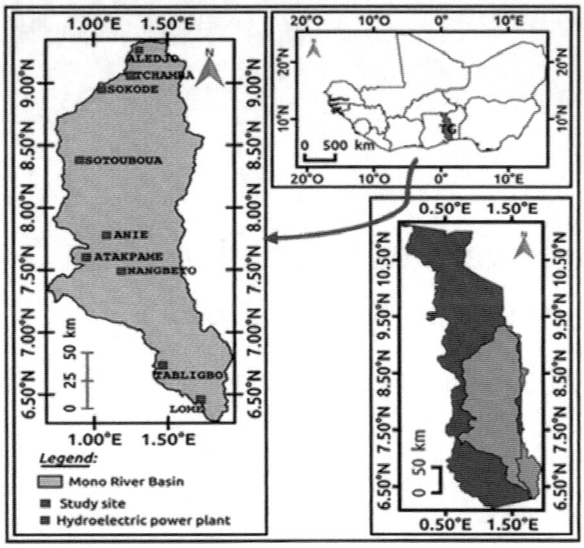

Figure 1. Study area location and considered stations (sources Batablinlè et al., 2018).

the north to the south. This transboundary watershed covers 15 680 km^2 and is shared by Benin and Togo. Mono basin is located more precisely between 06°16′ N and 09°20′ N and 0°42′ E and 20°25′ E. This watershed is home to the Nangbéto hydroelectric dam that has been providing electricity to Benin and Togo since September 1987.

Within the West African context, rainfall in the study area is characterized by two types of Rainfall regimes In southern basin, (from 6°16 to 7°30 N) there are two rainy seasons which to mid-July and from mid-August to October. In northern basin (from 7°30 to 9°20 N), there is one rainy season which extends from April to October. The population of the basin is more than two million, with an annual increase of 2.9 %. This population also distributed at high densities in the south of the basin has as main activities, agriculture (mainly rain fed); in the lower valleys, fishing and salt-farming are the major activities (Ernest et al., 2014). The study area is presented in Fig. 1 below.

2.2 Data used and Methodological Approach

Two sources of data have been used in this paper. The first one is CORDEX program. Models *data have been* bias *corrected by* Batablinlè et al. (2018). In this study, multi-model's ensemble only will be considered. The last source of data is the observed data provided by the National Meteorology Agency of Togo for the period 1960–2010 for some stations spatially located as shown by Fig. 1. The ET of the current and future climate was calculated. In order to estimate ETo considering only temperature measurements, we used the Hargreaves–Samani equation (Hargreaves and Samini, 1985).

GR is a family of lumped hydrological models designed for flow simulation at various time steps. The models are now available in a flexible two main R-packages called airGR. The GR4J model has been extensively tested in a large number of watersheds in France, Australia, Brazil (Perrin et al., 2001). It has been used in a context of climatic variability in the Upper Ouémé basin (Ague et al., 2014). The GR4J model was used to simulate daily flow of the Mono watershed. The model parameters were calibrated and validated on a daily scale with respect to the hydro-meteorological inputs observed from 1971 to 1987 and from 1988 to 2010, respectively. These sub-periods are homogeneous and belong to the dry or wet sequence obtained after the analysis of the stationarity break in the rainfall and flow chronicles. Statistical analyses produced Nash-Sutcliffe efficiencies (NSEs) of 0.76 and 0.61 for model calibration and validation, respectively, which were considered acceptable.

The changes from the reference period are assessed as shown by Eq. (1) for ET, rainfall or flow :

$$I_p = \left(\frac{\mathrm{FF} - \mathrm{HIST}}{\mathrm{HIST}}\right) \times 100 \tag{1}$$

Where FF and HIST represent respectively the mean for the far future (2011–2100) and the historical or reference period (1981–2010).

Sensitivity analysis is used to determine the influence of two main climatic variables on future flows. The daily average variables used in this study are: Future precipitation and evapotranspiration (ET). Flows sensitivity to climate variables was estimated by varying each variable individually, leaving the other variables constant.

A sensitivity coefficient (Eq. 2) was calculated to determine the degree of influence of each climate variable on the flow:

$$S_{\mathrm{Vi}} = \frac{\partial Q}{\partial \mathrm{Vi}} \frac{\mathrm{Vi}}{Q} \tag{2}$$

where S_{Vi} is the sensitivity coefficient, ∂Q is the variation of the future flows (Q) caused by the change of a variable ∂Vi, Vi is the considered variable. A positive (negative) sensitivity coefficient indicates that the variable increases (decreases) the future flows. The higher the absolute value of the sensitivity coefficient is, the more the variable has influence on flows (Li et al., 2018).

The free software R was used to compute all the statistical parameters such as the seasonal averages as well as the annual averages and all plots presented.

3 Results et discussions

3.1 Results

3.1.1 Projected changes in ET and the rainfall

The changes relating to the evapotranspiration (ET) and rainfall between the future and the historical period are shown in

Fig. 2. Under the RCP4.5 scenario, an increase in ET is projected by the mean-models that can fluctuate from 20 % to 45 % on average in some areas in the south. This projection varies from 5 % to 38 % northward and from 5 % to 10 % in the basin center. On this same projection period and under the scenario RCP8.5, the mean-model also predicts also a strong increase in ET the southward of the basin ranging from 10 % to 45 % against 3 % to 40 % and 5 % to 15 % the northward and the center, respectively (Fig. 2 to the left). If we consider the seasonal change (over the same projection period), under the RCP8.5 scenario, the mean-models predicts a decrease in rainfall from −15 % to −41 % on average excepted July, August and September (15 %), while it projected an increase in ET under the same scenario with 38 % (mean value). Over the same projection period, the scenario RCP4.5 predicts a sharp decrease in rainfall on average from −15 % to −40 %. The analysis of the ET shows an increase in the basin Mono excepted April, May and June (−14 %) (Fig. 2 to the right).

3.1.2 Changes in seasonal flows

The changes relating to the streamflows between the future and the historical period are calculated. Regarding the results of relative changes for monthly streamflows (figure is not represented), the lowest decrease is −54.3 % for the time period 2071–2100 (the far future) and is projected in the month of much while the highest increase is 42.4 % and is projected in the month of December for the time period 2011–2041 (the near future) under RCP4.5. Under RCP8.5 the lowest decrease is −58.3 % and is projected in the month of April for the time period 2041–2070 while the highest increase is 39.1 % and is projected in the month of September under RCP8.5 for the time period 2011–2040. In summary, mean models used have announced in general, a decrease of the streamflows over Mono basin.

Figure 3 depicts relative changes of Annual and seasonal streamflows (the near future and the far future were considered). Regarding these results, the streamflows (MAM and JJA) for the time period 2011–2040 (the near future) will increase between 2 % and 50.3 % under the RCP4.5 scenario and 12 % and 55 % under the RCP8.5 scenario while the streamflows (DJF, SON and An) will decrease between 10 % and 49 % under the RCP4.5 scenario and between 4 % and 56 % under the RCP8.5 scenario, respectively. In summary, mean models used have announced in general, a decrease of the streamflows over Mono basin (Fig. 3) for the time period 2071–2100 (the far future) for the both scenarios. Monthly streamflows characteristics reflect the monthly changes in rainfall characteristics: A shift towards the rainy season (particularly, Murch to July) and a decrease in the dry months (particularly, November to April) are observed for all climate scenarios in Mono basin.

3.1.3 Changes in extreme flows

Figure 4 (to the right) presents the low (Q95) and high (Q05) flows magnitudes for the baseline period and the percent change in the future. The Q95 and Q05 here were computed based on the streamflows mean for the multi model ensemble. The relative mean changes of high flow (Q5), low flows (Q95) under two RCPs in the future are presented in Fig. 4. Simulations using the RCP4.5 scenario indicate more moderate changes to streamflows (Q5 and Q95), whereas the RCP8.5 climate scenarios project larger changes. Under both RCPs, decreases in general of Q5, and Q95, are projected in all three future sub periods (A = 2051–2080, B = 2061–2090 and C = 2071–2100), but the distribution of decreasing median values is generally wider under RCP8.5 than RCP4.5 in the middle and long terms. Furthermore, the distribution of Q95 values in the near term is much larger under RCP4.5 than under RCP8.5. The greatest decreases of Q5 is projected to occur in the near term under RCP8.5, whereas the greatest decrease of Q95 is projected to occur in the long term under the same RCP.

Figure 4 (to the left) shows the relationship between future flows and two main climatic variables. The results show that future rainfall will have a positive influence on future flows. Indeed, the increase of this variable leads to an increase in futures flows. On the other hand, an increase in future evapotranspiration leads to a decrease in future flows. For example, at annual scale, an increase of 0 % to 30 % of the rainfall leads to an increase in flows from 1 to 5 m^3/s. Conversely, increasing the evapotranspiration from 0 % to 28 % favors a decrease in flows from −3.5 to −1.5 m^3/s under RCP8.5. Under RCP4.5, a decrease from −22 % to 0 % of the rainfall leads to a decrease in flows from −3.5 to 0 m^3/s.

3.2 Discussions

The ET and rainfall are an important climate variables that directs the change in flows. Evapotranspiration is the highest outgoing water flux in the hydrological cycle and is of vital importance in assessing the effects of climate change in water availability. Allen et al. (1998) show, Evapotranspiration changes, on their own or in combination with rainfall changes, can contribute to changes in hydrological indices such as mean monthly river flows. Our study predicts an increase of ET while the rainfall will decrease over Mono basin. Simulations of future streamflows in the Mono River project lowest increases and highest decreases of streamflows.

This can have a negative effect on future water availability in the area. the energy use is also dependent on water resources, and the availability and reliability of renewable sources are function of climate conditions, which can vary according to global climate changes. A possible and expressive loss in electricity production capacity could bring serious social and economic risks to the country. It was also

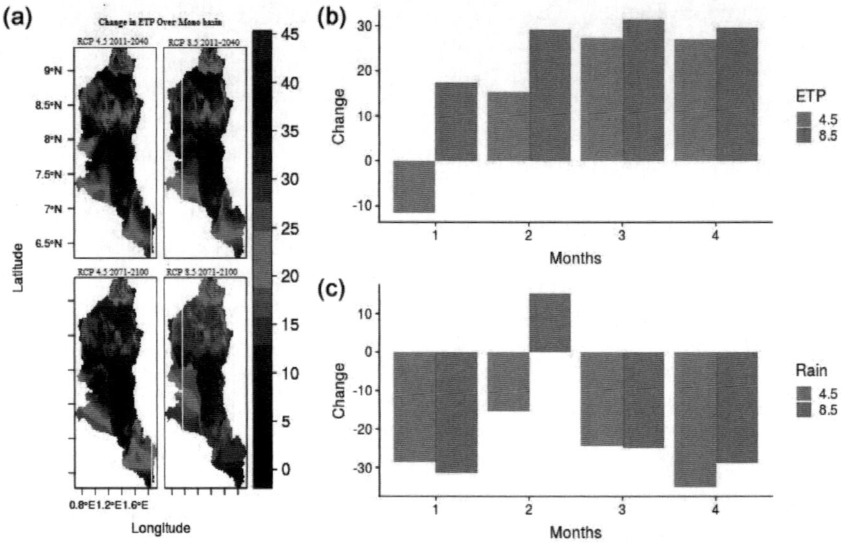

Figure 2. Spatial projected changes in ET (to the left, panel **a**) projected changes average ET and rainfall for each calendar month from the reference period (1981–2010) to (2011–2100); (1: April, May, June; 2: July, August, September; 3: October, November, December; 4: January, February, March), (to the right, panels **b** and **c**).

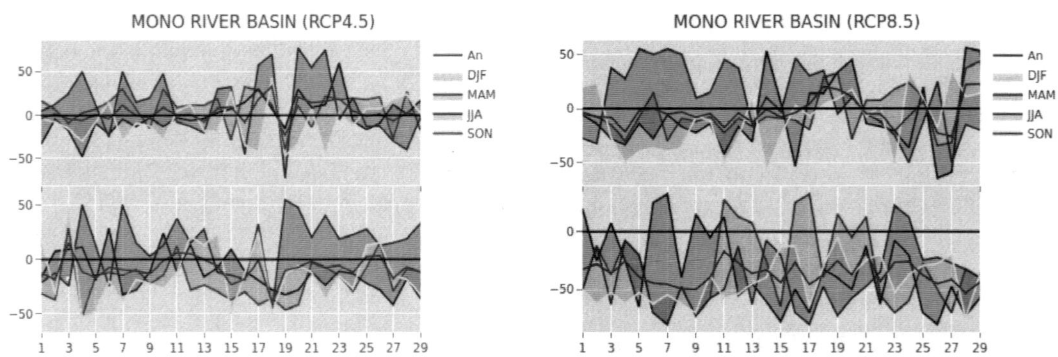

Figure 3. Annual and seasonal changes in streamflows between the futures (2011–2040 in top and 2071–2100 in below) and the historical period (1981–2010) for the average multi-models forced by RCP4.5 and RCP8.5 greenhouse gas scenarios in Mono Basin.

observed that the impact of rainfall reduction in flow production in the Mono Basin will be accentuated. Similar to previous studies in other region, results showed that the evapotranspiration is in?uenced, to a large extent, by precipitation (Liu et al., 2008). Morocco and Algeria appear to show an increase in ETref and a decrease in precipitation for future periods, meaning that water stress will become an even more severe problem in these countries in the future (TERINK et al., 2013). Jung et al. (2010) also found that the deficit of available water resources would lead to a negative global land surface evapotranspiration trend. Furthermore, a significant decrease in water availability (surface water and groundwater) due to a decrease in rainfall showed by Giertz and Diekkrüger (2003) will exacerbate following the scenario RCP8.5. Our work shows that such a change diagnosed on the present climate is likely to persist of more and more in the future. The decrease of the rainfall in the Mono basin appears to be consistent with the evolution of the ETP because a sharp increase in ETP could favor a decrease in the rainfall.

4 Conclusion

This study evaluated changes in climatic variables including precipitation, flows and evaporation due to climate change over the Mono basin in west Africa in the future periods (2011–2100) using regional climate models under middle and highest emission scenarios (i.e., RCP4.5 and RCP8.5). The rainfall, flows and evaporation over the historical period 1980–2010 have been analyzed at seasonal scales and interannual scales. The key findings of this study are summarized below:

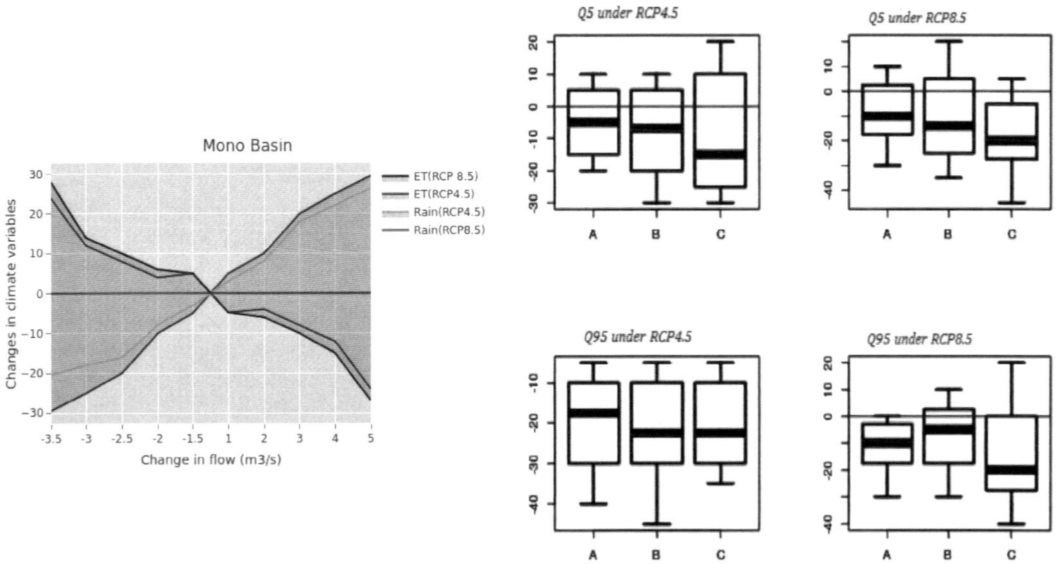

Figure 4. The relative mean changes of Q5 and Q95 under two RCPs (RCP4.5 and RCP8.5) in the future (A = 2051–2080, B = 2061–2090, C = 2071–2100). The changes from the reference period are assessed as shown by equation for Q5 and Q95 (to the right) and the sensitivity of flows to main Climatic Variables (to the left).

1. Compared to the base period, Simulations of future streamflows in the Mono River project lowest increases and highest decreases of streamflows. It predict an increase of ET while the rainfall will decrease.

2. The results of the Sensitivity of flows to climatic variables shows that there will be impacts on water availability from climate change, reducing the streamflows in the studied area.

3. These results can serve as a basis for the implementation of prevention and water resource management strategies in the Mono catchment.

Code availability. The software R is freely available (http://www.gnu.org/copyleft/gpl.html, GNU Operating System, 2021). The code used in this study is not publicly accessible. Please contact the authors if you are interested in this research code.

Data availability. All data can be accessed via the references: Models data have been used in this study. It is a set of eight climatic models outputs with daily time step and spatial resolution of 0.44° and available at: https://esgf-data.dkrz.de/projects/esgf-dkrz/ (DKRZ, 2017). The observed data used, are provided by the Togo National Meteorology Agency (Météo Togo). Please contact the authors if you are interested in this research data.

Author contributions. All authors developed the main research idea. LB, LEA and KDJ performed the analyses and evaluated the stress tests. LB wrote the manuscript with contributions and revisions from all authors.

Competing interests. The contact author has declared that neither they nor their co-authors have any competing interests.

Disclaimer. Publisher's note: Copernicus Publications remains neutral with regard to jurisdictional claims in published maps and institutional affiliations.

Special issue statement. This article is part of the special issue "Hydrology of Large River Basins of Africa". It is a result of the 4th International Conference on the "Hydrology of the Great Rivers of Africa", Cotonou, Benin, 13–20 November 2021.

Acknowledgements. All my gratitude to the International Association of Hydrological Sciences and local organizers who gave me financial support and took charge of my participation in 4th International Conference on the "Hydrology of the Great Rivers of Africa", Cotonou, Benin. We are also grateful to the anonymous reviewers whose thoughtful and constructive comments have improved the quality of the article.

Financial support. This research has been supported by the International Association of Hydrological Sciences (grant no. 13 SYSTA).

References

Ague, A. I., Afouda, A., and Lanhoussi, F. : Etude Comparative D'un Modele Conceptuel Global (gr4j) Et D'un Modele Semi-distribue (geosfm) Sur Le Bassin Versant De L'oueme A Save (benin, Afrique De L'ouest), Le Journal de l'Eau et de l'Environnement (LJEE), 13, 2–8, 2014.

Allen, R., Pereira, L., Raes, D., and Smith, M.: Crop evapotranspiration guidelines for computing crop requirements, FAO Irrig. Drain. Report modeling and application, J. Hydrol., 285, 19–40, 1998.

Amoussou, E., Tramblay, Y., Totin, H. S. V., Mahé, G., and Camberlin, P.: . Dynamics and modelling of floods in the river basin of Mono in Nangbeto, Togo/Benin, Hydrolog. Sci. J., 59, 2060–2071, https://doi.org/10.1080/02626667.2013.871015, 2014.

Batablinlè, L., Lawin, E., and Agnide, C. M.: Africa Cordex Simulations Projection of Future Temperature, Precipitation, Frequency and Intensity Indices Over Mono Basin in West Africa, J. Earth Sci. Clim. Change, 9, 490, https://doi.org/10.4172/2157-7617.1000490, 2018.

Deutsches Klimarechenzentrum (DKRZ): Homepage, available at: https://esgf-data.dkrz.de/projects/esgf-dkrz/, last access: 3 August 2017.

Giertz, S. and Diekkrüger, B.: Analysis of the hydrological processes in a small headwater catchment in Benin (West Africa), Physics and Chemistry of the Earth, Parts A/B/C, 28, 1333–1341, https://doi.org/10.1016/j.pce.2003.09.009, 2003.

GNU Operating System: GNU General Public License, available at: http://www.gnu.org/copyleft/gpl.html, last access: August 2021.

Hargreaves, G. and Samini, Z.: Reference crop evapotranspiration from temperature, Appl. Eng. Agric., 1, 96–99, 1985.

Intergovernmental Panel on Climate Change (IPCC): Climate Change 2007: The Physical Science Basis. Contribution of Working Group I to the Fourth Assessment Report of the Intergovernmental Panel on Climate Change, available at: http://www.ipcc.ch/publications_and_data/ar4/syr/en/contents.html (last access: 12 September 2019), 2007.

Jung, M., Reichstein, M., Ciais, P., Seneviratne, S., Sheffield, J., Goulden, M., Bonan, G., Cescatti, A., Chen, J., de Jeu, R., Dolman, A., Eugster, W., Gerten, D., Gianelle, D., Gobron, N., Heinke, J., Kimball, J., Law, B., Montagnani, L., Mu, Q., Mueller, B., Oleson, K., Papale, D., Richardson, A., and Roupsard, O.: Recent decline in the global land evapotranspiration trend due to limited moisture supply, Nature, 467 951–954, 2010.

Li, M., Chu, R., Shen, S., and Islam, A. R. M. T.: Quantifying Climatic Impact on Reference Evapotranspiration Trends in the Huai River Basin of Eastern China, Water, 10, 144, https://doi.org/10.3390/w10020144, 2018.

Liu, B., Ma, Z., Feng, J., and Wei, R.: The relationship between pan evaporation and actual evapotranspiration in Xinjiang since 1960, Acta Geogr. Sin., 63, 1131–1139, 2008.

Perrin, C., Michel, C., and Andréassian, V.: Does a large number of parameters enhance model performance? Comparative assessment of common catchment model structures on 429 catchments, J. Hydrol., 242, 275–301, 2001.

Terink, W., Immerzeel, W. W., and Droogers, P.: Climate change projections of precipitation and reference evapotranspiration for the Middle East and Northern Africa until 2050, Int. J. Climatol., 33, 3055–3072, 2013.

Proc. IAHS, 384, 289–293, 2021
https://doi.org/10.5194/piahs-384-289-2021

Open Access PIAHS

Seasonal flow forecasting in Africa; exploratory studies for large lakes

Kevin Sene and Wlodek Tych

Lancaster Environment Centre, Lancaster, LA1 4YQ, UK

Correspondence: Kevin Sene (k.sene1@lancaster.ac.uk)

Published: 16 November 2021

Abstract. For many applications, it would be extremely useful to have insights into river flows at timescales of a few weeks to months ahead. However, seasonal predictions of this type are necessarily probabilistic which raises challenges both in generating forecasts and their interpretation. Despite this, an increasing number of studies have shown promising results and this is an active area for research. In this paper, we discuss insights gained from previous studies using a novel combined water balance and data-driven approach for two of Africa's largest lakes, Lake Victoria and Lake Malawi. Factors which increased predictability included the unusually long hydrological response times and statistically significant links to ocean-atmosphere processes such as the Indian Ocean Dipole. Other lessons learned included the benefits of data assimilation and the need for care in the choice of performance metrics.

1 Introduction

Seasonal river flow forecasts aim to provide useful information for operations and planning from weeks to months ahead. Potential applications include water supply planning, hydropower generation and irrigation scheduling. This is a developing area both in terms of the technical approaches used and interpretation of the outputs provided.

The forecasting techniques potentially available include statistical approaches, ensemble streamflow prediction, and the direct input of ensemble rainfall forecasts into hydrological models. However, forecast skill depends on several factors, including seasonal influences and the catchment size, location and antecedent conditions (e.g. Greuell et al., 2019; Mendoza et al., 2017; Robertson and Wang, 2012; Sene, 2016). Often a key challenge is that the required forecast lead times exceed hydrological response times, requiring a greater reliance on long-range rainfall forecasts, with the many uncertainties that entails.

One situation which can improve predictability is when there is considerable storage in a catchment, such as from snowpack, groundwater or large lakes and reservoirs. The Rift Valley lakes of east and southern Africa provide one such example, and due to their huge size, offer considerable potential for deriving seasonal forecasts over operationally useful timescales.

Here we describe insights gained from previous studies by the authors into two of Africa's largest lakes, Lake Victoria and Lake Malawi. These flow into the White Nile and Shire River respectively. A novel water balance and statistical approach was used, building on stochastic transfer function modelling techniques previously applied in a wide range of environmental applications (e.g. Young, 2011; Tych and Young, 2012). The approaches used are only briefly outlined here, but are described in more detail in Sene et al. (2017, 2018) and Sene and Tych (2018).

In contrast, most previous studies have focussed on developing statistical links between river flows and external drivers such as climate indices. For example, Siam and Eltahir (2015) describe a Bayesian model for the Nile incorporating Indian and Pacific Ocean climate indices, while Elganiny and Eldwer (2018) used Artificial Neural Network and ARMA approaches. Gehad et al. (2017) have also compared forecast skill with ensemble approaches for the Blue Nile. For the Shire River, Jury and Gwazantini (2002) describe a seasonal forecasting model for Lake Malawi using a statistical model including climate indices, and Jury (2014) describes a detailed statistical analysis of links between climate and hydrology using atmospheric model reanalyses and satellite data.

In this paper, other topics discussed include the potential for real-time updating of model outputs using techniques similar to those used in real-time flood forecasting, and the choice of performance metrics to deal with a statistically non-stationary response over timescales of months or more. An analytical approach also provided useful insights. Finally, some priorities for future research are considered in this challenging area, including the prospects for seasonal forecasting for smaller lakes and river basins.

2 Methodology

2.1 The study area

Lake Victoria lies on the equator and its catchment spans six countries: Burundi, the Democratic Republic of the Congo, Kenya, Rwanda, Tanzania and Uganda (Fig. 1). It plays an important role in hydropower generation, irrigation and water supply in the region. The lake has some of the longest hydrological records in Africa starting with routine observations of levels in the 1890s and rainfall a decade later, albeit initially with a sparse network of rain gauges. Regular observations of outflows began in the 1940s, and flows have been regulated for hydropower production since 1953. The lake is the largest in Africa with surface and catchment areas of about 68 000 and 194 000 km^2.

Lake Malawi, also shown in Fig. 1, lies several hundred kilometres to the south and its catchment is mainly in Malawi and Tanzania, with a smaller contribution from Mozambique. Level observations also began in the 1890s with regular outflow observations since 1948. The lake has been regulated for hydropower production since 1965. Its surface and catchment areas are about 28 750 and 95 570 km^2.

For both lakes, regional rainfall is affected by the annual passage of the Intertropical Convergence Zone (ITCZ). For Lake Victoria, this results in two main rainfall seasons, which are typically between March and May and October and December. In contrast, Lake Malawi lies near the southernmost end of ITCZ excursions and much of the catchment experiences a single main rainfall season from November to April or May.

The datasets used for these studies are described in Sene et al. (2017, 2018) and included rainfall, lake level and lake outflow records. To provide the best possible data coverage, extensive use was made of previously validated and infilled values, many of which themselves relied on observations and record extension techniques developed during an unusually intensive period of monitoring in the 1970s and 1980s (WMO, 1982, 1983). This meant that monthly data availability and completeness was particularly good and allowed the focus to be on issues related to model structure and performance, rather than trying to resolve issues due to sparse data. Nevertheless, the records included some of the most extreme flood and drought periods on record. The periods chosen for analysis were 1925 to 1990 for Lake Victoria and 1954 to 1980 for Lake Malawi.

Several climate indices were considered for use in the analyses. Exploratory work typically suggested a weak correspondence with rainfall for El Niño indices at lag times of a few months and a rather stronger correspondence with an index for the Indian Ocean Dipole. Based on these studies, the indices chosen for use in the simulations were NINO34 (Trenberth, 1997) and JAMSTEC's estimates for the Dipole Model Index (DMI) and illustrative results are discussed later. More detailed studies into links to global and regional climate typically suggest lag times of about two months for Shire River flows (Jury, 2014) and values up to several months for east Africa at a seasonal timescale (Nicholson, 2017), although there are many factors and variables to consider when interpreting these findings.

2.2 Simulation techniques

The studies used an innovative approach to estimating lake outflows, based on a combination of water balance and transfer function techniques. Only brief details are given here while the full methodology is described in Sene et al. (2017, 2018).

The starting point for the analyses was the water balance for a lake, which can be expressed as:

$$\frac{\mathrm{d}h}{\mathrm{d}t} = N(t) - \frac{Q_\mathrm{o}(t)}{A(t)} \tag{1}$$

where h is the water level, N is the net inflow, Q_o is the outflow, A is the surface area, and t is time. Based on previous studies (e.g. Piper et al., 1986) the lake areas were assumed to be constant and the lake outflows were estimated from an equation of the form $Q_\mathrm{o} = ah^b$.

The net inflow is normally expressed in the following form:

$$N = P(t) - E(t) + \frac{Q_\mathrm{i}(t)}{A(t)} = \Delta h + \frac{Q_\mathrm{o}(t)}{A(t)} \tag{2}$$

where P and E are the lake rainfall and evaporation and Q_i is the tributary inflow. Any additional terms which are difficult to measure or estimate, such as groundwater seepage, were considered to contribute to the overall model uncertainty.

To simulate the water balance, a data-driven stochastic framework was adopted which has proven useful in many other environmental applications, such as real-time flood forecasting and assessing the long-term variability in climate records (Beven, 2009; Young, 2013). Some advantages of this approach are that few assumptions are required about the statistical characteristics of the datasets used and the relationships between them, and that estimates for the uncertainty in model outputs are intrinsic to the approach.

A transfer function solution was sought, inspired by the observation by Young (2011) that for the linear case ($b = 1$)

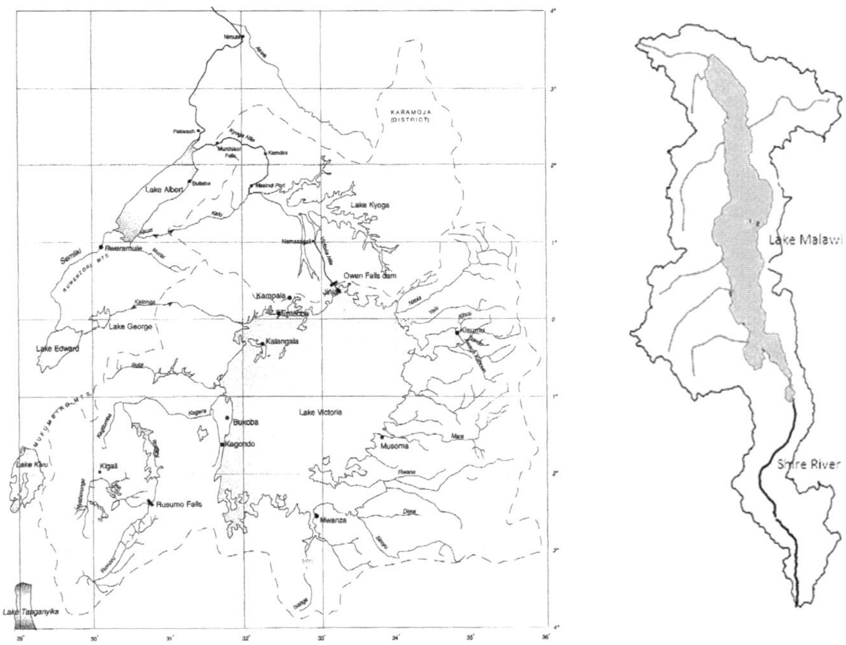

Figure 1. Map of the Lake Victoria region (Source: IAHS Press and Sutcliffe and Parks, 1999) (left) and Lake Malawi (right).

Eq. (1) can be solved in this form without further approximation. The following more generalised form was adopted, allowing for both serial dependence in model inputs and the influence of external variables (u_t), such as climate drivers:

$$y_t = \frac{\mathbf{B}(z^{-1})}{\mathbf{A}(z^{-1})} u_{t-\delta} + \frac{\mathbf{D}(z^{-1})}{\mathbf{C}(z^{-1})} e_t \tag{3}$$

Here, **A**, **B**, **C** and **D** are polynomial functions and z^{-1} is the backward shift operator $z^{-i} y_t = y_{t-i}$.

The parameter values were estimated using a stochastic recursive estimation approach, which also provides estimates for the uncertainty in parameter values and how they vary over time. Tych and Young (2012) and Young (2013) provide more details.

2.3 Results and discussion

Similar techniques were applied to both Lake Victoria and Lake Malawi and, again, the results are described in detail in Sene et al. (2017, 2018). Here we draw out some key insights which may be relevant to future studies of these two lakes and to other large lakes in Africa.

2.3.1 Net inflows

The net inflows were estimated from the outflow terms in Eq. (2) since, in principle, lake levels and outflows can be measured and infilled with greater precision than the lake rainfall and tributary inflows. This approach has been widely used in previous studies (e.g. Piper et al., 1986) and avoided the need to estimate the individual inflow terms for which there can often be many uncertainties, particularly when as here catchments are large and monitoring networks are sparse. As a further refinement, values were expressed in standardised form to focus on the underlying climate signals, thereby helping to reduce the influence of bias in observations.

For Lake Victoria, cross correlation analyses suggested that r^2 coefficients with lake rainfall were highest (> 0.8) for no time delay but still statistically significant with a 1 month delay, whilst autocorrelation coefficients were highest for a 1 month delay. This provides an indication of the potential forecasting lead times for net inflow from rainfall variability alone whilst, in contrast, linkages with tributary inflows were considerably less. Similar results were obtained for Lake Malawi and helped to inform the overall structure of the model in Eq. (3).

2.3.2 Data assimilation

When forecasting river flows at short timescales, such as in real-time flood forecasting, one widely used approach is to update model outputs based on telemetry data. This approach is called data assimilation and is effective over timescales comparable with the hydrological response time of a catchment. Clearly, this is unlikely to be the case in many seasonal forecasting applications, but for a large lake like Lake Victoria, given the huge storage influences it is potentially an option, and the use of this approach was a novel aspect of these studies.

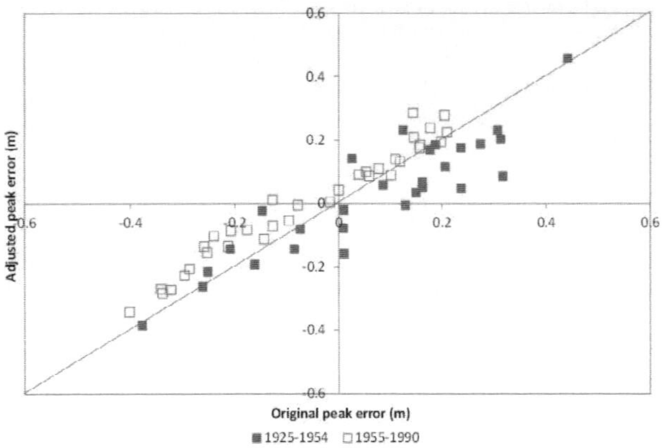

Figure 2. Illustration of the reduction in annual peak level errors for Lake Victoria at a lead time of 5 months when using a regression model for the forecast residuals linked to climate indices (adapted from Sene et al., 2018).

Various formulations were explored for the assimilation component, including an error prediction approach and multiple regression models for the forecast model residuals.

The best forecast skill was obtained by incorporating both El Niño and Indian Ocean Dipole indices into the regression models at lag times of 5–7 months for Lake Victoria, and 6–9 months and 3–5 months respectively for Lake Malawi. This gave better forecast skill than the more usual approach of developing statistical relationships between rainfall and climate indices. In contrast, an Autoregressive Regressive Moving Average (ARMA) approach for the residuals alone only provided benefits at lead times of 2–3 months.

2.3.3 Performance metrics

The data assimilation studies highlighted another important issue, which was the choice of performance metrics to use.

The reason for this was that, for large lakes such as Lake Victoria and Lake Malawi, one characteristic of the lake level and outflow series is that, due to the huge storage, there can be underlying longer term influences superimposed on the daily and seasonal variations due to rainfall. This non-stationary response requires special consideration and for this study it was considered most important to focus on the annual peaks in levels and outflows.

Figure 2 shows an example of this type of output, showing the improvements in peak level estimates when climate indices were included in the data assimilation component. This example was for Lake Victoria and in future studies, a further refinement would be to consider the timing errors in peaks as well. Many other measures could be considered and this is a worthwhile area for future research.

2.3.4 Analytical approaches

In some hydrological situations, an analytical approach can also give useful insights, and for a lake it is possible to solve Eq. (1) analytically for some integer values of b in the outflow term.

These analyses provided some useful additional insights into the key timescales for lake response; for example, suggesting that, following a sudden increase in net inflows, such as during heavy rainfall, for both lakes an approximately exponential decay in levels might be expected over timescales of 4–5 years, if inflows revert to their long-term mean values.

This provided further evidence of a non-stationary response, and an indication of lake response times due to storage, albeit for a highly idealized situation. Similar idealized solutions might be sought in other seasonal forecasting applications although normally, of course, a numerical solution is required for further insights.

2.4 Conclusions

This study has highlighted some methodological issues arising from exploratory studies into seasonal flow forecasting using a data-driven approach. Some issues to consider in developing models for other lakes include the potential role of climate indices in data assimilation, the choice of suitable performance metrics and the value of simple analytical solutions to explore the response.

Future studies might also consider the forecast skill gained from including seasonal rainfall and air temperature forecasts as model inputs for comparison with an ensemble streamflow prediction approach. The use of spatially varying (grid-based) climate indices might also be considered and, for smaller lakes, more consideration given to the spatial relationships between tributary inflows, perhaps at a daily time step. The modelling framework described here might also be developed further using a so-called State Dependent Parameter approach to integrate the dynamic input-output model and data assimilation components into a single modelling framework, using a unified State Space form and estimated (forecast) using Kalman Filter tools.

One key objective of studies such as these is to develop techniques that could be used operationally. For large lakes, perhaps the longest-established systems are for the Great Lakes in the USA and Canada. These combine empirical and physically based models, with precipitation and air temperature outlooks and ensemble forecasts (Gronewold et al., 2011; Bolinger et al., 2017). A Nile Basin Flow Forecasting System is also at the planning stage and the Eastern Nile Technical Regional Office (ENTRO) issues a Flood Preparedness and Early Warning Bulletin during the flood season, based on short to medium-range meteorological forecasts. The WMO Hydrological Status and Outlooks (HydroSOS) initiative (http://www.wmo.int, last access: 4 November 2021) is also considering Lake Victoria.

Data availability. No data sets were used in this article.

Author contributions. This paper draws on previous research by the authors on seasonal flow forecasting using modelling techniques developed jointly by the two co-authors (KS, WT), making use of a software framework described in Tych and Young (2012). The manuscript was prepared by KS with expert contributions from WT.

Competing interests. The contact author has declared that neither they nor their co-authors have any competing interests.

Disclaimer. Publisher's note: Copernicus Publications remains neutral with regard to jurisdictional claims in published maps and institutional affiliations.

Special issue statement. This article is part of the special issue "Hydrology of Large River Basins of Africa". It is a result of the 4th International Conference on the "Hydrology of the Great Rivers of Africa", Cotonou, Benin, 13–20 November 2021.

Acknowledgements. We would like to thank the Centre for Ecology and Hydrology for permission to use the datasets for Lake Victoria referred to in this paper; also Michael Kizza for providing additional independently derived records for comparison. The WMO (1983) study was led by Chris Kidd and provides a valued contribution to understanding of the hydrology of Lake Malawi.

References

Beven, K.: Environmental Modelling: An Uncertain Future?, CRC Press, UK, 2009.

Bolinger, R. A., Gronewold, A. D., Kompoltowicz, K., and Fry, L. M.: Application of the NMME in the development of a new Regional Seasonal Climate Forecast Tool, B. Am. Meteorol. Soc., 2017, 555–564, 2017.

Elganiny, M. and Eldwer, A.: Enhancing the Forecasting of Monthly Streamflow in the Main Key Stations of the River Nile Basin, Water Resour., 45, 660–671, 2018.

Gehad, N., Doaa, A., Shokry, A., and Tahani, Y.: Flow forecasting and skill assessment in the Blue Nile Basin, Nile Water Sci. Eng. J., 10, 29–37, 2017.

Gronewold, A. D., Clites, A. H., Hunter, T. S., and Stow, C. A.: An appraisal of the Great Lakes advanced hydrologic prediction system, J. Great Lakes Res., 37, 577–583, 2011.

Greuell, W., Franssen, W. H. P., and Hutjes, R. W. A.: Seasonal streamflow forecasts for Europe – Part 2: Sources of skill, Hydrol. Earth Syst. Sci., 23, 371–391, https://doi.org/10.5194/hess-23-371-2019, 2019.

Jury, M. R.: Malawi's Shire River Fluctuations and Climate, J. Hydrometeorol., 15, 2039–2049, 2014.

Jury, M. R. and Gwazantini, M. E.: Climate variability in Malawi, part 2: sensitivity and prediction of lake levels, Int. J. Climatol., 22, 1303–1312, 2002.

Mendoza, P. A., Wood, A. W., Clark, E., Rothwell, E., Clark, M. P., Nijssen, B., Brekke, L. D., and Arnold, J. R.: An intercomparison of approaches for improving operational seasonal streamflow forecasts, Hydrol. Earth Syst. Sci., 21, 3915–3935, https://doi.org/10.5194/hess-21-3915-2017, 2017.

Nicholson, S. E.: Climate and climatic variability of rainfall over eastern Africa, Rev. Geophys., 55, 590–635, https://doi.org/10.1002/2016RG000544, 2017.

Piper, B. S., Plinston, D. T., and Sutcliffe, J. V.: The water balance of Lake Victoria, Hydrolog. Sci. J., 31, 25–37, 1986.

Robertson, D. E. and Wang, Q. J.: A Bayesian Approach to Predictor Selection for Seasonal Streamflow Forecasting, J. Hydrometeorol., 13, 155–171, 2012.

Sene, K.: Hydrometeorology: Forecasting and Applications, 2nd Edn., Springer, Dordrecht, 427 pp., 2016.

Sene, K. and Tych, W.: Some challenges in seasonal forecasting for large lakes and reservoirs, Seasonal Forecasting: Meeting User Needs, British Hydrological Society National Meeting, Loughborough, UK, 2018.

Sene, K., Piper, B., Wykeham, D., McSweeney, R., Tych, W., and Beven, K.: Long-term variations in the net inflow record for Lake Malawi, Hydrol. Res., 48, 851–866, 2017.

Sene, K., Tych, W., and Beven, K.: Exploratory studies into seasonal flow forecasting potential for large lakes, Hydrol. Earth Syst. Sci., 22, 127–141, https://doi.org/10.5194/hess-22-127-2018, 2018.

Siam, M. S. and Eltahir, E. A. B.: Explaining and forecasting interannual variability in the flow of the Nile River, Hydrol. Earth Syst. Sci., 19, 1181–1192, https://doi.org/10.5194/hess-19-1181-2015, 2015.

Sutcliffe, J. V. and Parks, Y. P.: The Hydrology of the Nile, IAHS Special Publication no. 5, IAHS Press, Wallingford, ISBN 978-1-901502-75-6, 192 pp., 1999.

Trenberth, K. E.: The Definition of El Niño, B. Am. Meteorol. Soc., 78, 2771–2777, 1997.

Tych, W. and Young, P. C.: A Matlab software framework for dynamic model emulation, Environ. Model. Softw., 34, 19–29, 2012.

WMO: Hydrometeorological Survey of the Catchments of Lake Victoria, Kyoga and Mobutu Sese Seko, WMO Report, 1982.

WMO: A Water Resources Evaluation of Lake Malawi and the Shire River, WMO Report No. MLW/77/012, 1983.

Young, P. C.: Recursive Estimation and Time-Series Analysis: An introduction for the student and practitioner, 2nd Edn., Springer, 2011.

Young, P. C.: Hypothetico-inductive data-based mechanistic modeling of hydrological systems, Water Resour. Res., 49, 915–935, 2013.

Proc. IAHS, 384, 295–300, 2021
https://doi.org/10.5194/piahs-384-295-2021

Open Access PIAHS

Hydrology of Large River Basins of Africa

Perceptions communautaires du couple sécheresse/déboisement dans le sous-bassin versant de Koumagou à Boukoumbé (Nord-Ouest Bénin)

Ibouraima Yabi, Mahamoud N'tcha Tchanati, and Akindélé Akibou

Laboratoire Pierre PAGNEY Climat, Eau, Ecosystème et Développement (LACEEDE), Département de Géographie et Aménagement du Territoire (DGAT), Faculté des Sciences Humaines et Sociales (FASHS), Université d'Abomey-Calavi (UAC, Bénin), Bénin

Correspondence: Ibouraïma Yabi (yafid2@yahoo.fr)

Published: 16 November 2021

Résumé. Au-delà des analyses climatologiques et statistiques, la présente communication s'intéresse aux connaissances communautaires relatives au couple sécheresse-déboisement qui sont les principaux moteurs de la dynamique environnementale dans le sous bassin versant de Koumagou au Nord-Ouest du Bénin. Sur la base des investigations auprès d'un échantillon de 180 personnes, les perceptions communautaires relatives aux manifestations, causes et conséquences des deux fléaux sans oublier les mesures d'adaptation, sont analysées. Des paramètres de la statistique descriptive ont été utilisés pour le traitement des données. La sécheresse est causée par le déboisement massif et la colère des dieux alors que l'agriculture et l'exploitation forestière sont perçues comme la cause du déboisement. Les ressources en eau, les ressources végétales et les sols sont les plus affectés selon les répondants avec des conséquences socioéconomiques. Les mesures d'adaptation concernent l'ajustement des pratiques culturales, le reboisement, la création des aires protégées communautaires, la mobilisation des eaux souterraines.

Abstract. Beyond climatological and statistical analyzes, this communication focuses on community knowledge relating to the combination of drought and deforestation, which are the main drivers of environmental dynamics in the Koumagou sub-watershed in north-west Benin. Based on investigations of a sample of 180 people, community perceptions relating to the manifestations, causes and consequences of the two scourges, not to mention adaptation measures, are analyzed. Descriptive statistics parameters were used for data processing. Drought is caused by massive deforestation and the wrath of the gods while agriculture and forestry are seen as the cause of deforestation. Water resources, plant resources and soils are the most affected according to the respondents with socio-economic consequences. The adaptation measures concern the adjustment of cultural practices, reforestation, the creation of community protected areas, the mobilization of groundwater.

1 Introduction

La dégradation inquiétante des ressources naturelles et ses corolaires sont sans conteste une des préoccupations majeures actuelles de l'humanité. En effet, au-delà de leur importance écologique, les ressources naturelles contribuent également à améliorer et à diversifier les activités socio-économiques, participent à la création d'emplois et constituent une source de recettes pour les État (Ndiaye et Ndiaye, 2013). Ainsi, en termes économiques, le capital naturel représente un quart de l'ensemble des richesses des pays à faible revenu, et elles sont souvent la principale source de revenu pour les populations les plus démunies (OCDE, 2009). C'est au regard de cette réalité que la gestion durable de l'environnement et des ressources naturelles est essentielle à la réduction de la pauvreté dans les zones rurales des pays en développement (FIDA, 2011).

Malgré l'importance des ressources naturelles, elles font l'objet de pression aussi d'origine aussi bien naturelle qu'anthropique. Au nombre des facteurs naturels figure la sécheresse (OMM et GWP, 2016; Dardel, 2014). Or, les pays Ouest-Africains sont touchées par des graves sécheresses, notamment au cours des décennies 1970 et 1980 et la tendance n'est pas encore totalement normalisée par l'amélioration des cumuls pluviométriques à partir des années 1990. Ces sécheresses qui ont particulièrement touché les Pays pays sahéliens n'ont pas épargné les pays côtiers comme tel que le Bénin considéré comme plus humides. et Elles ont eu des conséquences sur les composantes écologiques, le couvert végétal en particulier. Les facteurs humains de dégradation des ressources naturelles quant à eux, concernent, l'agriculture, le pastoralisme, l'urbanisation, l'exploitation du bois, etc. En effet, une utilisation des terres au détriment des écosystèmes naturels a marqué les trois dernières décennies pour des motivations diverses : exploitation minière, étalement urbain, expansion agricole à travers le monde (Ndiaye et Ndiaye, 2013).

Le Bénin en général et le sous bassin de Koumagou en particulier (Fig. 1) n'est pas en marge de cette tendance à la dégradation des ressources naturelles en lien avec les causes naturelles (la sécheresse notamment) et les activités humaines.

La présente recherche s'intéresse particulièrement aux perceptions communautaires relatives aux effets du couple sécheresse/déboisement dans ce sous bassin à Boukoumbé. En effet, la connaissance et la prise en compte des perceptions, connaissances ou logiques communautaires sont capitales dans le processus de planification des mesures de gestion durable des ressources naturelles (Wicander, 2016) de cette entité hydro-écologique.

2 Données et méthodes

Les informations utilisées sont relatives aux perceptions ou connaissances des populations sur la sécheresse (ampleur et fréquence) et le déboisement. Il s'agit d'identifier, selon les populations locales, les causes et conséquences de ces phénomènes sur les composantes environnementales (eaux, végétation et sol). La taille minimale (n) de l'échantillon est déterminée par la formule de Dagnélie utilisée par Akouehou et al. (2014) :

$$n = \frac{U^2_{1-\frac{\alpha}{2}} \times (1-P)}{d^2}$$

Dans cette équation, $n =$ *taille minimale de l'échantillon considérée* ; $P =$ *proportion des personnes capables de fournir des informations* ; $U_{1-\alpha/2} =$ *valeur de la loi normale à la valeur de probabilité 1-α/2* (est de 1.96 avec $\alpha = 5\,\%$) ; *d : marge d'erreur de l'estimation* fixée à une valeur de 6 %. La valeur de P a été estimée à 0.8 au terme d'une enquête exploratoire qui a permis de sonder 60 personnes choisies au hasard. Sur la base de ces valeurs, l'application numérique de formule permet d'obtenir $n = 177$, arrondie à 180. Cet échantillon est réparti de façon proportionnelle dans les 4 arrondissements concernés par le sous bassin (Korontière : 36 ; Tabota : 53 ; Natta : 44 ; et Manta : 47). Le facteur genre est également pris en compte dans le choix des personnes. Ainsi, les personnes d'âges compris entre 25 et 55 ans représentent 65 %, contre 35 % pour les plus de 55 ans. Les hommes représentent 78 % des enquêtés contre 22 % pour les femmes.

Un questionnaire a permis de faire des entretiens individuels. Des possibilités sont également offertes aux répondants de faire des commentaires libres sur les aspects abordés. En outre, des séances d'observations individuelles et directes ont été organisées pour mieux apprécier les formes de déboisement et les problèmes de dégradation qui en découlent.

Les informations collectées ont été codifiées manuellement et les questionnaires ont été dépouillés dans une base de données. Le traitement des informations ethno-environnementales collectées a nécessité l'utilisation des paramètres de la statistique descriptive (fréquence, illustrations graphiques).

3 Résultats

3.1 Perceptions locales des causes de la sécheresse et du déboisement

Les résultats indiquent que les populations du bassin perçoivent une augmentation de la fréquence (85 %) et de l'ampleur de la sécheresse (65 %). Le déboisement est devenu massif selon 91 % des répondants. Ils ont indexé aussi bien les actions humaines que les facteurs naturels comme cause de cette tendance inquiétante (Fig. 2).

La Fig. 2 montre que la sécheresse est principalement causée par le déboisement (78 %), et la colère des dieux (62 %) alors que l'agriculture (83 %) et l'exploitations forestières (64 %) sont majoritairement perçues comme les causes du déboisement. Outre ces causes citées par la majorité, d'autres mobiles comme la variabilité naturelle du climat (sécheresse), les feux de végétation (déboisement), la forte chaleur (sécheresse) sont évoqués mais de façon minoritaire.

Pour expliquer la sécheresse, la variabilité climatique et le déboisement sont plus cités par les jeunes (ayant accès aux informations) alors que la colère divine est plus indexée par les vieux plus attachée à la tradition. Mais, les 2 catégories d'acteurs ont tous perçu l'agriculture et l'exploitation forestière (bois de chauffe, charbon de bois et bois d'œuvre) comme étant les principales causes des 2 fléaux qui sévissent dans le milieu.

Dans l'ensemble, les perceptions ont plus ciblé les causes liée directement ou non aux activités humaines. Ces résultats sont semblables aux conclusion de Hosonuma et al. (2012) et

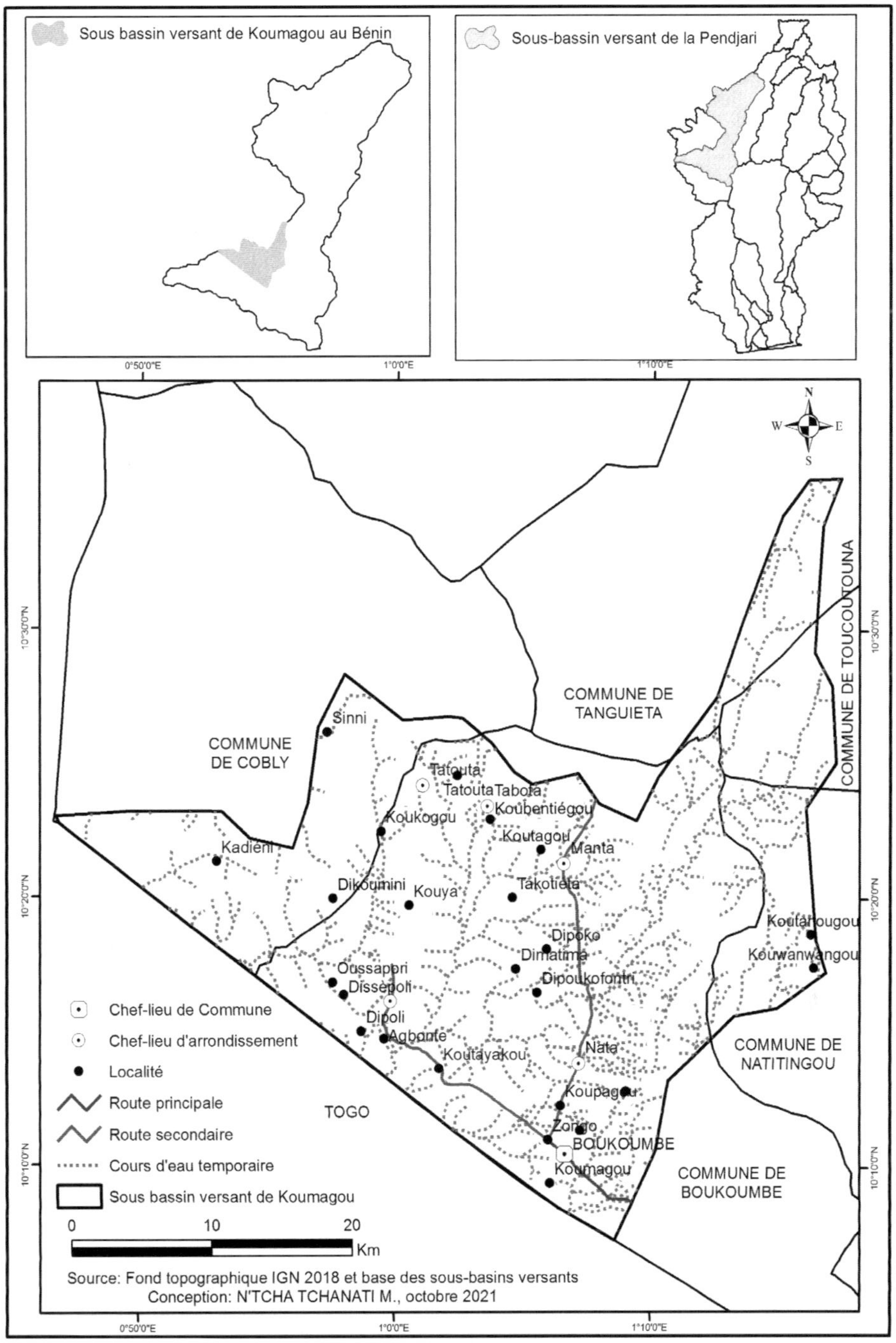

Figure 1. Situation géographique du sous-bassin de Koumagou à Boukoumbé.

de Gilet et al. (2016) qui ont mis en évidence l'agriculture extensive, l'exploitation du bois et l'extraction minière comme principaux facteurs directs de la dégradation du déboisement dans le monde. Au Bénin,

Arouan et al. (2016), Toko Imorou et al. (2019) ont obtenu des résultats similaires sur le rôle prégnant des actions humaines sur la dégradation du couvert végétal. Il en ressort que les perceptions paysannes sont globalement en phase avec les résultats d'analyses scientifiques. Il convient d'analyser les conséquences environnementales de ces phénomènes selon les perceptions.

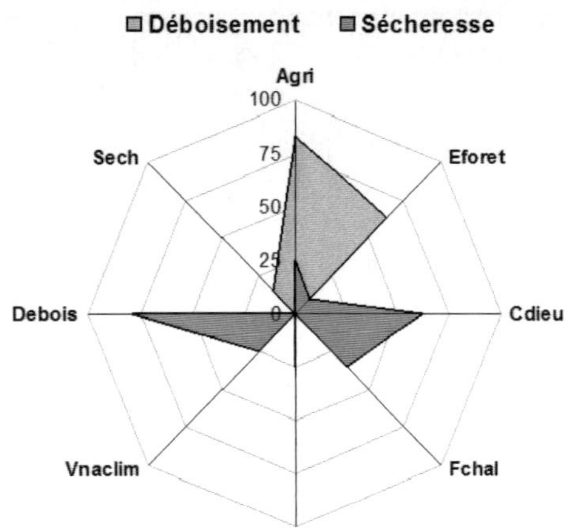

Figure 2. Causes du déboisement et de la sécheresse selon les populations locales. Légende : Agri = production agricole ; Apas = activité pastorale ; Eforet = exploitations forestières ; Cdieux = Colère des dieux ; Fchal = Forte chaleur ; Fvege = feux de végétation ; Vnaclim = variabilité naturelle du climat ; Debois = Déboisement ; Sech = sécheresse.

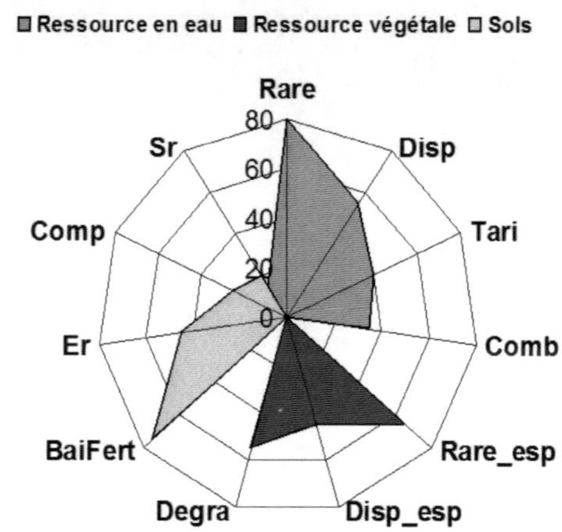

Figure 3. Effets du déboisement et de la sécheresse selon les perceptions. Légende : Rare = raréfaction ; Disp = disparition ; Tari = tarissement précoce ; Comb = Comblement ; Rare_esp = raréfaction d'espèces ; Disp_esp = disparition d'espèces ; Degra = dégradation d'écosystèmes ; BaiFert = baisse de fertilité ; Er = Erosion ; Comp = Compactage ; Sr = sans réponse.

3.2 Effets perçus de la sécheresse et du déboisement

Selon les répondants, les principales composantes environnementales, sources de moyens d'existence sont affectées par le déboisement et la sécheresse (Fig. 3). Ainsi, les ressources en eau (rivières, marigots, mares, etc.), sont devenues plus rares (80 %), disparaissent (55 %), tarissent tarissement précocement (40 %) et se comblent (35 %).

S'agissant des ressources forestières (espèces ligneuses, produits forestiers non ligneux), elles subissent des changements comme la raréfaction d'espèces (65 %), la disparition d'espèces (45 %), et la dégradation des écosystèmes caractéristiques (55 %).

Les sols quant à eux sont sujets à l'érosion (75 %), à la perte de fertilité (45 %) et au compactage (25 %). Dans l'ensemble, les femmes (ayant la charge de l'approvisionnement en eau du ménage et impliquées dans la production maraichères) ont plus cité la disparition et le tarissement précoce des sources d'eau alors que les agro-éleveurs ont plus perçu la dégradation des ressources forestières et la raréfaction/tarissement des ressources en eau. En effet, ils ont insisté sur le fait que la disponibilité de certaines espèces fourragères comme : *Daniellia oliveri*, *Pterocarpus erinaceus*, *Afzelia africana, Khaya senegalensis*, sont devenues rares. L'agriculteur quant à lui est plus sensible à la dégradation des sols qui sont le principal support de son activité. Ils pensent en effet que les sols sont plus dégradés et sont moins productifs, ce qui se traduit par la baisse des rendements agricoles.

Dans tous les cas, les activités et moyens d'existence des communautés sont touchées d'une manière ou d'une autre selon les enquêtes. Il s'en suit des retombées socioéconomiques négatives (pauvreté, instabilité du disponible alimentaire, tensions entre les acteurs, autour des ressources disponibles). Le croisement des différentes perceptions permet de constater l'existence d'un cercle vicieux avec comme maillons : activités humaines, déboisement/sécheresse, dégradation des ressources naturelles, altération des moyens d'existence des communautés dans le sous bassin. Plusieurs auteurs comme : Claude (1989), Al Hamndou et Requier-Desjardins (2008) ; Ndiaye et Ndiaye (2013), Ahononga et al. (2020), ont abouti aux conclusions identiques.

Pour tenter de limiter l'ampleur des effets néfastes, des mesures d'adaptation sont mises en œuvre par les paysans.

3.3 Mesures communautaires d'adaptation et perspectives

Les mesures d'adaptations citées concernent l'ajustement des pratiques culturales fondées sur les savoirs endogènes et les suggestions de l'encadrement rural. Il s'agit des pratiques agroforestières, des pratiques de Gestion Durable de Terre (GDT), la mise en valeur agricole de bas-fonds, etc. Il y a également des initiatives de reboisement qui permettent l'enrichissement des aires dégradées, des plantations agroforestières avec l'utilisation des plantes à valeur socioéconomique, etc. La création des aires protégés communautaires est également citée pour préserver les écosystèmes naturels sans oublier la mobilisation des eaux souterraines à travers les puits à vocation domestique et agricoles. A ces mesures, s'ajoutent l'organisation des prières et cérémonies en cas de sécheresse sévère qui menace la réussite de la campagne agricole. Ces mesures mobilisent plus les personnes âgées à faible niveau d'instruction.

Ces mesures visent à la fois à maintenir la couverture végétale, les services écosystémiques du milieu et l'amélioration des conditions de vie des populations. L'efficacité et la pertinence de ces mesures restent à évaluer. Elles témoignent toutefois de la proactivité des communautés à faire face aux défis environnementaux (Boko, 1988 ; Afouda et al., 2014 ; Yabi (2019). Il conviendrait de les analyser pour en ressortir celles qui sont durables et qui pourraient être vulgarisées. Les associations villageoises, les ONGs, les chefs traditionnels, les autorités locales devraient être associés à cet exercice dans le cadre d'une démarche participative et inclusive.

4 Conclusions

La présente recherche a permis de savoir que les communautés riveraines du sous bassin de Koumagou sont conscientes des dangers que constituent le déboisement et la sécheresse. Ces fléaux sont essentiellement causés par les activités humaines notamment l'agriculture et l'exploitation forestière selon les perceptions communautaires. Les principales composantes environnementales à savoir la végétation, les ressources en eau et les sols sont affectées par ces phénomènes d'après les répondants. Il s'ensuit un amenuisement de moyens d'existence des populations essentiellement tirés de l'agriculture. Face à l'ampleur de ces fléaux et de leurs effets, les populations mettent en œuvre des mesures d'adaptations qui concernent l'ajustement des pratiques culturales, les initiatives de reboisement, la création des aires protégées communautaires, la mobilisation des eaux souterraines.

La lutte contre la sécheresse et le déboisement passe par la mise en place des mesures participatives d'adaptation et de mitigation. Les bonnes pratiques communautaires actuelles pourraient être mises à contribution à cet effet.

Disponibilité des données. Acune donnée utilisée dans cet article n'est disponible au public.

Collaborateurs. IY a initié la recherche, a donné des orientations méthodologiques et a participé à la rédaction du manuscrit. MNT a participé à l'élaboration des outils de collecte des données, a participé aux enquêtes de terrain et a participé à la rédaction du manuscrit. AA a supervisé les enquêtes de terrain, a participé au traitement des données et a participé à la rédaction du manuscrit.

Intérêts concurrents. Les auteurs déclarent qu'ils n'ont aucun conflit d'intérêts.

Clause de non-responsabilité. Publisher's note : Copernicus Publications remains neutral with regard to jurisdictional claims in published maps and institutional affiliations.

Déclaration du numéro spécial. This article is part of the special issue "Hydrology of Large River Basins of Africa". It is a result of the 4th International Conference on the "Hydrology of the Great Rivers of Africa", Cotonou, Benin, 13–20 November 2021.

Remerciements. Les auteurs remercient sincèrement les évaluateurs qui ont contribué à l'amélioration du manuscrit et les organisateurs du colloque.

Références

Ahononga, F. C., Gouwakinnou, G. N., Biaou, S. S. H., et Biaou, S. : Vulnérabilité des terres des écosystèmes du domaine soudanien au Bénin de 1995 à 2015, Bois et Forêts des Tropiques, 346, 35–50, 2020.

Afouda, F., Salako, M. P., et Yabi, I. : Instabilité intra-saisonnière des pluies de la grande saison agricole dans la commune de Kétou au Bénin, Revue de Géographie du Laboratoire Leïdi, 2, 26–47, 2014.

Al Hamndou, D. et Requier-Desjardins, M. : Variabilité climatique, désertification et biodiversité en Afrique : s'adapter, une approche intégrée, VertigO [En ligne], 8, https://doi.org/10.4000/vertigo.5356, 2008.

Arouna, O., Eténé, C. G., et Issiako, D. : Dynamique de l'occupation des terres et état de la flore et de la végétation dans le bassin supérieur de l'Alibori au Bénin, Journal of Applied Biosciences 108, 10543–10552, 2016.

Boko, M. : Climat et communautés rurales du Bénin : Rythmes climatiques et rythme de développement, Thèse d'Etat ès lettres, Dijon, France, 607 pp., 1988.

Claude, J. : Rapport de synthès, in : Les Hommes face aux sécheresses : Nordeste brésilien, Sahel africain [en ligne], Paris : Éditions de l'IHEAL, https://doi.org/10.4000/books.iheal.1216, 1989

Dardel, C. : Entre désertification et reverdissement du Sahel : Diagnostic des observations spatiales et in situ, Thèse de Doctorat, Université Paul Sabatier, Toulouse III, France, 200 pp., 2014.

Fonds international de développement agricole (FIDA) : Gestion des ressources naturelles et de l'environnement : Moyens d'existence résilients pour une utilisation durable des biens naturels, Rome (Italie), 52 pp., 2011.

Gillet, P., Vermeulen, C., Feintrenie, L., Dessard, H., et Garcia, C. : « Quelles sont les causes de la déforestation dans le bassin du Congo ? Synthèse bibliographique et études de cas », BASE [En ligne], 20, 183–194, 2016.

Hosonuma, N. et al. : An assessment of deforestation and forest degradation drivers in developing countries, Environ. Res. Lett., 7, 044009, 2012.

Ndiaye, A. et Ndiaye, P. : Changement climatique, dégradation environnementale et quête d'utilisation des ressources naturelles. Occasional Paper No. 21, The African Capacity Building Foundation (ACBF), Hararé (Zimbabwé), 50 pp., 2013.

OCDE : Ressources naturelles et croissance pro-pauvres : Enjeux économiques et politiques, Paris (France), CEDEX 16, 189 pp., 2009.

OMM et GWP : Manuel des indicateurs et indices de sécheresse, Genève, Suisse, 52 pp., 2016.

Sanou, K., Amadou, S., Adjègan, K., et Tsatsu, K. D. : Perceptions et stratégies d'adaptation des producteurs agricoles aux changements climatiques au nord-ouest de la région des savanes du Togo, Agronomie Africaine, 30, 87–97, 2018.

Toko Imorou, I., Arouna, O. Zakari, S., Djaouga, M., et Thomas, O. : Évaluation de la déforestation et de la dégradation des forêts dans les aires protégées et terroirs villageois du bassin cotonnier du Bénin. Conférence OSFACO : Des images satellites pour la gestion durable des territoires en Afrique, Cotonou, Bénin, hal-02189556, 2019.

Wicander, S., Helfgott, A., Bailey, M., Munroe, R., Ampomah, G., Diouf, A., Devisscher, T., et Corrigan, C. : Planification de la résilience et de l'adaptation pour les communautés des aires protégées, Guide par étapes. Cambridge : UNEP-WCMC, 112 pp., 2016.

Yabi, I. : Paysannat vivrier face aux incertitudes pluviométriques de la seconde saison agricole dans la commune de Didja au Sud-Benin, Revue de Géographie de l'Université de Ouagadougou, 3, 145–164, 2019.

Yabi, I. : Sécheresse pluviométrique dans la zone soudanienne au Bénin : Manifestations et Implications pour une agriculture climato-résiliente, Dynamiques Spatiales et Développement : Revue semestrielle du Laboratoire d'Etudes des Dynamiques Urbaines et Régionales, 2, 115–134, 2018.

Yabi, I. et Afouda, F. Extreme rainfall years in Benin (West Africa), Quaternary International, 262, 39–43, https://doi.org/10.1016/j.quaint.2010.12.010, 2012.

Proc. IAHS, 384, 301–304, 2021
https://doi.org/10.5194/piahs-384-301-2021

Open Access
PIAHS

Hydrology of Large River Basins of Africa

Pratiques endogènes de gestion de l'eau au service des politiques d'adaptation aux changements climatiques dans la région du lac Tchad

Armel Sambo

Faculté des Arts, Lettres et Sciences Humaines, Université de Maroua, BP 644, Cameroun

Correspondence: Armel Sambo (samboarmel@yahoo.fr)

Published: 16 November 2021

Résumé. Dans un contexte-marqué par la variabilité climatique comme la région du lac Tchad, il est utile de prendre en compte et de revaloriser les stratégies d'adaptation aux phénomènes induits par les Changements Climatiques (CC), dans le but d'améliorer la gestion durable des ressources en eau. Cette étude vise donc à identifier et à interpréter les pratiques locales de gestion et d'accès à l'eau, des dynamiques de celles-ci, et son importance dans la cadre de mises en œuvre des projets d'adaptation aux changements climatiques dans la région du lac Tchad. La méthodologie adoptée a consisté d'abord à la collecte des données (par des entretiens, observation, etc.) auprès des populations riveraines du lac Tchad, ensuite à leur traitement analyse et enfin la rédaction de cet article. Ainsi, quelques techniques issues de savoir faire traditionnel et local, ont été identifiées et sont susceptibles d'être revalorisées dans un cadre stratégique. Il s'agit : l'existence des mares d'eau sacrée, la gestion collective des mares d'eau naturelles, la gestion communautaire des pêcheries, l'irrigation, etc. Toutes ces pratiques et techniques ont été inventées et ou améliorées par les populations pour répondre à des contraintes de production qu'ils connaissent bien.

1 Introduction

Le changement climatique risque d'amplifier les vulnérabilités socio-économiques auxquelles les populations font face, et particulièrement dans la zone du lac Tchad où l'économie est essentiellement basée sur l'agriculture, l'élevage et la pêche. C'est pourquoi, les travaux de terrain doivent focaliser l'attention non seulement sur les effets du changement climatique dans le bassin du lac Tchad mais aussi sur les pratiques d'adaptation aux variations environnementales qui peuvent être développées dans le cadre de la mise en place des projets de développement à base communautaire. Ainsi, l'objectif principal de ce travail est de savoir quelles sont les pratiques endogènes que les populations ont développé pour assurer une gestion durable des ressources en eau dans la zone du lac Tchad et comment peut- on les mettre en pratique à travers des projets communautaires dans le cadre des plans nationaux d'adaptation au Changement climatique en Afrique ? Ce retour d'expérience souligne l'intérêt – la nécessité – d'identifier, répertorier/recenser/classer ces pratiques, et d'en mesurer/d'évaluer l'impact/la valeur ajoutée sur la gestion durable et rationnelle des ressources en eau.

La méthode adoptée a consisté d'abord à la collecte des données (par des entretiens, observation, etc.) auprès des populations riveraines du lac Tchad, ensuite à leur traitement analyse et enfin la rédaction de cette communication. D'abord, elle présente le niveau et la vulnérabilité des populations aux effets du Changement Climatique, ensuite les stratégies endogènes de gestion rationnelle des eaux qu'elles ont développées et enfin, l'intérêt de l'intégration des ses savoirs locaux dans le processus du développement durable de la région du lac Tchad.

2 La région et les populations du bassin du lac Tchad affectées par le Changement Climatique

Les rapports du GIEC indiquent que la région du lac Tchad compte parmi les zones les plus affectées par le changement climatique (GIEC, 2007). L'on peut observer ce phénomène à travers la baisse de la pluviosité, l'assèchement du lac Tchad et la raréfaction des ressources en eau (Sambo, 2013).

Les populations riveraines sont de plus en plus vulnérables à ces modifications survenant dans leur environnement. L'évolution de la pluviométrie de ces 50 dernières années démontrent à suffisance de la baisse de la pluviosité. Elle est marquée par la chute la plus importante et la plus durable de sa pluviométrie. Une comparaison des isohyètes des années 1950, « considérées comme la décennie la plus humide, la plus sèche étant celles des années 1990, a montré un déplacement considérable vers le sud » (Olivry et al., 1996). La baisse de la pluviométrie a atteint près de 100 mm pour tous les 100 km de distance en dépit des variations annuelles et spatiales (CBLT, 2000).

Les variations de la pluviométrie se sont répercutées sur les écoulements les affluents du lac Tchad (Logone, Chari, etc.) et sur la quantité d'eau du lac. Le lac Tchad est donc en constante régression vu son évolution historique. Entre le XII*e* et le XVIII*e* siècles, le lac Tchad va connaître une baisse catastrophique des ressources en eau qui va conduire à son assèchement successivement au XV*e* et à la fin du XVIII*e* siècle (Fadimatou Dia, 2006, pp. 15–16). Depuis le XIX*e* et le XX*e* siècles, le bassin du lac Tchad a connu plusieurs épisodes de sécheresse. Le lac Tchad avait en 1963 une superficie d'environ 25 000 km^2 et suite à la sécheresse de 1973, sa superficie en juillet n'était que de 9000 km^2 (Beauvilain, 1989, p. 92). Depuis 2013, l'on observe une crue abondante en saison de pluies. Les images ci- dessous produites par la NASA sont contestées par certains chercheurs. Ils pensent que le lac Tchad, loin de s'inquiéter sur sa disparition est juste passé « à l'état de petit Tchad, comme il l'a été plusieurs fois au cours des 150 dernières années ». Depuis 2013, l'on constate que le lac a atteint une superficie inondée de 14 800 km^2 (Lemoalle et al., 2015). Toutefois, cette variabilité est influencée aussi ces dernières années par le changement climatique.

La dégradation du niveau de vie des populations, victimes des effets du changement climatique est une réalité dans la zone du lac Tchad. Toutefois les changements climatiques ne sont pas à la base de toutes les difficultés rencontrées dans cette zone écologique. En effet des facteurs anthropiques, comme la construction des barrages, la déforestation, ou les problèmes politiques liés à un encadrement efficace du monde rural, sont aussi responsables de la pauvreté des populations. Plus de 80 % des personnes interrogées établissent une relation entre ces transformations environnementales et la baisse des ressources en eau, et partant de la production agro – sylvo – pastorale ce d'autant plus que l'économie dans cette localité est essentiellement rurale (Sambo, 2013).

3 Les populations ont développé des stratégies de gestion rationnelle

Dans la majorité des cas, l'adaptation aux changements climatiques s'inscrit naturellement dans un cadre de développement puisque tous les pays s'efforcent d'une manière ou d'une autre de gérer les mêmes systèmes sociaux, écologiques et économiques en vue d'instaurer un développement durable (Mohamadou Sall et al., 2011). Il s'agit de répertorier, de classifier et des prioriser les stratégies d'adaptation qu'on peut soutenir financièrement. De ce fait donc, l'adaptation intègre les politiques environnementales des Etats africains. Il est important de noter que les communautés locales ont toujours adopté des stratégies et mesures afin de s'adapter aux conséquences des modifications des ressources en eau liées à celles du climat. Il s'agit entre autres selon ce tableau ci-dessous :

La prise en compte de ces pratiques dans un cadre stratégique comme on le constate dans ce tableau sera un atout énorme pour améliorer la résilience des populations du bassin du lac Tchad au changement climatique. C'est ainsi qu'il importe pour les Etats d'intégrer ces savoirs faire par la mise en place des projets de développement. Les pratiques les plus courantes observées dans la région et qui sont au centre des activités des ONG sont : la construction des mares d'eau, des infrastructures de rétention d'eau, l'organisation de la transhumance, l'irrigation, etc.

De plus en plus, l'on observe un changement dans les pratiques qui se modernisent. Ainsi, aux pratiques traditionnelles qui existent beaucoup plus en zones rurales à savoir entre autres l'aménagement des puits à ciel ouvert, l'usage des cours d'eau à des fins de lessive et bain, etc. se développent les aménagements modernes tels que les forages et les puits modernes. Ainsi, ces Pratiques et connaissances endogènes peuvent servir de mécanismes de gestion rationnelle et durables des ressources en eau, en partant des stratégies d'adaptation aux changements climatiques dans cette partie de l'Afrique.

4 La prise en compte de certaines des pratiques endogènes dans la politique environnementale est une nécessité

Les pratiques endogènes dans le domaine de gestion des ressources en eau sont donc des précieuses ressources qui peuvent contribuer à améliorer la base du processus de décision locale dans des domaines tels que la sécurité alimentaire, la gestion des ressources et par d'autres activités économiques. Sur cette base, il est possible de co-construire et de concevoir avec les populations du bassin du lac Tchad des systèmes techniques améliorés issus de les propres savoir-faire, afin de gérer rationnellement et durablement les ressources en eau sans bouleverser le monde rural.

Dans le processus de gestion des ressources en eau, la construction des mares d'eau, des puits, le dévelop-

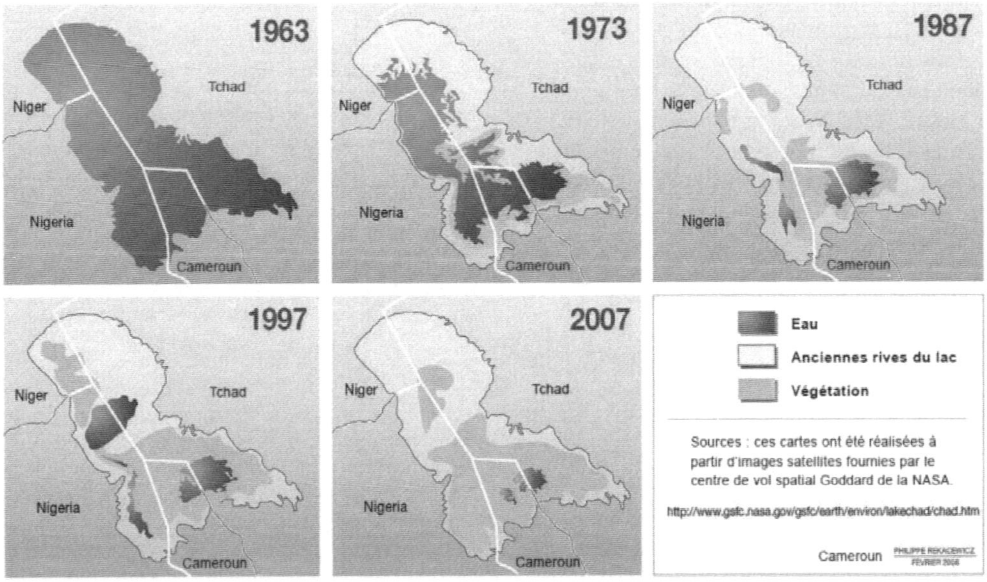

Figure 1. Evolution de l'étendue en eau du lac Tchad de 1967 à 2007 (source : Documentation de la CBLT, 2010).

Tableau 1. Les stratégies d'adaptation de gestion rationnelle des ressources en eau (Source : Enquête de terrain, 2012).

Systèmes	Impact du CC	Pratiques endogènes d'adaptation au CC	Les cadres stratégiques dans lesquels intégrer ces savoirs
Agriculture	Variabilité pluviométrique	réaménagement des calendriers agricoles	Mise à la disposition des paysans des données sur la pluviométrie. Sensibilisation sur l'évolution des calendriers de la pluviométrie
	Diminution des ressources en eau	L'irrigation pour la culture	Gestion rationnelle des mares d'eau, aménagements des ouvrages hydroagricoles, etc.
		Recharge artificielle des nappes souterraines	Développement des infrastructures de rétention d'eau pour la recharge des eaux souterraines
Élevage	Raréfaction de l'eau pour le bétail	transhumance	Organiser la transhumance
		Préservation des mares d'eau	Construction des mares d'eau artificielles (rétention des eaux de pluie). Renforcement des capacités des structures de gestion des mares
		Gestion des mares d'eau naturelles	Construction des mares d'eau artificielles
Pêche	Baisse de stock de poissons	gestion communautaire des pêcheries,	Appui financier des populations à travers des associations Renforcer les capacités des communautés
		la préservation des mares d'eau sacrée	Revaloriser de cette tradition

pement des systèmes d'irrigation dans la vallée du Logone, etc. contribuent à augmenter la production agro-sylvopastorales (Sambo, 2018). Quelques techniques issues de savoir faire traditionnel et local, ont été identifiées et sont susceptibles d'être revalorisées dans un cadre stratégique. Il s'agit : l'existence des mares d'eau sacrée (stratégie de préservation des ressources), la gestion collective des mares d'eau naturelles, la gestion communautaire des pêcheries, des systèmes de recharge artificielle des nappes souterraines, etc. Quelques exemples (gestion rationnelle des mares d'eau, développement des infrastructures de rétention d'eau pour la recharge des eaux souterraines, construction des mares d'eau artificielles, etc.) mis en œuvre par les Organisations paysannes et certaines associations locales ont prouvé leur importance dans le renforcement de résilience des populations riveraines du lac Tchad.

Il existe plusieurs programmes et projets en cours financés par des organismes bilatéraux et multilatéraux pour aider les pays africains à gérer les risques liés aux changements climatiques ainsi qu'à s'adapter à la variabilité climatique. Seuls quelques-uns de ces programmes et projets traitent des risques liés à la gestion des ressources en eau partagées par plusieurs Etats. Pourtant, la gestion rationnelle des ressources naturelles doit faire partie intégrante des ripostes aux changements climatiques.

5 Conclusions

En conclusion, ces résultats nous permettent de dire que les populations riveraines du lac Tchad ont développées des mécanismes endogènes de gestion de l'eau qui peuvent – ou pourraient – jouer un rôle primordial dans la gouvernance environnementale. C'est une contribution qui met à profit les « propres savoir faire » des communautés locales dans le choix des politiques environnementales des Etats africains. Ce faisant, les décideurs ont en main un outil important de prise de décision qui peut améliorer la gestion intégrée des ressources eau du lac Tchad, en partant l'adaptation au changement climatique pour le développement des activités économiques liés à l'eau. Il est important dans ce contexte d'encourager la mise en place des projets communautaires et favoriser des conditions permettant aux pratiques communautaires et locales innovantes d'être élevées à des niveaux de mise en œuvre régionaux et nationaux.

Intérêts concurrents. The contact author has declared that there are no competing interests.

Déclaration du numéro spécial. This article is part of the special issue "Hydrology of Large River Basins of Africa". It is a result of the 4th International Conference on the "Hydrology of the Great Rivers of Africa", Cotonou, Benin, 13–20 November 2021.

Références

Beauvilain, A. : Nord Cameroun : crises et peuplement, T.1 et T.2, thèse de Doctorat es Lettres et Sciences humaines, Université de Rouen, 1989.

CBLT : Gestion intégrée du bassin fluvial, les défis du bassin du lac Tchad, vision 2025, Audit préparé par Steveland Consult, Norway, 2000.

Fadimatou Dia : La Commission du Bassin du Lac Tchad : 1964–2004, Mémoire de Maîtrise d'Histoire, Université de Ngaoundéré, 2006.

GIEC : Climate change 2007 : The physical science basis, Cambridge University Press, Cambridge, 2007.

Lemoalle, J. and Magrin, G. (Dir.) : Le développement du lac Tchad : situation actuelle et futurs possibles, Marseille, IRD Editions, coll. Expertise collégiale, bilingue français-anglais, 2014.

Mohamadou Sall, Serigne Mansour Tall, Aly Tandian, et M Al Assane Samb : Changements climatiques : stratégies d'adaptation et mobilités. Evidence à partir de quatre sites au Sénégal, GIZ, accessible à l'adresse suivante : http://pubs.iied.org/10612IIED.html, consulté le 11 février 2011.

Olivry, J.-C. : Hydrologie du lac Tchad, ORSTOM, Paris, 1996.

Sambo, A. : Perceptions locales et pratiques d'adaptation au changement climatique dans la gestion rationnelle des ressources en eau du Lac Tchad, in : Geo-Eco-Trop (Revue internationale de géologie, de géographie et d'écologie tropicales), No. 37, Tome, 2, 293–302, 2013.

Sambo, A. : Les savoirs endogènes comme stratégies d'adaptation au changement climatique chez les agriculteurs, les éleveurs et les agropasteurs de l'Extrême Nord du Cameroun, in : Revue Science et Technique, Lettres, Sciences Humaines et Sociales, Spécial hors-série No. 4, Ouagadougou, Burkina Faso, 11–25, 2018.

Proc. IAHS, 384, 305–312, 2021
https://doi.org/10.5194/piahs-384-305-2021

Open Access
PIAHS

Hydrology of Large River Basins of Africa

Utilisation des ressources en eau dans un contexte de changement climatique : réalités écologiques, sociales et économiques dans les communes Ambositra I et Ambositra II Madagascar

Rindramampionona Fanambinantsoa Ankasitrahana Randriamifidison[1], Daniel Rakotoarisoa[1,2], Tsaralaza Jorlin Tsiavahananahary[3], Tantely Rakotovao[4], and Lily Arson Rene de Rolland[5]

[1]Institut Supérieur de Technologie d'Ambositra, Ambositra 306, Madagascar
[2]Laboratoire Engénierie et Géosciences, Ecole Supérieure Polytechnique, Antananarivo 101, Madagascar
[3]Faculté des Sciences de Technologie et Environnement, Université de Mahajanga, Mahajanga 501, Madagascar
[4]Direction de l'Energie de l'Eau et des Hydrocarbures Amoron'i Mania, Ambositra 306, Madagascar
[5]The peregrine Fund, Antananarivo 101, Madagascar

Correspondence: Rindramampionona Fanambinantsoa Ankasitrahana Randriamifidison (rindrarandriamifidison@yahoo.fr)

Published: 16 November 2021

Résumé. Les démarches de cette recherche ont été fondées au départ sur le fait que les ressources en eau diminuent avec le changement climatique, alors que les besoins de la population augmentent incéssament. Elle concerne la Commune urbaine d'Ambositra et la Commune rurale d'Ambositra, traversées par la rivière Isaha . Dans cette optique, cet article a pour objectif d'analyser le rapport entre les ressources et les besoins en eau pour diverses utilisateurs tenant compte des variabilités climatiques dans la zone. Pour ce faire, des variables d'entrée ont été utilisés pour modéliser de l'utilisation des ressources en eau par le biais du logiciel WEAP (Water Evaluation and Planning System), précédé par un traitement cartographique pour la délimitation de la zone par le logiciel QGIS. Les données sur la précipitation sont obtenues avec la médiane des valeurs issues des modèles climatiques du scenario RCP4.5. Les principaux résultats de cette étude concernent à la fois l'utilisation de l'eau à des fins agricoles et pour un usage urbain. Et les stratégies proposées concernent tous les acteurs de tous les niveaux, des utilisateurs à l'Etat central. Nous avons évoqué l'urgence de la concrétisation des recommandations émanant de cette étude vue les réalités actuelles ainsi que la prévision issue de la modélisation.

1 Introduction

La disponibilité en eau conditionne l'occupation humaine et toutes activités anthropiques. Cette ressource est devenue rare dans certains endroits à Madagascar (FAO, 2019). Cette étude a été conduite dans la zone d'Ambositra, située sur les hautes terres centrales de Madagascar, dans la Région Amoron'i Mania qui littéralement signifie au bord de la Mania, une grande rivière qui coule des montagnes centrales de Madagascar et qui se déverse dans le Canal de Mozambique en passant par le fleuve Tsiribihina (Chaperon et al., 1993). À Ambositra, la gestion de l'eau ne se fait pas d'une façon rationnelle. Traversé par la Route Nationale 7, Ambositra est une zone des hautes terres centrales de Madagascar. Une ville urbaine où règne toutefois un aspect rural. L'accès à l'eau potable y est encore très faible. Les contextes écologiques, économiques et sociales sont difficilement maîtrisables car ni la population, ni les autorités concernées ne sont assez conscientes de la réalité sur l'enjeu de cette ressource. Ce qui est à l'origine de cette recherche. Qui s'est proposée de (1) établir une base de donnée sur les ressources en eau disponibles selon les variabilités climatiques et saisonnières, (2) identifier, une typologie des principaux utilisateurs, et (3) proposer des stratégies adaptées aux réalités du site et aux paramètres de changement climatique

2 Matériels et méthodes

2.1 Présentation de la zone d'étude

Les deux Communes Ambositra I, Ambositra II se situent administrativement dans le District d'Ambositra, Région Amoron'i Mania ; la ville d'Ambositra étant le Chef Lieu de la Région Amoron'i Mania (Fig. 1), et se situe à une distance de 255 km de la Capitale de Madagascar en empruntant la RN7 vers le sud. La population d'Ambositra compte 33 276 habitants (INSTAT, 2019). Une espèce de plante actuellement disparue nommée vernaculairement « Somangana » dominait ces habitats ripicoles (Randriamifidison, 2015), ce qui était à l'origine de l'appellation plus ancienne et autochtone de *Sahasomangana*, actuellement Isaha ; avec *Sahasomangana iray* sur la partie sud de la ville (*iray* pour dire en amont) et *Sahasomangana iava* sur la partie nord de la ville (*iava* pour dire en aval).

2.2 La rivière Isaha

L'hydrologie de la Région Amoron'i Mania est dominée par le bassin versant de la Tsiribihina. La Fig. 2 montre le réseau hydrographique de la Région. Le réseau hydrologique de ce bassin versant prend sa source dans la région et se déverse dans le Canal de Mozambique, une fois récupéré par le fleuve Tsiribihina (Chaperon et al., 1993). La rivière de l'Isaha est un plan d'eau permanent issu du fleuve Mania.

2.3 Méthodes

Une descente sur terrain a été effectuée pour une observation de la hauteur d'eau dans la rivière en période d'étiage (octobre 2019). En effet, en référence avec la démarche méthodologique de Rakotoarisoa (2018) et Rakotondrabe (2007) l'apport en eau de surface disponible à été obtenue. Elle est considérée comme variable d'entrée dans le modèle.

Dans cette modèle, la priorité de la demande représente la priorité de l'allocation d'une ressource limitée entre plusieurs sites de demande. Pour cette étude, deux politiques de répartition seront étudiées pour chaque scénario :

- La politique 1 priorisant les demandes de la population, besoins en eau pour les infrastructures sanitaire et éducatif (Code de l'Eau de Madagascar, 1999) ;
- La politique 2 favorisant les besoins en eau pour le secteur Agricole (agriculture et élevages).

Dans cette recherche, la chaîne de modélisation intégrative de la balance demande et offre en eau avec WEAP[1] (Rakotoarisoa, 2018) a été adoptée et présentée dans la Fig. 3. DEP : demande en eau par la population ; DEinf-San : demande en eau par l'infrastructure sanitaire ; Deinf-Ed : demande en eau par l'infrastructure éducative ; DE-agr : demande en eau par l'agriculture DE-élg : demande en eau par l'elevageLe processus de création de scenarii se déroule comme suit :

- *Création de la zone d'étude* : il s'agit d'ajouter la carte de la zone d'étude sous forme vecteur.
- *Réglage de paramètres généraux* : Dans cette étude, l'année 1991 a été choisie comme l'année de base de modélisation en limitant toutes les informations sur le système introduit dans le compte actuel entre la période 1991–2018.
- *Création de site de demande* : après avoir numérisé la rivière de l'Isaha ; deux sites de demande ((Populations, agricultures) ont été créés.
- *Création des hypothèses clés* : ils sont des variables qui permettent de distinguer les scénarios proprement dits (climat variable, accroissement de population) aux scénarii de référence (situation imaginaire sans aucune variation interannuelle). Dans cet article, deux hypothèses clés ont été adoptées, le besoin de la population varie avec la croissance démographique et la variabilité climatique affecte la disponibilité de ressources pour les besoins de la population et de l'agriculture.
- *Création de Scénarii* : ils sont des variables construites à partir des hypothèses clés sur la base de l'état actuel. Trois scenarii ont été choisis dans cette étude pour la modélisation et comptée à partir de l'année 1991 jusqu'en 2050 (fin du scénario). : scenario de référence pris comme base avec une croissance continue de la population de 3.4 % ; le deuxième scénario considère un fort accroissement de population (5 %) ; et le troisième scénario représente la variation du climat.

Pour le scénario « Changement climatique », il a été attribué aux régimes climatiques (très sec, sec, normale, humide, très humide) des valeurs numériques (Tableau 1) pour que le logiciel puisse les reconnaître.

Le type de climat est défini à partir de la précipitation annuelle reçue par la zone d'étude. L'attribution des régimes climatiques pour chaque année a été basée sur l'analyse des données climatiques (précipitation moyenne mensuelle) issues du portail du World Bank Group sur le climat[2] selon les

1. Water Evaluation And Planning, est un outil convivial qui adopte une approche integrée dans la planification des ressources en eauUniversité de Stockholm.

2. https://climateknowledgeportal.worldbank.org/download-data (la date du dernier accès : 18 octobre 2021).

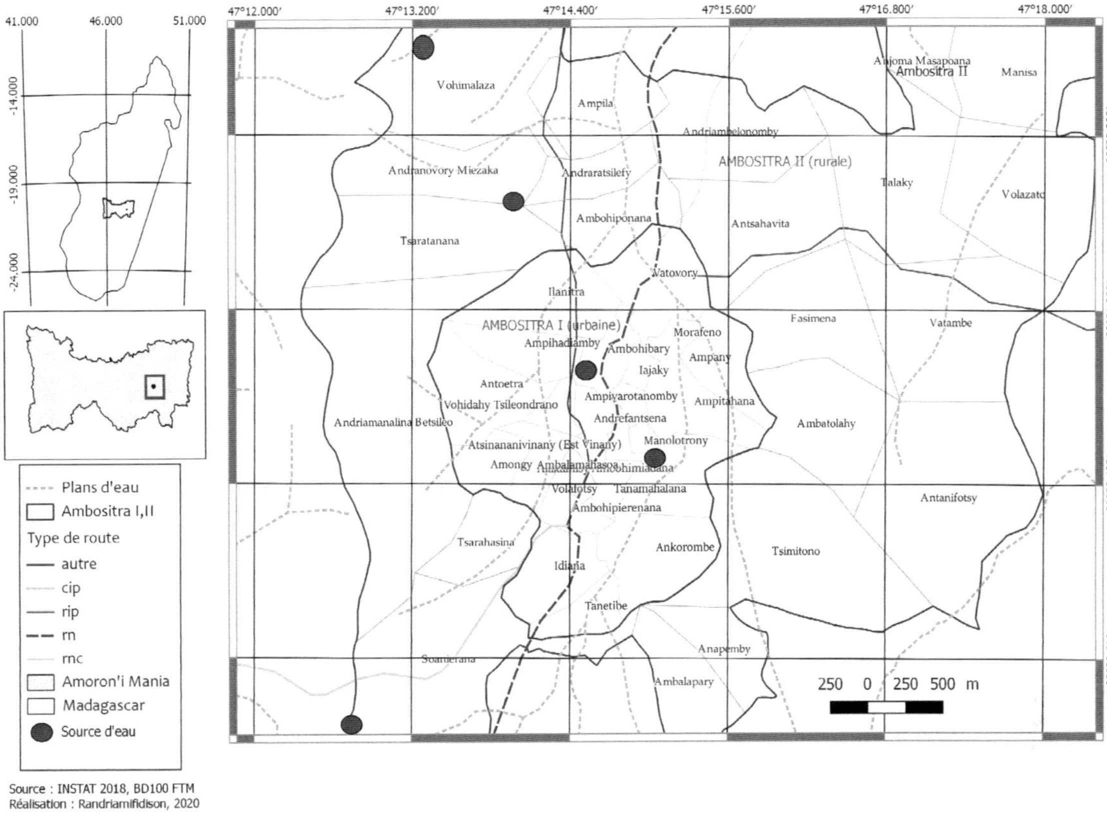

Figure 1. Carte de localisation de la zone d'étude (Réalisation : R. F. A – Rindramampionona Fanambinantsoa Ankasitrahana).

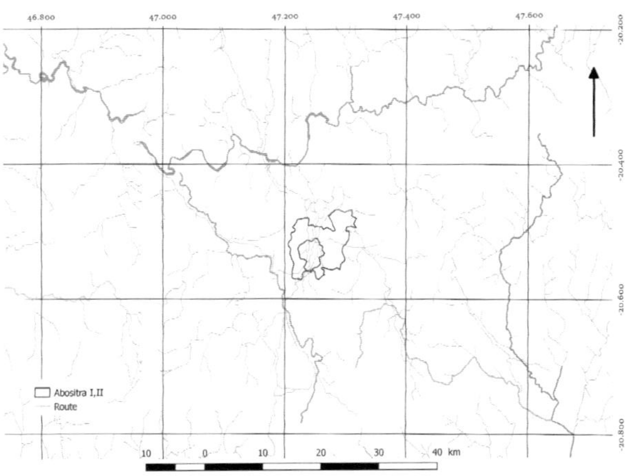

Figure 2. Aperçu hydrographique du site d'étude (Source : BD 100 FTM, Realisation : Randriamifidison, 2019).

modèles climatiques avec le scénario RCP4.5. Ainsi nous avons adopté ces figures (Figs. 4 et 5) pour classifier les années.

La valeur moyenne correspond à la valeur de référence « normal » puis en dessous « sec » et « très sec », et au dessus « humide » et « très humide ».

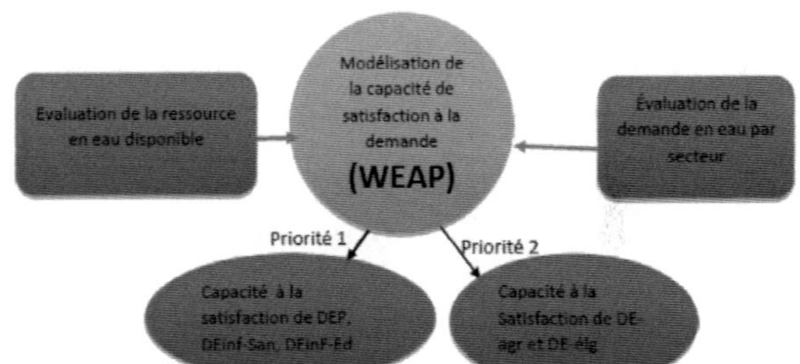

Figure 3. Modélisation avec le logiciel WEAP (Rakotoarisoa, 2018).

Ce qui nous amène à cette Fig. 6.

3 Résultats

3.1 Accès aux ressources en eau pour la population à Ambositra

Pour Ambositra urbaine, la distribution de l'eau domestique est assurée par la JIRAMA (Jiro sy Rano Malagasy ou Electricité et Eau de Madagascar). Toutefois, ce n'est pas accessible à tous. Ambositra II, dispose de quelques ressources

Tableau 1. Valeurs numériques relatives à chaque type de climat.

Types de climat	Valeurs correspondantes
Très sec	0.7
Sec	0.8
Normale	1
Humide	1.3
Très humide	1.45

Source : Rakotondrabe (2007) et Rakotoarisoa (2018).

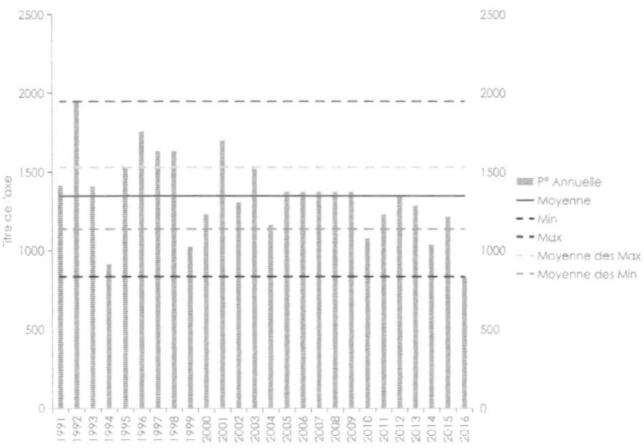

Figure 4. Précipitation annuelle depuis 1991 à 2015 (accessible à l'adresse suivante : https://climateknowledgeportal.worldbank.org/download-data, la date du dernier accès : 18 octobre 2021).

dont le captage d'Ambositra I par la JIRAMA. Actuellement, le captage se situe dans la Commune Rurale d'Ankazoambo, la source d'Anasana, au sud est de la ville.

Généralement, d'après les observations sur sites, toutes les sources hydrologiques de surfaces (rivières et fleuves) sont de nature simple et de types permanent et dont la largeur ne dépasse pas 10 m. Treize (13) sources hydrologiques sont répertoriées dont 3 sont de types temporaires. La production annuelle en eau potable par la JIRAMA est estimée en moyenne à 68 823 m^3, dont 65 088 m^3 utilisée (https://www.jirama.mg/ (la date du dernier accès : 28 octobre 2021), avec 2031 abonnés. Par rapport au nombre de population, ce chiffre est trop faible, moins de 3 % de la population sont abonnés à l'eau de la JIRAMA, seul distributeur.

Les figures suivant montrent la répartition de la population par types de ressources (Figs. 7 et 8).

Ces figures montrent seulement la provenance des eaux domestiques, et pas les moyens d'accès. Les « Non répertoriés » selon la Direction de l'eau et de l'assainissement de la Région Amoron'i Mania constituent 75 % de la population urbaine. Ils sont des abonnés de la JIRAMA (3 %) et le reste (97 %) utilise les bornes fontaines publiques pour leur usage domestique.

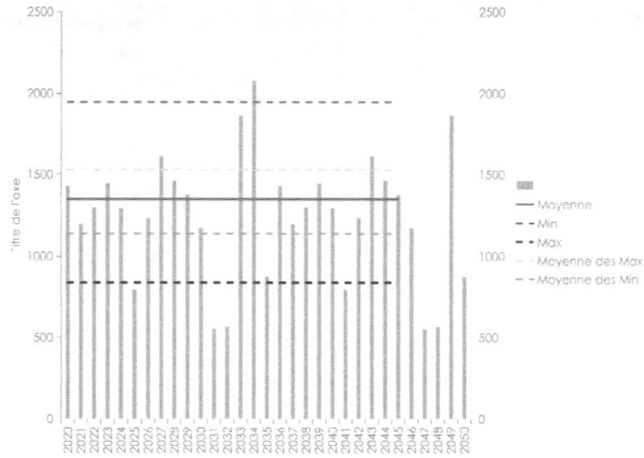

Figure 5. Précipitation annuelle depuis 2020 à 2050 (accessible à l'adresse suivante : https://climateknowledgeportal.worldbank.org/download-data, la date du dernier accès : 18 octobre 2021).

3.2 Évaluation quantitative de l'utilisation de l'eau

Au vu des données de la Fig. 9, force est d'admettre que la ressource en eau est largement suffisante pour Ambositra, la demande est satisfaite (Demande non satisfaisante = 0 m^3). Prenons en exemple pour le mois de Novembre (période de repiquage pendant laquelle le besoin en eau est le plus important pour l'agriculture), l'eau disponible est estimée à 8.9 millions de m^3, alors que l'agriculture n'utilisent que 1.63 millions de m^3.

Ces données semblent excellentes, mais la réalité n'est pas le cas. Ce qui nécessite l'analyse de l'aspect socio-économique dans la partie discussion.

3.3 Modélisation de l'utilisation de l'eau à l'horizon de 2050

La Fig. 11 donne la variation de la quantité de l'eau de surface qui n'est pas utilisée ni par la population, ni par l'agriculture. Comme la population a été priorisé dans l'analyse, la valeur du débit de retour (quantité non consommée après utilisation) étant nulle, donc seule l'utilisation à des fins agricoles présentent une quantité très importante d'eau non utilisée. La Fig. 12 qui suit montre la situation en 2050, avec les scenarii d'une forte croissance démographique de l'ordre de 5 % ainsi que des séquences hydrologiques sèches prolongées. Rappelons que la projection de la précipitation pour 2030 donne une baisse en moyenne mensuelle de la précipitation allant de 112.38 à 105.45 mm (−6.92 mm) (Randriamifidison, 2019). A partir de cette figure, il est démontré une augmentation des besoins totaux (population + agriculture) à partir de 2017–2018. Des variations mensuelles sont constatées, justifiés par une variabilité climatique qui perturbe les calendriers culturaux, dont les besoins en eau.

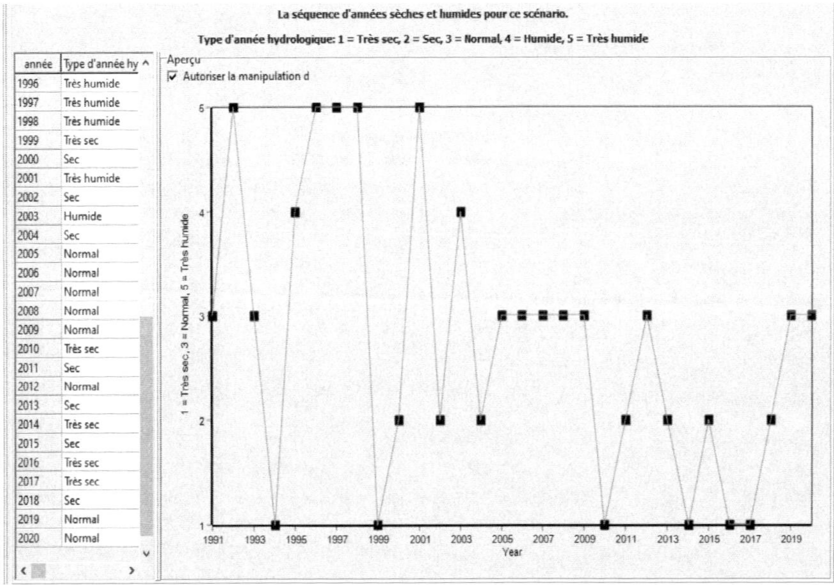

Figure 6. Attributions de séquences d'année sèches et humides pour identifier le scénario lié à la précipitation dans WEAP.

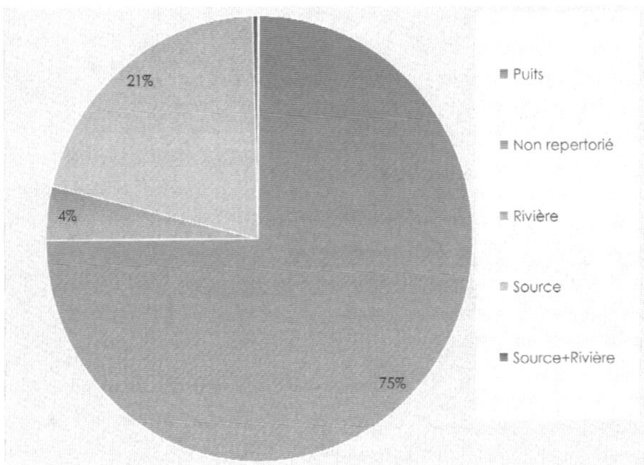

Figure 7. Répartition de la population par type de ressources dans la Commune Urbaine d'Ambositra (Direction de l'Eau Amoron'i Mania, données non publiées).

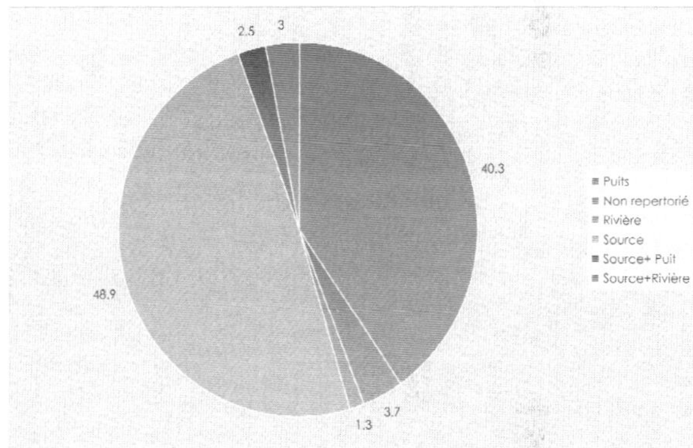

Figure 8. Répartition de la population par type de ressources dans la Commune Urbaine Rurale d'Ambositra (Direction de l'Eau Amoron'i Mania, données non publiées).

4 Discussion et recommandations

La pauvreté de la population affecte les moyens d'accès à l'eau potable (traitée). La plupart des ménages urbains sont en location d'appartement et n'envisage pas de s'abonner à la JIRAMA n'étant pas propriétaire. Pour y remédier, exiger une installation pour l'accès à l'eau courante pour chaque nouvelle construction, augmenter le nombre de bornes fontaines ou des puits à pompages manuelles ou électrique solaire au niveau de chaque quartier. Pour la zone rurale, l'utilisation de la source doit être améliorée, en ramenant l'eau par conduite forcée au village ou en aménageant le site et ses alentours pour le respect de l'hygiène et pour pérenniser la source, renforcer le reboisement avec des espèces à racines adventives pour faciliter l'infiltration et entretenir la nappe phréatique.

Les ressources en eau peuvent satisfaire les besoins de la population, toutefois, des problèmes d'irrigation se font ressentir de plus en plus au niveau des rizières de part et d'autres de la rivière Isaha. Cela résulte d'une mauvaise gestion de l'eau, le calendrier cultural est perturbé avec les variabilités du climat. Ce problème devient même une source de litige entre les paysans. Car l'eau peut être bloqué par un riziculteur en amont. Cela pourrait être résolu par la mise en place d'une association des usagers de l'eau pour respecter les exigences communes. La mise en place d'un bassin de décan-

	Jan	Fév	Mars	Avr	Mai	Juin	Juil	Août	Sept	Oct	Nov	Déc	Sum
Débits entrant dans l'aire (Million Mètre cube)													
En dessous de IsahaDébit de tête	29,38	26,47	29,73	16,30	9,91	6,74	6,43	5,92	4,38	4,18	8,09	22,42	169,95
Débits sortant de l'aire (Million Mètre cube)													
Demande agri	0,00	0,10	0,18	0,20	0,00	0,00	1,10	1,97	1,11	1,19	1,63	1,30	8,78
Demande pop	0,00	0,00	0,00	0,00	0,00	0,00	0,00	0,00	0,00	0,00	0,00	0,00	0,00
En dessous de Demande agri Retour	29,38	26,37	29,55	16,10	9,91	6,74	5,33	3,95	3,27	2,99	6,46	21,12	161,17
Sum	29,38	26,47	29,73	16,30	9,91	6,74	6,43	5,92	4,38	4,18	8,09	22,42	169,95
Demande non satisfaite (Mètre cube)													
Demande agri	0,00	0,00	0,00	0,00	0,00	0,00	0,00	0,00	0,00	0,00	0,00	0,00	0,00
Demande pop	0,00	0,00	0,00	0,00	0,00	0,00	0,00	0,00	0,00	0,00	0,00	0,00	0,00
Sum	0,00	0,00	0,00	0,00	0,00	0,00	0,00	0,00	0,00	0,00	0,00	0,00	0,00
Exigence distribution (avec pertes, recycl., GSD) (Million Mètre cube)													
Demande agri	0,00	0,17	0,30	0,33	0,00	0,00	1,83	3,29	1,85	1,98	2,71	2,17	14,64
Demande pop	0,00	0,00	0,00	0,00	0,00	0,00	0,00	0,00	0,00	0,00	0,00	0,00	0,00
Sum	0,00	0,17	0,30	0,33	0,00	0,00	1,83	3,29	1,85	1,98	2,71	2,17	14,64

Figure 9. Données de référence moyennes mensuelles sur l'eau disponible et utilisée (Source : Modélisation avec WEAP).

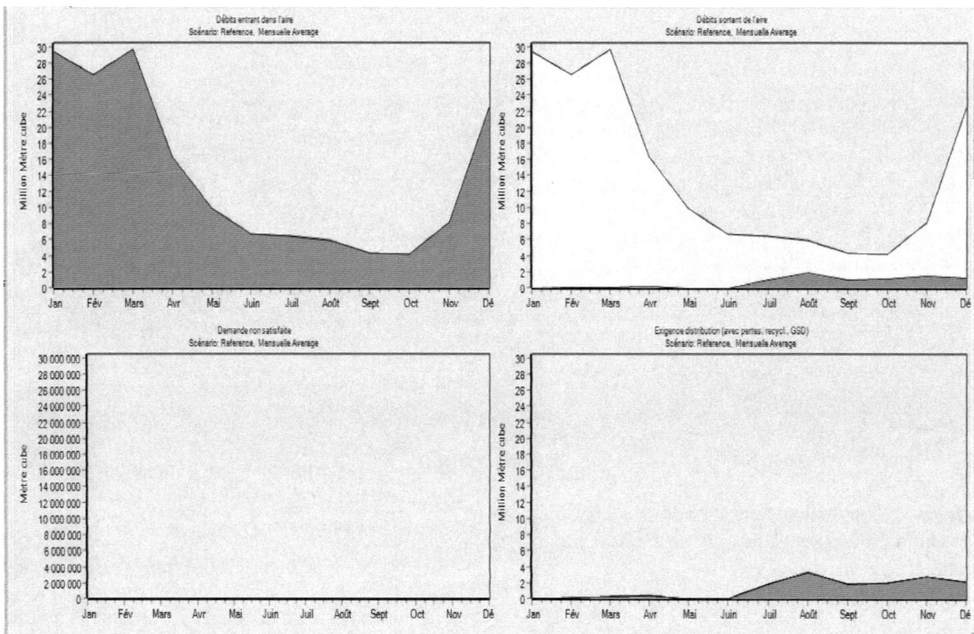

Figure 10. Débit entrant, débit sortant, demande non satisfaisante, exigence de distribution (Modélisation avec WEAP).

tation qui servira de stockage de l'eau de pluies au moment des crues pour l'utiliser après pour irriguer les rizières et les cultures vivrières. Promouvoir la riziculture pluvial. Et modifier la perception de la disponibilité en eau comme disponibilité de pluie.

5 Conclusion

Les ressources en eau de surfaces à Ambositra sont satisfaisantes, du point de vue quantitative. Cependant, tenant compte des réalités socio-économique et écologiques, il a été soulevé que la plupart de la population n'ont accès à l'eau potable au sein de leur habitations. Par ailleurs, la plupart de cette quantité élevée devient des eaux de ruissellement engendrant sur leur passage des érosions très fréquentes sur les

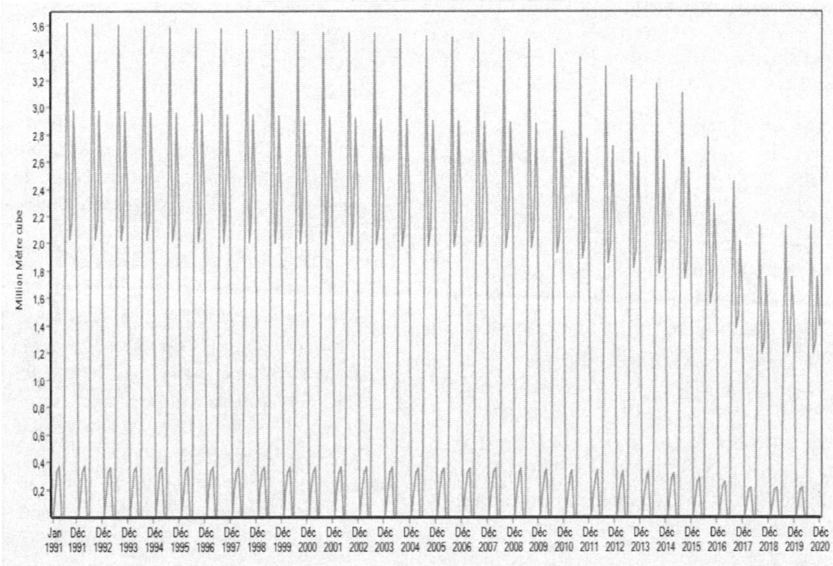

Figure 11. Débit de retour (non utilisée) (Modélisation avec WEAP).

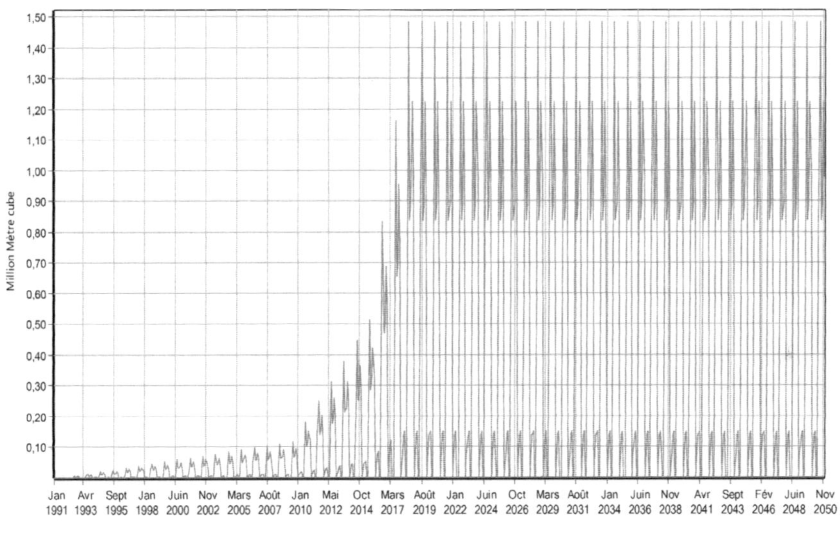

Figure 12. Modélisation de l'utilisation mensuelle de l'eau jusqu'en 2050 (Modélisation avec WEAP).

Hautes Terres de Madagascar. Ceci, du fait que la végétation ait disparu au profit des terrains dénudés, infertiles et hostiles à l'agriculture dans plusieurs cas, laissant place aux *tanety* [3] plein de *lavakas* [4]. Il n'est pas alors tard pour remettre en question les pratiques ancestrales et penser à la pérennisation de l'utilisation, en respect du principe d'équitabilité sociale et économique du développement durable, tout en tenant compte que sans couverture végétale, l'eau ne pourra être retenue. Cette étude est loin d'être complète, toutefois, elle ouvre une perspective assez large d'approfondissement de l'aspect des eaux souterraines, de la capacité d'infiltration, pour la résilience face aux crues et sécheresses.

3. Vaste étendue de collines dénudéss.

4. Fissures, grands trous causés par une forte érosion sur un tanety.

Disponibilité du code et des données. Codes and data are available upon request by contacting the correspondence author.

Collaborateurs. RFAR et DR ont assuré la conceptualisation, le montage du protocole expérimental ainsi que la redaction de l'article. TJT a apporté sa contribution dans les simulations. TR était l'administrateur du projet, LARDR pour la supervision.

Intérêts concurrents. Les auteurs déclarent qu'ils n'ont aucun conflit d'intérêts.

Clause de non-responsabilité. Publisher's note : Copernicus Publications remains neutral with regard to jurisdictional claims in published maps and institutional affiliations.

Déclaration du numéro spécial. This article is part of the special issue "Hydrology of Large River Basins of Africa". It is a result of the 4th International Conference on the "Hydrology of the Great Rivers of Africa", Cotonou, Benin, 13–20 November 2021.

Remerciements. We would like to thank all the co-authors, especially the Regional Director of Water, Sanitation and Hygiene of Amoron'i Mania. May this work find its place in the water management strategies of the Amoron'i Mania Region.

Références

Chaperon, P., Danloux, J., et Luc, F. : Fleuves et Rivières de Madagascar, Monographie hydrologique, IRD/DMH/CNREm IRD Éditions 2005, 548–557, accessible à l'adresse suivante : https://www.pseau.org/outils/ouvrages/ird_fleuves_et_rivieres_de_madagascar_1993.pdf (la date du dernier accès : 26 octobre 2021), 1993.

Code de l'Eau de Madagascar : J. O. no. 2557 E.S. du 27.01.99, p. 735, accessible à l'adresse suivante : http://www.droit-afrique.com/upload/doc/madagascar/Madagascar-Code-1999-Eau.pdf, (la date du dernier accès : 15 février 2020), 1999.

FAO : https://www.fao.org, la date du dernier accès : 20 Septembre 2019.

INSTAT : https://www.instat.mg (la date du dernier accès : 5 Janvier 2020), 2019.

Rakotoarisoa, D. : Gestion Integrée des Ressources en Eau du Bassin Versant de Soavina, Region Amoron'i Mania, Thèse de doctorat en Génie minéral et pétrolier, Ecole doctorale Ingénierie et Géosciences, ESPA, Université d'Antananarivo, Antananarivo, p. 46, 2018.

Rakotoarisoa, D., Rasolomanana, J., Ralaimaro, J., Andrimanaosoa, N., Rakoto, H., et Mandimbiharison, A. : Gestion intégrée des ressources en eau du Bassin versant de Soavina par télédétection et SIG, Région Amoron'i Mania Fianarantsoa, Revues Madamines, volume 1, ISSN 2220-068, accessible à l'adresse suivante : http ://madarevues.recherches.gov.mg/ ?Gestion-integree-des-ressources-en-eau-du-Bassin-versant (la date du dernier accès : 26 octobre 2021), 2018.

Rakotondrabe, F. : Etude de la vulnérabilité des ressources en eau aux changements climatiques, modélisation par le logiciel WEAP 21 : cas du bassin versant de Morondava (Sud-ouest de Madagascar), Mémoire de recherches pour l'obtention du diplôme d'etudes approfondies spécialisation : hydrogeologie, 70–83, accessible à l'adresse suivante : https://www.weap21.org/downloads/Madagascar.pdf (la date du dernier accès : 26 octobre 2021), 2007.

Randriamifidison, R. F. A., Rakotoarisoa, D., Tsiavahananahary, T. J., and Andriamampihantona, F. : La Région Amoron'i Mania face aux changements climatques : realités, adaptation, projection et orientations strategiques pour un développement durable, Acte du 32 ème Colloque de l'Association Internationale de Climatologie 2019, 45–50, accessible à l'adresse suivante : http://www.climato.be/aic/colloques/actes/Thessaloniki2019_actes.pdf (la date du dernier accès : 18 octobre 2021), 2019.

Proc. IAHS, 384, 313–318, 2021
https://doi.org/10.5194/piahs-384-313-2021

Open Access
PIAHS

A DPSIR and SAF Analysis of Water Insecurity in Lake Chad Basin, Central Africa

Lukman Adeboye Soboyejo[1], **Ahmad Mojisola Sakinat**[1], **and Abayomi Oluwatobiloba Bankole**[2]

[1]Department of Biological, Geological, and Environmental Science, University of Bologna, Ravenna Campus, Emilia-Romagna, 48123, Italy

[2]Department of Water Resources and Agricultural Meteorology, Federal University of Agriculture, 2240, Abeokuta, Nigeria

Correspondence: Lukman A. Soboyejo (lukeng200@gmail.com)

Published: 16 November 2021

Abstract. Lake Chad is a transboundary freshwater body located in the extreme south of the Sahara Desert. Many centuries ago, the synergies between nature and human activities in the basin were in harmony; and nowadays, the manifestation of unsustainable human activities and drier climate in the basin is now evident. This study assesses the water insecurity and associated environmental issues in the area using the combined Driver-Pressure-State-Impact-Response (DPSIR) and System-Approach-Framework (SAF) frameworks. In achieving this, we conducted literature review to establish the major effects and possible consequences of water scarcity in the area. The SAF defines the Lake boundaries and eventually links the active stakeholders involved. On the other hand, the DPSIR reveals that about 90 % of inhabitants depends on agriculture, with warming (temperature) significantly increasing in the basin (0.22 °C per decade) – socio-economic and natural Drivers respectively – as well as increase in population. Pressures include input from irrigation systems, river–flow modification, limited rainfall, and prolonged drought periods. These pressures have led to change in the state of Lake Chad, like freshwater shortages and loss of ecological status. Impacts on human welfares includes mass displacement, unemployment, paralyzed socio-economic activities, and social unrest. The societal-Response has prompted various measures like the launch of campaigns and awareness, planned Inter-Basin-Water-Transfer (IBTW), and several policy changes for better governance. Conclusively, the restoration of Lake Chad solely depends on climate conditions and management policies. However, this study recommends the prioritization of monitoring systems, water-allocation plans, ecological plans, and modelling tools for better decision processes.

1 Introduction: Issue Identification

Over 11 700 years ago, the surface area of Lake Chad covered approximately 350 000 km^2. At that time, the Lake's surface area was almost equal in size to the Caspian Sea, which is the largest Lake in the world (UNEP, 2004). The Lake Chad Basin (LCB) was a thriving socio-economic centre for the entire Sahel basin because of the lake (Vivekananda et al., 2019). However, in the last 50 years, Lake Chad and its river catchments are being withdrawn in an unsustainable way under unfavourable climate conditions (UNEP, 2004). So, the Lake was completely separated into two parts (after 1973), demarcated in the middle by a shoal barrier, and still fluctuating in size (Lemoalle et al., 2012). Inhabitants whose socio-economic activities depend on the Lake (>30 million people) started to witness a downturn in socio-economic activities, loss of ecological integrity, and loss of prominent fishing communities.

The change of state of Lake Chad prompted the coming together of the Lake Chad Basin Commission (LCBC) member states in 1964 to manage and rescue the shrinking freshwater Lake (Lemoalle et al., 2012). Despite the joint management, the waterbody still shrunk to 2500 km^2 (in 1987) but later increased to almost 14 000 km^2 (in 2013) due to slight increase in rainfall between 2000 and 2019 (Lemoalle and Magrin, 2014; Vivekananda et al., 2019). The significant rise in the ground and river water levels is still not evident in the

Published by Copernicus Publications on behalf of the International Association of Hydrological Sciences.

area (LCBC, 2016; International Conference on Lake Chad, 2018). Sadly enough, the future trend of rainfall in the Sahel region remains highly uncertain (Hoegh-Guldberg et al., 2018). This has generated misconceptions among the top regulators, influencers, researchers, and scientific communities as to whether Lake Chad will disappear or not. Measures (like dam projects, strategic action plans, and more) have been put in place, but have not yielded satisfactory outcomes, persuading the LCBC to re-visit the long-neglected project in 2018 (Inter-Basin Water Transfer – IBWT). This is seen as feasible and viable option by some Influencers and Regulators as a form of hope to revitalize Lake Chad (see Sayan and Nagabhatla, 2020 for detailed explanation).

2 Aim and Objectives

The purpose of this study is to analyse water insecurity in LCB by shedding more light on the causes and effects. The understanding of the problem is a significant prerequisite for tackling it. This study will go a long way in presenting a clearer picture of the problem to decision-makers, stakeholders, and co-researchers. The study objectives include (i) identifying the main actors responsible for the management of the shared water resources in the area, and (ii) determining the Drivers, the Pressures on Lake Chad, the present State, the Impact on human economic and social welfare, and the Responses.

3 Methodology

The methodology involves the use of SAF (explained in Newton and Elliott, 2016) and D(A)PSI(W)R framework (remodified by Elliott et al., 2017). These frameworks were combined to indicate water insecurity challenges in the LCB. In achieving this, we conducted literature review that included scholarly articles, regional and international assessment reports. We included climate change as crucial natural driver for this study, because Lake Chad's water resource is highly sensitive to climate anomalies. SAF was used to identify the main issues, define the boundary, and mapping out stakeholders. On the other hand, the D(A)PSI(W)R was used to identify the Driving forces, Human Activities, Pressures, physical and ecological States, Impacts (on Human-Welfare), and Responses.

Boundary Definition: Study Area

Lake Chad is located in West Central Africa and lies between latitude 6 and 24° N and longitude 7 and 24° E, covering about 2 434 000 km^2 (8 % of African landmass), but the hydrologically active land portion covers about 966 955 km^2 (Odada et al., 2005; UNEP, 2004) (Fig. 1).

4 Discussion on Water Insecurity Indicators

4.1 System Approach Framework (SAF)

According to Newton and Elliott (2016), stakeholder identification could be divided into six groups. In this study, five actors are considered: Extractors, Beneficiaries, Affectees, Regulators, and Influencers (see Table 1).

4.2 D(A)PSI(W)R Framework

4.2.1 Drivers (human activities, climate change, and population growth)

Water insecurity in the LCB is closely linked to climate change, migration, and socio-economic development. In Sub-Saharan Africa, agriculture as a major socio-economic driver of water scarcity accounts for about 60 % of the human occupation (Miletto et al., 2017). More so, in the LCB, about 90 % of agricultural jobs depend on the Lake for survival (FAO, 2017).

Another crucial factor is climate change, the inter-annual precipitation of LCB is slightly increasing in trend (Nagarajan et al., 2018) with future uncertainty (Hoegh-Guldberg et al., 2018). The first localised climate prediction study conducted by Rashid et al. (2019) showed that the annual temperature in the area has increased with a trend value of 0.22 °C per decade, while the precipitation trend has slightly decreased with a 14 mm per decade trend value. In the nearest future (2026–2035), the basin's temperature will increase by 0.6 to 1.6 °C, while precipitation will decrease by 5 % to 16 %.

Apropos to demographic growth, Sub-Saharan Africa will mark almost 3.8 billion (11 billion globally) by 2100. The growth statistics means that 1 out of 3 people across the globe will come from Sub-Saharan Africa (United Nations, 2019). This significant rise will certainly alter the way people advance water scarcity in the region.

4.2.2 Pressure

Intensive irrigation farming and other forms of water withdrawal practices were predominant in the basin between 1953 and 1979. The amount of water withdrawn from the Lake for irrigation alone quadrupled between 1983 and 1994 (Coe and Foley, 2001). In these periods, the volume of water dropped significantly (2500 km^2 in 1987) and water demands for irrigation doubled with expanding arable land (GWP, 2013), thus leading to direct human modification on tributaries.

4.2.3 State of the Environment

The aerial view of Lake Chad (Fig. 2) shows considerable change in the state and volume of water. Niger has access only to the northern part of the Basin and contains water

Table 1. Summary of the D(A)PSI(W)R and SAF frameworks, indicating water scarcity indicators and stakeholders, respectively.

	DRIVERS	ACTIVITIES	PRESSURES	STATES	IMPACTS	MAJOR RESPONSE
Human Driven	Agriculture (as socio-economic need)	Crop Farming (Irrigation) Livestock farming Domestic Use Navigation	Intensive Irrigation Water Diversion or Impoundment/ Dam Construction River Modification Population Growth	Loss of biodiversity The decline of landmark birds Loss of wintering or nesting areas for migratory birds Freshwater Shortages Disjointed water body Extinction of wildlife / livestock / fish species Invasive species Navigation problem Shoreline Retreat / Dune formation / Archipelago/ Wetlands	Disputes over emerging land and water resources Loss and change of job Loss of cultural heritage Loss of life arising from conflict Loss of source of income The low financial output from commerce and trade Loss of game reserves and national park Prone to affiliation with the terrorist group	Launching of campaigns to save Lake Chad Establishment of regional committees on critical matters Implementation of plans and projects Conferences, Seminars, Workshop/Training Sensitization and Awareness Strategic Action Programmes National Action Plans and Programmes Purported Inter Basin Water Transfer Governor's Forum Meeting
Climate Driven	Climate change (natural)	Rainfall Variation Drought and Desertification Temperature Rise Dry and Hot Wind Circulation (High Evapotranspiration)				
Others	Population growth					
SUMMARY OF SYSTEM APPROACH FRAMEWORK (SAF)						

Countries: Nigeria, Cameroun, Chad, Niger
Extractors, Beneficiaries, Affectees: Fishermen, Farmers, Herdsmen, Traders, Investors (Private, Public), Households, Water Vendors. (Tourists – Affectees)
Influencers: Media, NGOs, International Organization, Regional Bodies, Research Institutes, Colonial Countries, External Countries (like Government of Canada, Italy, France, and Germany)
Regulators: Local Actors (Local Government, Traditional rulers, Unions), National Actors (Ministries, State Ministries, Focal Ministries), Regional Actors (Lake Chad Basin Commission, LCBC)

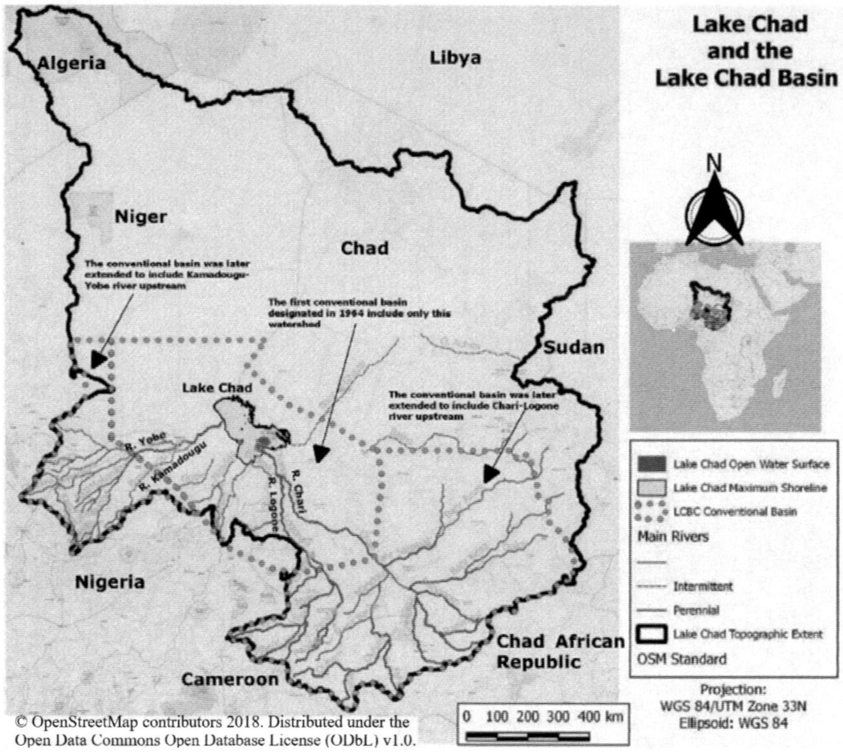

Figure 1. A map showing the conventional basin and the Lake Chad. Modified from International Conference on Lake Chad (© OpenStreetMap contributors 2018. Distributed under the Open Data Commons Open Database License (ODbL) v1.0).

every wet and dry season compared to the drought periods when both seasons often dry out completely (Binh et al., 2020; Vivekananda et al., 2019). Nigeria has a small portion of water points, while Cameroon and Chad still have more water within their territories. However, the extent and variation of surface water in the southern pool have maintained a stable trend since the 2000s. In contrast, the northern pool has slightly declined in extent due to annual and seasonal variability (Binh et al., 2020; Vivekananda et al., 2019). Furthermore, Binh et al. (2020), Lemoalle and Magrin (2014), Magrin (2016), and Vivekananda et al. (2019) claimed that the Lake will not disappear because the variability and extent of Lake Chad corroborates with historical and seasonal fluctuations. However, these researchers admit the negative effect of climate change on socio-economic activities, the lake water body extent and variability in space and time.

In reference to ecological status, the population of many indigenous flora and fauna species is threatened and endangered due to lowered water level (Hutchinson et al., 1992), leading to loss of ecological integrity and status.

4.2.4 Impacts (human welfare and ecological state)

In the Sahel region, irrigation is the only productive means by which inhabitants can settle down for a better livelihood (Miletto et al., 2017). Unfortunately, the Beneficiaries, Extractors, and Affectees in each riparian country have faced severe socio-economic problems due to freshwater shortages. Many households have consistently lamented about declined socio-economic activities. Also, they have seen the Lake retreated beyond its current shoreline. As a result, family members (especially the males) have migrated to possible job recruitment places for better lives (Odada et al., 2005; Vivekananda et al., 2019).

4.2.5 Responses

The sudden decline of Lake Chad and the resulting impact has raised concern among riparian member states. In the face of the increasing challenges, various management steps have been taken by the Regulators and Influencers to mitigate the effects and possible consequences (see Table 1). Unfortunately, the LCBC have been struggling with finance. To avoid this, the Commission adopted multi-donor approach; whereby Regulator's source for funds from Influencers (that is, external donors), member and conventional states, and through other means like levy systems, project management remuneration, and the sale of services (Odada et al., 2005).

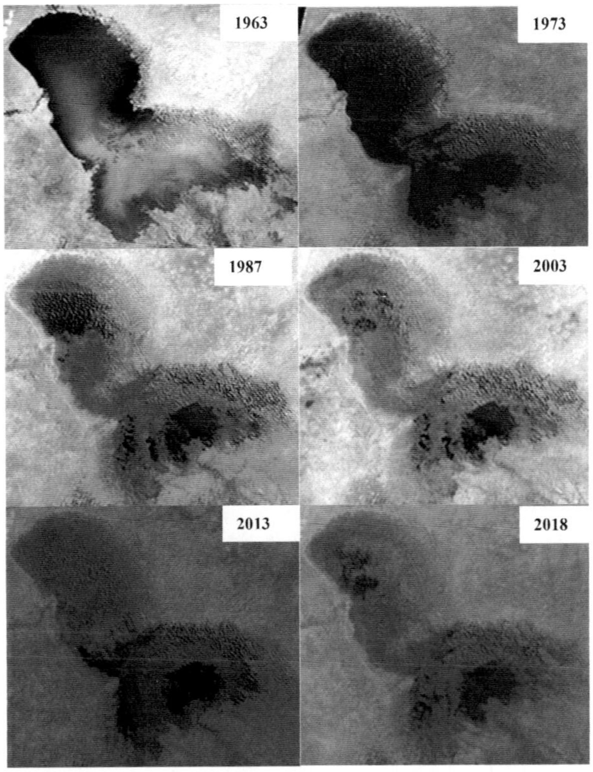

Figure 2. Chronological of changes in Lake Chad State from 1963 to 2018 (USGS, 2020).

5 Conclusion and Recommendation

The current surface area extent of Lake Chad is caused by the following set of Divers: population growth, human activities, and climate change, and management policies, leading to fluctuation and significant decline of the Lake. However, climate change is the primary determinant of Lake Chad's temporal and spatial water availability because of the imbalance in hydrological condition. The future recovery State of Lake Chad solely depends on future water management policies and climate change. Conclusively, we recommend the prioritization of monitoring systems, water-allocation plans and water release policy, ecological plans, and modelling tools for better decision processes.

Data availability. The manuscript is based on literature reviews and the use of assessment frameworks. SAF Framework – http://www.coastal-saf.eu/ (Coastal SAF, 2021). DPSIR Framework – see Elliott et al. (2017) for detailed explanation on usage.

Author contributions. Conceptualization of methodology, formal analysis, investigation, and writing (original draft preparation) were performed by the lead author LAS, while the final review and editing were carried out by LAS, AMS, AOB. All authors have read and agreed to the published version of the manuscript.

Competing interests. The contact author has declared that neither they nor their co-authors have any competing interests.

Disclaimer. Publisher's note: Copernicus Publications remains neutral with regard to jurisdictional claims in published maps and institutional affiliations.

Special issue statement. This article is part of the special issue "Hydrology of Large River Basins of Africa". It is a result of the 4th International Conference on the "Hydrology of the Great Rivers of Africa", Cotonou, Benin, 13–20 November 2021.

Acknowledgements. My [Lukman Adeboye Soboyejo] sincere gratitude goes to all Water and Coastal Management (WaCoMa) coordinators (https://wacoma.unibo.it/home.php, last access: 21 January 2021) as well as the European Union for the Erasmus Mundus Joint Master Degree (EMJMD) Programme and Scholarship provided as part of the 2018–2020 cohort at the University of Bologna – Italy, the University of Cadiz – Spain, and the University of Algarve – Portugal. More specifically, I am grateful to Alice Newton for exposing WACOMA students to DPSIR and SAF frameworks during the course of study, which has yielded this conference paper.

References

Binh, P. D., Sylvestre, F., Papa, F., Frappart, F., Bouchez, C., and Crétaux, J.-F.: The Lake Chad hydrology under current climate change, Scientific Reports, 10, 5498, https://doi.org/10.1038/s41598-020-62417-w, 2020.

Coastal SAF: Homepage, available at: http://www.coastal-saf.eu/, last access: 23 January 2021.

Coe, M. T. and Foley, J. A.: Human and natural impacts on the water resources of the Lake Chad basin, J. Geophys. Res.-Atmos., 106, 3349–3356, https://doi.org/10.1029/2000JD900587, 2001.

Elliott, M., Burdon, D., Atkins, J. P., Borja, A., Cormier, R., de Jonge, V. N., and Turner, R.: "And DPSIR Begat DAPSI(W)R(M)!" – A Unifying Framework for Marine Environmental Management, Mar. Pollut. Bull., 118, 27–40, https://doi.org/10.1016/j.marpolbul.2017.03.049, 2017.

Food and Agriculture Organization (FAO): Lake Chad Basin – Situation report July 2017, FAO – Food and Agriculture Organization of the United Nations, available at: https://www.fao.org/resilience/resources/resources-detail/en/c/1013218/ (last access: 21 October 2021), 2017.

Global Water Partnership (GWP): Transboundary Groundwater fact sheet, The Lake Chad Basin Aquifer System, Global Water Partnership, Global Secretariat, Sweden, 2013.

Hoegh-Guldberg, O., Jacob, D., Taylor, M., Bindi, M., Brown, S., Camilloni, I., Diedhiou, A., Djalante, R., Ebi, K. L., Engelbrecht, F., Guiot, J., Hijioka, Y., Mehrotra, S., Payne, A., Seneviratne, S. I., Thomas, A., Warren, R., and Zhou, G.: Impacts of 1.5 °C

Global Warming on Natural and Human Systems, in: Global warming of 1.5 °C. An IPCC Special Report on the impacts of global warming of 1.5 °C above pre-industrial levels and related global greenhouse gas emission pathways, in the context of strengthening the global response to the threat of climate change, edited by: Masson-Delmotte, V., Zhai, P., Pörtner, H. O., Roberts, D., Skea, J., Shukla, P. R., Pirani, A., Moufouma-Okia, W., Péan, C., Pidcock, R., Connors, S., Matthews, J. B. R., Chen, Y., Zhou, X., Gomis, M. I., Lonnoy, E., Maycock, T., Tignor, M., and Waterfield, T., World Meteorological Organization Technical Document, Geneva, Switzerland, 2018.

Hutchinson, C. F., Warshall, P., Arnould, E. J., and Kindler, J.: Development in Arid Lands Lessons from Lake Chad, Environ. Sci. Policy Sustain. Dev., 34, 16–43, https://doi.org/10.1080/00139157.1992.9931452, 1992.

International Conference on Lake Chad: Saving the Lake Chad to revitalize the Basin's ecosystem for sustainable livelihood, security, and development, available at: https://waterresources.gov.ng/lake-chad-conf/ (last access: 20 October 2021), 2018.

Lake Chad Basin Commission (LCBC): Report on the State of the Lake Chad Basin Ecosystem, GIZ Joint Report, Commission du Bassin du Lac Tchad (CBLT), Deutsche Gesellschaft für Internationale Zusammenarbeit (GIZ) GmbH, Bonn and Eschborn, Germany, 233 pp., 2016.

Lemoalle, J., Bader, J.-C., Leblanc, M., and Sedick, A.: Recent changes in Lake Chad: observations, simulations and management options (1973–2011), Glob. Planet. Change, 80–81, 247–254, https://doi.org/10.1016/j.gloplacha.2011.07.004, 2012.

Lemoalle, J. and Magrin, G.: Le développement du lac Tchad / Development of Lake Chad: Situation actuelle et futurs possibles / Current Situation and Possible Outcomes, Expertise collégiale, IRD Éditions, Marseille, France, 216 pp., 2014.

Magrin, G.: The disappearance of Lake Chad: history of a myth, J. Polit. Ecol., 23, 204–222, https://doi.org/10.2458/v23i1.20191, 2016.

Miletto, M., Caretta, M. A., Burchi, F. M., and Zanlucchi, G.: Migration and its interdependencies with water scarcity, gender, and youth employment, UNESCO Publishing, Paris, France, 34 pp., 2017.

Nagarajan, C., Pohl, B., Rüttinger, L., Sylvestre, F., Vivekananda, J., Wall, M., and Wolfmaier, S.: Climate-fragility profile: : Lake Chad basin, Adelphi research gemeinnützige GmbH, Berlin, Germany, 31 pp. 2018.

Newton, A. and Elliott, M.: A Typology of Stakeholders and Guidelines for Engagement in Transdisciplinary, Participatory Processes, Front. Mar. Sci., 3, 230, https://doi.org/10.3389/fmars.2016.00230, 2016.

Odada, E., Oyebande, L., and Oguntola, A. J.:. Lake Chad: experience and lessons learned brief, Lake Environment Committee Foundation, Kusatsu, Japan, 18 pp., 2005.

Rashid, M., Shaofeng, J., and Zhu, W.: Analysis of climate variability, trends, and prediction in the most active parts of the Lake Chad basin, Africa, Scientific Reports, 9, 6317, https://doi.org/10.1038/s41598-019-42811-9, 2019.

Sayan, R. C. and Nagabhatla, N.: Soft Power, Discourse Coalitions, and the Proposed Interbasin Water Transfer Between Lake Chad and the Congo River, Water Altern., 13, 752–778, 2020.

United Nation Environment Programme (UNEP): Lake Chad Basin, edited by: Fortnam, M. P. and Oguntola, J. A., Global International Water Assessment (GIWA – Regional Assessment-43), University of Kalmar, Kalmar, Sweden, 2004.

United Nations (Department of Economic and Social Affairs): Population Division: World Population Prospects 2019 – Highlights, ST/ESA/SER.A/423, United Nation Publisher, New York, USA, 2019.

United States Geological Survey (USGS): Lake Chad, West Africa, Earth shots: Satellite Images of Environmental Change, available at: https://earthshots.usgs.gov/earthshots/Lake-Chad-West-Africa#ad-image, last access: 15 December 2020.

Vivekananda, J., Wall, M., Sylvestre, F., Nagarajan, C., and Brown, O.: Shoring up Stability; Addressing Climate and Fragility Risks in the Lake Chad Region, adelphi research gemeinnützige GmbH, Berlin, Germany, 101 pp., 2019.

Proc. IAHS, 384, 319–324, 2021
https://doi.org/10.5194/piahs-384-319-2021

Open Access
PIAHS

Stratégies d'adaptation des producteurs agricoles aux risques hydroclimatiques dans le bassin Béninois de la pendjari à l'exutoire de Porga

Pierre Ouassa[1,3], Gervais Assaï Akanni Atchadé[1], Domiho Japhet Kodja[1], Fernand Ringo Avahounlin[2,3], and Expédit Wilfrid Vissin[1]

[1]Laboratoire Pierre Pagney : Climat, Eau, Ecosystèmes et Développement (LACEEDE), Université d'Abomey-Calavi, Bénin
[2]Ecole Normale Supérieure (ENS, UNSTIM, Rép. du Bénin), Bénin
[3]Chaire Internationale de Physique Mathématique et Applications (CIPMA CHAIRE-UNESCO, FAST, UAC, Rép. du Bénin), Bénin

Correspondence: Pierre Ouassa (ouaspeter@yahoo.fr)

Published: 16 November 2021

Résumé. West Africa's dependence on the increased impacts of extreme hydroclimatic hazards increases the vulnerability of food production to the resulting risks. This study analyses the adaptation strategies developed by agricultural producers in the Beninese Pendjari basin to cope with hydroclimatic risks. The methodological approach adopted consisted in the use of climatological data (height of daily and monthly rainfall) from the Tanguiéta and Natitingou stations over the period from 1965 to 2017 obtained in Météo-Bénin . Hydrological data consisting of the daily flows of the river calculation of the Pendjari indices at the Porga outlet, over the period 1965–2017 are extracted from the database of DG-Water and ltulization of socio-anthropological data. To better appreciate the adaptation strategies developed, a semi-structured survey that involved 121 people in 21 villages of the Beninese Pendjari basin is conducted according to the CAP approach (Knowledge, Attitudes, Practices). Direct observations in the field have also made it possible to better identify peasant adaptation strategies to the effects of hydroclimatic risks in the Beninese Pendjari basin. The information collected during the surveys was processed with Sphinx software. We studied socio-anthropological data, and we showed that Pendjari watershed in Benin is under the threat of hydroclimatic risks. To reduce their vulnerability, producers in the Benin Pendjari basin are developing several risk prevention and disaster management strategies. These include transplanting crops including sorghum by 100 % of the producers met, the disposal of logs, the enhancement of the shallows. In view of the effectiveness of the strategies identified and the limitations they present, it is important that the authorities develop a policy that is more sensitive to hydroclimatic risks, in order to improve the adaptation capacity of agricultural producers.

1 Introduction

Le Sahel et l'Afrique de l'Ouest sont, selon le GIEC (2007), parmi les régions les plus vulnérables aux fluctuations climatiques futures. Ainsi, la gestion des évènements extrêmes (inondations, sècheresses) est une des préoccupations majeures de la politique de l'eau (Hamdane, 2015). Dans le bassin de la Pendjari on observe une tendance à la baisse de la pluviométrie et un déficit pluviométrique pendant les décennies de 1971 à 2000 (Idiéti, 2012). Ce déficit a eu pour conséquence la baisse des débits allant de 30 à 50 % (Kimba, 2011).

Les conséquences de la variation du régime hydrologique sont entre autres la difficulté d'approvisionnement en eau, la chute de la production agricole, la diminution des surfaces irriguées (Vissin, 2007). Face à cette situation, les paysans du bassin Béninois de la Pendjari ont développé les stratégies pour faire face aux effets des risques hydroclimatiques. Ainsi, cette étude vise à analyser les stratégies d'adaptation développées par les producteurs agricoles du bassin Béninois de la Pendjari pour faire face aux risques hydroclimatiques.

2 Secteur d'étude

Le bassin Béninois de la Pendjari est situé entre les parallèles 10°21′ et 10°46′′ de latitude nord et les méridiens 1°12′ et 1°38′ de longitude est. Il est sous l'influence du climat soudanien (climat tropical chaud et humide) ou soudano-guinéen qualifié d'atacorien (Houndénou, 1999). En outre, il est caractérisé par l'alternance d'une saison sèche et d'une saison pluvieuse. La saison pluvieuse dure 5 à 6 mois (mai à octobre) est dominée par les flux de mousson, mais aussi ceux liés aux lignes de grains, génératrices de fortes averses sur la région. La pluviosité est de 1131.3 mm. Les mois de juillet, août et septembre sont les plus pluvieux avec une moyenne de 227.5 mm par mois. Les températures restent élevées toute l'année, mais elles ne sont jamais excessives. Les moyennes annuelles varient de 26′8°C à 27.7 °C, soit une augmentation de 0.9 °C sur la période 1965–2017. Ces moyennes, constamment élevées et presque homogènes pour l'ensemble de la région ne constituent pas un facteur limitant pour les activités agricoles. La Fig. 1 présente la situation géographique du bassin béninois de la Pendjari.

3 Données et méthodes

3.1 Données utilisées

Les hauteurs de pluies de la station de Tanguiéta et de Natitingou extraites de la base de données de Météo-Bénin et les débits mensuels de la station de Porga obtenu à la Direction Générale de l'Eau à Cotonou sur la période 1965–2017 ont été utilisées en plus des informations qualitatives (perceptions paysannes et effets des risques hydroclimatiques sur la production agricole, stratégies d'adaptation paysannes) recueillies lors des investigations socio-anthropologiques.

3.2 Méthodes utilisées

La variabilité hydro-pluviométrique a été déterminée à partir des anomalies centrées réduites. Le calcul des indices pluviométriques a permis d'identifier les années humides (excédentaires) au seuil de 1 et les années sèches (déficitaires) au seuil de −1. Ces indices sont obtenus par la formule suivante : $x_{i'} = \frac{x_i - \overline{x}}{\sigma(x)}$ (1) avec x représentant le total pluviométrique de l'année i, la moyenne de la série et σ représentant l'écart-type. Les tendances ont été mises en évidence par une droite de régression de type : $y = a \times +b$; elle est obtenue par le calcul de la pente a qui est un coefficient directeur.

Le choix des villages dans lesquels les enquêtes ont été menées s'est fait en privilégiant les villages riverains aux cours d'eau. Ils sont identifiés sur la base des cartes de relief et hydrographiques du bassin. Quant aux personnes enquêtées, le choix s'est fait sur la base de l'ancienneté dans la localité, vingt (20) ans au moins. L'enquêté est âgé d'au moins trente (30) ans et est acteur dans la production agricole. Au total 126 personnes dont 105 producteurs agricoles et 21 personnes ressources ont été enquêtées dans 21 villages dans le cadre cette étude. Les entretiens individuels ont été réalisés en utilisant les questionnaires pour recueillir les informations auprès des personnes ressources. Un focus groupe sur la base d'un guide d'entretien au niveau de chaque village est fait pour croiser les données recueillies et mieux identifier les meilleures stratégies d'adaptation aux risques hydroclimatiques. L'approche CAP (Connaissances, Attitudes, Pratiques) est utilisée pour collecter les informations relatives aux problèmes liés aux risques hydroclimatiques et appréhender les perceptions et les savoirs des communautés.

Les observations directes sur le terrain ont permis de mieux identifier les stratégies d'adaptation paysannes aux effets des risques hydroclimatiques dans le bassin. béninois de la Pendjari. Les informations recueillies au cours des enquêtes ont été traitées avec le logiciel Sphinx. L'ensemble de ces travaux réalisés a permis d'obtenir les résultats suivants.

4 Résultats

4.1 Tendance hydroclimatique dans le bassin Béninois de la Pendjari à Porga

La Fig. 2 présente la tendance d'évolution des hauteurs de pluies annuelles dans le bassin béninois de la Pendjari à Porga.

La moyenne annuelle des hauteurs de pluie sur la période 1965–2017 est de 1207 mm avec un écart de 186 mm. Cette figure 2 permet de distinguer trois principales séquences (1965–1969, 1970–1991 et 1992–2017). Ainsi, on retient que sur la période 1965 à 1969, la plupart des anomalies sont positives, ce qui correspond à la période humide. Inversement, sur la période 1970 à 1991, il y a une dominance des anomalies négatives traduisant ainsi les années déficitaires, ce qui correspond à la période sèche. La période 1992 à 2017 est marquée par une alternance d'années déficitaires et excédentaires, correspondant ainsi à la période de reprise des hauteurs pluviométriques.

Par ailleurs, les valeurs des débits au-dessus de la moyenne enregistrées à la station hydrométrique de Porga, ont permis d'identifier les années de crues qui sont les causes des inondations au cours de la période 1965–2017 (Fig. 3).

L'analyse de la Fig. 3 révèle que le bassin béninois de la Pendjari a connu 23 années de crues au cours de la période 1965–2017. Les débits les plus fortes ont été enregistrés en 1969 (588.79 m^3/s) et 1970 (625.2 m^3/s). Les crues les plus faibles ont été enregistrées en 2001 (286.29 m^3/s) et en 2005 (283.6 m^3/s). L'analyse de la Fig. 3 permet de constater également que le bassin enregistre ces dernières décennies de faibles débits et par conséquent des inondations moins fortes.

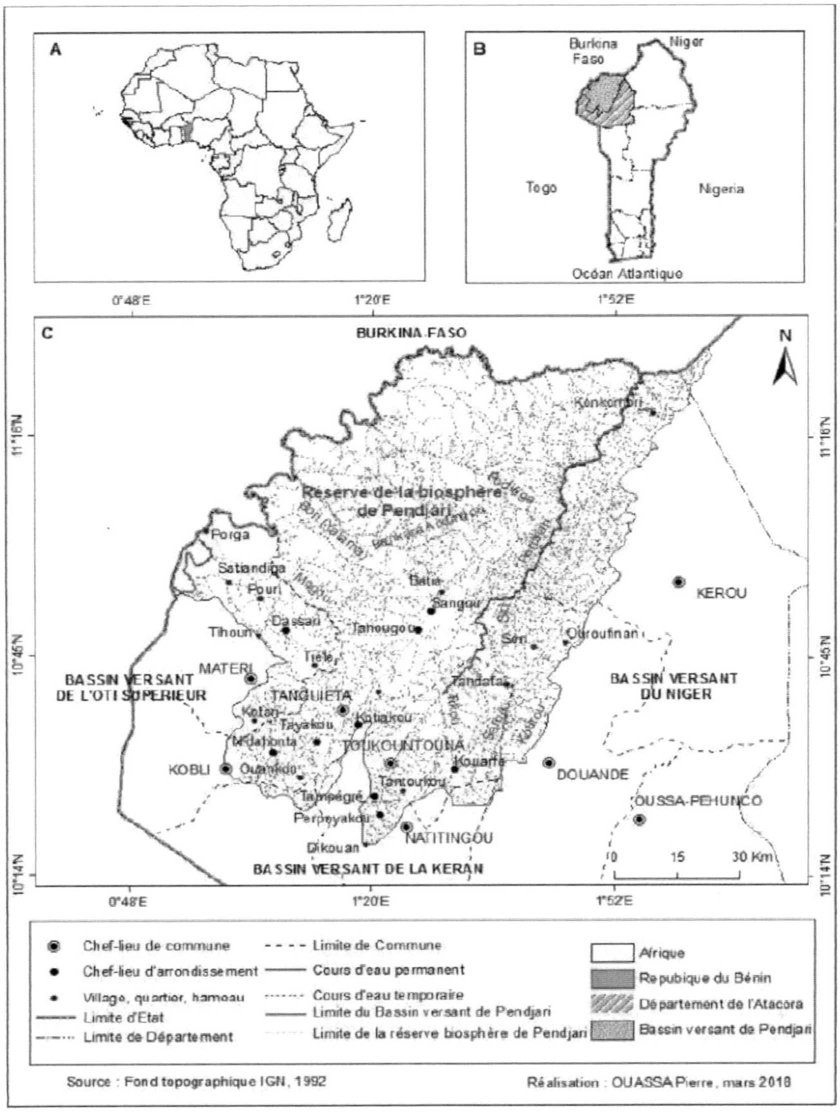

Figure 1. Situation géographique du bassin Béninois de la Pendjari à l'exutoire de Porga.

Ce qui se justifie par la tendance à la baisse des débits maximaux dans le bassin béninois de la Pendjari.

4.2 Stratégies d'adaptation paysannes face aux risques hydroclimatiques

4.2.1 Stratégies d'adaptation aux inondations

Dans le but de prévenir une éventuelle irrégularité des pluies au cours de la saison agricole, les paysans enquêtés orientent les billons de façon perpendiculaire ou parallèle à la pente selon le cas. La photo 1 montre la disposition des billons de façon parallèle à la pente dans le bassin.

Lorsque les cultures sont pratiquées sur un sol plus ou moins perméable, les paysans disposent les billons dans le sens contraire à la pente. Ce qui permet à l'eau tombée de stagner dans les sillons et de s'infiltrer progressivement. De plus, si les billons sont minces, cela assure une bonne alimentation des cultures en eau et les racines fournissent moins d'effort pour puiser les réserves d'eau. Lorsque, les cultures sont réalisées au bord des rivières, les billons réalisés sont de grandes tailles et parallèles à la pente. La disposition des billons de façon parallèle à la pente, favorise selon les paysans, l'évacuation des excès en eau en cas de pluviométrie exceptionnelle et empêche la destruction des billons, le pourrissement des racines lorsque les inondations surviennent.

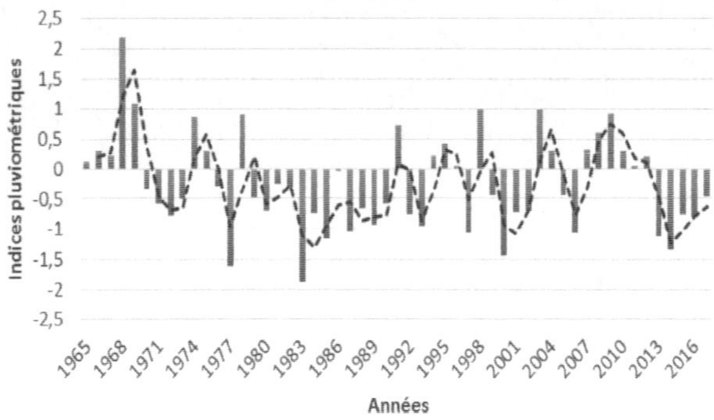

Figure 2. Indices pluviométriques au cours de la période 1965–2017 dans le bassin Béninois de la Pendjari à l'exutoire de Porga. Sources : Météo-Bénin, et traitement des données, 2020.

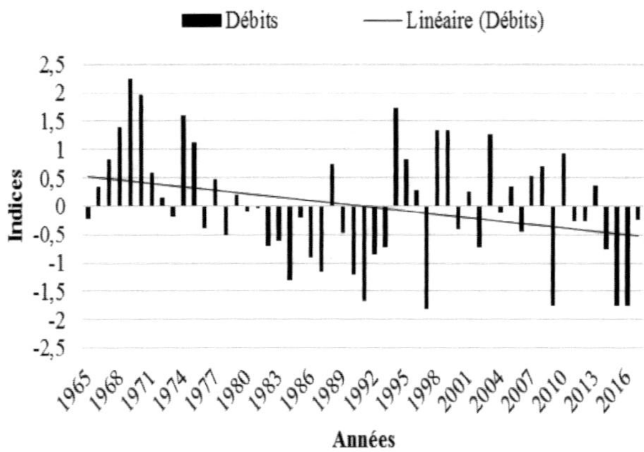

Figure 3. Variabilité interannuelle des débits dans le bassin Béninois de la Pendjari à Porga.

4.2.2 Stratégies d'adaptation à la sécheresse

Les stratégies les plus développées pour faire face à la sécheresse sont entre autres :

– *Repiquage des cultures*

Lorsque le calendrier agricole ne permet pas aux paysans de semer, ils procèdent au repiquage des cultures. Selon les producteurs interviewés, avant de procéder au repiquage il faut faire de nouveaux billons. En faisant ainsi, la culture plantée a plus de facilité à s'implanter dans le sol et donc a une meilleure croissance par rapport à celle plantée sans labourer, pouvant par contre rencontrer un sol plus ou moins compact qui ralentira sa croissance. Le repiquage (photo 2) concerne généralement les cultures de sorgho et de mil selon 78 % des enquêtés.

Cette pratique permet aux producteurs, de rattraper le calendrier agricole. En effet, cette stratégie permet aux paysans de se passer de la phase de semis pour permettre à la culture d'amorcer directement la phase de croissance.

Figure 4. Photo 1 : Disposition des billons de façon parallèle à la pente à Toucountouna. Prise de vue : P. Ouassa, Août 2015.

– *Mise en valeur des bas-fonds*

La sécheresse des années 1970 a révélé tout le profit qui pouvait être tiré des agro-systèmes de bas-fonds et les paysans ont spontanément commencé à les mettre en valeur (Idani, 2012). Les populations ne craignent plus le risque de voir leurs champs inondés. En effet, avec la baisse des pluies et la fin précoce de la saison des pluies, les bas-fonds constituent les meilleures terres de recours à cause de leurs fertilités et de la disponibilité en eau du sol. La photo 3 montre un bas-fond mis en valeur dans le bassin béninois de la Pendjari.

L'eau recueillie par ces bas-fonds après les pluies sert à alimenter le riz pendant les périodes de récession pluviométrique. Ceci permet de maintenir humide la zone cultivée pendant l'absence des pluies et d'évacuer les excès d'eau pour éviter les inondations. Somme toute, même si ces stratégies des producteurs agricoles sont peu efficaces dans leur mise en œuvre, par manque d'outils moderne, elles témoignent de la capacité des producteurs à s'adapter aux risques hydroclimatiques.

5 Discussion

Les changements climatiques se révèlent un phénomène unique sur l'ensemble de la planète, mais les causes sont réparties sur toute la surface du globe de façon inégale selon les régions ou les Pays ; ses conséquences connaîtront une répartition sans rapport avec les causes (Ogouwalé, 2006). De ce fait, pour réduire la vulnérabilité de leurs cultures vivrières à la survenance des catastrophes hydroclimatiques, les producteurs du bassin béninois de la Pendjari développent plusieurs stratégies de préventions des risques et de gestion des catastrophes. Il s'agit entre autres du repiquage des cultures no-

Figure 5. Photo 2 : Repiquage du sorgho dans le bassin Béninois de la Pendjari. Prise de vue : P. Ouassa, Septembre 2018.

tamment le sorgho, de la disposition des billons, de la mise en valeur des bas-fonds etc. Ce résultat montre que la population est réellement consciente de l'existence des risques climatiques pouvant impacter leurs activités. Ce qui corroborent avec les résultats de Nago et al. (2019), qui montrent que les prières et sacrifices sont des stratégies d'atténuation actuellement utilisées. Cependant Ouorou-Barrè (2014), indiquent que les paysans savent que leurs pratiques endogènes d'adaptation ne sont que des expérimentations compte tenu du degré d'incertitude qui entoure les prévisions sur le changement climatique. De ce fait, ces stratégies sont susceptible d'engendrer des pressions sur les ressources naturelles et de fragiliser davantage les conditions de vie des populations (Oloukoï et al., 2019). Il est donc nécessaire de développer des stratégies plus efficaces à l'instar de celles identifiées dans la présente étude. En effet, la création de variétés précoces et résistantes à la sècheresse prendra en compte la préoccupation de certains paysans qui estiment que les variétés précoces disponibles actuellement ne résistent pas aux longues séquences sèches. De même, l'exploitation efficace des bas-fonds exige de matériels appropriés et expose les paysans à des risques pathologiques (le paludisme, la bilharziose, l'onchocercose).

Figure 6. Photo 3 : Différentes formes de mise en valeur des bas-fonds dans le bassin Béninois de la Pendjari. Prise de vue : P. Ouassa, Août et Septembre 2015.

6 Conclusions

Au terme de cette étude, il faut retenir que pour réduire la vulnérabilité de leurs cultures vivrières à la survenance des catastrophes hydroclimatiques, les producteurs du bassin béninois de la Pendjari développent plusieurs stratégies de préventions des risques et de gestion des catastrophes. Ces différentes stratégies d'adaptation dans leur ensemble, présentent des limites et contraintes dans leur mise en œuvre.

Sur cette base, les pratiques à promouvoir doivent être proposées de façon concertée et en tenant compte des réalités agro-écologiques et socioéconomiques des bénéficiaires si la dimension de durabilité des stratégies veut être atteinte. Ainsi, la nécessité de rendre plus accessible et facilement réalisable ces stratégies d'adaptation contribueront certainement à rendre plus durable les effets de ces pratiques d'une part, et d'autres part faciliterait leur diffusion. Sur cette base, l'insécurité alimentaire des ménages vivant dans le bassin béninois de la Pendjari liée à la variabilité hydroclimatique pourrait être boutée.

Disponibilité du code. Pour la rédaction de cet article, nous avons pas utilisé de logiciel. Seulement Excel, a été utilisé pour la réalisation des graphiques.

Disponibilité des données. Les données sur les hauteurs pluviométriques de la période de 1965 à 2017 sont obtenus à l'Agence Nationale de Météorologie du Bénin. Les données sur les Débits du cours d'eau Pendjari sont obtenus à la Direction Générale de l'Eau. Ces deux données bien-qu'elles soient accessible au public ne sont pas disponible à ligne. Les données socio-économiques concernant les stratégies d'adaptation ont été directement collectées par nous mêmes auprès des populations.

Collaborateurs. PO a conçu l'étude, collecté et analysé les données, et a rédigé le manuscrit. GAAA, DJK, FRA et EWV ont contribué à l'analyse des données, l'amélioration de l'approche méthodologique et l'orientation scientifique et la relecture de l'article.

Intérêts concurrents. Les auteurs déclarent qu'ils n'ont aucun conflit d'intérêts.

Clause de non-responsabilité. Publisher's note : Copernicus Publications remains neutral with regard to jurisdictional claims in published maps and institutional affiliations.

Déclaration du numéro spécial. This article is part of the special issue "Hydrology of Large River Basins of Africa". It is a result of the 4th International Conference on the "Hydrology of the Great Rivers of Africa", Cotonou, Benin, 13–20 November 2021.

Remerciements. Les auteurs remercient les évaluateurs anonymes qui ont sensiblement contribué à l'amélioration de la qualité scientifique de cet article en consacrant leurs précieux temps à son évaluation critique.

Références

GIEC : Bilan 2007 des changements climatiques. Contribution des Groupes de travail I, II et III au quatrième Rapport d'évaluation du Groupe d'Intergouvernemental sur l'évolution du climat, GIEC, Genève, Suisse, 103 pp., 2007.

Hamdane, A. : Changement climatique et sécurité alimentaire en méditerranée et en Afrique de l'Ouest : Stratégies pour une agriculture, des territoires et une croissance durable-Etude du cas Tunisien., Séminaire SESAME, 25 pp., 2015.

Houndénou, C. : Variabilité climatique et maïsiculture en milieu tropical humide : l'exemple du Bénin, diagnostic et modélisation, Thèse de doctorat de géographie, UMR 5080, CNRS « climatologie de l'espace tropical », Université de Bourgogne, centre de recherche de climatologie, Didjon, 341 p.,1999.

Idiéti, M. E. : Les hydro-écorégions du Bassin de la Pendjari au Bénin : Analyse des déterminants socio-économiques et environnementaux de la dynamique des écosystèmes naturels, Thèse de doctorat de géographie, Université d'Abomey-Calavi, 229 pp., 2012.

Kimba, Y. G. : Impact de la variabilité pluviométrique sur les ressources en eau dans le bassin de la Pendjari, Mémoire de maîtrise de géographie, FLASH, UAC, 67 pp., 2011.

Nago, S. G. A., Gnohossou, P., Sagbo, R. R. S., et Bokonon-Ganta, E. : Perception du changement climatique et stratégies locales d'adaptation dans la pêcherie de la Réserve de Biosphère de la Pendjari, Bénin, Afrique SCIENCE, 15, 114–127, 2019.

Ogouwalé, E. : Changements Climatiques dans le Bénin méridional et central : indicateurs, scénarios et prospectives pour la sécurité alimentaire, Thèse de doctorat nouveau régime, EDP/FLASH, 301 pp., 2006.

Oloukoï, J., Yabi, I., et Houssou, S. C. : Perceptions et stratégies paysannes d'adaptation à la variabilité pluviométrique au Centre du Bénin, Int. J. Biol. Chem. Sci., 13, 1366–1387, 2019.

Ouorou-Barrè, I. F. : Contraintes climatiques, pédologiques et production agricole dans l'Atacora (Nord-Ouest Bénin), Thèse de doctorat de l'Université d'Abomey-Calavi ; Bénin., FLASH/ UAC., 241 pp., 2014.

Vissin, E. W. : Impact de la variabilité climatique et de la dynamique des états de surface sur les écoulements du bassin béninois du fleuve Niger, Thèse de Doctorat, Université de Bourgogne, Dijon, 310 pp., 2007.

Proc. IAHS, 384, 325–330, 2021
https://doi.org/10.5194/piahs-384-325-2021

Open Access
PIAHS

Hydrology of Large River Basins of Africa

Trajectoires de l'agriculture familiale face aux changements climatiques dans la zone agro-écologique 3 du Bénin

Biaou Ibidun Hervé Chabi, Akibou A. Akindele, Hervé Yabi, and Euloge Ogouwale

Laboratoire Pierre PAGNEY « Climat, Eau, Ecosystèmes et Développement » (LACEEDE), Abomey-Calavi, BP 15 Allada, Bénin

Correspondence: Biaou Ibidun Hervé Chabi (chabiherv@yahoo.fr)

Published: 16 November 2021

Résumé. La présente recherche décrit la trajectoire de l'agriculture familiale en lien avec les changements climatiques dans la Zone Agro-Ecologique 3 du Bénin (ZAE-3). Les données utilisées sont obtenues auprès des exploitants agricoles grâce à la Méthode d'Investigation Socio-Historique. L'analyse de la trajectoire de l'agriculture familiale a été faite suivant trois générations d'agriculteurs. Ainsi, la Classification Hiérarchique Ascendante et l'Analyse en Composantes Principales sont mises à contribution pour l'analyse des principales étapes d'évolution des exploitations agricoles familiales (EAF). Face aux effets des changements climatiques, les exploitants agricoles ont mis en œuvre diverses mesures ayant favorisé l'adaptabilité de l'agriculture familiale. Au-delà de leurs mérites, lesdites mesures présentent de limites qui se résument à leur inefficacité face aux événements climatiques extrêmes. Des mutations et transformations (techniques et sociales) profondes sont intervenues dans le processus d'évolution de l'agriculture familiale. Ainsi, dans le cadre de l'adaptation aux changements climatiques, quatre (4) grandes étapes ont marqué l'évolution des EAF à savoir le maintien du système traditionnel de production (13 %), l'agrandissement de la taille de l'exploitation (18 %), la tendance à l'intensification et l'innovation au sein des EAF (19 %) et la diversification des activités agricoles (49 %).

1 Introduction

Le système climatique dans lequel s'inscrit l'Afrique de l'Ouest en général, et le Bénin en particulier, subit des modifications à grande échelle qui restent amplifiées par les facteurs naturels et anthropiques tant régionaux que locaux. Ainsi, les climats ouest africains et béninois sont sujets à de fortes variabilités ou à des changements selon les échelles de temps et d'analyse dont les conséquences restent néfastes pour le développement durable (Boko, 1988 ; Afouda, 1990 ; Ogouwalé, 2006). Pourtant, au Bénin, l'agriculture, principalement celle de type familial, repose sur de petites exploitations familiales paysannes orientées vers la polyculture. Environ 34 % des exploitations couvrent moins de 1 ha (PAM, 2014).

Dans la Zone Agro-Ecologique 3 (ZAE-3) du Bénin, le caractère pluvial de l'agriculture familiale la rend vulnérable aux effets des changements climatiques. Mais, de par ses caractéristiques intrinsèques, elle s'adapte plus ou moins aux changements observés. Ainsi, malgré les changements climatiques révélés par de nombreuses recherches scientifiques, les petits producteurs continuent d'exister et d'exercer leur activité de production. Cela s'explique par le fait que ceux-ci ont su s'adapter aux contraintes qui s'imposent par des mesures mises en œuvre en fonction du temps. Ainsi, le contexte climatique dans lequel l'agriculture familiale a évolué, amène à se demander quelle est la trajectoire de l'agriculture familiale face aux changements climatiques dans la ZAE-3 du Bénin ? La présente recherche analyse les principaux facteurs favorisant l'adaptabilité de l'agriculture familiale ainsi que les grandes étapes de son évolution face aux effets des changements climatiques.

La ZAE 3 est située entre 9°11′ et 10°55′ de latitude nord et entre 1°24′ et 3°47′ de longitude est. Cette zone est composée de huit Communes à savoir : Bembèrèkè, N'Dali, Nikki, Kalalé, Kouandé, Péhunco, Pèrèrè et Sinendé (Fig. 1).

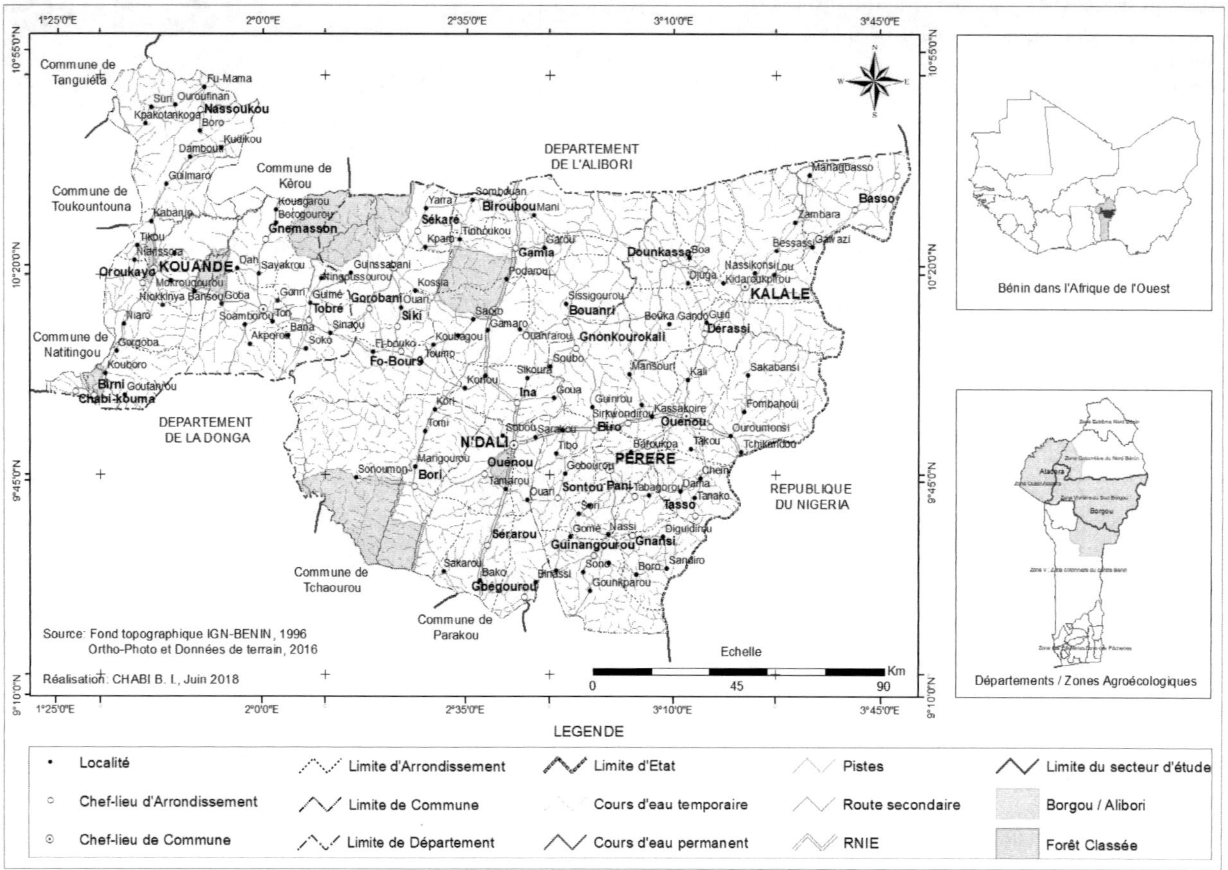

Figure 1. Situations géographique et administrative de la ZAE 3 du Bénin.

2 Données et méthodes

La Méthode d'Investigation Historico-Sociologique (MIHS) inspirée des travaux de Chabi et al. (2019), est utilisée pour la collecte des données sur la trajectoire de l'agriculture familiale. La période retenue pour l'analyse va de 1951 à 2015. En effet, pour apprécier la dynamique d'adaptabilité de l'agriculture familiale aux changements climatiques, la collecte de données socio-anthropologiques sur l'évolution de l'agriculture familiale a été faite auprès de trois générations d'agriculteurs. L'approche descriptive est utilisée pour l'analyse de la trajectoire de l'agriculture familiale. Celle-ci a permis de mettre l'accent sur ses caractéristiques pour mieux déterminer les facteurs de son évolution en lien avec les changements climatiques.

Ainsi, pour décrire les changements historiques intervenus dans la structure des exploitations agricoles, tout en minimisant les erreurs potentielles de données en raison de l'échec dans la mémoire des agriculteurs, les états successifs des exploitations et des ménages ont été décrits suivant des générations à l'aide d'un ensemble de variables qualitatives et quantitatives. Pour décrire l'évolution des caractéristiques des exploitations, un pas de temps de dix ans, permettant de capter les évolutions principales sur la période d'étude a été utilisé par Ryschawy et al. (2014). Contrairement à Ryschawy (2012), dans le cadre de cette analyse, le pas utilisé est relatif aux différentes périodes générationnelles retenues.

La typologie des trajectoires a été construite à partir des résultats obtenus selon une méthode d'analyse en deux étapes fréquemment utilisée pour l'élaboration des typologies (Garcia-Martínez et al., 2009) : (i) le tableau a été soumis à une analyse en composantes principales (ACP), qui a permis de retenir quatre facteurs selon le critère de Kaiser (valeurs propres > 1) et (ii) les coordonnées des exploitations sur les quatre premiers facteurs de l'ACP ont été utilisées pour effectuer une Classification Ascendante Hiérarchique (CAH) selon la distance euclidienne et le critère d'agrégation de Ward. La classification a ensuite été stabilisée en utilisant la méthode des K-means avec les centroïdes des classes obtenues par la CAH fixés comme barycentres. Les logiciels SPSS 21 et XLSTAT v 2008 sont utilisés pour les analyses statistiques.

 https://doi.org/10.5194/piahs-384-325-2021

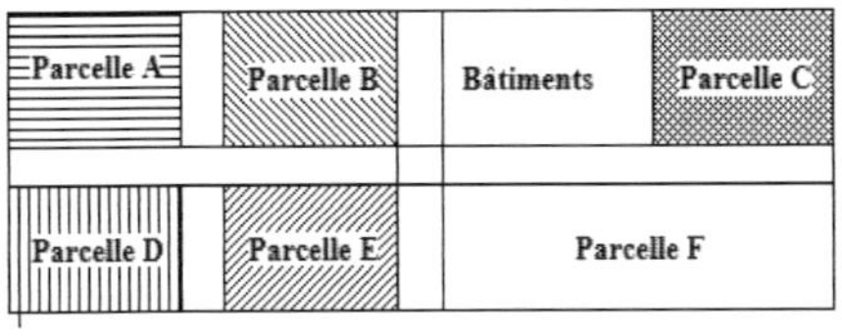

Figure 2. Configuration de l'espace d'une exploitation agricole familiale.

3 Résultats

3.1 Facteurs d'adaptabilité de l'agriculture familiale

Les mesures ayant permis l'adaptation de l'agriculture familiale aux changements climatiques sont diverses. Il s'agit principalement de l'usage de la main-d'œuvre familiale et externe, la diversification de cultures, le développement des activités secondaires (égrenage de maïs, d'arachide, du riz, etc.), l'accès des producteurs aux crédits agricoles, la mutation au niveau des outils de production, l'organisation du terroir des EAF, l'association et la rotation des cultures, l'adoption des variétés à cycle court, le semis échelonnés, l'usage des herbicides, l'association agriculture-élevage et la gestion de la surface agricole utile.

3.2 Changements techniques intervenus dans les EAF

3.3 Mutation au niveau des outils

Les producteurs agricoles abandonnent progressivement la houe au profit de la charrue et des motoculteurs. En effet, aujourd'hui, la houe est moins utilisée pour le billonnage (Fig. 2).

Certes, des changements qualitatifs sont intervenus pour améliorer les outils de production. Ceux-ci visent à réduire le temps mis pour labourer les champs. Aucune machine n'est utilisée pour la confection des buttes qui reste encore plus difficile que le billonnage. La Planche 1 illustre les outils utilisés pour le billonnage.

La photo 1.1 de la planche présente un motoculteur issu de la fabrication artisanale à Bembèrèkè. Ce motoculteur est fabriqué à l'aide d'un moteur à quatre temps. Il permet aux producteurs de labourer les surfaces à ensemencer plus vite que la charrue. La charrue, en activité sur la photo 1.2 prise à N'Dali, est la plus ancienne utilisée parmi ces trois outils. Le tracteur en pleine activité de labour à Nikki (1.3) est le moyen le plus rapide. Il est accessible aux producteurs qui disposent de moyens pour en louer.

3.3.1 Organisation du terroir des EAF d'hier à nos jours

L'adaptation de l'agriculture familiale aux changements climatiques est favorisée par les capacités que développent les producteurs qui réajustent des mesures développées après les évènements climatiques extrêmes. En effet, depuis le début des années 1970, le terroir des EAF a été morcelé en plusieurs parcelles cultivables. Ainsi, le nombre moyen de parcelles va de 2 à 6 parcelles dans le secteur. Pour les producteurs, « plus on pratique plusieurs cultures à la fois, plus on a la chance d'avoir un peu de récolte de chacune d'elles et moins on est vulnérable aux conséquences du raccourcissement de la durée de la saison culturale ». Aussi, dans l'organisation actuelle des EAF, une parcelle est réservée pour servir de zone de pâture aux animaux de trait (Fig. 3). Celle-ci est indirectement mise en jachère de courte durée.

Pour plus d'efficacité dans le fonctionnement des EAF, il a été mis en place un système d'association et de rotation des cultures sur l'exploitation. L'association a permis la diversification culturale. Tout ceci est favorisé par le système d'assolement (morcèlement de l'espace agricole en des parcelles).

3.4 Étapes d'évolution passées des exploitations agricoles familiales

L'agriculture familiale s'est profondément transformée au cours de son parcours d'adaptation aux changements climatiques. Cette métamorphose amène à dire que l'on est passé d'une agriculture paysanne traditionnelle à celle familiale. Face aux contraintes imposées par les changements climatiques en cours, les mesures d'adaptation autonomes et institutionnelles mises en œuvre par les agriculteurs, rendent compte de la diversité des trajectoires de l'agriculture familiale. Aussi, la complexité de l'organisation des EAF dépend-elle des étapes de trajectoire. La Fig. 4 présente les principaux types de trajectoire (T1–T4) de l'agriculture familiale identifiés après l'analyse des données.

Concernant le premier type de trajectoire, il faut noter que les EAF de cette catégorie ont conservé les pratiques traditionnelles de production surtout en ce qui concerne la population exploitante. Il s'agit des EAF très faiblement orientées vers le marché. Elles estiment que pour durer dans le temps et résister aux menaces climatiques, « il faut être nombreux à travailler ensemble pour constituer une force de production ». Par conséquent, face aux changements climatiques, la mobilisation de la main-d'œuvre familiale s'impose. Cette dernière est essentiellement utilisée dans ces exploitations pour les opérations culturales.

Le deuxième type de trajectoire est basé sur l'extension des superficies. Beaucoup d'exploitations ont opté pour cette mesure pour faire face aux changements climatiques. Il s'agit des EAF plus ou moins orientées vers le marché. Selon ces exploitants, « pour satisfaire les besoins de la famille et avoir des revenus agricoles pour investir dans l'exploitation, la diversification des activités est nécessaire ».

La diversification culturale et celle des sources de revenus est basée sur l'augmentation ou non des superficies. Pour mieux s'adapter aux effets des changements climatiques, les agriculteurs vont opter pour la valorisation de plusieurs cultures sur la même exploitation. Pour ces derniers, « plus

Plate 1. Outils les plus utilisés pour les opérations de labour.

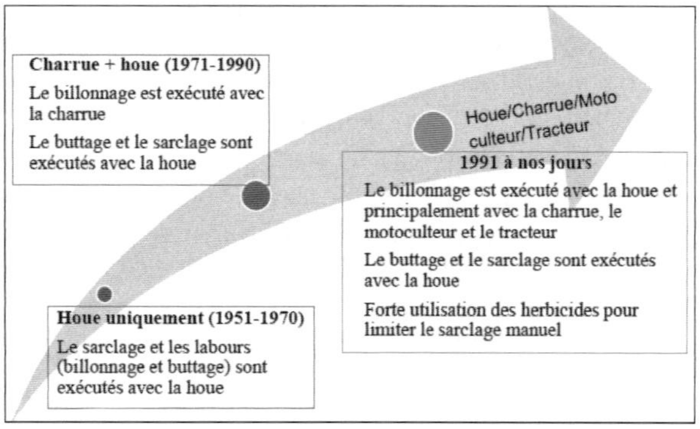

Figure 3. Changements intervenus dans l'outillage de la production agricole.

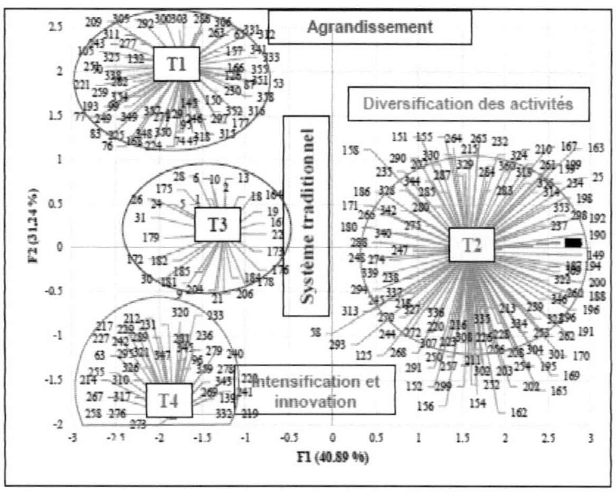

Figure 4. Typologie des trajectoires de l'agriculture familiale dans la ZAE-3.

les cultures sont nombreuses, moins on est vulnérable et plus on a la chance de récolter une petite quantité au moins au niveau de chacune des cultures ».

Le quatrième type de trajectoire est fondé sur l'intensification agricole et l'innovation. L'agriculture familiale est en ses débuts d'intensification. Pour améliorer les rendements culturaux, les agriculteurs ont apporté des innovations telles que l'association agriculture-élevage, l'usage des intrants chimiques (engrais et pesticides), l'utilisation des semences améliorées à cycle court, la faible adoption de l'association des cultures.

Les différentes analyses permettent de caractériser chaque génération par des qualificatifs propres du point de vue de la technicité de l'activité de production agricole. Ainsi, on a la « génération de culture manuelle », la « génération de culture attelée » et la « génération de culture motorisée ».

4 Discussion des résultats

La présente étude sur la trajectoire de l'agriculture familiale face aux changements climatiques a identifié quatre principaux chemins empruntés par les exploitations. Il s'agit de l'agrandissement des superficies, la diversification culturale, le maintien du système traditionnel et l'intensification et l'innovation. La méthode utilisée est identique à la démarche suivie par Ryschawy et al. (2013, 2014) et Ryschawy (2012). Ainsi, Ryschawy et al. (2013) ont analysé la diversité des trajectoires agricoles entre 1950 et 2005 dans les Coteaux de Gascogne afin d'identifier la diversité des parcours qui ont permis la survie des systèmes de polyculture-élevage et les forces motrices de cette survie. Les modèles individuels de trajectoire d'exploitation ont été réalisés par Ryschawy (2012) avec l'usage de 20 indicateurs pour l'étude des différences d'évolution des exploitations.

Plusieurs travaux similaires ont été réalisés sur les trajectoires et la dynamique des exploitations agricoles en particulier les exploitations d'élevage en Europe (Brédart et Stassart, 2017 ; Ruef et al., 2012 ; Ryschawy et al., 2013 ; Vall et al., 2017 ; Garcia-Martínez et al., 2009 ; Moulin et al., 2008). Ainsi, au terme de leurs travaux sur les trajectoires passées d'exploitations dans les coteaux de Gascogne, Ryschawy et al. (2014) identifient cinq types de trajectoire d'évolution. Selon ces derniers, pour durer dans le temps, les chemins empruntés ont consisté (i) à rechercher la maximisation de l'autonomie par une coordination forte entre culture et élevage au sein de l'exploitation ; (ii) à la recherche permanente de l'innovation technologique ; (iii) à la diversification des ateliers de production agricole pour bénéficier d'économies de gamme et se sécuriser par rapport aux fluctuations du marché des intrants et des produits agricoles ; (iv) à la sécurisation de l'exploitation par l'acquisition de capital *via* l'agrandissement et (v) aux adaptations progressives de pratiques « traditionnelles » en lien avec la main-d'œuvre familiale disponible.

De même, les travaux de Rueff et al. (2012) sur le changement de trajectoires des exploitations familiales depuis 1950 dans la région des montagnes des Pyrénées a identifié quatre (4) types de trajectoire agricole. Il s'agit du maintien au système traditionnel, la diversification des activités au sein de l'exploitation, la diversification des activités non agricoles et la spécialisation.

5 Conclusion

L'adaptabilité de l'agriculture familiale aux changements climatiques est due à la complémentarité des différentes mesures d'adaptation qui sont en grande partie tant autonomes qu'institutionnelles. L'utilisation des outils modernes et semi-modernes a révolutionné l'agriculture familiale. Face aux changements climatiques, elle fait partie des formes d'agriculture les plus adaptées. Malgré son niveau d'adaptabilité aux changements climatiques, elle est toujours dépendante des facteurs climatiques et susceptible d'être vulnérable à la survenance des phénomènes météorologiques extrêmes. Au-delà des changements climatiques, d'autres facteurs non climatiques ont contribué aussi à l'évolution de l'agriculture familiale. Les différentes tendances de trajectoire passée des exploitations agricoles ont permis d'adapter l'agriculture familiale aux effets des changements climatiques. Mais, il faut reconnaître que chaque type de trajectoire présente des limites.

Disponibilité du code. La publication d'un ouvrage vise pour objectif de le rendre accessible au grand public. A partir de cet instant, le document est accessible à tous sans condition aucune. Ainsi, je déclare que le manuscrit soit accessible à ceux qui le désirent.

Disponibilité des données. Je déclare que le manuscrit soit accessible à ceux qui le désirent. Les données utilisées dans ce texte proviennent de nos propres recherches.

Collaborateurs. BIHCH is the main author and initiator of the research. AAA is a co-author and he participated in the collection and processing of data. HY is a co-author and he participated in the data collection. EO is a co-author and he directed the work and made the internal evaluation of the article.

Intérêts concurrents. Les auteurs déclarent qu'ils n'ont aucun conflit d'intérêts.

Clause de non-responsabilité. Publisher's note : Copernicus Publications remains neutral with regard to jurisdictional claims in published maps and institutional affiliations.

Déclaration du numéro spécial. This article is part of the special issue "Hydrology of Large River Basins of Africa". It is a result of the 4th International Conference on the "Hydrology of the Great Rivers of Africa", Cotonou, Benin, 13–20 November 2021.

Remerciements. Nous remercions tous ceux qui ont contribué de quelconque manière à la réalisation de cet article. En particulier, nous témoignons nos reconnaissances aux agriculteurs qui ont favorisé la collecte des données lors de nos investigations.

Références

Afouda, F. : L'eau et les cultures dans le Bénin central et septentrional : étude de la variabilité des bilans de l'eau dans leurs relations avec le milieu rural de la savane africaine, Thèse de Doctorat nouveau régime, Institut de géographie, Université de Paris IV, Sorbonne, p. 428, 1990.

Boko, M. : Climats et communautés rurales du Bénin : Rythmes climatiques et rythmes de développement, Thèse de doctorat d'Etat ès Lettres et Sciences Humaines, 2 volumes, CRC, URA 909 du CNRS, Université de Bourgogne, Dijon, p. 601, 1988.

Brédart, D. and Stassart, P. M. : When farmers learn through dialog with their practices : A proposal for a theory of action for agricultural trajectories, J. Rural Stud., 53, 1–13, 2017.

Chabi, B. I. H., Akindélé, A. A., et Ogouwalé, E. : Dynamique d'adaptabilité de l'agriculture familiale aux changements climatiques dans la zone agro-écologique 3 du Bénin, 3ème année, no. 5, Vol. III, Baluki, Yolande, Berton-Ofoueme, 382–397, 2019.

Garcia-Martínez A., Olaizola, A., and Bernués, A. : Trajectories of evolution and drivers of change in European mountain cattle farming systems, Animal, 3, 152–165, 2009.

Moulin, C.-H., Ingrand, S., Lasseur, J., Madelrieux, S., Napoléone, M., Pluvinage, J., et Thénard, V. : Comprendre et analyser les changements d'organisation et de conduite de l'élevage dans un ensemble d'exploitations : propositions méthodologiques, in : L'élevage en mouvement : flexibilité et adaptation des exploitations d'herbivores, édité par : Dedieu, B., Chia, E., Leclerc, B., et Moulin, C.-H., Éditions Quae, Paris, 181–196, 2008.

Ogouwalé, E. : Changements climatiques dans le Bénin méridional et central : Indicateur, scénario et prospectives de la sécurité alimentaire, Thèse de doctorat unique, EDP/FLASH, UAC, Abomey-Calavi, p. 302, 2006.

PAM – Programme Alimentaire Mondial : Analyse Globale de la Vulnérabilité et de la Sécurité Alimentaire (AGVSA), République du Bénin, PAM Bénin/Togo, Cotonou, p. 146, 2014.

Rueff, C., Choisis, J.-P., Balent, G., and Gibon, A. : A Preliminary Assessment of the Local Diversity of Family Farms Change Trajectories Since 1950 in a Pyrenees Mountains Area, J. Sustain. Agricult., 36, 564–590, 2012.

Ryschawy, J. : Eclairer les conditions de maintien d'exploitations de polycultureélevage durables en zone défavorisée simple européenne, Une étude de cas dans les coteaux de Gascogne, Thèse de doctorat, INP-Université de Toulouse, Toulouse, 2012.

Ryschawy, J. Choisis, N., Choisis, J. P., and Gibon, A. : Paths to last in mixed crop–livestock farming : lessons from an assessment of farm trajectories of change, Animal, 7, 673–681, 2013.

Ryschawy, J., Choisis, N., Choisis, J.-P., Joannon, A., et Gibon A. : Quelles stratégies pour un maintien de la polyculture-élevage ? Une étude des trajectoires passées d'exploitations dans les coteaux de Gascogne in L'agriculture en famille : travailler, réinventer, transmettre, Edp sciences, France, 287–304, 2014.

Vall, E., Marre-Cast, L., et Kamgang, H. J. : Chemins d'intensification et durabilité des exploitations de polycultureélevage en Afrique subsaharienne : Contribution de l'association agriculture-élevage, Cah. Agric., 26, 25–36, 2017.

Proc. IAHS, 384, 331–336, 2021
https://doi.org/10.5194/piahs-384-331-2021

Open Access PIAHS

Ecohydrologie urbaine et changement climatique

Pascal Breil[1], Abdoulaye Faty[2], and Didier Orange[3]

[1]INRAE, RiverLy, 5 Rue de la Doua, CS 20244, 69625, Villeurbanne, France
[2]UCAD, Université Cheikh Anta Diop, Faculté de Géographie, UMI RESILIENCES, 10700, Dakar, Sénégal
[3]IRD Institut de Recherche pour le Développement UMR Unité Mixte de Recherche, Université Montpellier, CIRAD, INRAE, Institut SupAgro, UMR Eco&Sols, LMI IESOL, 18524, Dakar, Sénégal

Correspondence: Pascal Breil (pascal.breil@inrae.fr)

Published: 16 November 2021

Résumé. Due to global change, cities of the future will have to deal with more intense runoff and longer drought sequences, in addition to a growing urban and peri-urban population. French Mediterranean cities, such as Toulon, are already densely urbanised and exposed to the effects of global warming. The adaptation of their infrastructures is problematic. Cities with high development potential, such as Dakar, offer the opportunity to imagine other solutions for the management of water resources and its extremes in the context of global change. In particular, it is a question of managing the flows of water and substances linked to intense runoff events according to an ecohydrological logic that makes it possible to reduce environmental risks and increase social and economic benefits. To do this, we use a hydrologically-based geomatics model (IRIP) that produces predictive maps of areas of generation, transfer and accumulation of intense runoff and associated nutrients. This allows us to target effective intervention areas to reduce risks and increase water resources, for example by simulating land use change in appropriate locations and at the same time stimulating specific biological processes. The fundamental principle of ecohydrology is to balance energy flows with biological metabolic flows at the sub-catchment scale. The mapping of intense runoff processes is a first step illustrated in this article for the cities of Toulon and Dakar. This first step is part of the Dakar'2030 project, which aims to rethink urban development and adapt it to climate change.

1 Introduction

Pour le groupe d'experts méditerranéens sur le changement climatique et environnemental (Cramer et al., 2019), la région méditerranéenne présente une augmentation de sa température annuelle moyenne supérieure ($+1,5\,^{\circ}$C) à celle observée au niveau mondial ($+1,1\,^{\circ}$C). Les conséquences de ce réchauffement climatique « accéléré » en Méditerranée sont nombreuses, dont celles en lien avec le ruissellement intense pluvial. Ainsi, des sécheresses plus longues et des pluies torrentielles pourraient causer des problèmes majeurs pour l'agriculture, avec des pertes de production imprévues du fait de l'érosion des sols et de la perte en matière organique. En corolaire, les besoins en irrigation devraient augmenter ainsi que les incendies. Les zones humides, réservoirs de biodiversité et régulateurs des parasites seraient fortement menacées. Le risque d'inondation par ruissellement et de destruction par coulées de boue augmenterait aussi. Les inondations urbaines favoriseraient la transmission des maladies. Face à ces défis la gestion des espaces pour ralentir et conserver l'eau et les sols est une solution à envisager. L'écohydrologie propose le principe de la régulation duale (Zalewski, 2014) : les flux d'eau transfèrent les flux d'énergie associés (substances dissoutes et particulaires, thermie) vers les systèmes biotiques (microbes, plantes, chaine trophique) qui les transforment sous forme de biomasse ou de réactions biochimiques. Les flux d'eaux sont ainsi modifiés en quantité et en qualité par les interactions entre les systèmes biotiques et abiotiques. Si l'on souhaite réguler ces flux à l'échelle d'un bassin versant pour limiter les effets décrits du changement climatique, et promouvoir les services écosystémiques, il est nécessaire de représenter les circulations d'eau, en particulier lors des conditions de pluies intenses pour deux raisons : (I) localiser les pertes en sol, les destructions de cultures et d'infrastructures, les inondations ou encore des dépôts de matières en zone rurale ou urbaine ; (II) localiser les redistributions des nutriments produits et accumulés par les processus naturels ou par l'homme en temps normal. L'objectif est de minimiser les risques tout en utilisant les forces hydrauliques naturelles pour orienter les transferts de nutriments qui soutiennent les services écosystémiques.

1.1 Pluies plus intenses, sécheresses et réseaux d'assainissement

Parmi les risques dérivés, les processus d'érosion des sols par l'effet des sécheresses comme des pluies intenses sont de nature à créer de nombreux dysfonctionnements des réseaux d'assainissement unitaires et d'eau pluviales. En ce sens l'expérience africaine en climat tropical avec une longue saison sèche comme à Dakar peut s'avérer riche d'enseignement pour anticiper des problèmes de gestion du pourtour méditerranéen. En particulier, il a souvent été noté que la construction de réseaux d'assainissement unitaire impliquerait à Dakar un surdimensionnement pour absorber le ruissellement des pluies tropicales. Par ailleurs les longues périodes sèches sont néfastes au bon fonctionnement d'un réseau qui aura tendance à s'encombrer faute de capacité d'auto-curage, renforcé par des dimensions importantes et par l'apport d'eau ruisselée très chargée en flottants et sédiments divers lors des pluies orageuses. Les nuisances sont olfactives et sanitaires par débordement ou refus des réseaux. Enfin, le développement urbain sur plusieurs décennies se traduit systématiquement par un dépassement des capacités des réseaux anciens situés profondément dans les centres historiques des villes. Ce sont des goulots d'étranglement qu'il est difficile d'adapter aux flux gravitaires en provenance des zones périurbaines qui s'étendent avec l'accroissement urbain. Les raisons historiques, sociologiques et économiques de cette situation sont connues et posent aujourd'hui un problème majeur d'adaptation pour les gestionnaires des pays industriels. En effet, le coût de la réhabilitation des réseaux anciens qui se dégradent en vieillissant peut être insurmontable. Et le changement climatique n'est pas de nature à améliorer la situation car les pluies intenses ne pourront être gérées par des réseaux déjà défaillants.

1.2 La gestion alternative au « tout au tuyau » ... mais pas seulement en zone urbaine

Une stratégie de gestion alternative au tout tuyau pour les zones urbaines semble dès lors s'imposer pour gérer au mieux les flux d'eau, de sédiment et de substances transportés par le ruissellement. La gestion dite « à la source » consiste à stocker temporairement ou à infiltrer l'eau de ruissellement à proximité de sa zone de production. Cette possibilité est cependant contrainte par la disponibilité de l'espace en zone urbaine dense. Et il s'agit toujours dans ces solutions de gérer l'auto-inondation urbaine qui résulte du ruissellement généré par la ville elle-même. Dans le cas du changement climatique, les pluies plus intenses vont étendre les aires de connexions entre les zones périurbaines de l'amont et les centres urbains. Dans ce tissu rural et urbain non encore fortement imperméabilisé, le ruissellement intense peut résulter d'une intensité de pluie supérieure à la vitesse d'infiltration dans le sol ou encore par effet de saturation du profil de sol suite à des pluies répétées qui laissent peu de temps au ressuyage naturel. Les zones périurbaines sont donc stratégiques pour réguler les flux de ruissellement dans une logique écohydrologique qui permet de réduire les risques et d'accroitre les bénéfices. Le recyclage des eaux usées peut être intégré à cette logique d'ensemble pour ne pas surcharger les réseaux d'assainissement en usage. Cela nécessite cependant de mettre en place une planification de l'usage des sols entre naturel, agricole, urbain mais aussi dédiés à l'éco-ingénierie. Une première étape consiste à comprendre comment les processus du ruissellement intense s'organisent dans l'espace d'un bassin versant. La régulation duale des flux est regardée sur des exemples locaux tirés des cartographies du ruissellement intense produites par un modèle. La représentation cartographique des effets du ruissellement permet de préciser où des actions de régulation des flux d'eau et d'énergie seraient les plus efficaces pour fournir des services écosystémiques et réduire des risques potentiels. Dans les axes de transfert il est possible de casser l'énergie pour réduire l'érosion mécanique et les pertes en sol. Dans les zones d'accumulation, il possible de développer de la biomasse végétale à haut rendement en recyclant les eaux pluviales et usées via des biofiltres naturels. Ainsi, la connaissance des lieux de production du ruissellement, de ses axes de transfert, et des lieux de son accumulation sont des informations importantes pour imaginer les stratégies de gestion à mettre en place. Aujourd'hui, le besoin de connaissance concerne des dizaines de kilomètres carrés en devenir d'urbanisation en périphérie des villes. Cette étude explore la faisabilité d'une telle gestion dans un cas d'urbanisation très dense comme à Toulon (France), et dans le cas d'une ville à fort potentiel de développement, comme à Dakar (Sénégal).

2 Méthode

Nous proposons à cette fin d'utiliser un modèle géomatique à base hydrologique pour décrire les zones de production (P) du ruissellement intense, les zones de transfert (T) et érosion liées aux vitesses d'écoulement et les zones d'accumulation (A) d'eaux et de matières. Ce modèle, siglé IRIP pour Indicateurs du Ruissellement Intense Pluvial, a été décrit dans plusieurs publications (Lagadec et al., 2018 ; Braud et al., 2020) et fait l'objet d'évaluation de terrain pour en valider les facteurs. Le modèle IRIP est appliqué ci-après au bassin versant principal de la ville de Toulon et au territoire étendu de la péninsule de Dakar. Les scores varient de 0 à

5 pour chacun des processus P, T et A. Ils sont calculés en chacun des pixels définis par le modèle numérique de terrain.

2.1 Les couches d'information en entrée du modèle IRIP

Le modèle IRIP utilise 3 données de base : le modèle numérique de terrain (MNT) qui permet d'extraire le plan de drainage des écoulements de surface, mais aussi les mise en vitesse et les ralentissements selon les courbures topographiques concaves ou convexes ; l'occupation du sol dont les caractéristiques sont traduites en facteurs favorables ou non au ruissellement, en particulier les couverts imperméables ; et les caractéristiques pédologiques comme la texture, la densité apparente, dont sont dérivées des propriétés hydrodynamiques, la battance et l'érodibilité des sols. Les sources des données sont nationales (ISRIC data hub, 2020) et internationales dans les deux cas d'étude (voir Tableau 1).

Pour calculer la sensibilité des sols à produire du ruissellement hortonien, il a été considéré la pluie d'une heure de période de retour 5 ans pour Dakar (Bodian et al., 2016) et de 10 ans pour Toulon car elles sont comparables. Pour le ruissellement par saturation il est considéré la pluie de 24 h pour les mêmes périodes de retour.

3 Résultats

3.1 La péninsule de Dakar

La vue d'ensemble des cartes IRIP (Fig. 1a) avec les scores les plus forts (3, 4, 5) montre la contribution des zones urbanisées à produire du ruissellement intense (en rouge).

Certains sols à nu sont aussi des zones de production. Les zones d'accumulation du ruissellement sont bien visibles sous la forme de tâches bleues dans le tissu urbain. On retrouve aussi la grande zone humide qui collecte le ruissellement des alentours. La Fig. 1b est un agrandissement sur un quartier situé près de la zone humide (voir localisation sur Fig. 1a). Les zones de production et de transfert sont présentes sur la périphérie. La courbe de niveau d'altitude $z = 12$ m est marquée en blanc. Elle montre qu'il est possible de réduire l'accumulation en la transférant par gravité à l'aide d'un drain (flèche blanche) en direction des terrasses agricoles situées en bordure de la zone humide. Cela permet de rediriger les nutriments vers des zones de production agricole.

La Fig. 1c est une image aérienne localisée sur la Fig. 1b. Elle montre des cultures réalisées sur les pentes qui plongent vers la zone humide. C'est sur ces pentes qu'il est stratégique d'aménager des cultures en terrasse (ce qui n'est pas le cas pour l'instant) pour capter le ruissellement pluvial et les flux de substances associés, cela afin de maintenir l'humidité des sols, d'accroitre la fertilité et la production agricole. Les cultures devraient être généralisées sur ces pentes encore peu exploitées, ce qui favorise actuellement l'érosion et la perte des sols. Elles devraient être aménagées en terrasses, comme les « fanya juu » très pratiqués en Afrique de l'Est. Les eaux domestiques peuvent alimenter ces cultures après passage dans les terrasses supérieures dédiées à leur autoépuration. Cela maintiendrait l'humidité et le flux de nutriments.

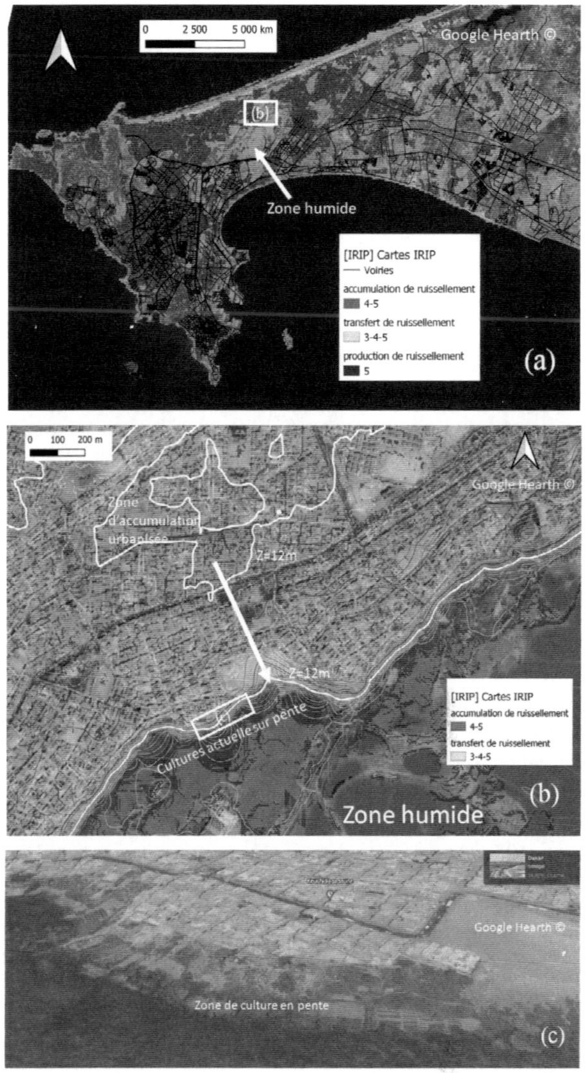

Figure 1. (a, b, c) Localisation des zones de production, transfert-érosion et dépôt-inondation dans la péninsule de Dakar.

3.2 Le bassin versant de la baie de Toulon

Dans la Fig. 2a, il est noté la présence de massifs montagneux forestiers qui expliquent l'absence de ruissellement intense au nord-ouest et en partie au sud-est du bassin versant. Cela car les sols forestiers favorisent l'infiltration et que la présence des arbres réduits considérablement les processus érosifs. Le reste du bassin descend progressivement en plaine continue vers la mer Méditerranée. Le développement

Tableau 1. Caractéristiques des données utilisées dans la modélisation IRIP.

	Dakar		Toulon	
	Résolution	Source	Résolution	Source
Modèle numérique de terrain	5 m	SRTM	5 m	IGN
Occupation du sol	250 m	ISRIC	20 m	THEIA
Nature du sol	250 m	Basegeo	250 m	ESDAC
Pluie	régionale	Bodian et al. (2016)	régionale	Météo-France
Occurrence & durée de pluie	5 ans, 1 h	5 ans, 24 h	10 ans, 1 h	10 ans, 24 h
Pluie mm	45.0	64.8	48.5	118.1

urbain est dense à Toulon et dans l'habitat périurbain. Deux zones humides d'importance sont visibles (aplats bleu). La plus grande a un rôle majeur de protection de la ville de Toulon contre les inondations. Elle est aussi très appréciée et fréquentée par les citadins pour son rôle d'ilot de fraicheur en période chaude. Elle améliore la qualité de l'eau des cours d'eau urbains qui l'alimente, avant de se déverser dans la baie de Toulon où l'activité économique est basée principalement sur le tourisme balnéaire.

La qualité des eaux de baignade est donc une contrainte forte. La grande zone humide fournit donc des services importants sur le plan économique et sanitaire. Il ne faut pas négliger son rôle de réservoir de biodiversité qui permet l'autorégulation des espèces, dont celles nuisibles aux cultures par exemple. Des zones de transfert érosif (jaune) sont présentes à flanc de massifs et en proximité de zones de très forte production (rouge) de ruissellement. Ces positions amont des aléas forts de ruissellement intense sont des facteurs de risque pour les habitations situées juste en aval. La pression urbaine est cependant très forte et des solutions d'aménagement contraintes sont à imaginer dans les secteurs exposés. La Fig. 2b montre comment il est possible de protéger des habitations situées dans une zone d'accumulation à l'aide d'un bourrelet de terre érigé le long d'une courbe de niveau pour créer une zone de stockage transitoire du ruissellement urbain provenant de l'amont. La zone aval, qui est alors protégée, est figurée en jaune. La cartographie IRIP sert ici à identifier des situations critiques vis-à-vis des inondations pluviales et des coulées boueuses. La zone de stockage transitoire peut être conçue comme une mini zone humide, cela afin de favoriser les différents services écosystémiques déjà évoqués. Cela implique une étude d'éco-ingénierie prenant en compte les fréquences des apports d'eaux de ruissellement et leurs natures.

La Fig. 2c montre un système de terrasses agricoles appelées « restanques » qui apparait comme des zones d'accumulation fortes du ruissellement intense. Ces terrasses sont limitées par des lignes de transfert fort du ruissellement. Ces lignes correspondent aux murs de pierres sèches, non érodibles, qui soutiennent les terrasses. Le rôle des terrasses agricoles est ainsi bien restitué par la modélisation IRIP, ce qui autorise a priori la construction de scénarios d'aménage-

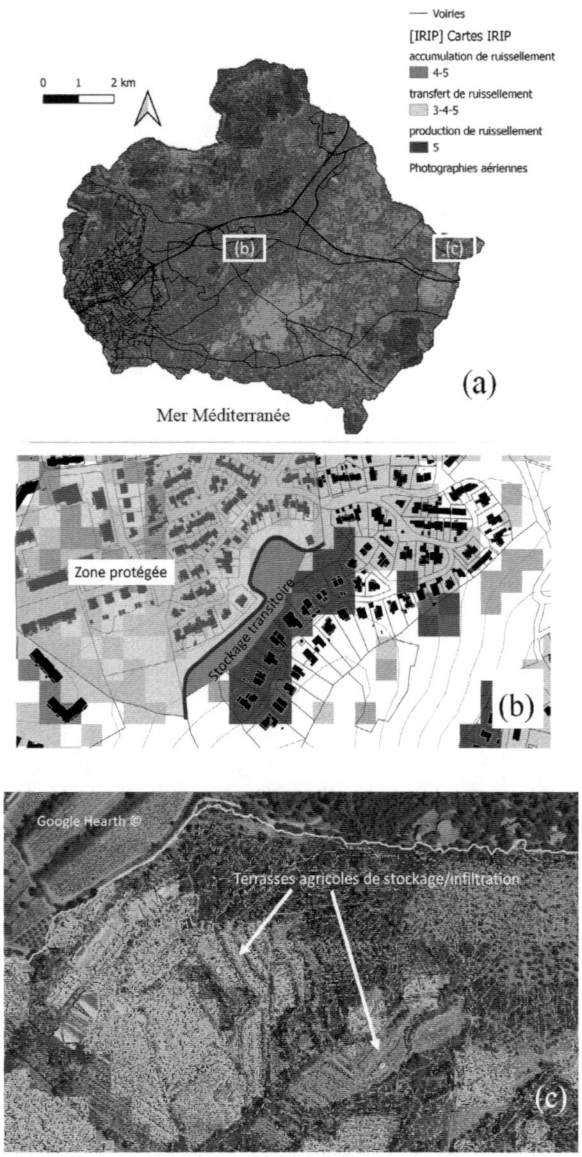

Figure 2. **(a, b, c)** Localisation des opportunités d'action pour réduire l'effet du ruissellement en périphérie de Toulon.

ment pour évaluer la réduction des risques en aval et l'augmentation des services écosystémiques associés.

4 Perspectives

Dans les exemples traités, la cartographie du ruissellement intense ouvre des perspectives pour définir les zones d'intervention qui permettraient de réguler les flux d'eau et d'énergie en utilisant de manière équilibré les services écosystémiques. Ces services sont multiples, directs, indirects, visibles ou cachés. A titre d'exemple, les terrasses agricoles, « restanques en France ou fanya juu en Afrique de l'Est » réduisent l'érosion, le transfert des flux d'eaux et de substances, la propagation des incendies, et favorisent l'augmentation de l'infiltration, la rétention de la matière organique et la biodiversité. Une étude détaillée (Hammad et al., 2004) montre que la productivité agricole est pratiquement doublée sur les parcelles en terrasse en comparaison avec des parcelles en pente. La maintenance, voire même le développement de nouvelles terrasses agricoles dans les versants en pente modérée est une solution écohydrologique d'intérêt pour atténuer les effets négatifs du changement climatique et accroitre des services écosystémiques et aussi économiques. La recherche d'un équilibre écohydrologique dans un versant ou bassin versant implique une cohérence entre les dynamiques des processus naturels (temps de séjours nécessaires aux processus d'assimilation, de biotransformations) et l'usage des bénéfices (production agricole, autoépuration de l'eau) réalisés. Pour les milieux anthropisés cet équilibre repose dans un premier temps sur la réduction des facteurs de déséquilibre, comme les pertes en sol, l'usage de susbtances nocives pour les processus naturels de bio-transformation. Le modèle IRIP participe des outils qu'il convient de développer pour sectoriser l'usage des sols dans les zones en devenir d'urbanisation et pour conserver des fonctions écologiques en des points clés comme les zones d'accumulation. Ce modèle peut être utilisé en simulation de changement d'occupation du sol. Le changement de paradigme proposé ici consiste à limiter l'utilisation de solutions grises basées sur l'évacuation rapide des excès d'eau, comme cela a longtemps été pratiqué dans les pays développés. L'idée est au contraire de favoriser la rétention dynamique et la répartition des excès d'eau et de substances associées pour accroitre la productivité des services écosystémiques.

Le projet d'étude Dakar'2030 a été initié par les trois instituts cosignataires de cet article avec l'idée de réfléchir sur ce que pourrait devenir cette ville en 2030 si l'on pouvait penser autrement le développement urbain.

Disponibilité du code. Le modèle est un plugin pour QGIS3. Il est disponible librement auprès de l'animateur du projet Pascal Breil (pascal.breil@inrae.fr), avec un mode d'emploi et un jeu de données test en français et en anglais. Il utilise QGIS version 3 avec Grass, sous Windows, Linux et MAC-OS.

Disponibilité des données. Toutes les données utilisées sont accessibles gratuitement à partir des sources du Tableau 1. Il semble que le site basegeo du senegal soit actuellement hors d'usage.

Collaborateurs. Le modèle utilisé est développé par PB qui l'a appliqué aux deux cas d'étude présentés. Les données françaises ont été recueillies par PB, les données Sénégalaises par AF et DO. La rédaction résulte d'un travail collaboratif.

Intérêts concurrents. Les auteurs déclarent qu'ils n'ont aucun conflit d'intérêts.

Clause de non-responsabilité. Publisher's note : Copernicus Publications remains neutral with regard to jurisdictional claims in published maps and institutional affiliations.

Déclaration du numéro spécial. This article is part of the special issue "Hydrology of Large River Basins of Africa". It is a result of the 4th International Conference on the "Hydrology of the Great Rivers of Africa", Cotonou, Benin, 13–20 November 2021.

Remerciements. Les résultats présentés pour la ville de Toulon ont été financés par l'Agence de l'Eau Rhône Méditerranée Corse via l'action 65 de son accord cadre avec INRAE.

Financement. La participation à cette conférence et l'article associé ont bénéficié du soutien financier du département AQUA de l'INRAE dans le cadre de ses actions incitatives 2021.

Références

Braud, I., Lagadec, L.-R., Moulin, L., Chazelle, B., and Breil, P. : A method to use proxy data of runoff-related impacts for the evaluation of a model mapping intense storm runoff hazard : application to the railway context, Nat. Hazards Earth Syst. Sci., 20, 947–966, https://doi.org/10.5194/nhess-20-947-2020, 2020.

Bodian, A., Dacosta, H., Ndimag Diouf, R., Ndiaye, E. H. O., and Mendy, A. : Contribution à la connaissance de l'aléa pluvial au Sénégal grâce à la valorisation des données pluviographiques historiques, Journal de l'association internationale de Climatologie, 13, 38–46, accessible à l'adresse suivante : http://lodel.irevues.inist.fr/climatologie/index.php?id=1194 (la date du dernier accès : 20 mai 2021), 2016.

Cramer, W., Guiot, J., and Marini, K. : Les risques liés aux changements climatiques et environnementaux dans la région Méditerranée, MedECC, accessible à l'adresse suivante : https://ufmsecretariat.org/wp-content/uploads/2019/10/MedECC-Booklet_FR_WEB.pdf (la date du dernier accès : 21 mai 2021), 2019.

Hammad, A. A., Haugen, L. E., and Borresen, T. : Effects of Stonewalled Terracing Techniques on Soil-Water Conservation and Wheat Production Under Mediterranean Conditions, Environ. Manage., 34, 701–710, https://doi.org/10.1007/s00267-003-0278-9, 2004.

ISRIC data hub : Africa SoilGrids – Texture , ISRIC [data set], accessible à l'adresse suivante : https://data.isric.org/geonetwork/srv/fre/catalog.search#/home (la date du dernier accès : 21 mai 2021), 2020.

Lagadec, L.-R., Moulin, L., Braud, I., Chazelle, B., and Breil, P. : A surface runoff mapping method for optimizing risk analysis on railways, Safety Science, 110, 253–267, https://doi.org/10.1016/j.ssci.2018.05.014, 2018.

Zalewski, M. : Ecohydrology and Hydrologic Engineering : Regulation of Hydrology-Biota Interactions for Sustainability, J. Hydrol. Eng., 20, A4014012, https://doi.org/10.1061/(ASCE)HE.1943-5584.0000999, 2014.

Proc. IAHS, 384, 337–342, 2021
https://doi.org/10.5194/piahs-384-337-2021

Open Access PIAHS

Impact of climate change on hydropower potential of the Lagdo dam, Benue River Basin, Northern Cameroon

Rodric M. Nonki[1], André Lenouo[2], Clément Tchawoua[3], Christopher J. Lennard[4], and Ernest Amoussou[5]

[1]Laboratory for Environmental Modeling and Atmospheric Physics, Department of Physics, Faculty of Sciences, University of Yaounde, P.O. Box 812, Yaounde, Cameroon
[2]Department of Physics, Faculty of Science, University of Douala, Douala, Cameroon
[3]Laboratory of Mechanics, Department of Physics, Faculty of Sciences, University of Yaounde, Yaounde, Cameroon
[4]Climate System Analysis Group, University of Cape Town, Cape Town, South Africa
[5]Laboratoire Pierre Pagney: Climat, Eau, Ecosystème et Développement (LACEEDE), University of Parakou, Parakou, Benin

Correspondence: Rodric M. Nonki (norome31@yahoo.fr)

Published: 16 November 2021

Abstract. Nowadays, special attention is paid to hydroelectric production because it is an efficient, reliable, and renewable source of energy, especially in developing countries like Cameroon, where hydropower potential is the main source of electricity production. It also represents a useful tool to reduce the atmospheric greenhouse gas concentrations caused by human activities. However, it is the most sensitive industry to global warming, mainly because climate change will directly affect the quality, quantity of water resources (streamflow and runoff), which are the important drivers of hydropower potential. This study examined the response of hydropower potential to climate change on the Lagdo dam located in the Benue River Basin, Northern Cameroon. Hydropower potential was computed based on streamflow simulated using HBV-Light hydrological model with dynamically downscaled temperature and precipitation from the regional climate model REMO. These data were obtained using the boundary conditions of two general circulation models (GCMs): the Europe-wide Consortium Earth System Model (EC-Earth) and the Max Planck Institute-Earth System Model (MPI-ESM) under three Representative Concentrations Pathways (RCP2.6, RCP4.5 and RCP8.5). The results suggest that, the combination of decreased precipitation and streamflow, increased PET will negatively impact the hydropower potential in the Lagdo dam under climate change scenarios, models and future periods.

1 Introduction

The Benue River Basin (BRB) is the Cameroonian part of the Niger River Basin (NRB) and occupies 4.4 % of the overall basin. The BRB lies between latitude 7 and 11° N and longitude 12 and 16° E (Fig. 1), and constitutes the second-largest river in Cameroon. It is the only perennial river in northern Cameroon and the neighboring areas where most of the rivers are seasonal (Olivry, 1986). The basin provides huge potential for development of water resources services that include hydropower, navigation and irrigation. In 1982, the Lagdo dam was built. The dam is 308 m long, 40 m in height and 9 m thick with an area of 586 km^2. It has an installed capacity of 72 MW and the entire northern part of Cameroon is currently supplied with electricity from Lagdo. Given its capacity on hydropower, there is a proposal to increase hydropower and irrigation capacity of the Lagdo dam with the objective to supply electricity to other countries such as the Chad Republic, the northern Nigeria and part of the Central African Republic (IRAP, 2015).

However, BRB experienced water related disasters such as floods and droughts (Molua and Lambi, 2006; Sighom-

Published by Copernicus Publications on behalf of the International Association of Hydrological Sciences.

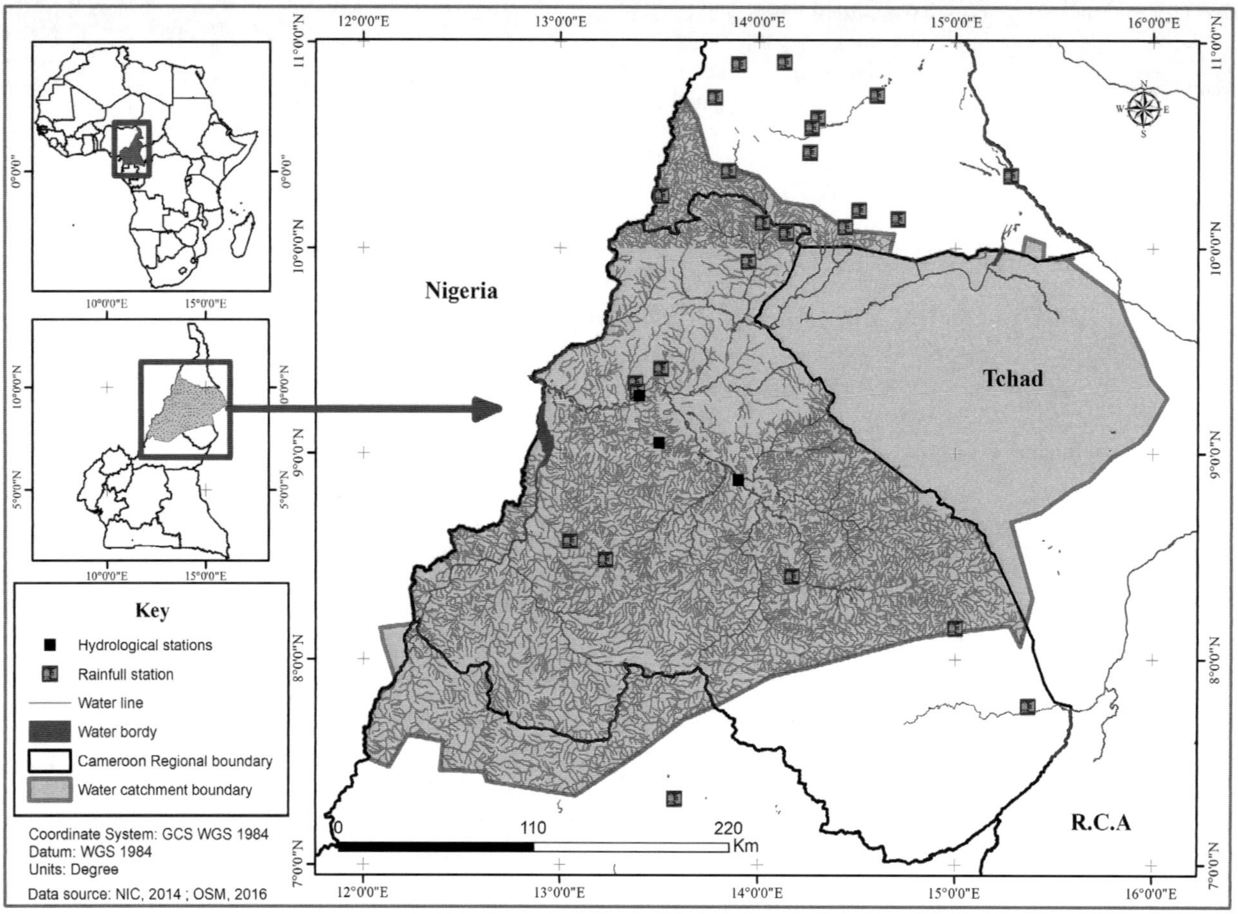

Figure 1. Study catchment: basin drainage area and rainfall and hydrological stations.

nou et al., 2013). In addition, recent studies showed the negative trend on rainfall and positive trend on temperature in the Northern Cameroon (Cheo et al., 2013; Dassou, 2019), while another studies revealed the increase of drought magnitude and intensity in the Northern Cameroon and in NRB (Guenang and Mkankam, 2014; Oguntunde et al., 2018) under changing environmental conditions. Nonki et al. (2019) also found that the BRB will move to extreme environmental drier conditions due to a decrease in excess water and an increase in evaporative demand under climate change. As key components of the hydrological cycle, any change in climate that affects rainfall and temperature will have proportionate impacts on the hydrological system in the area especially at the basin scale. According to Pal and Al-Tabbaa (2011), any change in the hydrological cycle due to global warming is likely to affect the distribution of water resources and consequently agriculture and hydropower production in the region as well as the availability and quality of freshwater. Therefore, climate change impact assessment on hydropower potential of this important watershed is highly important to support projects and develop new strategies for adaptation in this watershed.

The only existing study on this watershed (Grijsen et al., 2014) was based on General Climate Models (GCMs). According to Chou et al. (2014), climate change impact assessment at the regional and local scale needs higher resolution spatial data and GCMs with a coarse horizontal grid resolution do not capture these local and regional effects (Deb et al., 2015). In this context, the COordinated Regional climate Downscaling EXperiment project (CORDEX; Giorgi et al., 2009; http://www.cordex.org, last access: 17 October 2021) equipped African continent with many downscaled climate data from regional climate models (RCMs) that have downscaled many GCMs from phase 5 of the Coupled Model Intercomparison Project (CMIP5). Although several studies have assessed the impacts of CC on hydropower potential around the World and especially in Africa by using downscaled climate data, none of the above-mentioned studies assessed the climate change impacts on hydropower potential on BRB. Then, this study uses the downscaled precipitation and temperature data to assess the impact of climate change on hydropower potential of the Lagdo dam, BRB.

2 Materials and methods

2.1 HBV-Light hydrological model and data

The HBV-Light hydrological model (Seibert, 2005) used in this study is the modified version of HBV (Hydrologiska Byråns Vattenavdelning) hydrological model, initially developed in the Swedish Meteorological and Hydrological Institute (SMHI) by Bergstrom and Forman (1973). It is a conceptual-lumped model working in the daily time step and simulating discharges using daily precipitation and PET as input. This model was chosen due to his flexibility and robustness in simulating discharge in African context (Kwakye and Bardossy, 2020; Nonki et al., 2021a). More explanations of the model structure can be found in different studies (Seibert, 2005; Nonki et al., 2019, 2021b).

Daily measured weather data (precipitation) and potential evapotranspiration (PET) computed with Penman formula (Penman, 1948) from 1961–1978, used to calibrate and validate the model was obtained from the Direction of the National Meteorology of Cameroon (DNM), while the daily streamflow data measured at the available hydro-metric stations (Riao) located in the basin, were obtained from the Hydrosciences Montpellier – Système d'Informations Environnementales database (SIEREM; Boyer et al., 2006; http://hydrosciences.fr/sierem, last access: 17 October 2021).

In order to predict the climate change impact on hydropower potential, historical and projected scenarios of dynamically downscaled temperature and precipitation from the REMO regional climate model (RCM) forced by the boundary conditions data of the Europe-wide Consortium Earth System Model (EC-ESM) and the Max Planck Institute-Earth System Model (MPI-ESM) general circulation models (GCMs). REMO is integrated over a continuous period of 151 years, from January 1950 to December 2100 with 31 vertical levels and a horizontal resolution of 50 km (0.44°). REMO showed a strong ability to simulate the present climate in Central Africa in general (Fotso-Nguemo et al., 2017) and in the Northern Cameroon in particular (Nonki et al., 2019).

2.2 Calibration, validation and HBV-Light model performance assessment

In this study, the split sample test (Klemeš, 1986) and the Monte-Carlo simulations (MCS) (Robert and Casella, 2013) were used for model calibration and parameter optimization. Our data time series were separated into two sub-periods (1961–1970 for calibration and 1971–1979 for validation).

The model performance was then assessed by using different statistical techniques such as Nash and Sutcliffe Efficiency (NSE), Percent Bias (PBIAS), the ratio of the root mean square error (RMSE) and standard deviation of measured data (STDEVobs) (RSR) and the coefficient of determination (R^2). More details and descriptions of those statistical techniques can be found in Moriasi et al. (2007).

2.3 Assessment of the hydropower potential

The hydropower potential is the total energy from all natural runoff at stream gradient over the entire domain. It is achieved by converting the potential and kinetic energy of the water into electrical energy by electro-mechanical means (Turbines and Generators). Hydropower potential is estimated based on the streamflow, hydraulic head i.e. elevation gradient in this case and the total plant efficiency, as shown in the following equation (de Oliveira et al., 2017):

$$N_\mathrm{p} = Q \times H \times \rho_\mathrm{w} \times g \times \eta \qquad (1)$$

where N_p is the hydropower potential (W); Q is the stream flow ($\mathrm{m^3\,s^{-1}}$), H is the water head of the turbine (m), ρ_w is the water density ($\mathrm{kg/m^3}$), g is the gravitational acceleration and η is the total efficiency of the turbine (%).

Through that equation, reference and future streamflow simulated with the calibrated HBV-Light model were used to estimate the reference and future hydropower potential. The potential changes of hydropower potential by the near (2041–2065) and late (2071–2095) of the twenty-first century under RCPs 2.6, 4.5 and 8.5 relative to the baseline period (1981–2005).

3 Results and discussion

3.1 Hydrological model evaluation

Figure 2 shows the measured and modeled hydrographs as well as the flow duration curves at the Riao outlet during the calibration and validation periods. The model reproduces the timing and magnitude of the observed streamflow well although some small bias (PBIAS less than 15 %). The flow duration curves also revealed that the model reproduces the low flows more accurately than the high flows. Based on the statistical criteria analysis, the model performance is very good according to Moriasi et al. (2007) classification (with NSE of 0.86 and 0.78, RSR of 0.38 and 0.47 obtained during the calibration and validation periods respectively). There is also a strong relationship between modeled and observed streamflow (with $R^2 = 0.90$).

3.2 Projected changes in monthly, seasonal and annual hydropower potential

Figure 3 shows relative changes in monthly, seasonal and annual hydropower potential under the scenarios, models and time periods. The impact of climate change on hydropower potential is similar to the pattern of changes in streamflow (Nonki et al., 2019). In general, hydropower potential of the

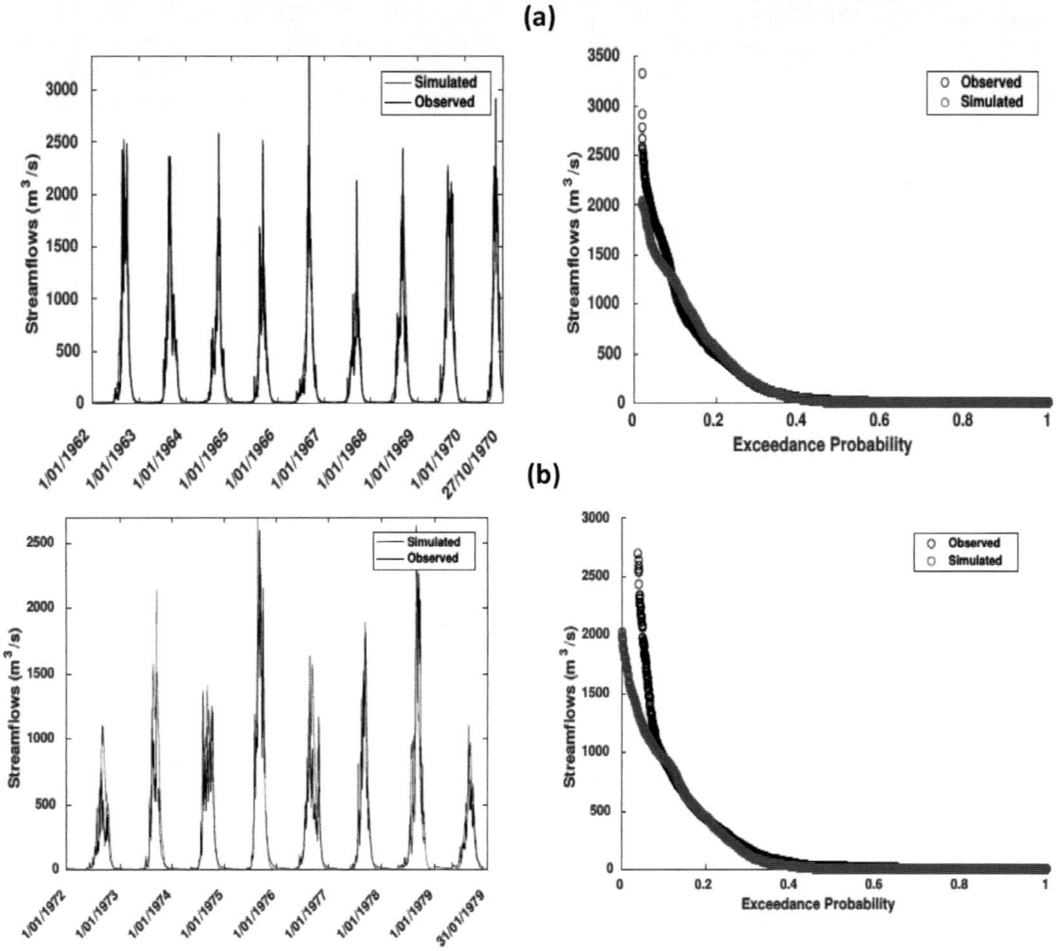

Figure 2. Comparison between observed and simulated hydrographs and flow duration curves at the Riao outlet during the calibration period **(a)** and validation period **(b)**.

Lagdo dam is projected to decrease. During September, the signal is strong under all scenarios, models and time periods, with the maximum decrease observed in the REMO-MPI combination during the late of the twenty-first century. The dry months do not exhibit a change signal which can be explained by absence of rainfall, thus very few streamflow. The seasonal and annual hydropower potential will decrease under both scenarios and time periods. The magnitude of the signal should vary according to the scenarios and time periods. This finding is consistent with that of Grijsen (2014). The similar negative trend was also reported in several studies in African catchments (Cole et al., 2014; Lumbroso et al., 2015; Falchetta et al., 2019).

4 Conclusions

The main focus of this work was to evaluate the influence of the projected temperature and precipitation change on hydropower potential of the Lagdo dam in the BRB, Northern Cameroon. Streamflow used to compute the hydropower potential was produced by coupling dynamically downscaled precipitation and temperature from the REMO model and the HBV-Light hydrological model under three (GHGs) concentration scenarios (RCP2.6, RCP4.5, and RCP8.5) during the future and baseline periods. The results showed that the optimized model parameters highly performed the HBV-Light model in the BRB during the calibration and validation stage. Based on climate scenarios, models and future time periods, the hydropower potential of the Lagdo dam will decrease, which is the result of the reduced precipitation and increase of PET, thus decrease in streamflow in the BRB. One major caveat of this study is the use of only one RCM with two GCMs models. Therefore, further works with a multi-model ensemble from CORDEX-Africa matrix are needed to quantify the range of uncertainty in this signal.

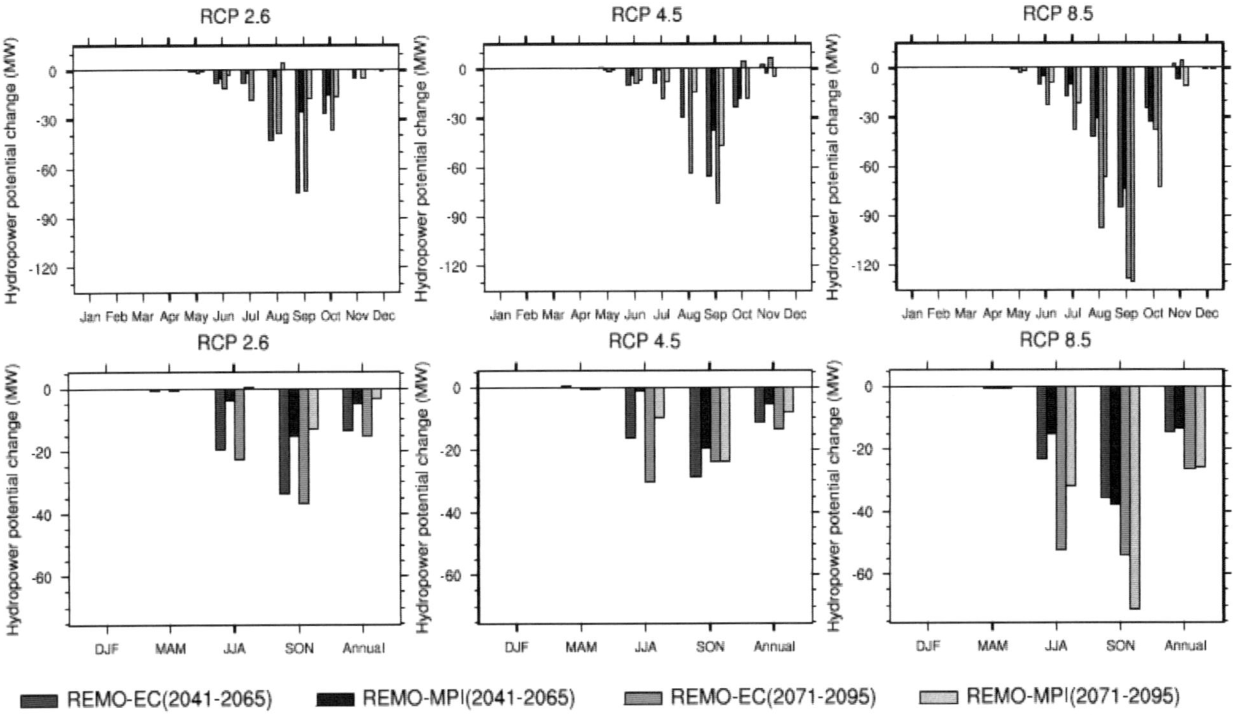

Figure 3. Projected monthly (top), seasonal and annual (bottom) changes in hydropower potential of the Lagdo dam under the three scenarios (RCP2.6, RCP4.5, RCP8.5) for the two future periods (2041–2065 and 2071–2095) relative to the baseline period (1981–2005).

Data availability. Meteorological and hydrological data are the properties of the Direction of the National Meteorology of Cameroon (DNM) and the SIEREM/HSM database respectively.

Author contributions. RMN: Conceptualization, Methodology, Software, Formal analysis, Validation, Writing – original draft. AL: Conceptualization, Software, Writing – original draft, Supervision. CT: Conceptualization, Methodology, Validation, Writing – review and editing. CJL: Data curation, investigation Visualization, Writing – review and editing. EA: Writing – review and editing.

Competing interests. The contact author has declared that neither they nor their co-authors have any competing interests.

Disclaimer. Publisher's note: Copernicus Publications remains neutral with regard to jurisdictional claims in published maps and institutional affiliations.

Special issue statement. This article is part of the special issue "Hydrology of Large River Basins of Africa". It is a result of the 4th International Conference on the "Hydrology of the Great Rivers of Africa", Cotonou, Benin, 13–20 November 2021.

Acknowledgements. The authors would like to thank the organizers of the 4th Conference on the hydrology of African Rivers, the data providers and the anonymous reviewers of this manuscript. We also thank the IAHS through SYSTA award for funding the first author to attend this conference.

References

Bergstrom, S. and Forman, A.: Development of a conceptual deterministic rainfall-runoff model, Nordic Hydrol., 4, 240–253, 1973.

Boyer, J. F., Dieulin, C., Rouche, N., CRES, A., Servat, E., Paturel, J. E., and Mahé, G.: SIEREM: an environmental information system for water resources, IAHS Publ., 308, 19–25, 2006.

Cheo, A. E., Voigt, H. J., and Mbua, R. L.: Vulnerability of water resources in northern Cameroon in the context of climate change, Environ. Earth Sci., 70, 1211–1217, https://doi.org/10.1007/s12665-012-2207-9, 2013.

Chou, S. C., Lyra, A., Mourão, C., Dereczynski, C., Pilotto, I., Gomes, J., Bustamante, J., Tavares, P., Silva, A., Rodrigues, D., Campos, D., Chagas, D., Sueiro, G., Siqueira, G., and Marengo, J.: Assessment of climate change over South America under RCP4.5 and 8.5 downscaling scenarios, Am. J. Clim. Change, 3, 512–525, 2014.

Cole, M. A., Elliott, R. J., and Strobl, E.: Climate change, hydrodependency, and the African dam boom, World Dev., 60, 84–98, 2014.

Dassou, E. F.: Impact de la variabilité climatique sur la réponse hydrologique du bassin versant supérieur de la Bénoué (Nord-Cameroun), Thèse de Doctorat/PhD, Université de Maroua, 332 pp., 2019.

Deb, D., Butcher, J., and Srinivasan, R.: Projected hydrologic changes under mid-21st century climatic conditions in a sub-arctic watershed, Water Resour. Manag., 29, 1467–1487, 2015.

de Oliveira, V. A., de Mello, C. R., Viola, M. R., and Srinivasan, R.: Assessment of climate change impacts on streamflow and hydropower potential in the headwater region of the Grande river basin, Southeastern Brazil, Int. J. Climatol., 37, 5005–5023, https://doi.org/10.1002/joc.5138, 2017.

Falchetta, G., Gernaat, D. E., Hunt, J., and Sterl, S.: Hydropower dependency and climate change in sub-Saharan Africa: A nexus framework and evidence-based review, J. Clean. Prod., 231, 1399–1417, https://doi.org/10.1016/j.jclepro.2019.05.263, 2019.

Fotso-Nguemo, T. C., Vondou, D. A., Tchawoua, C., and Haensler, A.: Assessment of simulated rainfall and temperature from the regional climate model REMO and future changes over Central Africa, Clim. Dynam., 48, 3685–3705, https://doi.org/10.1007/s00382-016-3294-1, 2017.

Giorgi, F., Jones, C., and Asrar, G. R.: Addressing climate information needs at the regional level: the cordex framework, World Meteorol. Org. (WMO) Bull., 58, 175–182, 2009.

Guenang, G. M. and Mkankam, K. F.: Computation of the Standardized Precipitation Index (SPI) and its use to assess drought occurrences in Cameroon over recent decades, J. Appl. Meteorol. Climatol., 53, 2310–2324, https://doi.org/10.1175/JAMC-D-14-0032.1, 2014.

Grijsen, J.: Understanding the impact of climate change on hydropower: the case of Cameroon, World bank Report, Washington, DC, 2014.

IRAP (International Rivers Africa Program): Hydropower in Africa: African Dams Briefing, 1250 Pretorius street, Suite G9 Ground Floor, East wing Pro Equity Court, Hatfield 00083, Pretoria, 2015.

Klemeš, V.: Operational testing of hydrologic simulation models, Hydrol. Sci. J., 31, 13–24, 1986.

Kwakye, S. O. and Bardossy, A.: Hydrological modelling in data scarce catchments: black Volta basin in West Africa, SN Appl. Sci., 2, 1–19, https://doi.org/10.1007/s42452-020-2454-4, 2020.

Lumbroso, D. M., Woolhouse, G., and Jones, L.: A review of the consideration of climate change in the planning of hydropower schemes in sub-Saharan Africa, Climatic Change, 133, 621–633, 2015.

Molua, E. L. and Lambi, C. M.: Climate, hydrology and water resources in Cameroon, Discussion Paper No. 33 Special Series on Climate Change and Agriculture in Africa, University of Pretoria, Pretoria: Centre for Environmental Economics and Policy in Africa [CEEPA], 2006.

Moriasi, D. N., Arnold, J. G., Van Liew, M. W., Bingner, R. L., Harmel, R. D., and Veith, T. L.: Model evaluation guidelines for systematic quantification of accuracy in watershed simulations, Trans. ASABE, 50, 885–900, https://doi.org/10.13031/2013.23153, 2007.

Nonki, R. M., Lenouo, A., Lennard, C. J., and Tchawoua, C.: Assessing climate change impacts on water resources in the Benue River Basin, Northern Cameroon, Environ. Earth Sci., 78, 1–18, https://doi.org/10.1007/s12665-019-8614-4, 2019.

Nonki, R. M., Lenouo, A., Lennard, C. J., Tshimanga, R. M., and Tchawoua, C.: Comparison between dynamic and static sensitivity analysis approaches for impact assessment of different potential evapotranspiration methods on hydrological models' performance, J. Hydrometeor., 22, 2713–2730, 2021a.

Nonki, R. M., Lenouo, A., Tshimanga, R. M., Donfack, F. C., and Tchawoua, C.: Performance assessment and uncertainty prediction of a daily time-step HBV-Light rainfall-runoff model for the Upper Benue River Basin, Northern Cameroon, J. Hydrol., 36, 100849, https://doi.org/10.1016/j.ejrh.2021.100849, 2021b.

Oguntunde, P. G., Lischeid, G., and Abiodun, B. J.: Impacts of climate variability and change on drought characteristics in the Niger River basin, West Africa, Stoch. Environ. Res. Risk Assess., 32, 1017–1034, https://doi.org/10.1007/s00477-017-1484-y, 2018.

Olivry, J. C.: Fleuves et rivières du Cameroun, Monographies hydrologiques ORSTOM, Thèse de Doctorat es Sciences, Université de Montpellier, 1986.

Pal, I. and Al-Tabbaa, A.: Regional changes of the severities of meteorological droughts and floods in India, J. Geogr. Sci., 21, 195–206, https://doi.org/10.1007/s11442-011-0838-5, 2011.

Penman, H. L.: Natural evaporation from open water, bare soil and grass, Proc. Roy Soc. A, 193, 120–145, https://doi.org/10.1098/rspa.1948.0037, 1948.

Robert, C. and Casella, G.: Monte Carlo statistical methods, Springer Science & Business Media, New York, 2013.

Seibert, J.: HBV-Light version 2, user's manual, Department of Earth Sciences, Uppsala University, Uppsala, 2005.

Sighomnou, D., Descroix, L., Genthon, P., Mahé, G., Moussa, B. I., Gautier, E., Mamadou, I., Vandervaere, J. P., Bachir, T., Coulibaly, B., Rajot, J. L., Issa, M. O., Abdou, M. M., Dessay, N., Delaitre, E., Maiga, F. O., Diedhiou, A., Panthou, G., Vischel, T., Yacouba, H., Karambiri, H., Paturel, J. E., Diello, P., Mougin, E., Kergoat, L., Hiernaux, P.: The Niger River Niamey flood of 2012: The paroxysm of the Sahelian paradox?, Secheresse, 24, 3–13, 2013.

Proc. IAHS, 384, 343–347, 2021
https://doi.org/10.5194/piahs-384-343-2021

Multiyear rainfall variability in the Mono river basin and its impacts on Nangbeto hydropower scheme

Salomon Obahoundje[1], **Ernest Amoussou**[2], **Marc Youan Ta**[3], **Lazare Kouakou Kouassi**[4], and **Arona Diedhiou**[1,5]

[1]LASMES – African Centre of Excellence on Climate Change, Biodiversity and Sustainable Development, Université Félix Houphouët Boigny, Abidjan, Côte d'Ivoire
[2]Département de Géographie et Aménagement de Territoire (DGAT/FLASH), Université de Parakou, Parakou, Bénin
[3]Centre Universitaire de Recherche et d'Application en Télédétection (CURAT), Université Félix Houphouët Boigny, Abidjan, Côte d'Ivoire
[4]UFR-Environnement, Université Jean Lorougnon GUEDE, 12 BP V 25 Daloa 12, Côte d'Ivoire
[5]Univ. Grenoble Alpes, IRD, CNRS, Grenoble INP, IGE, 38000 Grenoble, France

Correspondence: Salomon Obahoundje (obahoundjes@yahoo.com)

Published: 16 November 2021

Abstract. Hydropower energy, the main renewable energy source in West Africa, contributes to more than half of the Togo and Benin National electrification. This resource highly depends on water availability in rivers or reservoirs. The water availability heavily relies on climate patterns of the area. In the climate change context, the sustainability of hydropower plants is at risk. This work aims to assess the sensitivity of the Nangbeto hydropower plant to multiyear climate variability using statistical analysis. The results show that energy generation at Nangbeto hydropower is more modulated by four main variables namely inflow to reservoir, water level, rainfall of the actual and the previous year. The energy generation is found to be strongly and significantly correlated to inflow to reservoir, water level, and rainfall. Overall, the Nangbeto hydropower generation is more sensitive to inflow which is controlled by climate variables (rainfall, temperature) and land use/cover change. Therefore, the probable future change in these variables is suggested to be deeply investigated.

1 Introduction

The Mono basin plays a vital role in the socio-economic development of Togo and Benin countries. This basin is subjected to different water uses activities namely agriculture, livestock breeding, as well as small dam and Nangbeto hydropower generation. Nangbeto dam was constructed in 1987 to power Togo and Benin in electricity. This source has recorded rapid growth over the world especially over West African countries (IHA, 2019). In Togo, hydropower energy represents more than 80 % of the total electric energy consumption of the country which occupies more than 98 % share of renewable energy generation. The main challenge is that this source depends on water availability which is a function of climate variables. Thus, any change or variability in climate factors could affect hydropower plant (IHA, 2019; Rim, 2019) generation.

The water availability relies on precipitation and potential evapotranspiration which is a function of temperature. Therefore, any change in rainfall and temperature could affect the water availability for power generation (Hamududu and Killingtveit, 2012, 2016). It has been proven that West Africa's temperature has significantly increased over the past ranging from 0.3 to 1 and 0.2 to 0.5 °C over the Gulf of Guinea and this could be amplified in the future (Riede et al., 2016). The rising in temperature consequently leads to in increase the PET. However, the past and future changes of precipitation are uncertain and this change occurring faster than ever before (Riede et al., 2016). This could make the prediction of hydropower uncertain and may vary depending

Published by Copernicus Publications on behalf of the International Association of Hydrological Sciences.

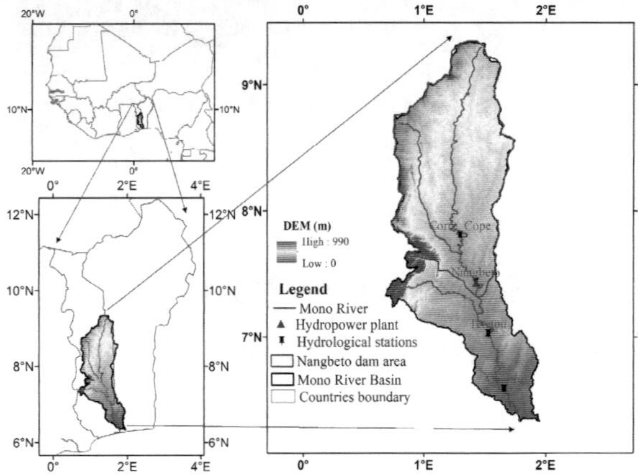

Figure 1. Presentation of Mono basin.

on the geographical location of the dam (Cole et al., 2014; Kling et al., 2016). Knowing the past could help us to learn the lesson and develop measures to adapt for the future.

This study has the goal of assessing the sensitivity of Nangbeto energy generation to change and variability of past temperature and precipitation over the 1988–2016 period. The observed rainfall, energy generation, inflow to reservoir, outflow as turbined water, and reservoir water level dataset as well as temperature were analysed. Aside from the general introduction which gives the state of the art concerning the topic, the study area, as well as the data and method used are described. Afterward, the results are presented and discussed following by a conclusion and some recommendations.

2 Study area, data and methods

2.1 Study area

Mono basin, transboundary river basin sharing between Togo and Benin, covers an area of 24 282.26 km^2. In Mono basin is constructed Nangbeto hydropower plant with an installed capacity of 65 MW (Fig. 1). The population living in the basin are mainly rural and their main activity is agriculture which exposes the basin under a rapid land cover/use dynamic (Obahoundje et al., 2018).

The cumulative annual precipitation in the basin is estimated around 1168 mm (FAO, 2005) with the mean monthly discharge at downstream (Athieme station) varies from 261 $m^3 s^{-1}$ (September) at the peak of the rainy season to about 0 $m^3 s^{-1}$ in the dry season (February to April). During the rainy season, the rainfall is greater than evapotranspiration (which varies from 4.5 to 6 mm d^{-1}; Amoussou, 2014), and vice versa, during the dry season. The basin mean, highest and lowest temperatures are, respectively, 26, 44, and 15 °C. The relative humidity varies between 20 %–30 % during the Harmattan period in the dry season and 70 %–80 % during the rainy season (Obahoundje et al., 2018).

2.2 Data and methods

The observed rainfall (1961–2016) data for four stations namely Corre-cope, Nangbeto, Tetetou, and Athieme as well as energy generation, inflow, outflow, and water level were obtained from Mono River Authority for 1988–2016. Due to a lack of temperature data, the reanalysis Climate Research Unit (CRU) air temperature data is used. CRU dataset is a month-by-month variation (from 1901 to the present) in climate over the last century and produced by the Climatic Research Unit at the University of East Anglia. CRU data has a spatial resolution of $0.5 \times 0.5°$ grids purposely made for climate variation study (New et al., 2002). However, the Thornthwaite (Thornthwaite, 1948) method was used to compute the potential evapotranspiration (PET) based on this CRU temperature data.

An ordinary Kriging method was used to interpolate the observed rainfall from four stations into basin average. Next, the method of Ali and Lebel (2008) was applied to compute the Standardized Precipitation Index (SPI) for the Mono basin. The formula is given by:

$$\mathrm{SPI}_{\mathrm{b}}^{i} = \frac{P_{\mathrm{b}}^{i} - \overline{P_{\mathrm{b}}}}{\sigma_{\mathrm{b}}} \tag{1}$$

With P_{b}^{i} the rainfall over basin b at year i, $\overline{P_{\mathrm{b}}}$ and σ_{b} the mean and standard deviation of the precipitation over basin b for the considered period.

The trend of all variables over the 1988–2016 period was computed using the Mann Kendall test for the Mono basin after an autocorrelation test, all with a significance at 95 % confidence level. The non-parametric Mann–Kendall test which is a rank-based non-parametric test for assessing the significance of a trend (Kendall and Gibbons, 1938) recommended by the World Meteorological Organization (WMO) for trend detection in hydro-meteorological time series (WMO, 2009) is used in this study. The description of the different factors guiding this test is presented in our previous work (Obahoundje et al., 2018).

Finally, the Nangbeto hydropower energy generation's sensitivity analysis was performed using the random forest algorithm. The random forest is proved to be a more reliable sensitivity analysis tool in matching production histories (Aulia et al., 2019) as it gives a robust, internally cross-validated estimation of the most important variables (Antoniadis et al., 2021). The Mean Decrease Accuracy (%IncMSE) is then used to determine the most important variables, i.e. those that influence the energy generation the most.

3 Results and discussion

3.1 Sensitivity analysis and correlation

Figure 2 presents the Mean Decrease Accuracy of all involved variables namely, inflow, water level, rainfall of the current (precip) and the previous (precip_lag1) year, the

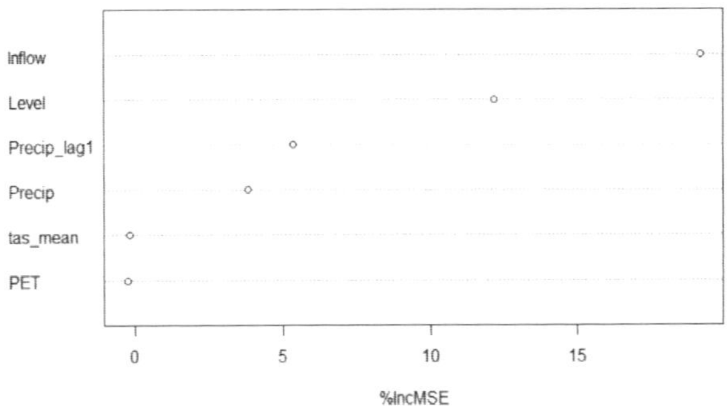

Figure 2. Random Forests-based sensitivity plot of energy generation at Nangbeto dam from 1988 to 2016.

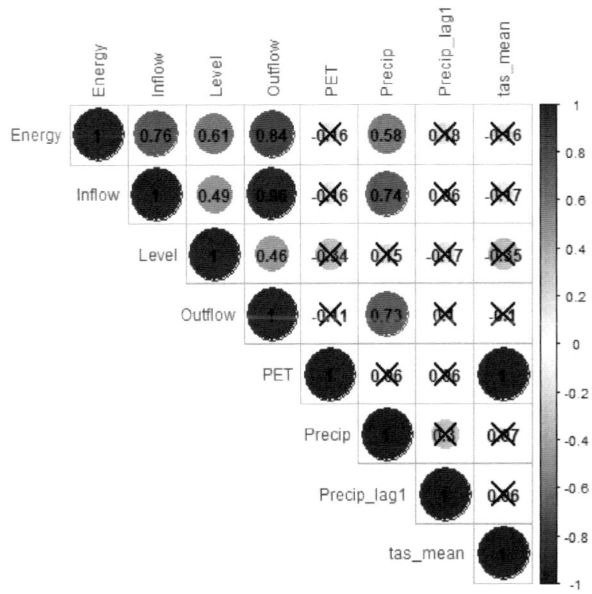

Figure 3. Correlation matric between energy generation and hydroclimatic variables in the Mono river basin. NB: The circle with × indicates that the correlation is not statistically significant at 95 % confident level while the values are the Pearson correlation coefficient *R*.

mean temperature, and the potential evapotranspiration. In this figure, the greater the %IncMSE of a variable, the more the energy is sensitive to that. Thus, the energy generation is found to be more sensitive to inflow to reservoir following by water level. The inflow is found to be strongly and significantly correlated with rainfall, water level, and outflow (Fig. 3). It is also important to highlight that the inflow is derived from river discharge which is a function of hydrological balance (rainfall, PET, runoff) modulated by basin characteristics mainly land use/cover change. The hydropower generation has been shown to be sensitive to both climate variables and land use/cover change in West Africa (Obahoundje et al., 2017, 2021). The water level is controlled by outflow (Fig. 3). The outflow is shown to be strongly and significantly correlated with rainfall, inflow and water level (Fig. 3) and water management policy in place at the dam. For instance, when focusing on rainfall, it was noted that the energy generation is more sensitive to the rainfall of the previous year than rainfall of the current year (Fig. 2). This exhibits the role that water management plays in dam operationalization.

Further analysis confirmed that the energy generation is strongly and significantly correlated with inflow to reservoirs, water level, and rainfall (Fig. 3). The rainfall is also shown to be strongly and considerably correlated with energy, outflow, and inflow. Therefore, the energy generation at the Nangbeto dam is dependent on rainfall.

Furthermore, the Mann Kendall test reveals that only water level (downward) and mean temperature and PET (upward) show a significant trend. It was also shown that water level and inflow are weakly and inversely correlated (not significant) with mean temperature and PET. This increase in PET may contribute to the observed decreasing trend in water level (not significant) and inflow despite the (also not significant) increase in precipitation. Addititionally, the downward trend in inflow could be due decreasing in occurrence of rainfall over Guinea region of West Africa (Bichet and Diedhiou, 2018). This is therefore a consequence of low outflow leading to a (non-significant) decrease in energy production.

3.2 Interannual variability of rainfall and energy

The long-term interannual rainfall variability was examined for observed rainfall data over the Mono basin. The 1961–1995 and 2011–2016 periods were marked by a steady decrease in rainfall while 1996–2011 was affected by an increase in rainfall (not shown). The total annual precipitation amount is greater within the 1961–1980 period than the 1981-2016 period. Generally, two decades (1961–1968 and 2002–2011) exhibit rainfall excess, and the last three (1969–1980, 1981–1991, and 1992–2001) show a rainfall deficit that affected West Africa in the early 1970s. In addition, the beginning of this last decade and (2012–2016) also exhibit a deficit in rainfall (Fig. 4). However, the 1968–1978 period recorded the lowest rainfall amount over the 1961–1980 period. The basin has recovered from this 1970s drought, but the amount of total rainfall recorded over the last decades as of 1980 is lower than the amount over the 1961-1980 period. This has been proved by different studies in West Africa (Ali and Lebel, 2008; Le Barbé et al., 2002; Diawara et al., 2014; Nicholson, 2013).

The standardized precipitation index (SPI) displays a period merely humid for 1988–2001 and a period globally humid for 2002–2016 (Fig. 4). The SPI analysis reveals that over 29 years (1988–2016) of study, 48.28 % (14 years), 20.69 % (6 years), and 31.03 % (9 years) were respectively normal, dry, and wet years leading to 27.59 % (8 years) of energy generation excess and deficit each. Both periods 1988–2001 and 2002–2016 were made of more normal years. For

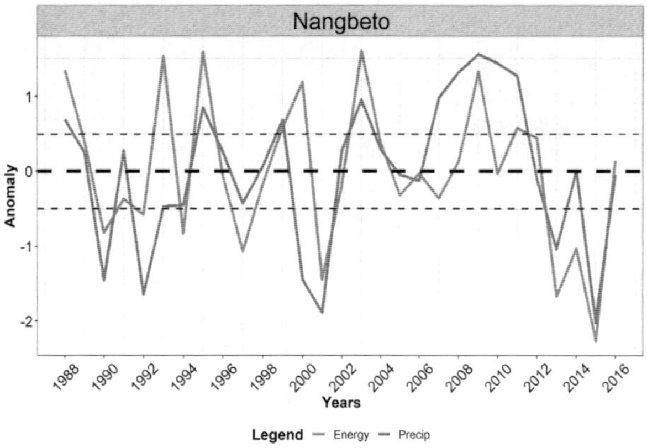

Figure 4. Interannual variability of rainfall and energy anomaly.

instance, 26.67 % (13.33 %) were with a deficit of rainfall and 20 % (40 %) were with excess rainfall over the 1987–2001 (2002–2016) period. The years 2001 and 2015 were the years with the highest deficit while the period 2007–2011 was wet with the highest excess rainfall record. It was noted that energy generation at a given year (y) depends also on the rainfall of the current (y) and/or previous ($y-1$) year. For instance, the dry year of 1990 led to deficit energy generation in the same 1990 year. The excess energy generation in 2000 despite the dry year 2000 is due to the previous wet year 1999. A higher deficit of energy generation in 2001 is due to consecutive dry years 2000 and 2001. This finding is in the agreement of previous studies in Akosombo in Ghana (Kabo-bah et al., 2016) and Kossou in Ivory Coast dams (Kouame et al., 2019).

4 Conclusion

This study evaluates the sensitivity of Nangbeto hydroelectric generation to multi-year hydroclimatic variability using statistical analysis and a random forest model. Energy is noted to be strongly and significantly correlated with inflow; water level is controlled by the climate variable. On the one hand, Nangbeto's hydroelectric generation is more sensitive to inflow and water levels. It was noted that inflow is controlled by precipitation, water level and outflow and also depends on basin characteristics, namely land use and cover change. Water level depends on runoff and water management. In addition, power generation at Nangbeto Dam is sensitive to rainfall in the current and previous years. Therefore, it is suggested that the likely future impacts of climate, basin features and socio-economic development changes on hydropower generation be thoroughly investigated while considering the incidences of climate extreme events as well as compounds extremes.

Code availability. The script written to analyse data during the current study are available from the corresponding author on reasonable request. The statistical analyse was performed used R language.

Data availability. The datasets generated during and/or analysed during the current study are available from the corresponding author on reasonable request.

Author contributions. SO conducted this research. AD, EA and MYT were the advisors: Conceptualization, SO and AD; Methodology, SO, AD, MYT and EA; Data Curation, LKK and EA; Writing – Original Draft Preparation, SO and AD; Writing – Review and Editing, EA, LKK and MYT; Funding Acquisition, AD.

Competing interests. The contact author has declared that neither they nor their co-authors have any competing interests.

Disclaimer. Publisher's note: Copernicus Publications remains neutral with regard to jurisdictional claims in published maps and institutional affiliations.

Special issue statement. This article is part of the special issue "Hydrology of Large River Basins of Africa". It is a result of the 4th International Conference on the "Hydrology of the Great Rivers of Africa", Cotonou, Benin, 13–20 November 2021.

Acknowledgements. The authors thank the Institute of Research for Development (IRD, France), Institute of Geosciences for Environment (IGE, University Grenoble Alpes) and LAPAMF – African Centre of Excellence on Climate Change, Biodiversity and Sustainable Development (CEA-CCBAD) for providing the facility (the Regional Climate Modelling Platform) to perform this study at the University Felix Houphouët Boigny (Abidjan, Côte d'Ivoire).

Financial support. This research has been supported by the Département Soutien et Formation, Institut de Recherche pour le Développement (UMR IGE Imputation (grant no. 252RA5)) and the Dan David Prize (2019 Dan David Prize Award for Young Researcher).

References

Ali, A. and Lebel, T.: The Sahelian standardized rainfall index revisited, Int. J. Climatol., 29, 1705–1714, https://doi.org/10.1002/joc.1832, 2008.

Amoussou, E.: Analyse hydrométéorologique des crues dans le bassin-versant du Mono en Afrique de l'Ouest avec un modèle conceptuel pluie-débit, available at: http://www.fmsh.fr (last access: 24 September 2018), 2014.

Antoniadis, A., Lambert-Lacroix, S., and Poggi, J. M.: Random forests for global sensitivity analysis: A selective review, Reliab. Eng. Syst. Saf., 206, 107312, https://doi.org/10.1016/j.ress.2020.107312, 2021.

Aulia, A., Jeong, D., Saaid, I. M., Kania, D., Shuker, M. T., and El-Khatib, N. A.: A Random Forests-based sensitivity analysis framework for assisted history matching, J. Pet. Sci. Eng., 181, 106237, https://doi.org/10.1016/j.petrol.2019.106237, 2019.

Bichet, A. and Diedhiou, A.: Less frequent and more intense rainfall along the coast of the Gulf of Guinea in West and Central Africa (1981–2014), Clim. Res., 76, 191–201, https://doi.org/10.3354/cr01537, 2018.

Cole, M. A., Elliott, R. J. R., and Strobl, E.: Climate change, hydrodependency, and the african dam boom, World Dev., 60, 84–98, https://doi.org/10.1016/j.worlddev.2014.03.016, 2014.

Diawara, A., Yoroba, F., Kouadio, K. Y., Kouassi, K. B., Assamoi, E. M., Diedhiou, A., and Assamoi, P.: Climate variability in the Sudano-Guinean transition area and its impact on vegetation: The case of the Lamto region in Côte D'Ivoire, Adv. Meteorol., 2014, 11, https://doi.org/10.1155/2014/831414, 2014.

FAO: L'irrigation en Afrique en chiffres – Enquête AQUASTAT 2005, Food & Agriculture Organisation, Rome, 2005.

Hamududu, B. and Killingtveit, A.: Assessing climate change impacts on global hydropower, Energies, 5, 305–322, https://doi.org/10.3390/en5020305, 2012.

Hamududu, B. H. and Killingtveit, Å.: Hydropower production in future climate scenarios; the case for the Zambezi River, Energies, 9, 1–18, https://doi.org/10.3390/en9070502, 2016.

IHA: Hydropower Status Report 2018, International Hydropower Association, Benjing, 2019.

Kabo-bah, A., Diji, C., Nokoe, K., Mulugetta, Y., Obeng-Ofori, D., and Akpoti, K.: Multiyear Rainfall and Temperature Trends in the Volta River Basin and their Potential Impact on Hydropower Generation in Ghana, Climate, 4, 49, https://doi.org/10.3390/cli4040049, 2016.

Kendall, M. and Gibbons, J. D.: Rank correlation methods, 5th edn., edited by: Arnold, E., Oxford University Press, New York, NY, London, C1990, available at: https://trove.nla.gov.au/work/3478976?q&sort=holdings+desc&_=1532491936434&versionId=21204985 (last access: 22 August 2018), 1938.

Kling, H., Stanzel, P., and Fuchs, M.: Regional Assessment of the Hydropower Potential of Rivers in West Africa, Energ. Proced., 97, 286–293, https://doi.org/10.1016/j.egypro.2016.10.002, 2016.

Kouame, Y. M., Obahoundje, S., Diedhiou, A., François, B., Amoussou, E., Anquetin, S., Didi, R. S., Kouassi, L. K., N'guessan Bi, V. H., Soro, E. G., and Yao, E. K.: Climate, Land Use and Land Cover Changes in the Bandama Basin (Côte D'Ivoire, West Africa) and Incidences on Hydropower Production of the Kossou Dam, Land, 8, 103, https://doi.org/10.3390/land8070103, 2019.

Le Barbé, L., Lebel, T., and Tabsoba, D.: Rainfall Variability in West Africa during the Years 1950–90, J. Climate, 15, 187–202, https://doi.org/10.1175/1520-0442(2002)015<0187:RVIWAD>2.0.CO;2, 2002.

New, M., Lister, D., Hulme, M., and Makin, I.: A high-resolution data set of surface climate over global land areas, Clim. Res., 21, 1–25, https://doi.org/10.3354/cr021001, 2002.

Nicholson, S. E.: The West African Sahel: A Review of Recent Studies on the Rainfall Regime and Its Interannual Variability, ISRN Meteorol., 2013, 1–32, https://doi.org/10.1155/2013/453521, 2013.

Obahoundje, S., Ofosu, E., Akpoti, K., and Kabo-bah, A.: Land Use and Land Cover Changes under Climate Uncertainty: Modelling the Impacts on Hydropower Production in Western Africa, Hydrology, 4, 2, https://doi.org/10.3390/hydrology4010002, 2017.

Obahoundje, S., Diedhiou, A., Ofosu, E. A., Anquetin, S., François, B., Adounkpe, J., Amoussou, E., Kouame, Y. M., Kouassi, K. L., Nguessan Bi, V. H., and Ta, M. Y.: Assessment of Spatio-Temporal Changes of Land Use and Land Cover over South-Western African Basins and Their Relations with Variations of Discharges, Hydrology, 5, 56, https://doi.org/10.3390/hydrology5040056, 2018.

Obahoundje, S., Youan Ta, M., Diedhiou, A., Amoussou, E., and Kouadio, K.: Sensitivity of Hydropower Generation to Changes in Climate and Land Use in the Mono Basin (West Africa) using CORDEX Dataset and WEAP Model, Environ. Process., 8, 1073–1097, https://doi.org/10.1007/s40710-021-00516-0, 2021.

Riede, J. O., Posada, R., Fink, A. H., and Kaspar, F.: What's on the 5th IPCC Report for West Africa?, in: Adaptation to Climate Change and Variability in Rural West Africa, edited by: Yaro, J. A. and Hesselberg, J., 7–23, Springer International Publishing, Switzerland, 2016.

Rim, B.: Energies renouvelables en Afrique: Enjeux , défis et opportunités, Policy Center for the New South, Maroc, 2019.

Thornthwaite, C. W.: On the deformation mechanisms in sap single crystals, Geogr. Rev., 38, 55–94, https://doi.org/10.1016/0022-3115(71)90076-6, 1948.

WMO: Extreme Values Analysis in The Guide to Hydrological Practices, vol. II, in: Management of Water Resources and Application of Hydrological Practices, World Meteorological Organization, Switzerland, 2, 302 pp., 2009.

Proc. IAHS, 384, 349–354, 2021
https://doi.org/10.5194/piahs-384-349-2021

Open Access PIAHS

Assessing climate change impact on the hydropower potential of the Bamboi catchment (Black Volta, West Africa)

Yacouba Yira[1,2], Tariro Cynthia Mutsindikwa[3,2], Aymar Yaovi Bossa[4,2], Jean Hounkpè[4,2], and Seyni Salack[2]

[1]Natural Substances Department, Applied Science and Technology Research Institute–IRSAT/CNRST, P.O. Box 7047, Ouagadougou, Burkina Faso
[2]West African Science Service Center on Climate Change and Adapted Land Use-WASCAL, Ouagadougou 06, P.O. Box 9507, Burkina Faso
[3]Water Engineering Department, Pan Africa University institute of Water and Energy Sciences, P.O. Box 119, 13000 Tlemcen, Algeria
[4]Department of Hydrology and Water Resources Management, National Institute of Water, University of Abomey–Calavi, P.O. Box 526, Cotonou 01, Benin

Correspondence: Yacouba Yira (yira_y@uni-bonn.de)

Published: 16 November 2021

Abstract. This study evaluates the impact of future climate change (CC) on the hydropower generation potential of the Bamboi catchment (Black Volta) in West Africa using a conceptual rainfall-runoff model (HBV light) and regional climate models (RCMs)–global climate models (GCMs). Two climate simulation datasets MPI-ESM-REMO (CORDEX) and GFDL-ESM2M-WRF (WASCAL) under RCP4.5 were applied to the validated hydrological model to simulate the catchment runoff. Based on reference and future simulated discharges, a theoretical 1.3 MW run of river hydro power plant was designed to evaluate the hydropower generation. Hydrological and hydropower generation changes were expressed as the relative difference between two future periods (2020–2049 and 2070–2099) and a reference period (1983–2005). The climate models' ensemble projected a mean annual precipitation increase by 8.8 % and 7.3 % and discharge increase by 11.4 % and 9.735 % for the 2020–2049 and 2070–2099 periods respectively (for bias corrected data). On the contrary an overall decrease of hydropower generation by −9.1 % and −8.4% for the 2020–2049 and 2070–2099 periods was projected respectively. The results indicate that projected increases in discharge should not solely be considered as leading to an increase in hydropower potential when prospecting climate change impact on hydropower.

1 Introduction

Hydropower is the leading source of renewable energy in West Africa (Hamududu et al., 2012). This source is climate sensitive, while West Africa is a region that is projected to be severely affected by CC (Sylla et al, 2016). Several studies (e.g. Obahoundje, 2015) have already assessed the impacts of CC on water resources in West Africa although not much work was done on its impacts on hydropower potential. Both Oyerinde et al. (2016) and Obahoundje (2015) undertook researches on the potential impact of CC on hydropower generation at the Kainji dam (the Niger Basin) and the Bui dam (Black Volta basin) respectively; and results showed that while CC is projected to decrease the hydropower potential of Bui, the opposite is expected for Kainji.

A strong reduction in the rainfall occurrence has been experienced in the Bamboi catchment (Black Volta) since 1960s, and notably during the 1973/1974 and 1983/1984 droughts (Boadi et al., 2017). These reductions have led to strong fluctuations in the river discharge with generally a negative trend from 1960–2000. The aim of this research is to assess the impacts of CC on hydropower potential of Bam-

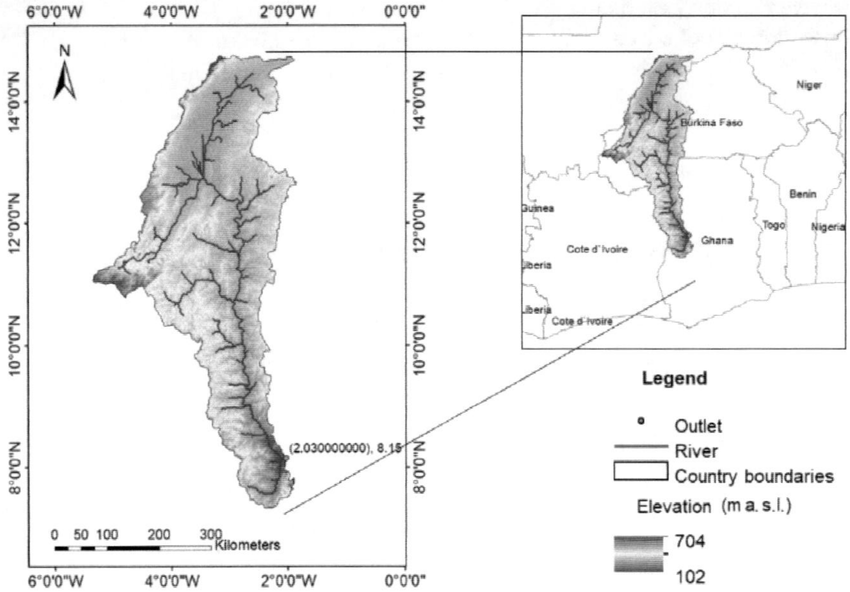

Figure 1. Location of the Bamboi catchment.

boi catchment. Accordingly, it has the following three objectives: (i) to set up the conceptual hydrological simulation model HBV light for the Bamboi catchment; (ii) to assess the impact of future CC on the hydrological regime of the catchment through scenario (RCP4.5) application comparing a reference period (1983–2005) to two future periods (2020–2049 and 2070–2099); and (iii) to assess the impact of CC on hydropower potential of the catchment.

2 Materials and methods

2.1 Study area

This study was carried out in the Bamboi catchment covering a total area of 134 200 km^2 (Fig. 1). The catchment spans across Mali, Burkina Faso, Ghana and Ivory coast. The main river in the basin is the Black Volta which originates as Mouhoun in Burkina Faso and as Sourou in Mali and it has a length of 1325 km (Atulley, 2013). Temperature in the basin ranges from 15 to 44 °C and March and April are the hottest months whereas the coolest month is August. Rainfall ranges from 400 to 1400 mm and it increases from North to South likewise humidity (Obahoundje, 2015).

2.2 Hydrological modelling

The hydrological modeling was carried out with the HBV-light model (Seibert and Vis, 2012). Compared to many other hydrological models, HBV has a limited number of parameters and requires few inputs for a classic application. Only three modules of the model structure were applied: (i) soil, (ii) routing, and (iii) response. The built-in Genetic Algorithm and Powell optimization (GAP) were adopted for the calibration of the model. Although the retrieved observed discharge data span from 1979–2007, the series of data over the 1979–2001 featured important gaps and was therefore only use as warm-up data. During the calibration phase, the period of January 1979–January 2001 was used for warming, and 1 January 2001–1 January 2004 was applied for parameters optimization. Statistical criteria were reported by considering *NSE*, R^2 and *KGE*. However, only *KGE* was set as objective function during parameters optimization. For the validation process the warm up period was taken from 1 January 1979–1 January 2004 and the validation period ranged from 1 January 2004–28 February 2007. Table 1 provides the type and source of applied datasets in the study.

2.3 Climate impact assessment on the hydrology and hydropower generation potential of the catchment

For climate impact assessment two RCM GCM climate datasets developed by WASCAL at 12 km spatial resolution using WRF and the CORDEX climate product (0.44° spatial resolution) with the same concentration pathway RCP4.5 were applied in this study. It is worth noting that the WASCAL dataset is only available for the RCP4.5 scenario. A climate data bias correction was applied to both RCMs–GCMs following the non-parametric quantile mapping using the empirical quantiles method. The bias corrected climate data (reference_1983–2005 and future_2020–2049 and 2070–2099) was used as inputs for the hydrological model to simulated both historical and projected discharges. Power and Energy production were calculated from the simulated

Table 1. Applied datasets.

Data set	Resolution	Source	Processing
Observed climate	Daily	Climate Forecast System Reanalysis-CFSR	Spatial average
Simulated climate	Daily	WASCAL (Heinzeller et al., 2018)	Bias correction
Simulated climate	Daily	CORDEX	Bias correction
Observed discharge	Daily	Global Runoff Data Center-GRDC (station 153 1100)	

discharges using the hydropower Eqs. (1) and (2).

$$P = e\rho Q g H \quad (1)$$

$$E = P \cdot \Delta t \quad (2)$$

where P is power (W), E is Energy (W h^{-1}), e is the efficiency of the turbine (set to 80 %), ρ is the density of water (kg m^{-3}), Q is the water released to the turbine (set to Q_{80} of the mean simulated historical and projected discharges, m^3 s^{-1}), g is the gravitational force (m s^{-2}), H set to 20 m is the water level above the turbine (mm), and Δt is time (h).

3 Results

3.1 Hydrological model calibration and validation

Goodness of fit for the Calibration and validation periods are shown in Fig. 2. The quality measures achieved during calibration equal 0.66, 0.66 and 0.73 for R^2, NSE and KGE respectively, indicating a satisfactory to good agreement between observed and simulated discharges. As for the validation phase, the KGE efficiency is within the same range as during the calibration. NSE and R^2 however dropped by 0.07 and 0.09 points compared the calibration phase. The visual inspection of the hydrographs shows a broad agreement between simulated and observed discharges.

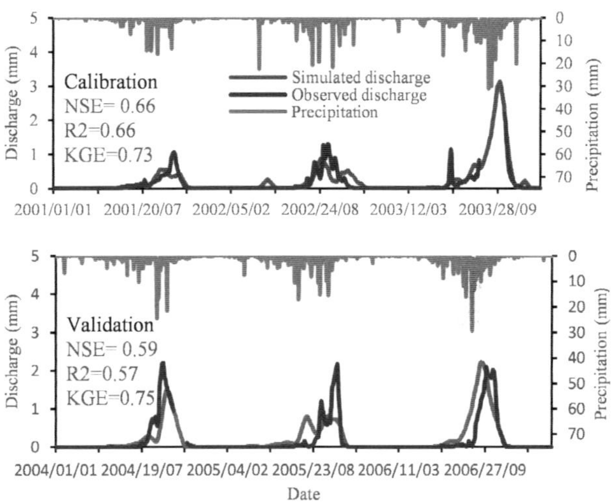

Figure 2. Calibration and validation results of the HBV for Bamboi catchment for the period 2001–2004 and 2004–2007 respectively.

3.2 Hydrological and hydropower potential changes

3.2.1 Discharge change

Future discharges simulated with GFD LESM2M–WRF and MPI–ESM REMO with both bias corrected and non-bias corrected climate data, were compared to the reference discharge. The comparison indicates an increase in the annual discharge by both climate models (Table 3). For the mid-century (2020–2049), GFD LESM2M–WRF projects a 14.4 % increase in annual discharge, which is a consequence of relative increase in precipitation of around 11 %. However, for the end of century (2070–2099) there is a slight decrease in discharge as compared to mid-century (2020–2049). These trends are similar with non-bias corrected climate data, and also apply to the MPI–ESM REMO climate dataset.

3.2.2 Hydropower generation change

Figure 3 indicates that the maximum amount of hydropower power will be generated in wet season and minimum generation will occur in the dry season. This is due to the fact that runoff increases with precipitation in the wet season so does discharge thereby increasing the hydropower production. Both climate datasets indicate that the average hydropower generation will reduce in the future as compared to the reference period (1983–2005) as more hydropower is produced in the historical period in the dry season compared to future periods.

This trend is consistent with the projected change in the annual discharge pattern (Fig. 4).

The projected hydropower potential trends by CORDEX climate product is in agreement with the WASCAL cli-

Table 2. RCM–GCM products and the corresponding label used in the study.

Framework	RCM	Driving GCM	Label used in the study
CORDEX	REMO	MPI-ESM	MPI-ESM-REMO
WASCAL	WRF	GFDL-ESM2M	GFDL-ESM2M

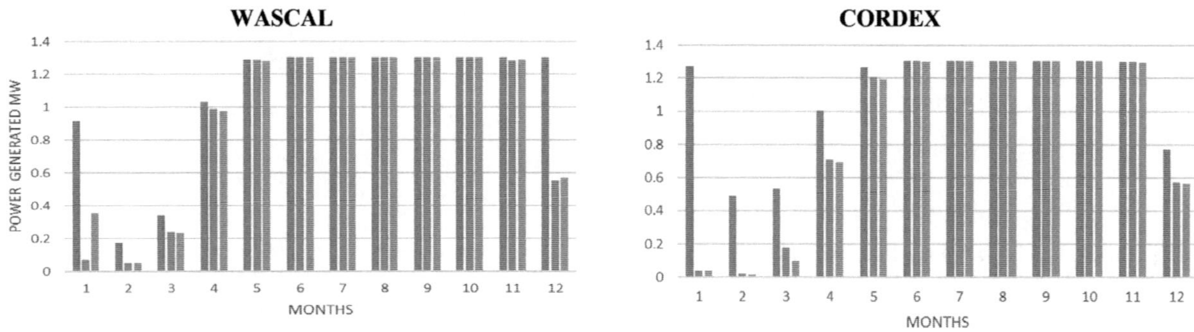

Figure 3. Projected and historical power by WASCAL and CORDEX climate products (corrected data). Blue refers to the reference period (1983–2005), orange is the future period of 2020–2049, and grey represents the future period 2070–2099.

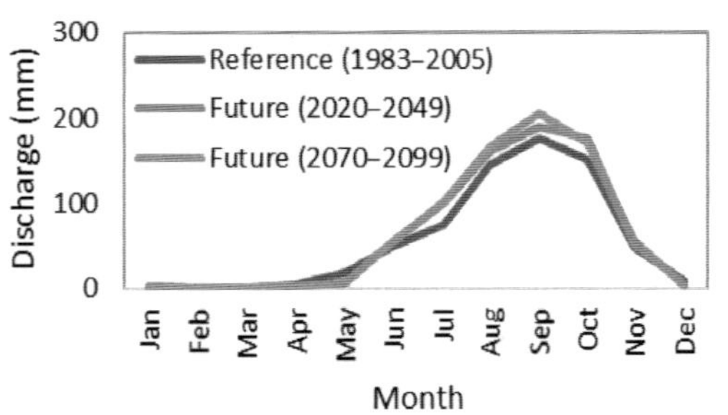

Figure 4. Change in the annual discharge pattern (CORDEX dataset).

mate product, only slight differences in the magnitude of the changes can be noticed.

It can be noted from that Table 4 that the average annual hydropower generation will decrease in the future for both climate datasets. For the mid-century (2020–2049) period a decrease in the average annual hydropower production by −9.1 % for bias corrected data and −7.6 % for uncorrected data climate ensemble is projected. By end of century (2070–2099), the average annual power production shows a slight increase as compared to the 2020–2049 period, however the amount is still lower compared to the reference period. These changes are not consistent with the projected annual discharges increase.

4 Discussion

Fair model statistical quality measures were achieved during the calibration and validation phases. However, several discrepancies can be observed; notably, (i) some peaks are not well reproduced in 2001 and 2002, (ii) peaks are underestimated in 2001, 2002, 2004 and 2005 and (iii) peaks are overestimated in 2003. Beside model imperfection, important gaps in the GRDC runoff time series combined with the use of reanalysis climate data (CFSR) can reasonably explain these discrepancies (Poméon et al., 2017). These authors also reported, the suitability of the HBV model to simulate the hydrology of some west African catchments. The often-mentioned misrepresentation of peak flow by the model is usually attributed to measurement errors during high discharge events as overbank flow characterizes many rivers in the region.

Both WASCAL and CORDEX climate datasets projected an increase in precipitation and discharge in the rainy season for the future periods. However, the increase in discharge occurs during the wet season where the power plant already operates at its full potential. As a consequence, the discharge increase does not translate in additional energy production. The annual discharge increase is overshadowed by a decrease in monthly discharge during the dry season. Indeed, the decrease of discharge over the dry season will lead to a loss of hydropower. A decrease in average annual hydropower generation in the future is also reported by previous studies in the region (e.g. Obahoundje, 2015). Such a decrease is however attributed to a reduction in precipitation.

5 Conclusions

GFDL ESM 2M of WASCAL and MPI ESM REMO of CORDEX were applied to the HBV to assess the impacts of CC on the hydropower generation of the Bamboi catchment comparing two future periods 2020–2049 and 2070–

Table 3. Projected discharge change between the reference between the reference (1983–2005) and future periods reference (2020–2049) and future (2070–2099).

Climate model	Reference discharge (mm)	Projected change_2020–2049 (%)	Projected change_2070–2099 (%)
Bias corrected			
GFD LESM2M–WRF	200.9	+14.4	+14.2
MPI–ESM REMO	219	+8.4	+5.3
Mean of the ensemble	210	+11.4	+9.7
Non bias corrected			
GFD LESM2M–WRF	314.8	+54.2	+35
MPI–ESM REMO	673.4	+14.3	+15
Mean of the ensemble	494.1	+34.2	+25

Table 4. Annual average projected energy generation.

		Energy (MWh)			Change (%)	
Period		1983–2005	2020–2049	2070–2099	2020–2049	2070–2099
CORDEX	Corrected	7917.1	7726.5	7636.3	−2.4	−3
	Uncorrected	8667	7726	8582	−10.9	−0.9
WASCAL	Corrected	9440.5	8050.6	8257.2	−15.4	−12.5
	Uncorrected	84 000	8050.6	8314	−4.2	−1.0
Average	Corrected	8678.8	7888.6	7946.7	−9.1	−8.4
	Uncorrected	8533.5	7888.3	8448	−7.6	−1

2099 to a reference period (19836–2005). The results indicated that precipitation will increase in both future periods, however the hydropower generation will decrease for both future periods. Hydropower is projected to decrease in the dry season while the discharge increase in the wet season. Both climate datasets, are agreeing that hydropower production is projected to decrease in the end of the century 2070–2099, while the decrease in the mid-century is even higher. These results underline that the projected increases in discharge do not necessarily mean an increase in hydropower generation.

Code availability. WaSIM code is not publicly accessible. The applied Richards version of the software is available for download at http://www.wasim.ch/en/products/wasim_richards.htm (last access: 10 October 2021) (WaSiM-ETH, 2021).

Data availability. The CORDEX-Africa data applied in this study are publicly accessible at https://esg-dn1.nsc.liu.se/search/esgf-liu/ (last access: 10 October 2021). The WASCAL climate data applied in this study are publicly accessible at https://doi.org/10.1594/PANGAEA.880512 (Heinzeller et al., 2018).

Author contributions. YY, AYB and TCM developed the concept. All authors contributed to implementing the concept and writing the manuscript.

Competing interests. The contact author has declared that neither they nor their co-authors have any competing interests.

Disclaimer. Publisher's note: Copernicus Publications remains neutral with regard to jurisdictional claims in published maps and institutional affiliations.

Special issue statement. This article is part of the special issue "Hydrology of Large River Basins of Africa". It is a result of the 4th International Conference on the "Hydrology of the Great Rivers of Africa", Cotonou, Benin, 13–20 November 2021.

Acknowledgements. The authors are grateful to Luc Olivier Crépin SINTONDJI for supervising this work.

References

Atulley, J.: Impact of small reservoirs and dugouts in ghana on hydrology and water allocation in the Black Volta Basin. Dissertation, Kwame Nkrumah University of Science and Technology, available at: https://hdl.handle.net/10568/42166 (last access 10 octobre 2021), 2013.

Boadi, S. A. and Owusu, K.: Impact of climate change and variability on hydropower in Ghana, Afr. Geogr. Rev., 6812, 1–15, https://doi.org/10.1080/19376812.2017.1284598, 2017.

Hamududu, B. and Killingtveit, A.: Assessing climate change impacts on global hydropower, Energies, 5, 305–322, https://doi.org/10.3390/en5020305, 2012.

Heinzeller, D., Dieng, D., Smiatek, G., Olusegun, C., Klein, C., Hamann, I., Salack, S., Bliefernicht, J., and Kunstmann, H.: The WASCAL high-resolution regional climate simulation ensemble for West Africa: concept, dissemination and assessment, Earth Syst. Sci. Data, 10, 815–835, https://doi.org/10.5194/essd-10-815-2018, 2018.

Obahoundje, S.: Potential Impacts of Land Use, Land Cover Change and Climate Change on Hydro power Generation in West Africa: The case study of Bui Dam (Ghana), MSc Thesis, Université Abdou moumouni, Niamey, Niger, 2015.

Oyerinde, G. T., Wisser, D., Hountondji, F. C. C., Odofin, A. J., Lawin, A. E., Afouda, A., and Diekkrüger, B.: Quantifying Uncertainties in Modeling Climate Change Impacts on Hydropower Production, Climate, 4, 34, https://doi.org/10.3390/cli4030034, 2016.

Poméon, T., Jackisch, D., and Diekkruger, B.: Evaluating the Performance of Remotely Sensed and Reanalysed Precipitation Data over West Africa Using HBV Light, J. Hydrol., 547, 222–235, https://doi.org/10.1016/j.jhydrol.2017.01.055, 2017.

Seibert, J. and Vis, M. J. P.: Teaching hydrological modeling with a user-friendly catchment-runoff-model software package, Hydrol. Earth Syst. Sci., 16, 3315–3325, https://doi.org/10.5194/hess-16-3315-2012, 2012.

Sylla, M. B., Nikiema, P. M., Gibba, P., Kebe, I., and Klutse, N. A. B.: Climate Change over West Africa: Recent Trends and Future Projections, in: Adaptation to Climate Change and Variability in Rural West Africa, edited by: Hesselberg, J. and Yaro, J. A., 25–40, https://doi.org/10.1007/978-3-319-31499-0_3, 2016.

WaSiM-ETH: WaSiM (Richards) [code], available at: http://www.wasim.ch/en/products/wasim_richards.htm, last access: 10 October 2021.

Proc. IAHS, 384, 355–361, 2021
https://doi.org/10.5194/piahs-384-355-2021

Integrated water resources management approach in mitigating the potential impacts of climate change on hydrology in Gurara reservoir catchment, Northwest Nigeria

Ifie-emi Francis Oseke[1], **Geophery Kwame Anornu**[2], **Kwaku Amaning Adjei**[2], and **Martin Obada Eduvie**[1]

[1]Department of Training, National Water Resources Institute, Mando-Road, Kaduna, Nigeria
[2]Department of Civil Engineering, Kwame Nkrumah University of Science and Technology, Kumasi, Ghana

Correspondence: Ifie-emi Francis Oseke (engroseke@gmail.com)

Published: 16 November 2021

Abstract. The strategies and actions in the management of African River Basins in a warming climate environment have been studied. Using the Gurara Reservoir Catchment in North-West Nigeria as a case study, summations were proposed using hypothetical climate scenarios considering the Global Climate Models prediction and linear trend of the data. Four (4) proposed scenarios of temperature increase (1 % and 2 %) coupled with a decrease in precipitation of (−5 % and −10 %) were combined and applied for the study area. The Water Evaluation and Planning Tool was used to model and evaluates the impact of the earth's rising temperature and declining rainfall on the hydrology and availability of water by investigating its resilience to climate change. Modelling results indicate a reduction in available water within the study area from 4.3 % to 3.5 % compared to the baseline with no climate change scenario, revealing the current water management strategy as not sustainable, uncoordinated, and resulting in overexploitation. The findings could assist in managing future water resources in the catchment by accentuating the need to put in place appropriate adaptation measures to foster resilience to climate change. Practically, it is pertinent to shape more effective policies and regulations within catchments for effective water resources management in reducing water shortage as well as achieving downstream water needs and power benefit in thefuture, while also allowing flexibility in the operation of a reservoir with the ultimate goal of adapting to climate change.

1 Introduction

The hydrology of large rivers in the world has been experiencing changes in their hydrological characteristics and variations in their morphological processes (Awotwi et al., 2015). Although these observable changes and variations, have widely been attributed to the impact of climate change (Wang and Hejazi, 2011). Climate change in itself can be best described as the average change in the weather. Although the climate is intertwined and closely related to weather, there are important differences (Lehmann et al., 2017; Salvati et al., 2017). A common distinction is a difficulty encounter by scientists in predicting the weather a few weeks while confidently predicting the climate in 50 years (Lehmann et al., 2017; Awotwi et al., 2015). According to Intergovernmental Panel on Climate Change (IPCC, 2017; Salvati et al., 2017), the past two centuries have seen a continuous increase in the surface temperature of the earth resulting in the occurrence of extreme weather and variability in precipitation leading to a reduction in runoff (Lehmann et al., 2017). On the other hand, anthropogenic activities construction of reservoirs, irrigation canals, alternate land use causing an increase in industrial water consumption have also resulted in changes in hydrological characteristics especially runoff (Kundzewicz, 2004; Lehmann et al., 2017).

Table 1. Land use and land cover around GRC (Source: modified from WEAP model outcome).

	Land Use in upper Gurara catchment		
S/N	Category	Area (km^2)	Percentage
1	Agricultural land and wet land	339.7	15.8 %
2	Forest land and wood land	722.4	33.6 %
3	Wet land	666.5	31.0 %
4	Urban land	303.2	14.1 %
5	Water body and grass land	116.1	5.40 %
6	Urban land	2.150	0.10 %
	Total	2,150	100 %

According to Abdullahi et al. (2016), in water basins, the impact of climate change is felt in the alteration of the amount and timing of runoff, thereby increasing the challenges in the coping capacities of large river basins especially in the wake of increasing water demands thus bringing higher risks of water shortages and food security. Meanwhile, the impact of climate change can go well beyond the water sector since many of the most serious effects of climate change on non-water areas are mediated via water (Salvati et al., 2017; Lehmann et al., 2017). These concerns have been the challenges among water resources managers in making efforts to protect the ecosystem especially the Gurara Reservoir Catchment (GRC) by quantifying the cumulative impact from the changing trend of climate and other anthropogenic activities on runoff using a long-term hydrological data-driven approach.

The GRC is Nigeria's pioneer and only reservoir catchment performing a water transfer operation (3 m wide and 75 km long pipeline). In the catchment, water is transferred from the Gurara Reservoir to Lower Usuma Reservoir in the Federal Capital Territory Abuja, the administrative capital of Nigeria, serving over 2.7 million people. The need to transfer water was initiated when from the fact that in 1999, Abuja had a population of 2.1 million and this number has been projected to increase to about 7 million in the next 50 years. This rapid population growth and accompanying developments brought about the tremendous surge in demand for water which was essential for sustaining its inhabitants. Before this study, there is limited converge on assessing the impact of climate change on hydrology using integrated water resources management approach within the (GRC) to the best knowledge of the authors, as such there is the need to fill this identified gap using multiple climate change scenarios while considering plausible socio-economic developments. This is because freshwater resources at GRC are considered very important by stakeholders in the basin. In this regard, the aim of this paper is to analyze the impact of climate change on hydrology within the GRC for estimating water availability and use. In particular, the focus is on estimating the runoff into the Gurara reservoir managed by the Department of Dams and Reservoir, Federal Ministry of Water Resources, Abuja Nigeria. Besides the water transfer operation, the multipurpose reservoir also supplies water to the Azare–Jere irrigation farms as well as generating 30MW of hydroelectric power. Thus, there is a high priority placed on the reservoir due to its strategic importance

2 Material and methods

2.1 Study area

The GRC is a transboundary catchment located in West Africa (Nigeria). The catchment is shared by Kaduna State in the West, Niger State in the North, Plateau State in the East and Abuja in the South. Remotely, the catchment cuts across Kachia and Kagarko Local Government Areas in Kaduna State and Bwari Area Council of Nigeria's FCT, covering an area of approximately 150 km^2. The catchment falls within Latitude 090:32 N to 090:48′ N and Longitude 070:29 E and 070:49′ E.

2.2 Data

The major data types are hydrological time series data, water demand data. However, due to the lack of sufficient hydrological data, the Soil Conservation Service Runoff Curve Number (SCSRCN) method was used for computing the runoff. The Curve Number (CN) utilized in this study as input for the SCSRCN method to compute the runoff was adopted from a previous study by Coyne et Bellier and de Crown (2008). The characteristics of Land cover within the GRC are listed in Table 1.

2.3 Model setup and simulation based on the WEAP platform

A WEAP model was set-up for the catchment consisting of meteorological and other relevant input data used for the hydrological modeling based on the WEAP platform. The model builds and analyzes different future scenarios using hypothetical climate assumptions about the impacts of cli-

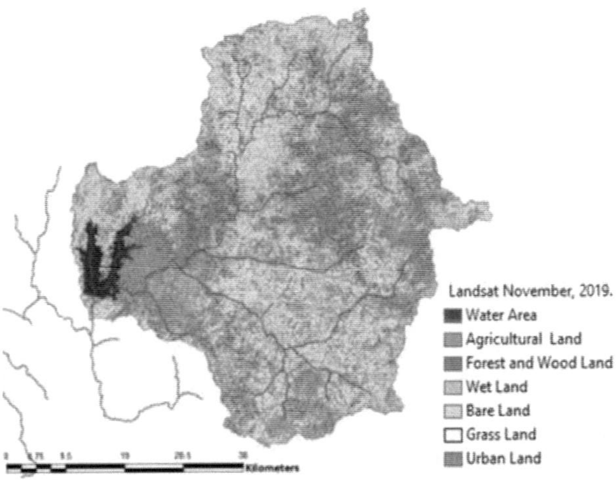

Figure 1. Land cover map showing changes within the GRC from 2010, the period after dam construction.

Table 2. Hypothetical climate change scenarios.

Scenario	Change in Temperature	Change in Precipitation
Baseline	No Change	No Change
Scenario 1	Temp (+0.01)	Precip (−0.05)
Scenario 2	Temp (+0.02)	Precip (−0.10)

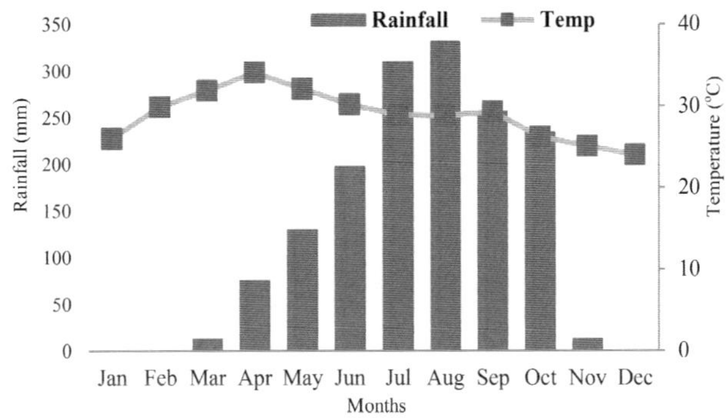

Figure 2. Mean monthly rainfall and temperature relationship for baseline scenario from 1988 to 2019 in GRC.

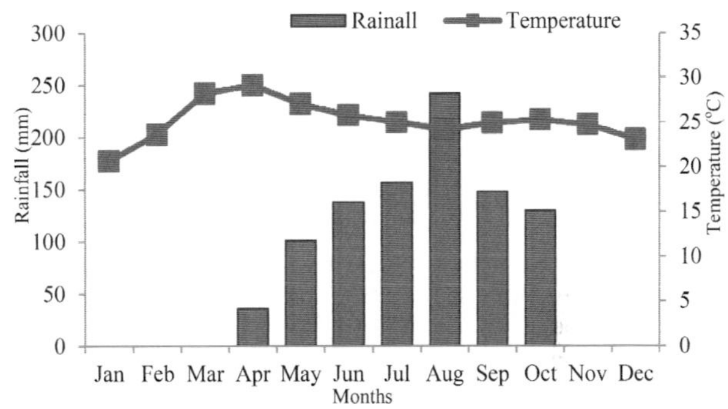

Figure 3. Mean monthly rainfall and temperature relationship for scenario 1 from 1988 to 2019 in GRC.

mate change on hydrology for water availability. This is to ensure that informed policies are made on the available water and water use dynamics.

2.4 Performance and evaluation of calibrated and validated model

The reliability of the simulation was assessed by two indices; the Nash-Sutcliff Efficiency (NSE) and the Coefficient of Determination (R^2) for performance, as proposed by Moriasi et al. (2007). The missing dataset was filled using the inverse distance weighting method based on Awotwi et al. (2015).

2.5 Climate change scenarios

This study relied on purely hypothetical scenarios derived from the outputs of General Circulation Models to assess the potential impacts of climatic change on runoff in GRC. The hypothetical scenario approach is well documented according to Abdullahi et al. (2014). The choice of using hypothetical temperature increases of 1 and 2 °C reflects the consensus that greenhouse emission will produce rises in this range, given an equivalent doubling of atmospheric CO_2 (Gebre and Ludwig, 2015). The climate scenarios are listed in Table 2.

Subsequently, a reduction of precipitation of 5 % and 10 % was used owing to greater uncertainty surrounding estimates of changes in regional precipitation. Both the increases and decreases in average temperature and precipitation respectively are modelled in this study.

2.6 Runoff computation at Gurara reservoir catchment

The general form of SCS-CN equation for computing the runoff is given as follows

$$Q = \frac{(P - l_a)^2}{(P - l_a) + S} \tag{1}$$

where Q = runoff, P = rainfall, S = potential maximum retention after runoff begins and l_a = Initial abstraction and is highly variable but generally is correlated with soil. l_a is estimated using:

$$l_a = 0.2S \tag{2}$$

S is related to the soil and covers conditions of the catchment through the runoff CN by the Eq. (2) and is estimated using:

$$S = \frac{1000}{\text{CN}} - 10 \tag{3}$$

A composite runoff CN value of 75 % developed by Coyne et Bellier and de Crown (2008) from previous was adopted. According to Coyne et Bellier and de Crown (2008). The

Table 3. Different climate warming impact on scenarios to water availability based on simulation of the sources for Gurara River.

	Amount of water available (mcm)		
	Baseline No Change	Scenario one +1.0 °C, −10 % P	Scenario two +2.0 °C, −10 % P
Available water (mcm)	4777	4458	4379
Difference (mcm)	n/a	319	398
(%) of reduction	n/a	3.5	4.3

n/a: not applicable.

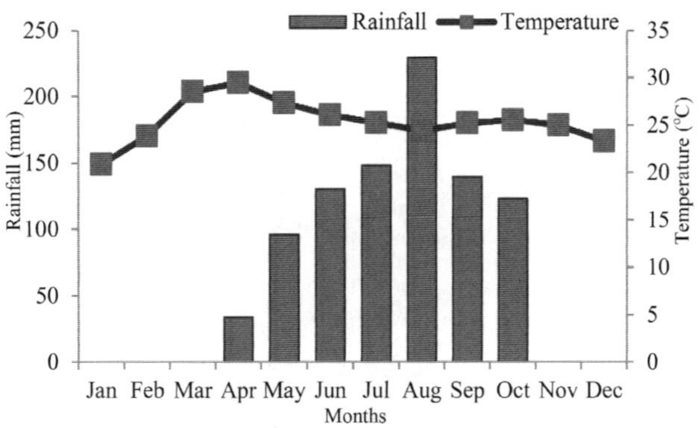

Figure 4. Mean monthly rainfall and temperature relationship for scenario 2 from 1988 to 2019 in GRC.

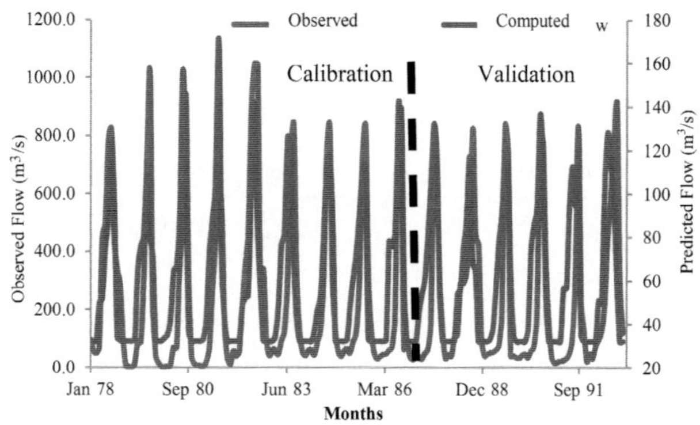

Figure 5. Computed runoff fit on the observed runoff for the period 1978 to 1991.

value is because of the development level within the catchment and bearing in mind that the catchment is also highly forested (Coyne et Bellier and de Crown, 2008). Shown in Fig. 1 is the land cover map obtained from the hybrid classification of the 2019 Enhanced Thematic Mapper (ETM+) Landsat image within the GRC. Thus Eqs. (1), (2), and (3) becomes:

$$Q = \frac{(P - 0.6667)^2}{(P + 2.6666)} \quad (4)$$

2.7 Analysis of climate change impact on hydrology

The analysis to quantify the available water in all scenarios from the hypothetical increase in the indicators of climatic change within the catchment was evaluated based on Eq. (5).

$$\% \text{ Impact} = \frac{\text{available Water Under a Scenario}}{\text{Available Water from observation}} \times 100 \quad (5)$$

3 Results

3.1 Temperature and rainfall

The mean monthly temperature and precipitation data obtained from the Nigerian Meteorological Agency (FMWR, 2013), is illustrated in Fig. 2 (baseline scenario), while the hypothetical climate change Scenarios is shown in Fig. 3 (Scenario 1) and Fig. 4 (Scenario 2). The hottest period occurring in the months of March and April, with temperature values between 30.6 to 36.5 °C. December is the coldest month with an average temperature ranging from 23.5 to 24.0 °C. The mean monthly precipitation reveals a varying trend between 350 mm in the dry season to 1412 mm in the wet season, with about 80 % occurring between July and September. The heaviest rainfall occurs in the month of August with a value of 294.8 mm.

3.2 Runoff

Illustrated in Fig. 5 is the comparison of observed and computed hydrographs for calibration and validation of the runoff in a monthly time step (1978 through 1991). A near excellent correlation was achieved and the computed percentage change in runoff followed the net supply and fits reasonably well with the observed. This fit statistically justifies the acceptability of our computed for hydrological modelling within the GRC. The calibration and validation using the SC-SRCN method for the observed and computed runoff resulted in NSE: calibration 0.72: validation 0.69 and R^2: calibration 0.72: validation 0.69.

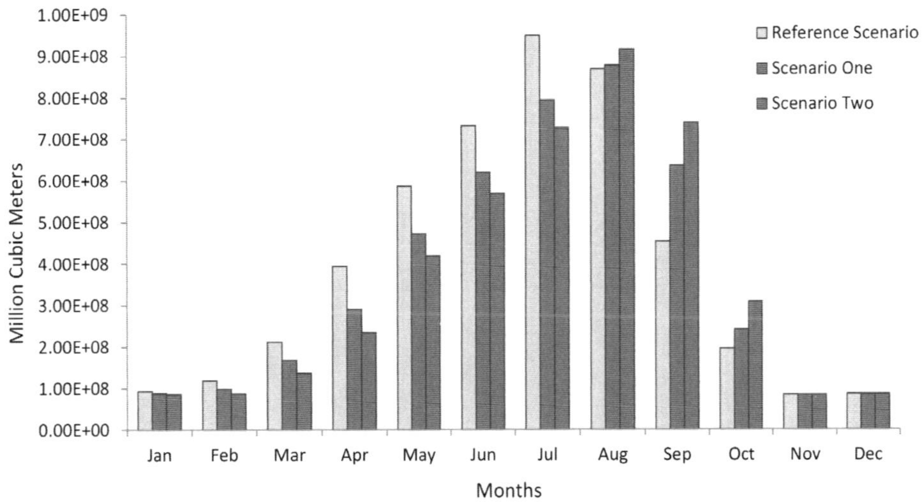

Figure 6. Simulated inflow (mcm) under current and future climate change for all scenarios.

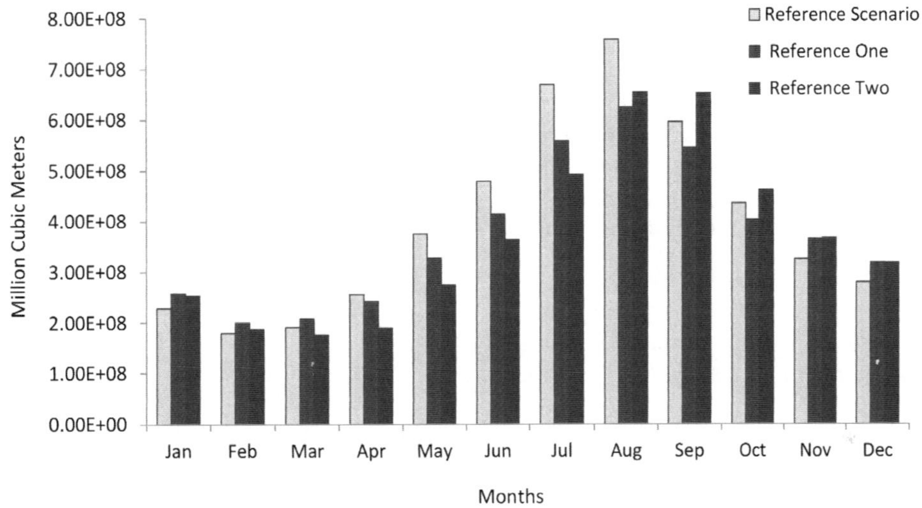

Figure 7. Simulated outflow (mcm) under current and future climate change for all scenarios.

3.3 Water availability

The result from the analysis of climate change impact is listed in Table 3. The cumulative inflow to the catchment is illustrated in Fig. 6, while the outflow is shown in Fig. 7 and the water balance for inflow, outflow, consumption, and evaporation is shown in Fig. 8, respectively.

The result reveals the different amounts of water available based on the simulated model due to the different hypothetical climate scenarios.

4 Discussion

The impact of climate change on hydrology is not uniform hence the water availability is season-dependent for all scenarios. The projected mean annual runoff for the simulation period in the WEAP model shows a reduction in available water compared to the baseline values. These results indicate that under scenario 2 regarded as an increase in temperature by 2° and a decrease in precipitation by 10 %, the available water within the catchment is estimated at 4379 mcm, a reduction from 4458 mcm in scenario 1. These translate into significant decreases in mean annual runoff relative to the baseline value, the mean annual runoff values for scenario 1 and scenario 2 showed a decrease of 3.5 % and 4.3 % respectively compared to the baseline (Table 3). Accordingly, the statistics of the available water within the catchment under the baseline and the hypothetical scenarios (see Table 3) show sensitivity to variability under climate change, compared to baseline. Abdullahi et al. (2014) reported a similar

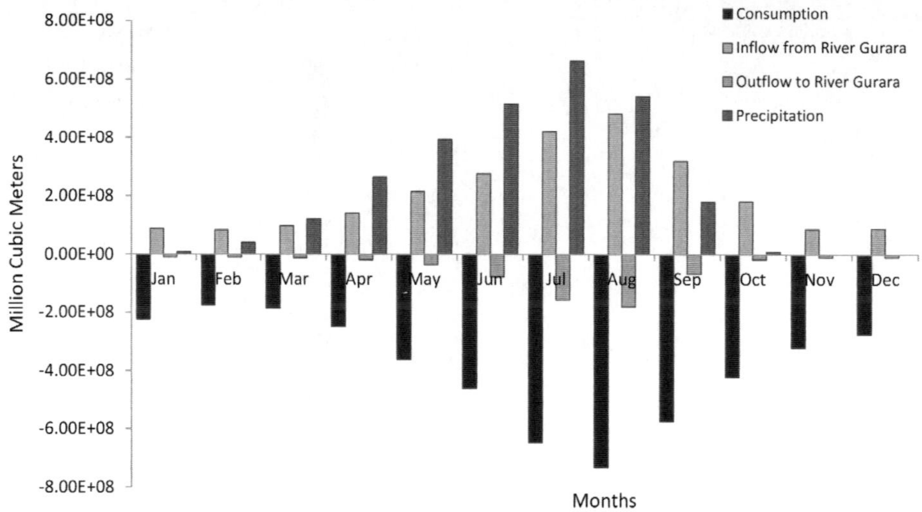

Figure 8. Water balance showing consumption, inflow, outflow and evaporation all scenarios with respect to baseline scenarios.

observation for the entire Sokoto Rima River Basin. The reduction in the percentage of available water under the different scenarios is more significant for the annual runoff, as it is the basis for the catchment's relevancy. The WEAP simulated model adopted for this study offers a simplified representation of the complex workings of the catchment's hydrology in estimating water availability by representing the actual conditions within the catchment while using various scenarios and key assumptions.

5 Conclusions and recommendations

The analysed projected temperature and rainfall hypothetical-based climate change scenarios show an increase in temperature and decrease in rainfall in the future within the catchment. The sensitivity analysis identifies runoff as the most sensitive input parameter and was much attention in the model calibration process. The runoff data were computed using the SCSRCN method while the impact analysis was modelled in WEAP. The WEAP model was successfully set up, calibrated, and validated and the climate change impact analysis was performed. The statistical correlation at calibration and validation carried out revealed monthly runoff was well predicted, and adequate for hydrological modelling during the calibration and validation periods.

Practically in Nigeria like many developing nations, a need to prepare adequately for the negative impact of climate change cannot be over emphasised, the development of models predicting projected climate change scenarios is recommended to serves as a platform that enables stakeholders better understand our climate system. In addition offering a means to access, plan and implement sustainable programs that will assist in combating these changes, thereby making our nation less vulnerable. Besides, as more climatic data are available, further research needs to be carried out to model variations based on decades, yearly, monthly and daily changes as a strategy for implementing integrated water resources management considering uncertainties and increasing warming climate for sustainability.

Data availability. The underlying research data can be accessed from the Federal Ministry of Water Resources Abuja Nigeria on request.

Author contributions. All the authors IFO, GKA, KAA, and MOE were involved in the concept development, data collection, data analysis and manuscript write-up.

Competing interests. The contact author has declared that neither they nor their co-authors have any competing interests.

Disclaimer. Publisher's note: Copernicus Publications remains neutral with regard to jurisdictional claims in published maps and institutional affiliations.

Special issue statement. This article is part of the special issue "Hydrology of Large River Basins of Africa". It is a result of the 4th International Conference on the "Hydrology of the Great Rivers of Africa", Cotonou, Benin, 13–20 November 2021.

References

Abdullahi, S. A., Muhammad, M. M., Adeogun, B. K., and Mohammed, I. U.: Assessment of Water Availability in the Sokoto Rima River Basin, Res. Environ., 4, 220–233, 2014.

Awotwi, A., Yeboah, F., and Kumi, M.: Assessing the impact of land cover changes on water balance components of White Volta Basin in West Africa, Water Environ. J., 29, 259–267, 2015.

Coyne et Bellier and De Crown Lot, A. (Eds.): Dam and Associated Works – Design and Works Progress April 2001–March 2007, Gurara Water Transfer Project to Federal Capital Territory, The Regent Printing and Publishing Ltd, Abuja, 2008.

Federal Ministry of Water Resources FMWR (Eds.): Gurara water transfer project to federal capital territory LOT A: Dam and associated works, Design and works progress April 2001–March 2007, Abuja, FMWR, 2008.

Gebre, S. L. and Ludwig, F: Hydrological Response to Climate Change of the Upper Blue Nile River Basin: Based on IPCC 5th Assmnt Rept, J. Climatol. Weather Forecast., 3, p. 121, 2015.

Kundzewicz, Z. W.: Searching for change in hydrologic data, Hydrol. Sci. J., 49, 3–6, 2004.

Lehmann, D., Brinkmann, K., Diogo, R. V., and Buerkert, A.: Temporal and spatial changes of land use in rare metal mining areas of Rwanda, Int. J. Mining. Reclam. Environ., 31, 519–529, 2017.

Moriasi, D. N., Arnold, J. G., Van Liew, M. W., Bingner, R. L., Harmel, R. D., and Veith, T. L.: Model evaluation guidelines for systematic quantification of accuracy in watershed simulations, Trans. Am. Soc. Agr. Bio. Engrs., 50, 885–900, 2007.

Salvati, L., De Zuliani, E., Sabbi, A., Cancellieri, L., Tufano, M., Caneva, G., and Savo, V.: Land-cover changes and sustainable development in a rural cultural landscape of central Italy: classical trends and counter-intuitive results, Int. J. Sustain Dev. Wld. Ecol., 24, 27–36, 2017.

Wang, D. and Hejazi, M.: Quantifying the relative contribution of the climate and direct human impacts on mean annual streamflow in the contiguous United States, Water Resour. Res., 47, 1–16, 2011.

Proc. IAHS, 384, 363–366, 2021
https://doi.org/10.5194/piahs-384-363-2021

Open Access
PIAHS

Calage et validation de SWAT sur le bassin versant du Bafing (Fleuve Sénégal) en amont de BAFING MAKANA : vers une application à la gestion du barrage de Manantali

Soussou Sambou[1], Moussé Landing Sane[1], Issa Leye[1], Didier Maria Ndione[1], Seydou Kane[1], and Mamadou Lamine Badji[2]

[1]Department of Physics, Cheikh Anta DIOP University, Dakar, 5005, Senegal
[2]Unité de Formation et de Recherche Science et Technologie, Laboratoire d'Océanographie, des Sciences de l'Environnement et du Continent, Université Assane Seck de Ziguinchor, Ziguinchor, Senegal

Correspondence: Soussou Sambou (sousamb@hotmail.fr)

Published: 16 November 2021

Résumé. En matière de Gestion des Ressources En Eau, il devient de plus en plus nécessaire de remonter à l'information pluviométrique. La modélisation est devenue ainsi un outil incontournable. Les modèles hydrologiques reposent sur le cycle hydrologique. Ils transforment les pluies en débits sur le réseau hydrographique. L'objectif de cet article est de décrire le fonctionnement hydrologique du bassin versant du Bafing (principal affluent amont du fleuve Sénégal) en amont de Bafing Makana. Le modèle hydrologique semi distribué à base physique SWAT a été choisi comme modèle de base pour calculer les hydrogrammes de crue à la station de Bafing Makana qui contrôle les entrées au barrage de Manantali. Pour représenter la topographie de la zone d'étude, un MNT de résolution 12.5 m × 12.5 m a été retenu. Les observations de pluies et débits sur la période 1979–1986 (pour le calage) et 1988–1994 (pour la validation) ont été utilisées comme entrées. L'analyse de sensibilité a permis de faire apparaître les paramètres les plus significatifs du modèle. Les valeurs de ces paramètres ont été déterminées dans la phase calage puis validées dans la phase validation. La valeur du critère de Nash est de 0.71 pour la phase de calage et de 0.65 pour la phase de validation. Les résultats validés à partir du critère de Nash montrent que les valeurs prises par le critère entrent dans l'intervalle des valeurs satisfaisantes retenues pour ce critère. L'ensemble des résultats obtenus montre que le modèle hydrologique SWAT peut être utilisé dans la gestion du barrage de Manantali à l'horizon 2090.

1 Introduction

Avec le changement climatique et la variabilité climatique qui s'en est suivie, mais également avec l'accroissement de la population, il devient plus que nécessaire d'utiliser de façon aussi rationnelle que possible les ressources en eau. Cette contrainte nécessite une connaissance approfondie de la physique du bassin versant et une très bonne compréhension de la façon dont les processus hydrologiques liés au cheminement de l'eau dans le bassin versant s'y produisent. Le modèle hydrologique qui permet de calculer les flux (débits, sédiments, nutriments, substances chimiques, ...) au sein du bassin versant est devenu incontournable de nos jours, aussi bien pour la recherche fondamentale en hydrologie que pour la planification et la gestion des ressources en eau (Xi Cheng et al., 2007). Les modèles hydrologiques peuvent être classés suivant la représentation spatiale du bassin versant ou selon la description des processus hydrologiques. Du point de vue spatial, on distingue les modèles globaux et les modèles distribués. Les modèles globaux sont unidimensionnels. Ces modèles ne tiennent compte ni des hétérogénéités spatiales du bassin versant ni des variations spatiales des variables hydrologiques et des processus hydrologiques. Les modèles distribués ont par contre une structure spatiale. Le bassin versant est divisé en mailles de résolution plus faible pouvant capturer explicitement la variabilité spatiale des processus hydrologiques et des caractéristiques du bassin versant. Du point de vue de la description des processus, le modèle hy-

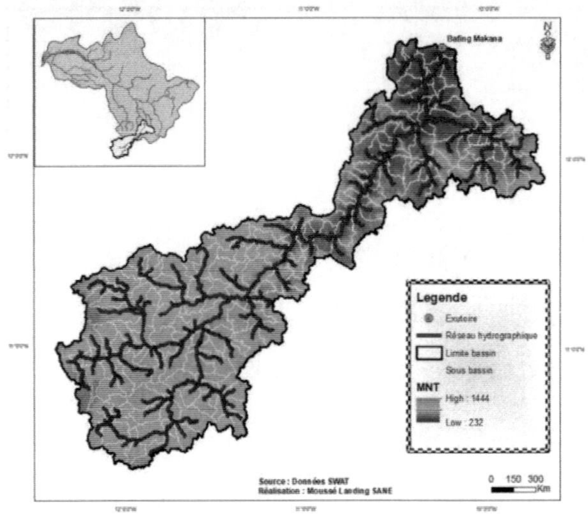

Figure 1. Bassin versant du Bafing Versant en amont de Bafing-Makana.

drologique peut être conceptuel ou à base physique. Dans les modèles conceptuels, les différents processus du cycle hydrologique sont représentés par des éléments de stockage. Les transferts entre ces différents réservoirs sont représentés par des fonctions mathématiques (Noandis and Mimikou, 2013). Les modèles à base physique représentent généralement les différents processus du cycle hydrologique à l'aide d'équations aux dérivées partielles non linéaires issues des lois de conservation de la mécanique des fluides. Les modèles distribués spatialement et à base physique fournissent une information détaillée sur le fonctionnement du bassin versant. Ils sont ainsi utilisés pour approfondir la compréhension de la dynamique des processus hydrologiques et ainsi simuler à l'échelle spatiale la réponse hydrologique d'un bassin versant. Ils permettent ainsi de faire des prévisions hydrologiques (Tashikawa Yasuto, 2011).

L'objectif de cet article est d'utiliser un hydrologique semi-distribué à base physique SWAT pour comprendre la dynamique du bassin versant du Bafing en amont de Bafing Makana. Après une analyse de sensibilité, les paramètres les plus sensibles ont été calibrés sur la période 1979–1986 et validés sur la période 1988–1994.

2 Matériels et méthodes

2.1 Zone d'étude

Le sous bassin versant du fleuve Sénégal en amont de Bafing Makana couvre une superficie de 22 000 km^2. Il se répartit entre la Guinée et le Mali. Le climat est subguinéen au Sud du sous bassin et soudanien au sud par soudanien au Nord. Les précipitations décroissent du Sud au Nord.

2.2 Sol et couvert végétal

La Fig. 1 représente le bassin versant du Bafing en amont de Bafing Makana. Le bassin versant comprend 4 différents classes de sol à savoir Acrisols, Cambisols, Leptosols and Regosols. Le couvert végétal est composé essentiellement de FRSD (37.2 %), RNGS (57.7 %), et dans une moindre proportion WWGR et CWGR.

2.3 Présentation du modèle hydrologique SWAT

SWAT est un modèle hydrologique continu, spatialement semi-distribué, à base semi-physique et à pas de temps journalier, téléchargeable gratuitement sur le site https://swat.tamu.edu/software/ (la date du dernier accès : 5 Novembre 2021). SWAT offre la possibilité de simuler un grand nombre de processus physiques dans un bassin versant avec une haute résolution spatiale. Pour cela, le bassin versant est divisé en sous bassins, eux-mêmes subdivisés en Unités de Réponses Hydrologiques (URH). Les URH sont des unités spatiales homogènes caractérisées par le même sol, le même couvert végétal, et la même pente. Les bilans d'eau calculés sur les URH permettent de déterminer les débits à l'exutoire des sous-bassins versants. Ces différents exutoires sont reliés entre eux par le réseau hydrographique. Le bilan d'eau simulé par SWAT utilise l'équation ci-dessous (Peng Shi et al., 2011) :

$$\mathrm{SW}_t = \mathrm{SW}_0 + \sum_{i=1}^{t} \left(R_{\mathrm{day}} - Q_{\mathrm{surf}} - E_{\mathrm{a}} - W_{\mathrm{seep}} - Q_{\mathrm{gw}} \right) \quad (1)$$

où t est le temps en jours, SW_t (mm) et SW_0 (mm) représentent respectivement l'humidité contenue dans le sol à l'instant t et à l'instant initial, R_{day} (mm) la lame précipitée à l'instant i, W_{seep} (mm) la lame infiltrée dans le sol à partir de la surface à l'instant i, Q_{surf} (mm) est la lame disponible pour le ruissellement à l'instant i, Q_{gw} (mm) le débit de base provenant de la nappe phréatique à l'instant i, et E_{a} (mm) est l'évapotranspiration à l'instant i. Dans ce travail, la méthode de Penmann-Monteith qui est la plus répandue est utilisée. La lame disponible pour le ruissellement est déterminée par la méthode du « Curve Number » mise au point par le SCS (Soil Conservation Service). L'écoulement de la surface du bassin versant vers le réseau hydrographique est simulé à l'aide de l'équation de l'onde cinématique. L'écoulement en rivière est calculé à l'aide de la méthode Muskingum (Peng Shi et al., 2011).

2.4 Données hydroclimatiques et topographiques utilisées

Les données utilisées dans cette étude sont les données hydrométriques, les données pluviométriques, les données climatiques et les données topographiques. Les données hydrométriques proviennent de la base de l'OMVS (l'organisation pour la Mise en Valeur du fleuve Sénégal). Il s'agit des débits moyens journaliers observés à la station de Bafing Makana sur la période 1961–2013. Les données pluviométriques observées s'étendent sur les périodes 1963–1986 et 1986–1994. Elles ont été extraites de la base ORSTOM. Les données climatiques journalières s'étendent sur la période 1979–2013. Il s'agit des précipitations, des températures maximales et minimales, du rayonnement solaire, de l'humidité relative, de la vitesse du vent tirées du site de SWAT https://globalweather.tamu.edu/ (la date du dernier accès : 5 Novembre 2021). Les données topographiques sont constituées par le MNT obtenu à partir du SRTM 12.5 × 12.5 m (https://vertex.daac.asf.alaska.edu/\penalty\@M\hskip.5\fontdimen2\font\relax?#, la date du dernier accès : 5 Novembre 2021).

2.5 Calage du modèle et analyse de sensibilité

Le calage est le processus qui permet d'estimer les valeurs des paramètres les plus significatifs du modèle en comparant pour un jeu de paramètres fixé les grandeurs calculées par le modèle aux grandeurs observées. SWAT comporte un grand nombre de paramètres dont le spectre de variation spatiale et temporelle est très étendu. Son calage est très complexe et consomme un temps de calcul qui peut être très élevé. Il est donc nécessaire d'éliminer les paramètres qui ne sont pas sensibles et de rendre ainsi le calage plus aisé. Cet exercice est appelée analyse de sensibilité. L'analyse de sensibilité fournit une liste finale de paramètres classés en ordre décroissant de sensibilité. Ce sont ces paramètres qui seront utilisés pour le calage. Le calage comme l'analyse de sensibilité peuvent être effectués manuellement ou automatiquement à l'aide de l'algorithme SUFI-2 du SWAT-CUP (Jinkang Du et al., 2013 ; Manoj et al., 2014).

2.6 Evaluation des performances du modèle

L'évaluation du modèle calibré permet de vérifier l'aptitude des paramètres estimés du modèle à reproduire ultérieurement la réponse du bassin versant de façon réaliste en utilisant des indicateurs de performance qui sont soit graphiques soit statistiques. Les indicateurs graphiques consistent principalement à comparer les hydrogrammes calculés par le modèle avec les hydrogrammes observés. Parmi les indicateurs statistiques existant, nous avons choisi le critère de Nash qui est le plus utilisé par les hydrologues.

3 Résultats et discussions

3.1 Analyse de sensibilité – Calage

La période 1979–1986 a été retenue pour l'analyse de sensibilité et le calage du modèle. A la suite de l'analyse de sensibilité, sur les 22 paramètres testés, 19 ont été reconnus suffisamment sensibles par le modèle. Ces paramètres ont été estimés lors de l'étape de calage en prenant comme critère de qualité le critère de Nash. La valeur de 0.71 obtenue pour ce critère peut être considérée comme bonne.. Nous présentons à titre d'illustration de la qualité du calage l'hydrogramme de la crue de l'année 1985 (Fig. 2) qui est restitué de façon satisfaisante par le modèle.

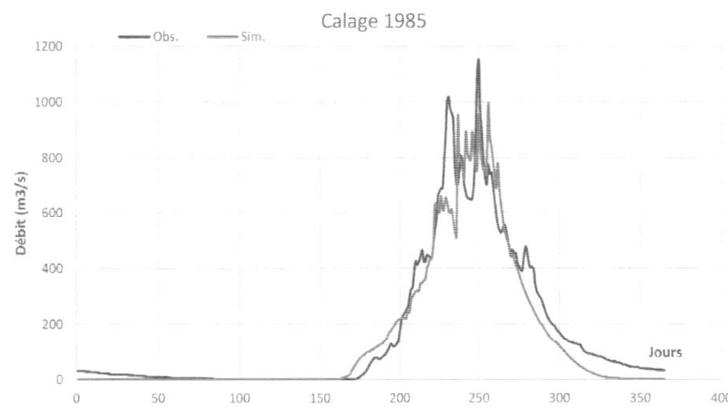

Figure 2. Hydrogrammes Bafing Makana : Calage année 1985.

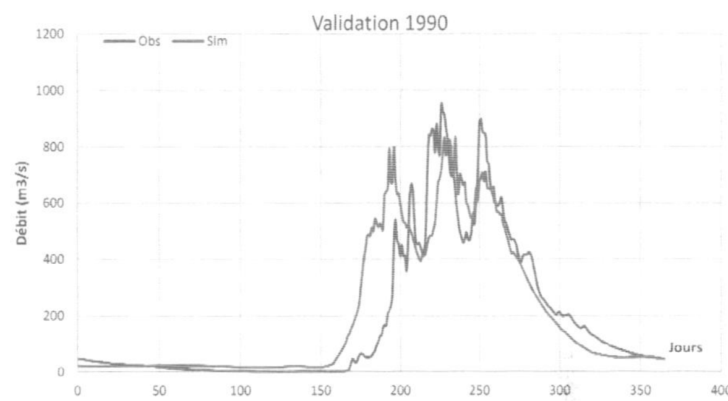

Figure 3. Hydrogramme Bafing Makana : Validation année 1990.

3.2 Validation

Les paramètres retenus après analyse de sensibilité et estimés après calage ont été appliqués sur la période 1988–1994 pour validation. Les mêmes indices de performance graphiques et statistiques ont été appliqués. Le critère de Nash calculé a une valeur de 0.65 sur l'ensemble de la période de validation. Bien qu'inférieure à celle obtenue lors du calage, il n'en demeure pas moins qu'elle est satisfaisante. Nous indiquons à titre d'illustration le graphe calculés-observés pour l'année 1990 (Fig. 3). Il y a un décalage important au début de la crue, alors que la décrue est mieux restituée.

4 Conclusion

La modélisation hydrologique joue un rôle très important dans la gestion des ressources en eau et la prise de décision. L'objectif de ce travail est de caler et valider le modèle hydrologique semi-distribué SWAT. L'application a été faite sur le bassin versant du Bafing en amont de Bafing Makana en vue de la gestion du barrage de Manantali. Les périodes de 1979–1986 et 1988–1994 ont été choisies respectivement pour le calage et la validation. La qualité du calage tout comme celle de la validation a été mise en évidence par le critère de Nash. Les valeurs obtenues aussi bien pour le calage (0.71) que pour la validation (0.65) montrent que le modèle hydrologique SWAT peut être utilisé pour reproduire le fonctionnement du bassin versant du Bafing en amont de Bafing Makana et faire une gestion du barrage de Manantali.

Disponibilité du code. Le code est en accès libre et a été téléchargé dans le site https://swat.tamu.edu/software/ (la date du dernier accès : 5 Novembre 2021).

Disponibilité des données. Les données hydrologiques proviennent de bases des services nationaux dans le cadre d'un projet sous condition de confidentialité et sans possibilité de divulgation. Les données climatiques proviennent du site de SWAT https://globalweather.tamu.edu/ (la date du dernier accès : 5 novembre 2021).

Collaborateurs. SS est l'auteur principal. Il a assuré la coordination, le montage et la relecture de l'article. MLS et MLB ont fait le calage et la validation de SWAT, et effectué tous les calculs correspondants. Ils ont également participé à la rédaction. DMN a participé à la collecte et à la mise en forme des données nécessaires au calage et à la validation de SWAT. SK a participé à la rédaction et à la relecture de l'article.

Intérêts concurrents. Les auteurs déclarent qu'ils n'ont aucun conflit d'intérêts.

Clause de non-responsabilité. Publisher's note : Copernicus Publications remains neutral with regard to jurisdictional claims in published maps and institutional affiliations.

Déclaration du numéro spécial. This article is part of the special issue "Hydrology of Large River Basins of Africa". It is a result of the 4th International Conference on the "Hydrology of the Great Rivers of Africa", Cotonou, Benin, 13–20 November 2021.

Remerciements. Les auteurs remercient le projet WEFE/AICS pour avoir appuyé le travail et les services techniques de l'Omvs pour avoir mis leur base de données à notre disposition.

Références

Jinkang Du, Hanyi Rui, Tianhui Zuo, Qian Li, Dapeng Zheng, and Ailin Chen : Hydrological simulation by SWAT Model with fixed and varied parameterization approaches under lad use change, Water Resour. Manag., 27, 2823–2838, 2013.

Manoj Jain, Survey Daman, and SharmaVan Sadhara : Hydrological modeling of river basin using SWAT. Internatiobal Conference on Emerging Trends in Computational and Imaging Processing (ICETCIP'2014), 15–16 December 2014, Pattaya, Thailand, 2014.

Noandis, L. I. and Mimikou, M. A. : Intercomparison of the lumped versus semi-distributed HEC-HMS hydrological model in the Kalamas River basin, Proceeding of the 13th International Conference of Environmental Scoence and Technology, 5–7 September 2013, Athens, Greece, 2013.

Peng Shi, Chao Chen, Ragawan Srinivasan, Xuesong Zhan, Xiuqin Fang, Simin Qu, Xi Chen, and Qiongfang Li : Evaluating the SWAT Model for hydrological modeling in the Xixian watershed and a comparison with the XAJ Model, Water Resour. Manag., 25, 2595–2612, 2011.

Tachikawa Yasuto : Distributed rainfall modeling, CE74.55 Modeling of water resources systems, 21 22 and 23 March 2012, Distributed flow routing model : 1K-FRM-event Demonstration and Project, disponible à : https://hywr.kuciv.kyotou.ac.jp/publications/lectureNote/2012_MWR-DHM-Demo-Tachikawa.pdf (la date du dernier accès : 5 novembre 2021), 2011.

Xi Chen, Yong Qin, David Chen, and Chong-yu Xu : A distributed monthly hydrological model for integrating spatial variation of basin topography and rainfall, Hydrol. Process. 21, 242–252, 2007.

Proc. IAHS, 384, 367–373, 2021
https://doi.org/10.5194/piahs-384-367-2021

Open Access PIAHS

Hydrology of Large River Basins of Africa

Variabilité saisonnière et intra-saisonnière de la pluviométrie en milieu forestier dans le Sud-ouest centrafricain

Jean-Marie Djebata[1], **Cyriaque R. Nguimalet**[2], **and Pierre Camberlin**[3]

[1]Ecole Normale Supérieure de Bangui, Bangui, République Centrafricaine
[2]Département de Géographie, Faculté des Lettres et Sciences Humaines, B.P. 1037, Bangui, République Centrafricaine
[3]Département de géographie/Centre de Recherches de Climatologie, UMR 6282 Biogéosciences, Université de Bourgogne Franche-Comté, Dijon, France

Correspondence: Cyriaque R. Nguimalet (cyrunguimalet@gmail.com) and Jean-Marie Djebata (djebata30@gmail.com)

Published: 16 November 2021

Résumé. Ce travail présente la variabilité intra-saisonnière de la pluviométrie dans le Sud-ouest centrafricain. Les données pluviométriques journalières utilisées couvrent la période 1981–2017. Elles ont été extraites aux points de grilles les plus proches des quatre stations représentatives de la zone d'étude (Bangui, Berberati, Boukoko et Nola) sur la base de données CHIRPS (Climate Hazards Group InfraRed Precipitation with Station data). Un contrôle de qualité de ces données a été effectué à partir d'une inter-comparaison entre les produits d'estimations et les données in-situ sur différentes sous-périodes : 1998–2011 à Nola, 1998–2012 à Berberati, 1998–2014 à Bangui et 2002–2017 à Boukoko. Le coefficient de corrélation entre les données CHIRPS et les observations au pas annuel est faible à Bangui ($r = 0,46$), moyen à Nola ($r = 0,57$) et Berberati ($r = 0,60$), et bon à Boukoko ($r = 0,73$). Les dates de début de la saison de pluies varient entre le 13 février et le 2 avril et celles de fin entre le 31 octobre et le 4 décembre. Des épisodes secs et humides sont mis en évidence dans le Sud-ouest centrafricain. A Bangui et Berberati, la période du 15 au 22 mai 1999 a été sèche soit 23 jours sans pluie. A Boukoko et Nola, la période du 26 juin au 24 juillet de la même année était caractérisée par des épisodes secs. Ces résultats établissent que la répartition intra-saisonnière des pluies n'est pas uniforme dans le Sud-ouest centrafricain.

1 Introduction

La connaissance sur l'évolution du climat est devenue l'un des défis majeurs des pays en développement, du fait des impacts négatifs et probables du climat sur l'homme et les écosystèmes (Abossolo et al., 2015).

Le système climatique est sans équivoque et, depuis les années 1950, beaucoup de changements observés sont sans précédent (GIEC, 2014). L'Afrique fait partie des régions du globe les plus vulnérables aux effets des changements climatiques en raison de son faible niveau de développement économique et de sa faible capacité de résilience.

En Afrique centrale, la République Centrafricaine (RCA) n'est pas épargnée de l'impact de la variabilité pluviométrique sur son économie. Ainsi, dans le Sud-ouest du pays, les populations du secteur forestier ont constaté que les pluies deviennent de plus en plus imprévisibles et varient considérablement d'une année à une autre. Les saisons pluvieuses sont parfois marquées par des épisodes secs et des sécheresses intercalaires.

De nombreux travaux ont permis de comprendre les modalités et facteurs de la variabilité pluviométrique en Afrique de l'Ouest, mais ces travaux restent beaucoup plus rares en Afrique centrale (Janicot et Sultan, 2001 ; Ardoin-Bardin, 2004 ; Camberlin, 2007). En particulier, la variabilité des

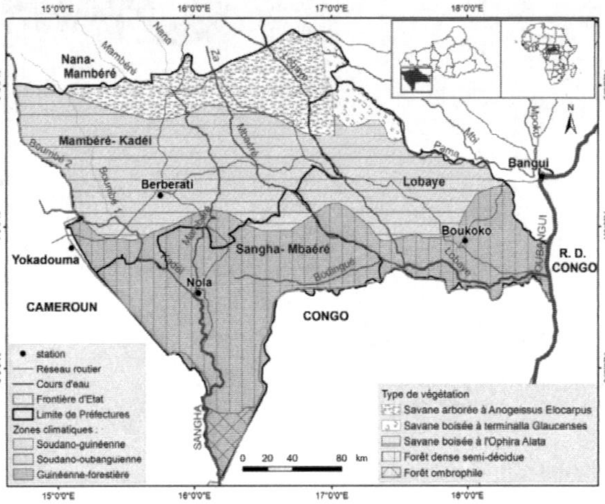

Figure 1. Localisation du Sud-ouest centrafricain. © Atlas of the Central African Republic (1984)

pluies à des pas de temps inférieurs à la saison (l'intra-saisonnier) est peu documentée (Camberlin, 2007), exceptés les travaux de Ndjéndolé (2001) dans le Nord-Ouest de la RCA et de Massouangui-Kifouala (2017) en République du Congo.

Toutefois, Zhou et al. (2014) ont détecté à partir de données satellitaires un assèchement de la petite saison pluvieuse (avril–juin) au cours des dernières décennies en Afrique centrale. Il a été aussi récemment démontré que l'intensité et la fréquence des orages avaient augmenté en février (Taylor et al., 2018) et en avril-mai-juin entre 1982 et 2016 dans le bassin du Congo (Raghavendra et al., 2018). Est-ce que les résultats obtenus sur la dynamique saisonnière, voire intra-saisonnière, de la pluviométrie en Afrique centrale ou dans le bassin du Congo sont-ils vérifiables dans le Sud-ouest centrafricain ? A cet effet, cette étude examine la variabilité des précipitations au pas de temps journalier et intra-saisonnier dans le Sud-ouest centrafricain.

2 Présentation de la zone d'étude

Le Sud-ouest centrafricain couvre une superficie de 94 200 km^2 sur les 623 000 km^2 que couvre la RCA, soit 15,12 % du territoire national. Il regorge aussi 41,70 % de forêts denses humides du pays (COMIFAC, 2015). Compris entre 2°13′ et 7°24′ de latitude Nord et entre 14°01′ et 19°02′ de longitude Est, il est limité au nord par la Préfecture de la Nana-Mambéré, au Sud par les Républiques Démocratique du Congo et du Congo, à l'Est par la Préfecture de l'Ombella-Mpoko et à l'Ouest par le Cameroun (Fig. 1).

Trois domaines climatiques caractérisent cet espace géographique : la zone guinéenne-forestière plus humide au sud, caractérisée par une pluviosité annuelle supérieure à 1700 mm, la zone soudano-oubanguienne au centre avec 1600 mm et la zone soudano-guinéenne au nord, marquée par un cumul pluviométrique annuel variant de 1200 à 1500 mm (Fig. 1).

3 Données et méthodologie

3.1 Données

Deux types de données de pluie au pas de temps journalier et annuel ont été utilisés pour caractériser la variabilité des précipitations : il s'agit des données pluviométriques réellement observées et des estimations de pluies du produit CHIRPS (Climate Hazards Group InfraRed Precipitation with Station data) produits par Earth Resources Observation and Science (EROS), un organisme américain) d'une résolution spatiale de 0,05. Les données d'observation sont incomplètes dans les quatre stations de la zone d'étude, c'est pourquoi, les données CHIRPS combinant les estimations satellitaires et les observations in-situ, couvrant la période 1981–2017 ont été utilisées. En Afrique centrale, ce produit d'estimation des pluies a été jugé comme l'un des plus fiables à l'échelle interannuelle (Camberlin et al., 2019).

3.2 Méthodes

Les données CHIRPS ont été choisies à partir d'une intercomparaison entre les données observées dans les stations principales de la zone d'étude et les points de grille les plus proches, sur différentes sous-périodes : 1998–2014 à Bangui, 1998–2012 à Berberati, 1998–2011 à Nola et 2002–2017 à Boukoko.

Le coefficient de corrélation entre les données CHIRPS et celles observées à l'échelle annuelle est faible à Bangui ($r = 0,46$), moyen à Nola ($r = 0,57$) et fort à Berberati ($r = 0,60$) et Boukoko ($r = 0,73$).

Pour valider le nombre d'événements pluvieux, nous avons confronté le nombre de jours de pluies des deux produits (CHIRPS et Observation) sur les différentes sous-périodes mentionnées ci-dessus. Avec CHIRPS, le nombre annuel de jours de pluies est de 184 jours à Bangui, 180 jours à Boukoko et 175 jours à Berberati. Cependant, avec les données d'observation, le nombre de jours de pluies est de 145 jours à Bangui, 140 jours à Boukoko et 138 jours à Berberati. On constate que CHIRPS surestime le nombre d'événements pluvieux, caractérisés par de faibles intensités (8,1 à 9 mm) contre (10,8 à 14,3 mm) pour les données in-situ.

L'analyse intra-saisonnière de la pluviométrie a reposé premièrement sur la détection des dates de début et de fin de la saison des pluies et la recherche des épisodes secs et humides au sein de la saison des pluies. Dans cette étude, nous avons choisi la méthode simple « d'anomalies cumulées de pluies » basée exclusivement sur les pluies journalières (Barraud, 2017). Avec la méthode d'anomalies cumulées, on a calculé tout d'abord, la moyenne journalière des pluies sur l'ensemble des années. Ensuite, cette moyenne est générale-

 https://doi.org/10.5194/piahs-384-367-2021

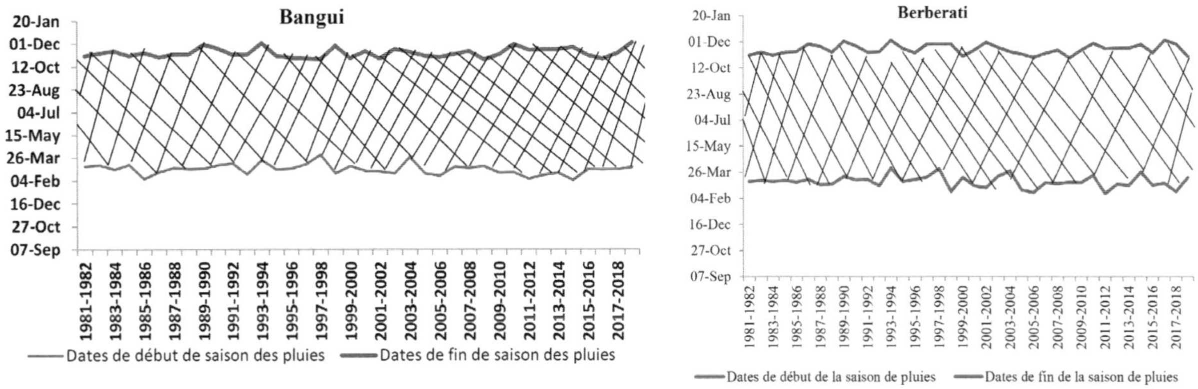

Figure 2. Variation interannuelle des dates de début et de fin de saisons des pluies à Bangui et Berberati de 1981 à 2018.

ment affectée d'un coefficient, lequel a permis d'obtenir une valeur de référence nommée (Pref) à partir de laquelle ont été calculées les anomalies journalières des pluies. Pour ce travail, un coefficient de 0,5 a été utilisé pour calculer la valeur de référence (Liebmann et al., 2012). Pour cette étude, la période de référence est comprise entre le 40ème jour (J1) à partir du 9 août et le 330ème jour (J2) et délimite la date de début et de fin de l'analyse. Le principal critère de choix de cette période est d'être centrée sur le cœur de la saison des pluies en débordant largement sur la Saison sèche. A l'intérieur de la période considérée, sont déterminées les dates de début et de fin de la saison des pluies. On calcule année par année, les anomalies cumulées de précipitations journalières. La date de la valeur minimale de ces anomalies cumulées, pour l'année A, est la date de début de saison des pluies de cette année. La date de la valeur maximale de ces anomalies cumulées est la date de fin de saison des pluies. L'utilisation de cette méthode a été justifiée par le fait qu'elle a déjà été appliquée à des zones similaires à celle de l'étude comme l'Amérique du Sud (Liebmann et al., 2008) ou l'Afrique dans son ensemble (Liebmann et al., 2012).

L'analyse interannuelle de la pluviométrie a reposé d'une part, sur une analyse en composantes principales (ACP), qui permet de détecter les structures spatiales de la variabilité (Sanders, 1989), et d'autre part, sur des tests appliqués aux séries centrées-réduites selon la formule proposée par (Lamb, 1983) qui s'écrit :

$$X_{ij} = \frac{(x_{ij} - \overline{x}_i)}{\sigma_i} \quad (1)$$

Où X_{ij} = hauteur de pluie totale pour une station i et une année j ; $\overline{x}_i$ = moyenne annuelle de pluie à la station i pendant la période de l'enregistrement ; σ_i = écart type de la série pluviométrique annuelle. Le test de Pettitt (1979) a été utilisé pour la détection des ruptures et tendances climatiques des séries temporelles.

Une analyse en composantes principales a été réalisée pour décrire la structure saisonnière ou la répartition des pluies selon la saison DJF, MAM, JJA et SON. La première composante principale de chaque saison, cartographiée sur Google (Climate explorer) à partir des coordonnées géographiques de la zone d'étude (latitude et longitude), de la période de référence (1981–2017) et de l'échelle (saisonnière) de la carte est retenue (Fig. 3) L'évolution des séries temporelles à l'échelle saisonnière a été présentée à partir des données issues des premières composantes principales cartographiées (Fig. 4).

4 Résultats et discussion

4.1 Dates de démarrage et de fin de saison des pluies

Les dates de début de saison de pluies à Bangui oscillent généralement entre le 5 février et le 2 avril (Fig. 2), et celles de fin de saison de pluies sont comprises entre le 28 octobre et le 3 décembre. A Berberati, les dates de démarrage et de fin de la saison humide sont comprises respectivement entre le 15 février et le 3 avril et entre le 19 octobre et le 3 décembre. Comparativement à la zone guinéenne-forestière (Bangui), la zone soudano-guinéenne (Berberati) connait un retard de 10 jours dans la mise en place de la saison humide.

Cependant, avec les données in-situ, la date de démarrage de saison de pluies se situe entre le 25 février et le 9 mars à Bangui et entre le 10 et le 25 mars à Berberati. La date de fin de saison des pluies est comprise entre le 1 et le 15 décembre à Bangui et entre le 1 et 30 novembre à Berberati. Par rapport aux données d'observation, on constate un début précoce de la saison de pluies avec les produits d'estimation, soit un écart de 20 jours à Bangui et 24 jours à Berberati. A l'inverse, on note un prolongement des dates de la fin de saison de pluies avec les données in-situ qu'avec les produits d'estimation. On note aussi un écart d'environ 33 jours à Bangui et 42 jours à Berberati.

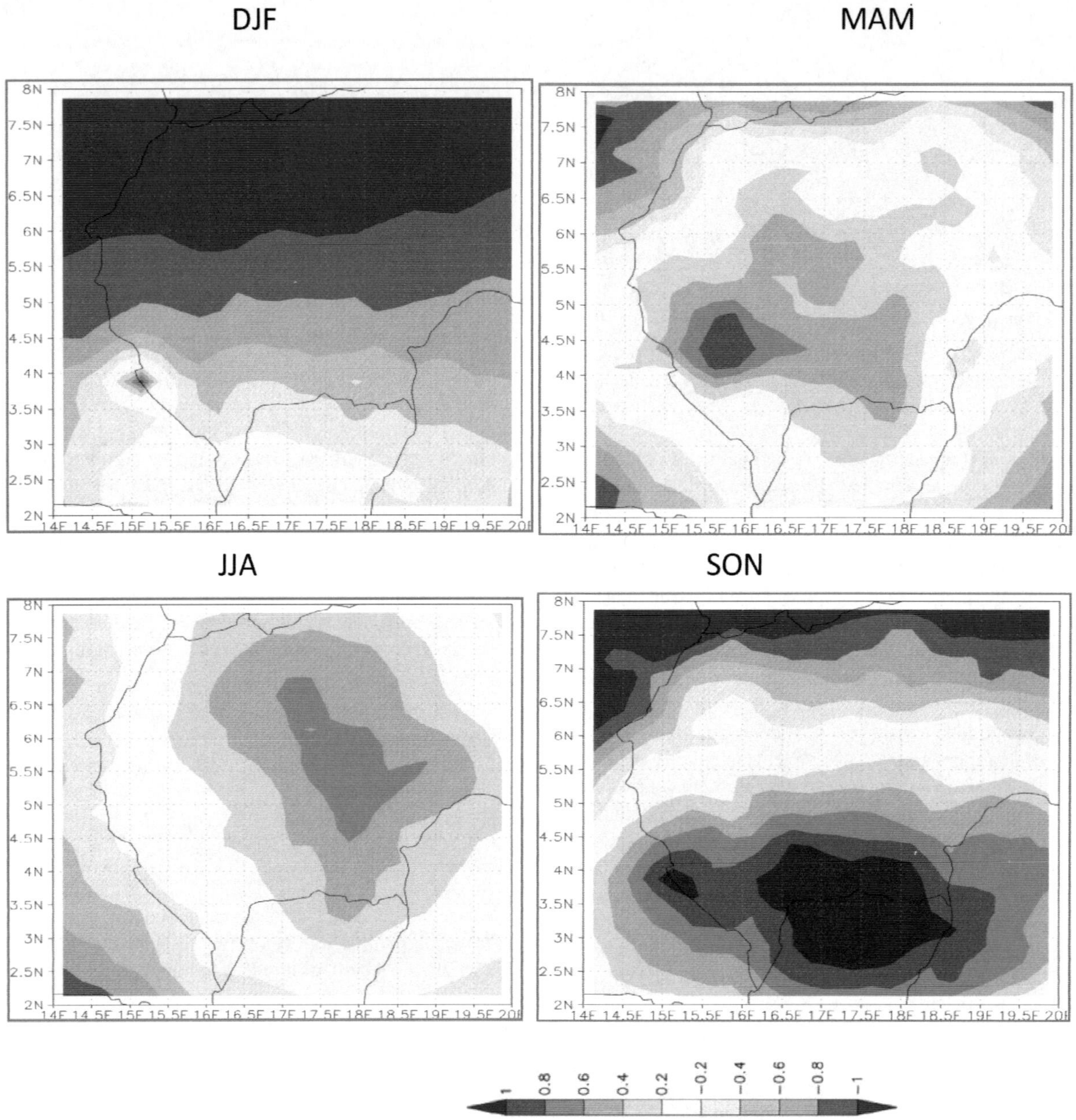

Figure 3. Structures spatiales des pluies associées aux composantes principales de 1981–2017.

4.2 La répartition spatiale des pluies associées aux composantes principales

L'analyse en composantes principales consiste à décrire la variabilité spatiale des pluies à l'échelle saisonnière.

La tendance des pluies est décroissante au trimestre décembre–janvier–février (DJF), croissante en mars–avril–mai (MAM), stationnaire en juin–juillet–août (JJA) et croissante en septembre–octobre–novembre (SON) (Fig. 4).

A l'échelle saisonnière, la répartition des précipitations dans le Sud-ouest centrafricain correspond au mouvement saisonnier de la Zone de Convergence Inter-Tropicale (ZCIT) vers le Nord du pays (Fig. 3). Les moyennes trimestrielles de décembre–janvier–février (DJF = 66 mm), mars–avril–mai (MAM = 309 mm), juin–juillet–août (JJA = 505 mm) et SON (420 mm) représentent respectivement 5,23 %, 23,64 %, 38,43 % et 32,68 % de la quantité totale des pluies sur la période considérée.

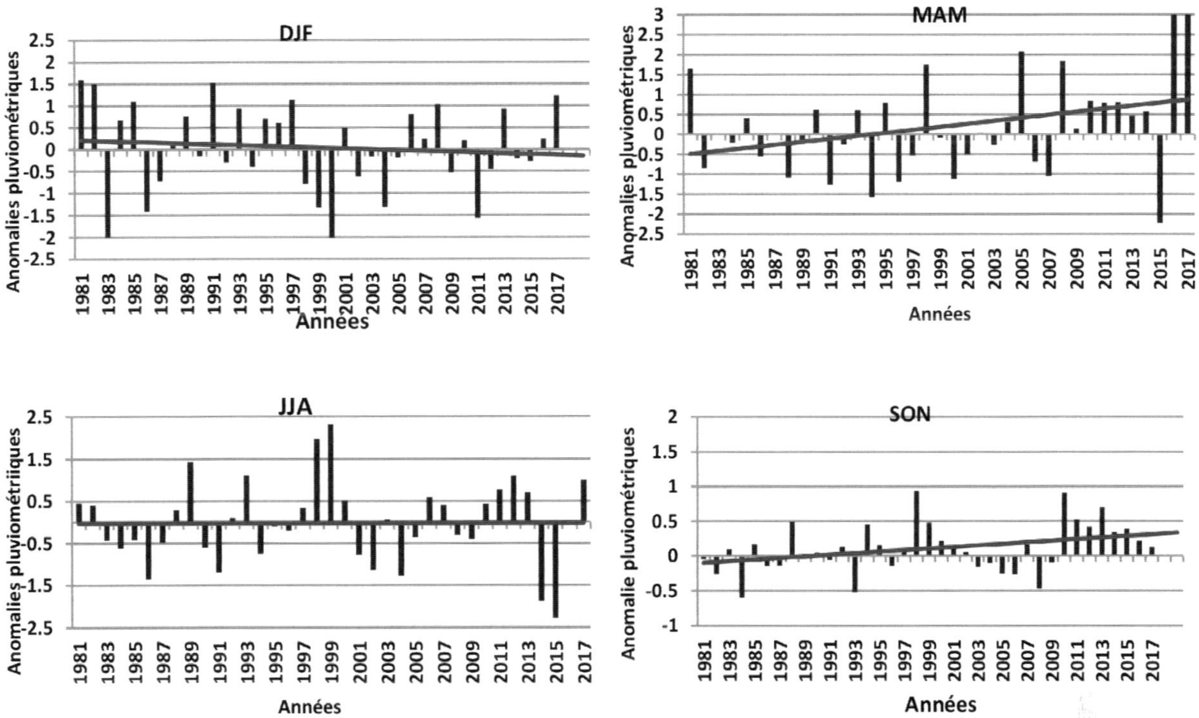

Figure 4. Evolution des séries temporelles associées à chacune des composantes principales cartographiées de 1981 à 2017.

4.3 Episodes secs et humides au sein de la saison des pluies

Dans le Sud-ouest centrafricain, la saison de pluies s'étend de mars à novembre, soit 9 mois, marquée par l'alternance des épisodes secs et pluvieux sans cohérence. Deux ou trois épisodes pluvieux sont séparés par une longue période durant laquelle les précipitations journalières peuvent être très faibles (inférieures à 20 mm) ou carrément nulles au sein même de la période normalement pluvieuse. Ainsi, une alternance des épisodes secs (4 à 6 jours sans pluie) et phases arrosées (3 à 5 jours de pluies consécutives) est notée dans les quatre stations représentatives de notre zone d'étude, exclusivement pour l'année 1999 (Fig. 5). En effet, l'année 1999 a été particulièrement humide sur l'ensemble du territoire centrafricain. Elle était caractérisée par des crues et inondations des parcelles agricoles, voire urbaines, dans le Nord du pays où ce phénomène est inhabituel. C'est pour cette raison qu'elle a été choisie ici pour nous permettre de mettre en évidence des épisodes secs au sein de cette année particulièrement humide. En d'autres termes, les sècheresses intercalaires ne s'observent pas seulement pendant les années déficitaires mais également excédentaires sur le plan pluviométrique. On note aussi qu'en 1999, la période du 15 au 22 mai a été sèche à Bangui et Berberati. Les sècheresses intercalaires, surtout en début de la campagne agricole, pourraient être préjudiciables pour les récoltes. En Afrique de l'Est par exemple, Barron et al. (2003) ont estimé qu'un épisode sec survenant durant la phase végétative de floraison peut entraîner une diminution de l'ordre de 72 % à 75 % des rendements de maïs sur sol sableux. A Boukoko et Nola, la période du 26 juin au 24 juillet était caractérisée par des épisodes secs (Fig. 5).

5 Discussion

La variabilité intra-saisonnière de la pluviométrie dans le Sud-ouest centrafricain est analysée sur les dates de début et de fin de saison de pluies, la saisonnalité et tendances des précipitations, et les épisodes secs et humides au sein de la saison humide. En domaine guinéen-forestier, les dates de début de saison des pluies oscillent entre le 5 février et le 2 avril. Les dates de fin de saison de pluies sont comprises entre le 28 octobre et le 3 décembre. Cependant, en domaine soudano-guinéen, les dates de démarrage et de fin de la saison humide sont comprises respectivement entre le 15 février et le 3 avril et entre le 19 octobre et le 3 décembre. Ces résultats mettent en évidence une installation plus précoce de la saison humide dans la zone guinéenne-forestière, en marges septentrionales de la grande forêt du bassin congolais, que dans la zone soudano-guinéenne. L'examen des précipitations à l'échelle saisonnière montre que DJF, MAM, JJA et SON représentent respectivement, 5,23 %, 23,64 %, 38,43 % et 32,68 % de la quantité totale des pluies enregistrées sur la période considérée. Nos résultats sont similaires à ceux de

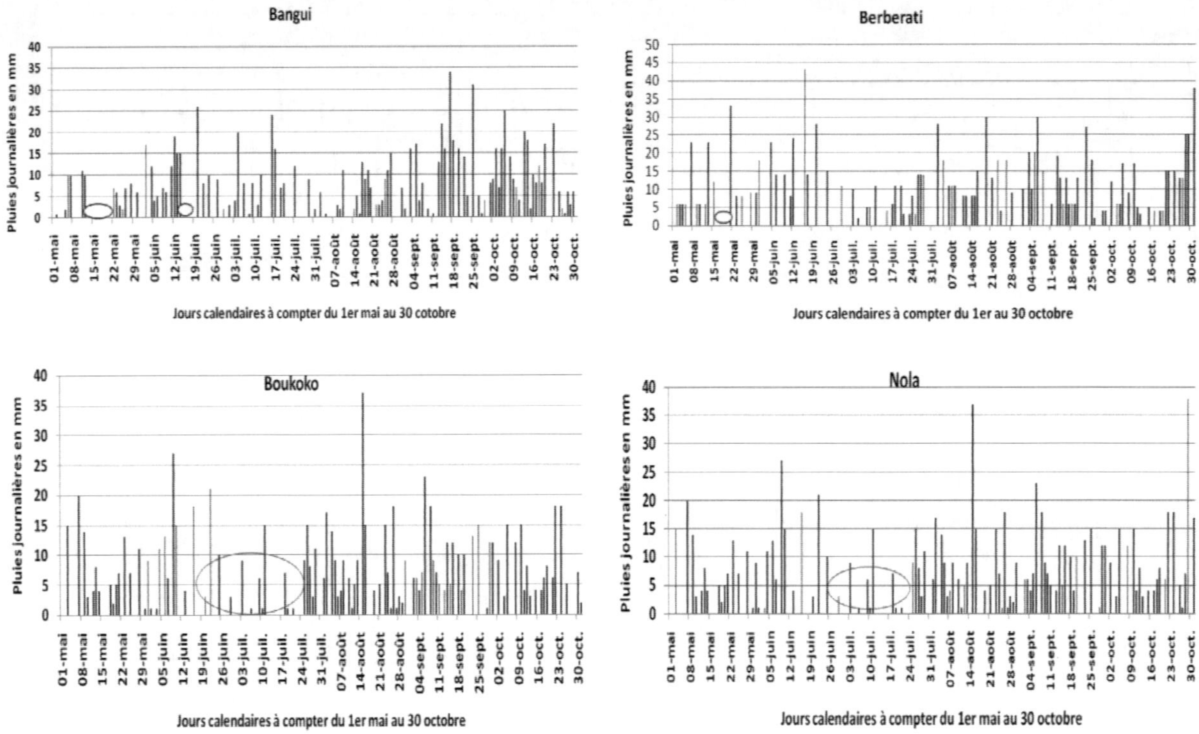

Figure 5. Variation intra-saisonnière des précipitations journalières dans les quatre stations représentatives de la zone d'étude à compter du 1 mai au 30 octobre 1999.

Ndjéndolé (2001). D'après cet auteur, le trimestre JJA constitue le cœur de la saison des pluies en RCA.

Concernant l'évolution des pluies saisonnières, les précipitations sont décroissantes en DJF, stationnaires en JJA, et croissantes en MAM et SON. Ce résultat est similaire à ceux de Raghavendra et al. (2018) et Taylor et al. (2018) qui ont démontré que l'intensité et la fréquence des orages avaient augmenté dans le bassin du Congo en avril-mai-juin et en février entre 1982 et 2016. Par contre, ce résultat ne confirme pas ceux de Zhou et al. (2014) qui ont détecté, à partir de données satellitaires, un assèchement de la petite saison pluvieuse (avril–juin) au cours des dernières décennies en Afrique centrale.

Les fluctuations intra-saisonnières des pluies dans cette zone seraient liées à des facteurs externes, notamment l'Oscillation de Madden-Julian, de l'ordre de 30 à 60 jours, ce qui peut renforcer ou diminuer les précipitations en Afrique centrale (Camberlin, 2007).

6 Conclusion

Cette étude révèle un caractère relativement homogène de la pluviométrie dans le Sud-ouest centrafricain, avec néanmoins quelques hétérogénéités locales liées aux conditions géographiques. Elle montre aussi que les dates de début de saison des pluies dans le Sud-ouest du pays oscillent généralement entre le 13 février et le 2 avril, et les dates de fin entre le 31 octobre et 4 décembre. Par ailleurs, cette étude indique que comparativement aux données d'observation, la date de début de la saison de pluies est précoce avec les produits CHIRPS. Alors cela ne pose-t-il pas un souci pratique pour la conduite des activités agricoles par exemple, vu qu'une modification du calendrier cultural est notée avec les données d'observation ?

Disponibilité des données. Les données pluviométriques utilisées dans ce travail proviennent de l'ASECNA (RCA), et les données CHIRPS disponible sur https://climexp.knmi.nl/ (la date du dernier accès : 14 mars 2019, World Meteorological Organization, 2019).

Collaborateurs. Trois auteurs ont contribué équitablement à la rédaction de ce travail.

Intérêts concurrents. Les auteurs déclarent qu'ils n'ont aucun conflit d'intérêts.

Clause de non-responsabilité. Publisher's note : Copernicus Publications remains neutral with regard to jurisdictional claims in published maps and institutional affiliations.

Déclaration du numéro spécial. This article is part of the special issue "Hydrology of Large River Basins of Africa". It is a result of the 4th International Conference on the "Hydrology of the Great Rivers of Africa", Cotonou, Benin, 13–20 November 2021.

Remerciements. Les auteurs remercient le comité d'organisation du Conférence international sur les grands fleuves africains de nous avoir donné l'opportunité de publier cet article. Merci pour leurs contributions.

Références

Abossolo, S. A., Amougou, J. A., Tchindjang, M., Mena, M. S., et Batha, R. A. S. : Analyse des précipitations annuelles à la station de Yaoundé de 1895 à 2006, Afrique Science, 11, 183–194, 2015.

Ardoin-Bardin, S. : Variabilité hydroclimatique et impacts sur les ressources en eau de grands bassins hydrographiques en zone soudano-sahélienne. Thèse de doctorat, option mécanique, génie mécanique et génie civil, Université Montpellier II, Montpellier, 440 pp., 2004.

Barraud, G. : Caractérisation des saisons sèches dans les zones forestières d'Afrique Centrale : approche multi-base de données combinant télédétection et observations, Mémoire de Master SEME, Université de Bourgogne, Dijon, 50 pp., 2017.

Barron, J., Rockström, J., Gichuki, F., and Hatibu, N. : Dry spell analysis and maize yields for two semi-arid locations in east Africa, Agric. For. Meteorol., 117, 23–37, 2003.

Camberlin, P. : L'Afrique Centrale dans le contexte de la variabilité climatique tropicale interannuelle et intra-saisonnière, Presses Universitaires d'Orléans, L'Afrique Centrale, le Cameroun et les changements globaux, Yaoundé, 25–39, 2007.

Camberlin, P., Barraud, G., Bigot, S., Dewitte, O., Makanzu Imwangana, F., Maki Mateso, J.C., Martiny, N., Monsieurs, E., Moron, V., Pellarin, T., Philippon, N., Sahani, M., and Samba, G. : Evaluation of remotely sensed rainfall products over Central Africa, Q. J. Roy. Meteor. Soc., 145, 2115–2138, 2019.

COMIFAC : Forêt d'Afrique centrale et changement climatique, Publication de l'Observatoire des forêts d'Afrique centrale, 2015.

GIEC : Rapport de synthèse. Contribution des Groupes de travail I, II et III au cinquième Rapport d'évaluation du Groupe d'experts intergouvernemental sur l'évolution du climat, Sous la direction de l'équipe de rédaction principale, édité par : Pachauri, R. K. et Meyer, L. A., GIEC, Genève, Suisse, 161 pp., 2014.

Janicot, S. et Sultan, B. : Intra-seasonal modulation of convection in the West African monsoon, Geophys. Res. Lett., 28, 523–527, 2001.

Lamb, P. : Subsaharan rainfall update for 1982 : continued drought, J. Climatol., 3, 419–422, 1983.

Liebmann, B., Bladé, I., Bond, N. A., Gochis, D., Allured, D., and Bates, G. T. : Characteristics of North American summertime rainfall with emphasis on the monsoon, J. Climate, 21, 1277–1294, 2008.

Liebmann, B., Bladé, I., Kiladis, G. N., Carvalho, L. M., Senay, D., and Allured, S. : Seasonality of African precipitation from 1996 to 2009, J. Climate, 25, 4304–4322, 2012.

Massouangui-Kifouala, M. : Caractérisation des saisons pluvio-agricoles dans les pays de Niari, République du Congo, en : Géographie du Congo, mélanges offerts au professeur Bonaventure Mengho, L'Harmattan, Paris, 225–246, 2017.

Ndjéndolé, S. : Contraintes pluviométriques sur les cultures annuelles : l'exemple de la culture cotonnière. Thèse de doctorat de Géographie, option agro-climatologie, Université de Bourgogne, Dijon, 227 pp., 2001.

Pettitt, A. N. A. : Non parametric approach to the change point problem, Appl. Stat., 28, 126–135, 1979.

Raghavendra A., Zhou, L., Jiang, Y., and Wenjian, H. : Increasing extent and intensity of thunderstorms observed over the Congo Basin from 1982 to 2016, Atmos. Res., 213, 17–26, 2018.

Sanders, L. : L'analyse statistique des données géographiques, G.I.P RECLUS, Montpellier, 269 pp., 1989.

Taylor, C. M., Fink, A. H., Klein, C., Parker, D. J., Guichard, F., Harris, P. P., and Knapp, K. R. : Earlier seasonal onset of intense mesoscale convective systems in the Congo Basin since 1999, Geophys. Res. Lett., 45, 458–467, 2018.

World Meteorological Organization : Climate Explorer, disponible sur : https://climexp.knmi.nl/ (la date du dernier accès : 14 mars 2019), WMO [data set], 2019.

Zhou, L., Tian, Y., Myneni, R. B., Ciais, P., Saatchi, S., Liu, Y. Y., and Hwang, T. : Widespread decline of Congo rainforest greenness in the past decade, Nature, 509, 86, 2014.

AUTHOR INDEX

International Association of Hydrological Sciences
Centre for Ecology and Hydrology
Wallingford, Oxfordshire OX10 8BB
United Kingdom

ISBN 978-1-7138-4503-4

CURRAN ASSOCIATES INC.
proceedings.com